S0-BWT-535

CONSTRUCTION REGULATIONS GLOSSARY

A REFERENCE MANUAL

CONSTRUCTION REGULATIONS GLOSSARY

A REFERENCE MANUAL

J. Stewart Stein, AIA, FCSI

A Wiley-Interscience Publication

JOHN WILEY & SONS

New York Chichester Brisbane Toronto Singapore

Library of Congress Cataloging in Publication Data:

Stein, J. Stewart.
Construction regulations glossary.

Companion manual to: Construction glossary.
"A Wiley-Interscience publication."
Includes index.
1. Building laws—United States—Terms and phrases.
2. Zoning law—United States—Terms and phrases.
3. Land subdivision—Law and legistration—United State—Terms and phrases. I. Title.
KF5701.A68S74 1983 343.73'07869'00321 83-5888
ISBN 0-471-89776-0 347.303786900321

Printed in the United States of America

10 9 8 7 6 5 4 3 2 1

To
my daughter
Diane

and my son
Barry

PREFACE

The *Construction Regulations Glossary* is the companion manual to the *Construction Glossary*. In the *Construction Regulations Glossary* the author has endeavored to complete the scope of definitions and meanings for words, phrases, and terms used by land developers, the construction industry, and all related and allied professions. The glossary was written after seven years of research, drawing from the author's experience in design and construction worldwide.

Approximately 500 zoning ordinances, subdivision controls, and building codes, dating from the late 1950s to the present, were used in the compilation of the glossary. Every possible definition for each word, phrase, or term is included in the text so that the reader will be able to compare the style, sentence structure, and nuances of meaning in various definitions. The glossary subjects are arranged in a format to represent most zoning ordinances and building codes.

The research and review conducted during the assembly of the glossary revealed that words that once had simple definitions and short explanations have now become complicated with legal meanings and lengthy, detailed explanations.

Construction regulations include a combination of zoning ordinances, subdivision controls, and building codes. Zoning ordinances have existed for several hundred years and set the pattern for the use and development of land. At the beginning of the twentieth century, it was recognized by the courts that some of the zoning laws which existed had exceeded constitutional authority and had invaded the rights of property owners. To establish fair and equitable use and development of land, the Department of Commerce, under Secretary Herbert Hoover, appointed an advisory committee in September 1921.

After several years of exhaustive study and review, the committee approved for publication a final document titled "A Standard State Zoning Enabling Act" in February 1924. (See the Appendix.) The act established a format that could easily be adopted by the individual states, who could make additions and revisions to suit local area conditions. The original act included nine sections that applied to land use and development in 1924, but since then the new updated zoning ordinances have included such subjects as pollution, flooding, fire safety, sanitation, health, noise abatement, earthquakes, hurricanes, sun shadows, and the ever-pressing problem of automobile parking. Each new subject added to the zoning ordinances requires new terminology and definitions to explain the advances in technology. The act, under the heading "Explanatory Notes in General," includes an item, #7 Definitions: "No definitions are included. The terms in the act are so commonly understood that definitions are unnecessary. Definitions are generally a source of danger. They give words to a restricted meaning. No difficulty will be found with the operation of the act because of the absence of such definitions." However, an examination of the act indicates that there are in fact 47 listed definitions and explanations, demonstrating the need for definitions.

The writing of building codes has occurred almost always because of an event that lead to the loss of life and to the destruction of property. Codes are constantly updated after a new event occurs, and when the Federal government invests in programs for the public, new materials, equipment, systems, and methods are used and are being tested for quality and endurance. Construction regulations are primarily written and enforced to protect the health, safety, and welfare of the general public. Code writing is a never-ending process, and so is the need for the definitions to interpret the regulations.

J. STEWART STEIN

Phoenix, Arizona
April 1983

ACKNOWLEDGMENTS

The author is indebted to thousands of men and women who daily serve the public in the art of code writing activities and to the organizations they represent. These individuals devote countless hours to research, question, and debate before final action is taken that results in the formulation of regulations. Without the experiences, frustrations, and accomplishments of these contributors, it would have been impossible to gather this collection of definitions into a recognizable format. I am indeed grateful to all of them, for their work will serve many people in the construction industry and those involved in the administration of construction regulations.

CONTENTS

Part One

ZONING ORDINANCES

1. CATEGORY A

ZONING REGULATIONS A-1 THROUGH A-19

A-1 BUILDING AND PROPERTY LINES

Abut. To touch or be contiguous. A building abuts upon a street when any part of the building touches or extends to the line of the street.

Abut, abuts, or abutting. The words "abut," "abuts," or "abutting" shall mean having a common lot line or zoning district line.

Abutting. The condition of two adjoining properties having a common property line or boundary, including cases where two or more lots adjoin only a corner or corners, but not including cases where adjoining lots are separated by a street or alley.

Adjacent property line. The dividing line of the lot or property from adjoining property, or of property facing or abutting upon a street, alley or publicway, for the purpose of exposure distance, the center of such street, alley or publicway.

Boundary line of zone lot. Any line separating a zone lot from a street, an alley, another zone lot or any other land not part of the zone lot.

Building line. A line between which line and any street or right-of-way line of a district, lot, tract, or parcel of land, no buildings or parts of buildings may be erected, altered, or maintained.

Building Line. A line established by law or agreement, usually parallel to property line, beyond which a structure may not extend. This generally does not apply to uncovered entrance platforms, terraces and steps.

Building line. A line established by the Zoning Code, generally parallel with and measured from the front lot line, defining the limits of a front yard in which no building or structure may be located above ground except as may be provided in the Code.

Building line. The front line of the building or the legally established line which determines the location of the building with respect to the street lot line.

Building line. That line, between which and the street line, no building or part thereof may be erected, except as provided in the building regulations.

Building line. A line beyond which the foundation wall of a building shall not project, nor beyond which shall a building or part thereof be erected, altered or maintained.

Building line. The line which establishes the minimum depth of front yard for the particular District as measured from the street line.

Building line. A line specified by Ordinance, beyond which the owner of a property may not erect a structure.

Building line. A line established beyond which no part of a building shall project, except as otherwise provided by the ordinance.

Building line. A line between any street right-of-way, either existing or future, and any building, parts of a building or structures which may be erected or altered on a lot, parcel or land or tract.

Building line. The line established by yard or setback requirements outside of which no principal building may be erected.

Building line. The right-of-way line of a public thoroughfare, or other line beyond which the construction of a building is prohibited by law.

Building line. The front line of the building or the legally established line which determines the location of the building with respect to the street line.

Building line. Line established by law, ordinance, or regulation, beyond which no part of a building, other than parts expressly permitted, shall extend.

Building line. The line, established by law, beyond which a building shall not extend, except as specifically provided by law.

Building line. Means the line, beyond which, property owners or others have no legal or vested right to extend a building or any part thereof, without special permission and approval of the proper authorities.

Building line. The line beyond which no building or part thereof may extend without special permission and approval of the proper authorities.

Building line. Is the line between the street line or lot line, which no building or other structure or portion thereof, except as provided in the Code, may be erected above the grade level. The building line is considered a vertical surface intersecting the ground on such line.

Building line. A line formed by the face of the building, and for the purposes of the code, a minimum building line is the same as a front setback line.

Building line. The line parallel to the street line at a distance equal to the depth of the front yard required for the district in which the lot is located.

Building line. The line established by law, beyond which a building shall not extend, except as specifically provided in the Ordinance.

Building line. Any private property line coterminous with a public way; or a building line established by City Ordinance.

Building line or setback line. A line or lines designating the area outside of which buildings may not be erected.

Building line setback. The minimum horizontal distance between the building line and the street or right-of-way line in a district, lot, tract, or parcel of land.

Building perimeter. The real or imaginary line formed on the surface of the ground by the intersection of the outer face of the enclosing walls of a building including porches and galleries whether roofed or not, but excluding stairs, terraces or steps when same are not over five (5) feet high (exclusive of railings and balusters) above established grade.

Building setback line. The distance between the building line and the street line.

Building setback line. Line establishing the minimum allowable distance between the nearest portion of any building, excluding the outermost 3 feet of any uncovered porches, steps, eaves, gutters, and similar construction, and the centerline or the property line of any street when measured perpendicularly thereto.

Building setback line. Line within a property defining the required minimum distance between any structure or building and the adjacent right-of-way line.

Building setback line. Line nearest the street and across a lot, establishing the minimum open space to be provided between buildings and specified structures and street lines.

Building setback line. The minimum horizontal distance between the front line of a building or structure and the right-of-way line.

Building site front. In the case of an interior lot, the side fronting the street; in the case of a triangular corner lot, any frontage and all yards other than the front yard, shall be considered side yards; in the case of a double frontage or through lot, both street frontages; in all other cases, the side selected by the Owner.

Common property line. Line dividing one lot from another when said lots are not of one ownership.

Contiguous. In contact with.

Curb line. The line coincident with the face of the street curb adjacent to the roadway, usually parallel to the property line.

Extended property line. Line, radial or perpendicular to the highway curbline, at each end of the frontage, extending from the right-of-way line to the curbline.

Exterior lot line. Boundary line between a lot and a street, alley, public way or railroad right-of-way.

Front. As applied to building location on a lot, shall mean the distance between lines drawn through the most remote points of the building perimeter, projected at right angles to a frontage space.

Frontage. The width of a lot measured at right angles to the depth on the front or street side of the lot.

Frontage. The property line of a site abutting on a street, other than the side line of a corner lot. Frontage shall be measured as the shortest distance, between the points, at which the side property lines intersect the street property line.

Frontage. That portion of all property abutting on a side of a street between 2 intersecting streets, or the end of such street if it does not meet another.

Frontage. Distance measured along an abutting public street right-of-way.

Frontage. Side of a structure which forms an angle of 45 degrees or less with a specified street and which has obstructed access to the street.

Frontage. All the property abutting on one side of a road or street between 2 intersecting roads, streets or railroads, or all of the property abutting on one side of a road or street, between an intersecting road or street, and the dead end of a road or street.

Frontage. The property line of a site abutting on a street, other than the side line of a corner lot. Frontage shall be measured as the shortest distance, between the points at which the side property lines intersect the street front property line.

Frontage. All the property on one side of a street between intersecting streets (crossing or terminating) measured along the property line of the street, or if the street is dead ended, then all of the property abutting on one side, between an intersecting street and the dead end of the street.

Frontage. All the property abutting on one (1) side of a street between intersecting streets. measured along the street line.

Frontage. That dimension of a lot measured along the front street line thereof, or if said front street line is curved, along the chord of the arc. The shorter street line of a lot abutting on more than one street shall be deemed to be the front street line thereof, regardless of the location of the principal entrance of a building on the lot.

Frontage. All property on one (1) side of a street between property lines measured along the line of the street.

Frontage. All the property on one side of a street between two (2) intersecting streets (crossing or terminating), measured along the line of the street, or if the street is dead ended, then all of the property abutting on one side, between an intersecting street and the dead end of the street.

Frontage. All the property on one side of a street, or place between two intersecting streets, or places measured along the line of the street or place, or if the street or place is terminated without intersecting another street or place, then all of the property abutting on one side, between an intersecting street or place and the terminus of the street or place.

Frontage. The length along the highway right-of-way line of a single property tract between the lateral property lines.

Frontage. All the property abutting on one side of a street between two intersecting streets (crossing or terminating) or if the street is dead ended, then all the property abutting on one side between an intersecting street and the point at which the street dead ends.

Frontage. All the property abutting upon one side of a street between two lot lines measured along the right-of-way line.

Frontage. All the property on one side of a street, between two (2) intersecting streets (crossing or terminating) or if the street is dead-ended, then all of the property abutting on one side, between an intersecting street and the dead end of the street.

Frontage. Means all the property fronting on one side of a street between an intercepting or intersecting street, or between a street and right-of-way, waterway, end of street or city boundary. An intercepting street shall determine only the boundary of the frontage on the side of the street which it intercepts.

Frontage lot. The distance for which the front lot line and the street line are coincident.

Frontage on corner lots and double frontage lots. On lots having frontage on more than one street, the minimum front yard shall be provided for each street in accordance with the provisions of the ordinance.

Frontage street. The distance along a street line from one intersecting street to another or from one intersecting street to the end of a dead-end street.

Front line of a zone lot. Any boundary of a Zone lot parallel to and within 20 feet of the right-of-way of a street or highway.

Front line of building. The line parallel to the street, intersecting the foremost point of the building, excluding steps.

Front line of building. Line of that part of the building nearest the front property line of the lot.

Front lot line. That boundary of a lot which adjoins a street line. On a corner lot it shall be the lot line having the shortest dimension adjoining a street.

Front lot line. That boundary of a lot which abuts a public street, or where no public street exists, abuts a private road.

Front lot line. The front street line; the shortest street line of a corner lot shall be the front lot line; both street lines of through lots shall be deemed front lot lines, except where no set-back requirement is applicable.

Front lot line. The boundary of a lot along a street or right-of-way; and for a corner lot the front lot line shall be the shorter lot boundary along a street or right-of-way. In no case shall a front lot line be less than twenty (20) feet in length.

Front lot line. A lot line which is a street lot line. Any street lot line of a corner lot may be established by the owner as the front lot line.

Front lot line. Where a lot abuts upon only one street, the lot line along such street shall be the front lot line. Where a lot abuts upon more than one street the assessment roll of the municipality shall determine the front lot line.

Front lot line. A line separating an interior lot from a street, or a line separating either the narrower or the wider street frontage of a corner lot from a street at the option of the owner.

Front lot line. In the case of an interior lot, that line separating said lot from the street. In the case of a corner lot, or double frontage lot, the "front lot line" shall mean that line separating said lot from that street which is designated as the front street in the plat and in the application for a zoning compliance permit.

Front lot line. That boundary of a lot which is along an existing or proposed right-of-way. The lesser dimension of the two, along rights-of-way in the case of corner lots.

Front lot line. In the case of an interior lot, a line separating the lot from the street or place; and in the case of a corner lot, a line separating the narrowest street frontage of the lot from the street, except in those cases where the latest tract deed restrictions specify another line as the front lot line.

Front lot line. Lot line separating an interior lot from the street upon which it abuts; or the shortest lot line of a corner lot which abuts upon a street. Unless the context clearly indicatés the contrary, it shall be construed as synonymous with street line.

Front lot line. That boundary of a lot which is along an existing or dedicated public street or where no public street exists, is along a private street or public way. The owner of a corner lot may select either street lot line as the front lot line.

Front lot line. The boundary line between a street and abutting property and in the case of a corner lot, a line separating the narrowest street frontage of the lot from the street.

Front lot line. The line separating the front of the lot from the street. When a lot or building site is bounded by a public street and one or more alleys or private street easements or private streets, the front lot line shall be the nearest right of way line of the public street.

Front lot line. The line abutting a street. On a corner lot, the shorter street line shall be considered the front lot line. Where new street lines are established by Ordinance, such lines shall be the front lot line.

Front lot line. That boundary of a lot which is along an existing or dedicated street. The owner of a corner lot may select either street lot line as the front lot line.

Front of lot. Front boundary line of a lot bordering on the street, and in case of a corner lot, may be either frontage.

Front of lot. Shortest street frontage.

Front of lot. The front boundary line of a lot bordering on the street, and in the case of a corner lot, the narrow portion of said lot will constitute the frontage.

Front open space required. Required open space extending the full width of the lot, and to equal to the required setback line, measured horizontally at right angles to the front lot line.

Front property line. Line dividing abutting property from a public right-of-way.

Front property line of a corner lot. Is the shorter of the 2 lines adjacent to the streets as platted, sub-divided or laid out. Where the lines are equal, the front line shall be that line which is obviously the front by reason of the prevailing custom of the other buildings in the block. If such front is not evident then either may be considered the front of the lot but not both.

Front property line of a corner lot. Shall be the shorter of the two (2) lines adjacent to the streets as originally platted or laid out. Where the lines are equal, the front line shall be that line which is obviously the front by reason of the prevailing customs of the other buildings in the block. If such front is not evident, then either may be considered the front of the lot but not both.

Front property line of an interior lot. Is the line bounding the street frontage.

Front property line of a through lot. That line which is obviously the front by reason of the prevaling custom of the other buildings in the block. Where such front property line is not obviously evident, the Board shall determine the front property line. Such lot over 200 feet deep shall be considered, for the purpose of this definition, as two lots each with its own frontage.

Future width lines. "Future width lines" means lines established adjacent to highways or streets for the purpose of defining limits within which no structure nor any part thereof shall be erected or maintained in order to insure the future acquisition of these limits as public rights-of-way.

Inside property line. A property line dividing lots (whether owned by the same person or not) or abutting a public way which is less in width than the separation required from such inside property line.

Interior lot line. A lot line other than a street line.

Interior lot line. Any lot line other than one adjoining a street or public space.

Interior lot line. Any lot line dividing one lot from another.

Interior lot line. Boundary line between a lot and an adjoining lot.

Interior lot line. Synonymous with common property line.

Interior lot line. Any lot line other than a street lot line.

Linear frontage. The length of the front lot line.

Line of building. The boundary of any side of a building, or projection therefrom, at its outermost vertical plane, excluding only such parts of the building as are located in and permitted as obstructions in one (1) or more required yards as described in the Ordinance.

Lot (front). The front boundary line of a lot bordering on a street; in the case of a corner lot, the shortest front boundary line.

Lot front. That portion of an interior lot between the front property line and the midline of the parallel to the front property line.

Lot frontage. That portion of a lot which abuts a public street. Each side of a lot so abutting a public street shall be considered as separate lot frontage.

Lot frontage. Distance for which the front lot line and the street line are coincident.

Lot line. Any boundary line of a lot on a legally recorded plat.

Lot line. A line of demarcation between a lot and a street or other public space or between two (2) lots.

Lot line. The boundary of a lot that separates it from adjoining land.

Lot line. A legally established line dividing one lot, plot of land or parcel of land from an adjoining lot or plot of land or parcel of land.

Lot line. Line which marks the boundary of a lot.

Lot line. The lot line or right-of-way line dividing a lot from a street or other public space.

Lot line. Any boundary line of a lot other than a street line.

Lot line. A line dividing one lot from another, or from a street or any public place.

Lot line. The boundary of a lot separating it from adjoining public, common or private land, including a public street.

Lot line. The property line, abutting right-of-way line, or any line defining the exact location and boundary of a lot.

Lot line. A line dividing one lot from another or from a street or other public place. In case any part of a lot lies adjacent to a public right-of-way or easement, the lot line of that part shall be the right-of-way line or easement boundary crossing such part.

Lot line. A property boundary line of any lot held in single and separate ownership, except that, in the case of any lot abutting a street, the lot line for such portion of the lot as abuts such street shall be deemed to be the same as the street line, and shall not be the centerline of the street, or any other line within the street lines even though such may be the deeded property boundary line.

Lot line. A line dividing one lot from another lot or from a street or alley.

Lot line. A line dividing one land unit from another, or from a street or other public space. A boundary line of a zoning lot.

Lot line. Line of demarcation between either public or private properties. A party line is the lot line between adjoining properties. When a lot line abuts on a street, avenue, park or other public property except an alley, it shall be known as a street line; when a lot line abuts on an alley it shall be known as an alley-line.

Lot line. Line dividing one lot from another, or from a street or other public space.

Lot line. Line dividing one premises from another, or from a street or other public space.

Lot line. Legally defined line dividing one parcel of property from another.

Lot line. Boundary of a lot separating it from adjoining land, public or private. It may or may not be the boundary line of a lot in a recorded subdivision.

Lot line. A property boundary line of any lot held in single or separate ownership, except that, where any portion of the lot extends into the abutting street or alley, the lot line shall be deemed to be the street or alley line.

Lot line. Line dividing one premise from another, or from a street or other public place.

Lot line. A line dividing one land unit from another, or from a street or other public space. A boundary line of a lot.

Lot line (rear). That boundary of a lot which is most distant from, and is or is approximately parallel to the front lot line. If the rear lot line is less than ten feet in length, or if the lot forms a point at the rear, the rear lot line shall be deemed to be a line ten feet in length with the lot, parallel to and at the maximum distance from the front lot line.

Lot line (side). Any boundary of a lot which is not a front or rear lot line.

Lot lines. Property lines bounding the lot. Where a lot abuts on a street or is partly contained in a street, the lot line shall be deemed to be a line drawn parallel to the center line of the street and thirty (30) feet distant therefrom, measured at right angles thereto, but the actual street line shall be construed as the lot line if such street line is established, and lies at a greater distance than thirty (30) feet from the center of such street.

Lot lines. The lines bounding a lot.

Lot lines. The property lines bounding the lot.

Lot lines. The lines outlining the boundaries of a lot or parcel of land.

Lot lines. The property lines bounding a lot or area.

Lot line wall. Wall adjoining and parallel to the lot line used primarily by the party upon whose lot the wall is located. Lot line walls may share common foundations.

Property. Lot or plot, including all buildings and improvements thereon.

Property line. Line establishing the boundaries of premises.

Property line. Demarcation limit of a lot dividing it from other lots or parcels of land.

Property line. Recorded boundary of a plot.

Property line. Legally established line dividing one lot, plot of land or parcel of land under one ownership from an adjoining lot or plot of land or parcel of land under another ownership.

Property line. A line separating parcels of real property having separate legal descriptions, but not including a building line.

Rear line. Of a lot is a dividing line between two (2) lots or between a lot and an alley or easement; provided it runs parallel to the shorter dimension of said lot. Any bounding line of a lot, which is not herein defined as a rear line or street line shall be deemed a side line.

Rear line of a zone lot. (1) If the zone lot has one front line, the boundary opposite that front line shall be the rear line; however, if there be in addition an alley abutting the zone lot and the boundary line opposite the front line and the alley are not coincidental, the rear line shall be fixed by the Zoning Administrator; (2) If the zone lot has two front lines the boundary opposite the shorter front line shall be the rear line; provided, however, if there be in addition an alley abutting the zone lot and the boundary line opposite the shorter front line and the alley are not coincidental, or if the zone lot is a corner zone lot not abutted by an alley on a block oblong in shape, and the longer dimensions of the lot coincide in direction with the longer dimensions of the block, or if the zone lot has two front lines of equal length, the rear line shall be fixed by the Zoning Administrator; (3) If the zone lot has two or more abutting alleys the rear line shall be fixed by the Zoning Administrator; (4) If the zone lot has three or more front lines there shall be no rear line; (5) Whenever the Zoning Administrator is required to fix the location of the rear line of a zone lot, he shall do so on the basis of the orientation of existing structures on the zone lot and on adjacent properties, so as best to protect such structures and permitted uses of such properties from any offensive effects of accessory structures and uses which may, under the Ordinance, be permitted in the rear portion of such zone lot.

Rear lines. Lines adjoining lots or alleys in the rear.

Rear lot line. The lot line that is most distant from and that is, or most nearly parallel to the front lot line.

Rear lot line. That boundary of a lot which is most distant from, and is or is most nearly parallel to the front lot line.

Rear lot line. The boundary of a lot at the opposite end of the lot from the front lot line.

Rear lot line. That boundary of a lot which is most distant from and most nearly parallel to the front lot line, or in the case of an irregular, triangular or gore-shaped lot, a line ten (10) feet long within the lot, parallel to and at maximum distance from the front lot line.

Rear lot line. That boundary of a lot which is most distant from and is, or is approximately, parallel to the front lot line. If the rear lot line is less than ten (10) feet in length, or if the lot forms a point at the rear, the rear lot line shall be deemed to be a line ten (10) feet in length within the lot, parallel to and at the maximum distance from the front lot line.

Rear lot line. A lot line which is opposite and most distant from the front lot line and, in the case of an irregular, triangular or gore-shaped lot, a line 10 feet in length within the lot, parallel to and at the maximum distance from the front lot line.

Rear lot line. Lot line that is generally opposite the lot line along the frontage of the lot; if the rear lot line is less than ten (10) feet in length, or if the lot comes to a point at the rear, the rear lot line shall be deemed to be a line parallel to the front lot line, not less than ten feet long; lying wholly within the lot, and farthest from the front lot line.

Rear lot line. Any lot line, except a street line, that is parallel or within 45 degrees of being parallel to, and does not intersect any street line bounding such lot.

Rear lot line. A lot line, not a front or side lot line, which is generally opposite the front lot line, and is not necessarily a straight line.

Rear lot line. The lot line opposite and most distant from the front lot line; in the case of irregularly shaped lot, such lot line shall be an imaginary line parallel to the front lot line, but is less than ten (10) feet long and measured wholly within said lot.

Rear lot line. The boundary opposite and most distant from the front lot line. The rear lot line of any irregular or triangular lot shall be a line not less than twenty (20) feet, lying wholly within the lot and parallel to the most distant from the front lot line.

Rear lot line. The record lot line or lines most distant from and generally opposite the front lot line, except that in the case of an interior triangular or gore-shaped lot, it shall mean a straight line ten (10) feet in length which (a) is parallel to the front lot line or its chord and (b) intersects the two other lot lines at points most distant from the front lot line.

Rear lot line. A lot line which is opposite and most distant from the front lot line.

Rear lot line. That boundary of a lot most distant from and most nearly parallel to the front line.

Rear lot line. A lot line parallel or within forty-five degrees of being parallel to the front lot line.

Rear lot line. For a triangular shaped lot line ten (10) feet in length within the lot and farthest removed from the front lot line and at right angles to the lot depth line.

Rear lot line for a pentagonal lot. Rear boundary of which includes an angle formed by two (2) lines, such angle shall be employed for determining the rear lot line in the same manner as prescribed for a triangular lot.

Rear lot line for a trapezodial lot. Rear line of lot which is not parallel to the front lot line, the rear lot line shall be considered to be a line at right angles to the lot depth line, and drawn through a point bisecting the recorded rear lot line.

Rear of building. The side of a building most nearly parallel with and adjacent to the rear lot line.

Rear open space required. Required open space extending the full width of the lot and to equal to the required setback line, measured horizontally at right angles to the rear.

Rear property line. That lot line opposite to the front property line.

Rear property line of a lot. That lot line opposite to the front property line. Where the side property lines of a lot meet in a point, the rear property line shall be assumed to be a line not less than ten (10) feet long, lying within the lot and parallel to the front property line. In the event that the front property line is a curved line then the rear property line shall be assumed to be a line not less than ten (10) feet long, lying within the lot and parallel to a line tangent to the front property line at its midpoint.

Residence lot frontage. Front boundary line of a lot that abuts on a street.

Setback. The minimum horizontal distance between the street line and the front line of the building. Where an official building line has been established, the normal setback for that area or the building line, whichever provides the greater setback from the center of the street, shall be used.

Setback. The distance required to obtain the front, side, or rear yard open space provisions of the Ordinance.

Setback. Minimum horizontal distance between the street line or the centerline of a highway and the nearest point of building or any projection thereof, excluding uncovered steps.

Setback. Lateral distance between the right-of-way line and the roadside structure, such as a building, pump island, display stand or other object.

Setback. Required distance between a structure and a property line or another feature of the lot.

Setback. The minimum horizontal distance between the street line and the nearest line of the building.

Setback. The minimum horizontal distance between a building and the street or lot line, disregarding encroachments permitted by the Ordinance.

Setback. The minimum horizontal distance between the front line or side line of the building and the street line, disregarding steps and unroofed porches.

Setback. The least horizontal distance from any existing or proposed building or structure to the nearest point in an indicated lot line or street line.

Setback building line. Is a building line back of the street, alley, bay, beach or side line.

Setback building line. Line established by Ordinance or official regulation for a specified distance from any lot line.

Setback building line. A building line back of the street line.

Setback line. Line drawn through the point of a building nearest to the street or lot line from which, the setback line is measured and lying parallel thereto.

Setback line. A line determined by measurement perpendicular to a property line, establishing a line beyond which no building or structure may be erected in the area thus created between a property line or right-of-way and the setback line.

Setback line or future right-of-way line. The line between which line and a street line no structure shall project, be erected, or placed.

Setback line required. Line marking the setback distance from the street or lot lines, which furnishes the minimum required front, side or rear open space of a lot.

Setback lines. Lines established along highways at specified distances from the center line, which prohibited building or structures shall be setback or outside of, and within which they may not be placed, except as hereinafter provided. "Within the setback line" means the setback line of highway.

Setback of building line. Line within a property defining the required minimum distance between any building and the adjacent right-of-way.

Setbacks. Refers to the open space between the property line or public thoroughfare and the nearest part of the building. Unenclosed terraces, slabs, or stoops without roofs or walls may project into this open space or setback.

Setbacks. A setback of a building is an open unoccupied space between the front or side street lot line and extending from such lot line to the face of a building wall or the face of a projecting bay, porch, piers or abutment.

Setbacks. No building shall hereafter be erected and no existing building shall be reconstructed or altered in such a way that any portion thereof shall be closer to the street line than twenty-five (25) feet. Along the side line of a corner lot, the minimum setback shall be twenty (20) percent of the width of the lot up to a maximum of fifteen (15) feet; provided, however, that no building permit shall be issued for a building or structure on a corner lot, when said building or structure is to be oriented in such a manner as to reduce the setback requirements on the street on which said corner lot as originally platted fronts.

Side line. Of a lot is a dividing line between two lots or a lot and a street; provided, it runs parallel to the longer dimensions of said lot.

Side lines. Are those lines between lots facing the same street or alley.

Side lot line. Any lot line which is not a front lot line or a rear lot line; a lot line separating from a side street is an exterior side lot line, while a lot line separating a lot from another lot, or lots, is an interior side lot line.

Side lot line. Any lot boundary line not a front lot line or a rear lot line.

Side lot line. Any lot line which is not a front lot line or a rear lot line.

Side lot line. Any boundary of a lot not a front or rear lot line.

Side lot line. Any lot line other than a front or rear lot line.

Side lot line. A lot line which is neither a front nor a rear lot line.

Side lot line. Any lot boundary not a front line nor a rear lot line. A side lot line separating a lot from another lot or lots is an interior lot line. A side lot line separating a lot from a street is a side street lot line.

Side lot line. Any lot line not a front line or a rear line. The "side line of a corner lot" is the line along the narrower or less important street, unless the lot is a sublot in a recorded subdivision and is clearly intended to front on the narrower street.

Side lot line. Any boundary of a lot which is not a front or rear lot line. On a corner lot, a side lot line may be a street lot line.

Side lot line. That boundary of a lot which is not a front lot line or a rear lot line.

Side lot line. Any lot line that is not a street line or a rear lot line.

Side property lines of a lot. Are those lot lines connecting the front and rear property lines of a lot.

Street line. A lot line separating a street from other land.

A-2 ENVIRONMENTAL STANDARDS

Air contaminant. The emission of an air contaminant into the open air such as smoke, soot, fly ash, dust, cinders, dirt, noxious or obnoxious acids, fumes, oxides, gases, vapors, odors, toxic or radioactive substances, waste, particulate, solid, liquid or gaseous matter, or any other materials in such place, manner or concentration as to cause injury, detriment, nuisance, or annoyance to the public, or to endanger the health, safety or welfare of the public, or as to cause or have a tendency to cause injury or damage to business or property.

Air pollutants. Matter in the air capable of creating or causing air pollution. Such matter may originate from any kind of combustion process, or industrial or laboratory processes, both chemical and physical, and may appear as, but is not limited to smoke, dusts, fumes, droplets, mists, vapors, gases, odors or a combination of them.

Air pollution. That condition of the air which results from the presence in the air of air pollutants in concentrations which may adversely affect the well-being of an individual or cause damage to property, animal or plant life.

Air pollution. The practice of allowing quantities of films, gases, vapors, dust, odors, etc., to escape to the open air in such a manner or quantity as to be detrimental or a nuisance to the public or to damage business or property.

Air quality standard. Ambient air quality goal established for the purpose of protecting the public health and welfare.

Animals and birds. The keeping of any animal or bird which by causing frequent or long-continued noise disturbs the comfort and repose of any person in the vicinity.

Asbestos. A non-inflammable natural mineral fiber.

Ashes. The residue from fires used for cooking foods and heating buildings. Ashes shall also include such floor sweepings as may accumulate in connection with the ordinary daily use of dwellings and stores. Ashes shall also include the bottles, glass, crockery and tin cans prohibited in the Code.

Attenuation. Sound reduction process in which sound energy is absorbed or diminished in intensity as a result of energy conversion from sound to motion or heat.

Blasting agent. Any material or mixture consisting of a fuel and oxidizer intended for blasting, not otherwise classified as an explosive, in which none of the ingredients are classified as explosives, provided that the finished product, as mixed and packaged for use or shipment, cannot be detonated by means of a No. 8 test blasting cap when unconfined. Materials or mixtures classified as nitro carbo nitrates by Department of Transportation regulations shall be included in this definition.

Carbon dioxide (CO_2). The colorless, odorless, electrically nonconductive inert gas.

Catalytic combustion system. An oven heater or any construction that employs catalysts to accelerate oxidization or combustion of fuel air or fume air mixtures for eventual release of heat to an oven process.

Chimneys. Shafts which contain one or more flues for transmitting the products of combustion from a stove, fireplace or heating device within the building to the outer air.

Coke. The solid fuel obtained by the carbonization of coal or the solid residue of petroleum product manufacture.

Cold boiler or furnace. A boiler or furnace in which fuel has not been consumed for a period of 24 hours or more.

Combustible rubbish. Rags, old clothes, leather, rubber, carpets, wood, excelsior, sawdust, tree branches, yard trimmings, wood furniture and other combustible solids not considered by the city to be of a highly volatile or explosive nature.

Combustible trade waste. Paper, rags, leather, rubber, cartons, boxes free of wire and metal scraps, wood excelsior, sawdust, garbage and other combustible solids except manure, and not considered by the city to be of a highly volatile or explosive nature.

Combustible waste matter. Magazines, books, trimmings from lawns, trees, or flower gardens, leaves, pasteboard boxes, rags, paper, straw, sawdust, packing material, shavings, boxes and all rubbish and refuse that will ignite through contact with flames or ordinary temperatures.

Combustion. A chemical process that involves oxidation sufficient to produce light or heat.

Corrosive liquids. Includes those acids, alkaline caustic liquids and other corrosive liquids which, when in contact with living tissue, will cause damage to such tissue by chemical action; or are liable to cause fire when in contact with organic matter or with certain chemicals.

Defect in vehicle or noisy load. The use of any automobile, motorcycle, street car or other vehicle so out of repair or loaded in such a manner, as to create loud or unnecessary grating, grinding, rattling or other noise.

Discrete pulses. Pulses which do not exceed one hundred (100) impulses per minute.

Dual-use fallout shelter. A dual-use fallout shelter is a space having a normal, routine use and occupancy as well as having an emergency use as a fallout shelter.

Dump. An area devoted to the disposal of refuse, including incineration, reduction, or dumping of ashes, garbage, combustible or noncombustible garbage or refuse, offal or dead animals.

Dump. Lot or land or part thereof used primarily for disposal by abandonment, dumping, burial, burning or any other means and for whatever purpose, of garbage, offal, sewage, trash, refuse, junk, discarded machinery, vehicles or parts thereof, or waste material of any kind.

Dump. Any lot or parcel of land on which garbage, refuse, trash or junk is deposited.

Dump. A place used for the disposal, by abandonment, discarding, dumping, reduction, burial, incineration or by other means of garbage, sewage, trash, refuse, waste material offal or dead animals.

Dust. Air-borne solid particles, fly-ash, cinders, soot and all other solid particles.

Dust. Particulate matter released into the air by natural forces, or by fuel-burning, combustion or process equipment or device, or by any other causes or work.

Dust. Processes that produce dust, lint, shavings, sawdust, or other fine particles of matter liable to spontaneous ignition or explosion; such as results from sawing, grinding, pulverizing, buffing, polishing and similar operations; and materials in dust form; such as sulphur, metal powders, and powdered plastics.

Dust free. Shall mean the property is maintained dust free by one (1) of the following methods: (1) asphaltic concrete, (2) cement concrete, (3) oil cake.

Dust free. Property which is maintained dust free by paving with one of the following methods; asphaltic concrete, cement concrete, penetration treatment of bituminous material and a seal coat of bituminous binder and mineral aggregate, or the equivalent of these methods.

Dust separating equipment. Any device for separating the solid products of any combustion process, i.e., dust, solids, particulate matter, fly-ash, or any combination thereof, from the gases in which they are carried.

Explosive. A chemical compound or mechanical mixture, that is commonly used or intended for the purpose of producing an explosion, that contains any oxidizing and combustible units, or other ingredients, in such proportions, quantities, or packing, that an ignition by fire, by friction, by concussion, by percussion or by detonator of any part of the compound or mixture may cause such a sudden generation of highly heated gases that the resultant gaseous pressures are capable of producing destructive effects on contiguous objects or of destroying life and limb.

Fallout shelter. A fallout shelter is any room, structure or space designated as such and providing its occupants with protection at a minimum protection factor of forty (40) from gamma radiation from fallout from a nuclear explosion as determined by an architect or engineer certified by the Office of Civil Defense as a Qualified Fallout Shelter Analyst.

Fly-ash. Solid particles resulting from combustion or incomplete combustion of coal, wood, or other solid fuels. Fly-ash does not include process materials.

Fly ash. Particulate matter capable of being gas-borne or air-borne and consisting essentially of fused ash and burned or unburned material.

Fuel. Any form of combustible matter such as liquid, solid, vapor or gas.

Fumigant. Includes any substance which by itself or in combination with any other substance emits or liberates a gas, fume or vapor used for the destruction or control of insects, fungi, vermin, germs, rodents, or other pests, and shall be distinguished from insecticides and disinfectants which are essentially effective in the solid or liquid phases. Examples are methyl bromide, ethylene dibromide, hydrogen cyanide, carbon disulphide and sulfuryl fluoride.

Garbage. The meat and vegetable waste solids resulting from the handling, preparation, cooking and consumption of foods. Garbage shall be considered to originate primarily in kitchens, stores, markets, restaurants, hotels and other places where food is stored, cooked and consumed.

Garbage. Animal and vegetable matter originating in houses, kitchens, restaurants and hotels, produce markets, etc.

Garbage. Putrescible animal and vegetable wastes.

Garbage. Solid wastes from the preparation, cooking and dispensing of food, and from the handling, storage and sale of produce.

Glare. Excessive illumination. Flickering or intense sources of light shall be controlled or shielded so as not to cross lot lines.

Glaring light near streets. Use of spotlights and other directional types of lights to illuminate buildings, displays or signs shall not be permitted unless the lights are directed or shielded in such a manner that they do not cause glare or other annoyance to pedestrians and drivers of vehicles on near-by streets and roads.

Horns—signaling devices and the like. The sounding of any horn or signal device on any automobile, motorcycle, bus, street car or other vehicle while not in motion, except as a danger signal that another vehicle is approaching apparently out of control; or if in motion, only as a danger signal or where the motor vehicle statutes requires the sounding of such horn or signal device.

Hurricane requirements. Lateral support securely anchored to all walls provides the best, and only sound structural stability against horizontal thrusts, such as winds of exceptional velocity.

Impact noise. A short-duration sound which is incapable of being accurately measured on a sound level meter.

Impact vibration. Vibrations occuring in discrete pulses separated by an interval of at least one (1) minute and numbering no more than eight (8) per twenty-four (24) hour period.

Impulse. Discrete vibration pulsation occurring no more often than one (1) per second.

Incineration. The burning of refuse or any other material.

Industrial waste. Liquid, gaseous or solid substances or a combination thereof, resulting from any process of industry, manufacturing, trade or business, or from the development or recovery of any natural resource.

Minimum landscaped open space. The percentage of lot area which must be maintained in grass or other living vegetation.

Noise. A disturbing sound of any kind caused by any circumstance.

Noises to attract attention. The use of any drum, loudspeaker, bell or other instrument or device for the purpose of attracting, by creation of such noise, to any performance, show or sale or display of merchandise, except where the director of police has given permission for the use of same on a given occasion, at a given place, which permission shall limit such uses to a period of not exceeding 48 hours.

Noncombustible rubbish. Metals, metal shavings, tin cans, glass, crockery and other similar materials, but not the wastes resulting from building construction or alteration work. It shall also include any small accumulation of cellar or yard dirt if stored as required in the Code.

Noncombustible trade waste. Metals, metal shavings, wire, tin cans, cinders, earth and other materials, but not the wastes resulting from building construction or alteration work.

Noxious matter. Material which is capable of causing injury or malaise to living organisms by chemical reaction, or is capable of causing detrimental effects upon the health, or the psychological, social or economic well-being of human beings.

Octave band. A prescribed interval of sound frequencies which classifies sound according to pitch.

Odor. Emission of noxious, odorous matter in such quantities as to be readily detectable at any point along lot lines, when diluted in the ratio of one volume of odorous air to four or more volumes of clean air or as to produce a public nuisance or hazard beyond lot lines is prohibited.

Odor nuisance. Any noxious odor in sufficient quantities and of such characteristics and duration as to be injurious to human, plant, or animal life, to health, or to property, or to unreasonably interfere with the enjoyment of life or property.

Odorous matter. Any material that produces an olfactory response among human beings.

Odor threshold. The lowest concentration of odorous matter in the air which will produce an olfactory response in a human being. Odor thresholds shall be determined in accordance with ASTM Method D-1391, "Standard Method for Measurement of Odor in Atmospheres (Dilution Method)."

Odor threshold. The concentration of odorous matter in the atmosphere necessary to be perceptible to the olfactory nerve of normal persons.

Open fire. Any fire wherein the products of combustion are emitted into the open air and are not directed thereto through a stack, chimney or flue.

Paper. Newspapers, periodicals, cardboard and all other wastepaper.

Particulate matter. Material other than water which is suspended in or discharged into the atmosphere in a finely-divided form as liquid or solid at outdoor ambient conditions.

Particulate matter. Finely divided liquid or solid material which is often but not always suspended in air or other gases at atmospheric temperature or pressure.

Performance standard. A criteria established to control noise, vibration, smoke and particulate matter, toxic matter, odorous matter, fire and explosive hazards, and glare and radiation hazards generated by or inherent in uses of land or structures.

Petroleum coke. The residue of various petroleum processes which may be handled and burned as a solid fuel.

Poisonous gas. Any noxious gas of such nature that a small amount of the gas in air is dangerous to life. Examples are chlorine, cyanogen, fluorine, hydrogen cyanide, nitric oxide, nitrogen tetraoxide and phosgene.

Poisonous gases or liquids. Includes gases which are highly poisonous even when present in the air in very small proportions; also liquids which give off highly poisonous vapors at ordinary temperatures.

Portable equipment. Equipment designed for the purpose of being readily transferred from one location to another.

Potable water. Water free from impurities present in amounts sufficient to cause disease or harmful physiological effects. Its bacteriological and chemical quality shall conform to the requirements of the Department of Health.

Power boiler. A boiler carrying more than 15 pounds per square inch (gauge) steam and of more than 10 boiler horse power.

Process furnace. Any furnace, kiln, still or combustion device, other than a boiler furnace used for the generation of heat or power.

Processing. Any operation changing the nature of material or materials such as the chemical composition or physical qualities. Does not include operations described as fabrication.

Processing of fuel. The washing, cleaning, screening drying and pulverizing, flotation, coking, carbonization, quenching, briquetting, bagging and packaging of solid fuels; the refining of liquid fuels; and the manufacture of gaseous fuels.

Pyrophoric dust. A dust in a finely-divided state that is spontaneously combustible in air.

Radiation hazards. The deleterious and harmful effects of all ionizing radiation, which shall include all radiation capable of producing ions in their passage through matter. Such radiations shall include, but are not limited to, electromagnetic radiations such as x-rays and gamma rays and particulate radiations such as electrons or beta particles, protons, neutrons and alpha particles.

Radios, TVs, phonographs, etc. The playing or permitting the playing of any radio, television set, phonograph, musical instrument or machine or device for the production or reproducing of sound, in such a manner or with such volume as to unreasonably annoy or disturb the quiet, comfort or repose of persons in any dwelling, hotel or any other type of residence particularly between the hours of midnight, and 8:a.m.

Refuse. Garbage, paper, rubbish, ashes, and trade waste. All as defined in the Ordinances.

Refuse burning equipment. Any destructor, incinerator, furnace, oven or other apparatus and appurtenances thereto used primarily for the purpose of destroying, reducing or consuming refuse as herein defined, or any other material by combustion. This shall also include crematories.

Repair. Any work which requires the equipment to be wholly or partially dismantled and which results in the restoration of the equipment to its original state.

Residual oil. Fuel oil having a viscosity heavier than 125 seconds Saybolt Universal at 100 degrees F., referred to as grades numbered 5 and 6 in Commercial Standard CS-12 U.S. Department of Commerce.

Ringlemann chart. The chart described in the U.S. Bureau of Mines Information Circular 7718, on which are illustrated graduated shades of gray for use in estimating the light obscuring capacity of smoke-smoke intensity.

Ringlemann number. The number of the area on the Ringlemann Chart that most nearly coincides with the visual density of emission or light-obscuring capacity of smoke.

Rubbish material. All miscellaneous matter, such as bottles, rags, mattresses, worn out furniture, old clothes, old shoes, broken glass, leather, carpets, crockery, metal, rubber, cut grass, leaves, tree branches, lumber or any materials that may accumulate as the result of building operations.

Salvage operation. Any business, trade or industry engaged in whole or in part in salvaging or reclaiming any product or material, including, but not limited to, metals, chemicals, shipping containers or drums.

Sealed source. A quantity of radiation so enclosed as to prevent the escape of any radioactive material, but at the same time permitting radiation to come out for use.

Septic tank. A watertight settling tank in which solid sewage is decomposed by natural bacterial action.

Serious hazard. Hazard of considerable consequence to safety or health through the design, location, construction, or equipment of a building, or the condition thereof, which hazard has been established through experience to be of certain or probably consequence, or which can be determined to be, or which is obviously such a hazard.

Sewage. Any liquid waste containing animal or vegetable matter in suspension or solution, and may include liquids containing chemicals in solution.

Smoke. The product resulting from the incomplete combustion of fuel or other burnable materials and is composed chiefly of finely-divided particles of unburned carbon. Included, but not limited to, in varying amounts, are other constituents such as tarry compounds, sulphur compounds, carbon dioxide, carbon monoxide and fine ash containing silica and iron compounds.

Smoke. The visible discharge from a chimney, stack, vent, exhaust or combustion process which is made up of particulate matter.

Smoke oven. Any piece of equipment which is used for smoking food products.

Smoke unit. The number obtained when the smoke density in the Ringlemann Number is multiplied by the time of emission in minutes.

Solid fuel. Any fuel that is not liquid or gaseous.

Soot. Means a dark substance, essentially carbon, resulting from the burning or heating of coal, wood, oil or other fuels, and burnable materials.

Sound level. The intensity of sound of an operation or use as measured in decibels.

Sound level meter. An instrument standardized by the American National Standards Institute for the measurement of the intensity of sound.

Steam whistles. The blowing of any steam whistle attached to any stationary boiler except to give notice of the time to begin, stop work or as a danger warning, is not permitted.

Thermal insecticidal fogging. The use of insecticidal liquids which are passed through thermal fog-generating units where they are, by means of heat, pressure and turbulence, transformed and discharged in the form of a fog or mist that is blown into an area to be treated.

Three-component measuring system. Instruments which measure simultaneously earthborn vibrations in horizontal and vertical planes.

Toxic matter or material. Those matters or materials which are capable of causing injury to living organisms by chemical means.

Trade waste. All material resulting from the prosecution of any business, trade or industry, conducted for profit. Trade waste shall be classified as combustible trade waste and noncombustible trade waste, as defined in the Code.

Vapor. Any material in a gaseous state which is formed from a substance, usually a liquid, by increase in temperature or release of pressure.

Vibration. The periodic displacement, measured in inches, of earth at designated frequency = cycles per second.

Vibration. All machinery shall be so mounted and operated as to prevent transmission of ground vibration exceeding a displacement of three thousandths (0.003) of one (1) inch measured anywhere outside the lot line of its source, or ground vibration which can be readily perceived by a person standing anywhere outside the lot lines of its source.

Volatile matter. The gaseous constituent of fuels as determined by standards of American Society for Testing and Materials.

A-3 FAMILY, PERSON, INDIVIDUAL AND OWNER

Agent. Any person who shall have charge, care or control of any building as owner, or agent of the owner, or as executor, executrix, administrator, administratrix, trustee or guardian of the estate of the owner. Any such person representing the actual owner shall be bound to comply with the provisions of the Code to the same extent as if he were the Owner.

Agent. Any person other than the subdivider who, acting for the subdivider, submits subdivision plans to the authorities for the purpose of obtaining approval thereof.

Agent of owner. Agent of owner shall mean any person who can show certified written proof that he is acting for the property owner.

Ambulatory person. One able to walk and physically able, when not restrained to leave the premises.

Disabled person. Any person who by reason of his or her physical or mental condition is not sufficiently ambulatory, or otherwise by reason of physical or mental incapability unable to reach or use the means of egress most accessible to his or her living quarters without assistance and with reasonable facility, or who is unable to attend to his or her daily personal and bodily needs.

Domestic servant. Person who lives in the family of another, paying no rent for such occupancy and paying no part of the cost of utilities therefor performing household duties and working solely within the house for the upkeep thereof and for the care and comfort and convenience of the family and occupants thereof.

Emancipated minor. Any person under the age of 21 who is gainfully employed and who is self-supporting or who is married to a spouse who is gainfully employed and who supports the said minor, or who is a student living away from home and in regular attendance in an institution of higher learning.

Family. One or more persons occupying a dwelling unit and living as a single housekeeping unit, whether or not related to each other by birth, adoption, or marriage, but no related group shall consist of more than five (5) persons, as distinguished from a group occupying a boarding house, lodging house, or similar group living quarters as defined in the Ordinance.

Family. Consists of one or more persons each related to the other by blood, marriage, or adoption, who are living together in a single dwelling and maintaining a common household. A "family" includes any domestic servants and not more than one gratuitous guest residing with such "family."

Family. One (1) person, his or her spouse, their offspring, legally adopted children or foster children.

Family. Any number of persons living together under one head as a single housekeeping unit, whether related to each other legally or not; and shall be deemed to include servants, but shall not include paying guests.

Family. Constitutes a man and wife or a father or a mother and their children by natural birth or adoption and the parents of either or both and may include 2 additional persons who occupy rooms for which compensation may or may not be paid; provided any group of persons, not so related, but inhabiting a single housekeeping unit, shall be considered to constitute one family for each 5 persons, exclusive of domestic employees contained in each group.

Family. One or more persons related by blood, marriage or adoption but not including persons other than the following: (a) nominal head of the household; (b) spouse of the nominal head of the household; (c) children or adopted children of the nominal head of the household or of the spouse of the nominal head of the household; (d) three (3) other additional persons related by blood, marriage or adoption to any of the persons enumerated in (a), (b) or (c).

Family. Any number of individuals living together as a single housekeeping unit, including domestic servants for whom, subject to the provisions of the Ordinance, separate living quarters may be provided, but excluding fraternities, sororities, residential clubs, homes of an institutional character, and housing for the elderly.

Family. One (1) or more persons immediately related by blood, marriage or adoption and living as a single housekeeping unit in a dwelling shall constitute a family. A family may include, in addition thereto, two (2) but not more than two (2) persons not related by blood, marriage or adoption.

Family. Individual or two (2) or more persons related by blood or marriage, or foster children living with family, or a group of not more than five (5) persons (excluding servants) who need not be related by blood or marriage living together in a dwelling unit.

Family. One (1) or more persons occupying a premises and living as a single housekeeping unit, as distinguished from a group occupying a boardinghouse, lodging house, hotel, club, fraternity or sorority house. A family shall be deemed to include servants.

Family. Family is an individual or two (2) or more persons related by blood, marriage, or adoption and usual servants, living together as a single housekeeping unit in a dwelling unit, or not more than three (3) persons, who need not be related, living together as a single, non-profit, housekeeping unit in a dwelling unit.

Family. One (1) or more persons occupying a single living unit. Such persons do not have to be related by birth or marriage to constitute a family unit.

Family. Household constituting a single housekeeping unit occupied by one (1) or more persons.

Family. One (1) or more persons related by blood or marriage or not more than three (3) persons not related by blood or marriage, living together as a single housekeeping unit.

Family. One (1) or more persons living together, whether related to each other by birth or not, and having common housekeeping facilities.

Family. Individual, or two (2) or more persons related by blood or marriage or legal adoption, or group of not more than five (5) persons (excluding servants) who need not be related by blood or marriage, but living together and subsisting in common as a separate non-profit housekeeping unit which provides only one (1) kitchen.

Family. Body of persons who live together in one (1) dwelling unit as a single housekeeping entity.

Family. One or more persons living privately as a single house-keeping unit and using cooking facilities in common.

Family. One or more persons occupying a dwelling unit and living as a single non-profit housekeeping unit, not including a group of five or more persons who are not within the second degree of kinship.

Family. One or more persons occupying a premise and living as a single housekeeping unit and using common cooking facilities, as distinguished from a group occupying a hotel, motel or club.

Family. An individual or two or more persons related by blood, marriage, or adoption and usual servants, living together as a single housekeeping unit in a dwelling unit, or not more than three persons, who need not be related, living together as a single, non-profit, housekeeping unit in a dwelling unit.

Family. An individual or two or more persons related by blood or marriage, or foster children living with a family, or a group of not more than five persons (excluding servants) who need not be related by blood or marriage living together in a dwelling unit.

Family. Any number of individuals related by blood or marriage, or not to exceed five (5) persons not so related, living together on the premises as a single house-keeping unit, including domestic servants for whom, subject to the provisions of the Ordinance, separate living quarters may be provided.

Family. Group of persons living together who share at least in part of their living quarters and accommodations.

Family. One or more persons living together, whether related to each other by birth or not, and having common housekeeping facilities, in a dwelling unit.

Family. Any number of individuals related by blood or marriage, or a group of not more than 5 individuals not so related, who live together in a single household unit.

Family. Individual or a group of 2 or more persons related by blood, marriage, or adoption, together with not more than 3 additional persons not related by blood, marriage or adoption, living together as a single housekeeping unit in a dwelling unit.

Family. Persons or any number of persons united by blood, marriage or legal adoption living together in a dwelling unit.

Family. One (1) or more persons living privately as a single housekeeping unit and using cooking facilities in common.

Family. One (1) or more persons occuping a premises and living as a single housekeeping unit, as distinguished from a group occupying a hotel, club, fraternity or sorority house.

Family. Individual or two (2) or more persons who are related by blood or marriage living together and occupying a single housekeeping unit with single culinary facilities, or a group of not more than four (4) persons living together by joint agreement and occupying a single housekeeping unit with single culinary facilities on a non-profit cost sharing basis. Domestic servants, employed and residing on the premises shall be considered as part of the family.

Family. One (1) or more persons who are related by blood, or marriage living together and occupying a single housekeeping unit with single culinary facilities, or a group of not more than four (4) persons living together by joint agreement and occupying a single housekeeping unit with single culinary facilities on a non-profit, cost-sharing basis.

Family. Number of individuals living together on the premises as a single non-profit housekeeping unit, including domestic servants.

Family. Individual or two (2) or more persons related by blood, marriage, or adoption, or a group of not more than five (5) persons (excluding servants) who need not be related by blood, or marriage, living together in a dwelling unit. Foster children living with a family as defined herein shall be considered to be part of the family.

Family. One (1) person living alone or a group of two (2) or more persons living together under one management.

Family. One (1) or more persons occupying a premises and living as a single housekeeping unit and using common cooking facilities, as distinguished from a group occupying a boarding house, a lodging house, hotel, club, or fraternity house.

Family. One (1) or more persons occupying a premises and living as a housekeeping unit, as distinguished from a group occupying a boarding house, lodging house, or hotel as herein defined. A family shall be deemed to include necessary servants.

Family. One (1) or more persons occupying a premises and living as a single, non-profit housekeeping unit.

Family. Individual or two (2) or more persons related by blood, marriage or adoption, or a group of not more than five (5) persons, not including servants, who need not be related, living as a single housekeeping unit.

Family. Any number of individuals living together as a single housekeeping unit and doing their cooking on the premises.

Family. One (1) person, or a group of two (2) or more persons, living together and interrelated by bonds of blood, marriage or legal adoption, occupying the whole or a part of a dwelling unit as a separate housekeeping unit. The persons thus constituting a family may also include gratuitous guests and domestic servants.

Family. Any number of individuals related by blood, marriage or legal adoption, and not more than four (4) persons not so related, living together as a single housekeeping unit. Foster children and domestic servants are considered part of a family.

Family. An individual or two (2) or more persons related by blood, marriage, or adoption and usual servants, living together as a single housekeeping unit in a dwelling unit or a group of not more than five (5) persons, who need not be related, living together as a single housekeeping unit in a dwelling unit.

Family. Single person occupying a dwelling unit and maintaining a household; two (2) or more persons related by blood or marriage, occupying a dwelling unit, living together and maintaining a common household, including not more than one (1) boarder, roomer or lodger; or, not more than three (3) unrelated persons occupying a dwelling unit, living together and maintaining a common household.

Family. One (1) or more persons occupying a single housekeeping unit and using common cooking facilities, provided that unless all members are related by blood or marriage, no such family shall contain over five (5) persons.

Family. Individual or two (2) or more persons related by blood or marriage living together; or a group of individuals, of not more than six (6) persons, not related by blood or marriage, but living together as a single housekeeping unit. In each instance the faimly shall be understood to include the necessary servants.

Family. Any number of individuals living together as a single non-profit housekeeping unit and having a single common cooking facility.

Family. One (1) person living alone, or a group of two (2) or more persons living together, whether related to each other by birth of not.

Family. One (1) or more persons who live together in one (1) dwelling unit and maintain a common household, and who are related by blood or marriage or adoption or who are foster children placed with such family by the state board of child welfare or a duly incorporated child care agency, for the purpose of this title, a family includes only a husband and wife, son, son-in-law, daughter, daughter-in-law, father, father-in-law, mother, mother-in-law, brothers, sisters, grandparents, grandchildren, stepchild, adopted children and foster children placed by the state board of child welfare or a duly incorporated child care agency, and bona fide family servants living in and working full time on the premises; provided, however, that in no event shall the total number of persons comprising a family exceed one (1) person per room in the dwelling, exclusive of sun porches. This limitation as to room requirement shall not apply where the premises are occupied solely by the parents or parent and the children thereof or foster children, as aforesaid.

Family. One (1) or more persons living together in one (1) dwelling unit and maintaining a common household, including domestic servants and gratuitous guests, together with boarders, roomers or lodgers not in excess of the number allowed by the ordinance as an accessory use.

Family. One (1) person or two (2) or more persons related by blood or marriage, or a group of not more than four (4) persons not necessarily related by blood or marriage; in any case living together as a single housekeeping unit.

Family. One (1) or more persons living together as a single housekeeping unit, as distinguished from a group occupying a boardinghouse, lodging house, club, fraternity, or hotel.

Family. Any number of individuals living together as a single non-profit housekeeping unit and having a single common cooking facility, excluding however, occupants of a club, fraternity house, lodge, residential club or rooming house.

Family. Family is one (1) person living alone or two (2) or more persons living together as a single housekeeping unit.

Family. Two (2) or more individuals who are related to each other by blood, marriage, adoption or legal guardianship. For purposes of the Code a group of not more than four (4) persons not necessarily related by blood or marriage, living together in a single living unit will be considered equivalent to a single family.

Family. One (1) or more persons, related by blood or marriage, living together and occupying a single housekeeping unit, with single culinary facilities, or a group of not more than four (4) persons living together by mutual agreement and occupying a single housekeeping unit with single culinary facilities on a non-profit cost sharing basis. The usual domestic servants resident on the premises shall be considered as part of the family.

Family. One (1) or two (2) persons or parents, with their direct lineal descendents and adopted children (and including the domestic employees thereof) together with not more than two (2) persons not so related, living together in the whole or part of a dwelling comprising a single housekeeping unit. Every additional group of two (2) or less persons living on such housekeeping unit shall be considered a separate family for the purpose of the Code.

Family. One (1) or more persons occupying a single dwelling unit, provided that unless all members are related by blood, marriage, or adoption, no such family shall contain over four (4) persons.

Family. Either one (1) individual, two (2) or more persons related by blood, marriage or adoption, or not more than three (3) persons not related by blood, marriage or adoption, who live together in one (1) dwelling unit and maintain a common household.

Family. One (1) or more persons, with their domestic servants, occupying a dwelling unit as a single, non-profit, housekeeping unit, but not including a group occupying a hotel, boarding house, club, dormitory, fraternity or sorority house.

Foster child. A child giving, receiving and sharing affection and care in a family as if related by blood or as if legally adopted.

Individual. Shall mean a single human being as distinguished from a group.

Individual. A particular being or group of beings, or a single human being.

Individual. A single member of a class or species. A particular person, animal or thing.

Individual. A single member of a family.

Individual. In accordance with the Ordinance, an "individual" shall mean a private or natural person as distinguished from a partnership, association, corporation or any organized group.

Lessee. The person in possession of a building under a lease from the owner thereof.

Multiple family. As in a building, meaning more than two (2) families or households living independently of each other and doing cooking within their living quarters; includes apartments, tenements and flats.

One ownership. Ownership of a parcel or contiguous parcels of property or possession thereof under a contract to purchase by a person or persons, firm, corporation or partnership, individually, jointly, in common, or in any other manner whereby such property is under single or unified control.The term shall include condominium ownership.

Owner. A person having title or interest in land or his agent or attorney.

Owner. Any person, firm or corporation controlling the property under consideration.

Owner. Any person, firm, corporation, association or organization holding the legal title to any dwelling or dwelling unit within the city, and any personal representative, trustee, committee or guardian having charge.

Owner. Includes his duly authorized agent or attorney, a purchaser, devisee, fiduciary and an individual having a vested or contingent interest in the property in question.

Owner. The owner of the land as recorded in the registry of deeds for the County, or as registered in the land Court, except as otherwise provided in the Zoning Ordinance.

Owner. Means a person recorded as such on the records of the City Assessor, and includes his duly authorized agent or attorney, a purchaser, devisee, fiduciary, and a person having a vested or contingent interest in the property in question.

Owner. Any person, firm or corporation owning or controlling property, including a duly authorized agent or attorney. Guardians or trustees shall also be regarded as owners.

Owner. Includes the owner or owners of the freehold of the premises or any lessor estate, area, a mortgages or validity of possession, or assigning for rent, receiver, executor, trustee, lessee, or other person, firm, or corporation in control of a building.

Owner. The term shall include his duly authorized agent, a purchaser, devisee, fiduciary, property holder or any other person, firm or corporation having a vested or contingent interest, or in case of leassed premises, the legal holder of the lease contract, or his legal representative, assign or successor.

Owner. The person holding record title to the property served or his duly authorized agent.

Owner. A person having legal title to premises; a mortgagee or vendee in possession; a trustee in bankruptcy; a receiver or any other person having legal ownership or control of premises.

Owner. Any person, who alone or jointly or severally with others shall have legal title to any premises or equipment, with or without accompanying actual possession thereof, or shall have charge, care, control of any premises or equipment, as owner or agent of the owner, lessee, or as executor, executrix, administrator, adminstratrix, trustee, or guardian of the estate of the owner, or as a mortgagee in possession either by virtue of a court order or by voluntary surrender by the person holding legal title, or a collector of rents.

Owner. A person who has an interest in the real property improved and for whom an improvement is made and who ordered the improvement to be made. "Owner" includes successors in interest of the owner and agents of the owner acting within their authority.

Owner. Includes his duly authorized agent or attorney, a purchaser, devisee, fiduciary, and a person having a vested or cantingent interest in the property in question.

Owner. A person recorded as such on the records of the city assessor.

Owner. Includes the owner or owners of the freehold of the premises or lesser estate therein, a vendee in possession, a mortgagee or receiver in possession, an assignee of rents, a lessee or joint lesses of the whole thereof, an agent or any other person, firm or corporation directly in control of such building and land.

Owner. Holder of the title whose interest is shown of record or who is in possession of a dwelling, or any person in control of dwelling, or an agent of any much person.

Owner. Owner of record of a parcel of land.

Owner. The person for the time being, receiving the rent of the land or premises in connection with which the word is used, whether on his account or as agent or trustee of any other person, or who would so receive the rent, if such land or premises were let.

Owner. The owner of the freehold of the premises or lesser estate therein, a mortgage or vendee in possession, assignee of rents, receiver, executor, trustee, lessee, or other person, firm or corporation in control of a building.

Owner. Person, firm, corporation or partnership exercising "one ownership" as defined in the Ordinance.

Owner. Any person, agent, firm, or corporation having a legal or equitable interest in the property.

Owner. Any person who alone, or jointly or severally with others shall have legal title to any building, structure of premises with or without accompanying actual possession thereof and shall include his duly authorized agent or attorney, a purchaser, devise, fiduciary and any person having a vested or contingent interest in the property in question.

Parties in interest. Persons, other than mortgagees or holders of vendor's liens who have an interest of record in or who are in possession of a dwelling.

Person Any natural person, firm, association or corporation.

Person. A natural person, his heirs, executors, administrators, or assigns, and also includes a firm, partnership, or corporation, its or their successors or assigns of the agent of any of the aforesaid.

Person. Any individual, firm, co-partnership, joint venture, association, social club, fraternal organization, corporation, estate, trust, receiver, syndicate, this and any other County, City and County, City, Municipality, District or other political subdivision, or any other group or combination acting as a unit.

Person. Any and all persons, including individuals, firms, partnerships, associations, public or private institutions, municipalities or political subdivisions, governmental agencies, or private or public corporations organized or existing under the laws of the State or any other State or Country.

Person. Natural person, his heirs, executors, administrators, or assigns; and shall also include a firm, partnership, or corporation, its or their successors or assigns. Singular includes plural; male includes female.

Person. Includes any individual, group of persons, firm, corporation, association, organization, and any legal public entity.

Person. Includes any firm, co-partnerhip, corporation any association of natural persons acting jointly or by a servant, agent, or otherwise, or school district, County, State, United States Government or any other public agency or instrumentality.

Person. Includes a corporation as well as an individual.

Person. An individual, firm, association, corporation, trust or any other legal entity, including his or its agents.

Person. Any individual, political subdivision, estates, trust, or body of persons, whether incorporated or not, acting as a unit.

Person. Includes a corporation and partnership as well as individual.

Person. Includes firms, corporations, and associations.

Person. Includes a corporation or a co-partnership as well as an individual.

Person. Includes any domestic or foreign corporation, association, syndicate, joint stock company, partnership of any kind, joint venture, club, business, or common law trust, society or individual.

Person. Individual, group of individuals, partnership, firm, corporation or association.

Person. An individual, partnership, corporation, or other legal entity.

Person. Every natural person, firm, co-partnership, association, or corporation.

Person. Any natural person, partnership, association, trust, and public or private corporation.

Person. Shall mean natural person, joint venture, joint stock company, partnership, association, club, company, corporation, business trust, organization or the manager, lessee, agent, servant, officer, or employee of any of them.

Person. Includes an individual, partnership, corporation, association, county, municipality, co-operative group, or other entity engaged in the business of operating, owning or offering the services of a hotel, motel, tourist home, retirement home or rooming house.

Person. Includes a corporation, firm, partnership, association, organization and any other group acting as a unit as well as individuals. It shall also include an executor, administrator, trustee, receiver or other representative appointed according to law. Whenever the word "person" is used in the Code prescribing a penalty or fine, as to partnerships or associations, the word shall include include the partners of members thereof, and as to corporations, shall include the officer, agents or members thereof who are responsible for any violation of such section.

Person. Any individual, firm, member of firm, partnership, corporation or any officer, director or stockholder of any corporation or any agent, or any employees of any such firm, partnership or corporation.

Person. Includes an individual, firm, corporation, partnership, limited partnership, association, or limited partnership association, or any other organized group of individuals or the legal successor or representative agent or servant of any of the foregoing, any department, bureau or agency of the city and any other public body or agency.

Person. Any individual, firm, member of a firm, partnership or member thereof, corporation or any officer, director or stockholder of such corporation, unless otherwise specified in the Ordinance.

Person. Any person, persons, partnership, firm, corporation, or cooperative enterprise of any kind.

Tenant. Any tenant, licensee, occupant, person in charge or in possession of any dwelling or part thereof.

A-4 FIRE PROTECTION—MATERIALS AND CONSTRUCTION

Area of fire division. Maximum horizontal projected area of the division within the property lines including exterior walls, one or more of which may be party walls, and in the case of separation walls within the property lines to the center of the separation wall.

Area of refuge. Room in a building or structure, another building or structure, or an open space outside of a building or structure affording escape and protection from fire and smoke.

Automatic. Constructed and arranged to operate other than manually; if open, it will close when subjected to a pre-determined temperature or rate of temperature rise.

Automatic. As applied to fire protection devices, is a device or system providing an emergency function without the necessity for human intervention and activated as a result of a pre-determined temperature rise, rate of rise of temperature, or combustion products; such as is incorporated in an automatic sprinkler system, automatic fire door, automatic fire shutter, or automatic fire vent.

Automatic. An automatic closing device is one which functions without human intervention, and is actuated as a result of the pre-determined temperature rise, rate of rise of temperature, combustion products or smoke density.

Automatic. Automatic as applied to a fire protective device is one which functions without human intervention and is actuated as a result of the pre-determined temperature rise, rate of rise of temperature, combustion products or smoke density such as an automatic sprinkler system, automatic fire door, automatic fire shutter, or automatic fire vent.

Automatic. As applied to a fire door or other opening protective, means normally held in an open position and automatically closed by a releasing device that is actuated by abnormal high temperature or by a predetermined rate of rise in temperature.

Automatic. Providing a function without the necessity of human intervention.

Brick for fireproofing. Brick shall be laid in Type M, S, N or O mortar. Solid clay and shale brick shall conform to the "Standard Specifications for Facing Brick" ASTM C216 or "Standard Specifications for Building Brick," ASTM C62. Hollow clay and shale brick shall conform to the "Standard Specification for Hollow Brick", ASTM C652. Concrete brick shall conform to the "Standard Specification for Concrete Building Brick", ASTM C55. Sand-lime brick shall conform to the "Standard Specification for Calcium Silicate Face Brick (Sand-Lime Brick", ASTM C73.)

Bureau or Bureau of Fire Prevention. Means the Bureau of Fire Prevention and Fire Safety Inspection in the fire department of the city.

Ceiling protection. Five protection membrane suspended beneath the floor or ceiling construction which, when included with the construction, develops the fire-resistive rating for the overall assembly.

Closure. Device for shutting off an opening through a construction assembly, such as a door or a shutter, and includes all components such as hardware, closing devices, frames and anchors.

Combustible. Material or combination of materials which will ignite and support combustion when heated at any temperature up to 1382°F.

Combustible. Capable of undergoing combustion.

Combustible. That property of a material by which it fails to conform to the criteria of the definition of noncombustible.

Combustible. Relative to materials which will burn under some conditions and not under others. The primary purpose of the definition is to provide a basis for the classification of construction for purposes of fire protection.

Combustible. Materials or assemblies capable of igniting and continuing to burn or glow at or below a temperature of 1200°F.

Combustible. Material or structure made of or surfaced with wood, compressed paper, plant fibers or other materials that will ignite and burn.

Combustible. Materials not included in the definition of noncombustible. When referring to assemblies of materials, means those assemblies which are made up in any part of combustible materials.

Combustible. Material or combination of materials which is not noncombustible.

Combustible. Capable of igniting and continuing to burn or glow with a flame at or below a temperature of 1200 degrees Fahrenheit.

Combustible construction. That type of construction that does not meet the requirements for noncombustible construction.

Combustible construction. Building or structure constructed in whole or part of combustible materials.

Combustible construction. Assembly such as a wall, floor or roof having components of combustible material.

Combustible fibers. Finely divided fibers, thin sheets or flakes of any materials or rapid combustion.

Combustible fibers. Fibrous materials which are highly flammable such as hay, straw, broomcorn, hemp, tow, jute, sisal, moss, paper, rags, excelsior, hair, kapok, oakum, cotton, henequen, ixtle, cocoa fiber, and wheat straw.

Combustible fibers. Includes readily ignitable and free burning fibers, such as cotton, sisal, henequen, ixtle, jute hemp, tow, cocoa fiber, oakum, baled waste paper, kapok, hay, straw, Spanish Moss, excelsior, certain synthetic fibers and other like materials.

Combustible liquids. Any liquid having a flash point at or above one hundred (100) degrees F., shall be known as Class II or III liquids.

Combustible material. Inflammable material which will ignite at or below a temperature of 1200°F and continue to burn and glow.

Combustible material. All materials not classified as non-combustible are considered combustible. This property of a material does not relate to its ability to structurally perform under fire exposure. The degree of combustiblity is not defined by standard fire test procedures.

Combustible material. Any material which will burn or is destoryed by heat.

Combustible material. A flammable material which will ignite at or below a temperature of 1200 degrees F., and continue to burn or glow.

Combustible mixture. Means any liquid or solid mixture or substance or compound which does not emit a flammable vapor at a temperature below 500 degrees F. when tested in a Tagliabue open cup tester, but which may be ignited and caused to burn.

Combustible plastic. A plastic material more than one twentieth (1/20) inches in thickness which burns at a rate of not more than two and one-half (21/2) inches per minute when subjected to ASTM D635, Standard Method of Test for Flammability of Self-Supporting Plastics, listed in the Code.

Combustion. Chemical process that involves oxidation sufficient to produce light or heat.

Conflagration hazard. The fire risk involved in the spread of fire by exterior exposure to and from adjoining buildings and structures.

Directly applied fire-resistive protection. Coating material applied directly to the structural element for the purpose of fire protection.

Draft curtain. Thin partition extending from the underside of the roof usually to the lower edge of the roof trusses, or a solid partition installed in attics extending from the underside of the roof to the ceiling of the top story. Draft curtains also are called curtain boards or draft stops, retard the spread of fire, hot gases and smoke along the ceiling or through an attic or suspended ceiling areas.

Draft curtain. Noncombustible curtain or baffle, suspended from the ceiling for the purpose of deflecting the flow of heated gases and smoke.

Draft curtain. A noncombustible curtain, suspended in a vertical position from a celing, for the purpose of retarding the lateral movement of heated air, gases, and smoke along the ceiling in the event of fire.

Draft curtain. A curtain or baffle, extending downward from a roof or ceiling, to stop drafts and bank up heat from a fire.

Exception to fireresistive requirements. In fireresistive construction, the fireprotective covering may be omitted from roof trusses, girders, beams and purlins when every part of the structural framework is 20 feet or more above the floor immediately below; and the roof slabs or other construction between purlins shall be entirely of incombustible materials of the required strength or of mill-type construction; or may be constructed of approved composite laminated type roof deck of materials which are not more combustible than wood.

Exposed. Usually with reference to a part of a building and its fire separation or fire exposure, means that said part is in or on a side of the building which is not at right angle to, and faces toward the line, wall, or other object, to which the fire separation under consideration is measured.

Fire area. Floor area enclosed and bounded by fire walls or exterior walls of a building to restrict the spread of fire.

Fire area. Floor area of a story of a building within exterior walls, party walls, fire walls, or any combination thereof.

Fire area. Adjacent buildings on the same premises shall be deemed to constitute one fire area, unless the exterior walls of such buildings have proper fire-resistance ratings for outside exposure, and openings in such exterior walls are equipped with opening protectives. A portion of a building which is cut off from the other portions of a building by fire-resistive walls or partitions and floors and ceilings, having an approved fire-resistant rating not less than 1 hour, with communicating and exposed openings protected with approved opening protectives, so as to resist the spread of fire, smoke, heat, and noxious gases.

Fire area. The floor area of a story of a building within exterior walls, party walls, fire walls, or any combination thereof, as listed in the Code.

Fire area. A floor area enclosed by fire divisions and/or exterior walls.

Fire area. Area of a building separated from the remainder of the building by construction having a fire resistance of at least 1 hour and having all communicating openings properly protected by an assembly having a fire resistance rating of at least 1 hour.

Fire barrier. A fire barrier is a continuous membrane, either vertical or horizontal, such as a wall or floor assembly, that is designed and constructed with a specified fire resistance rating to limit the spread of fire and which will also restrict the movement of smoke. Such barriers may have protected openings.

Fire compartment. Enclosed space in a building that is separated from all other parts of the building, by enclosing construction providing a fire separation having a required fire-resistance rating.

Fire compartment. A fire compartment is a space within a building that is enclosed by fire barriers on all sides, including the top and bottom.

Fire damper. A damper arranged to seal off air flow automatically through part of an air duct system, so as to restrict the passage of heat. The fire damper may also be used as a smoker damper if location lends itself to the dual purpose.

Fire damper. A damper arranged to seal off air flow automatically through part of an air duct system so as to restrict the passage of heat.

Fire damper. Approved automatic or self-closing noncombustible barrier designed to prevent the passage of air, gases, smoke or fire through an opening, duct or plenum chamber.

Fire damper operators. Designed to hold the damper in the open position under normal usage and release and automatically close the damper under fire conditions.

Fire dampers. Include single-blade, multiblade, or curtain types.

Fire department access. Every building more than one story in height which does not have windows opening directly upon a street in each story above the first, shall be provided with suitable access for fire department use. Such access shall be a window or door opening through the wall on each floor above the first story. The opening shall be at least 36 inches in width and not less than 48 inches in height, with the sill not more than 32 inches above the floor. The openings shall be so spaced that there will be one opening in each 100 feet of wall length in any accessible wall of the building. This requirement for access openings for fire department use shall not apply where a building is equipped throughout with an automatic sprinkler system approved for fire protection purposes.

Fire division. The interior means of separation of one part of a floor area from another part together with fireresistive floor construction to form a complete fire barrier between adjoining or superimposed floor areas in the same building or structure.

Fire drill. The organized procedure conducted with or without a private fire brigade for vacating the occupants of a building and for operating the first-aid fire appliances and equipment for the extinguishing of fire and safeguarding of life.

Fire exposure. Subjection of a material or construction to a high heat flux from an external source, with or without flame impingement.

Fire grading of buildings. All buildings and structures shall be graded in accordance with the degree of fire hazard of their use in terms of hours and fractions of an hour and as regulated by the building code.

Fire hazard. Any thing or act which increases or may cause an increase of the hazard or menace of fire to a greater degree than that customarily recognized as normal by persons in the public service regularly engaged in preventing, suppressing or extinguishing fire; or which may obstruct, delay, hinder or interfere with the operations of the Fire Department or the egress of occupants in the event of fire.

Fire hazard. The potential degree of fire severity existing in the use and occupancy of a building and classified as high, moderate or low hazard.

Fire hazard. The degree of severity of a potential fire in the use of a structure.

Fire hazard. Potential ease of ignition and potential degree of fire severity existing in the use of a building and classified as high, moderate, or low.

Fire limits. Boundary line establishing an area in which there exists, or is likely to exist, a fire hazard requiring special fire protection.

Fire load. Combustible contents of a room or floor area expressed in terms of the average weight of combustible materials per square foot, from which the potential heat liberation may be calculated based on the calorific value of the materials, and includes the furnishings, finished floor, wall and ceiling finishes, trim and contemporary and movable partitions.

Fire load. The combustible contents within a building during normal use.

Fire prevention. The preventible measures which provide for the safe conduct and operation of hazardous processes, storage of highly combustible and flammable materials, conduct of fire drills, and the maintenance of fire detecting and fire-extinguishing service equipment and good housekeeping conditions.

Fire prevention. Preventive measures which provide for fire-safe conduct and operations in buildings and includes the maintenance of fire-detection, fire-alarm, and fire-extinguishing equipment and systems, exit facilities, opening protectives, safety devices, good housekeeping practices and fire drills.

Fireproof. Fire resistive.

Fireproof. Term which has been discontinued by many agencies as misleading, since no material is immune from intense heat or duration of fire. Term is used when referring to types of fire-resistive construction in certain building and fire codes.

Fireproof buildings. Buildings or structures composed entirely of materials which will resist the action of fire.

Fire protection. Such fire hydrants and other protective devices as required by the Chief Engineer of the Fire Department.

Fire protection. The provision of safeguards in construction and of exit facilities; and the installation of fire alarm, fire-detecting and fire-extinguishing service equipment to reduce the fire risk and the conflagration hazard.

Fire protection. The provision and maintenance of safeguards in construction and of exit facilities; and the installations, maintenance, and provision for the use of fire alarm, fire-detecting and fire-extinguishing service equipment to reduce the conflagration hazard for the safeguard of human life and the protection of property.

Fire protection. Constructions which have been made fire retardant or have a fire resistance rating or approved fire extinguishing or fire alarm devices.

Fire protection appliance. The apparatus system or equipment provided or installed close at hand for immediate use in the event of fire.

Fire protection equipment. Apparatus, assemblies or systems either portable or fixed, for use to prevent, detect, control or extinguish fire.

Fire protection equipment. Includes apparatus, assemblies or systems either portable or fixed, for use to prevent, detect, control, or extinguish fire.

Fire protection system. A system including systems, devices and equipment to detect a fire, actuate an alarm or suppress fire or any combination thereof.

Fire resistance. As applied to building materials and construction means the ability to withstand fire or give protection from it for given periods under prescribed test conditions.

Fire resistance. The property of a material or assembly to withstand fire or give protection from it; as applied to elements of buildings, it is characterized by the ability to confine a fire or to continue to perform a given structural function, or both.

Fire resistance. The ability of certain materials and assemblies to resist exposure to fire and to prevent its spread. The measure of this ability is known as the Fire Resistance Rating.

Fire resistance. Ability of certain materials and assemblies to resist exposure to fire and to prevent its spread.

Fire resistance. Property of a material or of a construction type to resist failure because of high temperatures and to prevent or retard the passage of high temperatures, hot gases, or flames.

Fire resistance. The ability of materials and construction to withstand fire or give protection from it for given periods of time and under prescribed test conditions.

Fire resistance. That property of materials, construction or assembly of materials, which under fire conditions prevents or retards the passage of excessive heat, hot gases, or flames.

Fire resistance and fire-resistive material. Materials having the properties to withstand fire or give protection from it. As applied to the elements of buildings or structures, it is characterized by the ability to confine a fire or to continue to perform a given structural design function.

Fire resisting ceiling. Ceiling which is part of a floor and ceiling, or roof and ceiling assembly having a fire resistance rating of 1 hour or more.

Fire resistive. Term applies to properties of materials or design to resist the effects of any fire to which a material or structure may be expected to be subjected.

Fire resistive. Term applied to materials and assemblies in which the resistance to fire is of prime consideration.

Fire resistive. That quality of materials and assemblies to resist fire and prevent its spread.

Fire resistive. That quality of materials, assemblies, constructions, or structures to resist fire and prevent its spread; fireproof.

Fire resistive. Having fire resistance.

Fire resistive. In the absence of a specific ruling by the authority having jurisdiction, applies to materials for construction not combustible in the temperatures of ordinary fires and that will withstand such fires without serious impairment of their usefulness for at least one (1) hour.

Fire-resistive ceiling. Ceiling construction which is used as part of a floor assembly or roof system and has a fire resistance rating of 1 hour or more.

Fire-resistive materials. Materials which offer a degree of resistance to the passage or the effects of fire or heat sufficient to meet the minimum requirements of the code.

Fire-resistive protection. Thermal insulating materials applied directly, attached to, or suspended or sprayed to a structural assembly, to maintain the structural integrity of structural member or system for a specified time rating as established by testing procedures.

Fire-resistive separation. Every room containing a central heating plant in applicable occupancies as stated in the code shall be separated from the balance of the building or structure by not less than a 1 hour fire-resistive separation with all openings in such separation protected by a fire assembly having not less than a 1 hour fire-resistive rating.

Fire retardant. Materials, buildings, or structures which are combustible in whole or in part, but have been treated or have surface applications to prevent or retard ignition or the spread of fire under the conditions for which they were designed to be used.

Fire retardant. Having or providing comparitively low flammability or flame spread properties.

Fire retardant. The quality of any substance to prevent or retard ignition or the spread of fire in a combustible material.

Fire retardant. Construction that has a fire-resistance rating of at least one hour.

Fire retardant ceiling. The ceiling portion of a floor and ceiling or a roof and ceiling assembly having a fire resistance rating of one hour or more that is used to protect a floor or roof assembly other than that which it was tested.

Fire retardant ceiling. Ceiling used in a floor and ceiling assembly that has a fire resistance rating of 1 hour or more.

Fire retardant chemical. Chemical or preparation of chemicals or solutions used to reduce flammability or to retard spread of flame.

Fire retardant lumber. Lumber treated by pressure impregnation to reduce combustibility.

Fire retardant materials. Materials used in structures for which fire retardant ratings have been developed.

Fire retardant materials. Materials used in buildings and structures which will provide fire-resistive ratings of not less than 1 hour.

Fire separation. A construction of specific fire resistance separating parts of a building.

Fire shutter. Shutter capable of resisting a fire as required by code.

Fire stop. A solid, tight closure of a concealed space, placed to prevent the spread of fire and smoke through such a space.

Fire stop. A draft-tight barrier within or between construction assemblies that acts to retard the passage of smoke and flame.

Firestop. Fire-resistive closure, placed so as to restrict the spread of smoke and fire in concealed spaces.

Firestopping. A barrier effective against the spread of flames or hot gases within or between concealed spaces.

Firestopping. A barrier within concealed spaces which is effective against spread of flames or hot gases.

Firestopping. Effective prevention or retarding of horizontal or vertical spread of smoke or fire through hollow, concealed spaces in wall or floor assemblies or attic spaces.

Fire stopping. A barrier which prevents the spread of flames or hot gases within or between concealed spaces in a structure.

Fire treatment of materials. Surface-applied or pressure type treatment for control of fire or flame spread, when specified, such as: treatment by brushing, spraying, dipping, or by pressure application of chemical fire retardants to protect against combustion or flame spread.

Fire terrace. Level space or area at a setback of an exterior wall of a building and at approximately the same elevation as that of the curb or grade level of the higher street, to provide a safe termination for fire escapes from upper stories of the building.

Fire valve. An automatic self opening noncombustible device designed to permit the passage of gases, smoke, or fire through an opening or vented area.

Flameproof. As applied to decorations, curtains, draperies, scenery, tents, woodwork, or other normally combustible materials, means treated so that it will not propagate flame.

Flameproof. Materials which will not readily ignite and will not propagate flame under test conditions.

Flameproof. Materials treated so that they will not propagate flame, such as: decorations, draperies, curtains, scenery used on a stage, tents, woodwork, or other normally combustible materials of like use and purpose.

Flameproof material. Material which because of its nature or as a result of chemical treatment will not burst into flame at or below a temperature of 1200°F.

Flame resistance. Property of materials or combinations of component materials which restricts or resists the spread of flames, as determined by approved flame-resistance tests.

Flame resistant. That property of materials or combination of materials which restricts the spread of flames.

Flame resistant. That property of a material which is flame resistant by nature or has been made so by an accepted method.

Flame resistant material. Material which is flame resistant by nature or has been made flame resistant in conformity with generally accepted standards.

Flame spread. Propagation of flame over a surface.

Flame spread. Flaming combustion along a surface.

Flame spread. The rate at which flames spread over surfaces of various materials as tested by ASTM E-84.

Flammable. Capable of igniting within 5 seconds when exposed to flame, and continuing to burn.

Horizontal compartment. That portion of a building bounded by exterior walls and the required fire rated and smoke barriers for floors and roof assembly.

Incombustible. Wherever the word is used, the matter described shall be sufficiently fire-retarding or fire-resisting for its purpose.

Incombustible. Material that will not of and by itself ignite when its temperature and that of the surrounding air is 1200 degrees Fahrenheit. (649°C).

Incombustible. Does not apply to surface finish materials. Material required to be incombustible for reduced clearances to flues, heating appliances, or other materials shall refer to material conforming to the Code. No material shall be classified as incombustible which is subject to increase in combustibility or flame-spread rating beyond the limits herein established, through the effects of age, moisture or other atmospheric conditions.

Incombustible. Same as noncombustible; not combustible.

Incombustible. Has the same meaning as noncombustible, is subject to misunderstanding due to the prefix. Noncombustible is preferred.

Incombustible construction. Synonymous with "Non-combustible Construction."

Incombustible material. Synonymous with non-combustible material.

Incombustible material. Any material which will not ignite at or below a temperature of 1000° Farenheit during an exposure of five minutes and which shall not continue to burn or glow at that temperature.

Incombustible material. Any material which will not ignite nor actively support combustion in a surrounding temperature of twelve hundred degrees fahrenheit during an exposure of five minutes and which will not melt when the temperature of the material maintained at nine hundred degrees fahrenheit for a period of at least five minutes.

Incombustible material. Material having a structural base of incombustible material as defined in the code.

Incombustible material. Material of which no part will ignite and burn when subjected to fire.

Incombustible material. Material which will not ignite at or below a temperature of 1200°F and will not continue to burn or glow at that temperature.

Labeled fire dampers. Only fire dampers which have been tested, listed and labeled by Underwriters Laboratories, Inc., or an equivalent test and labeling by other accredited testing laboratories shall be deemed to meet the requirements of the Code and for the recommended locations and use as listed in the Code.

Non-combustible. Materials having a structural base of non-combustible material as defined in the Code, with a surfacing not over 1/8 inch thick which has a flame spread rating not higher than 50.

Non-combustible. Does not apply to the flamespread characteristics of interior finish or trim materials. No material shall be classed as a non-combustible building material which is subject to increase in combustibility or flamespread beyond the limits herein established through the effects of age, moisture or other atmospheric conditions.

Non-combustible. Materials or combination of material which when tested in accordance with the provisions of ASTM E 136 exposed to a furnace temperature of 1382° F during an exposure of 5 minutes, will not ignite and support combustion.

Non-combustible. Any material which will not ignite or actively support combustion in air at a temperature of 1200° F during exposure of five (5) minutes in a vented tube or vented crucible furnace.

Non-combustible. Does not apply to surface finish materials nor to the determination of whether a material is non-combustible from the standpoint of clearances to heating appliances, flues, or other sources of high temperature. No material shall be classed as non-combustible which is subject to increase in combustibility or flames spread rating beyond the limits herein established, through the effects of age, moisture, or atmospheric condition.

Non-combustible. No materials shall be classed as non-combustible which is subject to increase in combustibility or flame spread rating beyond the limits herein established, through the effects of age, moisture or other atmospheric condition, as for example, various types of treated wood. Flame spread rating as used in the Code refers to ratings obtained according to the method for fire hazard classifications of the Underwriters Laboratories Inc.

Noncombustible. Material or combination of materials which will not ignite and support combustion when heated at any temperature up to 1382°F (750°C), during an exposure of 5 minutes.

Noncombustible. Materials, no part of which will ignite when subjected to fire. Any material which liberates flammable gas when heated to any temperature up to 1380°F for five minutes shall not be considered noncombustible within the meaning of the Code.

Non-combustible material. Material which will not ignite when heated to a temperature of 1200 degrees Fahrenheit.

Non-combustible material. Non-inflammable material which will not ignite at or below a temperature of 1200°F and will not continue to burn or glow at that temperature.

Non-combustible materials. Having a surface flamespread rating no greater than twenty-five (25) without evidence of continued progressive combustion on any exposed surface that may be exposed by cutting through the material in any way.

Non-combustible materials. Materials having a structural base of non-combustible material with a surfacing not more than one-eighth (1/8) inch thick which in addition has a flame spread rating not greater than fifty (50) when tested in accordance with the "Method of Test for Surface Burning Characteristics of Building Materials, ASTM Designation E84."

Nonflammable. Not flammable; not combustible.

Opening protective. An assembly of materials and accessories, including any incidental frames, mullions, muntins, anchors, and hardware, which when installed in an opening in a wall, partition, floor, or roof, prevents the passage of flame, heat, fumes, and smoke through the opening for a specified time.

Opening protective. Assembly of materials and accessories, including frames and hardware, installed in a wall, partition, floor, ceiling, or roof opening to prevent, resist or retard the passage of fire, flame, smoke, excessive heat, or hot gases.

Protected construction. That in which all structural members are constructed, chemically treated, covered or protected so that the individual unit or the combined assemblage of all such units has the required fire-resistance rating specified for its particular use or application.

Protective assembly. An opening protective including its surrounding frame, casings and hardware attachments.

Protective assembly. Method of protecting an opening in a wall, partition, floor, or roof against the spread of fire, heat, and smoke.

Protective equipment. Any equipment, device, system or apparatus, whether manual, mechanical, electrical or otherwise, permitted or required by the Bureau to be constructed or installed in any hotel or multiple dwelling for the protection of the occupants or intended occupants thereof, or of the general public.

Separation. System of walls, floors or other construction serving to separate or cut off one unit of occupancy from another.

Smoke and heat vents. For the purpose of localizing the heat released by a fire to the immediate fire area, minimizing sprinkler operation, lowering building temperatures, exhausting smoke, and improving accessibility for fire-fighting personnel to permit close approach and direct action against the seat of the fire.

Smoke barrier. Partition in corridors designed and constructed to retard the passage of smoke.

Smoke barrier. A smoke barrier is a continuous membrane, either vertical or horizontal, such as a wall, floor, or ceiling assembly, that is designed and constructed to restrict the movement of smoke. A smoke barrier may or may not have a fire resistance rating. Such barriers may have protected openings.

Smoke barriers. Continuous from outside wall to outside wall, from a fire barrier to a fire barrier, from a floor to a floor, from a smoke barrier to a smoke barrier, or a combination thereof; including continuity through all concealed spaces such as those found above a ceiling, including interstitial spaces.

Smoke compartment. A smoke compartment is a space within a building enclosed by smoke barriers or fire barriers on all sides, including the top and bottom.

Smoke damper. Damper arranged to seal off air flow automatically through a part of an air duct system so as to restrict passage of smoke.

Smoke detector. A device which senses visible or invisible particles of combustion.

Smokeproof enclosure. An enclosed stairway, with access from the floor area of the building either through outside balconies or ventilated vestibules, opening on a street or yard or open court, and with a separately enclosed direct exitway to the street at the grade floor.

Smokeproof tower. Stairway enclosure so designed that the movement into the smokeproof tower of products of combustion produced by a fire occurring in any part of the building shall be limited.

Smoke stop. A partition in corridors, or between spaces, to retard the passage of smoke, with any opening in such partition protected by a door equipped with a self-closing device.

Smoke-stop door. A door or set of doors placed in a corridor to restrict the spread of smoke and to retard the spread of fire by reducing draft.

Unprotected opening. Doorway, window or opening other than one equipped with a closure having the required fire-protection rating, or any part of a wall forming part of the exposing building face that has a fire-resistance rating less than required for the exposing building face.

Vertical opening. Openings through floors, such as for stairways, elevators, ventilating shafts, etc., which if unprotected, may serve as channels for the spread of fire or smoke.

Vertical opening. Opening through a floor or roof.

Vertical opening. Opening in a floor or roof for giving access vertically from the story below or above for light, ventilation, the movement of persons or materials or for any other purpose.

Vertical opening. Openings through floors for stairways, elevators, conveyors, and the like or for purposes of light and ventilation.

Vertical service space. Shaft oriented essentially vertically that is provided in a building to facilitate the installation of building services including mechanical, electrical, plumbing installations and facilities such as elevators, refuse chutes, and linen chutes.

A-5 FLOODPLAIN REGULATIONS

Artificial lake. A man-made basin other than a swimming pool, designed or intended to permanently contain water, which shall have a depth of not less than two (2) feet or a minimum area of two hundred and fifty (250) square feet at the low water stage.

Artificial obstruction. Any obstruction which is not a natural obstruction, including any which, while not a significant obstruction in itself, is capable of accummlating debris and thereby reducing the flood-carrying capacity of the stream.

Base Flood. A flood having a one (1) percent chance of being equalled or exceeded in any given year.

Base flood elevation. An elevation equal to that which reflects the height of the base flood as defined in the Code.

Channel. A natural or artificial water-course with definite bed and banks to confine and conduct the normal flow of water.

Drainage channels. Channels used for interrupted flow of surface water.

Drainage facility. Any ditch, gutter, pipe, culvert, storm sewer or other structure designed, intended, or constructed for the purpose of diverting surface waters from carrying surface waters off streets, public rights-of-way, parks, recreational areas, or any part of any sub-division or contiguous land areas.

Drainage right-of-way. The lands required for the installation of storm water sewers or drainage ditches, or required along a natural stream or watercourse for preserving the channel and providing for the flow of water therein to safeguard the public against flood damage, in accordance with the Code.

Encroachment. Any fill, structure, building, accessory use, use or development in the floodway.

Encroachment floodway lines. Lines which are limits of obstruction to flood flows. These lines are on both sides of and generally parallel to the stream. The lines are established by assuming that the area landward (outside) of the encroachment lines will be ultimately developed in such a way that it will not be available to convey flood flows.

Federal flood insurance. Insurance available to property owners in limited amounts from private insurance carriers against casualty losses due to floods through the National Flood Insurance Program.

Fifty-year flood. A flood that has a two (2) percent chance of occurring in any one (1) year based upon criteria established by the State Water Commission.

Flood base elevation. That elevation of the highest flood on record, determined by the City Engineer's record of the elevations of the highest flood locations as indicated on the Flood Plain map of the City on file in the office of the City Engineer. Flood base elevations at intermediate locations shall be interpolated along the water course between the two (2) nearest flood base elevations, one (1) each upstream and downstream. The controlling flood base elevation for any building site shall be the same as the flood base elevation at the nearest point of the water course as measured on a line perpendicular to the direction on the water course.

Flood fringe. That portion of the floodplain outside of the floodway, which is covered by floodwaters during the regional flood; it is generally associated with standing water rather than rapidly flowing water.

Flood hazard. A hazard to land or improvements due to overflow water having sufficient velocity to transport or deposit debris, scour the surface soil, dislodge or damage buildings, or erode the banks of water courses.

Flooding. Land subject to flooding or other hazards to life, health or property deemed to be topographically unsuitable shall not be plotted for permanent residential occupancy or for such other uses as may increase danger to health, life, or property, or aggravate erosion or flood hazard until all such hazards have been eliminated or unless adequate safeguards against such hazards are provided by the development plans. Such land within the subdivision shall be set aside on the plot for uses as shall not be endangered by periodic or occasional innundation or shall not produce unsatisfactory living conditions.

Flood or flooding. A general and temporary condition of partial or complete inundation of normally dry land areas from the overflow of inland waters, or the unusual and rapid accumulation or runoff of surface waters from any source.

Flood or flood waters. A temporary overflow of water on lands not covered by water.

Floodplain. The relatively flat areas of lowlands adjoining the channel of a watercourse, or areas where drainage is or may be restricted by manmade structures which areas have been or may be covered partially or wholly by floodwater resulting from a 100 year flood.

Floodplain. The land which has been or may be covered by floodwater during the regional flood. The floodplain includes the floodway and the flood fringe.

Flood Plain. Continuous land area adjacent to a water course, the elevation of which is equal to or below the flood base elevation.

Floodplain regulations. The Codes, Ordinances and other regulations relating to the use of land and construction within the channel and floodplain areas, including zoning ordinances, sub-division regulations, building codes, housing codes, setback requirements, open area regulations and similar methods of control affecting the use and development of the areas.

Floodplain regulations. Includes the codes, ordinances, and other regulations relating to the use of land and construction within the channel and floodplain areas, including zoning ordinances, subdivision regulations, building codes, setback requirements, open area regulations and similar methods of control affecting the use and development of the areas.

Flood profile. A graph or a longitudinal profile showing the relationship of the water surface elevation of a flood event to locations along a stream or river.

Floodproofing. Involves any combination of structural provisions, changes or adjustments to properties and structures subject to flooding, primarily for the purpose of reducing or eliminating flood damage to properties, water and sanitary facilities, structures and contents of buildings in flood hazard areas.

Flood reservoir. A ponding area created for the purpose of impounding flood waters and alleviating flood damage which might result from man-made fills.

Floodway. An improved channel or an area designated for a future improved channel designed to pass the 100-year flood without raising the flood level more than one (1) foot at any point.

Floodway. That portion of the channel and floodplain of a stream designated to provide passage for the 100-year flood, without increasing the elevation of that flood at any point by more than one foot.

Floodway. The channel of a river or stream and those portions of the floodplain adjoining the channel required to carry and discharge the floodwater or flood flows associated with the regional flood.

Fully developed watershed. Planned or estimated intensity of development in the watershed or drainage area.

Hydraulic reach. A "hydraulic reach" along a river or stream is that portion of the river or stream extending from one significant change in the hydraulic character of the river or stream to the next significant change. These changes are usually associated with breaks in the slope of the water surface profile, and may be caused by bridges, dams, expansion and contraction of the water flow, and changes in stream bed slope or vegetation.

Inundation. Ponded water or water in motion of sufficient depth to damage property due to the presence of the water or to deposits of silt.

Irrigation. The controlled application of water to arable lands to supply water requirements not satisfied by rainfall.

Irrigation. The artificial application of water to the soil for the benefit of growing crops.

Irrigation. Natural, artificial, or mechanical watering of land.

Modified floodway. Any land area subject to inundation by slow moving water caused by natural forces.

One-hundred-year flood. A flood that has one percent chance of occurring in any one year based on criteria established by the State Water Commission.

Open floodway. Any land lying in the path of velocity flood waters caused by natural forces.

Permanent water level. The term "permanent water level" shall mean sea level unless special conditions exist. If special conditions exist, the term "permanent water level" shall mean such lower level as the superintendent in his opinion may deem to represent the permanent water level.

Regional flood. A flood determined to be representative of large floods known to have generally occured in the state and which may be expected to occur on a particular stream because of like physical characteristics. The flood frequency of the regional flood is once in every one hundred (100) years; this means that in any given year there is a one (1) percent chance that the regional flood may occur or be exceeded.

Retention basin. A holding pond or reservoir for surface water drained at a site to prevent flooding and fast accumulation of water.

Storage capacity of floodplain. The volume of space above an area of floodplain land that can be occupied by floodwater of a given stage at a given time, regardless of whether the water is moving.

Subject to flooding. Where any area within the proposed subdivision is known to be subject to flooding, such area shall be clearly marked "Subject to Periodic Flooding" on the lot plan and shall not be plotted in the streets and lots.

Land which normally will be inundated less frequently than once in five years may be used for recreational residential lots, or other recreation uses. In any event, easements must be reserved from normal flow line to the annual high water flow line of any water course or lake.

Watercourse. Any channel, creek, arroyo, lake, river, stream, or other body of water having natural banks and bed through which waters flow at least periodically. The term may include specifically designated areas in which substantial flood damage may occur.

Watercourse. A channel in which a flow of water occurs, either continuously or intermittently.

Watercourse. Any lake, river, creek, stream, wash, arroyo, channel or other body of water having banks and bed through which waters flow at least periodically. The term may include specifically designated areas in which substantial flood damage may occur.

Water course. Natural or constructed drainage way for water. Includes permanent streams and intermittent streams.

Waterway. A channel of water not less than fifty (50) feet wide and navigable by small boats. For the purpose of determining allowable floor areas, but not exit arrangement, of buildings, waterways shall be considered as streets.

Waterway. Any body of water, including any bayou, creek, canal, river, lake, or bay, or any other body of water, natural or artificial, except a swimming pool or ornamental pool located on a single lot.

Waterway lane. Line marking the normal division between land and a waterway as established by the appropriate governmental agency having jurisdiction.

A-6 HANDICAPPED USE OF AREAS, BUILDINGS AND STRUCTURES

Access aisle. A pedestrian space between elements such as parking spaces, seating and desks.

Accessible route. A continuous unobstructed path connecting accessible elements and spaces in a building or facility.

Automatic door. A door used for human passage and equipped with a power-operated mechanism and controls that open and close the door upon receipt of a momentary actuating signal.

Building or facility. All or any portions of buildings, structures, equipment, roads, walks, parking lots, parks, sites, or other real property or interest in such property.

Common areas. Those interior and exterior spaces available for use by all occupants and users of a building or facility, exclusively of any spaces that are made available for the use of a restricted group of people or the use of which is restricted to particular functions.

Controls. Switches and controls for light, heat, ventilation, elevators, windows, draperies, fire alarms, and all similar controls of frequent or essential use, shall be placed within the reach of individuals in wheelchairs.

Cross slope. The incline that is perpendicular to the direction of travel.

Curb ramp. A short ramp cutting through a curb or built up to it.

Developmentally disabled. Person having autism, cerebral palsey, epilepsy or mental retardation.

Disability. Any physiological disorder or condition, cosmetic disfigurement, or anatomical loss affecting one or more of the following bodily systems: neurological; musculloskeletal; special sense organs; respiratory, including speech organs; cardiovascular; reproductive; digestive; genito-urinary; hemic and lymphatic; skin; and endocrine.

Doors and doorways. Doors shall have a clear opening of not less than thirty-two (32) inches when open and shall be operable by a single effort.

Egress or means of egress. A continuous and unobstructed way of exit travel from any point in a building or facility to an exterior walk or out of a fire zone. It includes all intervening rooms, spaces, or elements.

Element. An architectural or mechanical component of a building, facility, space, or site, e.g. telephone, curb ramp, door, drinking fountain, seating and water closet.

Elevators. In a multiple-story building, elevators are essential to the successful functioning of physically disabled individuals. They shall conform to the following requirements:

Elevators shall be accessible to, and usable by, the physically disabled on the level which they use to enter the building, and at all levels normally used by the general public.

Entrances. At least one primary entrance to each building shall be usable by individuals in wheelchairs, avoiding abrupt changes in levels and with thresholds flush with the floor. At least one entrance usable by individuals in wheelchairs shall be on a level which shall make the elevators accessible.

Essential features. Those elements and spaces that make a building or facility usable by, or serve the needs of, its occupants or users. Essential features include but are not limited to entrances, toilet rooms, and accessible routes. Essential features do not include those spaces that house the major activities for which the building or facility is intended, such as classrooms and offices.

Exterior accessible routes. May include but are not limited to parking access aisles, curb ramps, walks, ramps, and lifts.

Grading. The grading of ground shall be such that it attains a level with a normal entrance or a gradient ramp thereto, and makes the facility accessible to individuals with physical disabilities.

Handicapped. Persons with reduced mobility, flexibility, coordination or perceptiveness due to age or physical conditions.

Handicapped persons use of public buildings. All buildings regulated by the Code and that are owned by the state or any political subdivision thereof, which are open to and used by the public, shall faciliatate the reasonable access and use by all handicapped persons as required under the Code. Lodging facilities shall be deemed in compliance with these regulations if 10 per cent of the units are accessible to handicapped persons. If all required units for the handicapped are located on one floor, other stories of the building need not comply. In these required units, switches and controls for lights, heating and ventilating equipment, windows, draperies, fire alarms and all similar controls of frequent or essential use required by the public, shall be placed within reach of persons in wheel chairs and of a type easily operated by handicapped persons. Utility receptacles shall be located at least 18 inches above the floor. Toilet rooms in these units shall have space to allow movement of persons in wheel chairs.

Interior accessible routes. May include but are not limited to corridors, floors, ramps, elevators, lifts, and clear floor space at fixtures.

Operable part. A part of equipment or an appliance used to insert or withdraw objects, to activate or deactivate equipment, or to adjust the equipment.

Parking. Where parking facilities are provided suitable parking spaces shall be provided and so identified for the use of the physically handicapped.

Physically handicapped person. Any person who has a disability which substantially limits one or more major life activity including but not limited to such functions as performing manual tasks, walking, seeing, hearing, speaking, learning and working.

Power-assisted door. A door used for human passage, and with a mechanism that helps to open the door, or relieve the opening resistance of a door, upon activation of a switch or a continued force applied to the door itself.

Principal entrance. A main entrance to a building or facility.

Public telephones. If public telephones are provided in the building, an appropriate number shall be made accessible to and usable by the physically handicapped.

Ramp. A walking surface that has a running slope greater than 1:20.

Ramps with gradients. Where ramps with gradients are necessary or desired, they shall conform to the following specifications:

Ramps shall not have a slope greater than one (1) foot rise in twelve (12) feet, or eight and thirty-three one hundredths (8.33) percent, or four (4) degrees - fifty (50) minutes.

Running slope. The slope that is parallel to the direction of travel.

Signage. The display of written, symbolic, tactile, or pictorial information.

Site improvements. Includes landscaping, paving for pedestrian and vehicular ways, outdoor lighting, recreational facilities, and similar site additions.

Space. A definable area, such as a toilet room, hall, assembly area, parking area, entrance, storage room, alcove, courtyard, or lobby.

Tactile. Perceptible through the sense of touch.

Tactile warning. A surface texture applied to or built into walking surfaces or other elements to warn visually impaired persons of hazards in the path of travel.

Toilet rooms. An appropriate number of toilet rooms, in accordance with the nature and use of a specific building or facility, shall be made accessible to and usable by the physically handicapped. Such toilet rooms shall have space to allow traffic of individuals in wheelchairs. A space of approximately 60″ × 60″ is required for turning. Each toilet room shall have at least one toilet stall which is or has:

(1). Three (3) feet wide.

(2). At least four (4) feet eight (8) inches preferably five (5) feet deep.

(3). A door, where doors are used, which is thirty-two (32) inches wide and swings out.

(4). Handrails on each side, thirty-three (33) inches high and parallel to the floor, one and one-half (1-1/2) inches in outside diameter, with one and one-half (1-1/2) inches clearance between rail and wall, and fastened securely at ends and center.

Walk. An exterior pathway or space with a prepared surface intended for pedestrian use and having a slope of 1:20 or less. It includes general pedestrian areas such as plazas and courts.

Warning signals. Audible warning signals shall be accompanied by simultaneous visible signals for the benefit of those with hearing disabilities.

A-7 HEALTH AND SAFETY

Air conditioning. The process by which the temperature, humidity, movement, cleanliness, and odor of air circulated through a space are controlled simultaneously.

Bathroom. Any enclosed space containing one or more bathtubs, or showers, or both and which also may contain water closets, lavatories, or fixtures serving similar purposes.

Community off-site sanitary sewage disposal. Sanitary sewage collection system in which sewage is carried from indivdual lots by a system of pipes to a temporary central treatment and disposal plant, generally serving a neighborhood area.

Contaminant. An undesirable substance or material in the form of poisonous or deleterious materials, dusts, vapors, gases, fumes or mists that are dispersed in the atmosphere in amounts in excess of the amount normally present in the atmosphere.

Corrosive liquid. Those acids, alkaline caustic liquids and other corrosive liquids which, when in contact with living tissue, will cause severe damage of such tissue by chemical action; or in case of leakage will materially damage or destroy other containers of other hazardous commodities by chemical action and cause the release of their contents; or are liable to cause fire when in contact with organic matter or with certain chemicals.

Dust. Solid particles generated by handling, crushing, grinding, rapid impact, detonation and decrepitation of organic or inorganic materials.

Environmental conditions. The aggregate of conditions and influences existing within the confines of a place of employment and in which a person is required to work.

Fire hazard. Any device or condition likely to cause fire and which is so situated as to endanger either persons or property. The creation, maintenance, or continuance of any physical condition by reason of which there exists a use, accumulation or storage for use of combustible or explosive material sufficient in amount, or so located or in such manner as to put in jeopardy, in event of ignition, either persons or property.

Fire protection. The provision of safeguards in construction and of exit facilities; and the installation of fire alarm, fire detecting and fire extinguishing service equipment to reduce the fire risk and the conflagration hazard.

Fire safety. Measure of protection of a building or structure against exterior and interior fire exposure hazards accomplished through fire-resistive construction and the provision of safe exit ways, fire detection, and extinguishing equipment, and access for fire-fighting purposes.

Fumes. Solid particles generated by condensation from the gaseous state.

Fumigation. The use within an enclosed space of a fumigant in concentrations which may be hazardous or acutely toxic to man.

Garbage. All waste, animal, fish, fowl or vegetable matter incident to the use and storage of food for human consumption, spoiled food, dead animals, animal manure, fowl manure, and septic tank effluent.

Gas. Normally formless fluids which occupy the space of enclosure and which can be changed to the liquid or solid state only by the combined effect of increases in pressure and decreased temperature.

Good industrial hygiene engineering practices. The most economical and practical engineering methods which are known to prevent the existence of a health hazard.

Habitable room. A room which is designed or may be used for living, sleeping, eating or cooking. Storerooms, bath rooms, toilets, closets, halls or spaces in attics are not habitable rooms.

Harmful concentrations. The existence of contaminants in working areas or occupied area in excess of the maximum allowable concentration specified in the Code.

Healthful and comfortable environment. A working area under conditions of effective temperature at which neither disease nor injury to health will result and as near the comfort level as good industrial engineering practices will permit. This does not apply to environments where abnormal temperature or humidities are encountered through the inherent nature of the industrial process and where individual protection is supplied through clothing, supplied air or other means.

Health hazard. An exposure to any contaminant or condition encountered in the environment or occupied area as a result of an industrial operation sufficient to injure any part of the body or reduce in efficiency the normal function of the body.

Highly toxic material. A material so toxic to man as to afford an unusual hazard to life and health during fire fighting operations. Examples are: parathion, TEPP (tetraethyl phosphate), HETP (hexaethyl tetraphosphate), and similar insecticides and pesticides.

Human excrement. The bowel and kidney discharge of human beings.

Infectious agent. Pathogenic micro-organisms capable of producing disease by entrance into the body.

Infestation. The presence of any insects, rodents or other pests. Infestation shall include breeding areas on the exterior of the premises so located that products thereof may spread to the interior of any building subject to these regulations.

Insecticide. Material or substance used to kill insects.

Livable room. The term "livable room" shall mean any room used for normal living purposes in a residence structure and shall not include kitchens, laundry rooms, bathrooms or storerooms.

Maximum allowable concentrations. The maximum allowable concentrations allowed to exist in an environment to which persons may be exposed for an 8-hour period.

Mist. Suspended liquid droplets generated by condensation from the gaseous to the liquid state or by breaking up a liquid into a dispersed state.

Noxious matter. Material which is capable of causing injury to living organisms by chemical reaction or is capable of causing detrimental effects upon the physical or economic well-being of individuals.

Nuisance. A discharge, dissemination, spreading, or emission into the open air of any air pollutant in quantities which may cause injury, detriment or damage or which may endanger, interfere with or disturb the comfort, repose, health or safety of an individual, or which causes injury or damage to business, property, plant life or animals.

Nuisance. Any nuisance which may prove detrimental to the health, or safety of children whether in a building, on the premises of a building, or upon an unoccupied lot. This includes, but is not limited to abandoned walls, shafts, basements, excavations, abandoned iceboxes, refrigerators, motor vehicles, any structurally unsound fences or structures, lumber, trash, fences, debris, or vegetation such as poison ivy, oak or sumac, which may prove a hazard for inquisitive minors.

Nuisance. Includes any building or premises, which are constructed or maintained poses a danger to the safety of occupants of adjacent property, or which unreasonably interferes with the peaceful use of adjacent property and shall include in the unlawful use of a building or premises for use which annoys, injures, or endangers the comfort repose, health, or safety of occupants of adjacent buildings or property, and shall include any attractive nuisance where a condition is permitted to exist which as existing may attract and be dangerous to children, such as abandoned wells, shafts, basements, cesspools, septic tanks, excavations, abandonded refrigerators, abandoned motor vehicles, structurally unsound fences, or abandoned and unclosed structures, lumber piles, trash piles, or similar hazards.

Nuisance. A nuisance is unlawfully doing an act, or omitting to perform a duty, or is any thing or condition which either, (1) annoys, injures or endangers the comfort, repose, health, or safety of others, (2) offends decency, (3) unlawfully interferes with, obstructs, or tends to obstruct or renders dangerous the use of any mode of transportation, or any public park, or other public property, (4) in any way renders other persons insecure in life or in the use of property, (5) affects at the same time an entire community or neighborhood or any considerable number of persons, although the extent of the annoyance or nuisance or damage inflicted upon individuals may be of unequal nature.

Occupational disease. Any pathological condition or affliction resulting from exposure to a harmful substance or condition in industry.

Outside air. Air that is taken from outside the building and is free from contamination of any kind in proportions detrimental to the health or comfort of the persons exposed to it.

Protection factor. A factor used to express the relation between the amount of fallout gamma radiation that would be received by an unprotected person and the amount that would be received by one in a shelter.

Protective equipment. A device, a permanent installation, clothing and other means for the adequate protection of the worker against health hazards defined in the Code.

Refuse. All putrescible solid wastes (except body wastes), including but not limited to garbage, rubbish, ashes, street cleanings, dead animals, abandoned automobiles, and solid market and industrial wastes.

Rubbish. Includes, but not limited to: miscellaneous waste material resulting from housekeeping, merchantile enterprises, trades, manufacturing and offices, including ashes, tin cans, glass, scrap metals, rubber, paper, rags, scrap lumber and scrap auto parts.

Rubbish. Non-putrescible solid waste consisting of both combustible and non-combustible waste, such as paper, wrappings, cigarettes, tin cans, yard clippings, leaves, wood, glass, bedding, crockery and similar materials.

Rubbish (thrash). Combustible and noncombustible waste materials including the residue from the burning of coal, wood, or coke or other combustible material, paper, rags, cartons, tin cans, metals, mineral matter, glass crockery, dust and discarded refrigerators, heating, cooking or incinerating appliances.

Sanitary. Applied to a building, means free from danger or hazard to the health of persons occupying or frequenting it or to that of the public, if such danger arises from the method or materials of its construction or from any equipment installed therein for the purpose of lighting, heating, ventilating, or plumbing.

Sanitary condition. The physical condition of working environments such as will tend to prevent the incidence and spread of disease.

Sanitary system. Either (a) the public system for disposal of sewage from the premises; or (b) in the absence of such system, any private system available to the premises for the disposal of sewage.

Sewage disposal system. A system for the disposal of sewage by means of a septic tank, cesspool, or mechanical treatment, all designed for use apart from a public sewer to serve a single establishment, building or development.

Trash. Includes, but without limitation upon other meaning, refuse, litter, ashes, leaves, debris, paper, combustible materials, offal, rubbish, waste, and useless or unused or uncared for matter of all kinds, whether solid or in liquid form.

Water supply. Such water system supply and distribution facilities as are necessary to provide a reliable and adequate water supply for private use and public fire protection purposes.

A-8 HEIGHT AND STORY REGULATIONS

Attic. Space between the ceiling beams of the top habitable story and the roof rafters, in which the area at a height of seven and one-third (7-1/3) feet above the attic floor is not more than one-third (1/3) the area of the floor next below. A habitable attic is an attic which has a stairway as a means of access and egress.

Attic. Finished or unfinished story situated within a sloping roof, the area of which at a height of four feet (4) above the level of its finished floor does not exceed two-thirds (2/3) of the area of the story immediately below it. There shall be only one attic in any building, and it shall be considered as a half story.

Attic. The space between the top of the ceiling framing of the top habitable story, or any flooring over such framing, and the roof framing and any walls constituting a part of the enclosure of such space.

Attic. Low story above the main entablature of a building, with walls vertical to ceiling; in some building codes defined as any story, in whole or part, which is situated in the roof.

Attic. The space between the ceiling beams of the top habitable story and the roof rafters.

Attic. Space not used for human occupancy located between the ceiling of the uppermost story and the roof.

Attic. A room built within the sloping roof of a dwelling. May be finished or unfinished.

Attic. Shall be taken to mean any space immediately under the roof rafters and above the ceiling joists of the story nearest to the roof.

Attic. Any space immediately under the roof of a building. Attics of business buildings when used for business or stores shall be counted as a story. Attics of residence buildings shall be counted as a story if their habitable area exceeds 50 percent of the habitable area of the floor below.

Attic (habitable). A habitable attic is an attic which has a stairway as a means of access and egress and in which the ceiling at height of seven and one-third (7-1/3) feet above the attic floor is not less than one-third (1/3) the area of the floor below.

Attic or attic story. A story which is wholly or partly in the roof space which does not have a usable floor area exceeding fifty percent (50%) of the floor below, and which is used only as a portion of a dwelling situated on the floors of the same building. An attic story shall be considered as a half story, but any story exceeding the above area, and any story occupied separately as a dwelling or used for any other purpose, shall be rated as a full story.

Attic or roof space. Space between the roof and the ceiling of the top story or between a dwarf wall and a sloping roof.

Attic room (finished attic). Attic space which is finished as living accommodations but which does not qualify as a half-story.

Attic story. Any story situated wholly or partly in the roof so designated, arranged or built as to be used for business, storage or habitation.

Basement. A portion of a building located partly underground but having less than half its clear floor-to-joist height below the average grade of the adjoining ground.

Basement. A story partly under ground, which, unless subdivided and used for tenant purposes, shall not be included as a story for purposes of height measurement.

Basement. Floor space in a building having part, but not more than one-half (1/2) of its ceiling height below grade. When a basement is used for storage, as a garage for use of occupants of the building or other facilities common to the rest of the building, it shall not be counted as a story.

Basement. A story having part but not more than one-half (1/2) of its height below grade.

Basement. A story which may have part but not more than one-half (1/2) of its height above grade and used for storage, garages for use of occupants of the building, or other utilities common to the building.

Basement. A story partly or wholly underground which, if not designed, arranged, or occupied for living purposes by other than the janitor and his family, shall not be included as a story for the purpose of height measurements.

Basement. That portion of a building between floor and ceiling, which is located that one-half (1/2) or more of the clear height from floor to ceiling is below grade.

Basement. That portion of a building between floor and ceiling which is partly below and partly above grade, but so located that the vertical distance from grade to the floor below is less than the vertical distance from grade to ceiling, provided, however, that the distance from grade to ceiling, shall be at least four (4) feet six (6) inches.

Basement. A space of full story height below the first floor wholly or partly below exterior grade and which is not used primarily for living accommodations. Space, partly below grade, which is used primarily for living accommodation or related commercial use is not defined as basement space.

Basement. A story partly underground, but having less than one-half its clear height (measured from finished floor to finished ceiling) below the curb level; except that where the curb level has not been legally established, or where every part of the building is set back more than 25 feet from a street line, the height shall be measured from the adjoining grade elevation. A basement shall be deemed a story when its ceiling is five (5) feet or more above curb level, unless otherwise specifically ordered.

Basement. A story of a building or structure having one-half or more of its clear height below grade.

Basement. The term "basement" shall mean a story partly under-ground, but having at least one-half of its height, measured from finished floor to finished ceiling, above the curb level at the center of the street front.

Basement. That portion of a building which is below the first story but more than 50 per cent above grade, and is used for storage, garages for use of occupants of building, janitor or watchman quarters, or other utilities (exclusive of rooms of habitation or assembly) common for the rest of the building. A basement used for the above purposes shall not be counted as a story, provided its height in the clear does not exceed 7 feet 6 inches.

Basement. Means that portion of a building partly below grade, but so located that the vertical distance from grade to the floor is not greater than the vertical distance from the grade to the ceiling. Provided, however, that if the vertical distance from the grade to the ceiling is five (5) feet or more, such basement shall be counted as a story.

Basement. A story, partly, but not more than one-half of its height below the level of a street grade or ground nearest the building. A basement shall not be counted as a story for the purpose of height regulation, unless it is subdivided, rented, sold or leased for dwelling purposes.

Basement. A story partly or wholly underground; where the basement floor is 6 feet or less below the average level of the adjoining ground, a basement shall be counted as a story for the purpose of height measurement.

Basement. That portion of a building between floor and ceiling, which is partly below and partly above grade, the ceiling in which is four feet (4′) above the grade.

Basement. A story partly underground and having at least one-half (1/2) of its height above average level of the adjoining ground. The basement shall be counted as a story for the purpose of height measurement if subdivided and used for dwelling or business purposes.

Basement. Any story or floor level below the main or street floor. Where, due to grade differences, there are two levels each qualifying as a street floor, a basement is any floor level below the lower of the two street floors.

Basement. A story having part but not more than one-half (1/2) of its height below grade. A basement is counted as a story for the purpose of height regulations.

Basement. A room or set of rooms located below the first floor joists and having one-half or more of its height below the average level of the adjoining ground. When designed for, or occupied by, dwellings, business or manufacturing, it shall be considered to be a story.

Basement. A story partly underground, but having less than one-half its clear height (measured from finished floor to finished ceiling) below the curb level; except that where the curb level has not been legally established, or where every part of the building is set back more than 25 feet from a street line, the height shall be measured from the adjoining grade elevation.

Basement. That portion of a building between floor and ceiling, which is partly below and partly above grade, but so located that the vertical distance from grade to the floor below is less than the vertical distance from grade to ceiling, provided, however, that the distance from grade to ceiling shall be at least four (4) feet six (6) inches.

Basement. A story having more than one-half (1/2) of its height below the average level of the adjoining ground. A basement shall not be counted as a story for purposes of height measurement.

Basement. A story partly underground which, if not occupied for living purposes by other than the janitor, or by the janitor and his family, shall not be included as a story for purposes of height measurements.

Basement. A portion of the building partly underground, but having less than half its clear height below the average grade of the adjoining ground. A basement which is used for a use other than storage shall be considered a story.

Basement. A space of full story height below the first floor which is not designed or used primarily for year-round living accommodations. Space, partly below grade, which is designed and finished as habitable space is not defined as basement space.

Basement. The space of a building where the floor level is more than four feet (4′) below the adjoining finished grade.

Basement. The story partly underground but having less than one-half (1/2) its clear height below grade. Also referred to as a "cellar".

Basement. Any area covered by a dwelling and having a clearance of more than 16 inches between the overhead structure and existing grade.

Basement. That portion of a building which is wholly below grade or partly below and partly above grade, the ceiling in which is less than 4 ft. 6 in. above grade.

Basement. A story of a building partly or wholly below grade.

Basement. A story partly underground and having at least one-half (1/2) of its height above the average level of the adjoining ground. A basement shall be counted as a story if subdivided and used for dwelling or business purposes.

Basement. Means a floor level partly or completely below grade. It shall be considered a story if more than 33-1/3 percent of the perimeter walls of a basement are 5 feet or more above grade.

Basement. That portion of a story partly below grade, the ceiling of which is four feet or more above the adjacent finished grade.

Basement. Basement means a story, partly underground, the greater part of which is above grade. A basement shall be counted as a story unless otherwise specifically stated, but shall exclude crawl space.

Basement. A story partly underground, which if occupied for living purposes, shall be counted as a story for purposes of height measurement.

Basement. Shall mean a story partly underground and having at least one-half (1/2) of its height, measured from its floor to its finished ceiling, above the average adjoining grade. A basement shall be counted as a story, if the vertical distance from the average adjoining grade to its ceiling is over five (5′) feet.

Basement. A story having part but not more than one-half (1/2) its height below finished grade. A basement is counted as a story for the purpose of height regulations, if subdivided and used for business or dwelling purposes, by others than a janitor employed on the premises.

Basement. A story wholly or partly underground. For purposes of height measurement a basement shall be counted as a story when more than one-half of its height is above the average level of the adjoining ground, or when subdivided and used for commercial or dwelling purposes by other than a janitor employed on the premises.

Basement. A story wholly or mainly below ground level.

Basementless space (crawl space). An unfinished accessible space below the first floor which is usually less than full story height.

Building height. The building height of a building or structure is the vertical distance between the mean level of the side yards immediately contiguous to the side of the building or structure and the mean level of its roof, except in the absence of the side yard, the vertical measurement shall be made from the mean original level of that portion of the lot built upon, provided that the measurement shall be made from not more than three feet (3′) below or six feet (6′) above the mean curb level, or in absence of curb level, the mean front street level.

Building height. The perpendicular distance measured in a straight line from top of the highest point of the roof beams in the case of flat roofs, and from the average height of the gable in case of roof having a pitch of more than 25 degrees with a horizontal plane, downward to the established grade in the center of the front of the building. Any penthouse or bulkhead covering less than 15 percent of the roof area need not be considered in determining the height of the building if used only for the enclosure of a staircase, water tank, elevator machinery, or other mechanical device.

Building height. Height of a building in stories does not include basements and cellars, except in school buildings of ordinary, light noncombustible or frame construction; the basement or cellar shall be deemed a story when used for purposes other than storage or heating.

Building height. Of a building in stories does not include basements and cellars, except in school buildings of ordinary noncombustible or frame construction, the basement or cellar shall be deemed a story when used for purposes other than storage or heating.

Building height. As applied to a building, means the vertical distance from grade to the highest finished roof surface in the case of flat roofs, or to a point at the average height or roofs having a pitch of more than (1′) foot in four and one-half (4-1/2′) feet.

Building height. Shall mean the vertical distance measured from the average level of the highest and lowest point of that portion of the building Site covered by the building to the highest point of the structure.

Building height. Vertical distance measured from curb or grade level to the highest level of a flat or mansard roof, or to the average height of a pitched, gabled, hip or gambrel roof, excluding bulkheads, penthouses and similar constructions enclosing equipment or stairs, providing they are less than 12 feet in height and do not occupy more than 30 percent of the area of the roof upon which they are located.

Building height. As applied to a building, the vertical distance from grade to the average elevation of the roof of the highest story. Does not include basements.

Building height. Is the vertical distance measured from the grade level to the average height of the coping of the street or outside wall for flat roofs, to the deck line of mansard roofs, and to the mean height between eaves and ridge for gable, gambrel, or hip roofs.

Building height. The vertical distance from the average natural grade of the building Site to the highest point of the coping of a flat roof, or to the deck line of a mansard roof, or the mean height level between eaves and ridge for gable, hip or gambrel roofs, except that for lots having a cross slope of ten percent (10%) or more, the vertical distance shall be measured from the natural grade at the lowest point around the perimeter of the building.

Building height. The vertical distance measured from the average elevation of the grade at the front of the building, to the highest point of the coping of a flat roof, to the mean height level between eaves and ridge for gable, hip and gambrel roofs and to the deck line of a mansard roof.

Building height. The vertical distance from the mean finished grade to the highest point of the building. The height of a wall is the vertical distance from the grade to the mean level of the top of the wall, including any dormers or gables on the wall. This does not include cooling towers, elevator equipment rooms and other structures not intended for human occupancy.

Building height. The vertical distance measured from the average elevation of the finished grade along the front of the building, to the highest point of the roof; and to the mean height level between eaves and ridge for gable, hip and gambrel roofs.

Building height. The vertical distance measured from the curb level to the highest point of the roof surface.

Building height. The vertical distance from the finished grade to the highest point of the coping of a flat roof or to the deck line of a mansard roof or to the average height of the highest gable of a pitch or hip roof.

Building height. In linear measure, the vertical distance of the highest point of the roof, excluding penthouses and roof structures, above the mean grade of the sidewalk at the line of the street or streets on which the building abuts; and, if the building does not abut on a street, above the mean grade of the ground around and contiguous to the building; and provided, further, that for the purposes of establishing said mean grade, the ground bounded by the lot lines and contiguous to the building and within twenty (20') feet of it, shall be considered to slant toward the building not more than one (1') foot upward or downward in two (2') horizontal feet. In stories, the number of stories above the floor of the first story.

Building height. The vertical distance measured from the average crown of the abutting road to the highest point of the roof surface of a flat roof, to the deck line of a mansard roof and to the mean height between eaves and the ridge of gable, hip and gambrel roofs.

Building height. Vertical distance measured from curb or grade level, whichever is the higher, to the highest level of a flat roof or to the average height of a pitched roof, excluding penthouse or other roof, excluding penthouse or other roof appendages occupying less than 30 percent of the roof area. Where a height limitation is set forth in stories, such height shall include each full story as defined herein.

Building height. The vertical distance measured from the natural grade level to the highest level of the roof surface of flat roofs, to the deck line of mansard roofs, or to the mean height between eaves and ridge for gable gambrel or hip roofs. The building height limitations of this ordinance shall not apply to church spires, belfries, cupolas, domes, monuments, water towers, chimneys, flues, vents, flag poles, radio towers, TV towers, fire lookout towers or airway beacons; nor to any bulkhead, elevator, water tank or similar structure extending above the roof and occupying an aggregate area of not greater than twenty-five (25) percent of roof area.

Building height. The vertical distance from the established average sidewalk grade, street grade, or finished grade at the building line, whichever is the highest, to the highest point of the building.

Building height. The vertical distance measured from the highest point of the established sidewalk grade to the point of intersection of the roof line with the face of the building wall, or where those would meet if produced.

Building height. The term "Building Height" shall mean the vertical distance measured from the established grade to the highest point of the roof surface for flat roofs; to the deck line of mansard roofs; and to the average height between eaves and ridge for gable, hip, and gambrel roofs. Where a building is located on sloping terrain, the height may be measured from the average ground level of the grade at the building wall.

Building height. The vertical distance from the grade to the top of the highest roof beams of a flat roof, or to the mean level of the highest gable or slope of a hip roof. When a building faces on more than one street, the height shall be measured from the average of the grades at the center of each street front.

Building height. The vertical distance from the grade to the ceiling of the topmost story.

Building height. The vertical distance from the grade to the highest point of the coping of a flat roof or to the deck line of a mansard roof; or to the mean height level between eaves and ridge for gable, hip and gambrel roofs.

Building height. The vertical distance measured from the natural grade level to the highest level of the roof surface of flat roofs, to the deck line of mansard roofs, or to the mean height eaves and ridge for gable, gambrel or hip roofs.

Building height. The vertical distance from grade to the highest point of the roof.

Building height in storys. The number of storys in a building or structure contained between the roof and the floor of the first story.

Building heights. Where portions of a building are of different heights, the height of the building shall be the greater height.

Building or wall height. Elevation of the street grade opposite the center of the wall nearest to and facing the street lot line.

Ceiling height. The clear vertical distance from the finished floor to the finished ceiling.

Ceiling height. The height from the finished floor to the finished ceiling, or when there is no finished ceiling, the height from the finished floor to the underside of the exposed secondary framing next above the floor.

Ceiling height. The clear vertical distance from the finished floor to the unfinished ceiling.

Ceiling height. Vertical distance between the floor and the ceiling. Where a finished ceiling is not provided, the underside of the joists shall determine the uppermost point of measurement.

Ceiling heights. Minimum ceiling height in any room exclusive of a room used only for storage purposes shall be 7 feet, 6 inches except under sloping roofs where the minimum shall be 7 feet, 6 inches for not less than 50% of the floor area.

Ceiling heights. Habitable rooms, other than a kitchen, shall have a ceiling height of not less than 7′-6″. Kitchens, storage rooms, laundry rooms, hallways, corridors, bathrooms, and water closet rooms shall have a ceiling height of not less than 7 feet measured to the lowest projection from the ceiling.

Court height. Vertical distance from the level of the floor of the lowest story served by that court to the level under consideration.

Court height. Vertical distance from the lowest level of the court to the mean height of the top of the enclosing walls.

Curb. When used in defining the height of a structure, shall mean the legally established level on the curb in front of the structure, measured at the center of such front. When a building faces on more than one street, the term "curb" shall mean the average of the legally established level of the curbs at the center of each front.

Curb. The term "curb" when used in fixing the depth of an excavation, shall mean the legal curb level at the nearest point of that curb which is nearest to the point of the excavation in question.

Curb elevation. The average elevation of a curb adjacent to a development from which the height of a building is determined.

Curb level. Used as the permanently established grade of the street curb in front of the lot.

Curb level. Elevation of the street curb as established in accordance with the law.

Curb level. The elevation of the street grade as established by governmental authority.

Curb level. A datum for the depth of an excavation or for grading, means the elevation of the street, as established by governmental authority, nearest to the point of excavation or grading. When the point of excavation is equally distant from 2 or more streets, the curb level is the average of the elevations of the streets, so established, nearest to said point.

Curb level. The mean level of the established curb in front of the building.

Curb level. The level of the established curb in front of the building measured at the center of the front lot line.

Curb level. Referring to a building, means the elevation at the point of the street grade that is opposite the center of the wall nearest to and facing the street line.

Curb level. The level of the established curb in front of the building measured at the center of such front. Where no curb has been established the City engineer shall establish such curb level or its equivalent for the purpose of the requirements of the Ordinance.

Curb level. The elevation at that point of the street grade that is opposite the center of the wall nearest to and facing the street line.

Curb level. The elevation of the curb opposite the center of the front of the building. If a building faces on more than one street, the curb level shall be the average of the elevations of the curbs at the center of each side or front of the building. Where no curb level or equivalent has been established by the municipal authority, the average elevation of the finished grade immediately adjacent to the front of the building shall be considered as the curb level. If a building faces on more than one street where no curb level has been established, the average of the elevations of the finished grade on each street side of the building shall be considered as the curb level.

Curb level at building. Elevation at that point of the street grade that is opposite the center of the wall nearest to and facing the street line.

Curb level in front of multiple dwellings. When an open unoccupied space in front of any multiple dwelling is above the curb level, and also extends along the entire street lot line on any street and is not less than five feet in depth, the level of such open unoccupied space shall be considered the curb level, provided it is not more than three feet above the level of the established curb in front of the building measured at the center of such front.

Curb levels. Where no curb has been established a curb level shall be established by the agency empowered to fix curb levels.

Curb line. Line of the vertical face of the curb nearest the sidewalk.

Established. Established as in the elevation of a street as established by a governmental agency.

Established datum. An elevation of a level as related to a known used datum or fixed location in height to measure other related levels or heights.

Established grade. Elevation of the street or street curb as established by a governmental agency.

Established street grade. Elevation established by the City, at the roadway centerline, or curb in front of the lot.

Exception to height regulations. Height limitations contained in the schedule of District regulations do not apply to spires, belfries, cupolas, antennas, water tanks, ventilating equipment, chimneys, or other appurtenances, excluding signs, usually required to be placed above the roof level and not intended for human occupancy.

Finished grade. The elevation of the finished surface of the ground adjoining the building after final grading and normal settlement.

Finished grade. Relative to floors above the basement is the finished grade at the lowest required grade exit serving the floor under consideration.

First floor. Floor of the first story.

First story. The lowest story or the ground story of any building, the floor of which is not more than twelve (12″) inches below the average basement or cellar used for residence purposes shall be deemed the first story; provided that a basement or cellar used purely for recreational purposes shall not be deemed the first story.

First story. Lowest story in a structure in which height of the story is more than 1/2 above the mean grade.

First story. Lowermost story that has at least half its total floor area designed and finished as living accommodations. For the purpose of determining this area, the area of halls, closets, and stairs is included. The area of storage, utility or heating rooms or spaces is not included. The location of the first story as defined in the code is based upon the use of the space rather than on the location of entrance doors or the finished grade.

First story. The story with its floor closest to grade and having its ceiling more than 6 feet above grade.

First story. The lowest story of which sixty-five per cent or more of the height is above the mean grade from which the height of the building is measured.

First story. The lowest story wholly above ground or the lowest story having more than two-thirds (2/3) of its entire wall area above ground.

First story. The lowermost story which is primarily above exterior grade on one or more sides, containing at least 50 percent living accommodations or related nonresidential uses, and from which there is direct egress to the exterior. Stories below grade used for storage, parking, mechanical equipment or other services are considered basement stories. Related nonresidential uses include laundry space, recreation or hobby rooms, commercial use and related corridor space.

Floor. Bottom or lower part of an enclosed space including any portions raised or depressed by not more than 3 feet from the designated principal level where the raised or depressed portion is treated architecturally as a part of the same principal level.

Floor level. Upper surface of a floor treated architecturally as the designated principal floor at a given elevation.

Floor level. The upper surface of the lower floor of a story, whether the floor is actually horizontal or has a slight slope to overcome differences in elevation. A floor level also includes any portions raised or depressed by not more than 3 feet from the principal floor level, where the raised or depressed portion is treated architecturally as part of the same story.

Grade. The average elevation of the ground adjoining the building or structure on all sides.

Grade. Average of the finished ground level at the center of all walls of a building. In case walls are parallel to and within five (5′) feet of a sidewalk, the ground level shall be measured at the sidewalk.

Grade. Required elevation of the ground at a building or building site as established by the Building Code of the City.

Grade. Elevation with reference to the city base, namely, a horizontal plane of reference established and used by the city.

Grade. The ground elevation established for the purpose of regulating the number of stories and the height of the building. The building grade shall be the level of the ground adjacent to the walls of the building if the finished grade is level. If the ground is not entirely level, the grade shall be determined by averaging the elevation of the ground for each face of the building.

Grade. With reference to a building means, when the curb level has been established, the main elevation of the curb level opposite those walls that are located on, or parallel with and within fifteen (15′) feet of, street lines; or, when the curb level has not been established, or all the walls of the building are more than fifteen (15′) feet from street lines, GRADE means the average of the finished ground level at the center of all walls of a building.

Grade. (as applying to the determination of building height) means the average level of finished ground adjoining a building at all exterior walls, as determined by the authority having jurisdiction.

Grade. The average of the finished ground level at the center of all walls of a building.

Grade. With reference to a building or structure, the average elevation of the ground adjoining the building or structure on all sides.

Grade. The average elevation of the ground, paved or unpaved, adjoining a building or structure, at the center of each exterior wall.

Grade. For buildings adjoining one street only, the elevation of the established curb at the center of the wall adjoining the street. For buildings adjoining more than one street, the average of the elevations of the established curbs at the center of all walls adjoining streets. For buildings having no wall adjoining the street, the average level of the ground adjacent to the exterior walls of the building. All walls approximately parallel to and not more than fifteen (15′) feet from a street line are to be considered as adjoining the street.

Grade. The mean elevation of the curb opposite those walls of the main building that are located on or within 5 feet of the street line. Where all the walls of the main building are more than 5 feet back from the street line, the grade is the mean elevation of the ground adjoining the building on all sides. The grade at any wall is the mean elevation of the ground adjoining that wall.

Grade. (1) As applied to a building less than ten feet from a street property line, grade shall be the established sidewalk elevation. (2) As applied to a building more than ten feet from a street property line, grade shall be the finished ground elevation at the building wall. (3) As applied to a building facing a two-level street, grade shall be determined by the upper level unless otherwise provided.

Grade. The elevation of the ground or paved surface adjoining a structure; also a term designating the quality of a material with respect to its mechanical or chemical properties.

Grade. The elevation of the public sidewalk at the center of the front wall, or the average elevations of the public sidewalks where the property abuts on more than one street, or the average level of the proposed ground surface at the center of walls which do not abut on, or are more than fifteen (15′) feet from a public sidewalk. In the absence of sidewalks or proposed sidewalks, the elevation of the center of the public street shall be used.

Grade. When used as a reference point in measuring "height of building" the "grade" shall be the average elevation of the finished ground at the exterior walls of the main building.

Grade. For buildings having walls adjoining one street only, the elevation of the sidewalk at the center of the wall adjoining the street.

Grade. Elevation of the ground at a building or building site, as established by the City Engineer.

Grade. A sidewalk elevation at the center of the building wall fronting on the sidewalk when such wall is located with five (5′) feet of the street lot line; or the average finished ground elevation along such wall when located more than five (5′) feet from the street lot line.

Grade. The surface of the ground, courts, lawns, yard or sidewalks adjoining the building. The established grade is the grade of the street curb lines, fixed by the City. The natural grade is the undisturbed natural surface of the ground, and the finished grade is the surface of the ground, court, lawn or yards, after filling or grading to desired elevation, or elevations, around a building or structure, but where the finished grade is below the level of the adjoining street, the established grade shall be deemed the finished grade.

Grade. (a) For buildings having walls adjoining one (1) street only, the elevation of the sidewalk, measured at one hundred (100) foot intervals for the length of the building. (b) For buildings having walls adjoining more than one (1) street, the average of the elevations of the sidewalks, measured at one hundred (100) foot intervals for the length of the walls facing the streets. (c) For buildings having no wall adjoining the street, the average level of the finished surface of the ground adjacent to the exterior walls of the building, measured at one hundred (100) foot intervals for the length of the building.

Grade. The established grade shall be prescribed by the City Engineer. Where no such grade has been established, the grade shall be the average elevation of the sidewalks at the property lines. Where no sidewalks exist, the grade shall be the average elevation of the street curbs adjacent to the property lines. Where no street curbs exist, the grade shall be the average elevation of the streets adjacent to the property lines.

Grade. The elevation established by the City Engineer for the sidewalk at the center of the wall adjoining the street for structures adjoining one (1) street only; the average of the established elevations of the sidewalks at the centers of walls adjoining streets for structures adjoining more than one (1) street; or the average level of the finished surface of the ground to the exterior walls of the structure for structures not adjoining any street. For these purposes any wall approximately parallel to and not more than twenty-five (25) feet from the street right-of-way is to be considered as adjoining the street.

Grade. For buildings adjoining one street only, the elevation of the side walls at the center of that wall adjoining the street. For buildings adjoining more than one street, the average of the elevations of the side wall at the centers of all walls adjoining streets. For buildings having no wall adjoining the street, the average level of the finished ground level at the center of all walls of the building. All walls approximately parallel to and not more than five (5′) feet from a street line are to be considered as adjoining a street.

Grade (adjacent ground elevation). The lowest point of elevation of the finished surface of the ground between the exterior wall of a building and a point five feet distant from said wall, or the lowest point of elevation of the finished surface of the ground between the exterior wall of the building and the property ground between the exterior wall of the building and the property line if it is less than five feet distant from said wall. In case walls are parallel to and within five feet of a public sidewalk, alley or other public way, the grade shall be the elevation of the sidewalk, alley or public way.

Grade at building. Elevation of surface of paved or unpaved ground adjacent to wall of a building.

Grade elevation of the established curb. For buildings adjoining one street only, the elevation of the established curb at the center of the wall adjoining the street.

Grade floor. Floor above and closest to the grade.

Grade (ground level). The average of the finished ground level at the center of all walls of a building. Where a wall is parallel or approximately parallel to and within 5 ft. of a sidewalk, the ground level shall be the elevation of the adjoining sidewalk, at the center of such wall.

Grade level. Elevation of the curb opposite a wall of a main building where the wall is located on or within 5 feet of the street line. Where all the walls of a main building are more than 5 feet back from the street line, the "grade level" is the elevation of the ground at the main entrance of the building.

Grade of average elevations of established curbs. For buildings adjoining more than one street, the average of the elevations of the established curbs at the center of all walls adjoining streets.

Grade of average elevations at exterior walls. For buildings having no wall adjoining the street, the average level of the ground adjacent to the exterior walls of the building. All walls approximately parallel to and not more than fifteen (15) feet from a street line are to be considered as adjoining a street.

Grades for building. Having walls adjoining more than one street, the average of the elevations of the sidewalk at the center of the walls adjoining the streets.

Grades for buildings. Having no walls adjoining the street, the average level of the finished surface of the ground adjacent to the exterior walls of the building.

Ground level. Horizontal plane passing through the average of the highest and lowest elevation of the ground along that facade of the building or structure which is nearest the street. In the absence of any building or structure, the points shall be located on the front setback line, between the two side setback lines (or between the side lot lines, if no side setback is specified in the district).

Ground story. The lowest story of a building, other than a basement or cellar as defined in the Building Code. The ground story of a dwelling shall not be counted as a story when devoted only to accessory uses.

Half story. A story under a sloping roof, the floor of which is not more than two (2) feet below the plate.

Half story. A story under a gable, hip or gambrel roof, the wall plates of which on at least two opposite exterior walls are not more than two feet above the finished floor of such story.

Half story. The uppermost story lying under a sloping roof. The usable floor area of which does not exceed 75 percent of the floor area of the story immediately below it and not used or designed or arranged or intended to be used in whole or in part as an independent housekeeping unit or dwelling.

Half story. A space under a sloping roof which has the line of intersection of roof decking and wall face not more than four (4) feet above the top floor level.

Half story. A story which is situated in a sloping roof, the floor area of which does not exceed two-thirds (2/3) of the floor area of the story immediately below it and which does not contain an independent apartment.

Half story. Story immediately under a sloping roof, which has the point of intersection of the top line of the rafters and the face of the walls not to exceed 3 feet above the top floor level, the floor area of which does not exceed 2/3rds of the floor area immediately below it, and which does not contain an independent apartment.

Half story. That part of a story under a gable, hip or gambrel roof, the wall plates of which on at least 2 opposite exterior walls are not more than 2 feet above the floor of such story.

Half story. A space under any roof except a flat roof which, if occupied for residential purposes shall be counted as a full story.

Half story. Portion of a building between a finished floor and the roof construction above, where the space thus enclosed has an average clear height of not more than 5 feet.

Half story. A partial story under a gable, hip or gambrel roof, the wall plates of which on at least two (2) opposite exterior walls are not more than three (3) feet above the floor of such story.

Half story. A space under a sloping roof which has the line of intersection for roof decking and wall not more than three (3) feet above the top floor level, and in which space not more than sixty (60) percent of the gross horizontal floor area is five (5) feet or more in height, measured from floor to rafters.

Half story. A story which is located within a sloping roof, the floor area of which does not exceed two-thirds (2/3) of the floor area of the story immediately below it and which does not contain an independent apartment.

Half story. That portion of a building between the eave and ridge lines of a sloping roof.

Height. Where the word "height" appears in the code it shall be interpreted to mean height above grade, surface or elevation of the structure inferred by the code unless otherwise specified or qualified.

Height. Vertical distance of a building is measured from the grade to the highest finished roof surface in the case of flat roofs or to a point at the average height of roofs having a pitch of more than 1 foot in 4-1/2 feet; height of a building in stories does not include basements and cellars, except as specifically provided otherwise in the Code.

Height. Vertical distance from one story to the next above is the top to top of 2 successive finished floor finishes.

Height. Vertical distance of an exterior wall is the top measured from the foundation wall, at grade, or from a girder or other immediate support of such wall.

Height. The vertical distance measured from the grade to the highest point of the coping of a flat roof or the deck line of a mansard roof, or to the mean level between the eaves and the ridge of a gable, hip, or gambrel roof, or to the highest point of a shed roof.

Height. The vertical distance of a wall, the top measured from the foundation wall, or from a girder or other immediate support of such wall.

Height. No building or structure shall exceed 40 feet in height, or 2-1/2 stories plus basement if any, which ever is the more restrictive, except as provided in the Ordinance.

Height. The height of a building, the vertical distance from the grade to the highest point of the coping of a flat roof or to the deck line of a mansard roof or to the average height of the highest gable of a pitch or hip roof.

Height. The term "height" of a structure shall mean the vertical distance from the curb level to the highest point of the roof beams in the case of flat roofs, or to a point at the average height of the gable in the case of roofs having a pitch of more than one foot in four and one-half, except that in the case of a structure where the grade of the street has not been legally established or where the structure does not adjoin the street, the average level of all the ground adjoining such structures shall be used instead of curb level.

Height above grade for theaters. The height of the sills of the principal entrance doors to any theater, as defined in the Code, shall be not more than 18 inches above the outside grade at that point. The floor level at the highest row of seats on the main floor shall not be more than 6 feet above the outside grade at the main entrance; the floor level at the lowest row of seats on the main floor shall be not more than 6 feet below, or above the grade at the nearest exit.

Height and area limitations. All buildings shall be constructed and protected to develop the fire-resistive ratings specified in the code; and the areas and heights of all buildings and structures between exterior walls or between exterior walls and fire walls shall not exceed the limitations fixed in the code subject to the fire limit restrictions provided in the code.

Height (building). The height of a building is expressed in both feet and stories.

Height limit. The limit of height as imposed in the ordinance for any structure or building or permitted use within the zoning district.

Height of a yard or court. The established height of a yard or court is the vertical distance from the lowest level of such yard or court to the highest point of any bounding wall.

Height of building. The perpendicular distance at the center of a building's principal front measured from the established grade to the high point of the roof for a flat roof, and to the mean height level for gable, hip or gambrel roofs. Chimneys and spires shall not be included in calculating the height. The height of any story of a building shall be excluded from the calculation of building height when seventy-five (75) percent or more of the gross floor area of said story consists of parking required for the building.

Height of building. The vertical distance measured, in the case of flat roofs, from the curb level to the level of the highest point of the roof beams adjacent to the street wall; and in the case of pitched roofs, measured from the curb level to the average height level of the gable. In the case of both flat roofs and pitched roofs, the measurement shall be made from the curb level to the center of the street facade. Where there are structures wholly or partly above the roof, the height shall be measured from the curb level to the level of the highest point of the building. Where a building stands or is to be erected on sloping ground, or will be set back from the street building line, the average level of the ground adjoining the walls of the building may be taken in measuring its height instead of the curb level.

Height of building. The vertical distance measured from the sidewalk level or its equivalent established grade opposite the middle of the front of the building to the highest point of the building for flat roofs; to the deck line for mansard roofs; and to the mean height level (between eaves and ridge) for gable and hip roofs. Where a building is located upon a terrace, the height may be measured from the average ground level of the terrace at the building wall. For accessory buildings the height shall be measured from the floor level.

Height of building. The vertical distance from grade to the top of the highest roof beams of a flat roof or to the mean level of the highest gables or slope of a hip, pitch or sloped roof. When a building faces on more than one street, the height shall be measured from the average of the grades at the center of each street front. The height of the building in stories does not include the basement and cellar.

Height of building. The vertical distance measured from the curb level to the mean level of the slope of the main roof. The height of a wall is the vertical distance from the curb level to the mean level of the top of the wall, including any dormers or gables on the wall.

Height of building. The vertical distance measured from the average elevation of the proposed finished grade at the front of the building, to the highest point of the roof for flat roofs, to the deck line of mansard roofs, and to the mean height between eaves and ridge for gable, hip, and gambrel roofs.

Height of building. The vertical distance measured from mean level of the ground surrounding the building to a point midway between the highest and lowest point of the roof, but not including chimneys, spires, towers, elevator penthouses, tanks, and similar projections.

Height of building. Height of a building or structure is the perpendicular distance measured in a straight line from the finished grade of the lot to the highest point of the roof beams at the front wall. The measurements in all cases to be taken through the center of the facade of the building or structure.

Height of building. Vertical distance from the finished grade to the highest point of the coping of a flat roof or to the deck line of a mansard roof or to the average height of the highest gable of a pitch or hip roof.

Height of building. Building's vertical measurement from the mean level of the ground surrounding the building to a point midway between the highest and lowest points of a roof, provided that chimneys, spires, signs, towers, elevator penthouses, tanks and similar projections shall not be included in calculating the height.

Height of building. A building's vertical measurement from the mean level of the ground abutting the building to the highest point in the roof line of a flat roof or of a roof having a slope of less than fifteen (15) degrees or more; provided that chimneys, spires, towers, elevator penthouses, tanks, and similar projections of the building shall not be included in calculating the height.

Height of building. The distance from the grade level, measured from the center of that face of the building having the principal entrance, to a line extended from the highest point of the building.

Height of building. The vertical distance from the grade to the top of the highest roof beams of a flat roof, or to the mean level of the highest gable or slope of a hip roof.

Height of building. The vertical distance from the average curb level in front of the lot or the finished grade at the building, to the building flat roof, to the deck line of a mansard roof, or to the average height of the gable or gambrel, hip or pitched roofs.

Height of building. The vertical distance as measured through the central axis of the building from the elevation of the lowest finished floor level to the highest point or ceiling of the top story in the case of a flat roof; to the deck line of a mansard roof; and to the mean height level between the eaves and ridge of a gable, hip, or gambrel roof.

Height of building. The vertical distance measured from the curb level to the highest point of the roof surface, if a flat roof; to the deck line of a mansard roof; and to the mean height level between eaves and ridge for a gable, hip, or gambrel roof; provided, however, where buildings are set back from the street line, the height of the building may be measured from the average elevation of the finished grade along the front of the building.

Height of building. The vertical distance measured from the average ground level at the sides of the building to the extreme high point of the building, exclusive of chimneys and similar fixtures.

Height of building. The vertical distance measured from the highest point of the coping of a flat roof, or the vertical distance measured from the ridge of a pitched roof to the average finished grade across the face of the building containing its principal entrance.

Height of building. The vertical distance from the established average grade to the highest point of the roof.

Height of building. The vertical distance from the average ground elevation around the foundation to the level of the highest point of the roof surface.

Height of building. The vertical distance measured from the grade of the average level of the highest and lowest point of the portion of the site covered by the building to the ceiling of the uppermost story.

Height of building. The vertical distance from the grade, to the highest point of the coping of a flat roof, or to the deck line of a mansard roof, or to a point midway between elevation of the eaves and elevation of the ridge, for gable, hip and gambrel roofs.

Height of building. The vertical distance from the grade to the highest point of the coping of a flat roof, or to the deck line or highest point of coping or parapet of a mansard roof, or to the mean height level between eaves and ridge for gable, hip shed, and gambrel roofs. When the highest wall of a building with a shed roof is within 30 feet of a street, the height of such building shall be measured to the highest point of coping or parapet.

Height of building. The vertical distance from the grade to the top of the highest roof deck of a flat roof, to the deck level of a mansard roof, and to the mean level of the highest gable or slope of a pitched roof.

Height of building. The vertical distance from the ground level adjoining the building to the highest point on the roof surface if a flat roof, to the deck line for mansard roofs, to mean height level between eaves and ridge for gabled, hip or gambrel roofs.

Height of building. Height of building is measured from the average of the exit discharge grade elevations of all required first story exits to the top of a level roof or to a point 1/2 of the distance between the intersection of the exterior wall surface (extended) with the roof surface, and the highest part of the roof but not to include penthouses.

Height of court. The vertical distance from the lowest level of such court to the highest point of any bounding wall.

Height of court. The vertical distance from the lowest level of the court as actually constructed, or from grade level, whichever is higher, to the top of the walls bounding the court, or to the level under consideration. In case the tops of such walls are at different elevations, the measurement shall be taken to the average elevation of the two highest walls that are opposite.

Height of court or yard. The vertical distance from the lowest level of such court or yard to the highest point of any building wall, forming a boundary of such court or yard.

Height of court or yard. The vertical distance from the lowest level of such court or yard to the highest point of any bounding wall.

Height of story. The vertical distance from the top surface of the floor to the top surface of the floor next above. The height of the topmost story is the distance from the top surface of the floor to the top surface of the ceiling joists. When any story exceeds twelve (12) feet in height, each twelve (12) feet or fraction thereof shall be deemed an additional story.

Height of story. Vertical distance from top to top of two successive tiers of floor beams or finished floor surfaces, and for the topmost story, from the upper limit of the story below to the top of the uppermost ceiling joists, or, where there are no ceiling joists, to the uppermost level to which the height of the building is measured.

Height of wall. Vertical distance from the foundation wall or other immediate support of the wall to the top of the wall.

Main floor. First or principal tier within a building, space, room used for any occupancy.

Mean grade. Shall be calculated as the average grade or elevations of the finished surface of ground adjacent to exterior walls surrounding a building, and within 20 feet of the exterior walls at the slope stipulated but not beyond property lines. Where a building abuts other buildings, only the grades proximate to exterior walls shall be considered in computing the Mean Grade. At abrupt change of grades exceeding one story in height, as at high retaining walls proximate or adjacent to the building, Mean Grade shall be computed excluding from a point adjacent to the top of the retaining wall and extending to the adjacent wall of the building.

Mean lot elevation. The average elevation of a lot.

Mezzanine. An intermediate story between the floor and ceiling of a main story and extending over not more than twenty-five (25) percent of the main floor.

Mezzanine. An intermediate floor placed in any story or room. When the total area of any such mezzanine exceeds 33-1/3 percent of the total floor area in that room or story in which the mezzanine occurs, it shall be considered as constituting an additional story. The clear height above or below a mezzanine-floor construction shall not be less than seven (7) feet.

Mezzanine. A low ceiling story between two (2) main stories of a structure, containing no more than one-third (1/3) of the floor area of either the floor directly above or below.

Mezzanine. An intermediate floor between the floor and ceiling of any room or story.

Mezzanine. An intermediate floor in any story occupying, not to exceed one-third (1/3) of the floor area of such story.

Mezzanine. An intermediate or fractional story between the floor and ceiling or a main story occupying and more than one third (1/3) of the floor area of such main story.

Mezzanine. An intermediate floor placed in any story of a building and limited in area as required elsewhere in the Code.

Mezzanine. An intermediate floor between the floor and ceiling and covering one-third or less of the floor area of such story.

Mezzanine. An intermediate floor in any story of a building, when such floor has an area of not more than 33-1/3 per cent of the area of the floor of the room or area enclosed by walls or partitions, in which the intermediate floor is placed.

Mezzanine. An intermediate floor between the floor and ceiling of any story, and covering less than thirty-three and one-third (33-1/3) per cent of the floor area immediately below.

Mezzanine. An intermediate floor placed within a story or room of a building. A mezzanine shall have a clear height of not less than 6 ft. 8 in. above and below the mezzanine floor construction. If the floor area of a mezzanine is more than 1/3 the floor area of floor below, it shall be considered as constituting an additional story.

Mezzanine floor. An intermediate floor placed in any story or room. When the total floor area of any such mezzanine floor exceeds 33-1/3% of the total area in that room or floor, it shall be considered as constituting an additional "story."

Mezzanine story. A story which covers one-third (1/3) or less of the story directly underneath it.

Minimum habitable room height. Clear height from finished floor to finished ceiling shall not be less than 8 feet, except under special conditions as provided in the code.

Minimum habitable room height. A clear height from finished floor to finished ceiling of not less than eight (8) feet in the basement, seven and one-half (7-1/2) feet in the first story and of not less than seven and one-third (7-1/3) feet in the second and third stories.

Minimum habitable room height. A clear height from finished floor to finished ceiling of not less than seven and one-half (7-1/2) feet in basement and upper stories and not less than seven and one-third (7-1/3) feet for attic and top halfstories over not less than one-third (1/3) the area of the floor when used for sleeping, study or similar activity.

Number of stories. Of a multi-story building includes all stories except the basement(s), ground floor(s), attic or interior balcony(ies) and/or mezzanine floor(s).

One-half story. Space under a sloping roof which has the line of the intersection of roof decking and wall face not more than 3 feet above the top floor level, and in which space not more than 2/3rds of the floor area is finished for use. A half story containing independent apartment or living quarters shall be deemed a full story.

One story building. Any building one story in height. It may or may not have a basement or cellar. It may be either a separate building or a one-story portion of a building separated from the remaining portion of the building by a fire wall.

Story. That portion of a building between any floor and the next floor above, except that the topmost story shall be that portion of a building between the topmost floor and the ceiling or roof above it. If the finished floor level directly above a basement or cellar is more than six (6) feet above grade such basement or cellar shall be considered a story.

Story. That portion of a building included between the surface of any floor above the average elevation of ground at the foundation wall and the surface of the next floor above, or if there is no floor above it, then the space between the floor and ceiling next above it.

Story. That portion of a building, excepting a cellar or basement, included between the upper surface of any floor and the upper surface of the floor next above, except that the topmost story shall be that portion of a building included between the upper surface of the topmost floor and the roof above. A cellar or basement is that portion of a building situated between the upper surface of any floor which is constructed below grade and the upper surface of the floor next above, except that if said upper surface of the floor next above is more than six (6) feet above grade such portion shall be deemed to be a story of the building rather than a cellar or basement.

Story. That portion of a building included between the upper surface of any floor and the upper surface of the floor next above, except that the topmost story shall be that portion of a building included between the upper surface of the topmost floor and the ceiling or roof above. If the finished floor level directly above a basement or cellar is more than six (6) feet above grade such basement or cellar shall be considered a story.

Story. That portion of a building included between the surface of any floor and the surface of the next floor above it, or if there be no floor above it, then the space between such floor and the ceiling next above it.

Story. That portion of a building included between the upper surface of any floor above (except basements), and that portion of a building included between the upper surface of the topmost floor and the ceiling or roof above. A floor of a building having a ceiling more than 4 feet 6 inches above the averaged finished exterior grade at its perimeter shall be considered a story.

Story. That portion of a building between a floor and the next floor above, or roof.

Story. That portion of a building included between the surface of any floor and the surface of the floor next above it, or the space between such a floor and the ceiling next above it.

Story. That portion of a building included between the surface of any floor and the surface of the next floor above it, or if there is no floor above it, then the space between the floor and the ceiling next above it. A basement, the ceiling of which is less than four feet six inches (4′6″) above the grade level, shall not be considered a story. A mezzanine floor shall be considered a story if it exceeds thirty-three and one-third percent (33-1/3%) of the area of the floor next above.

Story. That part of a building comprised between a floor and the floor or roof next above.

Story. That portion of a building between a floor and the next floor above.

Story. Portion of a building which is between one floor level and the next higher floor level or the roof. If a mezzanine floor area exceeds one third of the area of the floor immediately below, it shall be deemed to be a story. A basement shall be deemed to be a story when its ceiling is 6 or more feet above the finished grade. A cellar shall be deemed to be a story. An attic shall not be deemed to be a story if unfinished and without human occupancy.

Story. That portion of a building included between the upper surface of any floor and the upper surface of the floor next above, except that the topmost story shall be that portion of a building included between the upper surface of the topmost floor and the ceiling or roof above.

Story. That part of a building between the top of a floor and the top of the next floor above it, or if there is no floor above it, that part between the top of a floor and the ceiling above it, but does not include a penthouse that is not used by the public, and the story closest to grade having its ceiling more than six feet above grade shall be deemed to be the first story.

Story. That portion of a building between the surface of any finished floor and the surface of the finished floor next above it, or if there be no floor above it, then the space between any floor and the ceiling next above it.

Story. That part of a building comprised between a floor and the floor or roof next above. The first story shall be the one whose floor is not more than 5 feet above the average level of ground surrounding the building.

Story. A space between any two floors or between the topmost floor and the ceiling.

Story. That part of a building above the basement or cellar, between the top of any tier of floor beams, or floor slab if there are no beams, and the top of the tier of floor or roof beams, floor slab, or rafters next above.

Story. That portion of a building between floors and the floor or roof above.

Story. That part of a building comprised between a floor and the floor or roof next above.

Story. That part of a building between the surface of a floor and the ceiling immediately above.

Story. That portion of any building comprised between any floor and the floor or roof next above.

Story. That part of a building between a floor and a floor or roof next above.

Story. That portion of a building between a floor and the next floor above is considered a story.

Story. That portion of a building included between the surface of any floor and the surface of the floor next above it, or if there be no floor above it, then the space between such floor and the ceiling above it, provided that a basement shall not be considered a story.

Story. That portion of a building included between the surface of any floor and the surface of the next floor above it, or if there be no floor above it, then the space between such floor and the ceiling next above it. In computing the height of building, the height of basement or cellar if below grade shall not be included.

Story. That portion of a building, other than a cellar, included between the surface of any floor and the surface of the floor next above it, or, if there be no floor above it, then the space between the floor and the ceiling next above it.

Story. That portion of a building included between the upper surface of any floor and the upper surface of the next floor, except basements, and except that the topmost story shall be that portion of a building included between the upper surface of the topmost floor and the ceiling or roof above. If the ceiling over a basement is more than four (4) feet above grade, such basement shall be considered a story.

Story. That portion of a building, (other than a cellar or a basement used for dwelling purposes), included between the surface or any floor and the surface of the floor next above it; or if there be no floor next above it, then the space between such floor and the ceiling next above it. A cellar or basement being used for dwelling purposes shall be considered as a story.

Story. That part of a building comprised between a floor and the floor or roof next above. A mezzanine shall be considered a story if it exceeds 33-1/3 percent of the area of the floor immediately below. A penthouse shall be considered a story if it exceeds 1,000 square feet or 33-1/3 percent of the roof area. The basement of a building used for educational occupancy shall be considered a story if it is used for purposes other than storage or heating.

Story. That portion of a building wholly above the ground included between the upper surface of any floor and the ceiling or the upper surface of the floor next above; or that portion of a building immediately under the roof having a floor area that is 50% or more of the floor area of the story immediately below and with a ceiling height of at least eight (8) feet.

Story. The space between any finished floor of a building and the next finished floor above, excepting that a cellar or basement shall not be considered as a story.

Story. That portion of a building included between the surface of a floor and the surface of a floor next above it, or if there is no floor above it, then the portion of the building between the surface of a floor and the ceiling or roof above it. A basement shall be counted as a story for the purposes of height regulations if the vertical distance from grade to the ceiling is more than 7 feet, or if used for business or for dwelling purposes by other than a janitor or a caretaker, with or without his family.

Story. That part of a building except a mezzanine, as defined in the code, included between the surface of one floor and the surface of the next floor, or if no floor above, then the ceiling next above. A story, thus defined, shall not be counted as a story when more than 50 percent by cubic content is below the height level of the adjoining ground.

Story. That portion of a building other than a basement included between the structure of the floor next above it, or, if there is no floor above it, the space between the floor and the ceiling next above it.

Story. That portion of a building included between the upper surface of any floor and the upper surface of the floor above, or any portion of a building between the topmost floor and the roof having a usable floor area equal to at least fifty (50) percent of the usable floor area of the floor immediately below it, a top floor with less floor area is a half-story.

Story. That portion of a building which is between one floor level and the next higher floor level or, where there is no higher floor level, is the portion of a building which is between the highest floor level and the underside of the ceiling or roof surface directly above. If the ceiling over a basement or cellar is more than 4 feet 6 inches above grade level such basement or cellar shall be considered an above-grade-level story of the building.

Story. That part of any building comprised between the level of one finished floor and the level of the next higher finished floor, or if there is no higher finished floor, then the term "story" shall mean that part of the building comprised between the level of the highest finished floor and the top of the roof beams. A basement shall be counted as a story. A cellar shall not be counted as a story.

Story. (as applying to plumbing systems) An interval between 2 successive floor levels, or floor level and roof beginning at the lowest gravity soil-or-waste pipe.

Story. That portion of a building which is situated between the top of any floor and the top of any floor of the floor next above it, and if there is no floor above it, that portion between the top of such floor and the ceiling above it.

Story. That portion of a building, included between a floor which is calculated as part of a "building's floor area," and the floor or roof next above it.

Story. Space in a building between the surfaces of any floor and the floor next above or below, or roof next above, or any space not defined as basement, ground floor, mezzanine, balcony, penthouse or attic.

Story. Any horizontal portion through a building between floor and ceiling of which the ceiling is six feet or more above the average grade of the sidewalk or ground adjoining.

Story. That portion of a building included between the top surface of a floor and the top surface of the next floor or roof above, except that a space used exclusively for the housing of mechanical services of the building shall not be considered to be a story if access to such space may be had only for maintenance of such services.

Story. That portion of a structure between the upper surface of any floor and the upper surface of the floor or flat roof immediately above. In the case of sloping roofs and uninhabitable attics, the top story shall be from the surface of the topmost story to the ceiling above.

Story. That portion of a building included between the upper surface of a floor and the upper surface of the floor or roof next above.

Story. That portion of a building included between the surface of any floor and the surface of the floor next above it, or if there be no floor above it, then the space between such floor and the ceiling next above it, shall be considered a story.

Story. A space within a building included between the surface of any floor and the surface of the ceiling above.

Story. That portion of a building included between the upper surface of any floor and the upper surface of the floor next above, except that the topmost story shall be that portion of a building included between the upper surface of the topmost floor and the ceiling or roof above. If the finished floor level directly above a basement, cellar or unused underfloor space is more than six feet above grade as defined herein for more than 50% of the total perimeter, or is more than twelve feet (12′) above grade as defined herein at any point, such basement, cellar, or unused underfloor space shall be considered as a story.

Story. That portion of a building, except a mezzanine as defined in the Building Code, included between the surface of any floor and the surface of the next floor above it or if there is no floor above it, then the space between the surface of the floor and the ceiling next above it.

Story. That portion of a building included between the surface of any floor and the surface of the next floor above it, or if there is no floor above it, then the space between the floor and the ceiling next above it. A basement, the ceiling of which is less than four feet six inches (4′6″) above the grade level shall not be considered a floor. A mezzanine floor shall be considered a story if it exceeds forty per cent (40%) of the area of the floor next below it.

Story. That portion of a building, included between the surface of any floor and the surface of the floor next above it, or if there be no floor above it, then the space between the floor and the ceiling next above it.

Story (first). The lowest story in a building other than a basement.

Story height. Vertical distance from top to top of two successive finished floor surfaces.

Story height. The vertical distance from the surface of a floor to the surface of the next floor above, or to the ceiling of the top story.

Story height. As applied to a story, means the vertical distance from top to top of two (2) successive tiers of floor beams or finished floor surfaces.

Story height. The vertical distance from the top of one floor to the top of the next floor or roof beam above. Any floor or the common area of floors at any one level extending over less than 33-1/3 per cent of the horizontal area included within the outside walls at that level shall not be considered a floor for the purpose of determining story heights.

Story height. As applied to a story, means the vertical distance from top to top of two successive finished floor surfaces.

Story height. The height of a building in stories does not include basements and cellars, except as specifically provided in the Code.

Story height. The term "height" as applied to a story, shall mean the vertical distance from top to top of two successive tiers of floor beams.

Story height. Vertical distance from top to top of two (2) successive tiers of beams or finished floor surfaces; and, for the top-most story, from the top of the floor finish to the top of the ceiling joists, or, where there is no ceiling, to the top of the roof rafters.

Superficial floor area. Net floor area within the enclosing walls of the room in which the ceiling height not less than 5 feet, excluding built-in equipment such as wardrobes, cabinets, kitchen units, floor fixtures.

Top story. The story between the uppermost floor and the ceiling or roof above.

Upper story. Any story above the first story.

Wall height. The height as measured from its base line either at the grade or at the top of a girder to the top of coping or the center of the highest gable. In measuring the height of a wall, the height of the parapet above the top of the roof beams shall not be included.

Wall height. As applied to walls, shall mean the distance above the base of the wall or its means of support, but shall not include the parapet if the latter is four (4) feet or less in height.

Wall height. Vertical distance from the foundation wall or other immediate support of such wall to the top of the wall.

Wall height. Vertical distance to the top measured from the foundation wall, or from a girder or other immediate support of such wall.

A-9 LAND AND BUILDING AREAS

Access corridor. Portion of the site providing access from a street and having a minimum dimension less than the required site width, except that no portion of a site having side lot lines radial to the center of curvature of a street from the street property line to the rear lot line shall be deemed an access corridor. The area of an access corridor shall not be included in determining the area of the site.

Area. As applied to a building or structure, means the maximum horizontal projected area of the building or structure at or above grade.

Area. As applied to a form of construction, means an uncovered sub-surface space adjacent to a building.

Area. As applied to the dimensions of a building, the maximum horizontal project area of the building at grade.

Area. The surface of a building or site, in one plane, measured in square units such as square feet or square yards.

Area. The term "area" shall mean an open space below the ground level immediately outside of a structure, and enclosed by substantial walls.

Area. Maximum horizontal area of the building at finished grade, exclusive of unroofed porches, terraces, steps, and areaways.

Area. Area contained within the property lines of the individual parcels of land as shown on a subdivision plan, excluding any area within a street right-of-way, but including the area of any easement.

Area. The maximum horizontal projected area of the building or structure at or above grade.

Area (building). Maximum horizontal projected area of the building at or above grade, including all enclosed extensions.

Area (floor). The area included within surrounding walls of a building (or portion thereof), exclusive of vent shaft and courts.

Area (floor surface measurement). Horizontal projected floor area inside of exterior enclosure walls or between exterior walls and fire walls.

Area (gross). Maximum horizontal projected area within the perimeter of the outside surface of walls or supports of the building or structure. Exterior cantilever open balconies are not included.

Area (net). Occupied or usable floor area in a building but not including space occupied by columns, walls, partitions, mechanical shafts or duct spaces.

Area of a lot. Total horizontal area within the lot boundary lines of a zoned lot.

Area of a structure. The term "area of a structure" shall mean the total horizontal area including the exterior walls.

Area of a structure. The term "area of a structure" shall mean, except in the application of the building zone resolution, the horizontal area within the exterior walls or between fire walls. Premises between fire walls shall be considered as separate structures.

Area of house trailer spaces. Based upon the gross area of the park, the number of individual unit spaces shall be not more than ten (10) per gross acre. The minimum area of any space for a house trailer shall be not less than three thousand (3,000) square feet with no dimension less than forty (40) feet. No such space shall be located less than twenty-five (25) feet from street lot lines or interior lot lines. House trailers shall be located on each space so that there will not be less than fifteen (15) feet to any other house trailer or building within the park.

Area of principal building. Horizontal area bounded by the outside of the foundation walls and of the floors of roofed porches and roofed terraces inclusive.

Area of unit of occupancy. Maximum horizontal projected area of a unit of occupancy.

Area of zone lot. Area of land enclosed within the boundaries of a zoned lot.

Average width of lot. The quotient resulting from the division of the area of the lot in square feet by the maximum depth of lot in feet.

Basic floor area. Total amount of gross floor area a building contains, expressed as a percentage of the total area of the lot.

Basic floor area ratio (FAR). The basic floor area ratio shall determine the maximum floor area allowable for the building or buildings on a lot in direct ratio to the gross area of the lot. The basic floor area ratio of the building or buildings on any lot is the total area of all principal and accessory buildings on the lot divided by the area of such lot. Basic floor area ratio shall not include any applicable premiums.

Buildable area. The area of that part of the lot not included within the yards or open spaces required by the Ordinances.

Buildable area. Area of a lot not included within the yards or open spaces required.

Buildable area. The space remaining on a parcel after the minimum open-space requirements (maximum ground coverage, yards, setbacks) have been met.

Buildable area. The buildable area of a lot is the space remaining after the minimum open space requirements of the ordinance have been complied with.

Buildable area. The portion of a lot remaining after required yards have been provided.

Buildable width. The buildable width of a lot is the width of the buildable area, generally the distance between the inner boundaries of the two required side yards.

Buildable width. The maximum width on a parcel which may be occupied by a principal structure or use after the minimum open-space requirements (maximum ground coverage, yards, setbacks) have been met.

Buildable width or buildable depth. The width or depth respectively of that part of the lot not included within the open spaces required by the code.

Building area. The portion of the lot occupied by the main building, including porches, carports, accessory buildings, and other structures.

Building area. The greatest horizontal area of a building above grade within the outside surface of exterior walls, or within the outside surface of exterior walls and the center line of firewalls.

Building area. The total of areas taken on a horizontal place at the main grade level of the principal building and all accessory buildings exclusive of uncovered porches, terraces, and steps.

Building area. That area within and bounded by the building lines established by required yards and setbacks.

Building area. The maximum horizontal projected area of a building and its accessories.

Building area. The total areas taken on a horizontal plane at the mean grade level of the principal buildings and all accessory structures.

Building area. The maximum horizontal projected area of a building above the finished grade, including all enclosed appendages.

Building area. The maximum horizontal projected area of a building above ground, within the property lines, including exterior walls one or more of which may be party walls and including covered porches but excluding terraces, steps and cornices.

Building area. The total ground area taken on a horizontal plane at the mean grade level of each building and accessory building, but not including uncovered entrance platforms, terraces, and steps.

Building area. The maximum horizontally projected area of the building at or above grade, exclusive of court and vent shafts.

Building area. The aggregate of the maximum horizontal cross-section area, excluding cornices, eaves, and gutters, of all buildings on a lot.

Building area. The maximum horizontal projected area of a building and its accessory buildings, provided that the following shall be excluded in computing the building area occupied: open patios, steps and terraces below the first floor level (but not terraces, patios or porches which are permanently roofed over); temporary awnings and temporary supports, whether or not above open terraces or patios; detached structures (except carports) having no permanent roof; and fall-out shelters below ground level.

Building area. The horizontal area of a building including all projections from the building.

Building area. The maximum horizontal projected area of a building and its accessory buildings, excluding permitted obstructions in required courts.

Building area. The ground area in square feet under the roof of and bounded by the inside surfaces of the exterior walls of a building and including the areas under projections from the building unless excluded by the Code.

Building area. The total of areas taken on a horizontal plane at the main grade level of the principal building and all accessory buildings exclusive of uncovered porches, terraces, and steps.

Building area. The total areas, taken on a horizontal plane at the mean grade level, of the principal buildings and all accessory buildings exclusive of uncovered porches, terraces, and steps.

Building area. The maximum horizontal projected area of the building at or above grade, including all enclosed extensions.

Building area. The total of areas taken on a horizontal plane at the main grade level of the principal building and all accessory buildings or constructions exclusive of steps.

Building bulk. The term used to indicate the size and setbacks of buildings or structures and the location of a building or structure with respect to another building or structure, and includes the following: (a) size and height of buildings, (b) location of exterior walls at all levels in relation to lot lines, streets, and other buildings or structures, (c) gross floor area of buildings or structures in relation to lot area, (d) all open spaces allocated to buildings, and (e) amount of lot area provided per dwelling unit.

Building coverage. Ratio of the total ground floor area of all buildings and other structures on a lot to the total area of the lot on which they are located.

Building coverage. That percentage of the total plot or lot area covered by the principal building.

Building coverage. No more than 15% of the lot area shall be occupied by a building or buildings in the case of a building or buildings of six stories. For each additional story in height, the lot coverage shall be reduced by three-fourths of one percent. Where there are principal buildings of varying heights, the average height in stories will determine the lot coverage.

Building lot coverage. The ratio between the ground floor area of all buildings and structures on a lot and the total area of the lot.

Buildings and land areas. The area of buildings, and area at the ground level of the main building and all accessory buildings, excluding unenclosed porches, terraces and steps, measured from the outside surface of exterior walls.

Building site area. The ground area of a building or buildings, together with all open spaces, as required by the Ordinance.

Buildings on same lot. If more than one building is hereafter placed on a lot, or if a building is placed on the same lot with existing buildings, the several buildings may be treated as a single structure for the purpose of the code, provided equivalent uncovered lot area or other adequate sources of light and ventilation are provided for all habitable and occupiable spaces and rooms.

Bulk. The term used to define the size and setbacks of buildings or structures and location of same with respect to one another and includes the following: (a) Size and height of buildings; (b) Location of exterior walls at all levels in relation to lot lines, streets or to other buildings; (c) Gross floor area of buildings in relation to lot area (floor area ratio); (d) All open spaces allocated to buildings; and (e) Amount of lot area per dwelling unit, as defined in the Ordinance.

Bulk. A composite characteristic of a given building or structure as located upon a given lot, not definable as a single quantity but involving all of these characteristics: (1) location of exterior walls at all levels in relation to lot lines, streets, or to other buildings or structures, (2) size and height of building or structure, (3) floor area ratio, (4) all open spaces allocated to the building or structure, and (5) amount of lot area provided per dwelling unit.

Bulk. The three-dimensional space of a structure.

Bulk. A term used to indicate the size and setbacks of buildings or structures and the location of same with respect to one another and includes the following: (1) Size and height of buildings; (2) Location of exterior walls at all levels in relation to lot lines, streets, or other buildings; (3) Gross floor area of buildings in relation to lot area (floor area ratio); (4) All open spaces allocated to buildings; (5) Amount of lot area provided per dwelling unit.

Bulk. Term used to describe the size and shape of a building or structure and its relationship to the other buildings, to the lot area for a building, and to the open spaces and yards.

Bulk limitations (floor area ratio). The number of square feet of floor area as defined herein which is permitted for each square foot of lot area.

Coverage. That percentage of the total plot or lot area covered by the building area.

Coverage. That percentage of the plot area covered or occupied by buildings or roofed portions of structures.

Coverage. That percentage of the plot area covered by buildings including accessory buildings.

Coverage. The percent of the total site area covered by structures, open or enclosed, excluding uncovered steps, patios and terraces. The area covered by a structure shall be the area under the roof of the structure.

Coverage. The lot area covered by all buildings located thereon, including the area covered by all the hanging roofs.

Density. The numerical value obtained by dividing the total dwelling units in a development by the gross area of the tract of land upon which the dwelling units are located.

Depth. Horizontal distance between the front and rear property lines of a site measured along a line midway between the side property lines.

Depth of lot. Horizontal distance between the front and rear lot lines, measured in the general direction of its side lot lines.

Depth of lot. The mean distance between its front street line and its rear line. The greater frontage of a corner lot is its depth, and its lesser frontage, its width.

Depth of lot. The mean distance from the street line of the lot to its rear line measured in the general direction of the side lines of the lot.

Depth of lot. The mean horizontal distance, between the front and rear lot lines.

Depth of lot. The average horizontal distance between the front and rear lot lines.

Depth of lot. The horizontal distance between the front and rear property lines of a site measured along a line midway between the side property lines.

Extension. Increase in the amount of existing floor area used for an existing use in an existing building.

Floor area. The total floor area of one story within exterior enclosing walls of a building or between exterior walls and firewalls of a building.

Floor area. The space on any story of a building between exterior walls and required firewalls, including the space occupied by interior walls and partitions but not including exits and vertical service spaces that pierce the story.

Floor area. The area of all floors computed by measuring the dimensions of the outside walls of a building excluding attic and basement floors, porches, patios, terraces or breezeways, and carports, verandas and garages.

Floor area. The sum of the gross areas of the several floors of a building or buildings, measured from the exterior faces of exterior walls or from the center lines of walls separating 2 buildings.

Floor area. The total area of all stories devoted to residential use, including halls, stairways, elevator shafts and other related non-residential use, measured to outside faces of exterior walls.

Floor area. The sum of the gross horizontal areas of the several floors of a dwelling unit, exclusive of porches, balconies, garages, basements and cellars, measured from the exterior faces of the exterior walls or from the center lines of walls or partitions separating dwelling units. For uses other than residential, the floor area shall be measured from the exterior faces of the exterior walls or from the centerline of walls or partitions separating such uses, and shall include all floors, lofts, balconies, mezzanines, cellars, basements, and similar areas devoted to such uses.

Floor area. Sum of the gross horizontal areas of all floors of a building measured from the exterior faces of the exterior walls or from the center line of walls separating buildings.

Floor area. Shall mean, for the purpose of computing the minimum allowable floor area in a residential dwelling unit, the sum of the horizontal areas of each story of the buidling measured from the exterior faces of the exterior walls. The floor area measurement is exclusive of areas of basements, unfinished attics, attached garages, breezeways, and enclosed and unenclosed porches.

Floor area. A floor area within surrounding walls of a building or portion thereof.

Floor area. An area included within the outside dimensions of a building, which measurements shall include bays, porches, galleries, and other projections for the floor of which they are a part.

Floor area. The area included within surrounding walls of a building exclusive of vent shafts, elevator shafts, courts and fire towers.

Floor area. For the purpose of computing the minimum allowable floor area in a residential dwelling unit the sum of the horizontal areas of each story of a building shall be measured from the exterior faces of the exterior walls. The floor area measured is exclusive of areas of basements, unfinished attics, attached garages, or space used for off-street parking, breezeways, and enclosed and unenclosed porches, elevators, or stair bulkheads, common hall areas and accessory structures.

Floor area. An area included within exterior walls of a building exclusive of courts. Total or gross floor area is the sum of the floor areas of each story plus the floor areas of all basements and intermediate floors.

Floor area. As applied to area limitations: the floor space enclosed by exterior walls, firewalls or a combination of these structural elements. As applied to the capacity of a building or floor of a building: the floor space enclosed by the exterior walls of a building, excluding elevators, stairways or other shafts. As applied to a space or room: the net area within the enclosing walls or partitions.

Floor area. The horizontal projected floor area inside of the exterior enclosure walls, and extremities of the floor where there is no exterior wall, or between the inside of exterior enclosure walls, and extremities of the floor where there is no exterior walls, and the nearest side of fire walls.

Floor area. Where used as the basis for requirements for circuits and service in the standard, area is computed from the outside dimensions of the house and the number of floors, including unfinished spaces which are adaptable to future living use. Open porches and garage may be excluded from the calculation.

Floor area. The area included within the outside lines of the exterior walls of the main structure at the ground floor level, not including garages, breezeways, unenclosed porches, and not including attached utility or accessory rooms having three (3) or more exterior sides. Provided, however, utility rooms with no more than two (2) exterior walls, enclosed porches and breezeways, equipped with heating and ventilating facilities and finished on the inside, may be included. The livable floor area shall be the area on the the floors having a minimum clear height of five (5) feet and shall include interior partitions.

Floor area. The area included within the surrounding exterior walls of a building or portion thereof, exclusive of vent shafts and courts. The floor area of a building, or portion thereof, not provided with surrounding exterior walls shall be the usable area under the horizontal projection of the roof or floor above.

Floor area. The area at floor level, projected to a horizontal plane, within exterior and fire walls, exclusive of elevator shafts, ducts, stairways, walls, smokeproof towers, columns and similar parts of a structure not used for the permitted purposes under the Occupancy Classifications. Floor areas shall be the sum of the areas at each floor level, within the external dimension of the sturcture.

Floor area. Total area of all stories or floors finished as living accommodations. This area includes bays and dormers but does not include space in garages or carports or in attics. Measurements are taken to the outside of exterior walls.

Floor area. The floor area of a building or buildings shall be the sum of the gross horizontal areas of the several floors of such building or buildings exclusive of cellars or basements, except as to single family residences as provided in the Code, measured from the exterior faces of exterior walls or from the centerline of party walls separating two buildings.

Floor area. The sum of the gross horizontal areas of the several floors of a building or buildings measured from the exterior faces of the exterior walls or from the center lines of walls separating two buildings.

Floor area. The minimum floor area of individual units may be less than the requirement of the original district provided the City Council after being presented to the Planning Commission finds that such reduction of floor area is within the intent of the planned unit development ordinance.

Floor area. Floor area shall be determined by measuring the outside dimensions of all enclosed floor area under the roof, excluding basements, uninhabited attics, garages, carports and screen covered areas.

Floor area. Area included within surrounding walls of a building exclusive of vent shafts and courts.

Floor area. Floor space enclosed by exterior walls or fire walls or by a combination of them.

Floor area. Useable area of each story of a building or portion thereof, within surrounding walls.

Floor area. Total number of square feet of floor space within the exterior walls of a building, not including storage space in cellars or basements and not including space used for the parking of automobiles.

Floor area. The area in square feet within the exterior walls of a building, but not including the area of inner courts, shaft enclosures, or exterior walls.

Floor area. For the purpose of determining off-street parking and loading requirements, shall be the sum of the horizontal areas of the several floors of the building, measured from the interior faces of the walls, including accessory storage areas located within selling or working space, such as counters, racks or closets, and any basement floor area devoted to retailing activities, to the production or processing of goods, or to business or professional offices. However, floor area for the purpose for determining off-street parking and loading requirements shall NOT include: floor area devoted primarily to storage purposes (except as noted herein); floor area devoted to utility purposes, stairwells, or elevator shafts; floor area devoted to off-street parking and loading facilities, including aisles, ramps, and maneuvering space; or basement floor area other than that area devoted to retailing activities, to production or processing of goods, or to business or professional offices.

Floor area. The net floor area within the enclosing walls of any room in which the ceiling height is not less than five (5) feet, excluding floor area occupied by built-in equipment such as wardrobes and cabinets.

Floor area. The projected horizontal area enclosed inside of walls, partitions, or other enclosing construction.

Floor area for commercial, business and industrial uses. The sum of the gross horizontal areas of the several floors of a building measured from the exterior faces of the exterior walls, or from the centerline of walls separating two buildings but not including; (1) attic space providing less than seven (7) feet of headroom. (2) cellar space not used for retailing. (3) outside stairs or fire escapes, roof overhangs and balconies. (4) accessory water towers or cooling towers. (5) accessory off street parking spaces. (6) accessory off street loading area.

Floor area for residential dwellings. The gross horizontal areas of the several floors of the dwelling exclusive of garages, cellars and open porches, measured from the exterior faces of the exterior walls of a dwelling.

Floor area (gross). The sum of the gross horizontal area of the several floors of a structure measured from the exterior faces of the exterior walls. The "floor area" of a structure shall include basement (but not the cellar as defined in the Ordinance) floor area, elevator shafts and stairwells at each floor, penthouse and attic space having headroom of seven feet four inches (7′4″) or more, interior balconies and mezzanines, and mechanical equipment space (except equipment, open or enclosed, located on the roof).

Floor area (net). When used to determine the occupant load of a space, shall mean the horizontal occupiable area within the space, excluding the thickness of walls, and partitions, columns, furred-in spaces, fixed cabinets, equipment, and accessory spaces such as closets, machine and equipment rooms, toilets, stairs, halls, corridors, elevators and similar unoccupied spaces.

Floor area of a dwelling. Floor area of the main dwelling shall not be less than 1400 square feet, exclusive of garage, basement, and storage areas for outdoor equipment.

Floor area of a room. Number of square feet of floor space enclosed by walls.

Floor area or gross floor area. The sum of the gross horizontal areas of the several floors of all buildings on the lot, measured from the exterior faces of exterior walls and from the center line of walls separating two buildings. The term shall include basements, elevator shafts and stairwells at each story, floor space used for mechanical equipment, (with structural headroom of six feet, six inches or more), penthouses, attic space (whether or not a floor has actually been laid, providing structural headroom of six feet, six inches or more), interior balconies, and mezzanines. It shall not include stair and elevator penthouses or cellars unless said cellars are utilized for anything other than storage rooms, utility rooms, and mechanical equipment rooms.

Floor area ratio. The ratio of the gross floor area of all the buildings on a lot to the area of the lot. In cases in which portions of the gross floor area of a building project horizontally beyond the lot lines, all such projecting gross floor area shall also be included in determining the floor area ratio. If the height per story in a building, when all the stories are added together, exceeds an average of fifteen (15) feet, then additional gross floor area shall be counted in determining the floor area ratio of the building, equal to the gross floor area of one additional story for each fifteen (15) feet or fraction thereof by which the total building height exceeds the number of stories times fifteen (15) feet; except that such additional gross floor area shall not be counted in the case of a church, theatre, or other place of public assembly.

Floor area ratio. The numerical value obtained by dividing the aggregate floor area within a building on a lot by the area of such lot. The floor area ratio designated for each zoning district, when multiplied by the lot area in square feet, shall determine the maximum permissible square footage of floor area for the building on such lot.

Floor area ratio. The numerical value obtained by dividing the gross floor area of a building or buildings by the lot area on which such building or buildings are located.

Floor area ratio. The gross floor area of all buildings on a lot divided by the lot area on which the building or buildings are located.

Floor area ratio. The floor area of a building or buildings on any lot divided by the area of the lot.

Floor area ratio. Ratio of the total gross floor area of a building or buildings on one lot to the total area of the lot.

Floor area ratio. For purposes of site planning, the maximum square foot amount of total floor area (all stories) permitted for each square foot of land area. FAR is found by dividing the total floor area by the land area.

Floor area ratio. Ratio which is achieved by dividing the total floor area on a lot by the lot area of that lot. For example, a building containing 20,000 square feet of floor area on lot of 10,000 square feet has a floor area ratio of 2.0.

Floor area ratio. The floor area ratio of a building shall be the ratio of the gross floor area of the building, excluding those parts of the building specifically excluded, to the gross land area of the site which gross land area may include one-half of all abutting streets and alleys which are dedicated to public use as determined in accordance with the Ordinance.

Floor area ratio. The ratio of the total floor area of building expressed in square feet to the area of its lot expressed in square feet. The area of one or more floors which are at least seven (7) feet below grade and used for off-street parking shall not be included in floor area computation.

Gross acre. Includes any public rights-of-way which fall within the development and is to be measured to the center line of any bordering streets, alleys, or other rights-of-way.

Gross area. The maximum horizontal projected area within the perimeter of the outside surface of walls or supports of the building or structure.

Gross floor area. The sum of the total areas of the several floors of a building excluding mechanical equipment areas, janitorial service areas, rest rooms, stairwells, basement or attic storage areas, and other areas not normally considered rentable areas.

Gross floor area. For the purpose of determining the number of persons for whom exits are to be provided, or for purposes of classification of occupancy, gross floor area shall be the floor area within the perimeter of the outside walls of the building under consideration with no deduction for hallways, stairs, closets, thickness of walls, columns, or other features. Where the term "area" is used elsewhere in the Code, it shall be understood to be gross area unless otherwise specified.

Gross floor area. The sum of the gross horizontal area of the several floors of a building and its accessory buildings on the same site excluding: basement or cellar areas used only for storage; space used for off-street parking or loading; steps, patios, decks, terraces, porches, and exterior balconies, if not enclosed on more than three sides. Unless excepted above, floor area includes but is not limited to elevator shafts and stairwells measured at each floor (but not mechanical shafts), penthouses, enclosed porches, interior balconies, and mezzanines.

Gross floor area. The sum total of the gross areas of the several floors of a building or buildings measured from the exterior faces of exterior walls or from the center lines of walls separating two buildings; and including open land area used for service to the public as customers, patrons, clients, or patients. Gross floor area shall not include: underground parking space, uncovered steps, exterior balconies, and exterior walkways.

Gross floor area. The sum of the gross areas of the several floors of a building or buildings, measured from the exterior faces of exterior walls or from the center lines of walls separating two (2) buildings. Where columns are outside and separated from an exterior wall (curtain wall) which encloses the building space or are otherwise so arranged that the curtain wall is clearly separate from the structural members, the exterior face of the curtain wall shall be the line of measurement, and the area of the columns themselves at each floor shall also be counted.

Gross floor area. For the purpose of determining the ratio of the floor area of a building to the area of the lot, the "gross floor area" shall be the sum of the gross horizontal areas of the several floors of the building excluding areas used for accessory garage purposes and such basement and cellar areas as are devoted exclusively to uses accessory to the operation of the building. All horizontal dimensions shall be taken from the exterior faces of walls, including walls or other enclosures of enclosed porches. In computing the gross floor areas of buildings in Residence Districts, areas of floors with story heights greater than 10 feet shall be considered as H/10 times the gross area of the floor, where "H" is the height of the story in feet measured from the floor level to 1 foot above the ceiling level for top stories, and from floor level to floor level for other stories.

Gross floor area. The sum of the gross horizontal areas of the several floors of a building, including interior balconies and mezzanines, but excluding exterior balconies. All horizontal dimensions of each floor are to be measured by the exterior faces of walls of each such floor, including the walls of roofed porches having more than one wall. The floor area of a building shall include the floor area of accessory buildings on the same Zone Lot, measured the same way. In computing the gross floor area there shall be excluded the following: any floor area devoted to mechanical equipment serving the building, providing, that the floor area of such use occupies not less than 75% of the floor area of a story in which such mechanical equipment is located; any floor area in a story the ceiling whereof is less than four feet above grade at the nearest building line; any floor area used exclusively as parking space for motor vehicles.

Gross floor area. Floor area of any individual story, for the purpose of determining the capacity of the floor area in numbers of persons for whom exits are to be provided, or for purposes of classification of occupancy, should include the area within the perimeter of the outside walls of the building or that section of the building devoted to the occupancy under consideration, with no deduction for hallways, stairs, closets, thickness of walls, columns or other features.

Gross floor area. Aggregate area of all floors including the area of the outside walls and measured to exterior of such walls.

Gross floor area. Area of the plan projection of all floors of whatever nature within or attached to a building exclusive of pedestrian ways for general public use.

Gross floor area. Floor area included within the surrounding walls of a building, exclusive of vent shafts, elevator shafts, stairways and toilet rooms, except toilet rooms in apartment dwellings.

Gross floor area. Floor area included within the surrounding walls of a building, exclusive of vent shafts, elevator shafts, stairways and toilet rooms, except toilet rooms in apartment dwellings as specified in the Code.

Gross floor area. Sum of the gross horizontal areas of the several floors of a building excluding areas used for accessory garage purposes, and basements, and cellar areas. All dimensions shall be taken from the exterior faces of walls, including the exterior faces of enclosed porches.

Gross floor area. The sum of the gross horizontal areas of the several floors of the building measured from the exterior faces of the exterior walls or from the center line of walls separating two (2) buildings, The "gross floor area" at each floor, floor space used for mechanical equipment (except equipment, open or enclosed, located on the roof), ten (10) inches or more, interior balconies and mezzanines, enclosed porches, and floor area devoted to accessory uses. However, any space devoted to off-street parking or loading shall not be included in "gross floor area".

Gross floor area. The sum of the gross horizontal areas of the several floors of a building or apartment measured from the exterior faces of the exterior walls, or from the centerline of a wall separating two (2) buildings or two (2) apartments.

Gross lot coverage. The ratio between the ground floor area of all buildings and structures plus all area used for off-street parking facilities, loading areas, vehicular access-ways or driveways, and the total area of the lot.

Ground area. Total ground area between the outer lines of the exterior walls of a building, less an outer court.

Ground floor. That level of a building on a sloping or multilevel site which has its floor line at or not more than 3 feet above exit discharge grade for at least one-half of the required exit discharges.

Ground floor area. The sum of the gross horizontal area of the ground floor of a building, measured from the exterior faces of the exterior walls or from the center line of walls separating two (2) buildings. The ground floor area of a building shall also include recessed, unenclosed or partially enclosed areas under a floor above exterior stairways, porches and similar areas but excluding open terraces.

Ground floor area. The area contained within the outer plane of the enclosing walls of a building at grade, exclusive of garages, carports, unenclosed porches, and unenclosed breezeways.

Interior floor area. The gross floor area of an apartment measured from the interior faces of the exterior walls of an apartment unit or of the portion of an accessory building used for living quarters, disregarding any floor area occupied bv partition walls.

Intensity of use of lot. That portion of the area of a lot which is occupied or which may be occupied under the Ordinance, by buildings and their accessory buildings.

Intensity of use of lot. That proportion of the area of a lot which is occupied or which may be occupied under the Ordinance by a building and its accessory buildings and by major recreational uses of property.

Land coverage. Percentage of a lot covered by the main and accessory buildings.

Livable floor area. Area of the floors of a dwelling, excluding basements not designed for human occupancy, cellars, garages, breezeways, unenclosed or unheated porches or attics. It shall include only such floor area immediately under a roof for which the headroom is not less than five (5) feet, provided that at least sixty-five percent (65%) of such floor area has a ceiling height of at least seven (7) feet and if any such floor area is situated above another story, it has access to the floor below by a permanent, built-in stairway; and it meets the requirements of the building Code for light and ventilation. Measurements of livable floor area shall be made from exterior faces of exterior walls or from center lines of party walls, except for areas under a sloping roof.

Living area. Includes all areas included within the enclosing walls of the building except garages, outside utility rooms, carports, cabanas, porches, patios, and unroofed or unenclosed areas.

Lot area. Total number of square feet within the exterior lines of the lot, not including any area in a public or private street, nor any water area more than 10 feet from the shoreline.

Lot area. The total horizontal area within the lot lines of the lot.

Lot area. The area in square feet lying within the lot lines of the lot and not including any part of any abutting public or private street or alley.

Lot area. The total area exclusive of streets within the boundary lines of a lot.

Lot area. The total land area within the property lines of a lot.

Lot area. The area of a horizontal plane bounded by the front, side, and rear lot lines.

Lot area. The area of a horizontal plane bounded by the front, side and rear lot lines measured within the lot boundaries.

Lot area. The total horizontal area within the lot.

Lot area. The area of any lot within its exterior lot lines, not including any area lying within a street.

Lot area. The area of a horizontal plane bounded by the vertical planes through front, side and rear lot lines.

Lot area. Total horizontal area of the lot lying within the lot lines, provided that no area of land lying within any street line shall be deemed a portion of any lot area. The area of any lot abutting a street shall be measured to the street line only.

Lot area per family. Every building hereafter erected or structurally altered for occupancy by one family shall provide a lot area of not less than 20,000 square feet per family and no such lot shall be less than 100 feet in width. Every building hereafter erected or structurally altered for occupancy by more than one (1) family shall provide a lot area of not less than 5,000 square feet per family and no such lot shall be less than 1000 feet wide.

Lot area per family. Every building hereafter erected or structurally altered shall provide a lot area of not less than 1,200 square feet per family and six hundred (600) square feet for every lodger, roomer, boarder, child or patient requiring personal or limited nursing care, other than members of the immediate family. On an originally platted lot or on a lot subdivided prior to the revisions to the Ordinance, where the lot area divided by a number of square feet per family results in a surplus of seven hundred (700) square feet, one additional family may be permitted.

Lot coverage. That part or percent of the lot occupied by buildings, including accessory buildings.

Lot coverage. Area of the lot covered by a structure(s) exclusive of permitted overhang.

Lot coverage. Percentage of the lot area that is occupied by the Area of the Principal Building in Residence Zones; and the percentage of the lot area that is occupied by the Area of the Principal Building plus the area of accessory buildings in Business, Industrial, or any other zone.

Lot coverage. Lot coverage shall not exceed 10 percent, exclusive of accessory structures.

Lot coverage. Amount of lot area stated in terms of percentage that is covered by all buildings or structures located thereon. This shall be considered to include all buildings, porches, breezeways, patio roofs, eaves, awnings and the like whether box-type, lathe roof, or fully roofed, but shall not be considered to include fences, walls, swimming pools, or hedges used as fences.

Lot coverage. The part or percent of the lot occupied by buildings, including accessory structures.

Lot coverage. The percentage of the lot area that is occupied by the area of the principal building and accessory building.

Lot depth. Distance measured in the mean direction of the side point of the front lot line to the mid-point of the lot.

Lot depth. Horizontal length of a straight line drawn from the midpoint of the front lot line to the midpoint of the rear lot line.

Lot depth. Distance measured in a mean direction of the side lines of the lot from mid-point of the front line to the mid-point of the opposite rear line of the lot.

Lot depth. The mean horizontal distance of a lot measured between the front and rear lot lines.

Lot depth. The mean horizontal distance between the front and rear lot lines.

Lot depth. The depth of a lot is the mean distance between its front street line and its rear line.

Lot depth. The average depth from the front line of the lot to the rear line of the lot.

Lot depth. The mean distance between front and rear lot lines.

Lot depth. The horizontal distance between the front and rear lot lines measured along the median between the side lot lines.

Lot depth. The length (or depth) of a lot shall be: (1) If the front and rear lines are parallel, the shortest distance between such lines. (2) If the front and rear lines are not parallel, the shortest distance between the midpoint of the front line and the midpoint of the rear lot line. (3) If the lot is triangular, the shortest distance between the front lot line and a line parallel to the front lot line, not less than ten (10) feet long lying wholly within the lot.

Lot depth. The horizontal distance between the front and the rear lot lines measured in the mean direction of the side lot lines.

Lot depth. The mean horizontal distance between the front lot line and rear lot line of a lot measured within the lot boundaries.

Lot depth. Mean distance from the street line of the lot to its rear line measured in the mean direction of the side lines of the lot.

Lot depth. Horizontal length of a straight line connecting the bisecting points of the front and the rear lot lines.

Lot depth. The mean horizontal distance between the front lot line and the rear lot line. In the case of a corner lot, the lot depth is the greater of the horizontal distances between the front lot lines and the respective lot line opposite each.

Lot depth. The horizontal distance between the front and rear lot lines measured in the mean direction of the side lot lines.

Lot length. Length (or depth) of a lot shall be: If the front and rear lines are parallel, the shortest distance between such lines. If the front and rear lines are not parallel the shortest distance between the midpoint of the front lot line and the midpoint of the rear lot line. If the lot is triangular, the shortest distance between the front lot line and a line parallel to the front lot line, not less than ten (10) feet long lying wholly within the lot.

Lot measurements. Depth of lots shall be considered to be the distance between the midpoints of straight lines connecting the foremost points of the side lot lines in front and the rear-most points of the side lot lines in the rear. Width of lots shall be considered to be the distance between straight lines connecting front and rear lot lines at each side of the lot, measured across the rear of the required front yard.

Lot width. The width of a lot is its mean width measurement at right angles to its depth.

Lot width. The width, measured at a distance back from the front line equal to the minimum depth required for a front yard.

Lot width. The least horizontal dimension between side lot lines measured at the distance from the street line or street center line which establishes the minimum required front yard.

Lot width. The width of a lot along a line parallel to the frontage thereof and lying a distance therefrom equal to the required setback on said lot required by the Ordinance or of a greater setback line if established.

Lot width. The distance parallel to the front of a building erected or to be erected, measured between side lot lines at the building line.

Lot width. The straight line distance between the side lot lines, measured at the two points where the minimum building line, or setback, intersects the side lot lines.

Lot width. The horizontal distance of a lot measured along the building line at a right angle to the mean lot depth line. Width at the front lot line is measured along the street line.

Lot width. Horizontal distance between the side lot lines measured at the front "set-back" line.

Lot width. Mean horizontal distance between the side lot lines.

Lot width. Least horizontal distance across the lot between side lot lines, measured at the front setback of a main building erected or to be erected on such lot or at a distance from the front lot line equal to the required depth of the front yard.

Lot width. The horizontal distance between the side lot lines measured at right angles to the lot depth at a point midway between the front and rear lot lines.

Lot width. The minimum horizontal distance between the side lot lines of a lot measured at the narrowest width within the area between the front yard line and a line parallel to and thirty (30) feet immediately to the rear thereof.

Lot width. Average horizontal distance between the side lot lines measured at the required front yard line and parallel to the front street line, or measured at the street line if no front yard is required.

Lot width. Mean width measured at right angles to its depth.

Lot width. Horizontal distance between the side lot lines measured at right angles to the line comprising the depth of the lot at a point midway between the front and rear lot lines.

Lot width. Length of a straight line drawn between the points where the front required setback line cuts the side lot lines.

Lot width. The distance between the side lot lines measured at right angles to the lot depth line at a point midway between the front and rear lot lines.

Lot width. The width of a lot shall be: (1) If the side property lines are parallel, the shortest distance between these side lines. (2) If the side property lines are not parallel, the width of the lot shall be the length of a line at right angles to the axis of the lot at a distance equal to the front setback required for the district in which the lot is located. The axis of a lot shall be a line joining the midpoints of the front and rear property lines.

Lot width. For the purpose of the Ordinance, the width of a lot shall be measured at the front wall of the building.

Lot width. The width of a lot measured at the building line and at right angles to its depth.

Lot width. The distance between straight lines connecting front and rear lot lines at each side of the lot, measured between the midpoints of such lines.

Lot width. The mean horizontal distance between side lot lines measured at right angles to the lot depth, or the horizontal distance between side lot lines measured at the mean lot depth point, whichever is lesser.

Lot width. The horizontal distance between the side lot lines of a lot measured within the lot boundaries, and at the minimum required front setback line.

Maximum coverage. The maximum amount of land that may be covered by buildings on any lot.

Maximum depth of lot. The distance between the front base line of the lot and the farthest point of the lot therefrom measured along a line perpendicular to the front base line, or its extension, and passing through the farthest point therefrom of the lot. The front base line of a lot shall be a line passing through the two (2) termini of the front lot line.

Maximum net residential densities. For Garden apartments, 12 units per acre, for Townhouse uses, 8 units per acre, for semi-detached dwelling units, 6 units per acre, and for Detached housing, 4 units per acre.

Minimum lot area. That area of a lot in any zoning district, exclusive of the area of any street, road or access easement on or across such lot.

Net area. The occupied or usable floor area in a building but not including space occupied by columns, walls, partitions, mechanical shafts or ducts.

Net floor area. Aggregate area of all floors included within the outer walls of a building, measured at the interior of such walls, excluding basements, cellars, rooms for furnace equipment, garages, carports and unenclosed porches, breezeways, and including only such floor area under a sloping ceiling for which the headroom is not less than 5 ft. 6 in. and then only if at least 50% of such floor area has a ceiling height of not less than 7 ft. 4 in. and if any such floor that is situated above another story has access to the floor below by a permanent built-in-stairway.

Net floor area. For the purpose of determining the number of persons for whom exits are to be provided, net floor area shall be the actual occupied area, not including accessory unoccupied areas or thickness of walls.

Net floor area. The actual occupied area not including accessory unoccupied areas or thickness of walls.

Net site area. The net site area of the property shall be the remaining ground area of the gross site area after deleting all portions for existing and proposed perimeter rights-of-way and alleys.

Open space ratio. For the purpose of site planning, the minimum square foot amount of open space which shall be provided for each square foot of floor area. The OSR is found by dividing the total of the open space by the total of the floor area. The open space is the total land area minus the building area plus the usable roof area.

Residential density. The number of residential dwelling units occupying a given land area. Expressed in terms of either dwelling units per gross acre of land or dwelling units per net acre of land area.

Site area. The total area included within the property lines of a site, exclusive of the area of access corridors, streets, portions of the site within future street plan lines; provided however, all lots in subdivisions with acute angles less than 45 degrees formed by adjacent sides shall be discouraged in planning.

Site area. The total horizontal area included within the property lines of a site, exclusive of the area of access corridor, streets, portions of the site within future street plan lines. Provided, however, all lots in subdivisions with acute angles less than 45 degrees formed by adjacent sides shall be discouraged by the Planning Commission at the time of the Tentative Map Approval.

Site coverage. Permitted maximum site coverage in the planned development zone shall not exceed the maximum permitted site coverage in the original district; however, site coverage may be calculated on the total land involved in the planned development.

Superficial floor area. The net floor area within the enclosing walls of the room, excluding built-in equipment such as cabinets, closets, or fixtures which are not readily removable and excluding the floor area where the floor to ceiling height is less than 4-1/2 feet.

Usable floor area. For the purposes of computing parking, that area used for or intended to be used for the sale of merchandise or services or for use to serve patrons, clients, or customers. Such floor area which is used or intended to be used principally for the storage or processing of merchandise, or for utilities shall be excluded from this computation of "Usable Floor Area." Measurement of floor area shall be the sum of gross horizontal areas of the several floors of the building, measured from the interior faces of the exterior walls.

Width of lot. The horizontal distance between side lot lines measured at the street frontage.

Width of lot. The mean width measured at right angles to its depth.

Width of lot. Average horizontal distance between side lot lines.

Width of lot. The mean horizontal distance between its side lot lines.

Width of site. Horizontal distance between the side property lines of a site measured at right angles to the depth of a point midway between the front and rear property lines.

A-10 LOTS, PLOTS, SITES, PARCELS, AND TRACTS

Acre. A parcel of land containing 43,560 square feet in whatever shape.

Acre. Shall mean a land area measuring 43,560 square feet.

Acre (square). A parcel measuring 208.708 feet by 208.708 feet.

Acreage. Any parcel of land described by metes and bounds and not shown on a plot of a recorded subdivision legally admitted to record.

Acreage. Any tract or parcel of land which has not been subdivided and platted.

Acreage. The total number of acres in a piece of land or tract.

Block. The properties abutting on one (1) side of a street and lying between the two (2) nearest intersecting or intercepting streets, or nearest intersecting or intercepting street and railroad right-of-way, unsubdivided land, watercourse, or city boundary.

Block. A tract of land bounded by streets, or a combination of streets and public parks, cemeteries, railroad rights-of-way or other lines of demarcation. A block may be located in part beyond the corporate limits of the City.

Block. A tract of land bounded by streets, or by a combination of streets and public parks, cemeteries, railroad right-of-way, shore lines or waterways, or municipal boundary lines.

Block. A tract of land bounded by streets or by a combination of one or more streets and public parks, cemeteries, railroad rights of way, bulkhead lines or shore lines of waterways or corporate boundary lines.

Block. Tract of land bounded by streets, public parks, railroad rights-of-way, or corporate boundary lines.

Block. That property abutting on one side of a street between two (2) nearest intersecting streets, railroad right-of-way, or other natural barriers provided, however, that where a street curves, so that any two (2) chords thereof form an angle of one hundred twenty (120) degrees or less measured on the lot side, such curve shall be construed as an intersecting street.

Block. That property abutting on one side of a street and lying between the nearest intersecting or intercepting streets, or nearest intersecting or intercepting street, and railroad right-of-way, waterway, or other barrier to or gap in the continuity of development along such street.

Block. An area of land surrounded on all sides by streets, railroads or other rights-of-way, regardless of size or shape of such land or the number of such lots thereon.

Block. A block shall include the property having frontage on one side of a street and lying between the two nearest intersecting or intercepting streets, or nearest intersecting or intercepting street and railroad right-of-way.

Block. A block shall be deemed to be that property abutting on a street on one side of said street and lying between the two nearest intersecting or intercepting railroad rights-of-way or streets.

Block. Area fronting on the same side of a public street or road, situated between street intersections, except that where the distance between street intersection is greater than 1200 feet, the area fronting on the same side of a public street or road not more than 600 feet on either side of the parcel, lot, or tract of land being considered as a building site shall be considered to be a block.

Block. Shall mean all property fronting upon one side of a street between intersecting and intercepting streets, or between a street and a right-of-way, water way, terminus of dead end street, or city boundary. An intercepting street shall determine only the boundary of the block on the side of the street which it intercepts.

Block. The properties abutting on one side of a street and lying between the two nearest intersecting or intercepting streets, or nearest intersecting or intercepting street and railroad right of way, unsubdivided land, watercourse, or city boundary, delineated on Zoning Maps.

Block. That property abutting one side of a street and lying between the two nearest intersecting streets, or nearest intersecting street and railroad right-of-way, unsubdivided acreage, waterways, but not an alley, of such size as to interrupt the continuity of development on both sides thereof.

Block face. The properties abutting on one side of a street and lying between the two nearest intersecting or intercepting streets, or nearest intersecting or intercepting street and railroad right-of-way, unsubdivided land, or watercourse.

Border lot. Lot contiguous to a zone boundary.

Building site. The ground area of a building or buildings, together with all open spaces which are required.

Building site. Shall mean (1) the ground area of one (1) lot; or (2) the ground area of two (2) or more lots when used in combination for a building or permitted group of buildings, together with all open spaces as required by the ordinance.

Building site. Area occupied by a building or structure, including the yards and courts required for light and ventilation, and such areas that are prescribed for access to the street.

Building site. Ground area of a building or group of buildings together with all yards and open spaces as required by the ordinance.

Building site. The area of a building together with associated parking areas and open space required by the Ordinance. A building site may encompass more than one lot.

Cluster lot. One of a group of 3 or more lots, each of which must abut common or dedicated ground on one or more sides and does not necessarily front on a public street.

Corner lot. Lot abutting upon 2 or more streets at their intersection.

Corner lot. Lot abutting on 2 intercepting or intersecting streets, where the interior angle of interception or intersection does not exceed 135 degrees.

Corner lot. Lot situated at the junction of 2 or more streets, alleys, or open passageways of not less than 15 feet in width.

Corner lot. Lot with 2 adjacent sides abutting upon streets or other public spaces.

Corner lot. Lot abutting upon 2 or more streets at their intersection, the shortest side fronting upon a street shall be considered the front of the lot, and the longest side fronting upon a street shall be considered the side of the lot.

Corner lot. Lot at the junction of and fronting on 2 or more intersecting streets both of which are 20 feet or more in width.

Corner lot. Lot located at the intersection of 2 streets or a lot bounded on 2 sides by crossing or intersecting streets, any 2 corners of which have an angle of 120 degrees or less measured on the lot side.

Corner lot. Lot on a corner fronting not more than 60 feet on one street and not more than 120 feet on an intersecting street.

Corner lot. Parcel of land not over 60 feet in width at the junction of and fronting on, 2 intersecting streets, having an area not greater than 6000 square feet and a frontage on one of the intersecting streets not greater than 100 feet.

Corner lot. A site bounded by two (2) or more adjacent street lines which have an angle of intersection of not more than 135 degrees.

Corner lot. A lot abutting two (2) or more streets at their intersections. A lot abutting on a curved street or streets shall be considered a corner lot if straight lines drawn from the foremost point of the side lot lines to the foremost point of the lot (or an extension of the lot where it has been rounded by a street radius) at an interior angle of less than one hundred and thirty-five (135) degrees.

Corner lot. A lot situated at the intersection of two (2) streets, the interior angle of such intersection not exceeding one hundred and thirty-five (135) degrees.

Corner lot. A lot situated at the junction of two or more streets or places.

Corner lot. A lot which at least two (2) adjacent sides abut for their full length upon a street.

Corner lot. A lot situated at the junction of two or more streets.

Corner lot. A lot fronting on and at the intersection of two or more streets.

Corner lot. A lot situated at the intersection of and abutting two streets that have an angle of intersection of not more than 135°.

Corner lot. Any lot abutting upon 2 intersecting streets at their intersection or upon 2 parts of the same street; and, in either case forming an interior angle of not less than 135 degrees.

Corner lot. A lot abutting on 2 or more streets at their intersection, provided that the interior angle of such intersection is less than 135 degrees.

Corner lot. A parcel of land at the junction of and fronting on two or more intersecting streets.

Corner lot. A lot situated at the intersection of two (2) or more streets, which have an angle of intersection of not more than one hundred thirty-five degrees (135°).

Corner lot. A lot at the junction of, and having frontage on, two or more intersecting streets, or a lot bounded on two or more sides by the same street.

Corner lot. Any lot situated at the intersection of two streets and abutting such streets on two adjacent sides.

Corner lot. A lot at the point of intersection of and abutting on two intersecting streets, each more than 20 feet in width, the angle of the intersection being not more than 135 degrees. It is the land occupied or to be occupied by the corner building and its accessory buildings. For the purposes of the Ordinance, in computing the area of a corner lot or in applying the yard regulations, the width shall be considered as extending not more than 50 feet from the side street, line measured at right angles thereto, or to the actual lot line, whichever distance is less; and the depth shall be considered as 150 feet, or the actual depth, whichever is less.

Corner lot. A lot abutting the intersection of two or more streets. A lot abutting a curved street shall be considered a corner lot if the angle of the apex of a triangle is 135 degrees or less. Which triangle shall establish as follows; a base line shall be drawn between the two most extreme opposite side points of the lot which intersect with the abutting street; from this line a perpendicular line shall be extended to the most distant point on the front line which point shall be the said apex; thereafter the said points shall be connected to form this pertinent triangle.

Corner lot. A lot abutting on two or more intersecting streets where the interior angle of intersection does not exceed 135 degrees. A corner lot shall be considered to be in that block in which the lot fronts.

Corner lot. A lot or portion of a lot more than 50 feet wide at the junction of and fronting on two intersecting streets. Any portion of a lot more than 50 feet distant from the street with the greater frontage shall comply with all the provisions of the code respecting interior lots.

Corner lot. A lot where the interior angle of two adjacent sides at the intersection of two streets is less than one hundred thirty five (135) degrees. A lot abutting upon a curved street or streets shall be considered a corner lot for the purposes of the Ordinance, if the arc is of less radius than one hundred fifty (150) feet and the tangents to the curve, at the two points where the lot lines meet the curve or the straight street line extended, form an interior angle of less than one hundred thirty-five (135) degrees.

Corner lot. A lot located at the intersection of two (2) streets or a lot bounded on two (2) sides by crossing or intersecting streets, any two corners of which have an angle of one hundred twenty (120) degrees or less measured on the lot side as described herein.

Corner lot. The front property line of a corner lot shall be the shorter of the two lines adjacent to the streets. Where the lines are equal, the front line shall be that line which is obviously the front by reason of the prevailing custom of the block. If such front is not evident, then either may be considered the front of the lot but not both.

Corner lot. A lot abutting upon two (2) or more streets or roads (including platted or provided but unopened streets or roads) at the intersection. A lot abutting on a curved street or streets shall be considered a corner lot if straight lines drawn from the foremost points at the side lot lines to the foremost point of the lot meet at an angle of less than 135 degrees.

Corner lot. A lot at the junction of and abutting on two or more intersecting streets, or at the point of deflection in alignment of a single street, the interior angle of which does not exceed 135 degrees.

Corner lot. A lot abutting on two (2) or more intersecting streets where the interior angle of intersection does not exceed one hundred and thirty-five (135) degrees.

Corner lot. A lot situated at the junction of and abutting on two (2) or more intersecting streets; or a lot at the point of deflection in alignment of a single street, the interior angle of which does not exceed one hundred thirty five (135) degrees.

Corner lot. A lot located at the intersection of two (2) streets, the interior angle of which does not exceed 135 degrees.

Corner lot. A lot in one ownership located at the intersection of two (2) streets or rights-of-way or a lot bounded on two (2) sides by a curving street or right-of-way, any two (2) adjacent chords of which form an interior angle not exceeding one hundred and twenty (120) degrees.

Corner lot. Every corner lot in a residential District having on its side street an abutting interior lot shall have minimum setbacks from both streets equal to the minimum required front setback of the District in which it is located, provided, however, that this does not reduce the buildable width of any lot of record to less than twenty-five (25) feet. On corner lots where a rear lot line abuts a side lot line on the adjoining lot, accessory buildings on the corner lot shall have a rear setback from the rear lot line a distance equal to the smaller of the side yard setbacks required for the District.

Corner lot. A lot which has at least two (2) adjacent sides abutting on a street, provided that the interior angle at the intersection of such two (2) sides is less than one hundred thirty-five degrees (135°).

Corner lot. A lot situated at the intersection of two or more streets having an angle of intersection of not more than 135 degrees.

Corner lot. A lot abutting two (2) intersecting streets, where the interior angle of intersection does not exceed one hundred thirty-five (135) degrees.

Corner zone lot. A zone situated at the junction of two or more intersecting or intercepting streets where the angle of intersection of the lot lines coterminous with the street lines does not exceed 135 degrees.

Double frontage lot. Lot extending between and having frontage on a major traffic street and a minor street, and which vehicular access solely from the latter.

Double frontage lot. Lot having frontage on 2 parallel or approximately parallel streets.

Double frontage lot. Interior lot having frontage on 2 parallel streets. For the purpose of determining front yard requirements, each frontage from which access is permitted shall be deemed a front lot line.

Double frontage lot. Lot having frontage on two (2) nonintersecting streets.

Double frontage lot. Lot other than a corner lot with frontage on more than 2 streets.

Double frontage lot. Any lot having frontages on two or less parallel streets as distinguished from a corner lot. In the case of a row of double frontage lots, all sides of said lots adjacent to streets shall be considered frontage, and front yards shall be provided as required.

Double frontage lot. A lot which runs through a block from street to street and which has two (2) nonintersecting sides abutting on two (2) or more streets.

Double frontage lot. An interior lot having frontage on two (2) parallel or approximately parallel streets.

Double frontage lot. A lot having a pair of opposite lot lines along two (2) more or less parallel streets, and which is not a corner lot.

Double frontage lot. A lot having a frontage on two (2) streets, but which does meet the criteria for a corner lot. Also called a "through lot".

Double frontage lot. A lot having a frontage on two (2) nonintersecting streets, as distinguished from a corner lot.

Double-fronted lot. Interior lot bounded by a street on front and back.

Flag lot. A lot so shaped and designed that the main building site area is set back from the street on which it fronts and includes an access strip not less than 20 feet in width at any point connecting the main building site area to the frontage street.

Front lot. That boundary of the lot which abuts on a street. In the case of a corner lot, the narrowest boundary fronting on a street. In case the corner lot has equal frontage on two or more streets, the lot shall be considered to front on the principal street, or on that street which the greatest number of buildings have been erected within the same block.

Front lot. That portion of a lot abutting a street right-of-way. On corner lots the front shall be that portion of the lot having the least horizontal distance as measured along each of the abutting street rights-of-way. Through lots shall be considered as having two (2) fronts.

Front lot. Front boundary line of a lot bordering on a street, in the case of a corner lot, the shortest front boundary line.

Gore-shaped lot. Triangular piece of property.

Individual lots. Deviation from the applicable requirements for lot area, lot dimensions, yards, setbacks, location of parking areas, and public street frontage may be allowed, but only if such deviation is consistent with the total design of the planned development.

Interior lot. A lot other than a corner lot.

Interior lot. A lot other than a corner lot, or a reversed corner lot.

Interior lot. One which faces on one street with opposite sides on two (2) streets.

Interior lot. A lot other than a corner lot with only one frontage on a street.

Interior lot. A lot other than a corner lot or through lot.

Interior lot. A lot which faces on one street or with opposite sides on separate streets.

Interior lot. A lot having but one side abutting on a street.

Interior lot. A lot bounded by a street on one side only.

Interior lot. Property side lines which do not abut on a street.

Interior lot. Land occupied or which may hereafter be occupied by a building and its accessory buildings, together with such open spaces as are required under the code, and not having its principal frontage upon a street or officially approved place.

Interior lot. Lot other than a corner lot or reversed lot.

Interior lot. Any lot bounded on both sides by other lots.

Interior lot. Front property line of an interior lot shall be the line coterminal with the street frontage.

Interior lot. A lot adjacent to two (2) other lots, having common side lot lines, having frontage on one (1) abutting street, and a rear property line on an alley or easement.

Interior lot. A lot having the front side abutting on a street.

Key lot. The first interior lot to the rear of a reversed corner lot.

Key lot. An interior lot, one (1) side of which is contiguous to the rear line of a corner lot.

Key lot. The first interior lot to the rear of a reversed corner lot and not separated therefrom by an alley.

Key lot. Shall mean the first lot to the rear of a reversed corner lot whether or not separated by an alley.

Land measure.

1 mile—80 chains, 320 rods, 1,760 yards or 5,280 feet.
16-1/2 feet—1 rod, perch or pole.
1 chain—66 feet, 100 links or 4 rods.
1 link—7.92 inches.
25 links—1 rod.
4 rods—1 chain.
144 square inches—1 square foot.
9 square feet—1 square yard.
30-1/4 square yards—1 square rod.
160 square rods—1 acre.
10,000 square links—1 square chain.
10 square chains—1 acre.
1 acre—208.708 feet by 208.708 feet.
1 acre—43,560 square feet.
1 acre—4,840 square yards.
1 acre—160 square rods.
640 acres—1 square mile or section.
36 square miles or sections—1 township

Legal non-conforming lot of record. (a) was created by a plat or deed recorded at a time when, it came into ownership separate from adjoining tracts of land at a time when, the creation of a lot of such size, shape, depth and width at such location would not have been prohibited by any Ordinance or other regulation; and (b) has remained in separate and individual ownership from adjoining tracts of land continuously during the entire time that the creation of such a lot has been prohibited by any applicable Ordinance or other regulation.

Lot. A land occupied or to be occupied by a building and its accessory buildings together with such open spaces as required under the Ordinance and having its principal frontage upon a street or officially approved place or having vehicular and pedestrian access to a public street by private ways found adequate for such purposes by the City Council.

Lot. A portion or parcel of land considered as a unit, devoted to a certain use or occupied by a building or group of buildings that are unlimited by a common interest or use, and the customary accessories and open space belongs to the same.

Lot. Land occupied or to be occupied by a building and any buildings accessory thereto, or by a building group and any buildings accessory thereto, together with the open spaces appurtenant to such building or group, and either having its principal frontage on a street. A parcel of land shall be deemed to be a lot in accordance with this definition regardless of whether or not the boundaries thereof coincide with the boundaries of lots or parcels as shown on any map of record.

Lot. A parcel of land existing in common ownership, regardless of whether acquired in separate parcels or as a whole, occupied by or which may be occupied by one principal building or use of land or group of principal buildings and accessory buildings, including the yards and other open spaces designed to be used in connection with such buildings as referred to in the Ordinance.

Lot. A single or individual parcel or area of land legally recorded, or validated by other means acceptable to the Administrative Authority, on which is situated a building or which is the site of any work regulated by the Code, together with yards, courts, and unoccupied spaces legally required for the building or works, and which is owned by or is in the lawful possession of the owner of the building or works.

Lot. A parcel of land occupied, or intended to be occupied, by a main building or a group of such buildings and accessory buildings, or utilized for the principal use and uses accessory thereto, together with such open spaces as are required under the provisions of the Ordinance. A lot may or may not be specifically designated as such on public records.

Lot. For the purpose of the Ordinance, a lot is a parcel of land of at least sufficient size to meet minimum zoning requirements for use, coverage and area, and to provide such yards and other open spaces as are herein required. Such lot shall have frontage on a public street and may consist of: (a) A single lot of record; (b) A portion of a lot of record; (c) A combination of complete lots of record, of complete lots of record and portions of lots of record, or of portions of lots of record; and (d) A parcel of land described by metes and bounds, provided that in no case of division or combination shall any residual lot or parcel be created which does not meet the requirements of the Ordinance.

Lot. As used in the Ordinance, a lot is a lawful building site. Such building site may consist of all, portions or combinations of land parcels described by metes and bounds or lots as shown on a subdivision plat.

Lot. A parcel of land which is occupied, or is to be occupied by one principal building or other structures or uses together with any accessory buildings or structures or uses customarily incidental to such principal buildings or other structure or use, and any such open spaces as are arranged or designed to be used in connection with such principal building or other structure or use, such open spaces and the area and dimensions of such lot being not less than the minimum required by the Ordinance.

Lot. A tract, site or parcel of land occupied, or to be occupied, by a building, together with such open spaces as may be required under the Ordinance, having its frontage on a public street, and having not less than the minimum area and frontage required by the Ordinance.

Lot. A parcel of land occupied or intended for occupancy by a building together with its accessory buildings, including the open space required under the ordinance. For the purpose of the ordinance the word "lot" shall be taken to mean any number of contiguous lots or portions thereof, upon which one or more main structures for a single use are to be erected.

Lot. A piece, parcel, tract or plot of land occupied or to be occupied by one principal building and its accessory buildings and including the required yards and shall include all lots of record included in such piece, parcel, tract, or plot of land, and all lots otherwise designated.

Lot. A subdivision of a block or other parcel intended as a unit for the transfer of ownership or for development.

Lot. A piece or parcel of land occupied or to be occupied by a building and its accessory buildings, or by any other activity permitted thereon, and, including the open spaces required under the Ordinance and having frontage upon a public street. A lot may or may not be a lot of existing record.

Lot. A parcel of land or two or more contiguous parcels to be used as a unit under the provisions of the Ordinance, as shown in the County Assessor's office, and having its principal frontage on a street. In any district where a half street not less than one-half (1/2) of that width prescribed for that street by the Minimum Right-Of-Way standards map, and amendments thereto, has been dedicated, any lots facing or siding on such half street from which side the required width of dedication has been made shall be deemed to have frontage on a street.

Lot. A parcel of real property shown as a delineated parcel of land with a number or other designation on a final map of subdivision recorded in the office of the County Recorder; or a parcel of land the dimensions or boundaries of which are defined by a Record of Survey map recorded in the office of the County Recorder, in accordance with the law regulating the subdivision of land or a parcel of real property not delineated as stated above, and containing not less than the prescribed minimum area required in the zone in which it is located and which abuts at least one street and is held under one ownership.

Lot. Land occupied or to be occupied by a building and its accessory buildings, or by any other single activity permitted herein, together with such open spaces as are required under the Ordinance and having its principal frontage upon a street or official approved place.

Lot. Lot includes the word "plot" or "parcel".

Lot. A division of land separated from other divisions for the purpose of sale, lease or separate use, described on a recorded subdivision plat, recorded survey map, or by metes and bounds.

Lot. A parcel of land that is described by reference to a recorded plat or by metes and bounds.

Lot. A parcel of land, or two (2) or more contiguous parcels to be used as a unit under the provision of the Zoning Code, as shown in the record of the County Assessor's Office, and having its principal frontage on a street.

Lot. Land occupied or which may hereafter be occupied by a building and its accessory buildings, together with such open spaces as are required under the Code, and having its principal frontage upon a street or officially approved place.

Lot. A parcel of land considered as a unit.

Lot. A parcel of land occupied or which may be hereafter occupied by a building and its accessory buildings, together with such open spaces and parking spaces or area as are required under the Ordinance, and having its principal frontage upon an officially approved street or place.

Lot. A parcel of land having a width and depth sufficient to provide the space necessary for one main building and its accessory building, together with the open spaces required by the Ordinance and abutting on a public street or officially approved place.

Lot. A parcel of land on which a main building and any accessory buildings are or may be placed, together with the required open spaces. The area of a lot which abuts a street shall be measured to the side of the street right of way only.

Lot. A parcel of land occupied or capable of being occupied by one building and the accessory buildings or uses customarily incident to it, including such open spaces as are required by the Ordinance.

Lot. A piece, parcel, or plot of land occupied or intended to be occupied by one main building, accessory buildings, uses customarily incidental to such main building and such open spaces as are provided in the Ordinance, or as are intended to be used with such piece, parcel, or plot of land, except as otherwise permitted in the Ordinance.

Lot. Shall be considered as being a parcel of land surveyed or apportioned for sale or other purposes, as shown by the plat of the subdivision of which it is a part, filed in the office of the County Recorder, or as any tract of land the use of which is controlled or managed by any person or group of persons under a unified and specific plan. No parcel of land in a residence district shall be considered as being a lot having an area less than six thousand (6,000) square feet.

Lot. A parcel of land occupied or intended for occupancy by use permitted in the Ordinance including permitted buildings together with accessory buildings, the yard area and parking spaces required by the Ordinance, and having its principal frontage upon a publicly owned street.

Lot. A parcel of land occupied or to be occupied by one (1) building and accessory buildings and uses, including the open spaces required under the Ordinance. A lot may be land so recorded on the record of the Register of Deeds of the County, but it may include parts of or a combination of such lots when adjacent to one another, provided such land is used for one improvement. The area of the lot shall be measured to the street line only.

Lot. A parcel of land, or two or more contiguous parcels to be used as a unit under the provisions of the Ordinance, as shown in the records of the County Assessor's Office, and having its principal frontage on a street. In any district where a half-street has been dedicated not less than 25 feet in width, lots facing on such half-street shall be deemed to have frontage on a street.

Lot. A parcel of land in a single ownership occupied or to be occupied by one (1) principal building or use and the accessory buildings or uses customarily incident to the principal building or use, including such open spaces as are required by the Ordinance, except that where a parcel of land contains less than one (1) acre and not more than five (5) acres, more than one (1) principal building and the accessory buildings and uses.

Lot. A parcel of land exclusive of any adjoining street, the location, dimensions and boundary of which are determined by description as on a subdivision map of record, an official map or by metes and bounds, which is occupied or intended to be occupied by one building, and, if any, its accessory buildings; or by a group of buildings as permitted by the Ordinance and including such open spaces appurtenant to such building or group of buildings required by the Ordinance. In the event that more than one plot or lot as set forth on any map filed in the Office of the Clerk of the County, or on present or future assessment maps of the County is used in part or in full with one or more other such plot or lots for the erection of a building and its accessory buildings including yards, the aggregate of all such plots or lots shall, for the purpose of the Ordinance be deemed to be one lot.

Lot. Parcel of land devoted to a single principal use or land occupied or to be occupied by a building and its accessory buildings, together with such open spaces as are required under the ordinance, and having its principal frontage upon a street or officially approved place.

Lot. A parcel of land which is either a ''lot of record'' or a ''zoning lot.''

Lot. A parcel of land legally described and subdivided as a single lot, occupied by or intended for occupancy by one principal building together with its accessory structures and uses, including the yards required by the Ordinance and having frontage on a public street.

Lot. A parcel of land of at least sufficient size to meet minimum zoning requirements for use, coverage and area, and to provide such yards and other open spaces as are required by the Ordinance.

Lot. A parcel of subdivided or unsubdivided land of record in the office of the Recorder, or registered in the office of the Registrar of Titles, which is occupied or intended for occupancy by one (1) main building with accessory buildings, or by a group house or group houses, including such open spaces as are required by the Ordinance, and which parcel of land has frontage on a public street or on a private street connecting with a public street in which private street the owner of any such parcel of land, by virtue of a recorded or registered instrument, has the right to use the same for ingress to and egress from any such parcel of land.

Lot. A parcel of land having its principal frontage upon a street or an officially approved place and not divided by a public right-of-way.

Lot. A parcel of land occupied or to be occupied by a building or group of buildings, together with such yards, open spaces, lot width and lot area as required by the Code having frontage upon a street or upon a right-or-way approved by the planning commission. A lot may be land so recorded on a plat of record, or considered as a unit of property and described by metes and bounds, and which may include parts of or a combination of such lots, when adjacent to one another, providing such grounds are used for one improvement. All lots shall front on or have ingress or egress by means of an officially approved public right-of-way.

Lot. A parcel of land located within a single block, which at the time of filing for a building permit is designated by its owner or developer as a parcel to be used, developed or built upon as a unit under single ownership or control. A ''Lot'' may or may not coincide with a lot of record. ''Lot'' includes ''plot'' or ''site''.

Lot. Parcel of land occupied or capable of being occupied by a principal building or use, or a group of principal buildings or uses that are united by a common interest or customary accessory buildings or uses, and including such open spaces to be used in connection with such buildings or uses, lot may or may not be a lot of record.

Lct. Any piece of land described by metes and bounds in a recorded deed or on a subdivision plot of record which possesses or is in the process of being assigned a number for tax assessment identification purposes.

Lot. Parcel of land used or intended to be used as a unit. Each building and its accessory building, shall be on a separate lot.

Lot. Parcel of land occupied or capable of being occupied by one building or use and the accessory buildings or uses permitted by the Ordinance, including such open spaces as are required by the Ordinance and having its principal frontage upon a public street or highway.

Lot. Single and contiguous parcel of land under one ownership. Such a parcel is still a single lot even though the one owner may introduce interior lot lines.

Lot. Plot or parcel of land considered as a unit, devoted to a certain use, or occupied by a building or a group of buildings that are united by a common interest and use, and the customary accessories and open spaces belonging to the same.

Lot. Land occupied or to be occupied by a building and its accessory buildings together with such open spaces as are required under the Code and having its principal frontage upon a public street or officially approved place.

Lot. Parcel of land, or two or more contiguous parcels used as a unit; which lots are shown as such in the records of the County Assessor's Office, and having a frontage on a street.

Lot. Parcel of land occupied or to be occupied by a use, building or structure, and permitted accessory buildings, together with such open spaces, lot width and lot area as are required by the ordinance, and having access to a public street.

Lot. The smallest portion of parcel of land considered as a unit.

Lot. Portion or parcel of land considered as a unit.

Lot. A parcel of land defined by metes and bounds or boundary lines in a recorded deed fronting on a street. In determining lot area on boundary lines, no part thereof within the limits of the street shall be included.

Lot. A parcel of land shown as an individual unit on the most recent plat of record.

Lot. A parcel of land, the location, dimensions and boundaries of which are determined by the latest official tax assessors maps.

Lot. A tract or parcel of land intended for transfer of ownership, use or improvement.

Lot. A tract or parcel of land held in single and separate ownership.

Lot. A plot or parcel of land which is, or in the future may be offered for lease or sale, conveyance, transfer or improvement as one parcel, regardless of the method or methods in which title was acquired.

Lot. A parcel of real property shown on an approved final subdivision map or record of survey map or a recorded parcel described by metes and bounds.

Lot. A parcel of land which is or may be occupied by one main building or use and its accessories, including the open spaces required by the Ordinance.

Lot. A portion or parcel of land considered as a unit devoted to a certain use or occupied by a building or a group of buildings that are united by a common interest or use, and the customary accessories and open spaces belonging to the same.

Lot. Entire parcel of land occupied or to be occupied by a main building and its accessory buildings, or by a group such as a dwelling group or automobile court and their accessory buildings, including the yards and open spaces required therefore by the Ordinance and other applicable law.

Lot. A parcel of land, occupied or to be occupied by a use, building or unit group of buildings, and accessory buildings and uses, together with such yard, open spaces, lot width and lot area, as are required by the Ordinance, and fronting for a distance of at least 20 feet upon a street as defined herein, or upon a private street as defined in the Ordinance. The width of an access-strip portion of a lot shall not be less than 20 feet at any point.

Lot. Platted lot of a recorded subdivision, or a parcel of land, including, in addition to the land required to meet the regulations of the Code, all of the land area shown in a request for a Zoning Compliance Permit, occupied by a principal building or use and any accessory building or use.

Lot. Parcel of land occupied or intended for occupancy by a use permitted in the Code, including one main building together with its accessory buildings, the open spaces and parking spaces required by the Code, and fronting upon a street, as herein defined, except for lots of record as herein defined, which need not front on a public street.

Lot. Plot or parcel of land considered as a unit, devoted to a certain use, or occupied by a building or group of buildings that are united by a common interest and use, and the customary accessories and open spaces belonging to the same.

Lot. A parcel of land exclusive of any adjoining street, the location, dimensions and boundary of which are determined by description as on a subdivision map of record, an official map or by metes and bounds.

Lot. A parcel of land under one ownership which constitutes or is to constitute, a complete and separate functional unit of development, and which does not extend beyond the property lines along streets or alleys and has permanent access to a street or alley. A lot as so defined generally consists of a single Assessor's Lot, but in some cases consists of a combination of contiguous Assessor's Lots or portions thereof. In order to clarify the status of specific property as a Lot under the Code, the Zoning Administrator may, consistent with the provisions of the Code, require such changes in the Assessor's records, placing of restrictions on the land records and other actions as may be necessary to assure compliance with the Code.

Lot. A piece or parcel of land occupied or intended to be occupied by a principal building or a group of such buildings and accessory buildings, or utilized for a principal use and uses accessory thereto, together with such open spaces as required by the ordinance, and having access on a public street.

Lot of record. A lot which is a part of a subdivision, the plat of which has been recorded in the Office of the Register of Deeds of the County at the time of the adoption of the Ordinance, provided that said lot has a frontage of not less than forty (40) feet; or, an irregular tract lot as described by a deed recorded with the Register of Deeds of the County at the time of the passage of the Ordinance; provided such lot is numbered and described by the County Surveyor at the time of the passage of the Ordinance, and is not greater in area than one (1) acre at the time of the passage of the Ordinance.

Lot of record. A lot which is a part of a subdivision, the map of which has been recorded in the Office of the Registrar of Conveyances; or a parcel of land which became legally established and defined by deed or Act of Sale on or before the adoption of the Ordinance.

Lot of record. A lot which is part of a subdivision, the map of which has been recorded in the office of the Chancery Clerk of the County.

Lot of record. A lot which is a part of a plot, a map of which has been recorded in the office of the County Recorder.

Lot of record. A lot whose existence, location and dimensions have been legally recorded or registered in a deed or on a plat prior to the effective date of the Ordinance.

Lot of record. The land designated as a separate parcel on a plat map or deed in the records of the County.

Lot of record. A lot which is part of a subdivision, the map of which has been recorded in the Office of the Registrar of Conveyances; or a parcel of land which became legally established and defined by deed or Act of Sale.

Lot of record. Lot which is part of a platted sudivision or a parcel of land recorded in a County Deed Book or Official Record Book.

Lot of record. Lot which is part of a subdivision, the plat of which has been recorded in the office of the clerk of the circuit court of the County; or a parcel of land the deed of which has been recorded in the office of said clerk in the County.

Lot of record. Parcel of land for which the deed, prior to the adoption of the ordinance, is on record with the County Register of Deeds.

Lot of record. A parcel of land, the dimensions of which are shown on a recorded plat on file with the County Register of Deeds, at the time of inception of the Ordinance or in common use by Municipal or County Officials, and which actually exists as so shown, or any part of such parcel held in a record ownership at the time of inception of the Ordinance, separate from that of the remainder thereof.

Lot of record. A lot which is either part of a subdivision, the map of which has been recorded on the office of the Clerk of the District Court, or a parcel of land which has become legally established and defined by Deed or Act of Sale.

Lot of record. A tract, site or parcel designated as a separate lot on a legally recorded subdivision plat or in a legally recorded deed, prior to the effective date of the Zoning Resolution.

Lots of record. Herein designated as a separate and distinct parcel on a legally recorded subdivision plat or a legally recorded deed filed in the records of the County.

Lot of record. A lot which is part of a subdivision, a plat of which has been recorded, or a lot described by metes and bounds, the description of which has been recorded in the office of the Registrar of Deeds.

Lot (plot). Portion or parcel of land considered as a unit, devoted or to be devoted to a certain use or occupied by a building or group of buildings and accessory uses that are united by a common interest or use, and the open spaces belonging to the same. A lot may or may not be a platted lot.

Lot split. Any division of land by metes and bounds description into two (2) or more parcels for the purpose, whether immediate or future, of transfer of ownership, and which does not constitute a subdivision as defined in the Ordinance.

Lot zone. Single tract of land which, at the time of filing for a building permit, is designated by its owner or developer as a tract to be used, developed, or built upon as a unit under single ownership or control. A zone lot may or may not coincide with a lot of record.

Original tract. Contiguous body of land under the same ownership.

Outlots. Lots that do not meet the requirements of the ordinance as to minimum width and depth.

Parcel. A unit of land described by lot and block, by tract designation in a registered land survey, or by other legal descriptions.

Parcel. Any quantity of land capable of being described with such definiteness that its location and boundaries may be established.

Parcel of land. Contiguous quantity of land in the possession of, or owned by, or recorded as the property of, the same person.

Plot. Parcel of land consisting of one or more lots or portions thereof, which is described by reference to a recorded plat or by metes and bounds.

Premises. A distinct portion of real estate, land or lands with or without buildings or structures. It may or may not have the same meaning as "lot", "building" or "structure".

Premises. Land, including buildings or structures thereon, or any part thereof, except land occupied by streets, alleys, or public thoroughfares.

Premises. Lot and the buildings situated therein.

Premises. A lot, plot, parcel of land including the buildings or structures thereon.

Premises. Land including improvements or appurtenances or any part thereof.

Premises. A general term meaning part or all of any Zone Lot or part or all of any building or structure or group of buildings or structures located thereon.

Premises. Any land or any building, public or private, sailing, steam or other vessel, any vehicle, steam, electric or street railway car for the conveyancy of passengers or freight, any tent, van or other structure of any kind, any mine, or any stream, lake, drain, ditch or place, open, covered or enclosed, public or private, natural or artificial, and whether maintained under statutory authority or not.

Premises. The term "premises" shall describe a lot, parcel or plot of land including the buildings or structures thereon.

Private land. Land in a subdivision or development area which shall be adjoining, attached and assigned to a one-family, two-family or townhouse dwelling, to be held as an open space in ownership with the dwelling in the subdivision or development area, and which shall be identified on subdivision and development plans submitted to the City.

Rear lot. That portion of a lot lying at the opposite end to the front of a lot.

Rear of lot. Boundary opposite the front of the lot as herein defined, or that boundary most nearly parallel to the front of the lot.

Rear of zone lot. That portion of the Zone Lot farthest removed from the abutting street on any Zone Lot not a corner lot.

Reverse corner lot. Corner lot the side line of which is substantially a continuation of the front property line of the first lot to its rear.

Reversed corner lot. A corner lot, the rear lot line of which adjoins the side lot line of another lot, even though an alley may intervene.

Reversed corner lot. A corner lot where the street side lot line is substantially a continuation of the front lot line of the first lot to its rear.

Reversed corner lot. A corner lot, the side property line of which is substantially a continuation of the front property line of the first lot to its rear.

Reversed corner lot. A corner lot, the side street of which is substantially a continuation of the front lot line of the first lot to its rear.

Reversed corner lot. Corner lot having a lot or lots at its rear which fronts on the side street.

Reversed corner lot. Corner lot, the rear of which abuts the side lot line of another lot.

Reversed frontage lot. Corner lot, the side street line of which is substantially a continuation of the front lot line of the first platted lot to its rear.

Reverse frontage lot. Lot extending between and having frontage on a major traffic street and a minor street, and with vehicular access solely from the latter.

Section. A parcel of land containing 640 acres, and measuring one square mile. A mile measures 5,280 lineal feet.

Site. Area occupied by a building or structure, including the yards and courts required for light and ventilation, and such areas that are prescribed for access to the street.

Site or lot. Parcel of land or a portion thereof, considered a unit, devoted to or intended for a use or occupied by a structure or a group of structures that are united by a common interest or use. Site or lot shall have frontage on a street.

Split lot. Lot divided by a zone boundary.

Substandard lot. Any lot of less than five thousand (5000) square feet in area or less than fifty (50) feet in width at the front building line.

Through lot. An interior lot having frontage on two parallel, or approximately parallel streets and also known as a double fronted lot.

Through lot. The front property line of a through lot shall be that line which is obviously the front by reason of the prevailing custom of the other buildings in the block. Where such front property line is not obviously evident, the Board shall determine the front property line. Such a lot over 200 feet deep may be considered, for the purposes of this definition, as two lots each with its own frontage.

Through Lot. A lot having frontage on two (2) parallel or approximately parallel streets. A through lot having a depth of two hundred (200) feet or more shall; provide, front yard setbacks on both parallel frontages.

Through lot. "Interior lot" having frontage on two (2) parallel or approximately parallel streets.

Through lot. Lot having its front and rear lines on different streets, or having its front or rear line on a street and the other line on a river, lake, creek or other permanent body of water.

Through lot. An interior lot having frontage on 2 nonintersection streets.

Through lot. An interior lot having frontage on two parallel or approximately parallel, or converging streets.

Through lot. An interior lot having frontage on two streets.

Through lot. A lot abutting two parallel or approximately parallel streets.

Through lot. A lot in which the front lot line and rear lot line abut a street.

Through lot. A lot having frontage on two (2) streets (a corner lot, having frontage on two (2) parallel or approximately parallel streets, or two (2) streets, the center lines of which, if projected, would not make an angle of more than thirty (30) degrees).

Through lot. A lot other than a corner lot with frontage on more than one street. Through lots abutting two streets may be referred to as double frontage lots.

Through lot. A lot having two (2) opposite lot lines along two (2) more or less parallel streets, or along one (1) street and a body of water, and which is not a corner lot.

Through lot. A lot having a pair of opposite lot lines along two (2) or more or less parallel public streets, and which is not a corner lot. On a "through lot" both street lines shall be deemed front lot lines.

Through lot. An interior lot having a frontage on two (2) streets, which is not a corner lot.

Through lot. A lot having frontage on two parallel or approximately parallel streets, but not including those lots having frontage on a street and frontage on a navigable public canal or waterway parallel or approximately parallel to said street.

Through lot. A lot having a pair of opposite lot lines along two more or less parallel public streets and which is not a corner lot.

Through lot (double frontage). An interior lot having a street line for both the front lot line and the rear lot line.

Through lot (double frontage). A lot having a frontage of 2 approximately parallel streets or places.

Township. A geographical area containing 36 Sections or 36 square miles.

Tract. A parcel of land of no particular area, size or dimension.

Tract. A unit of land described by letter in a registered land survey.

Transitional lot. A lot not more than one hundred fifty (150) feet in width in a dwelling district, having one side lot line which is the side lot line of a lot in a district which is zoned, and which also has its front on the same street as a front in a district lot.

Transverse lot. Lot which is approximately at right angles to the general pattern of other lots in the same city block.

Wedged shaped lot. A lot situated so that the front is either wider or narrower than the rear of the lot.

A-11 NONCONFORMING LAND, BUILDING AND STRUCTURE USES

Abandoned. Discontinuance of a non-conforming use for a period of 6 months or more constitutes abandonment.

Abandonment. Land use shall be deemed abandoned where it has been discontinued and there is no clear indication from the property itself that it is to be resumed. Mere lack of occupancy, or vacancy, shall not, without more evidence be established as abandonment.

Damaged non-conforming building. No non-conforming building or structure which is hereafter damaged to an extent exceeding 50% of its then reproduction value, exclusive of foundation, by fire, flood, wind, explosion, earthquake, war, riot, or act of God, may be restored, reconstructed and used for any purposes other than a purpose permitted under the provisions of the zoning ordinance governing the district in which the building or structure is located.

Damages to nonconforming use. When a building containing a nonconforming use is damaged by explosion, fire, act of God or the public enemy to the extent of more than 50% of its current local assessed value, it shall not be restored except in conformity with the regulations of the district in which it is located. The total structural repairs or alterations in any non-conforming use shall not during its life exceed 50% of the local assessed value of the building at the time of its becoming a nonconforming use unless permanently changed to a conforming use.

Non-conforming. Not conforming to, or prohibited by, the provision of the Code.

Non-conforming. That state or condition of a building or other structure, use, or lot, which by reason of design, size or use, does not conform with the requirements of the district, or districts, in which it is located.

Non-conforming building. Any building which does not conform to the regulations for the district in which it is located.

Non-conforming building. A "non-conforming building" or other structure is any lawful building or structure which does not conform to one or more of the applicable area and bulk regulations of the district in which it is located either on the effective date of the Ordinance or as a result of a subsequent amendment thereto.

Non-conforming building. Building, or portion thereof, which was lawfully erected or altered and maintained, but which, because of the application of the ordinance to it, no longer conforms to the use, height, yard, setback or area regulations of the zone in which it is located.

Non-conforming building. Building or structure or portion thereof that does not conform to the height and area regulations of the zone in which it is located, or as a result of a subsequent amendment which may be incorporated into the Code.

Non-conforming building. Legally existing building which fails to comply with the regulations (for height, number of stories, size, area, yards and location) set forth in the ordinance applicable to the district in which this building is located.

Non-conforming building. A building, structure, or portion thereof which does not conform to the regulations of the Code, and which lawfully existed at the time the regulations with which it does not conform, became effective.

Non-conforming building. A building, structure or portion thereof, which does not conform to the regulations of the Ordinance and which lawfully existed at the time the Zoning regulations, with which it does not conform, became effective.

Nonconforming building. A lawfully established building which does not presently conform to all the applicable requirements of the Ordinance, governing height, bulk, location on a lot, off-street parking and off-street loading.

Nonconforming building. A building or portion thereof, existing at the effective date of the Ordinance or amendments thereto, and that does not conform to the provisions of the Ordinance, relative to height, bulk, area, yards, for the district in which it is located.

Non-conforming building or structure. Any building or structure which does not comply with all of the regulations of the Ordinance or of any amendments hereto for the Zoning District in which such use is located.

Non-conforming building or structure. The lawful use of a building or structure or portion thereof, existing at the time the Ordinance or amendments thereto take effect, and which does not conform to all the height, area and yard regulations prescribed in the zone in which it is located.

Non-conforming building or structure. Building or structure lawfully existing at the effective date of the ordinance or any amendment thereto affecting such building or structure, which does not conform to the Building Regulations of the ordinance for the Zone in which it is situated, irrespective of the use to which such building or structure is put.

Non-conforming building or structure. Any building or structure existing on the effective date of the ordinance or amendment thereto, which does not conform after the effective date of the ordinance or amendments thereto, with the building height, lot coverage, yard, open space, and/or interior living space regulations of the zone in which it is located.

Non-conforming buildings or use. A nonconforming building or use is one that does not conform with the regulations of a given use district.

Non-conforming buildings or use. A building or use lawfully existing at the time of adoption of the Ordinance and which does not conform with the use regulations of the district in which it is located.

Non-conforming land. Any lawful lot which does not conform to one or more of the applicable area regulations of the district in which it is located either on the effective date of the Ordinance or as a result of a subsequent amendment thereto.

Non-conforming lot. Lot existing lawfully at the time the Zoning Ordinance, or an amendment thereto, became effective, but which does not conform to the lot area, width, access or other requirements of the district in which it is located.

Non-conforming lot. A lot whose width, area or other dimension does not conform to the regulations of the Ordinance and which lawfully existed at the time the regulations with which it does not conform became effective.

Nonconforming lot. Any lot which does not meet the minimum dimensions, area, or other regulations of the district in which it is located.

Non-conforming lot of record. A lot of record which does not comply with the lot area, lot width, lot depth or lot shape requirements for the district in which it is located.

Non-conforming structure. A lawfully established structure which does not conform with the bulk regulations or parking regulations of the District in which it is located.

Non-conforming structure. A lawfully established structure which does not presently conform to all the applicable requirements of the Ordinance governing height, bulk, location on a lot, off-street parking and off-street loading.

Non-conforming structure. Any structure which existed lawfully at the time of the adoption of the Zoning Ordinance, or amendments thereto, all or substantially all of which is designed or intended for a use not permitted in the district in which it is located, or which does not comply with the applicable area, bulk, density, or setback requirements of the Ordinance.

Non-conforming structure. A building or structure, lawfully existing at the time of enactment of the Ordinance or a subsequent amendment thereto, that does not conform to the schedule of area, height, and placement regulations or to the supplementary regulations of the Ordinance for the Zoning District in which it is located.

Non-conforming structure. Structure which was lawfully erected, but which does not conform with the standards for yard spaces, height of structures, or distances between structures prescribed in the regulations for the district in which the structure is located, by reason of adoption or amendment of the Ordinance, or by reason of annexation of territory to the City.

Non-conforming structure. A structure or building which was lawfully erected, but which does not conform with the standards for yard spaces, heights of structures, or distances between structures prescribed in the Zoning regulations for the district in which the structure is located, by reason of adoption or admendment of the Ordinance, or by reason of annexation of territory to the City Corporate city limits.

Non-conforming use. Any building or land lawfully occupied by a use at the time of passage of the Ordinance and amendments thereto, which does not conform after the passage of the Ordinance or amendments thereto, with the use regulations of the district in which it is situated.

Non-conforming use. Use which was lawfully established and maintained, but which, because of the application of the Ordinance to it, no longer conforms to the use regulations of the zone in which it is located.

Non-conforming use. Any building or land lawfully occupied by a use on the effective date of the Ordinance, or any amendment thereto, which does not conform after the effective date of the Ordinance or amendments thereto, with the use regulations of the zone in which it is located.

Non-conforming use. Use of a building or structure or of a tract of land, lawfully existing at the time of enactment of the Code or a subsequent amendment thereto, that does not conform to the SCHEDULE OF USE REGULATIONS of the Code for the Zoning District in which it is situated.

Non-conforming use. A parcel of land lawfully occupied by a use that does not conform to the regulations of the district in which it is located.

Non-conforming use. Areas lawfully occupied by a building or land use at the time the Ordinance or amendments thereto take effect, and which does not conform with the use regulations of the zone in which it is located.

Non-conforming use. A structure or land lawfully occupied by a use that does not conform to the regulations of the district in which it is situated.

Non-conforming use. The use of a building or land that does not conform to the regulations of the Ordinance for the district in which it is located at the time of the adoption of the Zoning Ordinance, or at the time of the adoption of any amendments or supplements thereto, making it a non-conforming use.

Non-conforming use. Any lawful building or structure or any lawful use of land, premises, building or structure which does not conform to the regulations of the Ordinance for the district in which such building, structure, or use is located either on the effective date of the Ordinance or as a result of subsequent amendments hereto.

Non-conforming use. A legally existing use (of land or building) which fails to comply with "permitted uses" of the Ordinance, either at the effective date of the Ordinance, or as a result of subsequent amendments to the Ordinance.

Non-conforming use. A building or premise lawfully used or occupied at the time of the passage of the Ordinance or amendment thereto, which use or occupancy does not conform to the regulations of the Ordinance or any amendments thereto.

Non-conforming use. A use of a building, structure, or land existing at the time of enactment of the Ordinance and which does not conform to the use regulations of the district in which it is situated.

Non-conforming use. Any lawful use, in existence at the time of the adoption of the Ordinance, and not prohibited by the Zoning Ordinance, for a building or premises that does not conform with the regulations of the use district in which it is situated.

Non-conforming use. A building or land which does not conform with the height, area, or use regulations of the district in which it is located.

Non-conforming use. A building or the use of a building or land that does not conform to the regulations of the use district in which it is situated.

Non-conforming use. A building or premises occupied by a use which does not conform to the regulations of the district in which it is situated.

Non-conforming use. A use which is lawfully exercised within a structure or on land at the time of adoption of the Ordinance, or any amendment thereto, and which does not conform with the regulations of the zone district in which it is located.

Non-conforming use. Any use of a building or land that does not agree with the regulations of the district in which it is situated.

Non-conforming use. Use of a building or of land which existed previously that does not conform to the present regulations as to use for the district in which it is situated.

Non-conforming use. A building or land occupied by a use that does not conform with the regulations of the use district in which it is situated.

Non-conforming use. A use which lawfully occupies a building on a lot, or portion thereof, at the time of the effective date of the Ordinance, and which by virtue of the Ordinance does not conform to the use regulation of the zone in which it is located.

Non-conforming use. Any lawful use, whether of a building or other structure or of a tract of land which does not conform to the applicable use regulations of the district in which it is located, either on the effective date of the Ordinance or as a result of subsequent amendments thereto.

Non-conforming use. A use of a structure or land which was lawfully established and maintained, but which does not conform with the use regulations or required conditions for the district in which it is located, by reason of adoption or amendment of the Ordinance, or by reason of annexation of territory to the City.

Non-conforming use. Any building or land lawfully occupied by or used at the time of the passage of the Ordinance, which after passage does not conform to the use regulations of the District in which it is located. Improvements existing at the time of passage of the Ordinance not meeting required parking and loading regulations, height regulations, and area regulations for the District in which they are situated shall be considered as non-conforming use.

Non-conforming use. Any use of land or structure which existed lawfully at the time of the adoption of the Zoning Ordinance, or amendments thereto, which does not comply with the regulations of the District in which it is located.

Non-conforming use. A use of a building or premises that does not conform with the regulations of the District in which it is situated.

Non-conforming use. A building, structure or use of land, building or structure lawfully existing at the time of the enactment of the Ordinance, and which does not conform after the passage of the Ordinance to the regulations of the district in which it is located.

Non-conforming use. A use of land or of structures on the land which does not comply in every respect with every provision of the Ordinance.

Non-conforming use. Any building or land lawfully used at the time of the passage of the Ordinance or amendment thereto, which does not conform with the provisions thereof, shall be considered a non-conforming use.

Non-conforming use. A use of a building or premises occupied by, or if vacant, intended for a use that does not conform with the regulations of the use district in which such building or premises is located.

Non-conforming use. A building or portion thereof, or land or portion thereof, occupied by a use that does not conform with the regulations of use in the district in which it is situated.

Non-conforming use. A use of a building or premises that does not conform to the regulations of the Use District in which it is located.

Nonconforming use. A use of a building or land which does not conform to the regulations of the Ordinance and which lawfully existed at the time the regulations with which it does not conform became effective.

Nonconforming use. Any lawfully established use of a building or premises which on the effective date of the Zoning Code or amendment thereof does not comply with the use regulations of the zoning district in which such building or premises shall be located.

Nonconforming use. A legal use of a structure or tract of land in existence at the date of adoption of the ordinance which does not conform to the use regulations of the ordinance, or such legal use in existence at the date of annexation to the City, which does not conform with the use regulations of the ordinance as amended at the time of annexation, or such legal use in existence at the date of adoption of amendments to the ordinance which does not conform to the use regulations of the ordinance as amended.

Nonconforming use. Any use of a building, structure, lot or land, or part thereof, lawfully existing at the effective date of the ordinance or any amendment thereto affecting such use, which does not conform with the Use Regulations of the ordinance for the Zone in which it is situated.

Nonconforming use. Existing use of land or building which was legal prior to the effective date, but which fails to comply with the regulations set forth in the Ordinance, applicable to the district in which such use is located.

Nonconforming use. Any use, whether of a building, other structure, lot or tract of land, which does not conform to the use regulations of the Ordinance for the district in which such "nonconforming use" is located, or as a result of a subsequent amendment which may be incorporated into the Ordinance.

Nonconforming use. A use of any structure or land which, though originally lawful, does not conform with the provisions of the ordinance or any subsequent amendments thereto for the district in which it is located.

Nonconforming use. A structure or land lawfully used or occupied and which does not conform to the regulations of the use district in which it is situated.

Nonconforming use. A building or premises occupied by a use that does not conform with the Use Regulation of the district in which it is situated.

Nonconforming use. A use of a building or land, existing lawfully to the time the Zoning Code, or an amendment thereto became effective but which does not conform to the use regulations, off-street parking and loading requirements, performance standards or other use regulations of the district in which it is located.

Nonconforming use. A lawfully established use of land, buildings, structures or premises which is not presently listed permitted or special use under the Ordinance and which is not a use granted by variation subsequent to the adoption of the Ordinance.

Non-conforming use of building. Use of any building other than a use specifically permitted in the district in which the building is located.

Non-conforming use of land. Use of any land other than a use specifically permitted in the district in which the lot or parcel of land is located.

Non-conforming use of similar nature. For the purposes of the Ordinance a change from a non-conforming use to a non-conforming use of similar nature, means a change to any other use that would lawfully be permitted in the same zoning district in which the first use would lawfully be permitted.

Non-conforming uses. Building or land which does not conform with the height, area, or use regulations of the District in which it is located.

Non-conforming uses. Any use of a building or lot which does not comply with the provisions of the ordinance.

Non-conforming uses. Existing lawful use of a building or premises at the time of the enactment of the ordinance or any amendment thereto, may be continued although such use does not conform with the provisions of the ordinance for the District in which it is located, but such non-conforming use shall not be extended.

Non-conforming uses permitted. Except as herein specified the lawful use of any building or land existing at the time of the enactment of the Ordinance may be continued, although such use does not conform to the provisions of the Ordinance.

Non-conformity. Lot, structure, or use of land, or any combination thereof, which was lawful before the Ordinance was passed or amended, but which would be prohibited under the terms of the Ordinance.

Non-conformity. Any feature, such as location, size, bulk, height, or use of a building or premises, that does not conform to the regulations of the District in which it is located.

Privileged non-conforming use. Which was lawfully in existence on the effective date of the Ordinance, but only to the extent to which it had developed on said effective date.

Restoration. Non-conforming building wholly or partially destroyed by fire, explosion, flood, or other phenomenon, or legally condemned, may be reconstructed and used for the same non-conforming use, provided that (a) the reconstructed building shall not exceed in height, area, and volume the building destroyed or condemned, and (b) building reconstruction shall be commenced within one (1) year from the date the building was destroyed or condemned and shall be carried on without interruption.

A-12 OBSCENITY AND PORNOGRAPHIC MATERIALS AND DISPLAYS

Adult bookstore. An establishment: (a) having as a substantial portion of its stock in trade, books, magazines and other periodicals depicting, describing or relating to specified sexual activities or which are characterized by their emphasis on matter depicting, describing or relating to specified anatomical area; or (b) having as a substantial portion of its stock in trade, books, magazines and other periodicals and which excludes all minors from the premises or a section thereof.

Adult live entertainment establishment. An establishment which features dancers, go-go dancers, exotic dancers, strippers, or other similar entertainers, any of whom perform topless or bottomless.

Adult theater. An enclosed building or open-air drive-in theater regularly used for presenting any film or plate negative, film or plate positive, film or tape designed to be projected on a screen for exhibition, or films, glass slides or transparencies, either in negative or positive form designed for exhibition by projection on a screen and which regularly excludes all minors.

Adult theater. An enclosed building or open-air drive-in theater regularly used for presenting any film or plate negative, film or plate positive, film or tape designed to be projected on a screen for exhibition, or films, glass slides or transparencies, either in negative or positive form, designed to specified sexual activities or characterized by an emphasis on matter depicting, describing or relating to specified anatomical areas.

Article. Any picture, drawing, photograph, article or instrument. It does not include any of these articles used in or by any recognized religious, scientific or educational institution.

Filthy. Morally foul, polluted, nasty, dirty, vulgar, debasing, having tendency to corrupt or debauch.

Indecent. Morally offensive or depraving.

Lascivious. Wanton or lustful.

Lewd. Licentious, lecherous, dissolute, debauched, impure, salacious or pornographic.

Obscene. To the average person would mean, applying below the standards established by the community.

Pornographic. The portrayal of obscene scenes of persons in the nude.

Unfit. Obscene, lewd, lascivious, filthy or indecent.

A-13 OCCUPANCY AND OCCUPANCY CLASSIFICATIONS

Agricultural building. A building located on agricultural property and used to shelter farm implements, hay, grain, poultry, livestock, or other farm produce, in which there is no human habitation, and which is not used by the public.

Assembly building. A building or parts thereof, designed or used for the assembly of persons for civic, political, educational, religious, social, recreational or other similar activities.

Assembly occupancy. Buildings in which provision is made for the congregation or gathering of seventy-five (75) or more persons in one room or space shall be classified Assembly Occupancy. Such room or space shall include any occupied connecting room or space in the same story, or in story or stories above or below, where entrance is common to rooms or spaces. This occupancy includes buildings having an auditorium and a stage provided for the use of moveable scenery, or having an auditorium for viewing motion pictures, or for theatrical purposes.

Assembly occupancy. Occupancy or use of a building or structure or any portion thereof by a gathering of persons for civic, political, travel, religious, social, or recreational purposes.

Assembly occupancy. Occupancy or use of a building, or part thereof, by a gathering of persons for civic, educational, political, recreational, religious, social, or like purposes, or for the consumption of food or drink.

Business and personal services occupancy. Occupancy or use of a building or part thereof for the transaction of business or the rendering or receiving of professional or personal services.

Business buildings. A building or parts thereof, designed or used for the transaction of business or for the rendering of professional service or for other services that do not involve the storage of stocks of goods, wares, or merchandise, except such as are incidental for display purposes.

Business classification. Business classification shall include buildings, or parts thereof, which are occupied for the transaction of business, for the rendering of professional services, or for other commercial services that involve stocks of goods in limited quantities only, for uses incidental to offices or for sample purposes.

Business occupancy. The occupancy or use of a building or structure or any portion thereof for the transaction of business or the rendering or receiving of professional services.

Business occupancy. Buildings which are occupied for business or rendering of professional services shall be classified in the Code; Buildings which are occupied for the sale or display of merchandise, or the supplying of food or drink, are otherwise classified in the Code.

Commercial occupancy. Applies to that portion of a building used for the transaction of business; for the rendering of professional services; for the supply of food, drink, or other bodily needs and comforts; for manufacturing purposes or for the performance of work or labor, including among others, bake shops, fur storage, laboratories, loft buildings, markets, office buildings, professional buildings, restaurants, stores other than department stores, and similar occupancies.

Doubtful classification. In case a building is not specifically provided for, or where there is any uncertainty as to its classification, its status shall be fixed by a duly promulgated rule giving due regard to safety.

Extra hazardous occupancy. Includes occupancies involving highly combustible, explosive or unstable products or materials that constitute a special fire, life or toxic hazard because of the form, characteristics or volume of the materials used including every building or portion thereof used for the manufacturing, assembling, processing, or use, of highly combustible, flammable, explosive or unstable materials.

Fire hazard classification. A classification of occupancy or use of a building based on the fire load or danger of explosion therein.

General public occupancy. Applied to those who enter, use, or occupy a building, means persons other than the Owner of the building, a tenant in the building, and persons who are employed in the building.

Hazardous classifications. Includes structures, buildings, or parts thereof used for purposes that involve highly inflammable, or explosive products or materials which are likely to burn with extreme rapidity or which may produce poisonous fumes or gases, including highly corrosive, toxic, or noxious alkalies, acids, or other liquids or chemicals which involve flame, fumes, explosive, poisonous, irritant or corrosive hazards, also uses that cause division of material into fine particles or dust subject to explosion or spontaneous combustion, and uses that constitute a high fire hazard because of the form, character, or volume of the materials used.

Hazardous occupancy. Includes occupancies involving highly combustible products or materials that constitute an unusual fire and life hazard because of form, characteristics, and volume of the material used; including every building or structure or part thereof used for storage, warehousing, manufacturing, processing, use or sale of highly combustible products and materials.

High hazard industrial occupancy. Industrial occupancy containing sufficient quantities of highly combustible and flammable or explosive materials which, because of their inherent characteristics, constitute a special fire hazard.

High hazard occupancy. Occupancy or use of a building or structure or any portion thereof that involves highly combustible, highly flammable, or explosive material, or which has inherent characteristics that constitute a special fire hazard.

High hazard occupancy. Includes all buildings and parts of buildings used for purposes, processes, or storage involving highly combustible, highly flammable, or explosive products or materials, or products or materials that constitute a special hazard to life or limb because of the form, character, or volume of the materials used, or are hazardous because of special conditions incident to the processes or conditions of use or storage.

Light hazard. Where the amount of combustibles or flammable liquids present is such that fires of small size may be expected. These may include offices, schoolrooms, churches, assembly halls, telephone exchanges, etc.

Low hazard industrial occupancy. Industrial occupancy in which the combustible content is not more than 10 lbs. or 100,000 Btu per square foot of floor area.

Medium-hazard industrial occupancy. Industrial occupancy in which the combustible content is more than 10 lb. or 100,000 Btu per square foot of floor area and not classified as high hazard industrial occupancy.

Mercantile building. A building, or parts thereof, designed or used for the sale of goods, wares and merchandise and involving only incidental storage of such materials.

Mercantile occupancy. Occupancy or use of a building or structure or any portion thereof for the displaying, selling or buying of goods, wares, or merchandise, except when classed as a high hazard occupancy.

Mercantile occupancy. Occupancy or use of a building or part thereof for the displaying or selling of retail goods, wares or merchandise.

Minor occupancy. Occupancy subordinate to the major occupancy in a building of mixed occupancy.

Mixed occupancy. Where a minor portion of a building is used for an office, study, studio or other like similar purpose, the building shall be classified as to the occupancy on the basis of the major use. In some cases where a building is occupied for two (2) or more purposes not included in one class, the provisions of the code applying to each class of occupancy shall apply to each such parts of the building as come within that class; and if there should be conflicting provisions, the requirements securing the greater safety shall apply.

Mixed occupancy. Occupancy of a building in part for one use and in part for some other use not accessory to the first use.

Mixed occupancy. When a building is used for two or more occupancies, classified within different occupancy groups.

Occupancy. Purpose for which a building or part of a building is used or intended to be used.

Occupancy. Use of a building, structure, or premises.

Occupancy. Use or occupancy of a building, character of use, or designated purpose of a building or structure or portion thereof.

Occupancy. Use or uses of any building or structure for any of the following purposes: residential, instutional, assembly, business, mercantile, industrial, storage, hazardous or miscellaneous uses as defined in the ordinance.

Occupancy. Use or intended use of a building or part thereof for the shelter or support of persons, animals or property.

Occupancy. The purpose for which a building is used or intended to be used. Change of occupancy is not intended to include change of tenants or proprietors.

Occupancy. The purpose for which a building or part thereof is used or intended to be used. Change of Occupancy shall mean a change from one occupancy classification to another and shall not be construed as meaning a change of user or occupant. Mixed Occupancy shall mean the use of a building for more than one Occupancy classification.

Occupancy. The purpose for which a building is used intended to be used.

Occupancy. The purpose for which a building is used or intended to be used. The term shall also include the building or room housing such use. Change of occupancy is not intended to include change of tenants or proprietors.

Occupancy. Pertains to and is the purpose for which a building is used or intended to be used. Occupancy is not intended to include tenancy or proprietorship.

Occupancy. The purpose for which a room is used or intended to be used. The term "Occupancy" as used in the Code shall include the room housing such occupancy and the space immediately above a roof or structure if used or intended to be used for other than shelter.

Occupancy change. Change from one occupancy classification to another occupancy classification. Change of occupancy is not intended to include change of tenants or proprietors unless there is an incident change of occupancy classification.

Occupancy classification. A classification of buildings into occupancy groups based on the kind or nature of occupancy or use.

Occupancy classification or occupancy. The classification of a structure and the uses therein.

Occupancy content. Maximum number of persons occupying any building, floor, room, or space.

Occupancy factor. The square-feet-of-floor-area-per-occupant factor used in computing the number of occupants assumed to occupy a floor area. Such floor area shall be measured within exterior walls, deducting only the area occupied by stair, elevator, and other permanent shafts completely enclosed in fire-resistive enclosures.

Occupancy load. Number of individuals normally occupying the building or part thereof, or for which the exit facilities have been designed.

Occupancy load. Total number of persons occupying a building or structure at peak periods of the business hours, or the designed total load of the spaces.

Occupancy or tenant separation. Wall or floor assembly which separates occupancies, uses, or tenants, within a building and which has a fire resistance rating as required by code.

Occupancy or use. Purpose for which a building, structure, equipment, materials, or premises, or part thereof, is used or intended to be used as regulated in the code.

Occupant car ratio. For purposes of site planning, the number of parking and garage spaces, without time limits, required for each living unit.

Occupant load. Total number of persons that may occupy a building or portion thereof at any one time.

Occupant load. Number of persons for which a building or part thereof is designed.

Occupant load. Total number of persons that may occupy a building or portion thereof at any one time for a specified use.

Occupant load. The number of persons normally occupying a building, or the number of persons for which the exit ways are designed.

Occupiable room. Room or enclosed space designed for human occupancy in which large numbers of persons congregate for amusement, educational or similar purposes, or in which persons are in engaged in labor, and which is equipped with exits, light and ventilation facilities meeting the requirements of the code.

Occupied. Refers to any room or enclosure in a building or structure used b,y one or more persons for other than incidental maintenance.

Occupied space. Space within a building or structure and not otherwise defined in the code, wherein persons normally assemble, work, or remain for a period of time.

Occupied trailer coach or mobile home. Trailer coach or mobile home located on a site within the mobile home park when such mobile home is connected to any park facility such as sewerage collection system, water or electrical distribution system.

Occupier. Person in occupation or having the charge, management or control of any premises, whether on his own account or as an agent.

Posted use and occupancy. Posted classification of a building in respect to use, fire grading, floor load and occupancy load.

Private occupancy. As applied to part of a building, means that such part of the building is not normally subject to common use by those who occupy or enter the building.

Public assembly occupancy. Applies to that portion of the premises in which persons congregate for civic, political, educational, religious, social, or recreational purposes; including among others, armories, assembly rooms, auditoriums, ball rooms, bath houses, bus terminals, broadcasting studios, churches, colleges, court house without cells, dance halls, department stores, exhibtion halls, fraternity halls, libraries, lodge rooms, mortuary chapels, museums, passenger depots, schools, skating rinks, subway stations, theaters, and similar occupancies.

Public building. A building in which persons congregate for civic, political, educational, religious, social or recreational purposes.

Theater classification. Includes buildings or parts of buildings, containing an a assembly hall, having a stage which may be equipped with curtains or permanent or movable scenery, or which is otherwise adaptable to the showing of plays, operas, motion pictures or similar forms of entertainment.

Use group. Classification of a building or structure based on the character of usage or the purpose for which the building or structure is designed or used.

Use group. The classification of a building or structure based on the purpose for which it is used.

Use group. The classification of a building based on the purpose for which the building is used, intended, or designed to be used.

Use group classification. All buildings and structures shall be classified with respect to use in one of the following use groups; group (A) high hazard; group (B) storage; group (C) mercantile; group (D) industrial, group (E) business, group (F) assembly; Group (H) institutional; group (L) residential; and group (M) miscellaneous buildings.

Used or occupied. The words "used" or "occupied" as applied to any land or building shall be construed to include the phrase "intended, arranged or designed to be used or occupied".

Use-used. The purpose for which the building or structure is designed, used or intended to be used.

A-14 PRINCIPAL, MAIN, ACCESSORY BUILDINGS, STRUCTURES AND USES

Accessory. As applied to a use or a building or a structure, means customarily subordinate or incidental to, and located on the same lot with a principal use, building or structure.

Accessory. When applied to the use of a building or structure usually means subordinate or incidental to the principal use of the main building or structure on the same lot.

Accessory building. A subordinate building or portion of a principal building, the use of which is incidental to that of the principal building and customary in connection with said use.

Accessory building. A subordinate detached building, the use of which is incidental to that of the main building or to the main use of the premises.

Accessory building. A detached subordinate building clearly incidental to and located upon the same lot occupied by the main building. Any accessory building shall be considered to be a part of the main building when joined to the main building by a common wall not less than four (4) feet long, or when any accessory building and the main building are connected by a breezeway which shall be not less than ten (10) feet in width.

Accessory building. A detached subordinate building, the use of which is customarily incidental to that of the main building or to the main use of the land and which is located in the same or a less restrictive zone and on the same lot with the main building or use. The relationship between the more restrictive and the less restrictive zones shall be determined by the sequence of zones set forth in the Ordinance.

Accessory building. Any building on a lot other than a main building.

Accessory building. A subordinate building or structure on the same lot with the principal or main building, occupied or devoted exclusively to an accessory use.

Accessory building. A subordinate building detached from, but located on the same lot as the main building, the use of which is incidental and accessory to that of the main building or use.

Accessory building. No accessory building shall be erected in any required front yard, or in the portion of a required side yard between the required front yard and required rear yard, or within three feet of any lot line; a separate accessory building shall not be located within five feet of any other building.

Accessory building. A subordinate building containing an accessory use and situated on the same lot as the main building. An accessory building attached to the main building shall be considered to be a part of the main building and shall maintain any yards required for a main building.

Accessory building. A subordinate building or structure on the same lot with the principal building, or a part of the principal building, exclusively occupied by or devoted to an accessory use. Where an accessory building is attached to the front or side of a principal building, such accessory building shall be considered part of the principal building for the purpose of determining the required dimensions of yards; but if it is attached to the rear of the principal building in such a manner that it is completely to the rear of all portions of the principal building, it may be considered an accessory building for determining required yard dimensions.

Accessory building. Subordinate building or portion of the main building, the use of which is incidental to that of the main building on the same lot. Where an accessory building is attached to and made part of the main building, such accessory building shall comply in all respects with the requirements of the ordinance applicable to the main building. An accessory building, unless attached to and made a part of the main building, as above provided for, shall be not closer than seven (7) feet to the main building.

Accessory building. A Miscellaneous Use Unit located on the same lot and incidental to a main or principal occupancy.

Accessory building. A building detached from and subordinate to a main building on the same lot and used for purposes customarily incidental to the use of the main building or to any agricultural use shall be deemed to be an accessory building, whether situated on the same lot with a principal building or not.

Accessory building. A subordinate building the use of which is incidental and customary to that of the principal building.

Accessory building. A building, the use of which is customarily incidental to that of a dominant use of the main building or premises including bona fide household employees' quarters.

Accessory building. A subordinate structure, whether attached or detached or a subordinant adjunct to the principal building, the use of which is customarily incidental to the permitted use of the principal building.

Accessory building. A detached subordinate building situated on the same lot with the main building, and used for an accessory use.

Accessory building. A secondary building, the use of which is incidental to that of the main building and which is located on the same plot.

Accessory building. A subordinate building other than a garage on the same premises, the use of which is incidental to that of the main building.

Accessory building. A subordinate building or portion of main building for use customarily incident to the main use of the premises located on the same premises with the main building to whose use it is incidental.

Accessory building. A subordinate building or portion of main building, the use of which is clearly incidental to that of the main building.

Accessory building. A building or portion of a building subordinate to the main building and used for a purpose customarily incidental to the permitted use of the main building or the use of building, or is substantially attached there to, the side yard and rear yard requirements of the main building shall be applied to the accessory building.

Accessory building. A building subordinate to the main building on a lot and used for purposes customarily incidental to those of the main building.

Accessory building. A building customarily incident to a principal use of building and located on the same premises with such principal use or building.

Accessory building. A subordinate building, or portion of the main building on a lot, the use of which is customarily incidental to that of the main or principal building.

Accessory building. A subordinate building, the use of which is clearly incidental and secondary to that of the main building, on the same lot or parcel of land, which is used exclusively by the occupant of the main building.

Accessory building. A subordinate building, attached to or detached from the main building, the use of which is incidental to that of the main building and not used as a place of habitation except by domestic servants employed on the premises.

Accessory building. A subordinate building located on the same lot occupied by the main building, the use of which is incidental to that of the main building or to the use of the land. A trailer shall not be considered an accessory building.

Accessory building. A building the use of which is customarily incidental to that of the main building and which is located on the same lot as that occupied by the main building.

Accessory building. A subordinate building the use of which is purely incidental to that of the main building and which shall not contain living or sleeping quarters or storage space for commercial motor vehicles.

Accessory building. A subordinate building, the use of which is clearly incidental to that of the main building or the use of the land.

Accessory building. A subordinate building the use of which is incidental to that of the main building on the same lot. On any lot on which is located a dwelling, any building which is incidental to any agricultural use, shall be deemed to be an accessory building.

Accessory building. A subordinate building, or a portion of the main building, the use of which is incidental to that of the main building or to the use of the premises.

Accessory building. A subordinate building or structure on the same lot, or part of the main building, occupied by or devoted exclusively to an accessory use.

Accessory building. A building intended for a use that is accessory to the main use, subordinate to another building on the same lot, and having a total area of not over 1000 square feet.

Accessory building and use. Subordinate building located on the same lot with the main building, or subordinate use of the land, either of which is customarily incident to the main building or to the principal use of the land. Where a substantial part of the wall of an accessory building is a part of the wall of the main building or where an accessory building is attached to the main building in a substantial manner as by a roof, such accessory building shall be counted as part of the main building.

Accessory building and uses. Subordinate building or a portion of the main building, the use of which is incidental to that of the main building or land not used for a place of habitation or a living room, kitchen, dining room, parlor, bedroom or library. An accessory use is one which is incidental to the main use of the premises.

Accessory building or accessory use. A subordinate building or use customarily incident to and located on the same lot with the main building or use.

Accessory building or structure. A detached building or structure on the same site with, and of a nature customarily incidental and subordinate to, the principal building or structure, and the use of which is clearly incidental and subordinate to that of the dominant use of the principal building. The term "site" as used in this context shall mean a parcel or tract of land under single ownership containing one or more lots of record.

Accessory building or use. A building or use incidental and subordinate to the principal use or building and which: is located on the same lot with the use or building to which it is accessory; is reasonably necessary and incidental to the conduct of the principal use or building; does not alter the character of the premises on which it is located; is not detrimental to the neighborhood.

Accessory or auxillary use or structure. A use or structure customarily incidental, appropriate, and subordinate to the principal use of a building or to the principal use of land and which is located upon the same lot therewith.

Accessory building or use. A building or use which (a) is located on the same zoning lot as the principal building or use served, (b) is incidental to and subordinate in purpose to the principal building or use, (c) is operated and maintained solely for the comfort, convenience necessary, or benefit of the occupants, employees, customers, or visitors of or to the principal building or use, and (d) may be attached to a principal building or structure by means of a roof, or roof and walls. An accessory building may be detached from a principal building or structure.

Accessory structure. Structure, the use of which is incidental to that of the main building, and which is attached thereto, or is located on the same property.

Accessory structure. A structure, the use of which is incidental to that of the main building, and which is attached thereto, or is located on the same premises.

Accessory structure. A structure, the use of which is incidental to that of the main building or structure and which is located on the same lot.

Accessory structure. A building, the use of which is incidental to that of the main building and which is located on the same lot.

Accessory structure. A building or structure the use of which is incidental to that of the main building or structure and which is located on the same lot.

Accessory structure. A detached subordinate structure located on the same Zone lot with the main building, the use of which is customary to the main building.

Accessory use. A use, occupancy or tenancy customarily incidental to the principal use or occupancy of a building.

Accessory use. A use naturally and normally incidental to, subordinate to, and auxiliary to the permitted use of the premises.

Accessory use. A subordinate use which is customarily incidental to the principal use on the lot.

Accessory use. Is a use subordinate to the main use on a lot and used for purposes clearly incidental to those of the main use.

Accessory use. An activity, function or purpose existing on a lot as accessory or incidental to the principal use.

Accessory use. A use naturally and normally incident and subordinate to the main use of the premises or lot.

Accessory use. A subordinate use of a building, other structure, or use of land.

Accessory use. An accessory use is one which is incidental to the principal use of the premises.

Accessory use. A subordinate use, except a sign, drive-in facility or listed special use, which is clearly and customarily incidental to the use of the building or premises; and which is located on the same lot as the principal building or use, except that accessory parking facilities may be authorized to be located elsewhere.

Accessory use. A use which is incidental to the main use of the premises.

Accessory use. A building or structure, the use of which is incidental to the main building or structure, and is located on the same lot, or on a contiguous lot fronting on the same street as the lot or lots on which the main building is located and the use of which is manifestly incidental to that of the main building.

Accessory use. A subordinate use which is clearly and customarily incidental to the principal use of a building or premises and which is located on the same lot as the principal building or use.

Accessory use. A use subordinate to the principal use of land or a building or other structure on a lot and customarily incidental thereto.

Accessory use. A use customarily incidental and accessory to the principal use of a lot or building or other structure located upon the same lot as the accessory use.

Accessory use. A use which is customarily incidental and subordinate to the principal use of a lot or a building and which is located on the same lot therewith.

Accessory use. A land use which is subordinate to and incident to a predominant or main use, even though such land use, if standing alone as a predominant use or main use on its own lot, would not be permitted under the applicable district use regulations of the Ordinance.

Accessory use. A use which is incidental to the principal use of a building.

Accessory use. A use, located on the same zoning lot with the main use of the building, other structure or land, which is subordinate, and related to that of a main building or main use.

Accessory use. A use customarily incident and accessory to the principal use of lot or building or other structure, and located upon the same lot as the principal use.

Accessory use. A subordinate use which is incidental to and customary or necessary in connection with the main building or use and which is located on the same lot with such main building or use.

Accessory use. A subordinate use on the same lot with the principal use and incidental and accessory thereto.

Accessory use. A use subordinate in nature, extent or purpose to the principal use of a building or lot.

Accessory use. A use incidental to the principal use of a building or structure as defined or limited by the provisions of the local zoning laws.

Accessory use. A use customarily incident and accessory to the principal use of the land, building or structure located on the same lot or parcel of land as the accessory use.

Accessory use. A use subordinate to the main use on a lot and used for purposes customarily incidental to those of the main use.

Accessory use. Accessory use is either a subordinate use of a building, other structure, or tract of land, or a subordinate building or other structure; whose use is clearly incidental to the use of the principal building, other structure, or use of land, and which is customarily in connection with the principal building, other structure, or use of land, and which is located on the same zoned lot with the principal building, other structure or use of land, and which is not a use specifically permitted in a less restricted district.

Accessory use. A use customarily incidental to the principal use of a building, lot or land, or part thereof.

Accessory use. A use incidental to the principal use of a building as defined or limited by the provisions of the zoning law.

Accessory use. A use which is appropriate, subordinate, and customarily incidental to the main use of the site and which is located on the same site as the main use.

Accessory use. A subordinate use of a building, other structure, or use of land, which is clearly incidental to the use of the main building, other structure, or use of land, and which is located on the same zoned lot with the main building, other structure, or use of land.

Accessory use. A use conducted on the same lot as a principal use to which it is related; a use which is clearly incidental to, and customarily found in connection with a particular principal use.

Accessory use. A use, occupancy or tenancy customarily incidental to the principal use or occupancy of a building.

Accessory use. A use subordinate to the main use of land or of a building on a lot and customarily incidental thereto.

Accessory use. A use customarily incidental and subordinate to the principal use or building and located on the same lot with such principal use or building.

Accessory use. A use incidental to the principal use of a building as defined or limited by the provisions of the local zoning laws.

Accessory use, accessory building. A subordinate building or use which is located in its entirety on the same lot on which the main building or use is situated and which is reasonably necessary to the conduct of the primary use of such main building or main use and which, in the case of an accessory building, shall not be closer than 10 (ten) feet to any other building on the same lot, shall not be nearer to the street than one-half (1/2) of the depth of the lot, shall not occupy more than 6 (six) per cent of the area of the lot on which the main building or use is situated, and shall not be more than twenty-five (25) feet in height.

Accessory use or building. An accessory use or building is a subordinate use or building which is clearly incident to and located on the same premises with the main use or building.

Accessory use or buildings. Is a subordinate use or building customarily incident to and located on the same lot with the main use or building.

Accessory use or structure. Building or use which is subordinate to and incidental to that of the main building or use on the same lot.

Accessory use or structure. Use or structure on the same lot, and of a nature customarily incidental and subordinate to, the principal use or structure.

Accessory use or structure. Use or structure customarily incidental and subordinate to the principal use or building, and located on the same lot with such principal use or building.

Accessory use or structure. Use or structure on the same lot with, and of a nature customarily incidental and subordinate to the principal use or structure, but not including incomplete or inoperable motor vehicles.

Accessory uses of buildings or structures. Accessory uses of buildings or structures customarily incident to any use permitted by the Ordinance such as servant quarters, private garages; as hereinafter provided, or private work shops, provided, that none shall be conducted for gain or that no accessory buildings shall be inhabited by any one other than those employed by the owner or tenant of the premises, and, further provided, that in the event any such accessory structure or building is used in whole or part for living quarters, same shall be setback at least ten (10′) feet from the rear lot line of any lot.

Building accessory. The subordinate building, the use of which is customarily incidental to that of a principal building on the same lot.

Building (main). A building, or buildings, in which is conducted the principal use of the lot on which it is situated. In any residential district, any dwelling shall be deemed to be the main building of the lot on which the same is situated.

Building-principal. A building or buildings in which the principal use of the building site is conducted. In any residential district any dwelling shall be deemed to be the principal building on the building site.

Conforming use. Any lawful use of a building or a lot which complies with the provisions of the ordinance.

Erection of more than one principal building on a lot. In any district more than one building housing a permitted or permissible principal use may be erected on a single lot, provided that yard and other requirements of the ordinance shall be met for each building as though it were on an individual lot.

Main building. Any building having the predominant land use which is not an accessory building.

Main building. Building or structure occupied by the chief use or activity on the premises.

Main building. Principal building or buildings on a lot or tract which are used for, or within which is conducted, the principal use of the land.

Main building. Building occupied by the main use or activity on or intended for the premises, all parts of which building are connected in a substantial manner by common walls and a continuous roof.

Main building. Building devoted to the principal use of the lot on which it is situated. In any residential or suburban district, the dwelling shall be deemed to be a main building on the lot on which the same is situated.

Main building. Principal building on a lot or building site designed or used to accommodate the primary use in which such area is devoted; where permissible use involves more than one structure designed or used for the same primary purpose, as in the case of group houses, each such permissible building on one lot as defined by the ordinance shall be considered a main building.

Main building. A building or buildings, in which is conducted the principal use of the lot on which it is situated. In any residential district, any dwelling shall be deemed to be the main building of the lot on which the same is situated.

Main building. A building in which is conducted the principal use of the lot upon which it is situated.

Main building. The principal building or one of the principal buildings upon a lot, or the building or one of the principal buildings housing a principal use upon a lot.

Main building. One in which is conducted a principal of conditional use of the lot upon which it is situated. Every dwelling in specified Districts are main buildings.

Main or principal building. A building in which is conducted the principal use of the lot on which it is situated.

Main structure. Structure housing the principal use of a site or functioning as the principal use.

Main use. The principal use to which the premises are devoted and the principal purpose for which the premises exist.

Main use. Principal use of an activity conducted in a structure or on the land.

Principal building. Building in which is conducted the principal use of the lot on which it is situated. In a residence district any dwelling shall be deemed to be the principal building on the lot on which the same is situated. An attached carport, shed, garage or any other structure with one or more walls or a part of one wall being a part of the principal building and structurally dependent, totally or in part, on the principal building, shall comprise a part of the principal building and be subject to all regulations applicable to the principal building. A detached and structurally independent carport, garage, or other structure shall conform to the requirements of an accessory building.

Principal building. A building in which is conducted the main use of the lot on which said building is located.

Principal building. A structure in which is conducted the main use of the lot on which the structure is situated.

Principal building. A building in which is conducted the principal use of the lot on which it is situated.

Principal building. A building in which is conducted the principal use of the lot on which it is situated. In any residential district, any structure containing a dwelling unit shall be deemed to be the principal building on the lot on which the same is situated.

Principal building. Building which houses the main use or activity occurring on a lot or parcel of ground.

Principal building. Building in which is conducted the main or principal use of the lot on which said building is situated.

Principal building. Building or, where the context so indicates, a group of buildings in which is conducted the main or principal use of the lot on which said building is situated.

Principal building. A nonaccessory building in which is conducted the principal use of the lot on which it is located.

Principal building. A non-accessory building in which the dominant use of the lot is conducted.

Principal building or use. The main building or use of land or buildings as distinguished from an accessory building or use.

Principal or main building. A building in which is conducted the principal or main use of the lot on which said building is situated.

Principal structure or use. The main or primary structure or use on a parcel of land as distinguished from a secondary or accessory use. The uses permitted in the Ordinance are the principal uses.

Principal use. The main use of land or buildings as distinguished from a subordinate or accessory use. A "principal use" may be "permitted" or "special".

Principal use. The main use of land or structures as distinguished from a subordinate use.

Principal use. The main use of land, building or structure as distinguished from a subordinate or accessory use.

Principal use. Use which constitutes the primary activity, function or purpose to which a parcel of land or building is put.

Principal use. The main use of land or buildings as distinguished from a subordinate or accessory use.

Principal use. The primary and chief purpose for which a lot is used.

Principal use. Principal purpose for which a lot or building is designed, arranged, intended, occupied or maintained.

Principal use. Main use of land or structures as distinguished from an accessory use.

Principal use. Any main activity permitted by the Ordinance.

Principal use. Primary or predominant use of the premises.

Use. The purpose for which land or buildings thereon are designed, arranged, or intended to be occupied or used, or for which they are occupied or maintained.

Use. Activity which occurs in a building or on a site.

Use. The purpose for which a building is used, designed, or intended to be used.

Use. The specific purpose for which land or a building is designed, arranged, intended, or for which it is or may be occupied or maintained.

Use. The purpose for which land or a building is arranged, designed or intended or for which either land or a building is or may be occupied or maintained.

Use. The activity conducted in a structure.

Use. Any activity, occupation, business, or operation carried on, or intended to be carried on, in a building or structure, or on a tract of land.

Use. The purpose for which land or a building is arranged, designed or intended, or for which land or a building is or may be occupied.

Use. Synonymous with land use. The manner in which a parcel of land, or in which structures on the land, are used by parties in possession of the land.

Use. Any purpose for which buildings, other structures or land may be arranged, designed, intended, maintained or occupied; or any activity conducted in a structure or on the land.

Use. The purpose of which land or a building or other structure is designed, arranged, or intended, or for which it is or may be occupied or maintained.

Use. The purpose for which land or structures thereon is designed, arranged or intended to be occupied or used, or for which it is occupied, maintained, rented or leased.

Use. Specific purpose for which land or a building is designed, arranged, or intended, or for which it is or may be occupied or maintained. The term "permitted use" or its equivalent shall not be deemed to include any nonconforming use.

Use. Any purpose for which a building or other structure or a tract of land may be designed, arranged, intended, maintained or occupied, or any activity, occupation, business, or operation carried on in a building or other structure on a tract of land.

Use. Any activity, function, or purpose to which a parcel of land or building is put, and shall include the words used, arranged, or occupied, for any purpose including all residential, commercial, business, industrial, public or any other use.

Use. The purpose or activity for which the land or building thereon is designed, arranged, or intended or for which it is occupied or maintained.

A-15 PROPERTY — DEVELOPMENT

Abandonment. Land use shall be deemed abandoned where it has been discontinued and there is no clear indication from the property itself that it is to be resumed. Mere lack of occupancy, or vacancy, shall not, without more evidence be established as abandonment.

Access area. That land area which is an extended portion of a lot-in-depth or an easement area used to provide ingress and egress to one or more lots in a lot-in-depth subdivision.

Acre. An acre of land shall be deemed to be a commercial acre of forty thousand square (40,000) feet. Such area shall be exclusive of public streets or alleys or other public rights-of-way, lands or any portion thereof abutting on, running through or within a building site.

Active recreational areas. Usable common open space which is developed with active recreation facilities such as swimming pools, tennis courts, handball courts, golf courses, recreational buildings, clubhouses or other similar facilities.

Adjacent. The condition of being near to or close to but not necessarily having a common dividing line, i.e., two properties which are separated only by a street or alley shall be considered as adjacent to one another.

Apartment project. Group of two (2) or more apartment buildings.

Beneficial use. An interest, right, or use in a tract of land or the structures thereupon.

Block. In describing the boundaries of a district the word "block" refers to the legal description. In all other cases the word "block" refers to the property abutting on one side of the street between two (2) intersecting streets or a street and a railroad right-of-way or watercourse.

Block. A parcel of land, intended to be used for urban purposes, which is entirely surrounded by public streets; highways, railroad rights-of-way, public walks, parks or greenstrips, rural land or drainage channels, or a combination thereof.

Block Face. The properties abutting on one side of a street and lying between the two nearest intersecting or intercepting streets, or nearest intersecting or intercepting street and railroad right-of-way, unsubdivided land or watercourse.

Change of use. Alteration by change of use or occupancy in a building heretofore existing to a new use group which imposes other special provisions of law governing building construction, equipment or exits.

Change of use. Change in the use or occupancy of a building from one sub-group to another sub-group in the same general use group, according to the classification of buildings by use and occupancy in the applicable occupancy chapter of the code.

Change of use. An alteration by change of use in a building heretofore existing to a new use group which imposes other special provisions of law governing building construction, equipment or exits.

Change of use. An alteration by change of use in a building heretofore existing to a new use group or to another subgroup of the same use group which imposes other special provisions of law governing building construction, equipment or exits.

Change of use. Provisions of the code shall apply to every building, or portion of a building, devoted to new use for which the requirements under the code are in any way more stringent than the requirements covering the previous use.

Change of use. Change in the use or occupancy of a building from one use group to another use group, according to the classification of buildings by use and occupancy as listed in the ordinance.

Changes in use. If the conditions of use or occupancy of any building or premises or part thereof are substantially changed, or so changed as not to be in conformity with the conditions required by a certificate issued therefor or if the dimensions or area of the lot upon which a building is located or its yards or courts are reduced, said certificate shall be void and the owner shall notify the inspector who shall order an inspection of the buildings, premises, or lot. If the building, premises, or lot conforms to all the applicable requirements of the ordinance and the code, a new certificate shall be issued as herein provided.

Character. The distinguishing feature(s) of an area. In arriving at the determination of the character of an area the following shall be a minimum of the elements to be considered: the composite or aggregate of the characteristics of the structure, form, materials, and functions of the area under consideration.

Cluster development. Planned residential tract, subdivision of area where individual lots and/or buildings are arranged in clusters or groups of five (5) or more with open areas between such lot or building groups which areas are either publicly or privately owned and maintained in a parklike appearance.

Common ownership. Ownership of two (2) or more contiguous parcels of real property by one person or by two (2) or more persons owning such property jointly, as tenants by the entirety, or as tenants in common.

Common ownership: direct or indirect. Legal or beneficial ownership of more than one (1) lot by the same person or persons. For this purpose, ownership by the spouse of a person shall be considered the same as ownership by the person himself or herself.

Compatibility. The characteristics of different uses or activities that permit them to be located near each other in harmony and without conflict. In arriving at the determination of the compatibility of a particular use or activity the following shall be a minimum of the elements to be considered: intensity of occupancy as measured by dwelling units per acre; floor area ratio; pedestrian or vehicular traffic generated; volume of goods handled; and such environmental effects as noise, vibration, glare, air pollution, erosion, or radiation.

Covenant. A private or public written agreement relating to the use of, restriction of, interest in, or right to, real property.

Crosswalk or interior walk. Right-of-way or easement for pedestrian travel across or within a block.

Curb level. The level of the established curb in front of the building measured at the center of such front.

Dedication. Appropriation of land by the owner for any general and public use, with limitations reserved to himself for no further rights except that the dedicated land be used only for the purpose of the dedication.

Development. The performance of any construction or earth moving activity, the making of a material change in the use or appearance of any structure or land, the division of land into two (2) or more parcels, or the creation or termination of rights of access or riparian rights.

Easement. The right to use or benefit from the land of another, for a restricted use or purpose, without compensating the owner.

Easement. Right of a person to use common or private land owned by another for a specific purpose.

Easement. Vested or acquired right to use land other than as a tenant, for a specific purpose, such right being held by someone other than the owner who holds title to the land.

Easement. Limited right of use granted in private land for public or quasi-public purpose.

Easement. A right-of-way granted for limited use of private land for a public or quasi-public purpose.

Easement. A grant by the property owner to the public, a corporation or persons, of the use of a strip of land for specific purposes.

Easements. Easements for utilities shall be provided and shall conform in width and alignment to the recommendations of the appropriate utility company. Easements shall also be provided for all storm water drainage ditches, or sewers and water courses. All easements shall be shown on the final plan and the Township or its agents (to include the Planning Commission or other official representative) shall have the right to enforce the restrictive easements relative to water supply and sewerage disposal in the event the developer and/or lot owners association fail, or are unable to enforce them. They shall further have free access to all developments and lots at all times for the purpose of inspection and enforcement.

Easements. Easements with a minimum width of 20 feet shall be provided as necessary for utilities and drainage. The easement shall be centered on or adjacent to rear or side lot lines. Nothing shall be planted or constructed in the easement except lawn.

Fair market value. Price, between a seller who is willing but not forced to sell and a buyer who is willing but not forced to purchase.

Finished grade. The completed surfaces of lawns, walks, and roads brought to grades as shown on the official subdivision maps or recorded plats.

Finished grade. Natural surface of the ground, or surface of ground after completion of any change in contour.

Finished grade. The completed surfaces of lawns, walks, and roads brought to grades as shown on official plans or designs relating thereto.

Finish grade. Top surface elevation of lawns, drives, or other improved surfaces after completion of construction or grading operations.

Finish grade. Finish grade is the lowest required grade exit serving the floor under consideration.

Front of building. The side of a building most nearly parallel with and adjacent to the front lot line.

Front wall. That wall of the building facing a public street, alley, or public approved place. Where a lot abuts on more than one street, walls facing on any street shall be considered a front wall under the provisions of the code.

Grade. Slope in reference to a horizontal plane, usually expressed as a percentage; also known as gradient. For drainage piping, it is usually expressed as the fall in inches or fraction of an inch per foot of length of pipe.

Grade incline. Degree of inclination with respect to the horizontal.

Grading. Excavation or fill or any combination thereof and includes the conditions resulting from any excavation or fill.

Group housing development. Single-family, two-family or multiple-family dwellings built upon a minimum of a two (2) acre parcel of land, subject to an approved site plan. Density and layout are controlled without strict conformity to specific yard, lot coverage and lot size requirements, but the site plan must meet the density requirement of the pertinent use district.

Group housing project. One (1) or more residential structures on a site of at least 40,000 square feet in size where the building arrangement is such that the property cannot be subdivided into conventional streets and lots that meet the requirements of the subdivision chapter of the Code.

Housing project for elderly people. Planned multiple shelter of not less than 20 dwelling units under one (1) or more roofs, but upon one premises, designed and erected to be used exclusively for the housing of persons 60 years or more of age except as provided in the code and in which the normal operations of housekeeping and preparation of meals is customarilly performed by the tenants or occupants and provision is made therefore in each dwelling unit, excepting laundry, and heating equipment; provide further that one (1) dwelling unit may be occupied without restriction of age by a custodian or manager.

Housing unit. A housing unit for the purpose of the ordinance shall be any building whose function is to provide housing for over one hundred (100) families.

Incompatible use. A use or service which is unsuitable for direct association or contiguity with certain other uses because it is contradictory, incongruous, or discordant.

Integrated roadside development. A planned development including at least two of the following primary uses: restaurant, motel, and service station, and located within a reasonable distance of a freeway.

Land development. Improvement of one or more contiguous lots, tracts or parcels, of land for any purpose involving (1) a group of two or more buildings, or (2) the division or allocation of land between or among two or more existing or prospective occupants by means of, or for the purpose of streets, common areas, leaseholds, building groups, or other features.

Land development. Division of land into lots for the purpose of conveying such lots singly or in groups to any person, partnership, or corporation for the purpose of the erection of buildings by such person, partnership, or corporation.

Land use. The purpose or use made of a property. Developments such as a ball park, car sales lot, cemetery, drive-in-theater, fair ground, golf course, park, parking lot, play-ground, race track, shopping center, storage yard, swimming pool, trailer camp, wrecking yard, etc., the use of which land areas whether containing buildings or not, cause persons to congregate in large numbers are subject to land use restrictions. Roads adjacent to the right-of-way of a highway, power lines, pole lines, etc., adjacent to a controlled-access highway and pipe lines, public utilities, etc., are subject to land use restrictions and, if they are located on the right-of-way of a highway, they are also subject to encroachment restrictions.

Legal non-conforming. A non-conformity which is permitted to continue because the established amortization period has not expired or no amortization period is required for continuance.

Limiting distance. Distance from an exposing building face towards a property line, the centre line of a street, lane, public thoroughfare or an imaginary line between two buildings on the same property, measured at right angles to the exposing building face.

Livability space. Outdoor area, excluding parking and other service areas, which can be utilized for outdoor living, recreation, or passive use by the residential occupants of a parcel, and their guests. Such area may include space on the ground level and/or on roofs, decks, and balconies.

Mean grade. The average grade in the area within twenty (20) feet of the structure to which the term is applied or within the lot line, whichever is closer, provided that where the grade slopes down and away from the structure at a rate exceeding one (1) vertical unit or two (2) horizontal units, the average grade shall be computed as though the slope is one (1) vertical unit to two (2) horizontal units.

Mean grade. Average of the grades of midpoints of successive equal distances of not over 10 feet measured along a line or lines along which the mean grade is to be determined.

Metes and bounds. A system of land description by distances, terminal points, and angles (metes-terminal points; bounds-boundaries).

Mid point. Point of a boundary line equally distant from the 2 lot lines intersecting it at right angles or within 45 degrees of a right angle.

Model home. Residential structure used for demonstration purposes, not occupied as a dwelling unit and open to the public for inspection.

Natural grade. Elevation of the undisturbed natural surface of the ground prior to any excavation or fill.

Natural grade. The surface of the natural, undisturbed, and unfilled ground taken at any point.

Natural grade. The elevation of the original or undisturbed natural surface of the ground.

Natural grade. The surface of the ground prior to excavation, fill or grading.

New or finished. With respect to any specific project, the resulting level of the ground after the final grading where there is a cut, and after normal settlement where there is a fill.

Planned commercial development. Planned commercial development is an activity or group of activities in which the princpal activity is: concerned with consumer retail sales and consumer retail services and related to retail market or trade area; and located, planned, developed and maintained on and integrated with a site, which site has a continuing unity of management, maintenance and operation.

Planned development. A tract of land which is developed as a unit under single or unified ownership or control and which generally includes two or more principal buildings or uses, but which may consist of one building containing a combination of principal and supportive uses. Uses not otherwise allowed in the zoning district are prohibited within a Planned Development unless specific provision is otherwise made by Ordinance or by Resolution adopted pursuant to the provisions of the Zoning Ordinance.

Planned development. A tract of land which contains or will contain two (2) or more principal buildings, developed under a single ownership or control; the development of which may be of a substantially different character than that of surrounding areas. A planned development allows for flexibility not available under normal zoning district requirements.

Planned development. A tract of land of not less than eight thousand (8000) square feet which is developed as a unit under single ownership or control. One or more principal buildings may be located on a single lot, but no single building its required yards shall be located on more than one (1) lot of record.

Planned development unit. The purpose and intent of a Planned Development Unit is to provide a more flexible method whereby land may be designed and developed as a unit for residential, commercial or industrial purposes taking advantage of more modern site planning techniques than would be possible through the strict application of conventional zoining and subdivision regulations. It is intended that development will meet the broader objectives of the general plan and the Code and will exhibit excellence in design, site arrangements, integration of uses and structures and protection to the integrity of surrounding development, although such development may deviate in certain respects from the zoning regulations or subdivision. The regulations of the Code are intended to produce developments which meet standards of open space, light and air, pedestrian and vehicular circulation and a variety of land uses, which compliment each other and harmonize with existing and proposed land use in the vicinity.

Planned residential development. A group of residential buildings located on a lot having an area, including one-half the width of abutting streets, of five acres or more and arranged in accordance with a plan of development for the entire project with provision for adequate open spaces and conveniently located service facilities. When authorized by the Commission, a development may include churches, schools, hospitals, infirmaries, and recreational and commercial uses, which are an integral part of the development.

Planned residential development. Shall mean two or more dwelling units together with related land, buildings and structures, planned and developed as a whole in a single development operation or a programmed series of operations in accordance with detailed, comprehensive plans encompassing such elements as the circulation pattern and parking facilities, open space, utilities, and lots or building sites, together with a program for provisions, operation and maintenance of all areas of improvements, facilities and services provided for common use of the residents thereof.

Planned residential development. Contiguous area of land, controlled by a landowner, to be developed as a single comprehensively planned entity for a number of dwelling units, the plan for which does not necessarily correspond in lot size, bulk or type of dwelling, density, lot coverage and required open space to the Township Zoning District Regulations. Uses in a PRD may include dwelling units of various types and densities and those non-residential uses deemed to be appropriate for incorporation in the design of the PRD.

Planned unit development. Approved large-scale residential development or group of dwellings which is planned, constructed and maintained as a unified entity.

Planning agency. The official body designated by local ordinance to carry out the purposes of the Act and may be a planning department, a planning commission, the legislative body itself or any combination thereof.

Residential development or use. Buildings, structures, and uses which are permitted in a residential district, including dwellings, permitted residentially related public and semi-public facilities, and permitted accessory uses. All such uses being subject to applicable district regulations.

Service use. Use devoted to repair, maintenance, administration, teaching, or enhancement.

Single and separate ownership. Ownership of property by one or more persons, which ownership is separate and distinct from that of any adjoining property.

Single ownership. Ownership by one (1) person or by two (2) persons whether jointly, as tenants by the entirety, or as tenants in common, of a separate parcel of real property.

Tenancy. One or more occupants, tenants, lessees, owners, etc.

Tenant. A person or firm using a building, or part of a building, as a lessee or owner-occupant.

Terrace. A natural or artificial earthen embankment between a building and its street front. The "height of the terrace" shall be the difference in elevation between the average sidewalk level or its equivalent established grade opposite the front of the middle of the building, and the average elevation of the terrace at the building wall.

Through block connection. An ample passageway through or between any structures on a parcel, that allows the general public to walk from the street edge to the rear of the parcel.

Tract house. A house mass-produced on a tract development.

Use. The purpose for which land or a building is arranged, designed intended or for which either land or a building is or may be occupied or maintained.

Use. The purpose for which land or premises or a building there on is designed, arranged, or intended, or for which it is occupied or maintained, let or leased.

Use of property. The purpose or activity for which the land or building thereon is designed, arranged or intended, or for which it is occupied or maintained.

U.S. Rectangular survey system. The rectangular survey system adopted by Congress on May 20, 1785, and used outside the original 13 states (now 18 states) and Texas.

A-16 RODENT AND VERMINPROOFING

Extermination. The control and extermination of insects, rodents, and vermin by eliminating their harborage places, by removing or making inaccessible material that may serve as their food, by poisoning, spraying, fumigating, trapping, or by other approved means of pest elimination.

Extermination. The process of controlling and eliminating of insects, rodents or other pests by eliminating their harborage places, by removing or making inaccessible materials that may serve as their food, by poisoning, spraying, fumigating, trapping, or by any other recognized and lawful pest-elimination methods.

Extermination. The control and elimination of infestation, as defined in the code by eliminating harboring places, removing or making inaccessible any food, dirt, waste, or other materials that may stimulate increase infestation, and includes pest control by poisoning, spraying, trapping, fumigation by licensed fumigator, or other means of pest elimination procedure.

Extermination of Rats. The control and elimination of infestation by eliminating harboring places; removing and making inaccessible materials that may serve as food; and pest control by poisoning, spraying, trapping and fumigating by licensed persons, or any other recognized legal and effective pest elimination procedure.

Fumigant. Substance which by itself or in combination with any other substance emits or liberates a gas, fume or vapor used for the destruction or control of insects, fungi, vermin, germs, rodents or other pests and is distinguished from insecticides and disinfectants.

Fumigation. The use of any substance which emits or liberates a gas, fume, or vapor used for the destruction or control of insects, fungi, vermin, germs, rodents, or other pests, and is distinguished from insecticides and disinfectants.

Infestation. Presence of household pests, vermin or rodents.

Infestation. Household pests, vermin, rodents, insects, nesting places and conditions for nesting.

Infestation. The presence of insects, rodents, vermin, or other pests on the premises which constitute a health hazard.

Rat eradication. The elimination or extermination of rats within buildings so that the buildings are completely freed of rats or there is no evidence of rat infestation remaining, by any or all of the accepted measures, such as poisoning, fumigation, trapping, clubbing, and other accepted means.

Rat harborage. Any condition which provides shelter or protection for rats, thus favoring their multiplication and continued existence in, under or outside a structure of any kind.

Rat harborage. Any condition found to eixst under which rats may find shelter or protection, including any defective or inadequate construction which would permit the entrance of rats into a building.

Ratproof. Impervious to rodent infestation and propagation.

Rat proofing. Closing or protection of all openings, exterior walls and foundations, sidewalk or area gratings, of all buildings with rat proof materials installed in such manner as to prevent rats from gaining entrance to a building.

Ratproofing. A form of construction to prevent the ingress of rats into buildings from the exterior or from one building or establishment to another.

Rat-protection. Every food establishment shall be completely surrounded by a continuous exterior foundation wall not less than 12 inches below grade.

Rat stoppage. Any inexpensive form or ratproofing designed to prevent the ingress of rats into business buildings. It is essentially the closing or protecting of all openings in exterior walls and foundations or the grates in a sidewalk of business buildings with ratproof materials installed in such a manner as to prevent rats from gaining entrance.

Rat tunnels and nesting. Rats will tunnel under verical walls to depths of 4 feet to gain access to food or water while they will make no effort to burrow under a foundation that only penetrates the ground a few inches for the purpose of nesting. As a general rule foundations that go to a depth of 3 feet will prevent rats from gaining entrance to buildings if the rodents have not been feeding in them previously. In locations where rock formation is found at or near the surface the barrier need extend only to such formation.

Rodent and verminproofing. Beneath the exterior walls of every building or enclosed part thereof which is not supported on a continuous masonry wall, there shall be a tight masonry or concrete wall at least 3-3/4 inches thick if of brick, concrete or concrete blocks; or 2 inches thick if of reinforced concrete plates securely fastened together. Every such rat wall shall extend from a point at least 3 feet below the finished grade up to a point at least 6 inches above the finished grade, except in the case of dwellings having a concrete floor which is contiguous with the rat wall, in which case such rat wall need extend only 2 feet below the finished grade.

A-17 SUBDIVISIONS

Access Street. Any public street within a subdivision or along the boundaries of a subdivision which is located in a manner which would serve any properties outside the plat boundaries or provide a connection directly with a collector street.

Applicant. Any person who submits subdivision plans to the authority for the purpose of obtaining approval thereof.

Block. A tract or parcel of land established and identified within a subdivision which is surrounded by streets or a combination of streets and other physical features and intended to be further subdivided into individual lots or reserves.

Building setback restriction. A defined area designated on a subdivision plat in which no building structure may be constructed and is located between the adjacent street right-of-way line or other type of easement or right-of-way and the proposed face of a building.

Common land. Land in a subdivision or development area owned as private land or occupied by dwellings created for common usage by restrictions, easements, covenants or other conditions running with the land, and which is held for the use and enjoyment by or for the owners or occupants of the dwellings in a subdivision or development area.

Compensating open space. Those areas designated on a plat which are restricted from development, except for landscaping and recreational uses and which all owners of residential properties within the plat have a legal common interest or which is retained in private ownership and restricted from development, except for landscaping and recreational uses for the exclusive use of all owners of residential property within the plat. The terms compensating open space and common open space may be used interchangably and can be considered the same.

Correction plat. A plat, previously approved by the Planning Commissiion and duly recorded, which is resubmitted to the Commission for reapproval and recording which contains dimensional or notational corrections of erroneous information contained on the originally approved and recorded plat. A correction plat is not to be considered as a replat or resubdivision and may not contain any changes or additions to the physical characteristics of the original subdivision, but is intended only to correct errors or miscalculations.

Design plat. Plat indicating street alignment, grades and widths, alignment and widths of easements and rights-of-way for drainage and sanitary sewers and the arrangement and orientation of lots.

Developer. Person commencing proceedings under the ordinance to effect the development of a parcel of land for himself or for another.

Development area. Minimum area of tract of land permitted by the ordinance to be developed by a single owner or a group of owners, acting jointly which may consist of a parcel or assembled parcels, and includes a related group of one-family dwellings, townhouses and apartment dwellings planned and developed as an entity under the Subdivision unit plan development procedures.

Easement. A strip reserved by the subdivider for public utilities, drainage and other public purposes, the title which shall remain in the property owner, subject to the right of use designated in the easement reservation.

Easement. A right granted for the purpose of limited public or quasi-public uses across private land.

Extraterritorial jurisdiction. Refers to the unincorporated territory extended five miles beyond the city limits and which has been established as a result of the provisions of the Muncipal Annexation Act and the State subdivision acts.

Filing date. The date when a plat is formally presented to the Planning Commission for its approval and registered as part of the Commission's official meeting agenda. This date is to be considered as the inital date of the statutory thirty (30) day time period in which the Planning Commission is required to act upon a plat submitted to it under the provisions of the Ordinance.

Final map. A map prepared in accordance with the provisions of the Ordinance and with any applicable provisions of the Subdivision Map Act, designed to be recorded in the Office of the County Recorder.

Final plat. A map or drawing of a proposed subdivision prepared in a manner suitable for recording in the County records and containing accurate and detailed engineering data, dimensions, dedicatory statements and certificates and prepared in conformance with the conditions of preliminary approval previously granted by the Planning Commission.

Final plat. The final map of all or a portion of the subdivision which is presented to the central planning board for final approval in accordance with the regulations established by the Ordinance and which, if approved, shall be filed with the proper county recording officer.

Final plat. The formal layout of a proposed subdivision encompassing all requirements imposed by the City Council and prepared by a registered land surveyor.

Final plat. A complete and exact subdivision plan, prepared in form for official recording, to define property lines, proposed streets and other improvements.

Final plat. A map of a land subdivision prepared in a form suitable for filing of record with necessary affidavits, dedications, and acceptances, and with complete bearings and dimensions of all lines defining lots and blocks, streets and alleys, public areas, and other dimensions of land.

Final subdivision plan. Complete and exact subdivision plan, prepared for offical recording as required by State statute, to define propery rights and proposed streets and other improvements.

Frontage. That portion of any tract of land which abuts a public street right-of-way and where the primary access to said tract is derived.

Front lot. A "front lot" in a lot-in-depth subdivision is a lot which fronts on a public street and could be developed and provided access without the use of an access area.

General overall plan. A map or plat designed to illustrate the general design features and street layout of a proposed subdivision which is proposed to be developed and platted in sections. This plan, when approved by the Planning Commission, constitutes a guide which the Planning Commission will refer to in the subsequent review of more detailed sectional plats as they are presented to the Planning Commission covering portions of the land contained within the general overall plan and adjacent properties.

Improvement. Required installations, pursuant to the Act and subdivision regulations, including grading, sewer and water utilities, streets, easements, traffic control devices as a condition for the approval and acceptance of the final plat thereof.

Improvement. Such street work and utilities to be installed, or agreed to be installed by the subdivider on the land to be used for public or private streets, highways, ways, and easements, as are necessary for the general use of the lot owners in the subdivision and local neighborhood traffic and drainage needs and required as a condition precedent to the approval and acceptance of the Final Map. Such street work and utilities include necessary monuments, street name signs, guardrails, barricades, safety devices, fire hydrants, grading, retaining walls, storm drains and flood control channels and facilities, erosion control structures, sanitary sewers, street lights, street trees, traffic warning devices (other than traffic lights and relocation of existing traffic signal systems directly affected by other subdivision improvements) and other facilities as are required by the Bureau of Street Lighting or Bureau of Street Maintenance in conformance with other applicable provisions of the Code, or as are determined necessary by the Advisory Agency for the necessary proper development of the proposed subdivision.

Interior street. Any public street within a subdivision designed to serve only those properties within the boundaries of the subdivision in which it is dedicated and established. An interior street must be so designed and located as to form a closed circulation system. Cul-de-sacs and loop streets or street systems beginning from streets within a subdivision may be considered as interior streets. Interior streets may not, however, be any street which would allow access through the subdivision to other properties or directly connect with other streets outside the plat boundary.

Local street. Any public street not designated as a major thoroughfare, freeway or highway.

Lot. An undivided tract or parcel of land contained within a block and designated on a subdivision plat by numerical identification.

Lot. A portion of a subdivision or parcel of land, however designated, intended as a single building site or unit for transfer of ownership for development.

Lot. A parcel or portion of land separated from other parcels or portions by description, as on a subdivision or recorded map, or by metes and bounds description, for purpose of sale, lease or separate use.

Lot-in-depth. A lot which uses an access area for ingress and engress and which is located behind a lot fronting on a public street.

Lot-in-depth-subdivision. Any division of land which contains a lot-in-depth.

Lot of record. A lot which is part of a duly recorded plat of subdivision; or a parcel of land which has been conveyed by the identical description by a deed of record recorded meeting all of the requirements of the subdivision and the zoning ordinance then in effect.

Lot of record. A parcel of land that is a lot in a subdivision recorded on the records of the recorder of deeds of the County or that is described by a metes and bounds description in a deed which has been so recorded.

Lot of record. A lot which is part of a subdivision, the map of which has been recorded in the office of the Recorder of Deeds of the County; or a lot described by metes and bounds, the description of which has been heretofore recorded in the office of the Recorder of Deeds or in the office of the Registrar of Titles of the County prior to the effective date of the Ordinance.

Lot of record. Lot which is a part of a subdivision, the map of which has been recorded in the office of the register of deeds of the county, or a lot described by metes and bounds, the deed to which has been recorded in the office of the register of deeds of the county.

Major subdivision. All subdivisions are not classified as minor subdivisions.

Major thoroughfare. A public street designed for fast heavy traffic and intended to serve as a traffic artery of considerable length and continuity throughout the community and so designated on the latest edition of the Major Thoroughfare and Freeway Plan adopted by the Planning Commission.

Master plan. A composite of the mapped and written proposals as provided for in the Municpal Planning Act, recommending the physical development of the city which has been or shall be duly adopted by the central planning board.

Minor subdivision. Any subdivision containing adjoining lots, tracts, or parcels, all of which front on an existing improved street, and having an aggregate frontage on said street, of not more than two-hundred-fifty (250) feet.

Mobile home subdivision. A subdivision for residential use by mobile homes exclusively.

Parcel map. A map showing the division of land as described in the Subidivision Map Act and prepared in accordance with the provisions of the Subdivision Map Act.

Plat. A formal subdivision approved by the City Council, and filed with the appropriate county offices, which illustrates the organization of lots, blocks, streets, easements, and public properties.

Plat. A map of a subdivision.

Plot plan. A plan prepared primarily for title purposes which shows the location, dimensions, and bearings of parcels of ground.

Preliminary plat. A tentative subdivision layout, showing approximate proposed street and lot layout as a basis for consideration prior to preparation of a final plat.

Preliminary plat. A map of a proposed land subdivision showing the character and proposed layout of the tract in sufficient detail to indicate the suitability of the proposed subdivision of land.

Preliminary plat. A formal layout of the proposed subdivision prepared by a registered land surveyor.

Preliminary plat. A map or drawing of a proposed subdivision of land prepared to illustrate the features of the development for review and approval by the Planning Commission, but not suitable for recording in the County records.

Private street. A vehicular access way under private ownership and maintenance providing access to buildings containing residential dwelling units which buildings or any part thereof is located more than 300 feet beyond the nearest public street to which the private street intersects. A private street shall also include any vehicular access way under private ownership and maintenance which provides vehicular access to four (4) or more separate lots or building sites designed for individual sale the the general public for residential purposes. Parking lots and private driveways within shopping centers, institutions, commercial areas and industrial developments will not be considered as private streets.

Private street plat. A map or drawing of a proposed development required to be submitted to the Planning Commission for approval which contains a proposed private street or building arrangements and locations which require the provision of a private street.

Property data map. A map showing all existing and planned conditions affecting the property to be subdivided, required prior to the submission of an application for subdivision approval.

Public street. A public right-of-way, however designated, dedicated or acquired, which provides vehicular access to adjacent private or public properties.

Public utility easements. Easements established within a plat which are designed to accommodate public owned or controlled utility facilities necessary to provide various types of utility services to the individual properties within the plat boundaries. Public utility easements may be used for, but not limited to, facilities necessary to provide water, electrical power, natural gas, telephone, telegraph and sanitary sewer services. Storm sewers or open drainage ways must not be constructed within public utility easements unless specifically approved by the Director of the Public Works Department and where additional easement width is provided to conform to the standards established herein for drainage easements.

Recorded plat. A final plat bearing all of the certificates of approval required by the Act, any local applicable ordinance and other state statute.

Re-subdivision. Any subdivision or transfer of land, laid out on a plan whether or not approved previously by the Township which changes or proposes to change property lines and/or public rights-of-way not in strict accordance with the approved plan or the recorded plan.

Re-subdivision. Legal subdivision which has been altered by changing of a line, bearing, or other measurement which either reduced or enlarges the number of lots or size of lot(s) originally created, and which is subsequently platted and recorded in a legal manner.

Reverse frontage lot. A lot extending between and having frontage on a major street or highway and a residential street, with vehicular access solely from the latter.

Revised tentative subdivision map. A map involving a revised arrangement of the streets, alleys, easements or lots within property for which a tentative map has been previously approved, or a modification of the boundary of the property.

Right-of-way. Land reserved for use as a street, interior walk, or for other public purpose.

Right of way. A parcel of ground reserved by the subdivider, or deeded by the owner, for public use.

Sketch or preliminary plat. The proposed map of a subdivision of sufficient accuracy to be used for the purpose of discussion and classification and meeting the requirements of the Ordinance.

Sketch plan. An informal layout of proposed subdivision including location of existing structures.

Street dedication plat. A map or drawing illustrating the location of a public street only passing through a specific tract of land.

Stub street. A public street not terminated by a circular turnaround ending adjacent to undeveloped street or acreage and intended to be extended at such time the adjacent undeveloped property or acreage is subdivided. A "stub street" which has been dedicated, but cannot be extended into the adjacent property or terminated with a circular turnaround or cul-de-sac can then be considered to be a "dead-end street".

Subdivider. Any person seeking a subdivision of land under the Ordinance as owner or as agent for the owner of the land to be affected by a subdivision.

Subdivider. Any person, firm partnership, corporation or other entity, acting as a unit, subdividing or proposing to subdivide land as herein defined.

Subdivider. Any person, firm, corporation, partnership or association who causes land to be divided into a subdivision for himself or others.

Subdivider. Any individual, firm association, syndicate, co-partnership, corporation, trust or any other legal entity commencing proceedings under the Ordinance, to effect a subdivision of land hereunder for himself or for another.

Subdivider. Any landowner, agent of such landowner, or tenant with the permission of such landowner, who makes or causes to be made a subdivision of land or a land development.

Subdivider. Owner, or authorized agent of the owner, of a subdivision.

Subdivider (Developer). Any person or authorized agent thereof, proposing to divide or dividing land so as to constitute a subdivision or proposes to the terms and provisions set out in the Ordinance. The term "Developer" will mean the same as "subdivider" for the purposes of the Ordinance.

Subdivision. The division or redivision of land into two (2) or more lots, tracts, sites or parcels for the purpose of transfer of ownership or for development, or the dedication or vacation of a public or private right-of-way or easement.

Subdivision. The separation of land into two or more parcels, lots, or tracts; or any change in the property line or lines of a parcel, lot, or tract; or the establishment of the property lines of a parcel, lot, or tract not previously platted.

Subdivision. The division of a lot, tract or parcel of land into 2 or more lots, sites or other divisions of land for the purpose, whether immediate or future, of sale or building development.

Subdivision. (1) The division of land into one or more lots, sites, tracts or parcels or however otherwise designated, from a larger tract or parcel, for the purpose of transfer of ownership, leasing, or building development. (2) The dedication of a road, highway, street, alley or easement through or on a tract of land regardless of area. (3) The resubdivision of land heretofore divided or platted into lots, sites or parcels.

Subdivision. A lot, parcel or tract of land which has been legally subdivided, platted, and recorded according to the laws of the County and the State.

Subdivision. The division for lease or sale to the public of a tract or parcel of land into five (5) or more lots, tracts, or parcels of land, or, if a new street is involved, any division of a parcel of land or the division into more than two (2) parts of any residential lot, the boundaries of which have been fixed by a recorded plat, providing that a partitioning or division of land into tracts or parcels of land of five (5) acres or more and not involving a new street or the sale or exchange of parcels of land to or between adjoining property owners, where such sale or exchange does not create additional lots, shall not be deemed a subdivision. The partitioning of land in accordance with state statutes regulating the partitioning of land held in common ownership shall not be deemed a subdivision.

Subdivision. A division of any part, parcel or area of land by the owner or agent, either by lots or by metes and bounds, into lots or parcels three or more in number for the purpose of conveyance, transfer, improvement or sale with appurtenant roads, streets, lanes, alleys and ways, dedicated or intended to be dedicated to public use, or the use of purchasers or owners of lots fronting thereon. A subdivision as defined above includes division of a parcel of land having frontage on an existing street, into three or more lots each having frontage on an existing street. However, for the purpose of these regulations, division of land for agricultural purposes in parcels of more than ten (10 acres, not involving any new street or easement, shall not be deemed a subdivision. Any development of a parcel of land (for example, as a shopping center or a multiple dwelling project), which involves installation of streets and/or alleys, even though the streets and alleys may not be dedicated to public use and the parcel may not be divided immediately for purposes of conveyance, transfer or sale. The term "subdivision" includes "resubdivision" and, as appropriate in these regulations, shall refer to the process of subdividing land or to the land subdivided.

Subdivision. Division or redivision of a parcel of land into three (3) or more lots of parcels, either by plat or by metes and bounds description for the purpose of transfer or ownership or development, or, if a new street is involved, any division of a parcel of land.

Subdivision. Legally platted subdivision approved by the Planning commission.

Subdivision. Division or redivision of a lot, tract or parcel of land by any means into two or more lots, tracts, or parcels or other divisions of land, including changes in existing lot lines for the purpose, whether immediate or future, of lease, transfer, of ownership, or building or lot development; provided, however, that the division of land for agricultural purposes into parcels of more than ten (10) acres, not involving any new street or easement of access, shall be exempted.

Subdivision hearing and decision. As required by law, a public hearing shall be held by the Commission. A notice fifteen (15) days in advance, once in a newspaper, having a general circulation in the City, describing the location of proposed subdivision, the name of the applicant and the time and place of the hearing and a copy of said notice shall be mailed to the applicant on the day of the publication of the notice. The Commission shall either approve, modify and approve, or disapprove the application. The Commission may grant the applicant the right to submit a revised plan without extra cost for a second hearing which will not be advertised as stated above.

Subdivision Layout. Informal plan to scale, indicating salient existing features of a tract and its surroundings and the general layout of the proposed subdivision for discussion purposes only and not to be presented for approval.

Subdivision regulations. Regulations as promulgated and created by the Planning Commission and approved by the Zoning Board.

Sub-lot. Subordinate and integral part of a lot which lot is identified on a subdivision recorded in the maps and plats record of the County Recorder.

Submittal date. The date and time specified in the Ordinance when plats, related materials and fees must be received by the Planning Department in advance of the regular meeting of the Planning Commission. The "Submittal date" is not to be considered as the "Filing date" as herein defined or considered as the inital date of the statutory thirty (30) day time period in which the Planning Commission is required to act upon plats filed with it.

Tentative subdivision plan. Plan in lesser detail than a final plan, showing approximate proposed street and lot layout as a basis for consideration prior to preparation of a final plan.

Town. A political subdivision in a county having the same connotation as the term "township".

Township. A political subdivision in a county, having certain powers of self-government.

Township. A geographical area six (6) miles square, which is subdivided into thirty-six (36) sections, each one (1) square mile or 640 acres in area. An area used in government surveys.

A-18 WORD DEFINITIONS AND TERMS DEFINED

Action. The term "action" shall include suits, prosecution and all judicial proceedings.

Adequately. Effectively, securely and similar words, shall be interpreted as conditions subject to approval by the Building Official or other municipal authority having jurisdiction and the right of approval.

And. "And" indicates that all connected items, conditions, provisions or events shall apply.

Area. As applied to the dimensions of a building means the horizontal projected area of the building at grade.

Area. Surface square footage.

Building or structure. "Building or structure" includes any part thereof and the words are interchangeable.

City. As used in the Code, is any political subdivision which adopts a Code for regulation within its jurisdiction.

Cubic content of a building. The actual cubic space enclosed within the outer surfaces of the outside or enclosing walls and contained between the outer surfaces of the roof and six inches below the finished surfaces of the lowest floors. The above definition requires the cube of dormers, penthouses, enclosed porches and other enclosed appendages.

Definitions. For the purpose of the Ordinance, certain words in the singular number shall include the plural number and words in the plural number shall include the singular number unless the obvious construction of the wording indicates otherwise. The word 'shall' is mandatory. Unless otherwise specified, all distances shall be measured horizontally and at right angles to the line in relation to which the distance is specified. The word 'lot' includes the word 'plot'; the word 'used' shall be deemed also to include 'constructed', 'reconstructed', 'altered', 'placed', or 'moved'. The terms 'land use' and 'use of land' shall be deemed also to include 'building use' and 'use of building'. The word 'adjacent' means 'nearby' and not necessarily 'contiguous'.

Definitions. For the purpose of the ordinance, words used in the present tense include the future, the singular includes the plural and the plural the singular, the word "lot" includes the word "plot", the word "used" includes "designed" or "intended to be used", the word "building" includes the word "structure", and the word "shall" is mandatory and not discretionary. Unless otherwise specified, all distances shall be measured horizontally, in any direction.

Definitions. For the purpose of the code, certain words and phrases are defined and certain provisions shall be construed as herein setforth, unless it is apparent from the context that a different meaning is intended.

Definitions. In the interpretation of the code, all words other than terms herein specifically defined shall have their ordinarily accepted meanings as implied by the context or as customarily used in the construction industry.

Definitions. For the purposes of the Ordinance, the terms and words listed in the section shall have the meanings herein given. Terms and words not defined herein but defined in the Building Code, shall have, for the purpose of the Ordinance, the meanings given them in the Building Code, as the same now reads or may be amended.

Definitions. Words used in the present tense include the future tense; words used in the singular number include the plural, and words used in the plural include the singular. 'Map', means the 'Zoning map', the word 'person' includes a firm, partnership, trust, company, association, organization, individual, co-partnership or corporation. The word 'lot' includes the word 'plot' or 'parcel', the word 'building' includes the word 'structure', the word 'shall' is always mandatory, and not merely directory. The word 'used' or 'occupied' as applied to any land or buildings shall be construed to include the words 'intended', 'arranged', or 'designed to be used or occupied'.

Definitions. For the purposes of the Code, certain words and terms used herein are defined as set forth in this and the following sections of the Code. All words used in the present tense shall include the future. All words in the plural number shall include the singular number, all words in the singular number shall include the plural number, unless the natural construction of the wording indicates otherwise. The word (shall) is mandatory and not directory.

Definitions. For the purpose of the Zoning Code, certain terms and words are hereby defined as follows: Words used in the present tense include the future; words in the singular number include the plural, and words in the plural include the singular number; the word "shall" is mandatory and not permissive; the word "person" includes individuals, partnerships, corporations, clubs, or associations. The following words and terms, when applied in the Zoning Code, shall carry full force when used interchangeably: lot, plot, parcel, or premises; used, arranged, occupied or maintained; sold or dispensed; construct, reconstruct, erect, alter, structurally or otherwise, but not the term maintain or any form thereof.

Definitions and abbreviations. For the purpose of the code, certain abbreviations, terms, phrases, words, and their derivatives shall be construed as set out in the section of the code.

Developed distance. The shortest distance between two (2) points that free air would travel as measured horizontally, vertically or diagonally in a straight line or around corners.

Distances and areas. Terms which refer to measurements in a horizontal plane.

Effectively. Adequately, securely and similar words, shall be interpreted as conditions subject to approval by the Building Official or municipal authority having jurisdiction and the right of approval.

Either/or. Indicates that the connected items, conditions, provisions or events may apply singly but not in any combination.

Erect. To construct, reconstruct or excavate, fill, drain or conduct physical operations of any kind in preparation for, or in the pursuance of, construction or reconstruction, or to move a structure upon a lot.

Gender. Words importing the masculine gender shall include the feminine and neuter.

General terms. (a) The word "shall" is to be interpreted as having permission or being allowed to carry out a provision, "should" is to be interpreted as expressing that the application of such criteria or standard is desired and essential unless commensurate criteria or standards are achieved. (b) All words used in the singular shall be plural, and all words used in the present tense shall include the future tense, unless the context clearly indicates the contrary. (c) The phrase "used for" includes "arranged for," "designed for," "intended for" "maintained for" or "occupied for." (d) "Regulation" means a rule, restriction or other mandatory provision in the Zoning Code intended to control, require or prohibit an act.

Generally. Words used in the present tense include the future; the singular number includes the plural and the plural includes the singular. The word "lot" includes the word "plot;" the word "building" includes the word "structure;" the word "zone" includes the word "district;" the word "occupied" includes the words "designed or intended to be occupied" and the word "used" includes the words "arranged, designed or intended to be used."

Grammatical interpretation. "Genders," any gender includes the other genders.

Hereafter. After the time that the Code or Ordinance becomes effective.

Heretofore. Before the time that the Code or Ordinance became effective.

Inclusions. As used in the ordinance, words in the singular include the plural and those in the plural includes the singular; the word person includes a corporation, unincorporated association and a partnership, as well as an individual; the word building includes structure and shall be construed as if by the phrase or parts thereof; the word street includes avenue, boulevard, court, expressway, highway, lane, place, and road; the word watercourse includes channel, creek, ditch, dry run, spring and stream; the word may is permissive, the word shall and will are mandatory.

Interpretation of words. For the purpose of the ordinance, certain words shall have the meaning assigned to them as follows: words used in the present tense include the future; words used in the singular include the plural; the word "shall" or "must" is always mandatory; the word "building" includes "structure" and any part thereof; the phrase "used for" includes "arranged for", "designed for," "intended for", or "occupied for"; the word "person" includes an individual, corporation, partnership, incorporated association or any other similar entity; the word "includes" or "including" shall not limit the term to the specified example, but is intended to extend its meaning to all other instances of like kind and character.

In the city. The words "in the city" or "within the city" shall include all territory over which the city now has or shall hereafter acquire jurisdiction for the exercise of its police powers or other regulatory powers.

Lot. The word "lot" includes the word "plot" or "parcel".

Masculine Gender. Words used in the masculine gender include the feminine and neuter unless a contrary intention plainly appears.

May. Term giving permission but not, except in the negative, making a requirement. May is used in the code to emphasize that specified construction is not prohibited by the code when such prohibition might otherwise be implied or construed; or to limit the scope of a prohibition by excepting specified construction from its effect. A permission so expressed in the code in specific terms shall not be construed as a prohibition of other construction. "May not" is prohibited.

May. The word "may" is permissive.

Meaning of certain words. Whenever the words "accessory structure," "building," "premises," "room," "rooming unit," or "structure" are used in the Code, they shall be construed unless expressly stated to the contrary, to include the plurals of these words and as if they were followed by the words "or any part thereof." The word "shall" shall be applied retroactively as well as prospectively.

Meaning of words and phrases. The words and phrases defined in the Ordinance, and when used in the Ordinance, shall, for the purpose of the Ordinance, have the meanings ascribed to them in the Ordinance, except in those cases where the context clearly indicates a different meaning. Words used in the present tense include the future tense, words in the singular number include the plural number, words in the plural number include the singular number, and the word "shall" is always mandatory and not merely directory.

Month. The word "month" shall mean a calendar month.

Municipality. A city, town, county, district, or other public body created by or pursuant to State law, or any combinations thereof acting cooperatively or jointly.

New. That which is constructed, erected, or installed subsequent to the date at which the Code goes into effect.

New. Material never previously used in the construction work.

Nominal dimension. Dimension that may vary from an actual dimension in accordance with accepted engineering practice.

Nominal dimension. Dimension that may vary from the actual permissible dimension.

Nominal dimension. The dimension or size in which such material, part or unit is usually manufactured or supplied.

Nominally horizontal. At an agnle of less than 45 degrees with the horizontal.

Nominally vertical. At an angle of not more than 45 degrees with the vertical.

Number. Words importing the "singular" number may extend and be applied to several persons or things, and word importing the "plural" may include the singular.

Occupant or tenant. The words "occupant" or "tenant" applied to a building or land means any person who holds a written or an oral lease of or who actually occupies the whole or a part of such building or land, either alone or with others.

Occupied. Used, inhabited, or kept for use.

Occupied. Shall be construed as though followed by the words "or intended, arranged, or designed to be occupied."

Occupied. As applied to a building or structure, shall be construed as though followed by the words "or intended, arranged or designed to be occupied".

Or. Indicates that the connected items, conditions, provisions or events may apply singly or in any combination.

Or. The word provides an alternate of an option, unless the contrary is clearly indicated.

Owner. The word "owner" applied to a building or land shall include any part owner, joint owner, tenant in common, tenant in partnership or joint tenant of the whole or a part of such building or land.

Person. Includes an individual, firm, corporation, association or partnership.

Person. Includes natural persons, corporations, (private and public), partnerships and all other unincorporated organizations, trusts, estates, government agencies, and other legal entities, except when a contrary intention plainly appears.

Plan or plans. The word plan or plans shall be construed to mean drawing or drawings illustrating the work involved.

Plot. As used in the Ordinance shall mean the same as "LOT."

Post-code building. A building erected after the effective date of the Code and subject to the provisions thereof.

Pre-code. A term used to describe a structure in existence prior to the effective date of the adoption of the Code.

Pre-existing. In existence prior to the effective date of the Ordinance.

Present tense. Words used in the "present tense" include the "future tense."

Project. Any construction, alteration or demolition operation.

Property. The word "property" shall include real and personal property.

Public. General citizenry and/or the specific residents of a particular subdivision.

Public authority. Any officer who is in charge of any department of branch of the Government of the City or the State relating to health, fire, building regulations, or to other activities concerning dwellings in the city.

Public body. Any government or governmental agency, board, commission, authority or public body, of the City, County or State or the U. S. Government, or any legally constituted district.

Public enterprise. A community or public utility or development and the buildings, structures and land relevant thereto. An arena, ball park, cemetery, children's playground, church, fair ground, park, parking lot, pet cemetery, school, skating rink, swimming pool or tennis court, etc., may be classified as a public enterprise.

Regulations. Totality of text, charts, tables, diagrams, maps, notations, references, and symbols, contained or referred to in the Code.

Required. A mandatory provision of the Code.

Required. Shall be construed to be mandatory by provisions of the Code.

Semi-public body. Includes churches and organizations operating as a nonprofit activity serving a public purpose or service and includes such organizations as non-commercial clubs and lodges, theater groups, recreational and neighborhood associations, and cultural activities.

Shall. The word "shall" is always mandatory.

Should. Indicates that which is recommended, but not mandatory.

Single ownership. Ownership by one person or by two or more persons whether jointly, as tenants by the entirety, or as tenants in common, of a separate parcel of real property.

Single ownership. Holding record title, possession under a contract to purchase, or possession under a lease, by a person, firm, corporation or partnership, individually, jointly, in common, or in any other manner whereby the property is or would be under unitary or unified control.

Singular. The word used in the "singular" include the "plural", and words used in the "plural" include "singular."

Singular and plural. The singular number includes the plural, and the plural, the singular.

Singular number. The singular number includes the plural and the plural the singular, except where a contrary intention plainly appears.

Structure. The word "structure" as used in the Code, it shall not be deemed to include any coping, a fence not higher than three (3) feet, retaining wall, walk or stairway leading to a building.

Tense, gender and number. Words in the Ordinance used in the present tense include the future; words used in the masculine gender include the feminine and neuter; the singular number includes the plural and the plural the singular.

Tenses. Words used in the present tense include the past and future tenses and vice versa.

Terms. The following terms, when used in the Code, shall be construed to have the meaning here given. Words used in the present tense include the future as well as the present; the singular number includes the plural as well as the singular. Where terms are not defined in the Code, they shall have their ordinary accepted meanings or such as the context may imply.

Terms. The present tense shall include the future; the singular number shall include the plural; and the plural the singular. The word "shall" is always mandatory. The words "zone" and "district" are the same.

Terms. Terms and words used in the Ordinance are defined in the Ordinance. Words used in the present tense include the future; the singular number includes the plural and the plural the singular; the term "used" includes the words "arranged, designed or intended to be occupied"; the word "structure" includes the word "building."

Terms defined. For the purpose of the Code, certain terms and words are herewith defined, as set forth in the Code. All words used in the present tense shall include the future tense; all words in the plural shall include the singular number and all words in the singular number include the plural number; unless the natural construction of the wording indicates otherwise. The word "lot" includes the word "plot"; the word "building" includes the word "structure"; and the word "shall" is mandatory and not directory. Unless otherwise specified, all distances shall measured horizontally. Any words not herein defined shall be construed as defined in the Building Code of the City.

Terms not defined. Where terms are not defined, they shall have their ordinarily accepted meaning or such as the context may imply.

Terms not defined. Where terms are not defined, they shall have their ordinarily accepted meanings in accordance with Industry Standards or such as the context may imply.

Use. When used as a verb, shall mean that enjoyment of property which consists in its employment, occupancy, exercise, or practice. When used as a noun, the word shall mean that subdivision of a group of occupancies that refers to a specific purpose for which a building or lot is occupied or maintained. The word in every case shall include the terms "designed, arranged, or obviously intended."

Use of words and phrases. Words and phrases used in the Code and not specifically defined shall be construed according to the context and approved usage of the language.

Used or occupied. As applied to any land or building shall be construed to include the words "intended, arranged or designed to be used or occupied."

Used or occupied. The words "used" or "occupied" as applied to any land or building shall be construed to include the words "intended, arranged, or designed to be used or occupied."

Variance. Any modification of the terms of the ordinance.

Words. Used in the Code in the present tense include the future. Words in the masculine gender include the feminine and neuter. Words in the feminine and neuter gender include the masculine. The singular number includes the plural and the plural number includes the singular.

Words. Words used in the present tense include the future; words used in the singular number include the plural number; and words in the plural number include the singular number; the word "building" includes the word "structure" and the word "shall" is mandatory and not directory.

Words. Unless the context clearly indicates the contrary, the present tense shall include the future; the singular shall include the plural; the word LOT shall include the word PLOT; the word STRUCTURE shall include the word BUILDING; the word SHALL is always mandatory and not directory; the word MAY is permissive. The word USE and the word USED refer to any purpose for which a lot or land or part thereof is arranged, intended or designed to be used, occupied, maintained, made available or offered for use; and to any purpose for which a building or structure or part thereof is arranged, intended or designed to be used, occupied, maintained, made available or offered for use, or erected, reconstructed, altered, enlarged, moved or rebuilt with the intention or design of using the same.

Words. Whenever the words "dwelling," "dwelling unit," "habitable room," or "building" are used in the Code, they shall be construed so as to include the plural of these words, and be so interpreted as if they were followed by the words "or any part thereof."

Words and terms. Words used in the present tense include the future; words in the singular number include the plural and the plural the singular; "building" includes "structure" and "shall" is mandatory, not directory; words not included herein, but defined in the Building Code shall be construed as defined therein.

Workmanship. Shall conform to generally accepted good practice in the applicable trade.

Writing. Includes printing and typewriting.

Writing. The term includes printing, typewriting, or other forms of reproduction of legible symbols.

Writing. Includes handwriting, typewriting, printing, photo-offset or any other form of reproduction in legible symbols or characters.

Written. Include printed, typewritten mimeographed or multigraphed.

Written notice. A notification in writing delivered in person to the individual or to the parties intended, or sent by mail in an official United States Post Office depository.

Written notice. Shall be considered to have been served if delivered in person to the individual, or to the parties intended, or if delivered at, or sent by registered mail to, the last business address of the party given the notice.

Written notice. A notification in writing delivered in person to the individual or parties intended, or delivered at, or sent by certified or registered mail to the last residential or business address of legal record.

Written notice. Shall be any notice delivered or served in writing, either in person to the individual or to the parties intended, or forwarded by registered mail to the last known address of the party to be served.

Written or writing. The words "written" or "writing" may include printing and any other mode of representing words or letters, but when the written signature of any person is required by law to any official or public writing or bond required by law, it shall be in proper handwriting of such person, or in case he is unable to write, by his proper mark.

Year. The word "year" means a calender year unless otherwise expressed.

Zoning. The reservation of certain specified areas within a community or city for building and structures, or use of land, for certain purposes with other limitations such as height, lot coverage and other stipulated requirements.

A-19 ZONING AND DISTRICTS

Agricultural-rural residential district. The purpose of the district is to designate certain lands in the county which, because of soil conditions, location, natural resources, or economics, are best suited for farming, forestry or recreational and related uses, and to protect these lands from encroachment of unsuitable commercial and industrial activities and incompatible high density residential uses.

Appeal board. The City Planning Commission, for the purpose of hearing and making determinations upon appeals from actions of the Advisory Agency with respect to tentative subdivision maps, or the kind, nature and extent of improvement required in connection therewith.

Basic zoning regulations. Such zoning regulations as are applicable to the use district other than the regulations set forth in the Special zoning district.

Board of zoning appeals. For the purpose of hearing and making determinations upon appeals from actions of the Advisory Agency with respect to preliminary parcel maps or the kind with respect to preliminary parcel maps or the kind, nature and extent of improvements required in the connection therewith.

Boundaries of districts. Location and boundaries of districts shall be shown on the Zone map; provided that where the designation on the Zone map indicates a district boundary approximately on a street or alley line or on a lot line, such street or alley, or such lot line shall be construed to be the boundary.

Boundaries of districts established on "zoning district map." The boundaries of the districts shall be as shown and delineated on the "Zoning District Map" of the city, which map was declared to be part of the zoning ordinance.

Building zone maps. The building zone map consists of separate sectional maps, each identified by a sheet number, and a key map and chart, explaining symbols and indications which appear on the sectional maps. The key map and the sectional maps are part of the zoning ordinance.

Business. Includes the Commercial, Light Industrial and Heavy Industrial uses and Districts as defined in the Code.

Business district. The territory contiguous to and including a highway when within any six hundred (600) feet along such highway there are buildings in use for business or industrial purposes, including, but not limited to hotels, banks, office buildings, railroad stations, and public buildings, which occupy at least three hundred (300) feet of frontage on one side or three hundred (300) feet collectively on both sides of the highway.

City council review. Within 15 days following a decision of the Planning Commission or the Board of Appeals on a matter on which a public hearing is required by the Ordinance, or at its next regular meeting, whichever is later, the City Council may elect to review the action of the Commission or the Board. If the City Council elects to review an action and declines to confirm the decision, a public hearing shall be held by the City Council. The hearing shall be set and notice given as prescribed in the Ordinance.

City planning commission. It shall be the duty of the City Planning Commission to receive for the City Council and process, in accordance with the provisions of the Ordinance, petitions for zoning classifications, text revisions, text amendments, requests for large-scale planned developments, petitions for location of telephone exchanges and electric utility substations and transmission facilities, and petitions for conditional uses. When such proposals have been properly filed, the Commission shall hold a public hearing in the name of the Council relative to such proposals with the exception of proposals for telephone exchanges and electric substations in order to hear all proponents and opponents, after which a report and recommendation thereon shall be made and forwarded to the Council, except for petitions for telephone exchanges and electric utility substations.

Commercial district. District designated as a business or commercial district or special commercial district.

Conditional use. Uncommon or infrequent use which may be permitted in specific districts subject to the compliance with certain standards and explicit conditions set forth in the Zoning Code and the granting of a Conditional use permit.

Conditional use. A use listed in the zoning ordinance as a conditional use, requiring advertised public hearings and a Planning commission report before approval by the City Council.

Conditional use. A use which is generally not suitable in a particular Zoning District, but which may, under some circumstances and with the application of certain conditions be suitable.

Conditional use permit. Permit issued in accordance with stipulated conditions.

County. The largest political and administrative subdivision in a state, in area.

Date of non-conformity. The date at which a use or structure, which compiled to the applicable zoning regulations at the time of its initiation, was subsequently made non-conforming by the passage of a new Zoning Ordinance amendment.

Density. Permitted maximum residential densities in the planned development zone shall not exceed the permitted maximum densities in the original district. Minimum land area requirements for the various types of dwelling units shall be provided as required in the original district.

Development unit. That portion of a Planned Community District which is proposed for development at one time under one permit. Development units may consist of portions of a PCD or of the entire district.

District. A portion of the territory of the City within which certain regulations and requirements or various combinations thereof apply under the provisions of the Code.

District. One or more sections of the City for which the regulations governing height, area and use of buildings and premises are the same.

District. A section of the city for which the regulations governing the use, area, bulk, density, and setbacks of structures and property are uniform.

District. A section of the corporate area, in which all parts of the regulations of the Ordinance governing, provide for the height, area and use of buildings and buildings to be the same.

District. A geographical portion of the corporate area within which, on a uniform basis, certain uses of land and buildings are permitted, and certain other uses of land and buildings are prohibited, as set forth in the Ordinance, or within which certain lot areas, dwelling sizes and density requirements and other regulations are established, or within which combination of regulations are applied.

District. A portion of the city within which the use of land and structures and the location, height, and bulk of structures are uniform.

District. Section or sections of the City for which the regulations governing the use of buildings, the size of yards, and the intensity of use are uniform.

District. Any section of the City in which the zoning regulations are uniform.

District. Portion of the unincorporated territory of the Township which uniform regulations and requirements, or various combinations thereof, apply under the provisions of the ordinance.

District. Portion of the City within which the use of land and structures and the location, height, and bulk of structures are governed by the ordinance.

District. Section or sections of the City and the areas within 3 miles thereof for which regulations governing the use of buildings and premises, the height of buildings, the size of yards, and the intensity of use are uniform.

District. A district is a portion of the city within which certain uses of land and buildings are permitted or prohibited and within certain yards and other open spaces are required and certain limits are established, and certain offstreet parking areas are required, all as set forth and specified in the Ordinance concerning districts and the modification thereof by suffix or prefix.

District boundaries. Boundaries of the districts are established as shown on the official Zoning maps. Maps and all explanatory matter thereon is part of the ordinance as is fully written herein.

District boundary line. A line forming one of the boundaries of a given district.

Districts In order to classify, regulate, restrict and segregate uses of land, buildings and structures, and to regulate and restrict the height and bulk of buildings, the unincorporated territory shall be divided into a suitable number of districts which can be amended as required to meet the future growth of the area.

District use regulations. No structure hereafter shall be erected, reconstructed, structurally altered, enlarged, added to or moved, nor shall any structure or land be used for any purpose other than for a use originally permitted for the district in which such structure or land is located.

District use regulations. No Building or structure shall be erected or altered, nor shall any building or land be used for any purpose other than a use permitted in the district in which the building is located. No building or land shall be used so as to produce greater heights, smaller yards or less unoccupied area and no building shall be occupied by more families than hereinafter prescribed for such building for the district in which it is located. No lot which is now or may be hereafter built upon as herein required, shall be so reduced in area that the yards and open spaces will be smaller than prescribed by the ordinance, and no yard, court or open space provided about any building for the purpose of complying with the provisions of the ordinance shall be smaller than required.

Establishment of classes of districts. For the purpose of limiting and restricting to specified districts, and regulating therein buildings and structures according to their construction and the nature and extent of their use, and the nature and extent of the use of land, and to regulate and restrict the height, number of stories, and size of buildings and other structures, the percentage of lot that may be occupied. Also the size of yards, courts and other open spaces, the density of population, and the location and use and extent of use of buildings and structures and land, for trade, industry, residence or other purposes, the city is hereby divided into districts.

Establishment of districts. For the purpose of promoting the public health, safety, morals, and general welfare of the community, of the city, is hereby divided into districts as platted on the Zoning map.

Establishment of districts. In order to classify, regulate, restrict and separate the use of land, buildings and structures and to regulate and limit the type, height and bulk of buildings and structures in the various districts, and to regulate the area of yards and other open area abutting between buildings and structures and to regulate the density of population, the City is hereby divided into districts.

Establishment of districts. For the purpose of limiting and restricting to specified districts, and regulating therein buildings and structures according to their construction and the nature and extent of their use, and the nature and extent of the use of land, and to regulate and restrict the height, number of stories, and the size of buildings and other structures. The percentage of the lot that may be occupied, the size of yards, courts and other open spaces, the density of population, and the location and use and extent of use of buildings and structures and land, for trade and industry, residence or other purposes. The city is hereby divided into districts as indicated on the Zoning map which is a part of the ordinance.

Exceptions. A use permitted only after review and approval of an application by the Review Board. Such review being necessary because the provisions of the Zoning ordinance covering conditions, precedent or subsequent, are not precise enough to all applications without interpretations, and such review and approval is required.

Finding. A determination or conclusion based on the evidence presented and prepared by a hearings officer or Board in support of its decision.

Fire district. That area which contains closely built structures of business, mercantile or industrial occupancies having extraordinary density conditions created in areas or spaces, as well as the immediate surrounding blocks, which are exposed to such structures and density. District limits shall be indicated on the Zoning map.

Fire district. The territory defined and limited under the legal procedure of the governmental agency for creating and establishing fire districts.

Fire distirct. That area, within limits described by ordinance or by rule or regulation of the Board of Building Code Appeals, which contains closely built structures of business, mercantile or industrial occupancies, as well as the surrounding blocks, which are exposed to such structures or within which new construction of a business, mercantile or industrial character is developing.

Fire districts. Territories defined and limited by the provisions of the ordinance for the restrictions of types of construction.

Fire districts. The geographical territories established for the regulation of occupancy groups and construction classes within such districts.

Fire limits. Area in the City within boundaries, within which certain occupancies and types of construction may be limited or prohibited.

Fire limits. Boundary line establishing an area in which there exists, or is likely to exist, a fire hazard requiring special fire protection.

Fire limits. All areas as designated on the Fire Limit map, both inner and outer Fire Limits.

General plan. The comprehensive development plan for the City which has been officially adopted to provide long-range development policies for the area subject to urbanization in the foreseeable future and which includes, among other things, the plan for land use, land subdivision, circulation and community facilities.

General plan A General Plan is a comprehensive declaration of purposes, policies and programs for the development of the city, which includes, where applicable diagrams, maps and text setting forth objectives, principles, standards and other features, and which has been adopted by the City Council.

General purpose of a zoning plan. In order to provide for the public health, safety, morals, and general welfare, and to secure to the citizens in areas outside of municipal corporations, the social and economic advantages resulting from an orderly planned use of the land resources within the county; to regulate and restrict the location and use of buildings, structures and land for residence, trade, industry and other purposes; the height, number of stories and size of buildings and other structures; the size of yards, courts and other open spaces on the lot or tract; the density of population; to provide definite official land use plans for property publicily and privately owned in a district, outside municipal corporations; to guide, control and regulate the future growth and development of the district in accordance with such plan; and to provide for the administration and otherwise carrying out of such plan, there is hereby adopted and established an official zoning plan, pursuant to the authority of the State.

General regulations. No lot, building or structure shall hereafter be used, and no building or structure or part thereof shall be constructed, erected, raised, reconstructed, extended, enlarged or altered except in conformity with the regulations prescribed by the ordinance.

Governmental agency. City, municipality, village, town, township, county, state or federal governmental agency or subdivision thereof, exercising legal jurisdiction in the enactment, administration and enforcement of Codes, rules, regulations, laws or ordinances.

Hardship. The unusual situation or condition involving a particular property and making it impossible for the owner to use the property in a manner prescribed for the district by the Zoning Ordinance. A hardship exists only where the unusual situation or condition is not created by the owner of the property. A hardship as related to zoning is not to be confused with an economic hardship.

Higher classification. More restricted; a district or use subject, under the provisions of the Ordinance, to greater restrictions.

Legislative authority. By authority of Acts of the General Assembly of the State and by authority the laws of the State and the amendments thereto, and by power set forth in the Charter for the County, and by authority of the laws of the State and additions and amendments thereto, the County Council is empowered to regulate and restrict the height, number of stories, and size of buildings and other structures, the percentage of lot that may be occupied, the size of yards, courts, and other open spaces, the density of population, and the location and use of buildings, structures, and land for trade, industry, residence, or other purposes.

Local retail district. Business district adjacent to or surrounded on at least three (3) sides by Residence Districts in which such uses are permitted as are normally required for the daily local retail business needs of the residents of the locality only.

Method of regulation. The method to be used for carrying out the legislative intent shall be by ordinance of the Council dividing the corporate area into districts of such number, shape and area as may be deemed best suited to carry out the purpose of the legislative act. Within such districts, the Council may regulate and restrict the erection, construction, alteration, repair or use of buildings, structures, or land. All such regulations shall be uniform for each class or kind of buildings throughout each district, but the regulations in one district may differ from those in other districts. The regulations in the ordinance are intended to carry out the mandate of the Act as expressed in this paragraph.

Minimum hotel-motel units. In an interchange district, no hotel or motel shall have less than 20 lodging units containing living space.

More restricted—less restricted. The meaning and application of the terms "more restricted" and "less restricted" when used with reference to two or more zoning districts shall be determined by the City Council in each instance where applicable with reference to the uses, performance standards and building, lot and setback requirements and restrictions of the applicable zoning districts, provided that in all cases a residential zoning distict shall be deemed more restricted than a business or industrial zoning district.

Nature of the zoning ordinance. Ordinance shall consist of a zoning map designated certain districts and a set of regulations controlling the uses of land; the density of population; the bulk, locations, and uses of structures; the areas and dimensions of sites; the appearance of certain uses, structures, and signs; requiring provisions of usable open space, screening and landscaping and off-street parking and off-street grading facilities; and controlling the location, size, and illumination of signs. The zoning map shall be maintained on file with the Zoning Board.

Official zoning map adopted. Zoning map indicating the boundaries of each district, shown on a section map on file in the office of the Planning department, a reduced copy which is available to the public and can be purchased at the department office. the offical zoning map of the city is made part of the code by reference.

Open space zoning. Permitted reduction in lot sizes and lot area requirements in major subdivisions in the Residence District Zone in which the density requirement (dwelling units per acre) is maintained and where all resulting undeveloped land within said subdivision is deeded to the City for public use.

Original district. A zoning district described in the Code, other than a Planned Development District.

Pending prosecution. In the event that there are unremedied violations of any zoning ordinances replaled by the ordinance, which unremedied violations are also a violation of the ordinance, the City Council shall have the same rights and remedies as if the repealed ordinance were still in effect.

Permitted structure. Structure meeting all the requirements established by the Ordinance for the District, in which the structure is located.

Permitted use. Any use listed as Use by Right, a Use by Temporary Permit, a home occupation or an accessory use in any given District.

Permitted use. Use specifically permitted or analogous to those specifically permitted as set forth in the Ordinance.

Permitted use. Those uses specifically listed in the Ordinance as "uses permitted inherently" not to include uses defined in the Ordinance as "non-comforming use."

Permitted use. A permitted use in any district shall include any listed as a "Permitted Principal Use" or "Acessory Use", and shall include a "Conditional Use" as listed for the particular district provided a "Conditional Use Permit" is obtained.

Permitted use. Any use of land, buildings, structures or premises which is a listed permitted use in the zoning district in which such use is located or which is a use specified in the Ordinance.

Permitted use. A use which conforms with all requirements, regulations, and standards of a particular district.

Permitted uses. The permitted, accessory, and conditional uses set forth in the Code pertaining to the original district shall apply to and be permitted uses in the Planned Development District.

Property lines. The lines bounding a zoning lot.

Public agency. Any legally authorized body of the United States or the State which has taxing jurisdiction.

Purpose of ordinance. In their interpretation and application, the provisions of the Ordinance shall be held to be minimum requirement adopted for the promotion of the public health, safety, morals and general welfare. Among other purposes, the provisions of the Ordinance are intended to provide for adequate light, air and convenience of access; to lessen congestion in the streets; to secure safety from fire and other dangers; to avoid undue concentration of population by regulating and limiting the use of land, the height and bulk of buildings wherever erected to limit and determine the size of yards, and other open spaces; to regulate the density of population; and to conserve the value of property and encourage the most appropriate use of land in the city.

Purpose of ordinance. The ordinance is enacted for the purpose of promoting health, morals or general welfare of the people of the Township by protecting property values, reducing traffic hazards, congestion of population, and securing safety from fire, panic and other dangers; to provide adequate open spaces and regulate land development and land use as will enable to provide economical establishment of trasnporation service, drainage, sanitation, education, recreation, and other public improvements.

Purpose of the ordinance. The regulations and restrictions established herein have been made in accordance with a plan, called the Land Use Plan, which plan and these regulations and restrictions have been adopted by the City.

Purpose of the ordinance. For the purpose of promoting the health, safety and general welfare of the community, and to lessen congestion in the streets, to secure safety from fire, panic and other dangers, to provide adequate light and air, prevent the overcrowding of land, to avoid undue concentration of population, to facilitate the adequate provision of transportation, water, sewerage, schools, parks, and other public requirements, under the pursuant to the Ordinance as amended, the size of buildings and other structures, the percentage of lot area, that may be occupied, the size of yards, the density of populations, and the use of buildings, structures and land for trade, industry, residence or other purposes, are hereby restricted and regulated as herinafter provided in the Ordinance.

Purpose of zoning ordinance. The objective of the ordinance is for the purpose of promoting public health, safety, convenience, and general welfare of the community and of a serviceable and attractive municipality, by having regulations and restrictions that provide for the safety and security of home life; that preserve and create a more favorable environment in which to rear children; that stabilize and enhance property and civic values; that provide for a more uniform land-use pattern; that facilitate adequate provisions for increased safety in traffic, and for transportation, vehicular parking, parks, parkways, recreation, schools, public buildings, housing, light, air, water supply, sewerage, sanitation, and other public requirements; that lesssens congestion, disorder, and danger which often inhere in unregulated municpal development; that prevent overcrowding of land and undue concentration of population; that assist in carrying out the master plan, and to provide more reasonable and serviceable means and methods of protecting and safeguarding the economic structure upon which the good of all depends.

In order to more effectively protect and promote the general welfare and to accomplish the aims and purposes of this objective, the corporate area is divided into districts of such number, shape and area, and of such common unity of purpose, adaptability or use, that are deemed most suitable to provide for the best general civic use, protect the common rights and interests within each district, preserve use and occupancy of buildings, structures and land to be used for trade, industry, residence or other purposes, and the location, height, bulk of buildings and other structures, including the percentage of plot occupancy and coverage, street setback lines, size of yards, and other open spaces and most important to permit, protect, and promote the use of individual property in any way the owner may desire, consistant with the community goals embodied in the Ordinance.

Resident district. The territory contiguous to and including a highway not comprising a business district when the property on such highway for a distance of three hundred (300) feet or more is in the main improved with residences or residences and buildings in use for business.

Rezoning. Changes in zoning or zoning boundaries. The procedures for rezoning and conditional use permits shall in all cases be in accordance with the provisions of the Zoning Code.

Rural districts. All places not urban usually in the country, but in some cases within city limits.

Saving clause by court action. Should any section or provision of the ordinance or application of a provision under the Ordinance or application of a provision under the Ordinance be declared by the courts to be unconstitutional or invalid, such declaration shall not affect the validity of the Ordinance as a whole or any part thereof, other than the part or application so declared to be unconstitutional or invalid.

Special exception. Relaxation of the strict terms of the Ordinance to permit uses in use districts where such uses require additional controls and safeguards not required of principal uses.

Special exception. Deals with special permission, granted only by the Zoning Board to occupy land for specified purposes when such use is not permitted by right.

Special exception. Development or use, which though not the predominant type of development in a district, is nevertheless compatible, provided that specified conditions are met. A special exception is allowable where the facts and conditions prescribed and detailed in the Ordinance, as those upon which the special exception may be granted, are determined by the Zoning Board to exist.

Special exception. Permission or approval granted by the Board of Appeals in accordance with the Ordinance, in situations where provisions therefor is made by the terms of the Ordinance.

Special exception. Exception to the provisions of the Ordinance, when specifically mentioned in the Ordinance as one which may be granted by and at the discretion of the Board of Zoning Appeals.

Special exception use. Use in one or more Zones, for which the Zoning Board may grant a permit pursuant to the provision of the Ordinance.

Special use. A use, either public or private, which because of its unique characteristics, cannot be properly classified as a permitted use in a particular district or districts.

Special use. Any building, structure or use which on the effective date of the Ordinance complies with the applicable regulations governing special uses in the zoning districts in which such building, structure or use is located.

Strip commerical area. Developed business frontage along a street and no more than 200 feet in depth from the front line property.

Strip zoning. Similar to "spot zoning" except that it refers to the elongation of a use district along traffic arterials.

Transitional use. Use intended to permit a more gradual change of the character of uses at or near the boundaries of districts which have different use regulations and which may be permitted by the Zoning Board in accordance with the provisions of the Ordinance.

Urban district. The territory contiguous to and including any street which is built up with structures devoted to business, industry, or dwelling houses situated at intervals of less than one hundred (100) feet for a distance of a quarter of a mile or more.

Urban district. Thickly settled area, whether inside city limits or not, or where congested traffic often occurs. A highway, even though in the country, on which the traffic is often heavy, is considered urban.

Use by right. Use which is listed as a "Use by Right" in any given zone district in the Ordinance. "Uses by Right" are not required to show need for their location.

Use by temporary permit. Listed uses which may be permitted in any given district provided that the need for the use in the district can be established to the satisfaction of the Zoning Department.

Use district resolution. Whereas the Board of Supervisors and the State deems it necessary for the promotion of health, safety, morale and the general welfare to regulate and restrict therein the height, number of stories, use, size and location of buildings and other structures, the size and location of yards and other open spaces in relation to buildings and the use of land. Use Districts and Height and Area Districts are hereby created and regulations are hereby established to accomplish that purpose.

Variance. For the purposes of the Ordinance, a variance is a deviation from requirements of the Ordinance, granted by the Board of Zoning Appeals in cases of practical difficulty or unnecessary hardship under the provisions of and as limited in the Ordinance.

Variance. A modification or variation of the provisions of the Code as applied to a specific building or structure.

Variance. A modification of the literal provisions of the Zoning Ordinance.

Variance. A modification of the zoning regulations, permitted in instances where a literal application of the provisions of the Zoning Code would result in unnecessary hardships as a result of some peculiar or unique condition or circumstance pertaining only to the zoning lot in question in accordance with procedures and standards set forth in the Zoning Code.

Variance. A relaxation of the strict application of the terms of the ordinance. This definition shall not be construed to permit a use in any district which use is prohibited therein.

Variance. A relaxation of the terms of the zoning ordinance where such variance will not be contrary to the public interest and where, owing to conditions peculiar to the property and not the result of the actions of the applicant, a literal enforcement of the ordinance would result in unnecessary and undue hardship. As used in the ordinance, a variance is authorized only for height, area, and size of structure or size of yards and open spaces, establishment or expansion of a use otherwise prohibited shall not be allowed by variance, nor shall a variance be granted because of the presence of nonconformities in the zoning district or uses in an adjoining zoning district.

Variance. Permission or approval granted by the Zoning Board which constitutes modification of or a deviation from the exact provisions of the Ordinance as applied to the use of a specific piece of property or portion of the same.

Variance. Modification of the regulations of the Ordinance, granted on the grounds of practical difficulties or undue hardship, pursuant to the provisions of the Ordinance.

Variance. Variance deals with permissive waivers from the terms and conditions of the ordinance where literal enforcement would create hardship. A variance can be granted only by the Zoning Board of Appeals.

Variance. The granting of relief by the Zoning Board of Adjustment from the terms and conditions of a zoning ordinance, upon application of an individual, stating that (a) Special conditions and circumstances exist which are peculiar to the land, structure, or building involved and which are not applicable to other lands, structures, or buildings in the same district, and (b) That such conditions and circumstances have resulted in a hardship, in that a literal interpretation of the provisions of the Zoning Ordinance would deprive the applicant of rights commonly enjoyed by other property in the same district under the terms of the ordinance, and (c) That the special conditions and circumstances do not result from the actions of the applicant, and (d) That granting the variance requested will not confer on the applicant any special privilege that is denied by the ordinance to other lands, structures, and buildings in the same district.

Variance. A variance is defined to be a situation in which practical difficulties or unnecessary hardships would result from the carrying out of the strict provisions of the Ordinance and the waiver of such requirements should not unduly interfere with the general purpose and intent of the ordinance and the granting of such waiver would not adversely affect the health, safety, or general welfare of the residents of the City and would accomplish substantial justice.

Zero lot line district. District that allows the location of one wall of the building on a side lot line, while maintaining the sum of the two (2) normal side setbacks or yards on the opposite side lot line.

Zero lot lines. Lot which allows the location of the building on the lot lines. Original conception occurred where buildings were designed with large interior courts.

Zone. Section of the city in which the regulations governs the use, height, area, lot coverage, and size of buildings, structures and premises are uniform.

Zone. Zone shall mean a portion of the territory of the County within which territory certain uniform regulations and requirements, or various combinations thereof, apply pursuant to the provisions of the Ordinance. "Zone" shall include the word "district".

Zone height compliance. No building shall be erected, moved, reconstructed or structurally altered to exceed in height the limit established by the Ordinance for the zone in which such building is located.

Zone use compliance. No building shall be erected, moved, altered, enlarged, nor shall any land, building or premises be used, designed or intended to be used for any purpose or in any manner other than a use listed in the Ordinance as permitted in the zones in which such land, building or premises is located.

Zoning. The reservation of certain specific areas within a community or city for buildings or structures for use of land for certain specified purposes with other limitations such as height, lot coverage and other stipulated requirements.

Zoning. The reservation of certain specified areas within the City for buildings and structures for certain purposes with other limitations such as height, lot coverage and other stipulated requirements.

Zoning. The reservation of certain specified areas within a community or city for building and structures, or use of land for certain purposes with other limitations such as height, lot coverage, and other stipulated requirements.

Zoning. Reservation of certain specified areas for building and structures, or use of land, for certain purposes with other limitations such as height, lot coverage and other stipulated requirements.

Zoning. No building or land shall hereafter be used and no building or part thereof shall be erected, moved or altered unless in conformity with the regulations for the district in which it is located.

Zoning. Reservation of certain specified areas for specified uses, and the regulation of the height, bulk, location, percentage of lot occupancy, set back lines, area and dimensions of yards, courts, and other open spaces, of buildings and other structures within such areas.

Zoning. The reservation of certain specified areas within a community or city for buildings and structures, or use of land, for certain purposes with other limitations such as height, lot coverage, and other stipulated requirements.

Zoning. Zoning affects all structures, buildings, and land and the use thereof. No land or premises shall be used and no building or structure shall be erected, raised, moved, extended, enlarged, altered or used for any purpose other than a purpose permitted herein, for the zone district in which it is located, and all construction shall be in conformity with the regulations provided in the Zoning Ordinance and the Building Code.

Zoning district. Any district for which the regulations governing the height, area, use, structure, or size of buildings and premises are identical.

Zoning district dividing property. Where one (1) parcel of property is divided into two (2) or more portions by reason of different zoning district classifications, each portion shall be used independently of the other in its respective zoning classification, and for the purpose of applying the regulations of the ordinance, each portion shall be considered as if in separate and different ownership. Alternatively, the entire parcel may be used as permitted by the regulations applicable to the most restrictive zoning classification. However, nothing in the ordinance shall be construed as permitting residential use of any property within any industrial district classification or use of any property within any other district classification for any purpose not permitted by the ordinance.

Zoning districts. The districts into which the corporate area has been divided, as set forth on the Zoning District Map, for zoning regulations and requirements.

Zoning lot. A single tract of land located within a single block, which (at the time of filing for a building permit) is designated by its owner or developer as under single ownership or control. A zoning lot may or may not coincide with a lot of record.

Zoning lot. A single tract of land located within a single block, which (at the time of filing for a building permit) is designated by its owner or developer as a tract to be used, developed, or built upon as a unit, under single ownership or control. Therefore, a "zoning lot or lots" may or may not coincide with a lot of record. Every zoning lot must have access to a public street either by having frontage on a public street or by private road.

Zoning lot. Parcel of land abutting a dedicated street, occupied or intended to be occupied by a main and/or accessory use or a main or accessory building, as a unit together with such open spaces as required by the Zoning Ordinance. Unless the context clearly indicates to the contrary, the term lot is used synonymously with zoning lot in the Zoning Ordinance and it may or may not coincide with a lot of record.

Zoning map. The location and boundaries of the zoning districts established by the Zoning Ordinance are bounded and defined as shown on the map entitled "Zoning Map of the City", The said map consists of separate sheets numbered consecutively, and the same being attached to the Ordinance upon its introduction and passage is an effective and operative part thereof. The Zoning Map shall be kept and maintained by the Zoning Department, and shall be available for inspection and examination by members of the public at all reasonable times as any other public records.

Zoning ordinance. A municipal ordinance regulating the use of the land or structures, or both, as provided in the Act.

Zoning ordinance. No provisions of the Zoning Ordinance or any other legal statutes pertaining to the location, use or construction of buildings shall be nullified by the provisions of the Building Code.

Zoning ordinance. An Ordinance of the County to be known as "The Zoning Ordinance of the County" including a "Zone District Map" which shall be part of the Zoning Ordinance, to set forth the legislative authority; (1) To set forth standards for parking designed to lessen congestion in the streets; (2) To set forth standards and permissible uses designed to secure safety from fire, panic, and other dangers; (3) To promote health, aesthetics, and general welfare; (4) To provide adequate light and air; (5) To prevent the overcrowding of land; (6) To avoid undue concentration of population; (7) To facilitate the adequate provision of transportation, water, sewage, schools, parks, and other public requirements by dividing the County into districts of such size and shape as may be best suited to carry out the purposes of the legislative act and Ordinance, and by regulating and restricting the height and size of buildings and other structures, building lines and setbacks, the size of yards, the density of the population, and the location and use of buildings, structures, and land for trade, industry, residence or other purposes, in accordance with a comprehensive plan, to provide for off-street parking; and to provide for the administration and enforcement of the regulations and restrictions.

Zoning ordinance. An Ordinance to limit and restrict to specified Districts or Zones, and to regulate therein, buildings and structures according to their construction and the nature and extent of their use, and the nature and extent of the uses of land, in the Township and in the County and providing for the administration and enforcement of the provisions herein contained and fixing penalties for the violation thereof.

Zoning ordinance. The ordinance is adopted in order to promote, protect, and facilitate the public health, safety, morals, general welfare, coordinated and practical community development, proper density of population, the provisions of adequate light and air, police protection, vehicle parking and loading space, transportation, water, sewerage, public grounds, and other public requirements, as well as to prevent overcrowding of land, blight, danger and congestion in travel and transportation, loss of health, life or property from fire, flood, panic or other dangers. It establishes use of land and structures, area of lots, density of population, and similar regulations for the Township, and for such purposes divides the Township into districts. It provides for administration, enforcement, and amendments thereof in accordance with the provisions of the State Municipalities Planning Code.

Zoning ordinance. The Zoning Ordinance of the City is adopted for the following purposes: (1) To promote and to protect the public health, safety, morals, comfort, convenience and the general welfare of the people, (2) To maintain and promote pedestrial and vehicular circulation, (3) To secure safety from fire, panic, and other dangers, (4) To provide adequate standards of light, air, and open space, (5) To prevent the overcrowding of land and, thereby, ensure proper living and working conditions and prevent blight and slums, (6) To avoid undue concentration of population, (7) To facilitate the adequate provision of transportation, water, sewerage, schools, parks, and other public requirements, (8) To zone all properties with a view to conserving the value of buildings and encouraging the most appropriate use of land throughout the City.

Zoning ordinance. An ordinance to limit and restrict to specified districts or zones, and to regulate therein, buildings and structures according to their construction and the nature and extent of their use, and the nature and extent of the uses of land, in the Township and providing for the administration and enforcement of the provisions of the ordinance, and fixing penalties for the violation thereof.

Zoning ordinance. An ordinance in pursuance of the authority granted by the State, to provide for the establishment of Districts within the corporate limits of the City, to regulate within such districts, the height, number of stories, and size of buildings and other structures, the percentage of lot that may be occupied, the size of yards and open spaces, the density of population and the use of buildings, structures, and land; and to provide methods of administration of the Ordinance and penalties for the violations thereof.

Zoning ordinance. Adopted to protect and to promote the public health, safety, peace, comfort, convenience, prosperity, and general welfare. The Ordinance is not intended to abrogate, annul, impair, or interfere with any deed restriction, covenant, easement, or other agreement between parties, provided that where this Ordinance imposes a greater restriction on the use of the land or structures or the height or bulk of structures, or requires greater open spaces about structures or greater areas or dimensions of sites than is imposed or required by deed restriction, covenant, easement, or other agreement, the Ordinance shall control.

Zoning ordinance. Whereas in order to promote the health, safety, morals and general welfare of the inhabitants of the City to facilitate the adequate provision of transportation, sewerage, water, schools, parks and other public improvements and to regulate and restrict the location and uses of buildings, structures, land and water for trade, industry, residence or other purpose, to regulate and restrict the erection, reconstruction or alteration of buildings, and to regulate and restrict the height, number of stories, and size of all buildings and to structures, and the size of all yards and open spaces surrounding buildings; to regulate and restrict the density of population, and to divide the City into Districts of such number, shape and areas as may be best suited to carry out said purposes, it is desirable and necessary to adopt the Zoning Ordinance and plan for said City as hereinafter set forth.

Zoning ordinance. An Ordinance to provide for the establishment of Zoning Districts lying wholly within the unincorporated parts of the Township within the Zoning Districts the use of land, natural resources and structures, including tents and trailer, coaches, the height, the area, the size, and the location of buildings hereafter erected or altered, the light and ventilation of such buildings, the area of yards, courts and other open spaces and the density of population shall be regulated: To provide for amendments: To provide for administration, operation and enforcement of the Ordinance: To provide penalties for the violation of the Ordinance To provide for conflicts with other Ordinances and laws.

Zoning permit. A zoning permit shall be required prior to the erection, construction or alteration of any building, structure or any portion thereof, to be issued simultaneously with the required building permits where possible.

Zoning procedure. Procedure established by the Board which usually starts with an informal meeting to discuss the proposed project and the submission of preliminary drawings. Applications are filed and public hearings are held which finally results in a decision.

Zoning—vacated areas. Whenever any street or alley shall be vacated such street or alley or portion thereof shall automatically be classified in the same zoning district as the property to which it attaches.

Zoning variance. Usually a request by a petitioner to to seek relief from the requirements of an existing zoning ordinance.

Zoning variance. A modification of the literal provisions of the Ordinance, granted when strict enforcement of the Ordinance would cause practical difficulty or undue hardship owing to the circumstance unique to the individual property on which the variance is granted. The crucial points of a variance are practical difficulty, undue hardship and unique circumstances applying to the specific property involved. A variance is not justified unless all elements are present in each case.

1. Category B

SITE AND LAND USES B-1 THROUGH B-16

B-1 AGRICULTURE, FARMING AND LIVESTOCK

Agricultural building. A building located on agricultural property and used to shelter farm implements, hay, grain, poultry, livestock, or other farm produce or equipment, in which there is no human habitation, and which is not used by the public.

Agricultural building. A storage building located on agricultural property and used to shelter farm implements, hay, grain, poultry, livestock or other farm produce, in which there is no human habitation and which is not used by the public.

Agricultural building. An agricultural production or storage building or structure is a building located on agricultural property and used to shelter farm implements, forage crops, grain, poultry, livestock, or other farm produce, in which there is no human habitation and which is not used by the public.

Agricultural processing building. Agricultural processing building or structure is a building located on agricultural property and used to dehydrate, mill, pack, or otherwise process farm products. This type of agricultural building requires human occupancy to fulfill its intended use.

Agricultural purposes. Includes agriculture, farming, dairying, pasturage, agriculture, horticulture, floriculture, viticulture, ornamental horticulture, olericulture, pomiculture, and animal and poultry husbandry.

Agricultural tractor. A self-propelled vehicle designed or used for drawing other vehicles or wheeled machinery but having no provision for carrying loads, independently of such other vehicles, and used primarily for agricultural purposes.

Agriculture. The use of the land for agricultural purposes, including farming, dairying, pasturage, agriculture, horticulture, floriculture, viticulture, apiaries, and animal and poultry husbandry, and the necessary uses thereto; provided, however, the operation of any such accessory uses shall be secondary to that of the normal agricultural activities. For the purpose of the Ordinance "accessory use" shall mean supply, service, storage, and processing areas and facilities for any other agricultural land. The uses set forth in the Ordinance shall not include stockyards, slaughterhouses, hog farms, fertilizer works, or plants for the reduction of animal matter.

Agriculture. The use of land for farming, dairying, pasturage, horticulture, viticulture and animal and poultry husbandry.

Agriculture. Cultivating of the soil, and the raising and harvesting of the products of the soil, including, but not by the way of limitation, nurserying, horticulture, forestry and animal husbandry.

Agriculture. The art and science of cultivating the ground, the production of crops or livestock, and the processing only of milk produced on the farm on which the processing is located; excluding however, commercial greenhouses, the sale of nursery stock, riding stables, mink or fox or similar fur farms, hog or poultry farms using garbage as the principal source of feed and dairy processing operations.

Agriculture. Farms and general farming, including horticulture, floriculture, dairying, livestock, and poultry raising, farm forestry, and other similar enterprises, or uses, but no farm shall be operated as piggeries, or for the disposal of garbage, sewage, rubbish, offal or rendering plants, or for the slaughtering of animals except such animals as have been raised on the premises or have been maintained on the premises for at least a period of one year immediately prior thereto and for the use and consumption of persons residing on the premises.

Agriculture. The commercial production of field, row, or crops, but not including live stock other than those kept for the use of the family occupying the site.

Agriculture. The tilling of the soil, the raising of crops, horticulture and gardening, but not including the keeping or raising of domestic animals and fowl, except household pets, and not including any agricultural industry or business such as fruit packing plants, fur farms, animal hospital or similar uses.

Agriculture. The use of land for agricultural purposes including farming, dairying, pasturage, horticulture, animal and poultry husbandry provided, however, that the operation of accessory uses shall be secondary to that of normal agricultural activities. Accessory uses shall include storage and processing areas, including feed lots for farm animals.

Agriculture or agricultural. The bona fide use of a parcel of land of five (5) or more acres for the cultivation of land, raising of poultry and livestock or similar agrarian activity and the related buildings, structures and appurtenances necessary to carry out the aforementioned activities.

Animal waste processing. Processing of animal waste and by-products, including but not limited to animal manure, animal bedding waste, and similar by-products of an animal raising agricultural operation, for use as a commercial fertilizer or soil amendment and including composting operations.

Dairy. Any premises upon which three (3) or more cows or goats are kept for the commercial production or sale of milk and dairy products.

Farm. A parcel of land of five (5) acres or more on which bona fide agricultural and related uses are conducted.

Farm. An area which is used for the growing (but not selling) of the usual farm products such as vegetables, fruit, trees, and grain, and their storage on the area, as well as for the raising thereon of the usual farm poultry and farm animals. The term "farming" includes the operating of such an area for one or more of the above treatment or storing the produce; provided, however, that the operation of any such accessory uses shall be secondary to the normal farming activities, and provided further, that farming does not include the feeding of garbage or offal to swine or other animals.

Farm. Any parcel of land, five (5) acres or more, which is used for gain in the raising of agricultural products, livestock, poultry, and dairy products. It includes necessary farm structures within the prescribed limits and the storage of equipment used. It excludes the raising of furbearing animals, riding academies, livery or boarding stable, and dog kennels.

Farm. Tract of land in single ownership or single operation on which agriculture takes place.

Farm. Parcel of land which is worked as a single unit of not less than 5 acres in extent. A farm may be considered as greenhouses, nurseries, orchards, chicken hatcheries, apiaries and livestock.

Farm. An area of five (5) or more contiguous acres which is used for the production of farm crops such as vegetables, fruit trees, cotton or grain and their storage, as well as raising thereon of farm animals such as poultry or swine on a limited basis. Farms also include dairy produce. Farming does not include the commercial raising of animals, commercial pen feeding (feed lots) or the commercial feeding of garbage or offal to swine or other animals.

Farm-accessory buildings. Accessory buildings for bona fide farm uses are structures which are necessary in the execution of the agricultural processes. These buildings are not intended to include food processing or a manufacturing use of a related nature.

Farm and greenhouses in a single family residence district. Farms, truck gardens, and noncommercial greenhouses; provided that no greenhouse heating plant shall be operated within 25 feet of the lot line of any adjoining owner and that no fertilizer shall be stored within 50 feet of the lot line of any adjoining owner.

Farm building. Any building used for storing agricultural equipment or farm produce, housing livestock or poultry, and processing dairy products. Term "Farm Building" does not include dwellings.

Farm dwelling. A dwelling for permanent year-round residents of a farm, such as the owner, lessee, foreman, or others whose principal employment is the operation of the farm.

Farm employee housing. Includes living quarters, dwelling, boarding house, tent, bunkhouse, trailer coach, or other housing accommodations, maintained in connection with any farm work or place where farm work is being performed, and the premises upon which they are situated and/or area set aside and provided for camping of farm employees and their families.

Farming and farm animals. Growing of the usual farm products such as grain, and their storage, as well as the raising of the usual farm poultry and farm animals, and the operation of a dairy farm.

Farm labor camp. Any living quarters, dwelling, boarding house, tent, bunkhouse, trailer coach or other housing accommodations, maintained in connection with any farm work or place where farm work is being performed, and the premises upon which they are situated and/or the area set aside and provided for camping of five or more farm employees and their families.

Farm laborer's quarters. A building used to house laborers employed by the occupants of the main building on a farm and which shall not be rented or leased separate from the main building and the main use of the land.

Feed lot. Any tract of land or structure wherein any type of fowl or the by-products thereof are raised for sale at wholesale or retail; any structure, pen or corral wherein cattle, horses, sheep, goats, and swine are maintained in close quarters for the purpose of fattening such livestock before final shipment to market; the raising of swine under any conditions.

Hog ranch. Any premises where three or more hogs are kept.

Husbandry. Cultivation or production of plants and animals (livestock) and/or the by-products thereof.

Livestock—animal. Animals of any kind kept or raised for sale, resale, agricultural field production or pleasure, excluding fur-bearing animals.

Livestock feed yard. Lot or parcel of land improved with corrals, fences, buildings or improvements, and used primarily for the feeding and fattening of livestock for subsequent sale, and includes the feeding of garbage for disposal.

Miscellaneous definitions. *Agrarian:* Relating to the land. *Agriculture:* The cultivation of the ground. *Animal husbandry:* The business related to livestock and other animals. *Apiaries:* The keeping and housing of bees. *Apiculture:* The management of bees in hives. *Dairying:* The processing of milk and milk products. *Farming:* The business of agriculture. *Floriculture:* The cultivation of flowers. *Forestry:* The cultivation of trees and the management of growing timbers. *Horticulture:* The cultivation and management of gardens. *Nurserying:* The business of growing plants. *Olericulture:* The culture of edible vegetables. *Ornamental horticulture:* The cultivation and management of special types of ornamental flowers. *Pasturage:* The feeding and grazing of cattle. *Pomiculture:* The culture of fruit. *Poultry husbandry:* The business related to poultry. *Viticulture:* Cultivation and the growing of vine type plants.

Nursery. Any building or lot, or portion thereof, used for the cultivation or growing of plants, and including all accessory buildings.

Residence farm. Considered as a piece of platted or unplatted parcel of land of the aggregate size of one (1) acre or less, but not less than one-half (1/2) acre in area upon which is located a building used for a dwelling and which land is worked as a single unit by a single family, and/or with the assistance of members of the household or hired employees, except as provided in the Ordinance.

Small farm. Platted or an unplatted parcel of land containing not less than one (1) acre nor more than five (5) acres which is worked as a single unit by a single family, or with the assistance of members of the household or hired employees, except as provided in the Ordinance.

Small livestock farming. The raising or keeping of more than twelve (12) fowl of any kind, or twelve (12) rabbits or twelve (12) similar animals; any goats, sheep or similar livestock, but not hogs; or the breeding, raising or keeping of any cats or dogs; provided however, that "small livestock farming" as used in the Ordinance shall not include animal hospitals, commercial cat or dog kennels, hog raising or the breeding for commercial purposes of horses, cattle or similar livestock as determined by the Planning Commission.

Tenant dwelling. Residential structure located on a bonafide farm and occupied by a non-transient farm worker employed by the farm owner for work on the farm.

Tenant house. Accessory building on a lot used in whole or in part as dwelling quarters for one or more tenant farmers.

B-2 AIRPORTS, AIRCRAFT AND SERVICES

Adjacent to approach zones. Transition zones adjacent to approach zones shall be three thousand (3,000) feet for all personal and secondary-type airports; five thousand (5,000) feet wide each for continental, express trunk-line and feeder airports, and seven thousand (7,000) feet for intercontinental and intercontinental-express airports, measured at right angles to the centerline of the approach zone and shall be as long as the approach zone.

Adjacent to landing strip. Transition zones adjacent to a landing strip shall be one thousand fifty (1050) feet wide each, and shall be as long as the landing strip.

Adjacent to turning zones. A transition zone adjacent to a Turning Zone shall be three thousand (3,000) feet wide for personal and secondary-type airports and five thousand (5,000) feet wide for all larger-type airports.

Aircraft. A contrivance, now known or hereafter invented for use in or designed for navigation of or flight in air.

Aircraft hangar. A building, structure or space used for storage or servicing of aircraft in which gasoline, jet fuels, or other volatile flammable liquids, or flammable gases, are used, but shall not include such locations when used exclusively for aircraft which have never contained such liquids or gases, or which have been drained and properly purged.

Aircraft hangar. Building which is used for the housing, storage, maintenance, or repair of one or more aircraft.

Aircraft sales and display rooms. Rooms used for the display or sales of aircraft containing no flammable liquid during such sale or display.

Airfield. Any area of land or water utilized for the landing or taking-off of aircraft.

Airplane hangar. Building which is used to shelter or repair airplanes.

Airplane hangar (private). A hangar for the storage of four (4) or less single motor planes and in which no volatile or flammable oil is handled, stored or kept other than that contained in the fuel storage tank of the plane.

Airplane hangar (public). A hangar for the storage of aircraft.

Airplane hangers. Any building which is used to shelter or repair.

Airport. A landing area used regularly by aircraft for receiving or discharging passengers or cargo.

Airport. Any area of land which is used or is intended to be used primarily for the taking off and landing of aircraft, and any accessory areas which are used or intended to be used for airport buildings or facilities, including open spaces, taxiways and tie-down areas, hangars and other accessory buildings.

Airport. A tract of land which provides or may provide for the landing or taking-off of aircraft and related aviation activities conducted in the interest of the public.

Airport. The transportation buildings and the landing fields necessary for the landing and taking off of airplanes for receiving and discharging passengers and freight.

Airport. Any area of land which is used or intended to be used primarily for the taking off and landing of aircraft, and any accessory areas which are used or intended to be used for airport buildings or facilities, including open spaces, taxiways and tie-down areas, hangers and other accessory buildings.

Airport. Any landing area, runway or other facility designed, used or intended to be used either publicly or by any person or persons for the landing and taking off of aircraft, including all necessary taxiways, aircraft storage and tie-down areas, hangars and other necessary buildings and open spaces.

Airport approach zone. The area so designated on an officially approved Airport Plan for the use of aircraft approaching airport runway for landing purposes.

Airport approach zones. In an airport approach zone no building or structure shall be erected which is more than one (1) foot in height for each twenty (20) feet that the building or structure is in distance from a point two hundred (200) feet beyond the end of the landing strip for the outer forty thousand (40,000) foot section, for personal and secondary-type airports; one (1) foot in height for each forty (40) feet the building or structure is in distance from a point two hundred (200) feet beyond the end of the landing strip for a larger airport; provided that for an instrument runway, the ratio shall be not more than one (1) foot in height for each fifty (50) feet of distance from a point two hundred (200) feet beyond the end of the landing strip for the first ten thousand (10,000) feet and a ratio of one (1) foot for each forty (40) feet of distance for the next forty thousand (40,000) feet.

Airport building. Any building used as part of an airport.

Airport businesses and activities. Includes any person, subject to the provisions of a lease or permit, engaging in an activity which is supportive of or incidental to a airport. Such activities include, but are not limited to, aircraft storage facilities, restaurants, car rental agencies, novelty and gift shops, fresh flower booths, and any other concessions or shops which contribute to the operational and economic needs of the airport and as a convenience and service to the general public.

Airport hazard. Any structure or tree or use of land which obstructs the airspace required for the flight of aircraft in landing or taking-off of the aircraft.

Airport hazard area. Any area of land or water upon which an airport hazard might be established if not prevented as provided in the Code.

Airport or aircraft landing field. Any runway landing area or other facility designed, used, or intended to be used either publicly or privately by any person for the landing and taking off of aircraft including all necessary taxiways, aircraft storage and tiedown areas, hangars and other necessary buildings and open spaces.

Airport transition zones adjacent to approach zone. No building or structure within transition zones adjacent to approach zones shall be erected which exceeds the height limitations for an adjacent airport approach zone by more than one (1) foot for each seven (7) feet the building or structure is distant from the edge of the approach zone measured at right angles to the centerline of an approach zone.

Airport transition zones adjacent to landing strips. No building or structure within the transition zones adjacent to landing strips shall be erected more than one (1) foot in height for each seven (7) feet the building or structure is distant from an airport landing strip measured at right angles to the centerline of the landing strip.

Airport transition zones adjacent to turning zones. No building or structure within the transition zone adjacent to turning zones shall be erected which exceeds the height limitation for the airport turning zone by more than one (1) foot for each twenty (20) feet the building or structure is distant from the circumference of an airport turning zone measured radially from the circumference of the turning zone.

Airport turning zones. In any airport turning zone, no building or structure shall be erected to a height greater than one hundred fifty (150) feet.

Airport zones. Includes the following zones:

(a) APPROACH ZONE: The area leading from each end of a landing strip.

(b) TRANSITION ZONE: The area adjacent to each side of a landing strip, and the area outside of the circumference of a turning zone and adjacent thereto. Area adjacent to each side of an approach zone.

(c) TURNING ZONE. The circular area centered upon an airport reference point.

Express-type airport (continental and intercontinental). The approach zone of an Express-Type, Continental Type or Intercontinental-Type Airport shall have a width of five hundred (500) feet for a distance of two hundred (200) feet beyond the end of the landing strip, widening thereafter uniformly to a width of two thousand five hundred (2500) feet at a distance of ten thousand two hundred (10,200) feet beyond the end of the landing strip; its centerline being the continuation of the centerline of the landing strip.

Feeder-type airport. The approach zone of a Feeder-Type Airport shall have a width of three hundred (300) feet for a distance of two hundred feet beyond the end of the landing strip, widening thereafter uniformly to a width of two thousand three hundred (2300) feet at a distance of ten thousand two hundred (10,200) feet beyond the end of the landing strip; its centerline being the continuation of the centerline of the landing strip.

General aviation specialty shop operator. A person, firm or corporation subject to the provisions of a lease engaging in any one or more of the following: (1) Aircraft radio and accessories shop; (2) Aircraft instrument and accessories shop; (3) Flight school operator; (4) Aircraft and aircraft parts wholesaler; (5) Aircraft upholstery shop; (6) Aircraft maintenance shop; (7) Used aircraft sales operator; and (8) Airtaxi operator.

Hangar. A building in which aircraft are stored, serviced, or repaired.

Hangar. A building or part of a building designed or used for the shelter, storage or servicing of one or more aircraft.

Helicopter. Any rotocraft which depends principally for its support and motion in the air upon lift generated by one or more rotors that rotate on substantially vertical axes.

Helipad. An area on a heliport established for the landing or take-off of helicopters.

Heliport. A landing area solely for the use of helicopters. A heliport may include more than one helipad.

Heliport. Area of land or water or a structural surface which is used, or intended to be used, for the landing and takeoff of helicopters, and any appurtenant areas which are used, or intended for use, for heliport buildings and other heliport facilities.

Heliport. Area of land, water or structure or portion thereof or intended to be used for the landing and take-off of helicopters and having service facilities for such aircraft or providing for permanent basing of such aircraft.

Heliport. Landing field or area designed and adapted for the specific purpose of permitting helicopters to land and take off.

Helistop. Helistop is the same as a heliport, except that no refueling, maintenance, repairs or storage of helicopters is permitted.

Helistop. Area of land, water or structure or portion thereof used or intended to be used for the landing and take-off of helicopters providing no facilities for service or permanent basing of such aircraft are permitted.

Instrument runway. The approach zone to an Instrument Runway shall have a width of one thousand (1,000) feet for a distance two hundred (200) feet beyond the end of the landing strip, widening thereafter uniformly to a width of sixteen thousand (16,000) feet at a distance of fifty thousand two hundred (50,200) feet beyond the end of the landing strip; its centerline being the continuation of the centerline of the landing strip.

Landing area. Any locality, either land or water, including airports and landing fields, which is used or intended to be used for the landing and take-off of aircraft, whether or not facilities are provided for the shelter, servicing or repair of aircraft, or for receiving or discharging passengers and cargo.

Landing area. The area of the airport intended for use for the landing, taking-off, or taxiing of aircraft.

Landing threshold. A horizontal line to the runway centerline, established as the beginning of the usable landing runway.

Mean sea level. The United States Coast and Geodetic Survey zero datum plane, abbreviated MSL.

Personal-type airport. The approach zone of a personal-type airport shall have a width of two hundred (200) feet from the end of the landing strip for a distance of two hundred (200) feet, widening thereafter; uniformly to a width of two thousand two hundred (2200) feet at a distance of ten thousand one hundred (10,100) feet beyond the end of the landing strip; its centerline being the continuation of the centerline of the landing strip.

Private airplane hangar. Hangar for the storage of four (4) or less single motor planes and in which no volatile of flammable oil is handled, stored or kept other than that contained in the fuel storage tank of the plane.

Public airplane hangar. Hangar for the storage of aircraft.

Public airport. Any airport which complies with the definition contained in State Statutes, or any airport which services or offers to serve common carriers engaged in air transport.

Rotorcraft. Any aircraft deriving its principal lift or support in the air from one or more rotors or from the vertical component of the force produced by rotating airfoils.

Runway. That portion of the landing area intended for the landing and/or taking-off of aircraft.

Runway. A portion of the airport, having a surface especially developed and maintained for the landing and take-off of aircraft.

Secondary-type airport. The approach zone of a Secondary-Type Airport shall have a width of two hundred fifty feet for a distance of two hundred (200) feet beyond the end of the landing strip, widening thereafter uniformly to a width of two thousand two hundred fifty (2250) feet beyond the end of the landing strip; its centerline being the continuation of the centerline of the landing strip.

Trunk-line type airport. The approach zone of a Trunk-Line Airport shall have a width of four hundred (400) feet for a distance of two hundred (200) feet beyond the end of the landing strip, widening thereafter uniformly to a width of two thousand four hundred (2400) feet at a distance of ten thousand two hundred (10,200) feet beyond the end of the landing strip; its centerline being the continuation of the centerline of the landing strip.

Turning zones. The radius for the turning zones of personal-type and secondary-type airports shall be five thousand (5,000) feet; for feeder-type airports six thousand (6,000) feet; for trunk-line type airports seven thousand (7,000) feet for express-type airports ten thousand (10,000) feet; for intercontinental-type airports eleven thousand five hundred (11,500) feet; and for intercontinental-express airports thirteen thousand (13,000) feet.

B-3 AMUSEMENT PARKS AND AMUSEMENT DEVICES

Amusement attraction. A game of chance or skill or similar activity in which the public participates as a form of amusement.

Amusement device. The term "amusement device" shall mean a device used to convey persons in any direction as a form of amusement.

Amusement device. A mechanically operated device which is used to convey persons in any direction as a form of amusement.

Amusement device. A mechanically operated device which is used to convey persons in an unusual manner as a form of entertainment.

Amusement device. Is any manually or power operated device used to convey persons in any direction as a form of amusement.

Amusement device. A device or structure open to the public by which persons are conveyed or moved in unusual manner for diversion.

Amusement device. A device or structure open to the public, by which individuals are conveyed or moved in an unusual manner for diversion.

Amusement device. A mechanically operated device or structure, open to the public, used to convey persons in any direction as a form of amusement.

Amusement device approval. A detailed plan indicating the location of all amusement devices, and a description of the operation of each device shall be submitted to the Building department for approval before erection is started. A permit shall be issued designating the time period for the temporary operation of all of the devices which shall include roller coaster, scenic railway, water chute, ferris wheel, merry-go-round, or any other mechanical device used by persons such as for riding, sliding, swinging, turning, or the enclosure for these devices if used in the assembly.

Amusement devices. Before any mechanical amusement device, roller coaster, scenic railway, water chute, or other mechanical riding, sailing, sliding, or swinging device is erected, either in existing or new amusement parks, or places or sites where such devices are operated under carnival, fair, or similar auspices, a detailed plan shall be submitted to the Building department for review and approval.

Amusement park. A commercial amusement activity such as a carnival, circus, miniature golf course, or similar establishment which does not require an enclosed building.

Amusement park. Open or enclosed area which is used as a place of outdoor assembly, and includes all areas within the premises which are devoted to amusement devices, structures, concessions, activities, landscaping, access, and egress, which are accessory or incidental to such park.

Amusement parlor. Any place or premises wherein 10 or more coin operated amusement machines or devices are maintained for use and operation by the public.

Coin operated amusement devices. A machine or contrivance of the type commonly known and designated as bagatelle, baseball, or pin amusement game, or a similar machine or device operated, maintained or used, or to be operated, maintained or used, in any public or quasi-public place or in any building, store, or other place wherein the public are invited or wherein the public may enter, which provides no free play, no prizes, or return of any money.

Coin operated amusement machine or device. Any machine or contrivance operated by a coin, token or device of any nature whatsoever, or any machine or device which shall be automatic or semi-automatic in nature. This shall include, but not be limited to, all machines commonly known and designated as bagatelle, baseball, pin ball, or any other game or skill or table game; and shall also include all automatic or semi-automatic photographic machines, voice or music recording machines or such machines which shall reproduce photographs or motion pictures and any machine or electrical device which shall be used as a test of strength.

Coin operated amusement machine or device. Includes machines which deliver to the player any card or cards of printed material or photograph of any nature whatsoever; and shall also include all music boxes or other machines or devices coin operated, or of an automatic or semi-automatic nature which shall reproduce music.

Miscellaneous amusement business. Refers severally to any of the following businesses of keeping, conducting or operating a pool or billiard parlor, bowling alley, roller skating rink, outdoor miniature golf course, indoor miniature golf course, or place where the game of archery is conducted.

B-4 CEMETERIES AND RELATED SERVICES

Burial park. A tract of land for the burial of human remains in the ground or intended to be used, and dedicated for cemetery purposes.

Cemetery. A place for the interment of human remains. It includes, but not by way of limitation, a burial park for earth interments, a mausoleum for vault or crypt interments, a columbarium for cinerary interments, a crematory, or a combination of one or more thereof. A cemetery shall be deemed to be established or maintained or extended when the interment of any human remains is made in or upon any property, whether or not the same has been duly and regularly dedicated for cemetery purposes under the state laws, which at the effective date of the law was not included within the boundaries of a legally existing cemetery.

Cemetery. Tract of land designated by the Zoning ordinance and protected by the municipal laws, township or county regulations or the State statutes for the burial of human remains.

Cemetery. Place for the interment of human remains. It includes, but not by way of limitation, a burial park for earth interments, a mausoleum for vault or crypt interments, a columbarium for cinerary interments, a crematory, or a combination of one or more thereof. A cemetery shall be deemed to be established or maintained or extended when the interment of any human remains is made in or upon any property, whether or not the same has been duly and regularly dedicated for cemetery purposes under the laws of the State which at the date this part takes effect was not included within the boundaries of a legally existing cemetery. Any person who makes or causes to be made any interment in or upon such property, and any person having the right of possession of any such property who knowingly permits the interment of any human remains therein or thereupon, shall be deemed to have established or maintained, or extended, a cemetery within the meaning of the provisions of the ordinance.

Cemetery. Land used for the burial of the dead and dedicated for cemetery purposes, including columbariums, crematories, mausoleums and mortuaries when operated in conjunction with and within the boundary of such cemetery.

Cemetery. Any land used or intended to be used for the burial of the dead and may include columbariums, crematoriums, mausoleums and mortuaries when operated in conjunction with and within the boundary of such cemetery.

Cemetery. Land used or intended to be used for the burial of the human dead and dedicated for cemetery purposes.

Cemetery. Land used or intended to be used for the burial of the dead and dedicated for cemetery purposes, including columbariums, crematories, mausoleums and mortuaries when operated in conjunction with and within the boundary of such cemetery.

Cemetery permit. No person shall establish or maintain any cemetery, or extend the boundaries of any existing cemetery, at any location within the boundaries of the City, without a permit first having been applied for and obtained from the City Council.

Columbarium. A structure, room or other space in a building or structure containing niches for interment of cremated human remains in a place used, or intended to be used, and dedicated for cemetery purposes.

Crematory. A building or structure containing one or more furnaces for the reduction of deceased persons to cremated remains.

Crematory and columbarium. A building or structure containing both a crematory and columbarium providing functions as described in the code.

Funeral chapel. Room in an undertaking establishment where funeral services are conducted.

Human remains. The body of a deceased person and includes the body in any state of decomposition and cremated remains.

Human remains. Includes the body of a deceased person and includes also the body in any state of decomposition and the cremated remains.

Interment. The approved disposition of human remains by cremation, interment, entombment, or burial, in a place dedicated for cemetery purposes.

Interment. Disposition of human remains by cremation, interment, entombment, or burial.

Mausoleum. A structure or building for the entombment of human remains remains in crypts or vaults in a place used, or intended to be used, and dedicated for cemetery purposes.

Mortuaries. Mortuary, funeral parlor, or undertaking establishment is a building or portion thereof which is designated, arranged or let to be used or which is used for purposes connected with funerals.

Mortuary. Parts of an undertaking establishment, other than a "funeral chapel."

B-5 FENCES, SCREENING AND LANDSCAPING

Athletic back-stop fence. Fence erected for the specific purpose to act as a back-stop and the like for softball or baseball diamonds, tennis courts or other athletic grounds provided such back-stops are constructed of wire mesh or similar material regardless of whether the supports for same are of combustible or noncombustible material.

Barbed wire fence. Fence of woven wire enclosing commercial or industrial properties and may be topped with barbed wire provided the brackets supporting the barbed wire are securely fastened to fence posts and are angled toward the property and provided further that no strand of barbed wire may be closer than 7 feet to the ground. All other use of barbed wire for fencing and all electrified fencing is prohibited.

Blind fence. Fence so constructed that less than 50% of the superficial area thereof consists of regularly distributed apertures.

Bond. A bond or cash of an amount equal to $10.00 per lineal foot of required greenbelt shall be deposited with the municipality until such time as the greenbelt is planted. In the event that weather or seasonal conditions prevent transplanting, the petitioner shall be granted six months from date of issuance of certificate of occupancy to install said greenbelt or the township shall be authorized to use said funds to install said greenbelt.

Combustible blind fence. Fences enclosing athletic fields, amusement parks, airports, other open-air places of assembly and construction projects may be of height as great as 8 feet. Height of ordinary use combustible blind fences shall not exceed 6 feet unless used for protection as previously stated.

Combustible fence. Any fence not constructed entirely of noncombustible materials shall be classified as a combustible fence.

Combustible screen fence. Height of a combustible screen fence shall in no instance exceed 8 feet.

Construction and maintenance of fences. Every fence shall be so constructed and maintained as to safely resist all loads and forces to which it may within reasonable probability be subjected. Any fence subjected to lateral forces such as pressure of earth or stored materials shall be constructed to meet all applicable requirements of the ordinance or code for a retaining wall.

Fence. Independent structure forming a barrier at grade between lots, between a lot and a street or an alley, or between portions of a lot or lots. Fences shall be classified according to general form as screen fences or blind fences and according to material or their construction as combustible or noncombustible.

Fence. Barrier or partition having a combination of posts, wire, lumber, wood slats, brick, stone or a combination of these, or similar materials in which the principal dimensions are height and length and does not support or carry any loads, and is not attached to a ceiling or roof.

Fence. Includes any board, masonry, lawn or ornamental fence.

Fence. A structure, other than a building, which is a barrier and used as a boundary or means of protection or confinement.

Fence height. The vertical distance between the ground, either natural or filled, directly under the fence and the highest point thereof.

Fence or wall. A free-standing structure of metal, masonry, composition or wood, or combination thereof, including gates, resting on or partially buried in the ground, and rising above ground level, and used to delineate a boundary, or as a barrier or means of protection, confinement or screening.

Fences. Provisions of the ordinance shall not apply to fences or walls of less than 6 feet in height above the natural grade or to hedges, trees or shrubbery except as provided in the ordinance, or to terraces, steps of similar features of less than 3 feet in height above the level of the floor of the ground story.

Fences, walls, hedges. In any front yard, no fence or wall shall be permitted which materially impedes vision across such yard above the height of 30 inches, and no hedge or other vegetation shall be permitted which materially impedes vision across such yard between the heights of 30 inches and 10 feet.

Greenbelts. Greenbelt shall be a planting strip composed of deciduous or evergreen trees or a mixture of each, spaced not more than forty (40) feet apart and not less than one (1) row of shrubs, five (5) feet wide and five (5) feet or more in height after one (1) full growing season, which shall be planted and maintained in a healthy growing condition by the property owner.

Greenbelts. Prior to the commencement of construction of any structure or building in a Commercial District, Manufacturing District, Mobile Home Park District, or Travel Trailer Park District where such property abuts, adjoins, or is adjacent to a residential zone, a greenbelt shall be designated. Said greenbelt shall have a minimum width of six feet which shall be completed within six months from the date of issuance of a certificate of occupancy and shall thereafter be maintained with permanent plant materials. Specifications for spacing and plant materials are as indicated in the code. Materials to be used are merely suggestions and shall not be limiting, provided their equal in characteristics is used. All of the specified materials or their equal shall be used.

Height of fence. Vertical measurement from the topmost part thereof to the general elevation of the lowest adjacent ground or surface on either side of such fence.

Landscape. Any yard or other open space which is purposely designed to create an aesthetic environment composed of plant materials and/or other decorative elements such as fountains, ponds, sculptures, walls, fences and planters.

Landscaping. Act of providing landscaping or treatment of areas as defined in the ordinance.

Non-combustible blind fence. Height of a noncombustible blind fence shall not exceed eight (8) feet unless such fence is one for which greater height is permitted by the following: A non-combustible blind fence constructed of masonry, plain concrete or reinforced concrete may be of height as great as twelve (12) feet when such fence is employed to enclose an athletic field, amusement park, airdrome, place of open-air assembly or a yard of an institution or a commercial or industrial occupancy.

Non-combustible screen fence. Non-combustible screen fence shall not exceed ten (10) feet unless such fence is one for which greater height is permitted by the following: Non-combustible screen fence may be of height as great as fifteen (15) feet when such fence is employed to enclose a yard of an institution or a commercial or industrial occupancy.

Open fence. A fence which has over its entirety at least fifty (50) percent of its surface area in open space which affords a direct view through the fence, except that the required open space in louver-type fences may be viewed from any angle.

Prohibited material for fences. The uses of fragile, combustible material such as paper, cloth or canvas shall not constitute a part of any fence, nor shall any such material be employed as an adjunct or supplement to any fence.

Screen fence. One so constructed that at least fifty (50) percent of the superficial area thereof consists of regularly distributed apertures.

Screening. A structure erected or vegetation planted for concealing the area behind it from public view.

Screening. Structures, solid fences, or evergreen vegetation, maintained for the purpose of concealing from view the area behind such structures, solid fences or evergreen vegetation.

Solid fence. A fence, including solid entrance and exit gates, which effectively conceals from viewers in or on adjoining properties and streets, materials stored and operations conducted behind it.

Solid fence. A fence which is less than fifty (50) percent open.

Tree. Any object of natural growth.

Tree. Object of natural growth, with roots, trunk and branches.

B-6 GAS AND OIL EXPLORATIONS AND DRILLING

Controlled drilling site. That particular location within an oil drilling district in an "Urbanized Area" upon which surface operations for the drilling, deepening or operation of an oil well or any operation incident thereto, are permitted under the terms of the Code, subject to the conditions prescribed by written determination by the Administrator.

Offshore area. All property in the City which is between the high tide line and the outermost seaward City boundary.

Oil well. Any well or hole already drilled, being drilled or to be drilled into the surface of the earth which is used or intended to be used in connection with coring, or the drilling for, prospecting for, or producing petroleum, natural gas, or other hydrocarbon substances; or is used or intended to be used for the subsurface injection into the earth of oil field waste, gases, water or liquid substances; including any such existing hole, well or casing which has not been abandoned in accordance with the requirements of the Code.

Oil well class A. Any oil well drilled, conditioned, arranged, used or intended to be used for the production of petroleum natural gas or hydrocarbon substances.

Oil well class B. Any oil well drilled, conditioned, arranged, used or intended to be used only for the sub-surface injection into the earth of oil field waste, gases, water or liquid substances.

Temporary geological exploratory core hole. A seismic test hole or exploratory core hole used or intended to be used exclusively for geophysical or geological exploratory work preliminary to possible exploration for oil, natural gas or other hydrocarbons, to be drilled and abandoned within 30 days from the time of commencement of actual drilling operations.

B-7 JUNK YARDS, WRECKING AND STORAGE OF USED AND WASTE MATERIALS

Automobile wrecking. Shall mean the dismantling or wrecking of used motor vehicles or trailers, or the storage, sale or dumping of dismantled or wrecked vehicles or their parts. The presence on any lot or parcel of land of five (5) or more motor vehicles which for a period exceeding thirty (30) days have not been capable of operating under their own power, and from which parts have been or are to be removed for re-use or sale shall constitute prima facie evidence of an automobile wrecking yard.

Automobile wrecking. The dismantling or wrecking of used motor vehicles or trailers, or the storage, sale of dumping of dismantled, partially dismantled, obsolete or wrecked vehicles or their parts, but not including the incidental storage of damaged vehicles in connection with the operation of a repair garage.

Automobile wrecking. The dismantling or wrecking of used motor vehicles or trailers, or the storage, sale or dumping of dismantled or wrecked vehicles or their parts.

Auto wrecking yard. Any lot or parcel of ground where two (2) or more vehicles, not in running condition, are stored in the open; or any building or structure used principally for the wrecking and dismantling of obsolete or wrecked vehicles, or for the sale of the salvageable parts.

Auto wrecking yard. Is any land or building or other structure used primarily for the dismantling of motor vehicles for the purpose of converting such dismantled motor vehicles into scrap metal or salvageable parts.

Junk. Abandoned or dilapidated automobiles, trucks, tractors, and other such vehicles and parts thereof, abandoned or dilapidated wagons and other kinds of vehicles and parts thereof, scrap building material, scrap contractor's equipment, tanks, cans, barrels, boxes, drums, piping, bottles, glass, old iron, machinery, rags, paper, excelsior, hair, mattresses, beds or bedding or any other kind of scrap or waste material which is stored, kept, handled or displayed.

Junk business. Maintenance of a place where junk, waste, discarded or salvaged materials are bought, sold, exchanged, sorted, stored, baled, packed, disassembled, handled or abandoned; but not including pawn shops, antique shops, establishments for the sale, purchase or storage of used furniture, household equipment, clothing, used motor vehicles capable of being registered.

Junk classification. Any worn out, cast off, or discarded article or material which is ready for destruction or has been collected or stored for salvage or conversion to some use. Any article or material which, unaltered or unchanged and without further reconditioning, can be used for its original purpose as readily as when new shall not be considered junk.

Junk classification. Scrap iron, scrap tin, scrap brass, scrap copper, scrap lead or scrap zinc and all other scrap minerals and their alloy, and bones, rags, used cloth, used rubber, used rope, used tinfoil, used bottles, old or used machinery, used tools, used appliances, used fixtures, used utensils, used lumber, used boxes or crates, used pipe or pipe fittings, used automobiles or airplane tires, and other manufactured goods that are so worn, deteriorated or obsolete as to make them unusable in their existing condition, but are subject to being dismantled.

Junk classification. Abandoned or dilapidated wagons and other kinds of vehicles and parts thereof, trucks, tractors, and other such vehicles and parts thereof, scrap building material, scrap contractor's equipment, tanks, casks, cans, barrels, boxes, drums, piping, bottles, glass, old iron, machinery, rags, paper, excelsior, hair, mattresses, beds or bedding or any other kind of scrap or waste material which is stored, kept, handled or displayed.

Junk dealer. Any person who travels or intends travelling by foot or who uses or intends using or authorizing the use of any vehicle in the city for the purpose of engaging in the business of buying or selling or collecting old rope, old bottles, old iron, brass, tin, copper, lead, steel, rags, glass, paper or other metals or any discarded material of any kind, and not having a fixed place of business within the city for the aforesaid purposes.

Junk or salvage yard. A place where waste, discarded or salvage materials are bought, sold, exchanged, bailed, packed, disassembled or handled, or yards for storage of salvaged house wrecking and structural steel materials and equipment; but not including such places where such uses are conducted entirely within a completely enclosed building, and not including pawn shops and establishments for the sale, purchase, or storage of used furniture and household equipment, used cars in operable condition, or salvage material incidental to manufacturing operations.

Junk shop or junk yard. A place in which is conducted the business of purchasing, selling or storing junk, old metals, rubber, paper, rags, rope, bags or empty used bottles, in large or small quantities. This shall also include premises used for the wrecking of old motor vehicles and storing and selling of parts thereof.

Junk yard. Open area where waste, used or secondhand materials are bought and sold, exchanged, stored, baled, packed, disassembled, or handled including, but not limited to, scrap iron and other metals, paper, rags, rubber tires, and bottles. "Junk Yard" includes automobile wrecking yards and includes any area of more than two hundred (200) square feet for storage, keeping or abandonment of junk, but does not include uses established entirely within enclosed buildings.

Junk yard. Includes any lot or parcel of land on which is kept, stored, bought, or sold articles commonly known as junk, including scrap paper, scrap metal, and used automobile bodies and parts.

Junk yard. Use of a lot, or portion thereof, for the storage, keeping or abandonment of junk, dismantled automobiles, or other vehicles, or machinery, or parts thereof including scrap metals, rags, or other scrap materials.

Junk yard. Site or portion of a site on which waste, discarded, or salvaged materials are bought, sold, exchanged, stored, baled, cleaned, packed, disassembled, or handled, including used furniture and household equipment yards, house wrecking yards, used lumber yards and the likes; excepting a site on which such uses are conducted within a completely enclosed structure and excepting motor vehicle wrecking yards as defined in the Ordinance. An establishment for the sale, purchase, or storage of used cars or salvaged machinery in operable condition and the processing of used or salvaged materials as part of a manufacturing operation shall not be deemed a junk yard.

Junk yard. Premises upon which combustible or noncombustible waste is handled, stored or sold.

Junk yard. An open area where wastes, used or secondhand materials are bought, sold, exchanged, stored, processed or handled. Materials shall include, but not be limited to scrap iron and other metals, paper, rags, rubber tires and bottles. A junk yard shall include an automobile wrecking yard.

Junk yard. An open area where waste, used or second-hand materials are bought, sold, exchanged, stored, baled, parked, disassembled, or handled including but not limited to scrap iron and other metals, paper, rags, rubber tires, and bottles.

Junk yard. An open area where waste, used or second-hand materials are bought, sold, exchanged, stored, baled, parked, disassembled or handled; including, but not limited to, scrap iron and other metals, paper, rags, rubber tires and bottles. Junk yard includes a wrecking yard, but does not include uses carried on entirely within enclosed buildings.

Junk yard. An area of land, and any accessory structure thereon, which is used, except within completely enclosed buildings, primarily for buying, selling, exchanging, storing, baling, packing, disassembling or handling waste or scrap materials, including motor vehicles, machinery, and equipment not in operable condition, or parts thereof, and other metals, paper, rags, rubber tires and bottles.

Junk yard. Site or portion of a site on which waste, discarded, or salvaged materials are bought, sold, stored, baled, cleaned, packed, disassembled, or handled, not within a building. An establishment for the sale, purchase, or storage of used cars or salvaged machinery in operable condition and the processing of used or salvaged materials as part of a manufacturing operation shall not be deemed a junk yard.

Junk yard. Any land or building or other structure used for the storage, collection, processing or conversion of any worn out, cast off, or discarded metal, paper, glass or other material which is ready for destruction, or has been collected or stored for salvage or conversion to some use.

Junk yard. Open area where waste, used, or secondhand materials are bought, sold exchanged, stored, baled, packed, disassembled or handled, including but not limited to scrap iron and other metals, paper, rags, rubber tires and bottles. A "junk yard" also includes an auto wrecking yard, but does not include uses established entirely within enclosed buildings.

Junk yard. Any premise or building used for or in connection with the buying, selling, gathering, storing or shipping of old iron, rags, paper or other waste or salvage, material commonly included within the term junk or salvage, or the accumulating or wrecking of automobiles, trucks, tractors, or other motor vehicles or machinery shall be construed a junk or salvage yard.

Junk yard. Any place trading in or handling waste, discarded used or salvaged materials or articles.

Junk yard. Any area where waste, discarded or salvaged materials are bought, sold, exchanged, baled or packed, disassembled, kept, stored or handled, including house wrecking yards, used lumber yards and places or yards for storage of salvaged house wrecking and structural steel materials and equipment; but not including areas where such uses are conducted entirely within a completely enclosed building, and not including automobile, tractor or machinery wrecking and used parts yards, and the processing of used, discarded or salvaged materials as part of manufacturing operations.

Junkyard. The use of any lot, portion of a lot or tract of land for the storage, keeping, sale or abandonment of junk, including scrap metals or other scrap material; also, that which is used for the dismantling, demolition or abandonment of automobiles, other vehicles, machinery or parts thereof.

Junkyard or automobile-wrecking yard. Any space of two hundred (200) square feet or more of the area of any lot used for the business of storage, sale, keeping or abandonment of junk or waste material, including scrap metal or other scrap materials, or for the dismantling, demolition or abandonment of automobiles, other vehicles, machinery or parts thereof.

Junk yards. Use of a lot, or portion thereof, for the storage, keeping or abandonment of junk, dismantled automobiles, or other vehicles, or machinery, or parts thereof including scrap metals, rags, or other scrap materials.

Motor vehicle wrecking yard. Site or portion of a site on which the dismantling or wrecking of used vehicles, whether self-propelled or not, or the storage, sale, or dumping of dismantled or wrecked vehicles or their parts is conducted. The presence outside a fully enclosed structure of three (3) or more used motor vehicles which are not capable of operating under their own power shall constitute prima facie evidence of a motor vehicle wrecking yard.

Scrap metal recycling. Reclaiming ferrous and non-ferrous materials through cold processing.

B-8 OPEN AIR DRIVE-IN THEATERS

Aisle. Area of ground laid out in straight or curved rows, providing for the systematic and designated parking of cars.

Auditorium area. Auditorium area, where motion pictures are exhibited to the patrons.

Auditorium area. Separate area used exclusively for the actual exhibition of motion pictures to patrons seated in the interior of cars which are specially arranged and parked in this area.

Box office. Building used for the sale of admission tickets.

Bunkers. Concrete dividers which separate the ramp area into individual parking spaces.

Concession building. Enclosed structure located in the auditorium area used for the sale of food, soft drinks, candy, cigarettes, ice cream or popcorn.

Drive-in theater. Open lot or part thereof, with its appurtenant facilities, devoted primarily to the showing of moving pictures or theatrical productions, on a paid admission basis, to patrons seated in automobiles or on outdoor seats.

Drive-in theater. Theater so arranged and conducted that the customer or patron may view the performance while being seated in a motor vehicle.

Drive-in theater. Commercial enterprise devoted to the exhibition of sound motion picture films to be viewed by patrons seated in the interiors of automobiles parked in an open air auditorium.

Drive-in theater. Area, and all of the structures accessory thereto and on the premises, which is used for outdoor assembly of 100 or more persons, and into which area motor vehicles are driven and used for seating persons attending motion or televised pictures, or theatricals, religious services, or other similar activities.

Drive-in theater. An outdoor arena arranged with spaces for automobiles, whose occupants view motion pictures cast from a building housing a motion picture booth on an outdoor screen.

Drive in theatre. Open-air theatre designed for viewing by the audience from motor vehicles.

Entrance. Opening provided onto a roadway leading from the public highway into the entrance area.

Exit. Opening provided from the open air drive-in theater onto the public highway.

Exit area. Exit area where cars gather upon leaving the auditorium area preliminary to merging onto public highways.

Exit area. Separate area used exclusively for the parking or holding of cars exiting or breaking from on the auditorium area awaiting passage onto the public highway.

Hold-out area. Separate area used exclusively for the parking of cars awaiting admittance into the auditorium area of the theater.

Lighting. Entrances and exit driveways shall be adequately lighted and properly marked to avoid congestion and confusion and shall remain lighted throughout the performance and until the audience has left the area.

Lobby area. Entrance-way and Hold-out area, commonly known as the lobby area.

Occupancy rate. Established by allowing 2-1/4 persons for each vehicle accommodated, exclusive of vehicles parked in the waiting or hold-over area.

Outdoor theater. Place of outdoor assembly used for the showing of plays, operas, motion pictures and similar forms of entertainment in which the audience views the performance from self-propelled vehicles parked within the theater enclosure.

Outdoor theatre. Includes only those areas, buildings or structures designed and used for the commercial outdoor exhibit of motion pictures or live shows to passengers in parked motor vehicles.

Public toilet rooms. Located that the patrons must cross the ramp area in order to reach the toilet rooms, a suitable approach or passageway leading thereto shall be maintained. Such passageways shall be properly lighted and they shall be kept free from obstructions.

Ramp. Curved aisle of stalls with various elevations and inclines, which provides an elevated area where cars may be parked in individual stalls.

Ramps and speaker equipment. Ramps shall be spaced not less than 38 feet apart. The ramps shall be so designed that any vehicle can move from its parked position to the exit driveway without being required to back up.

Roadways. Passages lying between the ramps and aisles; also interconnecting passageways throughout the open air drive-in theater.

Running of engines. At each performance, an instructive trailer shall be shown on the screen informing the patrons of the danger of carbon monoxide poisoning when the engine is running and stating that when it becomes necessary to run the engine, the windows of the vehicle should be opened at least one inch.

Screen building. Structure on which one smooth flat surface of one side is used as the screen surface upon which the motion picture is exhibited.

Speaker posts. There shall not be less than 18 feet distance between speaker posts, measured parallel to the ramps. All electrical wiring and electrical equipment shall be installed in accordance with the provisions of the electrical Code. Each speaker post shall be wired with wire approved for underground use laid in trenches not less than 12 inches in depth.

Speaker stand. Pipe or post mounted into the concrete bunkers and connected with a carriage or junction box attached to the top of the post for use in the sound system.

Speed limit. In every outdoor theater, notices of a permanent character shall be prominently displayed designating the maximum speed limit permitted for cars driven within the area. Parking lights shall be used when cars are moving in the theater enclosure.

Surfacing of areas. All ramps, parking area, entrance and exit driveways shall be properly surfaced with a gravel surfacing or better, adequate to withstand the weight of the vehicles accommodated.

B-9 OUTDOOR ASSEMBLY

Amphitheater. Any building or outdoor arena having rising tiers of seats around a central area.

Common exit way. Exit way, such as a stairway, ramp, corridor, bridge tunnel, or similar exit way, which leads from a seating or other area in a grandstand or other place of outdoor assembly, and which affords a place for reasonably safe refuge or a safe exit way for persons who have vacated said areas and which leads to required exits from the structure or place of outdoor assembly.

Enclosed area. Area which is enclosed by a fence, wall, gates, doors, or similar barriers.

Enclosed space. Space which is enclosed by the ground or floor and walls, partitions, fencing, or railings, and a roof or similar enclosure.

Grandstand. Structure including bleachers, arranged in tiers, one above another at sides, ends or around a central space, intended primarily to support persons attending athletic events, parades, concerts, or other events.

Grandstand. Platform or other structure of one or more tiers, which is used to support persons for the purpose of outdoor assembly, except that grandstand does not include "movable seating" or "sectional benches".

Grandstand. Any structure, except movable seating and sectional benches, intended primarily to support individuals for the purposes of assembly, but, does not apply to the permanent seating in theatres, churches, auditoriums and similar buildings.

Grandstand. Unenclosed structure, constructed with other than movable seating, used for seating and supporting individuals assembled to observe a public event.

Minor place of outdoor assembly. Structure or enclosed area used for "outdoor assembly" as defined in the Ordinance and accommodating less than 200 persons.

Movable seating. Any form of seating which is not a fixed part of a structure or attached to the surface on which it rests.

Open-air assembly. Structures or enclosed areas designed or used for the assembly of persons in the open air.

Open air assembly unit. The structure, group of structures or part of a structure in or upon which persons assemble is open to the air on one or more sides for a horizontal distance equal to not less than 33-1/3% of its perimeter and for a height of not less than eight (8) feet in each seating level or story.

Open air grandstands and bleachers. Open air grandstands and bleachers shall refer to seating facilities which are located so that the side toward which the audience faces is unroofed and without an enclosing wall.

Outdoor arena. Place of outdoor assembly with an area for performances or exhibitions, such as boxing, wrestling, skating, musicals, or dramatic performances, and which area is substantially surrounded by seats arranged in tiers.

Outdoor assembly. Gathering of persons in or on a structure or in a fenced area for any of the purposes of a place of assembly; where the area of the structure used for such gathering is partly, or entirely unenclosed and is not equipped to be enclosed while occupied by such gathering; except that outdoor assembly includes such a gathering in a tent, but not in parts of a building in accordance with the Code.

Outdoor assembly shelter. One-story roofed structure designed, constructed, and used for outdoor assembly for dining, recreation, or shelter.

Outdoor places of assembly. Include grandstands, bleachers, coliseums, stadiums, drive-in theaters, tents, and similar structures, and enclosed areas, which are designed or used for outdoor gatherings of an aggregate of 200 or more persons.

Place of outdoor assembly. Premises used or intended to be used for public gathering or two hundred (200) or more individuals in other than buildings.

Place of outdoor assembly. Structure or enclosed area used for "outdoor assembly" as defined in the Code, and accommodating 200 or more persons.

Portable grandstand. Assembly of prefabricated units, which are readily erected, dismantled, and transported, and which is used as a temporary and movable grandstand.

Sectional benches. Seating benches made up for assembly in sections to accommodate, not to exceed 10 rows of seats and not to exceed 100 seat spaces per row, with the uppermost seats not more than 4 feet above the ground or floor level on which the sectional benches are supported.

Semi-enclosed place of outdoor assembly. Place of assembly which is railed, fenced, or otherwise enclosed, in a manner which prevents rapid egress to areas adequate for refuge, or to streets; or where means for escape are limited as in an indoor place of assembly.

Stadium. Area used primarily for athletic games and contests such as baseball or football and includes the grandstands and other structures, and areas accessory thereto which are used for outdoor assembly, and purposes incidental thereto.

Stadium. A structure providing seating for spectator events and which is not more than fifty (50) percent enclosed by walls.

Tent. A portable structure with walls of canvas or other like material, covered with similar materials, and supported by poles, stakes and roofs.

Tent. Shelter or structure, which is not an appendage to a building, nor a roof structure, the covering of which is wholly or partly of canvas, or other pliable material which is supported, and made stable by standards, stakes, and ropes.

Tent. A temporary shelter made principally of fabric or pliable matter.

Tent. Any structure used for living or sleeping purposes, or for sheltering a public gathering constructed wholly or in part from canvas, tarpaulin, or similar material, and shall include shelter provided for circuses, carnivals, side shows, revival meetings, camp meetings, and all similar meetings or exhibitions in temporary structures.

Tent. A pavilion or portable lodge or canvas house of skins, canvas, strong cloth, or other similar materials stretched and sustained by poles, guy wires, or ropes and used for shelter.

B-10 OUTDOOR PARKING

Accessory parking area. Open surfaced area on private property in a residential zone where permitted, immediately adjoining a commercial or industrial district, both fronting upon the same street surface to eliminate the dust nuisance, fenced, free of advertising signs, moderately illuminated, used for the parking of customers' and personnel automobiles.

Automobile parking space. Space within a public or private parking area of not less than one hundred and eighty (180) square feet (9 feet by 20 feet), exclusive of access drives or aisles, ramps, columns or office and work areas, for the storage of one (1) passenger automobile or commercial vehicle under one and one-half (1-1/2) ton capacity.

Automobile parking space. Space within a building or a private or public parking area, exclusive of driveways, ramps, columns, office and work areas, for the parking of one automobile.

Commercial parking area. Open surfaced area on private property in a commercial or industrial district other than a street or alley, used for parking of automobiles and trucks and available for public use at rental rates.

Commercial parking lot. Lot upon which cars are parked subject to renumeration.

Common parking area. Public or private parking area jointly used by two (2) or more establishments.

Customer parking area. Off-street parking designed and arranged to an approved standard established by the City Engineer, that provides parking for all standard passenger vehicles with a minimum of maneuvering and made available as an accommodation to occupants and patrons of the property.

Employee parking area. Off-street parking designed to minimum dimensions including arrangements of parking spaces which are accessible only through other spaces and made available only for parking of vehicles of the owner or employees associated with office or business use.

Loading berth. A space within the principal building or on the same lot as the principal building, providing for the standing, loading or unloading of trucks, and with access to a street or alley.

Loading space. A space accessible from a street, alley or way, in a building or in a lot, for the use of trucks while loading or unloading merchandise or materials.

Loading space. An area, other than a street or alley on the same lot with a building, or a group of buildings not less than ten feet (10′) wide, thirty-five feet (35′) long, and fourteen feet (14′) high which affords adequate ingress and egress for trucks from a public street or alley, and which is permanently reserved and maintained for the temporary parking of commercial vehicles while loading or unloading merchandise or materials.

Loading space. On the same premises with every building, structure, or thereof, erected and occupied for manufacturing, storage, warehouse, goods display, department store, wholesale store, market, hotel, hospital, mortuary, laundry, dry cleaning or other uses similarly involving the receipt or distribution of vehicles or materials, merchandise, there shall be provided and maintained on the lot adequate space for parking, loading and unloading services in order to avoid undue interference with public use of the streets or alleys. Such space shall include a ten (10) foot by twenty-five (25) foot loading space, with fourteen (14) foot height clearance for every twenty thousand (20,000) square feet or fraction thereof, of building floor use for the above mentioned purposes.

Loading space. Any off-street space available for the loading or unloading of goods.

Loading space. An off-street space or berth on the same lot with a building or contiguous to a group of buildings for the temporary parking of a commercial vehicle in order to load or unload merchandise or material, and which abuts upon a street, alley or other appropriate means of access.

Loading space. An open or enclosed space, other than a street used for the temporary parking of a commercial vehicle while its goods are being loaded or unloaded.

Loading space. A space within the main building or on the same lot, providing for the standing, loading or unloading of trucks.

Loading space. Space within a main building or on the same lot as a main building, providing for the standing, loading or unloading of trucks, having minimum area of 540 square feet, minimum width of 12 feet, a minimum depth of 35 feet, and a vertical clearance of at least 14.5 feet.

Loading space. Space, accessible from a street or way, in a building or on a lot, for the temporary use of vehicles, while loading or unloading merchandise or materials.

Location of parking areas. Off-street automobile parking facilities shall be located as specified in the Code, where a distance is specified, such distance shall be walking distance measured from the nearest point of the parking area to the nearest entrance of the building that said parking area is required to serve.

Off-street bicycle parking space. A paved and properly drained area, enclosed or unenclosed which is permanently reserved for parking one (1) bicycle, and which measures not less than two (2) feet in width and which shall provide minimum security and support structures permanently fastened to the paved area.

Off-street boat or recreational vehicle space. A paved and properly drained area enclosed or unenclosed which is permanently reserved for parking one (1) boat or recreational vehicle, and which is in the form of a rectangle measuring not less than ten (10) feet by twenty-four (24) feet and containing an area of not less than 240 square feet, exclusive of driveways and aisles.

Off-street loading berth. A space on privately owned property adequate for parking, loading and unloading service vehicles and trucks together with properly related access to a public street or alley. Required off-street loading space for three (3) or more vehicles shall have individual spaces marked and shall be designed, maintained and regulated so that no maneuvering incidental to parking, loading, or unloading shall be on or across any public way or right-of-way.

Off-street loading facilities. Site or portion of a site devoted to the loading or unloading of motor vehicles or trailers, including loading berths, aisles, access drives, and landscaped areas.

Off-street loading space. Space located on the premises for pickup and delivery at the premises. Required off-street loading space shall not be included as off-street parking space in computation of required off-street spaces.

Off-street loading space. An on-the-property-space for the standing, loading and unloading of vehicles to avoid undue interference with the public use of streets and alleys. Such space shall be not less than 12 feet in width, 14 feet in height and 45 feet in length, exclusive of access aisles and maneuvering space.

Off-street parking. The provision of space reserved exclusively for the parking of motor vehicles entirely off the public street and lying wholly within the property boundaries of the parcel of land affected.

Off-street parking. Parking spaces located in an area other than a street or public right-of-way and limited in use to vehicles not exceeding a net weight of three (3) tons and parked for periods of less than forty-eight (48) hours.

Off-street parking. Space adequate for parking standard passenger vehicles together with properly related access to a public street or alley.

Off street parking area. Space provided for vehicular parking outside the dedicated street right-of-way.

Off street parking area. Space located off any street, alley or other right-of-way which is adequate for parking an automobile with room for opening both doors on each side of the automobile and adequate maneuvering room on a parking lot with access to a public street or alley.

Off street parking area. An area of not less than one hundred sixty (160) square feet, exclusive of drives or aisles giving access thereto, located on private property, accessible from a public street or alley, and usable for the storage or parking of passenger vehicles. Such space may be open or enclosed.

Off-street parking facilities. Site or portion of a site devoted to the off-street parking of motor vehicles, including parking spaces, aisles, access drives, and landscaped areas.

Off-street parking space. Space of 200 square feet minimum for the parking of an automobile. Such space shall not be less than 10 feet in width, and in determining its dimensions access drives and aisles shall not be included. Minimum vertical clearance shall be 6-1/2 feet.

Off-street parking space. The standing storage space for one automobile, plus the necessary driveway access space. The standing storage space shall not be less than nine (9) feet by twenty (20) feet.

Off-street parking space. Paved area not in a public street or alley and having an area of not less than 180 square feet and a width of not less than 8 feet, exclusive of driveways, permanently reserved for the temporary storage of one vehicle and connected with a street or alley by a paved driveway which affords ingress and egress for a motor vehicle without requiring another motor vehicle to be moved.

Off-street parking space. Space on private land accessible from a usable street or alley, not less than nine (9) feet wide and twnety (20) feet long, with the necessary maneuvering room within the private property.

Off-street parking space. A paved and properly drained area, enclosed or unenclosed which is permanently reserved for parking one (1) motor vehicle and which is in the form of a rectangle measuring not less than nine (9) feet by twenty (20) feet, and containing an area of not less than 180 square feet, exclusive of driveways or aisles.

Open parking lot. A lot, or portion thereof, used for the storage or sale of more than four motor vehicles, but not used for the repair or servicing of such vehicles.

Parked vehicle. Any vehicle which is not in motion and which is not under control of the driver.

Parking. Temporary, transient storage of private passenger motor vehicles used for personal transportation while the operators of such vehicles are engaged in other activities. The term shall not include storage of new or used cars for sale, service, rental, or any other purpose except as specified herein.

Parking area. Open area, other than a street or alley, which contains more than four (4) parking spaces.

Parking area. Any area, other than a street, whether open, covered, or enclosed, used for the parking or storage of automobiles, boats, trucks, trailers or other wheeled vehicles, whether free or for compensation, or for accomodation of clients, customers, employees, members, visitors or for residents of structures of higher density than two (2) family.

Parking area. Open area, other than street or other public way, used for the parking of motor vehicles and available for public or private use whether for a fee or as a service or privilege for clients, customers, suppliers or residents.

Parking area. Required area set aside for the parking of motorized vehicles on the owners property.

Parking area. Lot or part thereof used for the storage or parking of motor vehicles, with or without the payment of rent or charges.

Parking area. Area of a lot used as an off-street parking facility, enclosed or unenclosed, including parking spaces and access drives and limited to the parking of vehicles used to transport students to and from schools and churches, automobiles, station wagons and pick-up trucks of no more than one ton capacitiy in residential areas except as allowed in private garages.

Parking bay. Parking space provided by means of identation of the street curb or other similar layout which would require egress from such space onto a public right-of-way by rearward movement of a vehicle.

Parking lot. A parcel of land upon which members of the general public may park their motor vehicles for the purpose of utilizing an adjacent use or facility.

Parking lot. Area or plot of land used for the storage or parking of vehicles.

Parking lot. Open area which is used for the temporary parking of motor vehicles, but is not a required off-street parking facility.

Parking lot. Premises, enclosure or other place where two or more motor vehicles are stored or parked for hire, not within a building, in a condition ready for use where rent or compensation is paid to the owner, manager or lessee of the premises for the storing, sheltering, keeping or maintaining of such motor vehicles.

Parking lot. Lot where automobiles are stored temporarily, but not including the wrecking of automobiles. Automobiles placed on a parking lot shall comply with the front yard and side yard regulations of the district in which said lot is located.

Parking lot. Open area which is used for the intermittent parking of automobiles, but is not a required off-street parking facility.

Parking lot. Parcel of land devoted to unenclosed parking spaces which may include partially enclosed one-story buildings, and where a charge is made for storage or parking of vehicles.

Parking lot. Area consisting of one or more parking spaces for the storage of automobiles, together with a driveway connecting the parking area with a street or alley and permitting ingress and egress for an automobile, provided that there shall be no storage of automobiles for the purpose of sale or resale.

Parking lot. Any outdoor area or space where more than three (3) motor vehicles may be parked or kept for a charge, fee or other consideration.

Parking lot. Impervious, open hard surfaced area used for temporary parking of motor vehicles.

Parking lot. A lot or plat of ground surfaced so as to be dust-proof and with adequate surface water drainage used for the purpose of standing, storing or parking motor vehicles, but where such motor vehicles may not be sold, offered for sale, repaired or services; and where no motor vehicles may stand, be stored or parked for a period exceeding forty-eight (48) hours.

Parking lot. An open lot or plot of ground used for the standing, storing or parking of motor vehicles.

Parking lot. An area reserved or used for parking motor vehicles, hauling trailers or boats on premises on which there is no principal building.

Parking lot. Any place, lot, parcel or yard, used in whole, or in part, for storing or parking motor vehicles, where a storage or parking fee is charged therefor and which is open to the general public.

Parking lot. An off-street open area or portion thereof solely for the parking of passenger automobiles. Such an area or portion shall be considered a parking lot whether or not on the same lot as another use, whether or not required by the Code for any building or use, and whether classified as an accessory, principal or conditional use.

Parking of vehicles. The assembling, or standing of vehicles for relatively temporary periods of time either with or without charge for such assembling and standing, but not for repair, sale or commercial storage thereof.

Parking space. A space within a public or private parking area or in a building, exclusive of driveways, ramps, columns, office and work area, for the parking use.

Parking space. Reasonably level space, available for the parking of one (1) motor vehicle, not less than ten (10) feet wide and having a total of 180 square feet to face of bumper.

Parking space. An accessible surfaced area, not less than eight (8) feet in width and twenty (20) feet in length, which can be used for parking a motor vehicle at all times.

Parking space. Permanently surfaced area, enclosed or unenclosed, having an area of not less than 180 square feet which will accommodate a car, minimum width 9 feet.

Parking space. Area, enclosed or unenclosed sufficient in size to store one automobile, together with a driveway connecting the parking space with a street or alley and permitting ingress and egress of an automobile.

Parking space. Land area of not less than one hundred eighty (180) sq. ft., exclusive of driveways and aisles, and adjacent to a driveway or aisle, with minimum dimensions of nine (9) feet by twenty (20) feet, designed so as to be usable for the parking of a private motor vehicle.

Parking space. Storage space for one motor vehicle which space is not less than ten feet (10′) by twenty feet (20′), plus the necessary access space. It shall not be located within any required yard.

Parking space. A permanently surfaced area of not less than two hundred (200) square feet, either within a structure or in the open, exclusive of driveways or access drives for the parking of motor vehicles.

Parking space. A permanently maintained dust-proof area in a building or in the open, in addition to the required front and side yards, not less than two-hundred (200) square feet exclusive of access or maneuvering area, ramps, columns, or other obstructions, kept available exclusively as a storage space for automobiles.

Parking space. Land area of not less than two hundred (200) square feet, exclusive of driveways and aisles, and adjacent to driveways and aisles, of such shape and dimensions, and so prepared as to be usable for the parking of a motor vehicle, and so located as to be readily accessible to a public street or alley.

Parking space. Permanently surfaced area within a structure or in the open, designed or used for the parking of a motor vehicle.

Parking space. Open space or a garage, on a lot, used for parking motor vehicles, the area of which is not less than two hundred (200) square feet and to which there is access from a street.

Parking space. Unobstructed space or area other than a street or alley, not less than nine feet (9′) wide and twenty feet (20′) long, provided with adequate ingress and egress, and which is permanently reserved and maintained for the parking of motor vehicles. Where more than four (4) parking spaces are grouped as a common facility, the area per parking space plus the area used for driveways shall total not less than two hundred and eighty (280) square feet per parking space, except where vehicular access to such spaces is directly from an abutting alley.

Parking space. Off-street space available for the parking of one motor vehicle and having an area of not less than 300 square feet (exclusive of access or maneuvering area, passageways and driveways appurtenant thereto) which has direct access to a public street.

Parking space. Permanent surfaced area within or outside of a building, sufficient in size to store one standard model automobile, and with an area of not less than 180 square feet, exclusive of driveways or access drives.

Parking space. A land area usable for the parking of one (1) motor vehicle, and having an area of not less than 160 square feet, exclusive of driveways and circulation areas, which are readily accessible to a public street or alley.

Parking space. Space enclosed or unenclosed sufficient in size to store one automobile, together with a driveway connecting the parking space with a street or alley and permitting ingress and egress of an automobile.

Parking space. Total space of not less than one hundred eighty (180) sq. ft., exclusive of driveways and aisles, and adjacent to a driveway or aisle, with minimum dimensions of nine (9) feet by twenty (20) feet, designed so as to be usable for the parking of a private motor vehicle.

Parking space. Impervious, hard surfaced area enclosed in the main building or in an accessory building or unenclosed, having a rectangular area of not less than one hundred sixty (160) square feet, with a minimum width of eight (8) feet when enclosed, or one hundred eighty (180) square feet with a minimum width of nine (9) feet when individually enclosed on two or more sides, exclusive of driveways, permanently reserved for the storage of one automobile, and connected with a street or alley by an impervious hard surface driveway at least eight (8) feet in width providing unobstructed ingress and egress for motor vehicles.

Parking space. Reasonably level space, available for the parking of one (1) motor vehicle, not less than ten (10) feet wide and having an area of not less than two hundred (200) square feet exclusive of passageways, or other means of circulation or access.

Parking space. Area of not less than 198 sq. ft., exclusive of access drives, or aisles, ramps, columns, or office and work areas, accessible from streets or alleys, or from private driveways or aisles leading to streets or alleys or to be usable for the storage of or parking of passenger automobiles or commercial vehicles under two ton capacity shall constitute a parking space. Parking space for trucks, tractors, and trailers of greater than two tons capacity shall be made adequate for the specific purpose.

Parking space. Space available for the parking of one motor vehicle and having an area of not less than 300 square feet (exclusive of access or maneuvering area, passageways and driveways appurtenant thereto) which has direct access to a public street.

Parking space. A permanent surfaced space within or outside of a building, sufficient in size to store one standard model automobile, and with an area of not less than 180 square feet, exclusive of driveways or access drives.

Parking space. A storage space for one motor vehicle which space is not less than ten (10) feet by twenty (20) feet, plus the necessary access space. It shall not be located within any required established yard.

Parking space. An off-street area, for the parking of a motor vehicle, of not less than eight (8) feet in width and eighteen (18) feet in length with at least seven (7) feet of vertical clearance, either within a structure or in the open, excluding driveways or access drives, but which abuts upon a street or alley or has other appropriate means of access thereto.

Parking space. A parcel of land or a stall in a garage to be used for the parking or storage of motor vehicles and not less than 160 square feet in area, exclusive of drives, lanes, or aisles, provided with unobstructed access lane thereto from a public street, alley, or other open space approved by the Building Official.

Parking space. Land area available for the parking of one motor vehicle, and having an area of not less than 160 square feet, exclusive of driveways and circulation areas, which is readily accessible to a public street or alley.

Parking space. Stall or berth which is arranged and intended for the parking of one motor vehicle in a garage or parking area.

Parking space. Space within a public or private parking area or a building, exclusive of driveways, ramps, columns, office and work area, for the temporary parking or storage of one (1) automobile.

Parking space. An accessible area used or intended for use for temporary storage of one (1) motor vehicle, hauling trailer or trailer mounted boat, which parking space may be located in a private or storage garage, a carport or in the open. Temporary storage shall be further limited to include only the storage of vehicles which are fully capable of legal operation on the public streets.

Parking space. An area, enclosed or unenclosed, reserved for the parking of one (1) motor vehicle, and which is accessible to and from a street or alley.

Parking space. A paved and imperviously surfaced area having access to a street, alley, or officially approved place for the temporary storage of one (1) motor vehicle.

Parking space. Off-street land area, nine (9) feet wide by twenty (20) feet in depth, having a total area of not less than 180 square feet, exclusive of drives or aisles giving access thereto accessible from streets or alleys or from private driveways or aisles leading to streets or alleys, and to be usable for the storage of self-propelled vehicles.

Park or parking. Means the standing of a vehicle, whether occupied or not, otherwise than temporarily for the purpose of and while actually engaged in loading or unloading merchandise or passengers.

Private parking area. Open area for the same uses as a private garage.

Private parking area. Area for parking vehicles which is not open for public use.

Private parking area. Open surfaced area surrounding buildings other than a street or alley used for the parking of automobiles of occupants of residential buildings and of personnel and citizens commuting with institutional and public buildings.

Private parking area. An open area located on the same lot with a dwelling apartment house, hotel, or apartment hotel, for the parking of automobiles of the occupants of such buildings.

Private parking lot. A lot upon which cars are parked without charge and in conformance with requirements of the Code.

Public parking area. An area, other than a street or other public way, used for the parking of automobiles and available to the public, whether for a fee, free, or as an accommodation for clients or customers, including all space needed for the movement of vehicles and people, screening or buffering space, and access drives.

Public parking area. Open or enclosed publicly owned area used for passenger automobile parking, with or without a fee.

Public parking area. Open area other than a street or alley used for the parking of automobiles and available for public use whether free, for compensation, or as an accommodation for clients or customers.

Public parking area. Any land area used or intended to be used for the parking of motor vehicles and for which a fee is charged.

Public parking area. Open area other than street, alley or other right-of-way used for the temporary parking of automobiles and available for public use whether free, for compensation or as an accommodations for clients or customers.

Public parking area. An open area, other than a street or other public way used for the parking of automobiles and available to the public whether for a fee, or free, or as accommodation for clients or customers.

Public parking area. Open usable area, other than a street or other public way, used for the parking and storing of automobiles and available to the public whether for a fee, or as an accommodation for clients or customers.

Public parking area. Open area other than street, alley or other right-of-way used for the storing and parking of automobiles and available for public use whether free, for compensation or as an accommodation for clients or customers.

Public parking area. Any land area used or intended to be used for the parking of motor vehicles and which is free or a fee is charged.

Public parking area. An open area, other than a street or a private parking area, used for the parking of more than four automobiles.

Public parking lot. An open area, other than a street, used for the parking of more than four (4) automobiles and available for public use, whether free, for compensation or as an accommodation for clients or customers.

Reservoir parking facilities. Off-street parking spaces allocated to automobiles awaiting entrance to a particular establishment.

Reservoir parking spaces. Off-street parking spaces allocated for the exclusive temporary standing of automobiles awaiting entrance to a drive-in establishment.

Semi-public parking area. Open area other than a street, alley or place, used for temporary parking of more than four (4) self-propelled vehicles and available for public use whether free, for compensation, or as an accommodation for clients or customers.

B-11 OUTDOOR SPORTS AND RECREATIONAL FACILITIES

Air-supported structure. Structure constructed of a single diaphragm lightweight flexible fabric or film, or any combination thereof, which derives its primary support and stability from inflation pressure.

Air-supported structure. A structural and mechanical system, which is constructed of high strength fabric or film and achieves its shape, stability and support by pretensioning with internal air pressure.

Baseball field. Required infield with bases 90 feet (27.45m) apart, pitcher's mound 60′–6″ from home plate. Except for the player's running path, the pitcher's mound and a small area at home plate all surfaces are usually natural grass areas. Approximately 3 acres is required for this field.

Basketball court. A hard surface finished court, 84 feet (26m) by 45′–9″ (14m).

Commercial recreation. Recreation facilities operated as a business and open to the general public for a fee.

Country Club. Membership organization formed for recreational purposes. It is normally located in a suburban or rural area and golf is most frequently the major sports activity.

Day camp. Organized recreational or educational enterprise for day use only and not overnight lodging except for owner or caretaker.

Football field. Recommended playing areas and end zones measuring 360 feet (109.80 m) by 160 feet (48.80 m). Natural grass or synthetic grass surfaces.

Golf course. Courses are designed for nine (9) or eighteen (18) hole courses. Depending on the length of the spacing of the holes the nine hole course could occupy as minimum as a 20 acre site and a 18 hole course as minimum as 30 acre site.

Hockey field. Frozen surface required by natural weather conditions or by the installation of a mechanical refrigeration system. Field is 300 feet (91.50m) by 180 feet (54.90m).

Major recreational equipment. Equipment having a dimension which exceeds eighteen (18) feet in length, seven (7) feet in width or five (5) feet in height.

Minor recreational equipment. Equipment having no dimension which exceeds eighteen (18) feet in length, seven (7) feet in width or five (5) feet in height.

Recreational ground. Any establishment operated as a commercial enterprise in which facilities are provided for all or any of the following: Camping, lodging, picknicking, boating, fishing, swimming, outdoor games and sports, and activities incidental and related to the foregoing, but not including miniature golf grounds, golf driving ranges, or any mechanical amusement device.

Recreational uses. Municipal recreational buildings, playgrounds, parks, swimming pools, municipal or private; provided, that no swimming pool shall be within twenty-five (25) feet of the lot line of any adjoining owner; athletic fields, polo fields, and golf courses.

Soccer field. Rectangular field, 110 yards (100m) by 80 yards (73m). Recommended natural grass surface.

Softball field. Required infield with bases 60 feet (18.30m) apart, pitcher's mound, 46 feet from home plate for men and 40 feet for women. Approximately one acre of land is required for this field.

Swimming pool. Structures designed to contain water to a depth in excess of twenty-four (24) inches or with a water surface area in excess of two hundred fifty (250) square feet or constructed with a nonportable water recirculation system.

Tennis club. A commercial facility for the playing of tennis at which there is a clubhouse including rest rooms.

Tennis court. The playing area and required surfaces for the engaging in the act of playing tennis by two or more persons in accordance with the established rules of the game.

Tennis court. Any temporary or permanent arrangement, grading or surfacing, together with the ordinary appurtenance thereto, intended for or used for playing tennis, when constructed or maintained in or on a lot, piece, parcel or tract of land. Only one (1) tennis court shall be considered an accessory use or a use customarily incidental to a detached single family dwelling.

Tent. A portable structure with walls of canvas or other like material, covered with similar materials, and supported by poles, stakes and assembled roofs.

Tent. Any structure, enclosure or shelter constructed of canvas or pliable material supported by any manner except by air or the contents it protects.

Track (running). Recommended track is usually the standard 400 meters distance, with eight (8) lanes. All weather tracks are finished with resilient synthetic surfaces, or a track surfaced with cinders, clay, or clay and cinders or a mixture of other natural materials.

B-12 PIERS AND WHARVES

Approach way. A structure used to gain access to a pier or wharf but not used to moor barges or vessels.

Bent. A main supporting framework consisting of a transverse row of piling with interconnecting pile cap and bracing.

Bulkhead building. A structure having a solid-fill type substructure and generally forming the land end of one or more piers.

Bulkhead wall. A retaining wall of timber, stone, concrete, steel or other material built along, or parallel to, navigable waters.

Dock. A natural open or artificial closed basin in which vessels may remain afloat when berthed at a wharf or pier.

Low water. In nontidal locations the normal low water level, in single tidal areas the mean low water level and in dual tidal areas the mean low water level as established in the tables of the National Ocean Survey.

Marina. A place located on or immediately adjacent to a body of water where boats and other watercraft are kept, moored, or stored for a consideration, and may include facilities for servicing, repairing, renting, or selling watercraft and watercraft equipment or accessories.

Marina. A place for docking or storage of pleasure boats or providing services to pleasure boats and the occupants thereof, including minor servicing and repair to boats while in the water, sale of fuel and supplies or provision of lodging, food, beverages, and entertainment as accessory uses. A yacht club shall be considered as a marina, but hotel, motel, or similar use, where docking of boats and provision of services thereto, is incidental to other activities shall not be considered a marina, nor shall boatdocks accessory to a multiple dwelling where no boat related services are rendered.

Marine installation. Includes the equipment for propulsion, power or heating on all types of marine craft and floating equipment.

Mean higher high water. Mean of the higher of the two high levels to which the tide rises daily.

Mean high water. The mean of all high tide water levels.

Mean lower low water. Mean of the lower of the two low levels to which the tide falls daily.

Mean low water. The mean of all low tide water levels.

Pier. A structure, usually of greater length than width, and projecting from the shore into navigable waters so that vessels may be moored alongside for loading and unloading or for storage.

Slip. An extension, artificial or otherwise, of a navigable water into the space between adjacent structures, within which vessels may be berthed or moored.

Substructure. That portion of the construction of a pier or wharf below the deck, but includes the deck.

Superstructure. That portion of the construction of a pier or wharf above the deck.

Tidal range. The range between mean lower low water and mean higher high water or, in places having only one tide daily, the range between mean low water and mean high water.

United States geological survey. Information and data provided by a government agency in reference to the topography of the land.

Wharf. A structure having a platform built along and parallel to navigable waters so that vessels may be moored alongside for loading and unloading, or for storage.

B-13 ROCK AND GRAVEL PITS

Borrow pit. An excavation of rock, stone, sand, soil or other mineral as it is found in its natural state as part of the earth for the purpose of disposition away from the immediate premises and which shall be in excess of two (2) feet in depth, or at least ten (10) cubic yards in volume that would be removed from one zoning lot in one year. Not included are excavations for buildings, structures, highways, streets, private roads, driveways, underground utilities, drainage improvements, or flood plain requirements which are authorized and controlled by other sections of the Ordinance, or other Ordinances of the City.

Borrow pit. Means any premises from which soil, sand, gravel, decomposed granite or rock is removed for any purpose.

Borrow pit. Any premises from which soil, sand, gravel, decomposed granite or rock is removed for sales purposes.

Completed. When all rock and gravel in commercial quantities is entirely excavated, produced and removed from a property within a district.

Conditions. All conditions prescribed for the development operation and maintenance of property within a district.

District. Any rock and gravel district established pursuant to the provision of the Code.

Excavations. Shall mean any act by which earth, sand, gravel, rock or any other similar material is cut into, dug, quarried, uncovered, removed, displaced, relocated, or bulldozed and shall include the conditions resulting therefrom.

Owner. The holder of the fee title to property in a district, whether a person, partnership, corporation or other entity recognized by law, and his or its lessees, permittees, assignees or successors in interest.

Production. The mining, quarrying, excavating, processing and stockpiling of rock and gravel within a district, but not including the removal of such materials from an area in order to prepare it for building or street construction.

Rock and gravel. Any rock, sand, gravel, aggregate or clay.

B-14 SWIMMING POOLS AND BATHING

Artificial swimming pool. A structure intended for wading, bathing, or swimming purposes, made of concrete, masonry, metal, or other impervious material, located either indoors or outdoors, and provided with a controlled water supply.

Bather. A person dressed for bathing or swimming and permitted to be in the pool area, on the pool deck or in the pool.

Bathing load. The number of persons using the pool at any given time.

Clean water. Water added to a swimming pool after treatment in the pool recirculation system.

Design load. The maximum number of persons permitted in the pool at any given time and to be determined on dividing the total square footage of the swimming pool water surface area by twenty-seven (27).

Diving area. That area on the deep side of the transition point in the swimming pool.

Diving board. A flexible board, having a non-slip surface, that is provided for diving, and is properly anchored.

Diving platform. A rigid platform having a non-slip surface, that is used for diving purposes.

Diving pool. A swimming pool intended for use exclusively by divers.

Face piping. The piping with all valves and fittings which is used to connect the filter system together as a unit.

Filter. Any material or apparatus by which water is clarified.

Filter element. That part of a filter device which retains the filter media.

Filter media. The fine material which entraps the suspended particles.

Filter sand. A type of filter media

Indoor pool. A swimming pool where the pool and the pool deck are totally enclosed within a building or structure covered by a roof.

Inlet. The fitting or opening through which filtered water enters the pool.

Lifeguard. A qualified, certified person appointed by the owner or operator to maintain surveillance during published and established time periods, over bathers or swimmers while they are in the pool area or enclosure to supervise and control behavior and safety.

Lifeguard service. The attendance, at all times that persons are permitted to engage in water contact sports, of one or more lifeguards who hold Red Cross or YMCA senior lifguard certificates or other equivalent qualifications and who have no duties to perform other than to superintend the safety of participants in water contact sports.

Main outlet. The outlet or outlets at the deep portion of the pool through which the main flow of water leaves the pool.

Main suction. The line connecting the main outlet to the pump suction.

Make-up water. Water added from an external source to the swimming pool system.

Overflow gutter. A device at the normal water level which is used as an overflow and to skim the pool surface.

Pool. Any constructed pool used for swimming, bathing or wading or as a fishpond or similar use.

Pool deck. The paved area around the pool.

Pool depth. The distance between the floor of the pool and the maximum operating water level when the pool is in use.

Potable water. Water fit for human consumption.

Private residential swimming pool. Any swimming or wading pool of over two (2) feet in depth or with a surface area exceeding two hundred and fifty (250) square feet, located on private residential property including portable or temporary type pools installed entirely above ground elevation.

Private residential swimming pool. Any swimming pool, located on private property under the control of the homeowner, the use of which is limited to swimming or bathing by members of his family or their invited guests

Private swimming pool. Any constructed pool, including portable and demountable above ground pools, which is used, or intended to be used, as a swimming pool in connection with a single family residence and available only to the family of the householder and his private guests shall be classified as a private swimming pool.

Private swimming pool. Any artificially constructed basin or other structure for the holding of water for use by the possessor, his family or guests, for swimming, diving and other aquatic sports and recreation. The term "swimming pool" does not include any plastic, canvas or rubber pool temporarily erected upon the ground holding less than one thousand (1,000) gallons of water an/or less than eighteen (18) inches in depth.

Private swimming pool. Pool established or maintained on any premises by an individual for his own or his family's use or for guests of his household.

Private swimming pool. Means a residential swimming pool located on private property under the control of the homeowner, the use of which is limited to swimming or bathing by members of his family or invited guests.

Private swimming pool. Any swimming pool located on private property under the control of the homeowner, the use of which is limited to members of his family and invited guests.

Public pool. Every swimming or wading pool, admission to which may be gained by the general public with or without the payment of a fee.

Public swimming pool. Any swimming pool, intended to be used collectively by numbers of persons for swimming or bathing.

Public swimming pool. Any swimming pool, other than a private residential swimming pool, intended to be used collectively by numbers of persons for swimming or bathing, operated by any person as defined herein, whether he be owner, lessee, operator, licensee, or concessionaire, regardless of whether a fee is charged for such use.

Public swimming pool. Swimming pool admission to which may be gained by the general public with or without the payment of a fee.

Public swimming pool. Any swimming pool, except a private swimming pool, which is intended to be used collectively by numbers of persons for swimming, diving, recreational, or therapeutic bathing. For purposes of these standards, public and semi-public pools shall be defined as listed in the following categories, based upon specific characteristics of size, usage, and other factors:

"A" Any municipal, community, school, athletic club or swimming club pool or pool for other similar usage and type.

"B" Institutional pools, such as for Girl Scouts, Boy Scouts, YMCA, YWCA, Campfire Girls, Boys' Camps, Girls' Camps, and for other similar type usage.

Recirculating piping. The piping from the pool to the filter and return to the pool, through which the water circulates.

Recirculating skimmer. A device connected with the pump suction used to skim the pool over a self-adjusting weir and return the water to the pool through the filter.

Residential swimming pool. Swimming pool installed on the same lot as a one (1) family dwelling.

Return piping. The piping which carries the filtered water from the filter to the pool.

Semi-private swimming pool. A swimming pool on the premises of, or part of, a hotel, motel, mobile home or travel trailer park, apartment house, private club, association, or similar establishment, where admission to the use of the pool is included in the fee or consideration paid or given for the general use of the premises.

Semi-public pool. A swimming or wading pool on the premises of, or used in connection with a hotel, motel, trailer court, apartment house, country club, youth club, school, camp, or similar establishment where the primary purpose of the establishment is the operation of the swimming facilities.

Shallow area. That area on the shallow side of the transition point in the swimming pool.

Special use pools. Pools designed and used primarily for a single purpose, such as wading, diving, or instructions.

Swimming pool. Any structure, basin, chamber or tank containing an artificial body of water for swimming, diving or recreational bathing and having a water depth of thirty (30) inches or more at any point.

Swimming pool. A contained body of water used for bathing or swimming purposes either above or below ground level with the container being eighteen (18) or more inches in depth and/or wider than eight feet (8) at any point measured on the long axis.

Swimming pool. Any temporary or permanent artificial pool or receptacle for water, installed, constructed or maintained in, on, or above ground, having a perimeter of more than twenty-five (25) feet and a depth of more than two (2) feet at any point.

Swimming pool. Any artificial structure, basin, chamber, or tank containing a body of water for the primary purpose of swimming, diving, recreational, or therapeutic bathing.

Swimming pool. Any structure, basin, chamber, or tank containing an artificial body of water for swimming, diving, or recreational bathing and having a depth of two (2) feet or more at any point.

Swimming pool. Any structure, basin, chamber, or tank containing an artificial body of water for swimming, diving, relaxation, or recreational bathing.

Swimming pool. Shall mean an artificial basin of water constructed primarily for the purpose of public swimming and auxiliary structures, including dressing and locker rooms, toilets, showers and other areas and enclosures that are intended for the use of persons using the pool, but shall not include pools and auxiliary structures and equipment at private residences intended only for the use of the owner and friends.

Swimming pool or pool. A structure of concrete, masonry, or other approved material and finish, located either indoors or outdoors, used or designed to be used for public bathing or swimming purpose by humans, and filled with a controlled water supply, together with buildings, appurtenances and equipment used in connection therewith.

Swimming pools. Any structure, portable or permanent, containing a body of water eighteen (18) inches or more in depth, intended for recreational purposes, including a wading pool, but not including an ornamental reflecting pool or fish pond or similar type pool, located and designed so as not to create a hazard or to be used for swimming or wading.

Training pool. A swimming pool not normally in excess of 3′-0″ deep at its point of maximum depth and usually reserved for use by persons learning to swim.

Transition point. That place in the floor of the pool between water depths of 4′-6″ and 5′-6″ where an abrupt change in slope occurs.

Turnover. The time required to recirculate the volume of water the pool contains through the filtration system and back to the pool.

Underdrain. An appurtenance at the bottom of the filter to assure equal distribution of water through the filter media.

Vacuum fitting. The fitting in the wall of the pool which is used as a convenient outlet for connecting the underwater suction cleaning equipment.

Vacuum piping. The piping which connects the vacuum fitting to the pump section.

Wading pool. Shall normally be a small pool for non-swimming children only for wading and shall have a maximum depth at the deepest part not greater than 18″.

Wading pool. Any pool used or designed to be used exclusively for wading or bathing and having a maximum depth of 24 inches.

Wading pool. A pool of water in a basin having a maximum depth of less than twenty-four (24) inches intended chiefly as a wading place for children.

B-15 VEHICULAR AND PEDESTRIAN TRAFFIC ACCESS FACILITIES

Adjacent. To lie near or close to; in the neighborhood or vicinity of.

Adjoining. Touching or contiguous, as distinguished from "adjacent."

Alley. A right of way with a width not exceeding twenty-four (24) feet which affords a secondary means of access to abutting property.

Alley. A public thoroughfare not more than twenty (20) feet wide.

Alley. Any public space, public park or thoroughfare twenty (20) feet or less, but not less than ten (10) feet, in width, which has been dedicated or deeded to public use.

Alley. A public right-of-way primarily designed to serve as secondary access to the side or rear of those properties whose principal frontage is on a street.

Alley. A public thoroughfare not over twenty (20) feet wide which affords only a secondary means of access to abutting property.

Alley. An officially designated public or private thoroughfare which affords a means of vehicular or utility access to the rear of abutting property.

Alley. A public right-of-way, other than a street, with a width of not less than ten (10) feet nor more than twenty (20) feet, which affords a secondary means of vehicular access to adjoining and adjacent properties.

Alley. Is any public space or thoroughfare 20 feet or less in width, but not less than 10 feet in width, which has been dedicated or deeded to the public for public travel and affords access to abutting property.

Alley. Any public thoroughfare less than 21 feet in width which has been legally dedicated or devoted to public use.

Alley. A narrow service way providing a secondary public means of access to abutting properties.

Alley. A street or thoroughfare less than 21 feet wide and affording only secondary access to abutting property.

Alley. A public or private way permanently reserved as a secondary means of access to abutting property is an alley to that property.

Alley. A permanent service right-of-way providing a secondary means of access to abutting properties.

Alley. Any public space or thoroughfare not less than ten feet or more than twenty feet in width which has been dedicated or deeded for public use.

Alley. A way which affords a secondary means of access to abutting property.

Alley. A public thoroughfare less than 30 feet in width which affords only secondary means of access to abutting property.

Alley. A public thoroughfare having a right-of-way not less than twenty (20) feet and serving the rear of properties.

Alley. A public thoroughfare which affords only a secondary means of access to abutting property and not intended for general traffic circulation.

Alley. Any public or private thoroughfare for the use of pedestrians or vehicles, not less than ten (10) feet nor more than thirty (30) feet in width, and which affords only a secondary means of access to abutting properties.

Alley. A minor right-of-way dedicated to public use, which gives a secondary means of vehicular access to the back or side of properties otherwise abutting a street, and which may be used for public utility purposes.

Alley. Any right-of-way dedicated to vehicular travel, being twenty (20) feet or more but less than fifty (50) feet in width.

Alley. Any right-of-way dedicated to set aside for vehicular traffic, entering or going through a city block. After the enactment of the Ordinance no alley shall be less than sixteen (16′) in width.

Alley. A minor right-of-way providing secondary vehicular access to side or rear of two or more properties.

Alley. A public way permanently reserved for vehicular service access to the rear or side of properties otherwise abutting on a street.

Alley. A public space or thoroughfare which is less than thirty (30) feet in width dedicated to public use.

Alley. A public way which affords only secondary access to abutting property.

Alley. A public way, other than a street, twenty (20) feet or less in width affording secondary means of access to abutting property.

Alley. A way affording a secondary means of access to property abutting thereon.

Alley. A public way which affords a secondary means of access to abutting property, twenty (20) feet or less in width.

Alley. Any dedicated public way providing a secondary means of ingress or egress to land or structures thereon, less than twenty four (24) feet in width.

Alley. Any dedicated public way affording a secondary means of vehicular access to abutting property, and not intended for general traffic circulation.

Alley. A narrow service way dedicated to public use providing a secondary public means of access to abutting properties and not intended for general traffic circulation.

Alley. A dedicated public way other than a street which provides only a secondary means of access to abutting property.

Alley. A dedicated right-of-way other than a street.

Alley. A main public thoroughfare other than a dedicated half street twenty (20) feet in width, permanently reserved as a secondary means of access at abutting properties.

Alley. A roadway which affords only a secondary means of access to abutting property and not intended for general traffic circulation.

Alley. Any roadway or public way dedicated to public use and twenty (20) feet or less in width.

Alley. A recorded roadway which affords only secondary means of access to abutting property and which is not intended for general traffic circulation.

Alley. A public passageway, 8′ to 25′ wide, affording a secondary means of access to abutting property.

Alley. A public passageway affording a secondary means of access to abutting property.

Alley. Any legally established public thoroughfare less than 30 feet in width but not less than 10 feet in width whether designated by name or number.

Alley. A public or legally established private thoroughfare, other than a street, which affords a secondary means of vehicular access to abutting property.

Alley. An area included between the right of way lines of a public thoroughfare 20 feet or less in width.

Alley. Any public space, public park or thoroughfare less than sixteen (16) feet but not less than ten (10) feet in width which has been dedicated or deeded to the public for public use.

Alley. A permanent public service way which affords only a secondary means of access to abutting property.

Alley. A public highway which is a narrow way, less in size than a street, and which is not designed for general travel, which is used primarily as a means of access to the rear of residences and business establishments and which, generally, affords only a secondary means of access to the property abutting along its length.

Alley. A narrow supplementary thoroughfare deeded and dedicated to the public for the public use of vehicles and pedestrians.

Alley. A public way, other than a street or highway, providing a means of vehicular access to abutting property.

Alley. A right-of-way less than thirty (30) feet in width permanently dedicated to common and general use by the public.

Alley. A public or private way not more than twenty five (25) feet wide, which affords only secondary access to abutting property.

Alley. Narrow supplementary thoroughfare for the public use of vehicles or pedestrians, affording access to abutting property.

Alley. Any public space or thoroughfare twenty (20) feet or less in width which has been dedicated or deeded for public use.

Alley. A secondary thoroughfare less than thirty (30) feet in width dedicated for the public use of vehicles and pedestrians affording access to abutting property.

Alley. Any public thoroughfare less than thirty (30) feet in width.

Alley. Any public space or thoroughfare less than sixteen feet (16′) but not less than ten feet (10′) in width which has been dedicated or deeded to the public for public use.

Alley. A public right-of-way designed for vehicular traffic and providing a secondary means of access only to the property which abut it. From and after the adoption of the Zoning Code, no alley shall be established less than sixteen (16) feet in width if serving residential lots, or less than twenty (20) feet in width if serving Commercial or Industrial lots.

Alley line. The center line of an alley right-of-way as determined by the city plats.

Alley or service drive. Minor right-of-way, privately or publicly owned, primarily for service access to the back or sides of property.

Arterial street. A street used for fast moving traffic, located between developed areas in the city.

Arterial street. A fast or heavy traffic street of considerable continuity and used primarily as a main traffic artery.

Arterial streets. A vehicular traffic facility used for heavy traffic flow.

Arterial streets. Are those streets which carry or are intended to carry a heavy volume of traffic.

Block. An area bounded by streets.

Block. A parcel of land bounded by streets or by streets and a natural or artificial barrier.

Business street. Street which services or is designed to serve as an access to abutting business properties.

Cartway. The portion of a street for vehicular use.

Cartway. Graded portion of a street or alley, including travelway or shoulders.

Cartway or roadway. Portion of a street or alley intended for vehicular use.

Center lane. Line connecting the points on highways from which setback distances shall be measured, at any point on the highway.

Clear sight triangle. An area of obstructed vision at street intersections defined by lines of sight between points at a given distance from the intersection of a street center line.

Clear sight triangle. Area of unobstructed vision at street intersections defined by the right-of-way lines of the streets and by a line of sight between points on their right-a-way lines at a given distance from the intersection of the right-of-way.

Collector street. Street supplementary to and connecting the major street system to local streets, designated as a minor street.

Collector street. Street which, in addition to giving access to abutting properties, intercept minor streets, provide routes to community facilities and to major traffic streets and serve or anticipated to serve 1,000 to 4,000 vehicles per day.

Collector street. A street supplementary to and connecting the major street system to the local streets.

Collector street. A street (including the principal access streets of a subdivision) which carries traffic from local streets either directly or via other existing or proposed collector streets to a major or secondary highway.

Collector street. A Minor street as designated on the Major Street Plan is a street which collects traffic from other Minor streets and serves as the most direct route to a Major street or a community facility.

Collector streets. Are those streets which carry traffic from minor streets to the major system of arterial streets and highways, including the principal entrance streets of a residential or industrial development and streets for circulation within such a development.

Commercial entrance. An entrance onto a highway from a commercial establishment of any kind or an entrance to an apartment house or multi-family dwelling serving more than four (4) families and includes an entrance to any property used wholly or in part as other than a private residence or farm.

Controlled-access highway. Every highway, street, or roadway in respect to which owners or occupants of abutting lands and other persons have no legal right of access to or from the same except at such points only and in such manner as may be determined by the public authority having jurisdiction over such highway, street, or roadway.

Controlled intersection. Any intersection at which traffic control signals are in operation or a traffic officer is directing traffic.

Control of access. Condition where the right of owners or occupants of abutting land or other persons to access, light, air, or view in connection with a State Highway is fully or partially controlled by the Department of Transportation.

Corner clearance. This is the distance from a projection of right-of-way lines perpendicular or radial to the curbline, thence along the curbline to the nearest edge of proposed curbline opening.

Cross walk. That part of a roadway at an intersection included within the connections of the lateral lines of the sidewalks on opposite sides of the highway measured from the curbs or, in the absence of curbs, from the edges of the traversable roadway. Any portion of a roadway at an intersection or elsewhere distinctly indicated for pedestrian crossing by lines or other markings on the surface.

Cul-de-sac. A street, one end of which connects with another street and the other end of which is a dead end but which allows for turning of vehicles.

Cul-de-sac. A street intersecting another street at one end and terminating at the other in a vehicular turnaround.

Cul-de-sac. A minor street intersecting another street at one end and terminating at the other by a vehicular turnaround.

Cul-de-sac. Dead ended street designed as such, offered for dedication, preferably shall not exceed 600 feet in length.

Cul-de-sac. Street having a closed end turn-around with a right-of-way having a minimum outside radius of not less than 50 feet and shall be paved to a radius of not less than 40 feet.

Cul-de-sac street. A Minor street having one end open to vehicular traffic and having one closed end terminated by a turnaround.

Cul-de-sac street. A minor street having but one vehicular access to another street and terminated by a paved vehicular turn-around.

Curb. Substantially vertical member along the edge of a street to form part of a gutter.

Curb. Finished shaped member at edge of paving usually of concrete, granite or stone.

Curb break. Any interruption, or break, in the line of a street curb in order to connect a driveway to a street, or otherwise to provide vehicular access to abutting property.

Curbline. Line which is the outer edge of the shoulder. It is also the gutter line.

Curbline opening. Overall opening dimension at the curbline, whether curbing exists or not, measured from the extreme outer edges of the radii.

Curb loading zone. A space adjacent to a curb reserved for the exclusive use of vehicles during the loading or unloading of passengers or materials.

Dead end street. Street or portion of a street with only one vehicular outlet but which has a turn-a-round and which is designed to provide ample maneuvering area.

Dedication. Appropriation of land by an owner for any general and public use.

Distance between driveways. Distance is measured along the curbline between the tangent projections of the inside edges of 2 adjacent driveways to the same frontage.

Driveway. An open space or private thoroughfare, other than a street or alley, providing vehicular access to one zoning lot.

Driveway. Private road, the use of which is limited to persons residing, employed, or otherwise using or visiting the property on which it is located.

Driveway. Private way for the use of vehicles.

Driveway. A minor way providing vehicular access into a private property or public way providing vehicular access to house groups containing more than eight (8) houses.

Driveway angle. The angle of 90 degrees or less between the driveway centerline and curbline.

Driveway width. Narrowest width of driveway, within the sidewalk area measured with the curbline.

Easement. The right of a person, utility or public agency to use common land or private land owned by another for a specific purpose.

Edge clearance. Distance measured along the curbline from the lateral property line extended to the beginning of driveway.

Expressway. Multi-lane divided highway for through traffic with full or partial control of access and with grade separations at some intersections and at all major railroad crossings.

Expressway. A highway or any portion thereof which has been laid out by planned delineation of outer limits of right-of-way to which the owners of abutting lands have no right or easement of access to or from their abutting lands or in respect to which such owners have or may have only restricted or limited right or easement of access, and includes all accessory highways and interchanges connected with and leading to or away from such expressway.

Feeder road. Street or road intersecting with a limited access highway and having traffic interchange facilities with such limited access highway.

Freeway. Divided arterial highway for through traffic to which access from the abutting properties is prohibited and all street crossings are made by grade separated intersections.

Freeway. A highway in respect to which the owners of abutting lands have no right or easement of access to or from their abutting lands or in respect to which such owners have only limited or restricted right or easement of access, and which is declared to be such in compliance with the Streets and Highways Code of the State.

Freeway. A highway, in respect to which the owners of abutting lands have no right or easement of access to or from their abutting lands or in respect to such owners have only limited or restricted right or easement of access, the precise route for which has been determined and designated as a freeway by an authorized agency of the State or a political subdivision thereof. The term shall include the main traveled portion of the trafficway and all ramps and appurtenant land and structures.

Freight curb loading zone. A space adjacent to a curb for the exclusive use of vehicles during the loading or unloading of freight or passengers.

Frontage. All property fronting on one side of a street between intersecting or intercepting streets, or between a street and right-of-way, waterway, end of dead-end street, or city boundary measured along the street line. An intercepting street shall determine only the boundary of the frontage on the side of the street which it intercepts.

Frontage or service street. A Minor street auxiliary to and located on the side of a Major street for service to abutting properties and adjacent areas and for control of access.

Frontage road. A street lying adjacent and approximately parallel to and separated from a freeway, and which affords access to abutting property.

Frontage space. A street; or an open space outside of a building, not less than 30 feet in any dimension, that is accessible from a street by a driveway, lane, or alley at least 20 feet in width, and that is permanently maintained free of all obstructions that might interfere with its use by the fire department.

Future right-of-way. The right-of-way width required for expansion of existing streets to accomodate anticipated future traffic loads; or a right-of-way established to provide future access to or through undeveloped land.

Future street or alley. Any real property which the owner thereof has offered for dedication to the City for street or alley purposes, but which has been rejected by the City Council subject to the right of said Council to rescind its action and accept by resolution at any later date and without further action by the owner, all or part of said property as a public street or alley.

Future width lines. Means lines established adjacent to highways or streets for the purpose of defining limits within which no structure nor any part thereof shall be erected or maintained in order to insure the future acquisition of these limits as public rights-of-way.

Half or partial street. Street, generally parallel and adjacent to a property line, having a lessor right-of-way width than normally required for satisfactory improvement and use of the street.

Highway. Any public thoroughfare, public street, or public road.

Highway—major. Any street, existing or proposed, designated on the Master Plan adopted after public hearing by the City Planning Commission as a Major Highway.

Highway—major. Any street designated as a major highway on the "Master Plan of Highways and Freeways."

Highway or street. The entire width between the boundary lines of every way publicly maintained when any part thereof is open to the use of the public for purposes of vehicular travel. The words "highway" and "street" are synonymous herein.

Highway—secondary. Any street designated as a secondary highway on the "Master Plan of Highways and Freeways."

Highway—secondary. Any street, existing or proposed, designated on the Master Plan adopted after public hearing by the City Planning Commission as a secondary highway.

Industrial street. A street designed and constructed to serve both truck and bus movements within an industrial area. Abutting property will have free access. On-street parking and loading is prohibited.

Inside radius. Inside smaller curve radius on edge of driveway.

Intersecting street. Any street which adjoins another street at an angle whether or not it crosses the other.

Intersection. The area embraced within the prolongation or connection of the lateral curb lines, or, if none, then the lateral boundary lines of the roadway of two (2) highways which join one another at, or approximately at, right angles, or the area within which vehicles traveling upon different highways joining at any other angle may come in conflict. Where a highway includes two (2) roadways thirty (30) feet or more apart, then every crossing of each roadway of such divided highway by an intersecting highway shall be regarded as a separate intersection. In the event such intersecting highway also includes two (2) roadways thirty (30) feet or more apart, then every crossing of two (2) road-ways of such highways shall be regarded as a separate intersection.

Landscaped freeway. Any part of a freeway that is now or hereafter classified by the State or a political subdivision thereof as a landscaped freeway, as defined in the Outdoor advertising Act; any part of a freeway that is not so designated shall be deemed a non-landscaped freeway.

Limited access highway. A trafficway, including toll roads, for through traffic, in respect to which owners or occupants of abutting property or lands and other persons have no legal right of access to or from same, except at such points only and in such manner as may be determined by the public authority having jurisdiction over such trafficway.

Local street. A street primarily for access to abutting residential properties and to serve local needs.

Local street. Any street other than a collector street, major or secondary highway, or freeway, providing access to abutting property and serving local as distinguished from through traffic.

Loop street. A street used primarily for access to interior lots in a block, beginning and terminating at different points on the same abutting street.

Main street. Street upon which the majority of lots within a block frontage are fronted; or any street officially so designated; or the commercial or business street with the most traffic.

Major arterial street. A public street which is primarily for moving fast or heavy traffic between large or intensively developed districts.

Major highway. A road used primarily for traffic between distant populated areas.

Major highway. Any surface thoroughfare existing or proposed, as shown on the Master Plan of Highways of the Commission or the "Official Highway Plan" of the City, having a width of 100 feet or more in non-hillside areas, and 60 feet or more in hillside or mountain areas, including primary and divided highways.

Major highway. A street used primarily for traffic not local in destination.

Major street. Street serving large volumes of comparatively high-speed traffic, and officially designated as such on the Major street plan of the City.

Major street. A street or highway shown as a major street upon the street plan of the City.

Major street. A street designated as such on the officially adopted Major Street Plan of the City.

Major street. A street or highway shown as a major street upon the Major Thoroughfare Plan of the City.

Major street. An arterial street which is designated on the Major Street Plan or Comprehensive Plan.

Major street. A street which carries traffic, generally local, to or from the system of major highways or which serves as main circulation for a large area.

Major thorofare. Arterial street which is intended to serve a large volume of traffic for the immediate area and the region beyond, and may be designated as a major thorofare, parkway, freeway, expressway, or equivalent term to identify those streets comprising the basic structure of the street plan. Any street with a width, existing or proposed, of 120 feet or greater shall be considered a major thorofare.

Major thoroughfare plan. The plan which shows the general location and extent of existing and planned streets and other transportation facilities for the City, duly adopted and officially accepted separately or as a part of the General Plan.

Major traffic streets. Are those serving large volumes or comparatively high-speed and long distance traffic and include facilities classified as main and secondary highways by the State Highway Department.

Marginal access road. Service road parallel to a feeder road; and which provides access to abutting properties and protection from through traffic.

Marginal access street. A local street providing access to lots which abut or are adjacent to a limited-access highway or major street.

Marginal access streets. Local streets providing access to lots which abut or are adjacent to a limited-access highway or major street.

Marginal access streets. Minor streets, parallel and adjacent to major traffic streets, providing access to abutting properties and control of intersections with the major traffic street.

Marginal streets. Are those streets which are parallel to and adjacent to arterial streets and highways, and which provide access to abutting properties and protection from through traffic.

Median. That portion of a divided highway separating the traveled ways of traffic proceeding in opposite directions.

Minor street. Any street not classified as a Major street on the Major Street Plan whose primary purpose is to provide access to adjacent properties.

Minor street. Any street not designated as a major or collector street and intended to serve or provide access exclusively to the properties abutting thereon.

Minor street. A street used primarily for access to abutting properties, or in some cases a connecting street between subdivisions not adjoining.

Minor streets. Are those streets which are used primarily for access to the abutting properties.

Minor streets. A street or highway not shown as a major street upon the street plan of the city.

Minor streets. Streets which are used primarily for access to abutting properties and serve, or are anticipated to serve less than 1,000 vehicles per day.

Minor streets. Those used primarily to provide access to abutting properties.

No-access strip. A strip of land within and along a rear lot line of a double-frontage lot adjoining a street which is designated on a recorded subdivision plat or a deed of conveyance as land over which motor vehicular travel shall not be permitted.

Official traffic-control devices. All signs, barricades, signals, markings, and devices not inconsistant with the Ordinance placed or erected by authority of a public body or official having jurisdiction, for the purpose of regulating, warning, or guiding traffic.

Outside radius. Outside or larger curve radius on edge of driveway.

Passenger curb loading zone. A place adjacent to a curb reserved for the exclusive use of vehicles during the loading or unloading of passengers.

Paved area. An area having a surface of masonry, concrete, or asphalt; except that for driveways and off-street parking spaces may include crushed rock with permanent curbing in referenced districts.

Paved area. An area which has been drained, graded, compacted, provided with adequate base, and surfaced with asphaltic concrete at least two (2) inches thick, or equivalent, so as to provide sufficient durable surface to render the area usable for the purpose specified under normal weather conditions.

Pedestrian. Any person afoot.

Pedestrian way. A public or private right-of-way solely for pedestrian circulation.

Permanent open space. Street, alley, permanent surface and air easement, waterway, public park, or railroad right-of-way, other than a siding for the loading, unloading, or storage of cars or motive power equipment.

Place. An open, unoccupied space other than a public street or alley permanently reserved to permit a means of access to abutting property. An officially approved place is one which was of record at the time of the adoption of the Ordinance or one that has since been approved by action of the Board.

Place. An open, unoccupied public or private space other than a street or alley, permanently reserved for purposes of joint access to abutting property.

Place. Open, unoccupied space other than a street or alley permanently reserved as the principal means of access to abutting property.

Plaza. Open area at ground level accessible to the public at all times, and which is unobstructed from its lowest level to the sky. Any portion of a plaza occupied by landscaping, statuary, pools, and open recreation facilities shall be considered to be a part of the plaza for the purpose of computing a floor area premium credit. The term "plaza" shall not include off-street loading areas, driveways, off-street parking areas or pedestrian ways accessory thereto.

Primary residential street. A street which serves the prime function of collecting or distributing intracommunity residential traffic.

Private driveways. Where provided, shall be located not less than forty (40) feet from the intersection corner of corner lots and shall provide access to the street of lower classification when a corner lot is bounded by streets of two (2) different classifications.

Private road. A private thoroughfare other than a public street or alley, permanently reserved in order to provide a means of access to more than one zoning lot.

Private road easement. A parcel of land not dedicated as a public street, over which a private easement for road purposes is proposed to be or has been granted to the owners of property continuous or adjacent thereto which intersects or connects with a public street, or a private street; in each instance the instrument creating such easement shall be or shall have been duly recorded or filed in the Office of the County Recorder.

Private road or driveway. Every way or place in private ownership and used for vehicular travel by the owner and those having express or implied permission from the owner, but not by other persons.

Private street. The area lying within the described limits of an easement or right-of-way, created by virtue of a recorded or registered instrument, for ingress and egress by one person, or any number of persons less than the public at large, over the land of another.

Private street. A street held in private ownership.

Private street. A parcel of land not dedicated as a public street over which a private easement for road purposes has been granted to the owners of property contiguous or adjacent thereto which intersects or connects with a public street, and the instrument creating same has been duly recorded or filed in the office of the Recorder and the County, and which has been determined by the Director to be adequate for access and for the purposes defined in the Code.

Private street. A private road easement as defined herein which has been determined by the Advisory Agency or the Director of Planning to be adequate for access and for the purposes set forth in the Code.

Public easement. A space other than a street or alley permanently reserved as a principal means of access to abutting property and as location for public utilities.

Public entrance. An entrance used by the public which entrance opens onto a highway from a public road, street, highway or other thoroughfare that is maintained by a municipality, or other public authority. A street entrance to a highway from a recorded subdivision is a public entrance.

Public place. As used in the Code, an unoccupied open space adjoining a building and on the same property, that is permanently maintained accessible to the Fire Department and free of all incumbrances that might interfere with its use by the Fire Department.

Public place. An unoccupied open space adjoining a building and on the same property, that is permanently maintained accessible to the Fire Department and free of all incumberances that might interfere with its use by the Fire Department.

Public place. A thoroughfare or open space over 21 feet wide which is dedicated to a governmental body maintaining accessibility to the Fire Department and other public services.

Public place. Unoccupied, open space outside of a building devoted to public use and permanently maintained accessible to the Fire Department and free of all encumbrances that might interfere with its use by the Fire Department. Where such public places are required, a notation to that effect shall be shown on the certificate of occupancy for the building.

Public street. The term "Public Street" means the land dedicated, accepted or condemned for use as a highway for the benefit of the public at large or established as such highway by the right of prescription or common user, however, any such right-of-way as may be created after the adoption of the Zoning Code shall not be considered to be a public street unless it is sixty feet (60′) or more in width except where streets are entirely local when the width of fifty feet (50′) may be declared to be a "Public Street" if approved by the Commission.

Public street. Any thoroughfare or public way not less than 60 feet in width, which has been dedicated to the public or deeded to the city for street purposes.

Public street. Any thoroughfare or public way not less than thirty (30) feet in width, which has been dedicated to the public or deeded to the City for street purposes; and also any such public way as may be created after enactment of the Ordinance, provided it is forty (40) feet or more in width.

Public street. Land dedicated and accepted or condemned for use as a highway for the benefit of the public at large or established as such highway by the right of description or common user; however, any such right-of-way as may be created after the enactment of the Ordinance shall not be considered to be a public street unless it is sixty (60) feet or more in width, except where streets are entirely local, when the width of fifty (50) feet may be declared to be a "Public Street" if approved by the City Zoning Commission.

Public thoroughfare. Any legally established street or alley as defined in the Code.

Public walkway. Any space designed or maintained solely for pedestrian use, without regard to ownership.

Public way. Any parcel of land more than 12 feet wide and appropriated to the free passage of the general public.

Public way. Any parcel of land unobstructed from the ground to the sky, more than 10 feet in width, appropriated to the free passage of the general public.

Public way. Public street, alley, sidewalk or park.

Public way. A parcel of land unobstructed from the ground to the sky, appropriated to the free passage of the general public.

Public way. Any street, alley or other parcel of land essentially open to the outside air, deeded, dedicated, or otherwise permanently appropriated to the public for public use having a clear width of not less than 10 feet.

Public way. A thoroughfare over 21 feet wide on privately owned, privately maintained property but designated for public use and which by agreement is kept accessible at all times to the fire department and other public services.

Public way. Any street, channel, viaduct, subway, tunnel, bridge, easement, right-of-way or other way in which a public agency has a right of use.

Public way. Any parcel of land unobstructed from the ground to the sky, more than twelve (12) feet in width appropriated to the free passage of the general public.

Public way. Any sidewalk, street, alley, highway, or other public thoroughfare.

Public way. Any street, alley or other similar parcel of land essentially open to the outside air, deeded, dedicated, or otherwise permanently appropriated to the public for public use and having a clear width and height of not less than 10 ft.

Public way. Any street, alley or other parcel of land open to the outside air, deeded, dedicated or otherwise permanently appropriated to the public for public use and having a clear width and height of not less than 10 feet.

Railroad right-of-way. Strip of land on which railroad tracks, switching equipment, and signals are located, but not including lands on which stations, offices, storage buildings, spur tracks, sidings, yards, or other uses are located.

Residence street. Is that portion of a street between its intersections with two other streets where the majority of the frontage is within a Residence District.

Residential entrance. An entrance onto a highway from a private residence or private farm and is used primarily by the persons living on the property, by the persons operating the farm or by persons providing service to the residents living on the property. Where a fruit or produce stand is operated adjacent to a highway, the entrance to the stand is classified as a commercial entrance.

Right-of-way. Property acquired by an agency or utility for access purposes.

Right-of-way. All of the land included within an area which is dedicated, reserved by deed or granted by easement for street purposes.

Right-of-way. Land reserved for use as a street, alley or other means of travel.

Right-of-way. Land reserved for use as a street, alley, interior walk, or for other public purpose.

Right-of-way. A tract or parcel of land used or assigned to be used in whole or in part for public or private roadway serving three (3) or more lots.

Right of way. The privilege of the immediate use of the roadway.

Right-of-way line. Outer edge of State Highway property, separating highway property from abutting properties of others.

Road. Public thoroughfare thirty feet or more in width.

Roadway. That portion of a right-of-way available for vehicular travel, including parking lanes.

Roadway. That portion of a public thoroughfare devoted to vehicular traffic, or that part included between curbs.

Roadway. That portion of a street which is used or intended to be used for the movement of motor vehicles.

Roadway. That portion of a right of way for a street or alley used or intended to accommodate the movement of vehicles.

Roadway. That portion of a public way appropriated to vehicular traffic.

Roadway. That portion of any street so designated for vehicular traffic and where curbs are normally placed, means that portion of the street between the curbs.

Roadway. That portion of a highway improved, designed, or ordinarily used for vehicular travel, exclusive of the shoulder. In the event a highway includes two (2) or more separate roadways, the term "roadway" as used herein refers to any such roadway separately, but not to all such roadways collectively.

Roadway. That portion of the street set apart for the use of vehicular traffic or as a parkway and which is included between the curb lines.

Safety zone or island. An area or space officially set apart within a roadway for the exclusive use of pedestrians and which is protected or is so marked or indicated by adequate signs so to be plainly visible at all times while set apart as a safety zone or island.

Secondary residential street. A street which is used primarily for residential access.

Service road. That part of a major or secondary highway, containing a roadway which affords access to abutting property and is adjacent and approximately parallel to and separated from the principal roadway.

Service street or alley. A minor way used primarily for vehicular service access to the back or the side of properties otherwise abutting on a street.

Shoulder. Graded part of the right-of-way that lies between the edge of the main pavement (main-traveled way) and curbline.

Shoulder. An improved portion of a street immediately adjoining the travelway, for parking and for access to abutting properties.

Side street. Street along the side line or a corner lot.

Side street. That street bounding a corner lot and which extends in the same general direction as the line determining the depth of the lot.

Sidewalk. A walkway at the side of a street. For the purpose of encroachment thereon, the width of a sidewalk shall mean the distance between the property line and the "Curb Line" as defined in the Code.

Sidewalk. That area of a street or other public way which is located between the curb line and the street line.

Sidewalk. That portion of a street between the curb lines, or the lateral lines of a roadway, and the adjacent property lines, intended for the use of pedestrians.

Sidewalk area. That portion of the right-of-way that lies between the curbline and right-of-way line, and within the limits of the extended property lines. This area varies greatly in width. Whether improved or unimproved, it is considered and controlled as sidewalk area.

Sidewalk space. Part of a public street provided or set apart as a walkway for pedestrians, including the planting strip when the same exists; as distinguished from the roadway of said street.

Sight distance. The length of street, measured along the center line, which is continuously visible from any point four and one-half (4-1/2) feet above the center line to an object four (4) inches above the road surface.

Sight distance. Maximum extent of unobstructed vision (in a horizontal or vertical plane) along a street from a vehicle located at any given point on the street.

Snow emergency. A state of street conditions that are hazardous and/or dangerous to vehicular and pedestrian traffic and so declared by the city as provided in the Ordinance.

Stand or standing. Means the halting of a vehicle, whether occupied or not, otherwise than temporarily for the purpose of and while actually engaged in receiving or discharging passengers.

Stop. When required, means complete cessation from movement.

Stop or stopping. When prohibited, means any halting even momentarily of a vehicle, whether occupied or not, except when necessary to avoid conflict with other traffic or in compliance with the directions of a police officer or traffic-control sign or signal.

Street. A public or private thoroughfare, affording the principal means of access to abutting property.

Street. Any public or private way set aside for common travel more than twenty-one (21) feet in width if such existed at the time of enactment of the Ordinance, or such right-of-way, fifty (50) feet or more in width if established thereafter.

Street. Any highway, public bridge, road, lane, footway, square, court, alley, or passage, whether a thoroughfare or not.

Street. A thoroughway which affords the principal means of access to abutting property, including avenue, place, way, drive, land, boulevard, highway, road and any other thoroughfare except an alley.

Street. A public right-of-way fifty (50) feet or more in width which provides a means of access to an abutting property; or any public right-of-way not less than thirty (30) feet in width which existed prior to the enactment of the Ordinance. The term "street" shall include avenue, drive, circle, road, highway or similar terms.

Street. Any public thoroughfare (street, avenue, boulevard) 30 feet or more in width which has been dedicated or deeded to the public for public use, and is accessible for use by the fire department in fighting fire. Enclosed spaces and tunnels, even though used for vehicular and pedestrian traffic, are not considered as streets for the purposes of the Life Safety Code.

Street. Thoroughfare dedicated and accepted by a municipality for public use or legally existing on any map of a subdivision filed in the manner provided by law.

Street. Public or private thoroughfare which affords principal means of access to abutting property.

Street. All property acquired or dedicated to the public and accepted by the appropriate governmental agency for street purposes. Property that has been commonly used or dedicated to be used for street purposes prior to the adoption of the code shall be considered a street.

Street. Way, alley, lane, court, sidewalk, public square, or other place that is part of a highway, including all of the rights-of-way between property lines that are shown on the Official Map of the City.

Street. A public right-of-way 30 feet or more in width.

Street. Any right-of-way which has been accepted by the city and is publicly maintained.

Street. The right-of-way of a public or private thoroughfare affording the principal means of access to abutting property.

Street. Any highway, road, boulevard, square or other improved thoroughfare, 30 feet or more in width, which has been dedicated or deeded for public use and is accessible to fire department vehicles and equipment.

Street. Any public thoroughfare (street, avenue, boulevard, park) or space more than twenty (20) feet in width which has been dedicated or deeded to the public for public use.

Street. A thoroughfare for vehicular traffic, generally includes everything found within the right-of-way.

Street. A thoroughfare right of way, dedicated as such or required for public use as such, other than an alley, which affords the principal means of access to abutting land.

Street. A public thorofare which affords the principal means of access to abutting property.

Street. A public way for purposes of vehicular travel including the entire area within the rights of way. The term includes, but is not limited to, avenue, alley, boulevard, drive, highway, road or freeway.

Street. A public or private thoroughfare which affords a primary means of access to abuting property is a street to that property for the purposes of the ordinance, except driveways to building.

Street. The principal means of access to abutting properties.

Street. A dedicated public passageway which affords a principal means of access to abutting property.

Street. Any thoroughfare or public park not less than sixteen (16) feet in width which has been dedicated or deeded to the public for public use.

Street. A strip of land, including the entire right-of-way, intended for use as a means of vehicular and pedestrian circulation.

Street. A right-of-way (or portion thereof) intended for general public use to provide means of approach for vehicles and pedestrians. The word "street" includes the words, road, highway, thoroughfare and way.

Street. Any public or private right-of-way set aside for public travel thirty (30) feet or more in width. The word "street" shall also include the words, road, avenue, boulevard, lane, drive, circle, thoroughfare, and highway.

Street. Any road, avenue, street, lane, alley or other way commonly used by the public for street purposes.

Street. Any legally established public thoroughfare 30 feet or more in width whether designated or not by name or number such as avenue, boulevard, circle, court, drive, lane, place, road or way. All-weather hard-surfaced areas 30 feet or more in width and extending at least 50 percent of the length of that side of building and accessible to fire-fighting equipment will be acceptable in lieu of streets.

Street. A dedicated and accepted public thoroughfare, or permanent, unobstructed, private easement of access, having a right-of-way of more than thirty (30) feet in width and a roadway suitable for vehicular travel at least ten (10) feet wide, which affords the principal means of vehicular access to abutting property.

Street. Any thoroughfare or way, other than a public alley, dedicated to the use of the public and open to public travel, whether designated as a road, avenue, highway, boulevard, drive, lane, circle, place, court, terrace, or any similar designation, or a private street open to restricted travel, at least sixty (60) feet in width.

Street. Wherever the word "street" is used in the Ordinance, it shall be construed as including any public thoroughfare thirty (30) feet or more in width.

Street. A thoroughfare 21 feet or more in width which has been legally dedicated or devoted to public use.

Street. Any thoroughfare or public space not less than sixteen (16) feet in width which has been dedicated or deeded to the public for public use.

Street. Any public thoroughfare (street, avenue, boulevard) 30 feet or more in width which has been dedicated or deeded to the public for public use and is accessible for use by the Fire Department in fighting fire. Enclosed spaces and tunnels, even though used for vehicular and pedestrial traffic are not considered as streets for the purposes of the Code.

Street. A public thoroughfare which affords the principal means of access to abutting property. This includes avenue, road, lane, drive, or other means of ingress or egress regardless of the term used to designate.

Street. Any roadway or public way dedicated to public use, except an alley. The street line is the property line.

Street. A paved, legally, opened thoroughfare or highway. For the purpose of the Code, a paced driveway fifty (50) feet or more in width and leading to a paved, legally opened thoroughfare or highway, shall be considered as a street.

Street. Any road, avenue, lane, alley or other way which is an existing public way, or which is shown on an approved plat, or any private right-of-way or easement approved by the Board.

Street. An area included between the right-of-way lines of a public thoroughfare more than 20 feet in width.

Street. A thoroughfare which affords a principal means of access to abutting property and which has been accepted by the City as a public street.

Street. A public street, road, or highway which is legally open or officially plotted by the Township, or a private street, road, or way over which the owners or tenants of two (2) or more lots held in single and separate ownership have the right-of-way.

Street. A public thoroughfare more than 14 feet in width which has been dedicated or deeded to the public for public use and which affords principal means of access to abutting property.

Street. All property dedicated or intended for public or private street purposes or subject, or public easements therefore and 21 feet or more in width.

Street. A public thoroughfare or road thirty feet, or more, in width.

Street. A public roadway which affords a principal means of access to abutting property, having a right-of-way of fifty (50) feet or more.

Street. A primary thoroughfare or highway thirty (30) feet or more in width as dedicated or devoted to public use by legal mapping use, or other lawful means.

Street. A thoroughfare which affords principal means of access to abutting property.

Street. A public thoroughfare which affords principal means of access to abutting property.

Street. A highway or thoroughfare dedicated or devoted to public use by legal mapping, or by the user, or by any other lawful procedure, which is thirty (30) feet or more in width and which affords the prescribed means of access to abutting property; and includes avenue, boulevard, concourse and similar public ways.

Street. A public thoroughfare, avenue, road, highway, boulevard, parkway, way, drive, lane, court or private easement providing, generally, the primary roadway to and egress from the property abutting along its length.

Street. A strip of land, including the entire right-of-way, intended primarily as a means of vehicular and pedestrian travel.

Street. A public or private thoroughfare, however designated, which affords the principal means of access to abutting property.

Street. A right-of-way municipally or privately owned, serving as a means of vehicular and pedestrian travel, furnishing access to abutting properties, and space for sewers and public utilities.

Street. A thoroughfare which affords the principal means of access to abutting property and includes road and highway.

Street. A public thoroughfare including public roads or highways which affords principal means of access to abutting property.

Street. Any street, avenue, boulevard, road, lane, parkway, viaduct, alley, or other way which is an existing state, county or municipal roadway, or a street or way shown upon a plat heretofore approved pursuant to law, or heretofore approved by official action, including all streets or ways on plats duly filed and recorded in the office of the county recording officer prior to the grant to such board of the power to review plats. The word "Street" includes the land between the street right-of-ways lines, whether improved or unimproved, and may comprise pavement, shoulders, gutters, sidewalks, planted strips, parking areas and other areas within such street lines.

Street. Thoroughfare dedicated and accepted by a municipality for public use or legally existing on any map of a subdivision filed in the manner provided by law.

Street. A right-of-way, thirty (30) feet or more in width, permanently dedicated to common and general use by the public, including any avenue, drive, boulevard, parkway, highway, freeway, or similar way.

Street. A public right-of-way at least 30 feet in width which affords a primary means of access to abutting property.

Street. A thoroughfare which has been dedicated or abandoned to the public and accepted by proper public authority, or a thoroughfare which has been made public by right of use and which affords the principal means of access to abutting property.

Street. A public right-of-way more than twenty (20) feet in width which provides a public means of access to abutting property and used primarily for vehicular circulation. The term street shall include avenue, drive, circle, road, parkway, boulevard, lane, place, highway, thoroughfare, and any other similar term.

Street. Wherever the word "street" occurs in the Code, it shall be held to include all streets, avenues, boulevards, highways or other public ways in the City and County, which have been or may hereafter be dedicated and open to public use.

Street. Any public thoroughfare other than an alley or walk, except that in those cases where a subdivision has been recorded containing lots which abut only on an alley or walk, said alley or walk, may be considered to be a street.

Street. A way for vehicular traffic.

Street. Any street, highway, sidewalk, alley, avenue, or other public way or grounds or public easements in the City.

Street. Public thoroughfare, not less than twenty feet (20') in width, which affords principal means of access to abutting property. The term "street" shall include the terms "place," "way," "boulevard," "parkway," "avenue," "circle," "court," and "drive."

Street. A thoroughfare right-of-way, dedicated as such or acquired for public use as such, other than an alley, which affords the principal means of access to abutting land.

Street. A public street, or a private street for vehicular traffic which the owners of three (3) or more parcels of land of separate ownership have the right to use for ingress and egress.

Street. All property dedicated or intended for public or private roadway, highway, or freeway purposes.

Street. A public land improved or unimproved, which affords a primary means of access to abutting property, whether designated as a street, avenue, highway, road, boulevard, lane, throughway, right-of-way or otherwise, but does not include private roads and driveways to buildings.

Street. A publicly dedicated right of way not less than thirty three (33) feet in width or a permanently reserved easement of access which affords a primary means of access to abutting property.

Street. A permanent right-of-way which affords the primary means of vehicular access to abutting properties.

Street. Any public thoroughfare such as, but not limited to, street, avenue, lane, place, terrace, or road, and which is more than twenty (20) feet in width and dedicated or deeded to the public for public use.

Street. A strip of land, including the entire right-of-way, whether dedicated or not, intended for use as a means of vehicular and pedestrian traffic. Street shall be deemed to include avenue, boulevard, court, expressway, highway, lane, road, and the like.

Street. Public or private way which affords principal means of vehicular access to properties which abut thereon.

Street alignment. Wherever street lines are deflected a total of seven and one-half degrees (7-1/2°) or more within five hundred feet (500′), connection shall be made by horizontal curves.

Street centerline. Line midway between the street right-of-way lines or the surveyed and platted centerline of a street which may or may not be the line midway between the existing right-of-way lines.

Street centerline. Line midway between street lines.

Street floor. Any story or floor level accessible from the street or from outside the building at ground level with floor level at main entrance not more than three risers above or below ground level at these points, and so arranged and utilized as to qualify as the main stories accessible from the street each is a street floor for the purposes of the Life Safety Code. Where there is no floor level within the specified limits for a street floor above or below ground level, the building shall be considered as having no street floor.

Street frontage. All of the property fronting on one side of a street between two (2) intersecting streets, or in the case of a dead-end street, all of the property along one side of the street between an intersecting street and the end of such dead-end street.

Street frontage. The distance along a street line from one intersecting street to another or from one intersecting street to the end of a dead-end street.

Street frontage. Distance the lot line abuts a public or private way.

Street grade curb level. Elevation of the street grade curb level as fixed by the municipal authorities.

Street intersection. The area common to two (2) or more intersecting streets.

Street intersection. The area common to two (2) intersecting streets.

Street intersections. Streets shall be laid out to intersect as nearly as possible at right angles. No street shall intersect another at an angle of less than sixty (60) degrees.

Street line. The dividing line between the lot and the front street.

Street line. Line dividing a lot, plot, tract, or parcel from a street.

Street line. The line dividing a lot, tract, or parcel of land and a contiguous street; also a street right-of-way line.

Street line. A street right-of-way line.

Street line. The division line between a lot, tract, or parcel of land and a contiguous street or right-of-way, including in such streets or rights-of-way, all property dedicated for street purposes or subject to public or private easements therefor.

Street line. The street right-of-way.

Street line. The boundary line between a street and abutting property.

Street line. A line separating a lot from a street. In any case where a future street line has been established or approved by the Board, such future street line shall be considered as the street line for the purposes of determining lot area and setback requirements.

Street line. A dividing line between a lot, tract, or parcel of land on a continuous street.

Street line. A lot line dividing a lot from a street.

Street line. Line dividing a lot, plot, or parcel from a street.

Street line. Any line dividing a street from a lot.

Street line. The dividing line between the street and the lot.

Street line. A line separating a lot, tract, or parcel of land and an abutting street right-of-way.

Street line. The line between the street and abutting property.

Street line. The line dividing a lot, plot, tract, or parcel of land and a contiguous street. Also, a street right-of-way line.

Street line. The right-of-way of a street or easement for ingress or egress.

Street line. The street right-of-way boundary line.

Street line. The right-of-way line of a street.

Street line. The dividing line between the lot and the street.

Street line. The dividing line between a lot and a street, provided that no street line shall be considered to be less than twenty-five (25) feet from the center line of the street, provided further, however, that on certain specified streets the street line shall be considered to be the distance from the center line of the street.

Street line. Dividing line between a lot and the outside boundary or ultimate right-of-way of a public street, road, or highway legally opened or officially plotted; or between a lot and a privately owned street, road, or way, over which the owners' or tenants of two (2) or more lots each hold in single and separate ownership have the right-of-way.

Street line. Line of demarcation between public and private properties on plotted or legally opened streets; defining the land reserved for use as a street.

Street line or highway margin. Dividing line between a lot and a public street, road or highway right of way.

Street lot line. A lot line dividing a lot from a street right-of-way.

Street lot line. That boundary line between the street right-of-way and abutting property; or, in the case where an additional yard is required by the Ordinance, the street lot line shall be that line parallel to the center line of the street and most distant from the center line as specified by the ordinance.

Street lot line. The lot line dividing a lot from a street or other public space.

Street lot line. The lot line dividing a lot from a street.

Street or alley line. Lot line dividing the lot from, respectively, a street, or an alley.

Street projection. Any part of a structure or material attached thereto extending or projecting beyond the street building line, including, but not limited to architectural features, marquees, fire escapes, signs, flag poles.

Street property line. For purposes of the Code, only street property line shall mean any line separating private property from either a street or an alley.

Street right-of-way line. Line which abounds the right-of-way set aside for use as a street.

Street right-of-way line. Lines separating private property from the street or alley, existing or dedicated in public ownership.

Street wall. Of a building, at any level, means the wall or part of the building (other than a one-story open porch), nearest to the street line.

Street width. Mean of the distance between the street lines within a block. Where a street borders a public park, or a navigable body of water, the width of such street may be taken as the width of such street, plus the width of such public park or body of water, provided that the maximum width of such street shall not be considered more than 100 feet, measured at right angles to the street line.

Street width. Width of a street in any block between two intersecting streets is the average distance between street lines on opposite sides of the street.

Tentative plat. A preliminary map, sketch, drawing or chart indicating to a reasonable degree the location and layout of the subdivision submitted for approval.

Tertiary residential street. A street which provides access to houses on lots in excess of 20,000 square feet, having 100 feet frontage at the building line, and having no house or garage located within 50 feet of such right-of-way; or one which serves not more than 6 lots or parcels.

Thoroughfare. Public street, road, way or other space customarily used for travel.

Thoroughfare-expressway. A primary thoroughfare with divided roadways, partial or full control of access in general with grade separations at intersections. A freeway shall mean an expressway with full control of access and meeting the standards of the Bureau of Public Roads, U. S. Department of Commerce.

Thoroughfare plan. The part of the comprehensive development plan referring to transportation development goals, principles and standards and also includes use of the words "major street plan" and "traffic-ways plan."

Thoroughfare-primary or secondary. An officially designated federal or state numbered highway or county or other road or street designated as a primary thoroughfare on the official thoroughfare or major street plan for the city or county or other road or street designated as a secondary thoroughfare on said plan, respectively.

Through highway. Every highway or portion thereof on which vehicle traffic is given preferential right of way, and at the entrances to which vehicular traffic from intersecting highways is required by law to yield right of way to vehicles on such through highway in obedience to either a stop sign or a yield sign, when such signs are erected as provided in the Code.

Traffic. Pedestrians, ridden or herded animals, vehicles, and other conveyances either singly or together while using any highway for purposes of travel.

Traffic-control signal. Any device, whether manually, electrically, or mechanically operated, by which traffic is alternately directed to stop and to proceed.

Traffic lane. Strip of roadway intended to accommodate a single line of moving vehicles.

Traffic signaling device. A sign, device of mechanical contrivance, used for the control of motor vehicular and pedestrian movement.

Travelway. That portion of a public street or road which is intended for vehicular movement.

Tree lawn. That portion of a right-of-way lying between the exterior line of the roadway and the outside right-of-way line.

Vehicular access rights. The right or easement for access of owners or occupants of abutting lands to a public way other than as pedestrians.

Visibility at intersections in residential district. On a corner lot in any residential district, nothing shall be erected, placed, planted, or allowed to grow in such a manner as materially to impede vision between a height of 2-1/2 and 10 feet above the center line grades of the intersecting streets in the area bounded by the street lines of such corner lots and a line joining points along said street lines 50 feet from the point of the intersection.

Vision clearance. An unoccupied triangular space at the street corner or a corner lot which is bounded by the street lines and a setback line connecting points specified by measuring from the corner of each street line.

Vision clearance at corners. In any residential district no fence, wall, hedge or other structure or planting or other obstruction above a height of three (3) feet shall be erected, placed or maintained within twenty (20) feet of the intersection of the right-of-way lines of two (2) streets or railroad, or of a street intersection with a railroad right-of-way.

Vision obstruction. On any corner lot, no wall, fence or other structure shall be erected or altered, and no hedge, tree, shrub, or other growth shall be maintained which may cause danger to traffic on a street or public road by obscuring the view.

Way. Street, alley or other thoroughfare or easement permanently established for passage of persons or vehicles.

Width of street. The shortest distance between the lines delineating the right of way of a street.

B-16 YARDS AND COURTS

Buffer yard requirements. Along each side or rear property line which adjoins existing single-family residence, no building shall be located closer than seventy-five (75) feet from such residence or district line in conformity with the Ordinance.

Common open space. Common open space shall mean any meaningful open space, other than private or frontage open space, intended for use by all occupants of a development. This space may include recreation oriented areas.

Common open space. Part of a parcel which is open and unobstructed except for natural features and permitted accessory uses and structures, and which is accessible and usable by all persons who occupy a principal use on the parcel.

Court. An open, uncovered, unoccupied space on the same lot with a building; inner court means any court other than an outer court or a yard, outer court means a court other than a yard having at least one side thereof opening on to a street, alley or yard or other permanent open space, yard means a court on the same lot with a building extending along the entire length of a lot line.

Court. An open, uncovered, and unoccupied space, unobstructed to the sky, bounded on three or more sides by exterior building walls.

Court. Any space other than a yard on the same lot with a building or group of buildings, and which is unobstructed and open to the sky above the floor level of any room having a window or door opening on such court. The width of the court shall be its least horizontal dimension.

Court. An open, uncovered and unoccupied space within the lot lines of a lot, and includes the yard.

Court. A space, open and unobstructed to the sky, located at or above grade level on a lot and bounded on three or more sides by walls of a building.

Court. An open space other than a yard on the same lot with the building; includes patio.

Court. An open, unoccupied space, other than a yard, bounded on two or more sides by the exterior walls and lot lines.

Court. Open unoccupied portion of a lot other than a front yard, side yard or rear yard.

Court. Open, uncovered unoccupied space partially or wholly surrounded by walls of the structure.

Court. Open, unoccupied space, bounded on 2 or more sides by walls of a building.

Court. Required open, unoccupied space on the same lot and fully enclosed on at least 3 adjacent sides by walls of the building.

Court. Open space which may or may not have access and around which is arranged a single building or group of related buildings.

Court. Open space from the ground upward, which may or may not have direct street access and around which is arranged a group of buildings.

Court. Any space, other than a yard, bounded on 2 or more sides by the walls of a building or buildings and which is unobstructed and open to the sky above the floor level of any room having a window opening on such court. The width of a court shall be its least horizontal dimension.

Court. Court as applied to exitways, shall mean an exterior open space, or properly enclosed corridor. Exit courts shall lead directly to a thoroughfare.

Court. Shall be an open space upon a lot containing a building, other than a front yard, side yard or rear yard. An inner court is any other than an outer court. The width of an outer court is its horizontal dimension substantially parallel with its principal open end. The width of an inner court is its lesser horizontal dimension.

Court. A required open, unoccupied space on the same lot and fully enclosed on at least three adjacent sides by walls of the building. An outer court is any court facing for its full required width on a street, or any other required open space not a court. An inner court is any other required court.

Court. An open, uncovered and unoccupied space, unobstructed to the sky; bounded on two or more sides by the exterior walls of a building. An inner court is a court bounded on all sides by the exterior walls of a building or exterior walls and lot lines on which walls are allowable.

Court. An open, unoccupied space on the same lot with a building and bounded on 2 or more sides by such building, or the open space provided for access to a dwelling group.

Court. An open unoccupied space, other than a yard, on the same lot with a building or group of buildings.

Court. Any portion of the interior of a lot or building site other than required front, side or rear yards which is wholly or partially surrounded by buildings.

Court. An open space on the same lot with a building which is bounded on two (2) or more sides by the exterior walls of buildings on the same lot.

Court. An open unoccupied space other than a yard, on the same lot with a building and which is bounded on two (2) or more sides by the building.

Court. An open, unoccupied, unobstructed space other than a yard on the same lot as a building.

Court. An open, uncovered, unoccupied space on the same lot with a building.

Court. An open, unoccupied space, bound on two or more sides by walls of the building. An inner court is a court entirely within the exterior walls of a building. All other courts are outer courts.

Court. When applied to exitways, an exterior open space, or a properly enclosed corridor. Exit courts shall lead directly to a thoroughfare.

Court. An open unoccupied space, other than a yard, on the same lot with two (2) or more buildings which is bounded on two (2) or more sides by such buildings.

Court. A space, open and unobstructed to the sky, located at or above grade level on a lot, and bounded on three or more sides by walls of a building or structure.

Court. An inner court or outer court.

Court. An open, unoccupied space, bounded on two or more sides by the walls of the building. An inner court is a court entirely within the exterior walls of a building. All other courts are outer courts unless otherwise defined in the Code.

Court. Any space other than a yard on the same lot with a building or group of buildings, and which is unobstructed and open to the sky above the floor level of any room having a window or door opening on such court. The width of a court shall be its least horizontal dimension.

Court. An unoccupied open space on the same site with a building, which is bounded on three (3) or more sides by exterior building walls.

Court. Any space on a lot other than a yard which, from a point not more than two (2) feet above the floor line of the lowest story in the building on the lot in which there are windows from rooms abutting and served by the court, is open and served by the court, is open and unobstructed to the sky, except for obstructions permitted by the Code.

Court (inner). An open uncovered, unoccupied space surrounded on all sides by the exterior walls of a building or structure or by such walls and an interior lot line of the same premises.

Court (inner). A court not extending to a street, alley, open passageway, or yard.

Court (inner-width). Least horizontal dimension.

Court (outer). An open, uncovered, unoccupied space which has at least one side opening on a legal open space.

Court (outer). A court extending to a street, alley, open passageway, or yard.

Court (outer-depth). Least horizontal dimension measured perpendicular to the width.

Court (outer-width). Least horizontal dimension measured across the open end of the court.

Courts. Courts not covered by a roof or skylight but the entire required area shall be open and unobstructed from the bottom thereof to the sky. No fire escape or stairway shall be constructed in any court unless the court is enlarged proportionately.

Court (vent). An inner court solely for the lighting and ventilating of water closets, bathrooms, kitchens, public halls, and stair halls.

Depth of court. The minimum horizontal dimension at right angles to the width.

Depth of rear yard. The mean horizontal distance between the rear line of the building and the center line of the alley, where an alley exists, otherwise the rear lot line.

Depth of rear yard. The mean horizontal distance between the rear of the main building and the rear lot lines.

Double frontage lot yard. In any residence district the front and rear yard requirements of a double frontage lot shall be the same as prescribed for any single lot in the district wherein the lot is located.

Enclosed court. A court bounded on all sides by the exterior walls of a building or exterior walls and lot lines on which walls are allowable.

Exception to yard conditions. Where the shape of a lot or other circumstances result in conditions to which the provisions of the Zoning ordinance governing yard requirements are inapplicable, the Building department shall prescribe such yard requirements.

Exit court. A yard or court providing egress to a public way for one or more required exits.

Exterior side yard. Side yard abutting a street line.

Exterior yard. Any yard within an industrial district along the property line and fronting on a street.

Front equivalent yard. That portion of a rear yard of a through lot extending along a street line and from the street line for a depth equal to a required front yard. Any front yard equivalent shall be subject to the regulations of the ordinance which applies to front yards.

Front yard. A yard extending across the full width of the lot, the depth of which is the minimum horizontal distance between the front lot line and a line parallel thereto on the lot.

Front yard. A yard extending across the full width of a site, the depth of which is the minimum horizontal distance between the front property line and a line parallel thereto on the site.

Front yard. An open space on the same lot with a building between the front line of the building and the front lot line or future width line, and extending across the full width of the lot. The depth of the front yard is the minimum distance between the front lot line and the nearest exterior wall of the building, the front of a bay window or the front of a covered porch, or other similar projections, whichever is nearest the front lot line.

Front yard. A yard extending across the front of a zoned lot.

Front yard. The area extending across the full width of a lot and lying between the front lot line and a line parallel thereto, and having a minimum distance between them equal to the required front yard depth as prescribed for each Zoning District, Front yards shall be measured by a line at right angles to the front lot line, or by the radial line in the case of a curved front lot line. On corner lots the primary front yard shall be parallel to the street frontage which has the least dimension, and the side front yard shall be parallel to the other street frontage.

Front yard. A yard extending along the full length of a front lot line and back to a line drawn parallel to the front lot line at a distance therefrom equal to the depth of the required set-back (including, where applicable the average set-back).

Front yard. A yard which is bounded by the side lot lines, the front lot line and the front yard line.

Front yard. A yard extending along the full width of a front lot line between side lot lines and from the front lot line to the front building line in depth.

Front yard. An open unoccupied space, on the same lot with a building, between the front of the building and the front line for the full width of the lot.

Front yard. Yard across the full width of the plot facing the street extending from the front line of the building to the front property line. Either yard facing a street may be selected as the front yard of a corner yard.

Front yard. Yard extending across the full width of the lot between the front lot line and the nearest line of the main building.

Front yard. Open space extending the full width of the lot between a building an the front lot line, unoccupied and unobstructed from the ground upward except as specified elsewhere in the Ordinance.

Front yard. Minimum horizontal distance between the street line and the building or any projection thereof other than steps, unenclosed balconies and unenclosed porches.

Front yard. An open unoccupied space on the same lot with a building situated between the nearest roofed portion of the building and the street line of the lot, and extending from side lot line to side lot line.

Front yard. A yard extending across the full width of a lot and having a depth equal to the shortest distance between the front line of the lot and the nearest portion of the main building, including an enclosed or covered porch, provided that the front yard depth shall be measured from the future street line for a street on which a lot fronts, when such line is shown on the official map or is otherwise established.

Front yard. A yard extending along the full length of a street line.

Front yard. Open space extending the full width of a lot and of a uniform depth measured horizontally between the front line and the foundation of the building.

Front yard. A yard across the full width of the lot, extending from the front line of the building to the front line of the lot, excluding steps and unenclosed porches.

Front yard. A required yard extending across the front of a lot between the side lot lines and being the minimum horizontal distance between the street line and the maximum permissible main building. On corner lots the front yard shall be considered as parallel to the street upon which the lot has its least dimension.

Front yard. A yard across the full width of the lot extending from the front line of the building to the front line of the lot.

Front yard. A yard extending across the full width of the lot adjoining the street on which the lot fronts, the depth of which shall be as specified in the dimensional requirements for each district.

Front yard. Every dwelling hereafter erected shall have a front yard not less than 30 feet in depth. In the case of a dwelling fronting on a street or highway on which there are existing neighboring buildings or structures having front yards less than 30 feet, the minimum allowable front yards of all dwellings hereafter erected and fronting on such street or highway shall be as determined by the Building Official.

Front yard. A yard extending across the full width of the lot and lying between the front line of the lot and the building or building group.

Front yard. Open unoccupied space on the same lot with a main building, extending the full width of the lot and situated between the street line and the front line of the building projected to the side line of the lot. The depth of the front yard shall be measured between the front line of the building and the street line. In the case of a corner lot, the front yard shall be on the same side as the front yards of the adjacent interior lots.

Front yard. A yard covering the area between the side lot lines extending from the front lot line to the nearest point of a principal use. The front yard depth of a lot located on a curve shall be measured from the chord connecting the arc of the front lot line or from the tangent of said arc, whichever measurements results in the lesser depth.

Front yard. A required open space, the full width of the lot, extending from the street line to the nearest structure on the lot, exclusive of overhanging eaves, gutters, or cornices.

Front yard. A yard extending the full width of the lot, between the front lot line and the nearest part of the main building excluding uncovered steps.

Front yard. A yard extending across the front of the lot between the inner side yard lines and measured between the front line of the lot and the front line of the building, and the front line of the lot and the nearest line of any open porch or paved terrace.

Front yard. A yard across the full width of the lot extending from the setback or specific building line or front yard line to the street line.

Front yard. A yard extending across the entire width of the lot between the front lot line and the front building line. The lot line of a lot abutting a public street shall be deemed the front lot line. The front yard of a corner lot shall be that yard abutting the street with the least frontage, unless otherwise determined on a recorded plat or in a recorded deed. The front yard of a lot existing between two (2) streets not intersecting at a corner of a lot, shall be that yard abutting the street on which adjoining properties face, unless otherwise determined on a recorded plat on in a recorded deed.

Front yard. A space extending the full width of the lot and situated between the front property line and the front line of the building projected to the side property line of the lot. The front yard of a corner lot shall be on the same side as the front yards of the adjacent interior lot.

Front yard. A yard extending across the front of a lot between the side lot lines, and being the required minimum horizontal distance between the street and/or building line and the buildable area. On corner lots the front yard shall be provided facing the street upon which the lot has its lesser dimension.

Front yard. An open space lying between the street line upon which a building or structure fronts and the front wall of said building, and running entirely across the lot to the two (2) side lot lines.

Front yard. A yard extending across the full width the front line of the lot and the front of the lot and the nearest line of the main building.

Front yard. A yard extending across the width of the lot parallel and adjacent to the front lot line, lying between the side lot lines and measured as the minimum horizontal distance between the building and the front lot line with a minimum width at the street line of not less than twenty feet.

Front yard. A yard across the full width of the lot extending from the front line of the building to the street frontage.

Front yard. A yard extending between side lot lines across the front of a lot adjoining a public street.

Front yard. A yard extending across the full width of a site, the depth of which is the minimum horizontal distance between the front property line and a line parallel thereto on the property.

Front yard. A yard extending from the front wall of the building to the front lot line across the full width of the lot.

Front yard. A yard extending across the front of the lot between the inner side yard lines and measured between the front line of the lot and the front line of the building. Covered porches, whether enclosed or un-enclosed, shall be considered as part of the main building and shall not project into a required front yard.

Front yard. A yard extending across the full width of the lot and measured between the front lot line and the building or any projection thereof, other than the projection of the usual steps. Legally non-conforming unenclosed porches extending into a required front yard, shall not be included in computing the average front yard setback requirements for adjoining buildings. On corner lots, the front yard shall be considered as the yard adjacent to the street upon which the lot has its least dimension.

Front yard. An open, unoccupied space unobstructed to the sky, extending across the full width of a lot, or plot of land between the street line and the base of a front building wall. Unenclosed terraces, slabs or stoops without roofs or walls may project into this open space.

Front yard. An open unoccupied space on the same lot with a principal building, extending the full width of the lot and situated between the street line and the front line of the principal building projected to the side lines of the lot.

Front yard. A yard across the full width of the plot facing the street extending from the front line of the building to the front property line. On a corner lot, both yards facing a street are considered front yards.

Front yard. A yard extending across the front of a lot between the side lot lines and measured between the front line of the lot and the nearest point of the principal building.

Front yard. A yard extending across the front of the lot between the inner side lines measured between the front line of the lot and the front line of the building.

Front yard. There shall be a front yard having a depth of not less than 50 feet, or not less than the average front yard depth of all buildings within 200 feet on the same block frontage, which ever is the greater. Where the dedicated width of the main street is less than 50 feet, the front yard shall extend back to a line not less than 75 feet from the center line of the main street and parallel thereto. Where a lot extends through from street to street the applicable front yard regulations shall apply on both street frontages.

Front yard. A yard across the full width of the lot, extending from the front line of the building to the front line of the lot. (The front line of the building is not the front line of the porch if the porch is an open porch.)

Front yard. A yard extending across the front of a lot.

Front yard. The required open space between the street line and the nearest part of any building on the lot, excluding cornices, eaves or gutters projecting not more than thirty inches, steps, one-story open porches, bay windows not extending through more than one-story and not projecting more than five feet, chimneys, open balconies and terraces.

Front yard. A yard extending across the full width of the lot, the depth of which is the distance between the front lot line and the main wall of the building.

Front yard (front set-back line). A yard extending across the front of a lot between the side lot lines and being the minimum horizontal distance between the front lot line and the nearest wall or column line of the building.

Front yard line. Line on or back of the street line between that and the street line no building or other structure or portion thereof, except as provided in the code, may be erected above grade level. The front yard line shall be the building line for the front of a building.

Inner court. Any court other than an outer court or yard.

Inner court. Open unoccupied space bounded by walls of a building, but located within the exterior walls of the building.

Inner court. Open, unoccupied space on the same lot with a building not extending to either street, alley or rear yard.

Inner court. A court which is bounded on all four (4) sides by building walls.

Inner court. An open unoccupied space surrounded on all sides by walls or by walls and a lot line or lines.

Inner court. An open space unoccupied and unobstructed from its lowest point to the sky, except as otherwise permitted in the Ordinance, surrounded on all sides by walls, or by walls and a lot line.

Inner court. Any court which is not an outer court.

Inner court. Any open area, other than a yard or portion thereof, that is unobstructed from its lowest level to the sky and that is bounded by either walls, or building walls and one or more lot lines other than a street line or building walls, except for one opening on any open area along an interior lot line that has a width of less than 30 ft., at any point.

Inner court. A court entirely within the area enclosed by the exterior walls of a building. An unroofed shaft shall be assumed to be an inner court.

Inner court. Any court other than an outer court.

Inner court. A court, enclosed on all sides by exterior walls of a building, or by exterior walls and lot lines other than street or alley lines.

Inner court. Open air shaft or court surrounded on all sides by walls.

Inner court. An open, outdoor space enclosed on all sides by exterior walls of a building or by exterior walls and property lines on which walls are allowable.

Inner court. An open, uncovered, unoccupied space surrounded on all sides by the exterior walls of a building or structure or by such walls and an interior lot line of the same premises.

Inner court. A court not extending to a street, alley, open passageway or yard.

Inner court. A court surrounded on all sides by the exterior walls of a structure or by such walls and an interior lot line.

Inner court. An open unoccupied space, other than a rear or front yard, on the same lot with a building and not extending to either yard or street.

Inner court. No inner court not on a lot line shall be less than fourteen (14) feet in width, nor less than 280 square feet in area for courts two (2) stories or less in height. For each additional story in height, every such court shall be increased by at least four (4) lineal feet in its length and three (3) lineal feet in its width.

Inner courts. No inner lot line court shall be less than eight (8) feet in width nor less than 100 square feet in area for courts two (2) stories or less in height. For each additional story in height, every court shall be increased by at least three (3) lineal feet in its length and two (2) lineal feet in its width.

Inner courts. No inner court on a lot line shall be less than six (6) feet in width nor less than sixty (60) square feet in area for courts two (2) stories or less in height. At each additional story height, every such court shall be increased by at least one and one-half (1-1/2) lineal feet in its length and one (1) lineal foot in width.

Inner courts. No inner lot line court shall be less than six (6) feet in width not less than sixty (60) square feet in area for courts two (2) stories or less in height, except that an inner lot line court one story high shall not be less than four

(4) feet wide and not less than forty (40) square feet in area. At each additional story height every such court shall be increased by at least one (1) lineal foot in its length and one (1) lineal foot in its width.

Inner courts. Shall be not less than 10 feet in width nor less than 150 square feet in area for courts 2 story or less in height; and for every additional story every such inner court shall be increased by at least one lineal foot in its length and one lineal foot in its width.

Inner court width. Court width is the least horizontal dimension of the court.

Inner court (width). As applied to an inner court, means its least horizontal dimension.

Inner lot line court. Court bounded on one side and both ends by walls and on the remaining side by a lot line.

Inner lot line court. A court bounded on three (3) sides by walls, and on the remaining side by a lot line or property line.

Inner lot line courts. Courts one story high shall be not less than four (4) feet wide and not less than forty (40) square feet in area. Inner lot line courts two (2) stories high shall be not less than six (6) feet wide and not less than sixty (60) feet in area. For every additional story every such inner lot line court shall be increased by at least one (1) lineal foot in length and one (1) lineal foot in its width.

Interior side yard. Side yard abutting a lot line of an adjoining lot.

Interior side yard. Side yard which is not a street yard.

Interior side yard. A side yard which adjoins another lot or alley separating such side yard from another lot.

Interior yard. Any yard within an industrial zoning district along the property line and not fronting on a street.

Legal open space. Open space on the premises, such as yards or courts, or an open space at least 25 feet wide permanently dedicated to public use which abuts the premises.

Length of an outer court. Length of an outer court is the horizontal distance between the end opening on a street or rear yard and the end opposite such street or rear yard.

Length of an outer court. Mean horizontal distance between the open and closed ends of the court.

Low level light area. Open area at ground level which is open and unobstructed to the sky, but which is not eligible to be a plaza, or a horizontal open area above ground level which is open and unobstructed to the sky.

Open court. Shortest horizontal dimension of an outer court measured in a direction substantially paralleled with the principal open end of such court.

Open inner court. Court bounded on all sides by building walls or other structures.

Open space. Part of a parcel which is open to the sky and unobstructed except for natural features.

Open space. Includes a street, alley, park, yard, court or other permanent unobstructed space open to the sky, providing safe access to a street.

Open space. Consists of: parks, common greens, other recreation space or generally open areas available to the public, or yards or other open areas or spaces provided in connection with residential buildings occupied by more than 2 families per lot which are intended for the sole use of the occupants of such buildings and their guests.

Open space. Land areas which are not occupied by buildings, structures, streets, alleys, excepting however approved landscape features and active recreation facilities, when developed in accordance with the Code.

Open space. Area included in any side, rear, or front yard, or any other unoccupied space on a lot that is open and unobstructed to the sky except for the ordinary projection of cornices and eaves of porches.

Open space required. Yard space of a lot which is established by and between the street, and side yard lines, and the required setback lines and which shall be open, unoccupied, and unobstructed by any building or structure or any part thereof from the ground to the sky, except as otherwise listed in the code.

Open spaces. Front, rear and side yards, exit courts, outer courts, and outer lot line as regulated by the code.

Open spaces. Include front, rear, and side yards or setbacks, exit courts, outer courts, and outer lot line courts on the same property with a building as regulated by the Ordinance.

Open use. Any use of a lot that is not conducted within a building.

Outdoors. On or above the surface of the ground but not within a building which has a roof composed of weather proof material and which is enclosed on at least 65% of its perimeter with exterior walls composed of weather proof materials.

Outer court. A court, one entire side or end of which is bounded by a front yard, a rear yard, a side yard, a front lot line, a street, or an alley.

Outer court. An open space unoccupied and unobstructed from its lowest level to the sky, except as otherwise permitted by the Ordinance, on the same lot with a building, extending to and opening upon a street, alley or yard.

Outer court. An open unoccupied space on the same lot with a building, extending to and opening upon a street, alley, or yard.

Outer court. A court which is open to the street, front yard or rear yard for its entire width.

Outer court. A court extending to and opening upon a street, public alley, or other approved open space, not less than (15) feet wide, or upon a required yard.

Outer court. An open space, drive or passage on the side or rear of a building, of sufficient area and width for egresses which connect thereto.

Outer court. An open, unoccupied space, opening onto a street, alley or yard.

Outer court. A court which opens on any yard on the lot, or which extends to any street line on the lot.

Outer court. A court having at least one side thereof opening onto a street, alley, or yard or other permanent open space.

Outer court. A court having at least one side open to a street, yard or other permanent open space.

Outer court. An open unoccupied space other than a yard on the same lot with a building extending to either the street, alley or the rear yard.

Outer court. A court enclosed on three (3) sides by exterior walls of a building, or by exterior walls and lot lines on which walls are allowable, with one (1) side or end open to a street, driveway, alley or yard.

Outer court. A court enclosed on not more than three sides by exterior walls of a building, or by exterior walls and lot lines, with one side or end open to a street or yard, or to an alley or other permanent open space not less than 20 feet wide.

Outer court. A court other than a yard having at least one side thereof opening on to a street, alley or yard or other permanent open space.

Outer court. No outer lot line court shall be less than three (3) feet wide for a court two (2) stories or less in height and forty (40) feet or less in length. At each additional story height the width of such court shall be increased one (1) foot, and for any additional length the width of such court shall be further increased at the rate of one (1) foot in ten (10) feet.

Outer court. No outer court not on a lot line shall be less than six (6) feet wide for a court two stories or less in height and forty (40) feet or less in length. At each additional story height the width of such court shall be increased one (1) foot, and for any additional length the width of such court shall be further increased at the rate of one (1) foot in ten (10) feet.

Outer court. Open, uncovered, unoccupied space which has at least one side opening on a legal open space.

Outer court. Court extending to a street or to front or rear yard.

Outer court. Open, outdoor space enclosed on at least 2 sides by exterior walls of a building or by exterior walls and property lines on which walls are allowable, with 1 side open to a street, driveway, alley, or yard.

Outer court. Court bounded on 3 sides with walls and on the remaining side by a street, alley or other open space not less than 16 feet.

Outer court. Court extending to a street, alley, open passageway, or yard.

Outer court. Any open area, other than a yard or portion thereof, that is unobstructed from its lowest level to the sky and that, except for an outer court opening upon a street line, a front yard, or a rear yard, is bounded by either building walls or building walls and one or more lot lines other than a street line.

Outer courts. Outer courts between wings or parts of the same building, or between different buildings on the same lot, shall be not less than 6 feet wide for a court 2 stories or less in height and 40 feet or less in length. For each additional story in height, the width of such court shall be increased one foot, and for each additional 10 feet or fraction thereof in length, the width of such court shall be further increased one foot. Where outer courts or outer lot line courts open at each end to a street or other open space not less than 15 feet wide, the above lengths may be doubled.

Outer court width. Shortest horizontal dimension measured in a direction substantially parallel with the principal open end of the court.

Outer court width. Least horizontal dimension measured across the open end of the court.

Outer court width. As applied to an outer court, means the shortest horizontal dimension measured in a direction substantially parallel with the principal open end of such court.

Outer lot line court. A court with one side on a lot line or property line and opening to a street or open space not less than fifteen (15) feet wide.

Place or court. Open, unoccupied space on the same lot with a building or group of buildings and bounded on three (3) or more sides by such building or buildings.

Public space. A plot or area of land outside of the building dedicated to or devoted to public use by legal mapping or by any other lawful procedure.

Public space. For the purpose of determining allowable floor areas and/or exit arrangement of buildings, such open spaces as public parks, right-of-ways, waterways, public beaches and other permanently unobstructed yards or courts having access to a street.

Public space. A legal open space on the premises, accessible to a public way or street, such as yards, courts or open spaces permanently devoted to public use which abuts the premises, and that is permanently maintained accessible to the Fire Department and free of all incumbrances that might interfere with its use by the Fire Department.

Rear yard. An open space extending across the full width of the lot between the most rear main building and the rear lot line. The depth of the required rear yard shall be measured horizontally from the nearest part of a main building toward the nearest point of the rear lot line.

Rear yard. A yard extending across the full length of a site, the depth of which is the minimum horizontal distance between the rear property line and a line parallel thereto on the site.

Rear yard. A yard extending along the full length of the rear lot line and back to a line drawn parallel to the rear lot line at a distance therefrom equal to the depth of the required rear yard.

Rear yard. An open, unoccupied space between the rear line of the building and the rear line of the lot, for the full width of the lot.

Rear yard. A yard extending across the rear of a lot.

Rear yard. A yard extending across the full width of the lot, the depth of which is the minimum horizontal distance between the rear lot line and a line parallel thereto on the lot.

Rear yard. A yard extending across the lot measured between the rear property line and the structure closest thereto; the depth of the rear yard is the minimum horizontal distance between the rear property line and a line parallel thereto touching any part of a structure, other than structures or parts thereof specifically excepted by the code.

Rear yard. A space on the same lot with a building situated between the nearest roofed portion of the building and the rear line of the lot, and extending from side lot line to side lot line.

Rear yard. A yard extending across the full width of the lot, measured between the rear lot line and the foundation wall of the building.

Rear yard. A yard extending across the full width of the lot adjoining the rear lot line, the depth of which shall be as specified in the dimensional requirements for each district.

Rear yard. An open space extending the full width of the lot, between a building and the rear lot line, unoccupied and unobstructed from the ground upward except as specified in the Ordinance.

Rear yard. A yard extending across the full width of the lot between the rear most main building and the rear lot line, the depth of which shall be the least distance between the rear lot line and the rear of such main building.

Rear yard. A yard extending across the rear of a lot between the side lot lines, and being the required minimum horizontal distance between the rear lot line and the rear of the buildable area. On both corner lots and interior lots, the rear yard shall in all cases be at the opposite end of the lot from the front yard.

Rear yard. A yard extending across the full width of a lot and having a depth equal to the shortest distance between the rear line of the lot and the main building.

Rear yard. A required yard extending across the rear of a lot between the side lot lines and being the minimum horizontal distance between a rear lot line and the rear of the maximum permissible main building. On all lots the rear yard shall be at opposite end of the building to the rear line of the lot.

Rear yard. A yard across the full width of the lot, extending from the rear line of the building, to the rear line of the lot.

Rear yard. A yard extending across the full width of the lot and lying between the rear line of the lot and the nearest line of the main building.

Rear yard. A yard, unoccupied except by an accessory building as hereinafter permitted, extending across the width of the lot parallel and adjacent to the rear lot line, lying between the side lot lines and measured as the minimum horizontal distance between the building and the rear lot line. In the case of irregular shaped lots, the rear yard may be measured as the average horizontal distance between the building and the rear lot line.

Rear yard. A yard extending across the rear of the lot between the inner side yard lines. Depth of a required rear yard shall be measured in such a manner that the yard established is a strip of the minimum width required by district regulations with its inner edge parallel with the rear lot line.

Rear yard. A yard extending the full width of the lot, being the minimum horizontal distance between the rear lot line and the nearest part of the building, excluding uncovered steps.

Rear yard. The required open space, the full width of the lot, extending from the rear property line of the lot to the nearest structure on the lot, exclusive of over-hanging eaves, gutters, or cornices, and exclusive of a private garage, which is not an integral structural part of a main building.

Rear yard. A yard extending across the full width of the lot and lying between the rear line of the lot and the building or building group.

Rear yard. A yard extending across the rear of a lot, between side lot lines.

Rear yard. The yard across the full width of the plot opposite the front yard, extending from rear line of building to rear property line.

Rear yard. The yard between the rear of the building and the lot lines most nearly parallel thereto.

Rear yard. A space extending the full width of the lot and situated between the rear property line of the lot and the rear line of the building projected to the side property lines of the lot.

Rear yard. A yard extending across the rear of a lot between the side lot lines and being the minimum horizontal distance between a rear lot line and the rear of the maximum permissible main building. On all lots the rear yard shall be at the opposite end of the lot from the front yard.

Rear yard. The yard across the full width of the plot opposite the front yard, extending from rear line of building to rear property line. The rear yard of a corner lot is the yard opposite the selected front yard.

Rear yard. A yard extending across the full width of the lot between the front lot line and the nearest line of the main building.

Rear yard. A yard extending across the rear of a lot between the side lot lines and measured between the rear line of the lot and the nearest point of the principal building, other than steps or unenclosed porches or terraces.

Rear yard. An open, unoccupied space, except for permitted accessory buildings or structures, on the same lot with a principal building, extending the full width of the lot and situated between the rear line of the lot and the rear line of the principal building projected to the side lines of the lot.

Rear yard. A yard extending across the full width of the lot between the most rear main building and the rear lot line. The depth of the required rear yard shall be measured from the nearest point of the rear lot line toward the nearest main wall of the building.

Rear yard. An open unoccupied space on the same lot with a main building, extending the full width of the lot and situated between the rear line of the lot and the rear line of the building projected to the side lines of the lot. The depth of the rear yard shall be measured between the rear line of the lot and that part of the building nearest to the rear line of the lot. The rear line of the lot shall mean that lot line which is opposite and most distant from the front lot line.

Rear yard. A yard extending across the rear of a lot between the side lot lines and being the minimum horizontal distance between the rear lot line and the rear of the main building or any projections other than steps, unenclosed balconies, or unenclosed porches. On corner lots, the rear yard shall be considered as parallel to the street upon which the lot has its least dimension. On both corner lots and interior lots the rear yard shall in all cases be at the opposite end of the lot from the front yard.

Rear yard. No rear yard shall be less than ten feet wide on an interior lot, nor less than five feet wide on a corner lot for a building two stories or less in height. At each additional story height the width of such rear yard shall be increased one foot.

Rear yard. Yard extending across the full width of a site, the depth of which is the minimum horizontal distance between the rear property line and line parallel thereto on the site.

Rear yard. Open space (unoccupied except by accessory buildings) on the same lot with a building, between the rear line of the building and the rear line of the lot for the full width of the lot.

Rear yard. Yard extending from the rear wall of the building to the rear lot line across the full width of the lot.

Rear yard. Yard extending across the full width of the lot and measured between the rear lot line and the building or any projections other than steps, unenclosed balconies or unenclosed porches. On both corner lots and interior lots the opposite end of lot from the front yard.

Rear yard. A yard which is bounded by side lot lines, rear lot line and the rear yard line.

Rear yard. The area extending across the rear of a lot and being the required minimum horizontal distance between the rear lot line and the rear of the main building or any projection thereof other than projections of uncovered steps, unenclosed balconies or unenclosed porches. On all lots, the rear yard shall be in the rear of the front yard.

Rear yard. The portion of the yard on the same lot with the principal building located between the building and the rear lot line and extending for the full width of the lot.

Rear yard. The area extending across the full width of a lot and lying between the rear lot line and a line parallel thereto, and having a minimum distance between them equal to the required rear yard depth as prescribed for each Zoning District. Rear yards shall be measured by a line at right angles to the rear lot line, or by the radial line in the case of curved rear lot line. On all lots but through lots the rear yard shall be at the opposite end of the lot from the front yard. On corner lots the rear yard shall be considered as parallel to, and opposite from, the street frontage which has the least dimension.

Rear yard. Yard extending across the rear of a lot measured between the rear lot line and rear of the main building, or any projection thereof, other than steps, unenclosed balconies or unenclosed porches except as otherwise provided in this title. The rear yard shall be at the opposite ends of the lot from the front yard and on corner lots it may extend across the narrowest part of the lot.

Rear yard. Yard extending the full width of the lot along the rear lot line and extending in depth from the rear lot line to the nearest point of any structure on the lot.

Rear yard. Yard extending across the full width of the lot and measured between the rear line of the lot (not a street line) and the extreme rear line or the main building, including an enclosed or covered porch.

Rear yard. Every dwelling hereafter erected shall have a rear yard no less than 25 deet in depth. Where it is impracticable to provide a rear yard of the depth specified above, the building inspector shall have discretionary authority to reduce the required depth of the rear yard, provided, however, that in no case shall the rear yard be less than 15 feet in depth and provided further, that the minimum requirement for the narrower side yard shall be increased by 6 inches for each foot of reduction in depth of the rear yard.

Rear yard. There shall be a rear yard with a depth of not less than 40 percent of the lot depth, or 75 feet, whichever is the lesser.

Rear yard. Yard immediately in the rear of the main building on the lot and across the full width of the lot. It may or may not be on the opposite side of the building from the front yard.

Rear yard. Yard extending across the full width of the lot and measured between the rear line of the lot and the rear line of the building.

Rear yard. Space unoccupied except by an accessory building or use as hereinafter specifically permitted extending across the full width of the lot between the rear line of any building, other than an accessory building, and the rear lot line.

Rear yard. Yard extending across the entire width of the lot between the rear lot line and the rear building line rear lot line shall be the lot line farthest removed from the front lot line.

Rear yard. Yard covering the area between the side lot lines extending from the rear lot line to the nearest point of a principal use. On a corner lot, a yard covering the area between the interior side lot line and the exterior side yard, extending from the rear lot line to the nearest point of a principal use.

Rear yard. (Rear set-back line) A yard extending across the rear of a lot between the side lot lines and being the minimum horizontal distance between the rear lot line and the nearest wall or column line of the building.

Required open space. Refers both to usable, open space and to other areas required by the Code.

Required setback line. Line beyond which a building is not permitted to extend under the provisions of the Ordinance establishing minimum depths and widths of yards.

Required yard. Minimum yard required between a lot line and building line or the line of any parking area or any other use requiring a yard in order to comply with the zoning regulations of the district in which the zoning lot is located. A required yard shall be open and unobstructed from the ground upward except for projections on buildings as permitted in the Zoning Code and except for walks, landscaping and other yard or site features.

Required yard. Yard having a depth or width set forth in the applicable district regulations. Such width or depth shall be measured perpendicular to lot lines.

Required yard lines. Lines whose location with respect to lot lines provide required yard spaces.

River frontage yard. (Set-back from river). Yard extending across the rear, or along the side of a lot that abuts the existing bulkhead, when such exists or that abuts the established bulkhead line and being the minimum horizontal distance between such bulkhead or bulkhead line and the nearest wall or column line of the building.

Rounded property corner-yards. Depth of required front yards shall be measured at right angles at a straight line joining the foremost point of the side lot line. The foremost point of the side lot line, in the case of rounded property corners at street intersections, shall be assumed to be the point at which the side and front lot lines would have met without such rounding. The front and rear lines of front yards shall be parallel.

Side open space required. Required open space extending from the required front open to the rear required open space and of a width equal to the side required setback line read horizontally at right angles to the side lot line.

Side street yard. Side yard along the side street the width of which extends from a setback building line to a street line.

Side yard. A yard extending between a side lot line and the nearest wall of the building, and from the front yard to the rear yard; provided, that for a corner lot, the side yards extends from the front yard to the rear lot line.

Side yard. A yard extending from the front yard to the rear yard and measured between the side lot lines and the nearest building.

Side yard. An open space extending along the side line of the lot between the side line and any building or structure on said lot.

Side yard. A yard between the side line of the lot and the building or building group and extending from the front yard to the rear yard, or, in the absence of either of such yards, to the front or rear lot line, as may be required by Code.

Side yard. A yard extending from the front yard to the rear yard between the side lot line and the nearest line of the main building.

Side yard. A yard measured between the side lot line and the nearest point of the principal building, extending from the front yard to the rear yard.

Side yard. A yard between the main building and the side line of the lot being the minimum horizontal distance between the building and the side yard line, and extending from the front lot line to the rear yard line.

Side yard. Any yard that lies between a front and rear yard.

Side yard. A yard between the building and the side line of the lot and extending from the front to the rear yard.

Side yard. A yard between the main building and the side lot line and extending from the required front yard to the required rear yard, and being the required minimum horizontal distance between a side lot line and the side line of the buildable area.

Side yard. A yard along the side line of a lot and extending from the front building line to the rear yard line or, in the case of a side street yard, to the rear lot line.

Side yard. A yard extending from the front yard to the rear yard and measured between the side line of the lot and the side line of the building or any projection thereof.

Side yard. That portion of a zoning lot lying between the side line of the lot and a line drawn through the nearest point of a main building extending from the front yard to the rear yard, or in the absence of either of said yards from the front to the rear lot lines respectively.

Side yard. An open unoccupied space on the same lot with a building between the building and the side line of the lot extending from the front building line to the rear building line or to the rear line of the lot, where no rear yard is required.

Side yard. A yard extending from the rear line of the required front yard to the rear lot line, or in the absence of any clearly defined rear lot line to the point of the lot farthest from the intersection of the lot line involved with the public street. In the case of through lots, side yards shall extend from the rear lines of front lines required. Width of a required side yard shall be measured in such a manner that the yard established is a strip of the minimum width required by the district regulations with its inner edge parallel with the side lot line.

Side yard. A yard on each side of the building between the side wall of the building at the finished grade line and the side line of the lot, extending from the street line of the lot to the rear yard.

Side yard. A yard between the side lot line and the nearest side line of the building and extending from the rear line of the building to front line of the building.

Side yard. A yard extending from the front yard to the rear yard, being the minimum horizontal distance between a building and the side lot line.

Side yard. The required open space, the full depth of the lot extending from the side line of the lot to the nearest structure on the lot, exclusive of over-hanging eaves, gutters, or cornices, and exclusive of a private garage which is not an integral structural part of a main building.

Side yard. A yard which is bounded by the rear yard line, front yard line, side yard line and side lot line.

Side yard. A yard extending along a side lot line between the front and rear yards, except that a side yard abutting a street shall be defined as the yard extending along a side lot line between the front yard and the rear lot lines.

Side yard. An open unoccupied space, on the same lot with a building, between the building and the side line of the lot, extending from the front yard to the rear line of the lot.

Side yard. The area between the main building and the side line of the lot, and extending from the required front yard to the required rear yard, and being the minimum horizontal distance between a side lot line and the side of the main buildings or any projections thereto.

Side yard. A yard extending along a side lot line and back to a line drawn parallel to the side lot line at a distance therefrom equal to the width of the required side yard, but excluding any area encompassed within a front yard or rear yard.

Side yard. The area extending along the side of a lot lying between the required front yard line, the required rear yard line, the side lot line, and a line parallel to the side lot line, and having a minimum distance between the side lot line and the line parallel thereto equal to the required side yard depth as prescribed for each Zoning District. Side yards shall be measured by a line at right angles to the side lot line, or by the radial line in the case of a curved side lot line.

Side yard. A yard more than six inches in width between a main building and the side lot line, extending from the front yard or front lot line where no front yard is required, to the rear yard. The width of the required side yard shall be measured horizontally from the nearest point of the side lot line toward the nearest part of the main building.

Side Yard. A yard between the building and the side line of the lot and extending from the street line of the lot to the rear yard.

Side yard. An open unoccupied space unless occupied by a use as hereinafter specified in the Code, on the same lot with the building between the building and the side lot line, extending from the front yard to the rear yard.

Side yard. A yard extending from the front yard to the rear yard and measured from the side lot line to the foundation wall of the building.

Side yard. A required yard between the main building and the side lot lines and extending from the required front yard to the required rear yard, and being the minimum horizontal distance between a side lot line and a side of the maximum permissible main building.

Side yard. A yard extending from the front yard or front lot line to the rear yard or rear lot line, and from the side line of the building to the side lot line nearest that side of the building, or of any accessory building attached thereto.

Side yard. An open space extending from the front yard to the rear yard between a building and the nearest side lot line, unoccupied and unobstructed from the ground upward except as specified elsewhere in the Ordinance.

Side yard. A yard extending from the front building line to the rear building line between the side line and the side building line.

Side yard. A yard between the side line of the lot and the main building extending from the front yard to the rear yard and having a width equal to the shortest distance between said line and the main building.

Side yard. An open unoccupied space on the same lot with a building between the building and the side line of the lot extending thru from the front building line to the rear yard or to the rear line of the lot, where no rear yard is required.

Side yard. A yard between the side of the building and the lot line most nearly parallel thereto and extending from the street line to the rear yard.

Side yard. A yard extending the full depth of the lot along a side lot line and extending in width from such side lot line to the nearest point of any structure on the lot.

Side yard. A yard between the building and the side line of the lot which shall be considered to extend from the required rear yard to the street line of the lot.

Side Yard. A yard between the building and the side line of the lot and extending from the street line to the rear yard.

Side yard. A yard between the main building and the side lot line, extending from the front yard or front lot line where no front yard is required, to the rear yard, the width of which shall be the least distance between the side lot line and the nearest point of the main building.

Side yard. A yard adjacent to any side lot line extending from the front yard to the rear yard. The width of the required side yards shall be as specified in the dimensional requirements for each district.

Side yard. An open, unoccupied space, except for permitted accessory buildings or structures, on the same lot with a principal building, situated between the side line of the principal building and the adjacent side line of the lot and extending from the rear line of the front yard to the front line of the rear yard.

Side yard. A yard, between a main building and the side lot line, extending from the front to the rear property line. The width of the required side yard shall be measured from the nearest point of the side lot line toward the nearest main wall of the building.

Side yard. A yard situated between the principal building and a side line and extending from the required front yard to the required rear yard. The depth of a side yard shall be measured from the building to the nearest building to the nearest point in a side line parallel to the front lot line.

Side yard. A yard between the side line of a building and the adjacent side line of the lot, extending from the front yard to the rear yard. If there be no front yard, the side yard shall be considered as extending to the front lot line and if there be no rear yard, the side yard shall be considered as extending to the rear line of the lot.

Side yard. A yard covering the area between the front yard and the rear yard extending from the side lot line to the nearest point of a principal use. In a corner lot where the side lot line abuts a street, a yard covering the area between the front yard and the rear lot line extending from such side lot line to the nearest point of a principal use.

Side yard. A yard extending from the rear line of the required front yard, or the front property line of the site where no front yard is required, to the front line of the required rear yard, or the rear property line of the site where no rear yard is required, the width of which is the minimum horizontal distance between the side property line and a line parallel thereto on the site.

Side yard. No side yard shall be less than three feet wide for a building two stories or less in height and 80 feet or less in length. At each additional story height the width of such side yard shall be increased one foot, and for any additional length the width of such side yard shall be further increased at the rate of one foot in 20 feet.

Side yard. A yard extending from the front yard to the rear yard, or from the front property line where no front yard is provided, to the rear property line where no rear yard is provided; the width of the side yard is the minimum horizontal distance between the side property line and a line parallel thereto on the lot touching any part of a structure, other than parts specifically excepted in the Code.

Side yard. A space situated between any side line of a building and the adjacent side property line of the lot and extending from the rear line of the front yard to the front line of the rear yard. If no front yard is required the front boundary of the side yard shall be the front property line of the lot and if no rear yard is required the rear boundary of the side yard shall be the rear property line of the lot. In the case of corner lots, the yard most nearly conforming to side yards of this adjacent lot shall be considered to be a side yard.

Side yard. Any yard that lies between a front and rear yard lot line.

Side yard. An open space between the main building and the side lot line and extending from the front yard to the rear yard.

Side yard. Yard between the side line of building and the adjacent side property line, extending from the front yard to the rear yard.

Side yard. Open unoccupied space on the same lot with a building situated between the nearest roofed portion of the building or of any accessory building and the side line of the lot, and extending through from the front yard or from the front street line where no front yard exists, to the rear yard.

Side yard adjoining a street. A yard which is bounded by a rear yard line, a front yard line, side yard line, and a side lot line adjoining a street line.

Side yards. There shall be two side yards on each lot. No side yard shall be less than five (5) feet wide for a building two stories or less in height and fifty (50) feet or less in length. For each additional story in height, the width of such side yard shall be increased one and one-half (1-1/2) feet; and for any additional length, the width of such side yard shall be further increased at the rate of one foot in 20 feet.

Side yards. Every dwelling hereafter erected on any lot or plot with side lines of record, shall be so located that one side yard shall conform to the requirements of the Zoning Ordinance. Every dwelling hereafter erected on a portion of a lot without side lines of record shall be located so that the clear space between it and another structure shall be not less than 20 feet or as required by the Zoning Ordinance or Subdivision and Land Development Regulations. The width of a side yard of a corner lot abutting on a or highway shall not be less than the minimum front yard required on an adjoining lot fronting on such street, but this shall not reduce the usable width for building purposes of any lot of legal record at the time the passage of the Ordinance to less than 30 feet measured at the foundation ground level.

Side yard. (Side set-back line). A yard extending along the side of a lot between the front yard and the rear yard and being the minimum horizontal distance between the side lot line and the nearest wall or column line of the building.

Through lot yards. In the case of through lots, unless the prevailing front yard pattern on adjoining lots indicates otherwise, front yards shall be provided on all frontages. Where one of the front yards that would normally be required on a through lot is not in keeping with the prevailing yard pattern, the administrative official may waive the requirement for the normal front yard and substitute therefore a special yard requirement which shall not exceed the average of the yards provided on adjacent lots.

Transitional yard. A yard that must be provided between zoning districts as prescribed by the Ordinance.

Usable common open space. Open space which is suitably located and improved for common recreational purposes and accessible to each lot or dwelling within a development through a system of public or private walkways; such walkways may abut a private or dedicated right-of-way.

Usable open space. Includes the aggregate area of side and rear yards, patios, and balconies and decks having a minimum horizontal dimension of not less than five (5) feet, on a building site or building, which is available and accessible to the occupants of the building for purposes of active or passive outdoor recreation. Usable open space does not include driveways, areas for off-street parking and services, and ground level areas with a width of less than five (5) feet.

Usable open space. The part or parts of a lot designed and developed for use by the occupants of the lot for recreation, gardens, or household service activities (such as clothes drying) which space is effectively separated from automobile traffic and parking, and readily accessible by all those for whom it is required. Open space shall be deemed usable only if all of the following conditions apply; at least seventy-five (75) percent of the area has a grade of less than eight (8) percent. Each dimension is at least twenty (20) feet. Balcony space may be credited toward open space requirements if said space has minimum dimensions of six (6) feet.

Usable open space. Land area meeting the qualifications and definitions of either usable common open space or usable private open space.

Usable open space. Ground area of a lot may qualify as usable open space provided that they are areas unobstructed from the ground to the sky.

Usable open space. Open space which has slopes of less than ten (10) per cent, has all dimensions a minimum of twenty (20) feet from any residential wall containing a window, and is easily accessible by all residents occupying the same parcel or related parcels.

Usable private open space. Open space which is designed and maintained for sole and exclusive use of the occupants of not more than one (1) dwelling unit and may include private patio areas.

Usable yard. One or more well drained open areas located on the same lot as the principal use for use by the residents thereon for outdoor activities. This definition does not include driveways, or off-street parking or loading areas; but does include private balconies containing at least 20 square feet, and roofs available for outdoor activities.

Width of a yard or court. The least horizontal dimension at its lowest level.

Width of a yard or court. Width of a yard or court is its least horizontal dimension at its lowest level.

Width of court. Minimum horizontal dimension parallel to the principal open side in the case of an outer court; and the least horizontal dimension in the case of an inner court.

Width of court. As applied to an inner court, is its least horizontal dimension. "Width" as applied to an outer court, is the shortest horizontal dimension measured in a direction substantially parallel with the principal open end of such court.

Yard. A court on the same lot with a building extending along the entire length of a lot line.

Yard. An unoccupied open space other than a court.

Yard. Open space on the same lot with a building or structure unoccupied and unobstructed by any portion of a structure from thirty-six (36) inches above the general ground level of the graded lot upward, but not including such things as yard recreational and laundry drying equipment, arbors and trellises, flagpoles, yard lights, statuary or other similar decorative things. In measuring a yard for the purpose of determining the depth of a front yard or the depth of a rear yard, the least distance between the lot line and the main building shall be used. In measuring a yard for the purpose of determining the width of a side yard, the least distance between the lot line and nearest permitted building shall be used.

Yard. Open space other than a court, on the same lot with the building, unoccupied and unobstructed from the ground upward, except as otherwise provided herein.

Yard. Open space on the same lot with a building, unoccupied and unobstructed from the ground upward, except as otherwise provided herein. In measuring a yard for the purpose of determining the width of a side yard, the depth of a front yard or the depth of a rear yard the minimum horizontal distance between the lot line and the main building shall be used.

Yard. Unoccupied area of a lot, open and unobstructed from the ground to the sky.

Yard. That portion of the open area on a lot extending between a building and the nearest lot line, or between an accessory use of building and the nearest lot line as established in the Zoning Code.

Yard. Required open space on the same lot with a principal use.

Yard. Open unoccupied space on the same lot with a building extending along the entire length of a street, rear or side lot line.

Yard. Open space on the same lot with a building unoccupied and unobstructed by any structure from the surface of the ground upward except for drives, sidewalks, lamp posts, open patios, retaining walls, entrance steps, fences and landscaping.

Yard. The open space existing on the same lot with a principal building; unoccupied and unobstructed by buildings from the ground to the sky, between the lot line and building line.

Yard. A space, other than a court, on the same lot with a building, extending along a street, rear, or interior lot line, and occupied by only such appendages and other construction as are specifically permitted by the provisions of the code, to project or be constructed in such space.

Yard. An open, unoccupied space other than a court unobstructed from the ground to the sky, except where specifically provided for in the Code, on the lot on which a building is situated.

Yard. An open, unoccupied space on the plot between the property line and the front, rear or side wall of the building.

Yard. An open space, other than a court, on the same lot with a building.

Yard. An open space, unoccupied and unobstructed from the ground upward, except by trees or shrubbery or as otherwise provided herein.

Yard. An open space other than a court on the same lot with a building, unoccupied and unobstructed from the ground upward.

Yard. Any open space on the same lot with a building, open and unobstructed from the ground to the sky.

Yard. A required open space on the same lot with a building, unoccupied and unobstructed by any structure or portion of a structure from ground level upward, except as otherwise provided in the code.

Yard. An open, unoccupied space, other than a court, on the same lot with a building, unobstructed from the ground to the sky, except as otherwise provided.

Yard. That portion of a lot extending open and unobstructed from the ground upward.

Yard. An open space on a lot, unoccupied and unobstructed from the ground upward, except as otherwise permitted in the Ordinance.

Yard. An open space on the same site as a structure, unoccupied and unobstructed by structures from the ground upward or from the floor level of the structure requiring the yard upward except as otherwise provided in the ordinance, including the front yard, side yard, rear yard, or space between structures.

Yard. A required open space other than a court, unoccupied and unobstructed by any structure or portion of a structure from 30 inches above the general ground level of the graded lot upward, provided, however, that fences, walls, poles, posts, and other customary yard accessories, ornaments, and furniture may be permitted in any yard subject to height limitations and requirements limiting obstruction of visibility.

Yard. A space open to the sky and unoccupied or unobstructed except by encroachment or structures specifically permitted by the Code, on the same lot with a principal building. All yard measurements shall be the minimum distances between the front, rear, or side yard line, as the case may be, and the nearest point of the building included enclosed or covered porches. Every part of every yard shall be accessible from every other part of the same yard.

Yard. A space on the same lot with a principal building, open, unoccupied and unobstructed by buildings or structures from the ground to the sky, except where encroachments and accessory buildings are expressly permitted.

Yard. An open space of uniform width, or depth on the same lot with a building or a group of buildings, which open space lies between the principal building or group of buildings and the nearest lot line and is unoccupied and unobstructed from the ground upward as permitted in the Code.

Yard. A court on the same lot with a building extending along the entire length of a lot.

Yard. An unoccupied space on the same lot with a building extending along the entire length of a street, or rear or interior lot line.

Yard. An open space on the same lot with a main building or structure, extending between the lot line and the extreme front, rear, or side wall of the main building or structure.

Yard. An open space at grade between a building and the adjoining lot lines unoccupied and unobstructed by any portion herein. In measuring a part to determine the width of a yard the minimum horizontal distance between the lot line and the maximum permissible main building shall be the yard dimension.

Yard. The required space other than a court on any lot, unoccupied by a structure and unobstructed from the ground upward except as otherwise provided herein, and measured as the minimum horizontal distance from a building or structure, excluding carports, porches and other permitted projections, to the property line opposite such building line.

Yard. A court that extends along the entire length of a lot line.

Yard. An open space other than a court on a lot unoccupied and unobstructed from the ground upward.

Yard. An open unoccupied space on the same lot with a building or other structure or use, open and unobstructed from the ground to the sky, except for public utility lines or facilities.

Yard. A space on the same lot with a main building, open, unoccupied and unobstructed by buildings or structures from the ground to the sky, except as otherwise provided in the Ordinance.

Yard. A space on a lot, on which a building is situated, unoccupied except where otherwise provided in the Ordinance, open and unobstructed from the ground to the sky.

Yard. A required space other than a court on any lot, unoccupied by a structure and unobstructed from the ground upward except as otherwise provided herein, and measured as the minimum horizontal distance from a building or structure, excluding carports, porches and other permitted projections, to the property line opposite such building line, provided, however, that where a future width line is established by the provisions of the ordinance for any street bounding the lot, then such measurement shall be taken from the line of the building to such future width line, determined by the Zoning Code.

Yard. That portion of a lot extending open and unobstructed from the lowest level to the sky along the entire length of a lot line.

Yard. An open space on the same site as the structure, unoccupied and unobstructed from the ground upward, except as otherwise provided, including a front yard, side yard, or rear yard. In any district the "required yard" shall mean that portion of a yard meeting the minimum dimensions for such a yard in that district.

Yard. An open space on the same lot with a building, unoccupied and unobstructed by any portion of a structure from the ground upward, except as provided in the Ordinance. In measuring a yard for purposes of determining the required width of a side yard, the required depth of a front yard or the required depth of a rear yard, the minimum horizontal distance between the lot line and the main building shall be used.

Yard. An open, unoccupied space other than a court, unobstructed from the ground to the sky, except where specifically provided by the Code, on the lot on which a building is situated.

Yard. An open space on the same lot with a building or a building group, lying between the building or building group and the nearest lot line, and unoccupied and unobstructed by any structure from the ground upward.

Yard. An open space on the same lot with a building and the adjoining lot lines, and unoccupied and unobstructed from the ground upward, except as otherwise provided in the code. The measurement of a yard is the minimum horizontal distance between the lot line and the building or structure.

Yard. An open space other than a court at existing ground level between a buildable area and the adjoining lot lines, unoccupied and unobstructed by any portion of a structure from the ground upward, except as otherwise provided in the code. For the purpose of determining yard measurements, the least horizontal distance between the lot line and the buildable area shall be used.

Yard. An open space on the same lot with a building, unoccupied and unobstructed from the ground upward, except by trees or shrubbery or as otherwise provided.

Yard. Yard means a court that extends along the entire length of a lot line.

Yard. An open unoccupied space other than a court, unobstructed from the ground to the sky, except where specifically provided by the Building Code.

Yard. An open space other than a court, on a lot, unoccupied and unobstructed from the ground upward, except as otherwise provided in the Ordinance.

Yard. An open space on a lot, other than a court, unoccupied and unobstructed from the ground upward, except as otherwise provided herein.

Yard. An open unoccupied space on the same lot, plot or parcel of land on which the building stands, which extends the entire length of the front or rear or interior lot line.

Yard. A required space other than a court on any lot, unoccupied by a structure and unobstructed from the ground upward except as otherwise provided herein, and measured as the minimum horizontal distance from a building or structure, excluding carports, porches, and other permitted projections, to the property line opposite such building line; provided, however, that where future width line is established by the provisions of the Zoning Code for any street bounding the lot, then such measurement shall be taken from the line of the building to such future width line determined by the Zoning Code.

Yard. An open space on the same site as a structure, or building, unoccupied and unobstructed by structures or buildings from the ground upward or from the floor level of the structure or building, requiring the yard upward except as otherwise provided in the Ordinance, including a front yard, side yards, rear yard or space between structures or buildings.

Yard. An open space measured from the nearest property line to the closest point of a structure which is unoccupied and unobstructed except for natural features and certain exceptions permitted by the Ordinance.

Yard. An open space at grade between a building and the adjoining lot lines, unoccupied and unobstructed by any portions of a structure from the ground upward, except as otherwise permitted by the Ordinance.

Yard. A required open space, on the same lot with a building or structure, unoccupied and unobstructed from its lowest point to the sky, except as otherwise permitted by the Ordinance.

Yard. An open space, on the same lot with a building unoccupied and unobstructed from the ground upward except as otherwise provided in the Ordinance.

Yard. An open unoccupied space on the same lot, plot, or parcel of land on which the building stands, which extends the entire length or width of the property.

Yard. An open space at grade between a building and adjoining lot lines, unoccupied and unobstructed by any portion of a structure or from a group of structures, upward except as otherwise provided in the ordinance. In measuring a yard to determine the width of a yard the minimum horizontal distance between the lot line and the maximum permissible main building shall be the yard dimension.

Yard. An open space, on a lot with a main building, left open, unoccupied and unobstructed by buildings from the ground to the sky except as otherwise provided in the ordinance.

Yard (front). A yard extending along the full width of the front lot line between side lot lines.

Yard (rear). The portion of the yard on the same lot with the building between the rear line of the building and the rear line of the lot for the full width of the lot. In those locations where an alley is platted in the rear of the lots; 1/2 of the width of the platted alley may be included as part of the rear yard equipment.

Yard (rear). A yard extending across the full width of the lot and measured between the rear line of the lot and the rear line of the building.

Yard (side). A yard extending along a side lot line between the front and rear yard.

Yard (side). A yard extending from the front yard to the rear yard and measured between the side line of the lot and the side line of the building or any projection thereof.

Yards; front, rear and side. (1) A yard which abuts a street or the waterway is a front yard. (2) All navigable waterfront yards other than Ocean or the Gulf are rear yards. (3) All yards which abut an alley are rear yards. (4) Rear yards, other than the above, occur only in the interior of a block and are parallel or nearly parallel to the front yard. (5) Yards other than above which are perpendicular or nearly perpendicular to a front yard are side yards. (6) Corner lots and double frontage lots which are not on an alley and are not on navigable waters, have two (2) front yards, two (2) side yards, and no rear yards. (7) A lot encircled by streets has only front yards. (8) In the event of irregular lot shapes, the Zoning Administrator shall determine yard locations, which shall as closely approximate the above rules as is practical.

1. Category C

BUILDING AND STRUCTURE USES C-1 THROUGH C-30

C-1 ANIMALS-BIRDS, HOUSING AND CARE

Animal. Any horse, mule, donkey, pony, cow, sheep, goat, swine, dog, cat, rabbit, chicken, goose, duck, turkey, or any other animal or fowl.

Animal at large. Any animal not securely confined by a fence or other means on premises under the control of or occupied by, the owner of the animal, and not under the control of the owner, a member of his immediate family twelve (12) years of age or older, or an agent of the owner, by leash or otherwise.

Animal clinic or animal hospital. A place where animals or pets are given medical or surgical treatment in emergency cases and are cared for during the time of such treatment. Use as a kennel shall be limited to a short time boarding period, and shall be only incidental to such clinic or hospital use and shall be enclosed in a soundproof building or structure.

Animal clinic or hospital. A facility for the impounding, treatment, and boarding of animals for a period of longer than twelve (12) hours.

Animal feed yard. A fenced area where livestock are fed a concentrated ration.

Animal hospital. A building wherein the care and treatment of sick or injured dogs, cats, rabbits, birds, and similar small animals are performed.

Animal hospital. An establishment in which veterinarian services are rendered to dogs, cats and other small animals and domestic pets.

Animal hospital. Any building or portions thereof, designated or used for the care, observation, or treatment of domestic animals.

Animal hospital. Establishment in which veterinarians service the needs of providing health care for dogs, cats and other small animals and pets.

Animal hospital. A place where animals or pets are given medical or surgical treatment and are cared for during the time of such treatment. The use of the premises as a kennel or a place where animals or pets are boarded for remuneration may be permitted only when incidental to the principal use.

Animal hospital. A building or portion thereof designed or used for the boarding, care, observation or treatment of domestic animals.

Animal hospital. A building or portion thereof designed or used for the care, observation or treatment of domestic animals as a business, including the boarding of domestic animals when conducted as an accessory activity to the foregoing listed activities.

Animal hospital or animal boarding place. Building, lot or plot of ground in or on which sick, injured or well animals are housed or confined, except that this definition shall not apply to rooms in hotels, private dwellings, or rooming houses not intended to be regularly occupied by sick, injured, or well animals.

Animal shelter. A building, structure or facility operated, owned, maintained, or used by a duly incorporated humane society, animal welfare society or other not-for-profit organization whose purpose is to provide for and promote the welfare, protection and humane treatment of animals, including animals impounded for rabies observation.

Aviary. Any lot, building, structure, enclosure or premise whereupon or wherein are kept more than 25 ornamental or song birds, in any combination whether such keeping is for pleasure, profit, breeding or exhibiting but not including poultry or birds kept for production and sale of meat and/or eggs.

Aviary. A house, large cage or enclosure for keeping and rearing of four (4) or more birds in confinement.

Bird. A warm-blooded, winged, feathered, egg-laying vertebrate, of the aves class of animals, primarily adapted for flying.

Commercial domestic animal kennel. The keeping of any dogs, cats or other domestic animals, regardless of number, for sale, breeding, boarding or treatment purposes except in an animal hospital, pet beauty parlor, or pet shop, as permitted by these regulations, or the keeping of 5 or more domestic animals or fowl, 6 months or older, on premises used for residential development, or the keeping of more than one domestic animal on vacant property or on property used for business or commercial purposes.

Commerical kennel. Use, for gain or profit, of all or a portion of any lot or premises for the sale, breeding, and/or the permanent or temporary boarding of dogs, cats or other household pets.

Commercial stable. An area of ten (10) or more contiguous acres containing a stable for horses, mules or ponies which are hired, bred, shown or boarded on a commercial basis. Commercial stables must meet the requirements of the Building Code.

Corral. Space other than a building, less than one (1) acre, or less than one hundred (100) feet in width, used for the confinement of animals.

Dog. Any dog, bitch or spayed bitch.

Dog kennel. Lot, building, structure, enclosure, or premises whereon or wherein three (3) or more dogs over four months of age are kept and maintained for any purpose whatsoever.

Dog kennel. Any premises where three (3) or more dogs over four (4) months of age are owned, boarded, bred and/or offered for sale.

Dog of licensing age. Any dog which has attained the age of 7 months or which possesses a set of permanent teeth.

Dog run. A fenced-in area for the containment of dogs or similar animals. For purposes of the Ordinance a "dog run" is considered a structure.

Dog shop. Any room or group of rooms, cage, or exhibition pen, not part of a kennel, wherein dogs for sale are kept or displayed.

Domestic pet service. A business establishment where clipping, bathing and related services, except boarding and housing of a veterinary nature, are rendered to dogs, cats and other domestic pets.

Horse ranch. A facility in which the breeding and raising of horses for sale is conducted.

Household pets. Animals or fowl ordinarily permitted in the house and kept for company or pleasure, such as dogs, cats, canaries and parakeets, but not including a sufficient number of dogs to constitute a kennel, as defined in the Code, unless the dogs are licensed under the provisions of the Code, relating to "Dog Fanciers" or a sufficient number of fowl to constitute an aviary, as defined in the code.

Kennel. Any lot, building, structure or premises on which four (4) or more dogs more than four months old are kept, but this term shall not include any dogs licensed under the provisions of the Code relating to "Dog Fanciers."

Kennel. Any establishment at which dogs and cats are bred or raised for sale, or boarded, cared for, commercially, exclusive of dental, medical or surgical care.

Kennel. Any lot or premises on which four (4) or more dogs, more than six (6) months of age, are kept.

Kennel. Any structure or premises where three (3) or more dogs over four (4) months of age are kept.

Kennel. A kennel is any lot, building, structure, enclosure, premise or place, whereon or wherein, three (3) or more dogs are kept or maintained for the purpose of breeding, training, raising, boarding or as pets.

Kennel. Any place in, or at which, more than three (3) dogs more than six (6) months of age or any number of dogs that are kept for the purpose of sale or in connection with the boarding, care, or breeding of which any fee is charged.

Kennel. Keeping and breeding of more than three (3) dogs that are more than six (6) months old.

Kennel. Any premises, except veterinary hospitals, where four or more dogs, six (6) months of age or older, are kept.

Kennel. Any lot or premises used for the sale, boarding or breeding of dogs, cats or other household pets, Kennel shall also mean the keeping of three (3) or more dogs over the age of six (6) months.

Kennel. Lot, building, structure, enclosure or premises whereupon or wherein are kept seven (7) or more dogs, cats or similar small animals in any combination, for more than ten (10 days, whether such keeping is for pleasure, profit, breeding, or exhibiting, and including places where dogs or cats or similar small animals in any combination are boarded, kept for sale, or kept for hire.

Kennel. Any lot or premises on which four or more domestic animals, at least four (4) months of age, are kept.

Kennel. Any premise, land or buildings, enclosed or unenclosed, wherein or whereon more than three (3) dogs or three (3) cats are housed or kept. When such animals are not raised or bred for sale, then in determining the number, for the purposes of the ordinance, dogs and cats under the age of four (4) months shall not be considered.

Kennel. Keeping of more than two (2) dogs or cats that are more than six (6) months old.

Kennel. Building or property where dogs and/or other small animals and/or pets are kept, sheltered, boarded or medically treated either for or without compensation.

Kennel. Any premises, except where accessory to an agricultural use, where five (5) or more dogs or other domestic animals which are not sick or injured and are ten (10) weeks in age or older are boarded for compensation, cared for, trained for hire, kept for sale, or bred for sale.

Kennel. Any premises where six (6) or more dogs or cats are bred, boarded and/or trained.

Kennel. Any premises where animals are boarded for remuneration or kept primarily for sale.

Kennel. Any structure or premises on which five (5) or more dogs over four (4) months of age are kept.

Kennel. Any lot or premises or portion thereof on which more than four dogs, cats or other household domestic animals, over six months old are kept or on which more than two such animals are boarded for compensation or kept for sale.

Kennel. Any establishment wherein or whereon the business of boarding or selling dogs or breeding dogs for sale is carried on, except a pet shop.

Kennel. Any lot or premises on which four or more dogs, at least four months of age are kept.

Kennel. Any lot, building, structure, enclosure, premise or place, whereon or wherein, three (3) or more dogs are kept or maintained for the purpose of breeding, training, raising boarding or as pets.

Kennel. Any lot or premises on which three (3) or more dogs are kept, either permanently or temporarily boarded.

Owner. Any person, firm, or corporation owning, harboring or keeping an animal. The occupant of any premises on which a domesticated or tamed animal remains, or to which it customarily returns, for a period of ten (10) days or more, is deemed to be harboring or keeping the animal.

Pen. Small enclosure used for the concentrated confinement and housing of animals or poultry; as a pig pen, a place for feeding and fattening animals; a coop; an enclosure within an enclosure. A pen is not to be construed as a pasture or range.

Pound. Any establishment for the confinement of dogs.

Private stable. A detached accessory building for the keeping of horses owned by the occupants of the premises, and not kept for remuneration, hire or sale.

Private stable. Stable is any building located on a lot on which a residence is located, designed, arranged, used or intended to be used for not more than four horses for the private use of the residence, but shall not exceed 6000 square feet in area.

Private stable. Accessory building with capacity for not more than two (2) horses provided, however, that the capacity of a private stable may be increased if the lot whereon such stable is located contains area of not less than twenty-five-hundred (2500) square feet for each horse stabled.

Private stable. Accessory building for the housing of not more than 2 horses or mules owned by a person or persons living on the premises and which horses or mules are not for hire or sale.

Private stable. Building or portion thereof for the housing of horses, ponies, or mules owned by the occupants of the premises, and not kept for remuneration, exhibition, hire, or sale.

Private stable. Stable with capacity for not more than two (2) horses, provided, however, that the capacity of a private stable may be increased if the premises whereon such stable is located contains an area of not less than two-thousand-five-hundred (2500) square feet for each horse accommodated.

Private stable. A detached accessory building for the keeping of horses, mules or ponies owned by the occupants of the premises and not kept for remuneration, hire or sale.

Private stable. A detached accessory building for the shelter of horses or similar hoofed animals for the use of the residents of the premises and their guests.

Private stable. A stable with a capacity of housing not more than four (4) horses.

Private stable. Stable, corral or paddock used or designed to shelter horses belonging to the occupants of a dwelling and where no horses are kept for hire or sale.

Private stable. Separate accessory building with a capacity for not more than one (1) horse for each three thousand (3,000) square feet of lot area whereon such stable is located and where such horses are owned by the owners or occupants of the premises and not kept for remuneration, hire or sale.

Private stable. Stable with a capacity for not more than two (2) animals.

Private stable. Stable with a capacity for not more than four (4) horses or mules.

Private stable. Stable with capacity for not more than two (2) horses, provided, however, the capacity of a private stable may be increased if the premises whereon such stable is located contains an area of not less than five-thousand (5,000) square feet for each horse accommodated.

Private stable. Stable wherein all quadrupeds housed are the property of the owner or lessee or of his immediate family.

Private stable. Stable with capacity for not more than two (2) horses; provided, that the capacity of a private stable may be increased if the lot whereon such stable is located is in accordance with the Ordinance.

Private stable. Accessory building for the housing of not more than two (2) horses or mules owned by a person or persons living on the premises and which horses or mules are not for hire or for sale.

Private stable. Stable with a capacity of not more than one (1) horse for each thirty-five hundred (3,500) square feet of lot area, whereon, such stables are located, and where such horses are owned by the owners or occupants of the premises, and are not kept for remuneration, hire or sale.

Public riding stable. A stable, other than a private stable, generally where horses are available for hire.

Public stable. Stable with a capacity for more than two (2) horses.

Public stable. Stable wherein some or all of the quadrupeds are housed for hire or by other than the owner or the lessee of such stable.

Public stable. Stable wherein some or all of the horses are housed for hire or by other than the owner or the lessee of such stable.

Public stable. Stable with a capacity for more than two (2) horses or mules.

Public stable. Any stable for the housing of horses or mules, operated for remuneration, hire, sale, or stabling, or any stable with a capacity for more than two (2) horses or mules, whether or not such stable is operated for remuneration, hire, sale or stabling.

Public stable. Stable with a capacity for the housing of more than two (2) horses or mules which stable may be operated for remuneration, hire, sale or stabling.

Public stable. A stable with a capacity of housing more than four (4) horses.

Quadruped. A four-footed mammal, for the purpose of the Code.

Riding academy. Any establishment where horses are kept for riding, driving, stabling for compensation, or incidental to the operation of any club, association, ranch, or similar establishment.

Riding stable. Any stable where horses are kept for hire.

Riding stable. Any place at which horses or ponies are kept for hire, either with or without instruction in riding.

Shelter. An establishment where dogs are received, housed and distributed without charge.

Small-animal hospital or clinic. Any institution where animals other than livestock are given medical or surgical treatment and are cared for during the period of such treatment. Use of such facilities as a kennel shall be limited to short term boarding of animals awaiting or convalescing from treatment. All kennel and treatment facilities shall be confined to completely enclosed soundproof buildings.

Stable. A structure for housing a horse or horses.

Stable. A structure designed, intended, or used for the keeping of one (1) or more horses.

Stable. The same meaning as "garage," a draft animal being considered the equivalent as one (1) self-propelled vehicle.

Stable. A building occupied or used for the housing of quadrupeds.

Stable. Building, structure or portion thereof, which is used for the shelter or care of horses, cattle or other similar animals either permanently or transiently.

Stable. Building occupied or used for the housing of one or more quadrupeds.

Stable (private). A stable other than a public stable with capacity for not more than 2 horses or mules; however, on lots having an area of 4,000 square feet or more, the capacity of a private stable may be increased, if the premises whereon such stable is located contain not less than 2,000 square feet for each horse or mule.

Stable (public). A stable other than a private stable, which is operated for remuneration, hire, sale, or stabling, with a capacity for more than 2 horses or mules.

Stable (public). A stable where horses are kept for remuneration, hire or sale.

Veterinary hospital or clinic - animal hospital. Institution for the treatment and care of illnesses and injuries of animals.

Vicious dog. Any dog which has attacked or bitten any human being or which habitually attacks other dogs or domestic animals.

C-2 AWNINGS, CANOPIES, MARQUEES, AND PORTE COCHERES

Awning. Term sometimes used to describe a marquee; any covering intended as a screen from the sun, or protection from the rain.

Awning. Any movable rooflike structure, cantilevered, or otherwise supported from a building, so constructed and erected as to permit its being readily and easily moved within a few minutes time to close an opening, or rolled or folded back to a position flat against the building or a cantilevered projection thereof, or is detachable.

Awning. Fragile shelter or shade of canvas or other woven material wholly suspended from a wall of a building or structure.

Awning. Roof like cover entirely supported by and extending from a building for the purpose of protecting openings therein from the elements.

Canopy. Same use as an awning.

Canopy. Structure or framework entirely or partly covered with canvas, cloth, or other similar fabric, extending from the entrance of a building over the major portion of the width of a sidewalk and supported in whole or in part by vertical supports to the ground or to the sidewalk, being in whole or part on public property, and erected for the major purpose of protecting exits and entranceways of a building from the elements.

Canopy. Rooflike structure serving the purpose of protecting pedestrians from rain and sun, which structure projects from a building and the width of which (width being taken as the dimension parallel to the face of the building) is not greater than 1/4th the width of the face of the building or 20 feet, whichever is lesser. Such structure must be open on 3 sides and, if ground supported, supports must be confined in number and cross-section area to the minimum necessary for actual support of the canopy.

Canopy. Projection from a building, cantilevered, suspended, or supported on columns intended only for shelter or ornamentation.

Canopy. Detachable, rooflike cover, supported from the ground, or deck, floor or walls of a building, for protection from sun or weather.

Canopy. Ornamental covering over a niche; name usually given to a marquee.

Drop awnings. Device so constructed which can be raised or lowered and supported entirely on the face of a building by a metal or wood frame suitably constructed and properly attached to the building so as to leave the area underneath it entirely unobstructed thereby. Drop awnings may be built of any approved material.

Entrance canopy. Fixed or stationary canopy or hood constructed to provide protection at the entrance of a building, either supported entirely from the building or supported partly from a building or posts either outside or within the street or alley line, and located 2 feet or more from the outer face of the curb.

Fire canopy. A solid horizontal projection, extending beyond the exterior face of a building wall, located over a wall opening so as to retard the spread of fire through openings from one story to another.

Fixed awning. Fixed structure supported by cantilevering or bracketing from the face of a building, and constructed to provide shelter or shade.

Fixed awning. Rigid, roof-like structure supported entirely from a wall of a structure. The roof shall be fire retardant material supported by a noncombustible framework.

Fixed awning. A device which is attached to a building and supported by suitable means which in effect shades that portion of the building and shall be securely attached and cannot be raised or lowered and shall leave the area underneath it entirely unobstructed.

Marquee. A permanent roof-like structure, movable or stationary, which projects from the wall of a building to afford shelter from the elements over a platform, stoop or sidewalk.

Marquee. A cantilevered or otherwise supported projection from a major building constructed to be, or appear to be, an integral part thereof by being of similar material and intended for the weather protection of the main entrance and extending on each side of the opening a distance not greater than the projection from the building.

Marquee. Permanent roofed structure attached to and supported by the building. Marquees projecting over public property are regulated by the code.

Marquee. Rigid shelter or shade wholly suspended from the walls or framing of a structure and located directly above the entrance thereto.

Marquee. Canopy which requires additional loading for displays, signs, or other advertising devices.

Marquee. A fixed structure constructed at the ground story level of a building to provide shelter for sidewalks or walks; supported by cantilevering outward from the face of a building, or by ornamental chains or rods, and not supported by posts or columns.

Marquee. A hood or cover projecting from the face of a building wall for the major purpose of protecting the entranceway of a building from the elements, and being wholly and rigidly attached to and supported by the walls of a building.

Marquee. A permanent roofed structure attached to and supported by the building and projecting over public property.

Marquee. A permanent roofed structure attached to and supported by a building.

Marquee. A permanent roofed structure attached to and supported by the building and projecting beyond the face of the building.

Marquee. Exterior, approximately horizontal overhang, permanently attached at a structure above the street entrance to provide shelter or to advertise events taking place within the structure.

Marquee. Any hood or awning of permanent construction projecting from the wall of a building above an entrance.

Non-retractable awning. Structure or framework used for major purpose of protecting windows of a building from the elements, wholly and rigidly attached to and supported by the walls of a building and being of such construction that may not be retracted or pulled back to the exterior face of a building wall.

Porte-cochere. A one-story porch under which vehicles may be driven for the purpose of providing shelter for either the vehicle or persons and which is open, full width, front and rear in the direction of vehicle travel, and open not less than fifty (50) percent on the outer side.

Porte-cochere. Awning or open shed type structure attached to a building to afford temporary shelter to vehicles at a point of entrance.

Porte-cochere. An awning or shed attached to a building to afford shelter to vehicles at a point of entrance.

C-3 BOARDING, LODGING, ROOMING AND TOURIST HOUSES

Accessory sleeping quarters. Space or building which is accessory to another occupancy and used as sleeping quarters for no more than 4 persons.

Accessory sleeping quarters for transients. Space or building which is an accessory to another occupancy and which is used as sleeping quarters for five or more transients.

Apartment and lodging or rooming house combined. A building designed, constructed or altered in such manner that the sum of the number of rooms in the apartment units (excluding bathrooms and kitchens) plus the number of lodging rooms shall not exceed 15.

Boarder (roomer, lodger). An individual living within a household who pays a consideration for such residence and does not occupy such space as an incident of employment therein.

Boarding home for the aged. Any building, however named, which is operated for the express or implied purpose of providing service or domiciliary care for three or more elderly people who are not ill or in need of nursing care, and in which there is no agreement that such service shall include personal care or special attention.

Boarding house. Building or premises where meals are regularly served by prearrangement for definite periods for compensation for four more persons, but not exceeding twenty persons, not open to transient guests, in contradistinction to hotels or restaurants open to transients.

Boarding House. Building other than a hotel or a motel, where, for compensation and by prearrangement for definite periods, meals, or lodging and meals, are provided for three or more persons, but not exceeding twenty persons, including nursing homes with less than twenty-one patients, and tourist homes accommodating not more than twenty persons.

Boarding house. Any building, not a hotel, inn, or tavern, in which persons are lodged or received for a consideration for a single day or night or longer period.

Boarding house. Building other than a hotel where meals are provided for compensation for four or more persons.

Boarding house. Building other than a hotel, where lodging and meals for five or more persons are served for compensation. A Boarding House may also include the dwelling unit occupied by the owner or operator.

Boarding house. Dwelling in which, for compensation, lodging or meals, or both, are furnished to not more than nine guests.

Boarding house. Any building where sleeping accommodations or meals are provided for compensation for less than eleven persons.

Boarding house. A dwelling in which three (3), four (4) or five (5) rooms are occupied as guest rooms and in which food may be served to the occupants thereof. Any dwelling in which more than five (5) rooms are occupied as guest rooms shall be deemed to be a hotel. A boarding house shall not include institutions for persons requiring physical or mental care.

Boarding house. Building other than a hotel, where, for compensation and by prearrangement for definite periods, meals, or lodging and meals, are provided for three or more persons, but not exceeding twenty persons.

Boarding house. A building arranged or used for lodging more than five (5) and not more than twenty (20) individuals, with or without meals for compensation.

Boarding house. Boarding House is a lodging house in which meals are provided.

Boarding house. A building designed, constructed, or altered to contain not more than 15 sleeping rooms, where lodging and meals are served for compensation, but not to anyone who may apply.

Boarding house. Means any building, not a hotel, inn, or tavern, in which persons are lodged or received for a consideration for a single day or night or longer period.

Boarding House. A building other than a hotel where meals, or lodging are furnished for compensation for 3 or more persons not members of a family.

Boarding house. Shall mean a building where lodging and meals are provided for compensation for not more than fifteen (15) persons, but shall not include rest homes.

Boarding house. Any structure or portion thereof originally designed for use as a single family dwelling or a two-family dwelling which is occupied by, or offered for occupancy by, more persons than consistent with the definition of the word "family."

Boarding house. A building where lodging and meals are provided for compensation by prearrangement for a definite period to five (5) or more, but not exceeding twelve (12) individuals.

Boarding house. A building where meals are regularly served to two (2) or more persons, not members of the operator's family, by prearrangement for definite periods for compensation.

Boarding house. A building or portion thereof where meals are provided to five (5) or more persons who are not members of the operator's family, and by prearrangement for definite periods of time and for compensation, whether direct or indirect.

Boarding house. A building other than a hotel where, for compensation and by prearrangement, lodging and meals are provided for at least three (3), but not more than twelve (12) persons who are not members of the principal family which occupies the building as a dwelling.

Boarding house. Any building, except a hotel, inn, or tavern, in which persons are lodged, for a day or night or longer period, for consideration.

Boarding house. A building where lodging or both meals and lodging, without individual cooking facilities, are provided for compensation and by prearrangement for a week or more at a time, for six (6) or more persons. The term includes a guest house, a rooming house, a lodging house, or a college or university dormitory. The term does not include a hotel or motel as defined in the Code, or any building where the accommodation of guests involves custodial care or supervision.

Boarding house. Any dwelling other than a hotel where meals or lodgings and meals for compensation are provided for five or more persons, pursuant to previous arrangement and not to any one who may apply.

Boarding house. Any house or building or portion thereof which is designed, arranged, or let to be used, or which is used to accommodate for compensation, more than four boarders or one-family in addition to the members of the family occupying said building.

Boarding house. Building where for compensation, lodging, or lodging and meals are provided for not more than 30 persons.

Boarding house. Lodging house, where for compensation and by prearrangement for definite periods, by the week or greater term, tableboard is provided to guests and others, but wherein, meals are not furnished to occasional or transient customers.

Boarding house. Dwelling in which there is no more than one (1) dwelling unit and more than two, but not exceeding five (5) rooming units or guest rooms. Meals may or may not be provided to the occupants thereof. A boarding house shall not include homes for persons not members of the family requiring professional or semi-professional care by reason of physical or mental infirmity or disease or by reason of age.

Boarding house. Dwelling in which 3, 4, or 5 rooms are occupied as guest rooms and in which food may be served to the occupants thereof. Any dwelling in which more than 5 rooms are occupied as guest rooms shall be deemed to be a hotel. A boarding house shall not include institutions for persons requiring physical or mental care by reason of age, infirmity or disease.

Boarding house. Residential establishment other than a hotel or motel where sleeping and eating accommodations are offered to the public on a non-transient basis.

Boarding house. Building other than a hotel where meals and lodging are provided for compensation for 4 or more persons.

Boarding house. Lodging house in which meals are provided.

Boarding house. A building other than a hotel, where for compensation and by prearrangement for definite periods, meals or lodging and meals, are provided for three (3) or more persons, but not exceeding five (5) persons.

Boarding house. A residential establishment other than a hotel or motel where sleeping and eating accommodations are offered to the public on a non-transient basis for compensation.

Boarding house. Building other than a hotel were meals, or lodging are furnished for compensation for 3 or more persons not members of a family.

Boarding house. Building where lodging and meals are provided for compensation for not more than fifteen (15) persons, but shall not include rest homes.

Boarding house. Building other than a hotel, where for compensation and by prearrangement for definite periods, meals or lodging and meals, are provided for three or more persons, but not exceeding five persons, but does not include the operator.

Boarding house. Dwelling where, for compensation for four or more persons on a weekly or monthly basis or with reference to longer periods, excluding transient guests or any patron not regularly residing on the property and paying a weekly or monthly rate therefore. "Boarding House" shall be interpreted to mean a traditional type of establishment, accommodating competent patrons, not a social welfare establishment.

Boarding house. Building other than a hotel, where lodging or meals or both are served for compensation.

Boarding house. Building other than a hotel where lodging and meals for five or more persons are served for compensation.

Boarding house. A building other than a hotel where lodging and meals for five (5) or more persons are served for compensation.

Boarding house. A building with not more than three (3) guest rooms where, for compensation, lodging and/or meals are provided for at least three (3) but not more than ten (10) persons.

Boardinghouse. A building other than a hotel where lodging and meals for five (5) or more guests are served for compensation.

Boarding house and rooming house. Where meals or lodging are provided for persons other than the family or their relation excluding facilities for transient persons such as hotels, motels, inns and other such facilities.

Boarding house, lodging house, tourist house. A building arranged or used for lodging, with or without meals for compensation, more than three (3) and not more than twenty (20) individuals.

Boarding house, lodging house, tourist house. Building arranged or used for lodging, with or without meals, for compensation, more than five not more than twenty individuals.

Boarding house, lodging house, tourist house. Building arranged or used for lodging, with or without meals for compensation, more than 6 and not more than 20 individuals, unless otherwise prescribed by zoning or other laws.

Boarding house—tourist house. Building arranged or used for lodging, with or without meals, for compensation by more than five and not more than twenty individuals.

Boarding-lodging house. Dwelling wherein lodging or meals for 3 or more persons, not members of the principal family therein, is provided for compensation, but not including a building in which 10 or more guest rooms are provided.

Boarding or rooming house. A dwelling containing a single dwelling unit and not more than five guest rooms or suites of rooms, where lodging is provided with or without meals, for compensation.

Boarding or rooming house. Dwelling, other than a hotel or motel, wherein furnished rooms with or without cooking facilities are rented for profit to 4 or more persons, not related to the owner or proprietor.

Boarding or rooming house. Dwelling, other than a hotel or motel, wherein furnished rooms with or without cooking facilities are rented for profit to four or more persons, not related to the owner or proprietor.

Boarding or rooming house. Building in which are provided sleeping accommodations for rent to more than four persons on either a transient or permanent basis, with or without meals, but without providing kitchens for individual units.

Boarding or rooming, or lodging house. Dwelling other than a hotel having one kitchen and used for the purpose of providing meals or lodging or both for compensation to persons other than members of the family occupying such dwelling.

Cabin camp. Any multiple dwelling or group of dwellings, other than bungalow courts, hotels, or apartment houses, which is designed or intended for the temporary residence of motorists or travelers. This term shall include "Cabin Camps", "Auto Courts", "Auto Camps", and "Motor Camps".

Dormitories. Buildings or spaces in buildings where group sleeping accomodations are provided for persons not members of the same family group in one room or in a series of closely associated rooms under joint occupancy and single management, as in college dormitories, fraternity houses, military barracks; with or without meals, but without individual cooking facilities.

Dormitory. A room having separate sleeping accommodations for more than four (4) persons.

Dormitory. Building intended or used principally for sleeping accommodations where such building is related to an educational or public institution, including religious institutions and hospitals.

Dormitory. Building or portion thereof in which group sleeping accommodations are provided for more than four persons not members of the same family group in one room or in a series of closely associated rooms under joint occupancy and single management including, but not limited to college dormitories, fraternity houses, barracks, and convents.

Dormitory. As applied to a building, means a building whose principal use is "accessory sleeping quarters for transients" as defined in the code.

Dormitory. As applied to a room, means a room which is used as "accessory sleeping quarters" as defined in the code for transients.

Dormitory. Building arranged or used for lodging six, but not more than twenty individuals and having common toilet and bathroom facilities.

Dormitory. Room occupied by more than two guests.

Dormitory. A building or portion thereof which contains living quarters for unmarried students, staff or members of an accredited college, university, boarding school, theological school, hospital, religious order or comparable organization; provided that said building is owned and managed by said organization and contains not more than one (1) cooking and eating area; and further provided that said building complies with the Rooming House Ordinance.

Dormitory. Any building or any portion of which, used as lodging or rooming quarters. The term includes any dormitory used in conjunction with another institution.

Dormitory. Any building or portion thereof used and maintained to provide sleeping accommodations for more than five (5) unrelated persons whether for compensation or not, including "cothouses", "flophouses" and similarly designated uses, but not including hotels, lodging houses, boardinghouses, hospitals or other approved institutions or similar uses. Each one hundred (100) square feet or fraction thereof of floor area used for sleeping purposes shall be considered to be a separate guest room.

Dormitory. Facilities used for housing students, which are owned and controlled by the university and which are distinguished from hotels, motels, and rooming houses.

Dormitory. Space in a unit where group sleeping accommodations are provided, with or without meals, for persons not members of the same family group, in one room, or in a series of closely associated rooms under joint occupancy and single management, as in college dormitories, fraternity houses, military barracks and ski lodges.

Dormitory. Sleeping room occupied by more than three persons.

Dormitory. Building or that portion thereof other than a hotel, motel, boarding house, fraternity house, or sorority house containing three or more rooming units or guest rooms or sleeping facilities for more than five persons. Such rooming units or guest rooms shall be for residential purposes only.

Dormitory. Room which is used as "accessory sleeping quarters" for transients.

Dormitory. Building arranged or used for lodging 6 but not more than 20 individuals and having common toilet and bathroom facilities.

Dormitory. Building whose principal use is "accessory sleeping quarters for transients."

Dormitory. Room in a residential building occupied by more than 2 guests.

Dormitory. A guest room designed, intended or occupied as sleeping quarters by more than two persons. Every 100 square feet of superficial floor area in a dormitory shall be considered as a separate guest room.

Dormitory. A space in a building where group sleeping accommodations are provided for persons not members of the same family group, in one (1) room, or in a series of closely associated rooms.

Family boarding home. Home operated by any person who receives therein for pay children, not to exceed six in number, who are not related to such person and whose parents or guardians are not residents of the same house, for supervision, care, lodging and maintenance.

Furnished rooming house. Building or part thereof, not a hotel or inn, in which sleeping rooms are available for hire as lodging with or without meals. Where equipment for cooking or provisions for the same are included in a sleeping room, such room shall be deemed to be a dwelling unit.

Guest. Any person who is in the building under consideration for the purpose of sleeping, and any person who rents therein a room or suite of rooms which has sleeping facilities; except that the owner of the building and his regular employees therein are not guests.

Guest. Any person hiring or occupying a room for living or sleeping purposes.

Guest. Each occupant of any unit or any hotel, motel, tourist home or related establishment included in the definition of a hotel, motel or tourist home contained herein.

Guest. Any person who occupies a unit of dwelling space either (a) as a temporary occupant or transient in an establishment holding itself out as serving transients or (b) on a temporary or permanent basis in an establishment providing housekeeping or dining services on a regular basis to occupants.

Guest car ratio. For purposes of site planning, the number of parking spaces required for each living unit for use of occupants.

Guest house. Accessory dwelling unit which includes cooking facilities, which is incorporated in, attached to or detached from a principal dwelling, and which is used exclusively by occupants of the principal dwelling and/or for the non-commercial accommodation of persons visiting the occupants of the principal dwelling.

Guest house. Living quarters within an accessory building for the use of persons employed on the premises, or for the temporary use by guests of the occupants of the main building, and having no kitchen facilities and not rented or otherwise used as a dwelling unit.

Guest house. A dwelling containing not more than five guest rooms or suites of rooms, but with no kitchen facilities.

Guest house or cottage. Means a separate dwelling structure located on a lot with one or more main dwelling structures and used for the housing of guests or servants of the occupant of the premises; and such building shall not have a kitchen and shall not be rented, leased or sold separately from the rental, lease or sale of the main dwelling.

Guest room. Any room or rooms used, or intended to be used, by guests for sleeping purposes. Every 100 square feet of superficial floor area in a dormitory is a guest room.

Guest room. Rooming unit of only one room. If a guest room contains sleeping accommodations for more than two persons, each unit of accommodations for two persons shall be considered as a separate guest room for purposes of calculating density.

Guest room. Any room in a hotel, dormitory, boarding or lodging house, or home for the aged, used and maintained to provide sleeping accommodations for not more than two (2) persons. Each one hundred (100) square feet or fraction thereof of floor area used for sleeping purposes shall be considered to be a separate guest room.

Guest room. Any habitable room except a kitchen, designed or used for occupancy by one or more persons and not in a dwelling unit.

Guest room or sleeping room. Room which is designed or intended for occupancy by, or which is occupied by, not more than two (2) persons, for compensation, not members of the family, but in which no provision is made for cooking and not including dormitories for sleeping purposes.

Guest room or unit. Any room for occupancy, meetings, display, conference or dining and shall include all other rooms integral to the hotel, motel or tourist home unit. In the case of a central toilet and/or shower room, the term "guest room" or "unit" shall also apply.

Guest (tourist) home. Private dwelling in which transient sleeping accommodations are provided for compensation, especially motor tourists or travelers.

Light housekeeping room. Any room which is designed and used both as a sleeping room and for the cooking or preparation of food in conformance with the provisions of the Code.

Light housekeeping room. Any guest room which is designed and used both as a bedroom and for the cooking and preparing of food, in conformance with the provisions of the Code. For the purpose of applying the lot area and automobile parking space requirements of the various zones, each light housekeeping room shall be considered as a separate guest room.

Lobby. An enclosed vestibule, directly accessible from the main entrance.

Lobby. Public lounge or space for waiting which is connected with an entrance or exit way which serves as a principal entrance or exit.

Lodger. Transient, temporary, or permanent paying guest.

Lodger. Transient or temporary paying guest in the home of another person.

Lodger. Transient or temporary paying guest in the residence of another person.

Lodging house. Building in which five or more rooms or suites of rooms are used by lodgers, or a one, two, or three-family dwelling in which five or more lodgers reside in any of such dwelling units.

Lodging house. Dwelling or that part of a dwelling in which living space, without kitchen facilities, is let to five or more persons not within the second degree of kinship. "Lodging House" shall include a rooming house, boarding house, or tourist home, but not a hotel or motel.

Lodging house. Dwelling other than a hotel where lodging for compensation is provided for five or more persons.

Lodging house. Building where lodging only is provided for compensation to three or more, but not exceeding twenty persons, in contradistinction to hotels open to transients.

Lodging house. Building other than a hotel, where lodging, with meals, for five or more persons is provided for compensation.

Lodging house. Building other than a one-family dwelling or a two-family-dwelling, which is occupied, or is intended, arranged or designed for occupancy compensation.

Lodging house. Establishment accommodating patrons paying on a weekly or monthly basis with reference to longer periods and also accepting overnight guests.

Lodging house. A house, building or portion thereof in which persons are lodged for hire for any term less than a week.

Lodging house. A building other than a hotel with not more than four (4) guest rooms where lodging for compensation is provided for not more than eight (8) persons who are not members of a family occupying such building.

Lodging house. A building where lodging is provided to two (2) or more persons not members of the operator's family, by prearrangement for definite periods for compensation.

Lodging house. Any building containing less than ten (10) rooms, intended or designed to be used or which is used, rented or hired out, or which are occupied for sleeping purposes by two (2) or more paying guests.

Lodging house. A building other than a hotel, motel, or motor hotel, where for compensation and by prearrangement, lodging is provided for three (3), but not more than twelve (12) persons who are not members of the family occupying the building as a dwelling.

Lodging house. A building or place where lodging (but not meals) is provided for compensation by prearrangement for a definite period to five (5) or more, but not exceeding twelve (12) individuals.

Lodging house. Building in which 5 or more rooms or suites of rooms are used by lodgers, or a one, two, or three-family dwelling in which 5 or more lodgers reside in any of such dwelling units.

Lodging house. A boarding house.

Lodging house. Building or place where lodging is provided for compensation for four or more, but not exceeding twelve individuals.

Lodging house. Building other than a hotel where lodging for three but not more than twenty persons is provided for definite periods for compensation pursuant to previous arrangement.

Lodging house. Dwelling in which lodging or lodging and meals are provided for compensation for more than three, but not more than 15 persons other than members of the resident family, excepting a nursing home as defined in the code.

Lodging house. Building, other than hotel, where for compensation, lodging is provided for four or more persons not members of a family.

Lodging house. Building or place where lodging is provided (or which is equipped to provide lodging regularly) by prearrangement for definite periods, for compensation, for three or more persons in contradistinction to hotels open to transients.

Lodging house. Building, other than a hotel, where for compensation, lodging is provided for four or more persons not members of a family, occupying such building.

Lodging house. Any building or portion thereof containing not more than five guest rooms which are used for not more than five guests where rent is paid in money, goods, labor or otherwise.

Lodging house. Multiple dwelling used primarily for the purpose of furnishing lodging, with or without meals, for compensation.

Lodging house. Building other than a hotel where lodging only is provided for compensation for not more than 3 persons not members of the family.

Lodginghouse. Building, other than a hotel, where lodging for five or more persons is provided for compensation.

Lodging room. A room rented as sleeping and living quarters, but without cooking facilities, and with or without an individual bathroom. In a suite of rooms without cooking facilities, each room which provides sleeping accommodations shall be counted as one (1) "lodging room" for the purpose of the Ordinance.

Lodging-rooming house. Any house, or other structure, or any place or location kept, used, maintained, advertised or held out to the public to be a place where living quarters, sleeping or housekeeping accommodations are supplied for pay to transient or permanent guests or tenants, whether in one or adjoining buildings.

Multiple dwelling. Apartment house, hotel, lodging house, tenement house, flat, dormitory, convent, monastery, rooming house for not more than six roomers, a one- or two-family dwelling when either of which has not more than 6 roomers in one dwelling unit, and similar residential occupancies, other than institutional occupancies.

Nontransient. Person who resides in the same building for a period of thirty days or more.

Occupied area. Any room, area or enclosure used by one or more persons.

One-family dwelling. A building containing one (1) dwelling unit with not more than four (4) lodgers or boarders.

One-family dwelling. A building containing one (1) dwelling unit with not more than five (5) lodgers or boarders.

Public hallway. A public corridor or space separately enclosed which provides common access to all the exitways of the building in any story.

Residential occupancy. The occupancy or use of a building or structure or any portion thereof by persons for whom sleeping accommodations are provided but who are not harbored or detained to receive medical, charitable or other care or treatment or are not involuntarily detained.

Roomer. Other than member of the family one who rents one or more rooms in a dwelling from the resident family.

Rooming and/or boarding house. Furnishing of lodging with or without meals for compensation to permanent guests.

Rooming and/or boarding house. Building containing guest rooms in which lodging is provided with or without meals for compensation and which is open to permanent guests only and where no provision is made for cooking in any guest room.

Rooming house. Any dwelling in which three or more persons either individually or as families are housed or lodged for hire with or without meals. A boarding house, lodging house, tourist home or a furnished rooming house shall be deemed a rooming house.

Rooming house. Building where a room or rooms are provided for compensation to three or more persons.

Rooming house. Residential establishment other than a hotel or motel where sleeping accommodations only are offered to the public, but on a non-transient basis.

Rooming house. Building designed, constructed, or altered to contain not more than 15 sleeping rooms, where lodging only is provided for compensation.

Rooming house. Lodging house.

Rooming house. Structure containing one or more rooming units, in which space is let to three or more persons.

Rooming house. Building, or portions thereof, other than a commercial or residential hotel, which contains sleeping rooms and which is regularly used or available for permanent occupancy only, by three (3) or more persons not members of the same family.

Rooming house. Any dwelling or that part thereof in which space is let by the owner, operator, or occupant of one or more rooming units to three or more persons who are not husband or wife, son or daughter, mother or father, grandparents, grandchildren, sister or brother or niece or nephew of the owner or operator or tenant or the spouse of any of these, but any child lawfully under the care of any of the above members of a family is not to be deemed a roomer.

Rooming house. The terms rooming house, boarding house, cooperative house and lodging house are used synonymously in the Ordinance. A building, other than a hotel or a dormitory, for compensation and by prearrangement for definite periods, lodging or lodging and food are provided for more than three (3) persons.

Rooming house. Any dwelling or that part of any dwelling containing 1 or more rooming units, in which space is let for compensation by the owner or operator to 5 or more persons, none of whom are related to the owner or operator by blood or marriage, but not including a hotel. Dwellings in which space is let to less than 5 persons is not to be deemed a rooming house.

Rooming house. Building operated for compensation by a resident family, in which a room or rooms are provided for living and sleeping facilities to one or more persons.

Rooming house. Residential building used or intended to be used, as a place where sleeping or house-keeping accommodations are furnished or provided for pay to transient or permanent guests or tenants in which less than ten and more than three rooms are used for the accommodation of such guests or tenants, but which does not maintain a public dining room or cafe in the same building, nor in any building in connection therewith.

Rooming house. Any building or portion thereof which contains not less than three or more than nine guest rooms which are designed or intended to be used, let, or hired out for occupancy by individuals for compensation whether paid directly or indirectly.

Rooming house. Building other than an apartment hotel, a hotel, or motor lodge where, for compensation and by prearrangement for definite periods, lodging, meals, lodging and meals are provided for three or more persons, but containing less than 20 sleeping rooms.

Rooming house. Building which has five (5) or more rooms or suites of rooms which are offered for use, or are permitted to be used by roomers; or a building containing one, two, or three dwelling units, with five (5) or more paying guests, the majority of whom are roomers, residing with a family in any one of such dwelling units.

Rooming house. Dwelling where for compensation and by prearrangement, four or more persons are provided with rooming accommodations on a weekly or monthly basis with reference to longer periods–overnight guests are excluded. Rooming House shall be interpreted to mean a traditional type of establishment, accommodating competent patrons, not a social welfare establishment.

Rooming house (furnished room house). Building or part thereof, not a hotel or inn, in which sleeping rooms are available for hire as lodging with or without meals. Where equipment for cooking or provisions for the same are included in a sleeping room, such room shall be deemed to be a dwelling unit.

Rooming house or lodging house. Any dwelling, or any part thereof in which space is rented or let to five or more persons or in which three or more "rooming units" are rented or let.

Rooming house or retirement home. Buildings in which separate sleeping rooms are rented providing sleeping accommodations for three (3) or more persons on a weekly, semi-monthly, monthly or permanent basis; with or without meals, but without separate cooking facilities for individual occupants.

Rooming unit. Any room or group of rooms forming a single habitable unit, used or intended to be used, for living and sleeping, but not for cooking or eating purposes. A room occupied by a person who is permitted to prepare meals anywhere in a dwelling is a family unit and not a rooming unit unless it is an institutional unit, or in a convent, monastery, a bona fide not-for-profit club, or in a dormitory, fraternity or sorority affiliated with an educational or other charitable institution.

Rooming unit. Single habitable space used or intended to be used for living and sleeping, but not for cooking or eating purposes.

Sleeping room. Room rented for sleeping or living quarters, but without cooking facilities and with or without an individual bathroom. In a suite of rooms not part of a dwelling unit, each room which provides sleeping accommodations is construed as one sleeping room.

Suite. Series of connecting rooms with one or more bedrooms and a parlor, suites occasionally include additional rooms like a dining room.

Suite of rooms. Two or more rooms which are arranged to be used as a unit and each room of which is accessible from within the unit without the use of a public exit way.

Temporary guest. A person who occupies or has the right to occupy a hotel or motel accommodation, boarding house, lodging house, or guest as a transient and not as his or her domicile or place of permanent residence.

Temporary residence. A property consisting of a tract of land and all tents, vehicles, buildings or other structures pertaining thereto, any part of which may be occupied by people who are provided with at least some part or portion of the sleeping facilities by the operator, owner, lessee or occupant thereof, with or without stipulated agreement as to the duration of their stay, whether or not they are supplied with meals, but who are supplied with such services or facilities as are necessary for their use of such property. It shall include, but shall not be limited to: a property occupied by adults, children, or both, primarily for educational, recreational or vacation purposes; a group of three (3) or more cabins or houses; a property used as a labor camp except a farm labor camp as defined in the Code, a tourist camp, motel, tourist home, hotel, boarding house or lodging house, or other establishment comparable or equivalent thereto, or, notwithstanding the provisions aforesaid to respect to some part or portion of the sleeping facilities, a property providing ground areas for the parking of occupied house trailers or the erection of tents or other shelters for overnight occupancy, or a summer day camp as defined in the Code.

Tourist court. An area containing one (1) or more building designed or intended to be used as temporary sleeping facilities of one or more transient persons.

Tourist home. Dwelling occupied in part by the owner or tenant in which sleeping accommodations for less than ten (10) persons are provided or afforded primarily for automobile travelers for compensation.

Tourist home. Building other than hotel where lodging is provided and offered for compensation for not more than twenty individuals and open to transient guests. A tourist home shall be considered as a dwelling use.

Toursit home. Tourist home shall be construed to mean a dwelling occupied in such a manner that rooms in excess of those used by member of the family, as herein before provided, and occupied as a home or family unit, are rented without cooking facilities, to the public for compensation and catering primarily to the public travel by motor vehicle.

Tourist home. Any establishment or premise where sleeping accommodations are furnished to transient guests for hire or rent on a daily or weekly rental basis in a private home when such accommodations are offered for hire or rent for the use of the traveling public.

Tourist home. A home which has been converted into premises offering rooms to transient guests for remuneration.

Tourist home. Building containing not more than fifteen (15) rooms with sleeping facilities where lodging or lodging meals are provided and offered to the public for compensation for one or more, but not exceeding 20 persons, and open to transient guests.

Tourist home. Rooming house in which transients are lodged for hire.

Tourist home. Dwelling in which overnight sleeping accommodations are provided or offered for not more than five (5) transient guests for compensation.

Tourist home. Building or portion thereof in which board or lodging, or both, are offered to the traveling public for compensation, open to transient guests.

Tourist home. Dwelling in which less than ten (10) guest rooms are provided for occupancy for compensation by transient guests.

Tourist home. One (1) or more buildings, each building containing only one (1) or more guest rooms.

Tourist house. Dwelling in which rooms for overnight sleeping accommodations are provided for or offered to transient guests for compensation.

Tourist house. Boarding or rooming house accepting transients.

Transient. A person who requests accommodations for a price, with or without meals, for a period of less than twenty-one (21) days.

Transient. Person who resides in the same building for a period of less than thirty days.

Transient guest. A tenant who does not have a lease and occupies an apartment, lodging room or other living quarters on a daily or weekly basis.

Transient guest. A guest for only a brief stay, such as the traveling public.

Transient lodging facility. Single or multiple family dwelling in which each dwelling unit contains sleeping and bathroom facilities, and which may or may not contain food refrigeration, cooking and/or dining facilities. Hotels and Motels are typical transient lodging facilities.

Transient occupancy. Rental of housing accommodations or rooms on a day to day basis including a daily change of linen and towels with no change in rate based on the period of occupancy.

Transient purposes. Intent to use and/or the use of a room or a group of rooms for the living, sleeping and housekeeping activities of persons on a temporary basis of an intended tenure of less than one month.

Transient residential buildings. Residential buildings usually occupied by transients.

C-4 BUILDINGS, STRUCTURES AND APPENDAGES

Addition. Extension or increase in area, height or equipment of a building.

Addition. An extension or increase in floor area or height of a building or structure.

Addition. As applied to a building or structure means any construction which increases the area or the height of any portion of the building or structure.

Addition. Any construction which increases the size of a building or adds to the building.

Addition. Any construction which increases the area or cubic content of a building or structure.

Addition. Any extension or enlargement of a building.

Addition. Any construction which increases the size of a building or adds to the building such as a porch or an attached garage or carport.

Addition. Extension or increase in area or height of a building.

Addition. An extension or increase in floor area or height of a building that increases its exterior dimensions.

Addition to a building. The result of any work that increases the volume of an existing building or replaces a demolished portion.

Air-supported structure. Structure consisting of a pliable membrane which achieves and maintains its shape and support by internal air pressure.

Air-supported structure. A structure consisting of skin diaphragms made of flexible material, which achieves its shape, support, and stability from internal air pressure.

Air supported structure. A structural and mechanical system which is constructed of high strength fabric or film and achieves its shape, stability, and support by pretensioning with internal air pressure; air structures may be used for temporary applications.

Alter. Any structural change in the supporting or load-bearing members of a building, such as bearing walls, columns, beams, girders or floor jointing.

Alter. To make a change in the supporting members of a structure, such as bearing walls, columns, beams, or girders, which will prolong the life of the structure.

Alter-alteration. Any change, addition or modification in construction or occupancy.

Alteration. Change in or addition to a building which reduces the means of exit or fire resistance or changes its structural supports, use or occupancy.

Alteration. A change or rearrangement in the structural parts of a building, or a change in required window or exit facilities, or in the building service equipment.

Alteration. Any change, rearrangement, or modification of an existing building or lot, other than additions or extensions thereto.

Alteration. A change from one major occupancy class or division to another, or a structural change such as an addition to the area or height, or the removal of part of a building, or any change to the structure such as the construction of, cutting into or removal of any wall, partition, column, beam, joist, floor or other support, or a change to or closing of any required means of egress or a change to the fixtures, equipment, cladding or trim where they are regulated by the code.

Alteration. As applied to a building or structure, means a change or a re-arrangement in the structural parts or in the exit facilities; or an enlargement, whether by extending on a side or by increasing in height; or the moving from one location or position to another.

Alteration. Construction which may change the structural parts, mechanical equipment or location of openings but which does not increase the size of the building.

Alteration. As applied to a building or structure, a change or rearrangement in the structural parts extending on a side or by increasing in height; or the moving from one location or position to another which does not involve a change in grade of Occupancy: the term "alter" in its various moods and tenses and its participial forms, refers to the making of an alteration.

Alteration. Any change, rearrangement, addition or modification in construction.

Alteration. Any change in the type of occupancy, or any rearrangement, addition or modification in construction, equipment, appurtenances, or facilities.

Alteration. As applied to a building or structure means any change or modification in construction, exit facilities, or permanent fixtures or equipment which does not include an addition to the building or structure.

Alteration. As applied to a building or structure, means a change or rearrangement in the structural parts or in the exit, lighting or ventilating facilities; or an enlargement, whether by extending on a side or by increasing in height; or the moving from one location or position to another; the term alter in its various moods and tenses and its participial forms, refers to the making an alteration.

Alteration. Any change, modification or deviation in construction or occupancy.

Alteration. Any change, rearrangement, or addition to a building, other than repairs; any modification in construction or in building equipment.

Alteration. As applied to a building or structure, shall mean any change or rearrangement in the structural parts or in the [exit facilities] egress facilities of any such building or structure, or any enlargement thereof, whether by extension on any side or by any increase in height, or the moving of such building or structure from one location or position to another.

Alteration. The term "alteration", as applied to a building or structure, shall mean any change or rearrangement in the structural parts or existing facilities of any such building or structure, or any enlargement thereof, whether by extension on any side or by any increase in height, or in the moving of such building or structure from one location or position to another.

Alteration. Any change, addition or modification in construction or occupancy; or the moving from one location or position to another.

Alteration. As applied to a building or structure, or their service equipment means a change or rearrangement in the structural parts or in the exit facilities or a vital change in the service equipment; or an enlargement by an increase in area or volume; or the moving from one location or position to another; or the change in use or occupancy from one use group to another as defined in the code.

Alteration. As applied to a building or structure, means a change or rearrangement in the structural parts or in the exit facilities; or an enlargement, whether by extending on a side or by increasing in height; including work, other than repairs, that would affect safety or a vital element of an elevator, plumbing, gas piping, wiring, ventilating or heating installation; the term "alter" in its various moods and tenses and its participial forms, refers to the making of an alteration.

Alteration. A change in size, location, shape, occupancy or use of a structure.

Alteration. Any change or modification of construction, space arrangement and/or occupancy of a building, or decreasing or not increasing the area or cubic contents thereof.

Alteration. Any change, rearrangement, or addition to a building, other than repairs; any modification in construction, or in building equipment, refers to an alteration.

Alteration. As applied to a building or structure, or their service equipment means a change or rearrangement in the structural parts or in the exit facilities or a vital change in the service equipment; or an enlargement by an increase in area or volume; or the moving from one location or position to another; or the change in use of occupancy from one use group to another as defined in the Code, creates an alteration.

Alteration of a building. Any change in supporting members of a building except such changes as may be required for its safety, or any addition to a building, or any change in use from one Zone classification to another, or removal of a building from location to another.

Alteration of building. Any change or rearrangement in the supporting members (such as bearing walls, beams, columns or girders) of a building, any addition to a building, or movement of a building from one location to another.

Alteration of building. Any change in the supporting members of a building, such as bearing walls, columns, girders, beams, joists and foundations, except such changes as may be required for its safety; any addition to a building; any change in use from that of one district classification to another; or of a building from one location to another.

Alter or alteration. Any change or modification in construction or occupancy.

Alter or alteration. Any change, rearrangement, addition, or modification in construction or occupancy.

Alteration or altered. Any addition to the height or depth of a building or structure, or any change in the location of any of the exterior walls of a building or structure, or any increase in the interior accommodations of a building or structure.

Alterations. Any change, addition or modification in construction or type of occupancy, any change in the structural members of a building, such as walls or partitions, columns, beams or girders, the consumated act of which may be referred to as altered or reconstructed.

Alterations. Any change in the occupancy classification or any change or modification of construction or space arrangement in any existing building or structure not increasing the area or cubic content thereof or any change which decreases the area and the cubic content of the building or structure.

Alterations. Changes, improvements, and replacement of parts, in a building or structure not affecting the supporting members of such buildings or structures.

Alterations. As applied to a building or structure, means a change or rearrangement in the structural parts or in the exit facilities; or an enlargement, whether by extending on a side or by increasing in height; including work, other than repairs, that would affect safety or a vital element of an elevator, plumbing, gas piping, wiring, ventilating or heating installation; the term "alter" in its various moods and tenses and its participial forms, refers to the making of an alteration.

Alterations. Includes, but are not limited to the following: all incidental changes in or replacements to the nonstructural parts of a building or other structure; minor changes or replacements in the structural parts of a building or other structure.

Alterations. Any change, rearrangement or addition to, or any relocation of, a building or structure; any modification in construction or equipment.

Alterations. Construction which may change the structural parts, mechanical equipment or location of openings but which does not increase the size of the building.

Alterations. Any change, addition or modification in construction or type of occupancy; any change in the structural members of a building such as bearing walls or partitions, columns, beams, or girders, or any substantial change in the roof or in the interior walls.

Alterations. As applied to a building or structure, a change or rearrangement in the structural parts or in the exit facilities, or an enlargement whether by extending on a side or by increasing in height.

Alterations. A change or rearrangement in the structural parts of a building, or a change in required window or exit facilities, or in the building service equipment regulated by the Building code.

Alteration to building. Any work on a building or structure that does not result in any addition to the building or structure.

Appurtenant structure. A device or structure attached to the exterior or erected on the roof of a building designed to support service equipment or used in connection therewith or for advertising or display purposes, or other similar uses.

Arcade. Continuous area at ground level open to a street or plaza, which is open and unobstructed to a height of not less than 12 feet, open at both ends, and which is accessible to the public at all stated times. Any portion of an arcade occupied by columns, landscaping, statuary, or pools shall be considered to be a part of an arcade for the purpose of computing a floor area premium credit. The term arcade shall not include off-street loading areas, driveways, off-street parking areas or pedestrian ways accessory thereto.

Arcade. An avenue or passageway, or roofed over and enclosed, except at the ends and serving as a common entrance and exit for shops, stores and similar places of business located thereon.

Arcade. Range of arches, supported either on columns or on piers, and detached or attached to a wall.

Arcade. An arched or covered passageway, generally lined with retail shops.

Arch. Arrangement of building materials in the form of a curve which preserves its given form when resisting pressure, and enables it, supported by piers or beams, to carry loads.

Arch-buttress. Sometimes called a flying buttress; an arch springing from a buttress.

Areaway. An uncovered sub-surface space adjacent to and considered a part of a building, but not within its perimeter, including but not limited to such spaces for the purpose of admitting light and air to a basement or cellar, or to provide access thereto.

Areaway. An open subsurface space adjacent to a building used to admit light and air or as a means of access to a basement or crawl space.

Areaway. An uncovered subsurface space adjacent to a building.

Areaway. An unroofed subsurface space adjacent to a building.

Areaway. Exterior area whose grade is below the grade (at building) and having at least one side consisting of the exterior wall of a building.

Areaway. Open subsurface space adjacent to a building wall used to admit light or air, or as a means of access to a basement or a cellar.

Areaway. An open, subsurface space, adjacent to a building for access to, or for lighting or ventilating purposes.

Areaway. A space below grade, adjacent to a building, open to the outer air and enclosed by walls.

Attached building. A building which has two party walls in common with adjacent buildings.

Attached building. Building which has any part of its exterior or bearing wall in common with another building or connected to another building by a roof.

Attic. Any area under a roof, with or without a finished floor, which does not meet other requirements for livable floor area.

Attic. Accessible space between top of uppermost ceiling and underside of roof. Inaccessible spaces are considered structural cavities.

Attic. A room built within the sloping roof of a dwelling. May be finished or unfinished.

Attic. A habitable room entirely within the roof space of a building.

Attic. Space between top of uppermost floor construction and underside of roof.

Balcony. The elevated floor space in a story or that portion of the seating space of an assembly room, the lowest part of which is raised four feet (4′) or more above the level of the main floor.

Balcony. Balcony within an auditorium, is a floor, inclined, stepped, or level, above the main floor, the open side or sides of which shall be protected by a rail or railings. Where a balcony of an auditorium has exits at two or more levels opening into separate foyers, one above another, each portion thereof served by such a foyer shall be considered a separate balcony for the purpose of the Code.

Balcony. That portion of the seating space of an assembly room, the lowest part of which is raised four (4) feet or more above level of the main floor.

Balcony. The first seating tier above a main floor within a place of assembly.

Balcony. A gallery or mezzanine floor above the main floor of an auditorium, theatre or church. Where such a balcony extends over 33 per cent or more of the horizontal area within the outer walls of the building it shall be considered a story.

Balcony. Partial floor in an assembly room.

Balcony. A railed platform projecting from the face of a building above the ground level with an entrance from the building interior. In a theater or auditorium, a partial upper floor with seats.

Balcony. Exterior auxiliary floor space projecting from the exterior wall of an enclosed structure and unenclosed by other than a railing or parapet, or as applied to places of assembly, a seating level located above the main floor of an auditorium.

Balcony. A railing-enclosed platform projecting from and supported by an outer wall of a building but not supported directly from the ground.

Balcony. That portion of a building which projects into the required yard and where the floor height of said projection is not less than four (4) feet above grade.

Balcony. A projection from the wall of a building, supported by columns, consoles or other cantilevers, and usually covered at its extremity by a balustrade.

Balcony. The exterior auxiliary floor space projecting from the outside of an exterior wall of an enclosed structure and unenclosed by other than a railing or parapet.

Balcony. An unroofed platform, enclosed by a railing or parapet, projecting from the wall of a building for the private use of tenants or for exterior access to the above grade living units. When a balcony is roofed and enclosed with operating windows, it is considered part of the room it serves.

Balcony (exterior). An appurtenance to a structure supported on the structure members of the building extending beyond the main wall of the building more than four (4) feet above grade.

Balcony (exterior). A narrow, open platform on the outside of a building wall and surrounded by a rail or balustrade.

Balcony (interior). That portion of seating for assembly space of an assembly room, the lowest part of which is raised four (4) feet or more above the level of the main floor.

Basement. A portion of the building partially underground, but having less than half its clear height below the grade plane.

Basement. That portion of a building or structure which is partly underground and which has one half or more of its ceiling height above the average finished grade of the ground adjoining the building.

Basement. A story partly underground, but having less than one-half in clear height (measured from finished floor to finished ceiling) below the curb level; except that where the curb level has not been legally established, or where every part of the building is set back more than 25 feet from a street line, the height shall be measured from the adjoining grade elevation.

Basement. That space of a building that is partly below grade which has more than half of its height, measured from floor to ceiling, above the average established curb level or finished grade of the ground adjoining the building.

Basement. That portion of a building between floor and ceiling, which is partly below and partly above grade but so located that the vertical distance from grade to the floor is less than the vertical distance from grade to ceiling.

Basement. A portion of the building partly underground but having less than half its clear height below the average grade of the adjoining ground.

Basement. A story of a building or structure having one-half or more of its clear height below grade.

Basement. That portion of a building which is below the first story, but more than 50 percent above grade, and is used for storage, garages for use of occupants of building, janitor or watchman quarters, or other utilities (exclusive of rooms of habitation or assembly) common for the rest of the building. A basement used for the above purposes shall not be counted as a story, provided its height in the clear does not exceed 7 feet 6 inches.

Basement. The basement floor is that level below the first or ground floor level with its entire floor below exit discharge grade.

Basement. The level in a structure which is partially or completely below the mean grade.

Basement. That portion of a building or structure located partly below the average adjoining grade and below the lower living quarters.

Basement. The lower part of a building, usually part of the basement is below the grade of the lot.

Basement. Any room or rooms built partially or wholly below ground level.

Basement. That portion of a building between floor and ceiling, which is partly below and partly above grade (as defined in the Code), but so located that the vertical distance from grade to the floor below is less than the vertical distance from grade to ceiling, provided, however, that the distance from grade to ceiling shall be at least four (4) feet six (6) inches.

Basement. Any story below the first story of a building.

Basement. That portion of a building which is partly below and partly above grade, and having at least one-half (1/2) its height above grade.

Basement. A story partly underground. A basement shall be counted as a story for the purpose of height measurement if its height is one-half (1/2) or more above grade.

Basement. A basement is that portion of the building between the floor and the ceiling, which is partly below and partly above grade, but so located that the vertical distance from the grade to the floor below is less than vertical distance from grade to ceiling.

Basement or cellar. That portion of a building between the floor and the ceiling which is either totally or partly below grade.

Basement parking garage. A portion of the basement area of a structure that is used for the storage of private vehicles.

Basements. The basement of a building shall not count as a story if the upper surface of the first floor above such basement is less than seven (7) feet above average grade.

Bay window. Any window space projecting outward from the walls of a building, either square or polygonal in plan.

Bay window. A window projecting beyond the wall line of the building and extending down to the foundations.

Bay window. A rectangular, curved or polygonal window supported on a foundation extending beyond the normal perimeter walls of the building in which it occurs.

Bay window. Any window projecting outward from the wall of a building and commencing from the ground. If they are supported on projecting corbels, they are called "oriel windows".

Bay window. Window projecting beyond the wall of a building and extending down to or below the ground.

Bay window. A window structure which projects from a wall. Technically, it has its own foundation. If cantilevered, it should be an oriel window, however, this is not common usage.

Bay window. A rectangular, curved or polygonal window extending beyond the main wall of the building.

Bay window. A rectangular, curved, or polygonal window, supported on a foundation extending beyond the main wall of the building.

Breezeway. Any structure connecting two (2) buildings and with a roof in keeping with the design and construction of the main building.

Breezeway. Roofed over space between buildings, usually between the garage and the residence.

Breezeway. A structure open to the outdoors consisting of a roof, roof supports, and floor, connecting a garage or other accessory building with a dwelling.

Building. A structure or part thereof, enclosing any occupancy including residential, institutional, assembly, business, mercantile, industrial, storage, hazardous and miscellaneous uses. When separated by fire walls, each unit so separated shall be deemed a separate building.

Building. Any structure such as but not limited to those having a roof supported by columns, piers, or walls, including tents, mobile homes, trailers, dining cars, camp cars, or other structures on wheels, or having other supports; and any unroofed platform, terrace or porch having vertical face higher than three (3) feet above the level of the ground over or upon which said structure is located.

Building. Any structure which is permanently affixed to the land, has one or more floors or stories, and is bounded by either lot lines or yards. A building shall not include such structures as billboards, fences, or structures with interior surfaces not normally accessible to human use, such as gas tanks, grain elevators, coal bunkers or similar structures. A building may accommodate more than one family and have more than one dwelling unit and be used for residential, or commercial or manufacturing purposes.

Building. A structure forming a shelter for persons, animals, or property, and having a roof. The word "building" shall be construed, where the context allows, as though followed by the words "or part thereof". The word "building" shall not include such frame-works and tents as are customarily used exclusively for outdoor carnivals, lawn parties or like activities.

Building. Any edifice, factory, house, residence, store, warehouse, etc., or any part thereof regardless of the use or purpose to which these are put.

Building. Any structure having a single or common roof supported by columns or walls. "Existing Building" means any building erected in conformance with legal permit issued therefor.

Building. Any structure having a roof supported by columns or walls.

Building. A structure enclosed within a roof and within exterior walls or fire walls, designed for the housing, shelter, enclosure, and support of individuals, animals or property of any kind.

Building. Any covered structure intended for the shelter, housing or enclosure of any person, animal or chattel.

Building. Anything constructed or erected, the use of which requires more or less permanent location on the soil, or attached to something having permanent location on the soil. The building area of any building shall be that area bounded by the exterior dimensions of the outer walls at the ground line.

Building. Any covered structure averaging more than six feet (6′) in building height which is intended for the shelter, housing or enclosure of persons, animals, chattels or property of any kind.

Building. Any structure having a roof supported by columns or by walls and intended for the shelter, housing or enclosure of any person, animal or chattel.

Building. A structure having a roof supported by columns or walls and when separated by a division wall without openings, each portion of such building shall be deemed a separate building.

Building. Any structure or enclosure devoted to human occupancy.

Building. An enclosed structure.

Building. A structure having a roof supported by columns or walls for the shelter, support, or enclosure of persons, animals or chattels, but not to include any trailer coach or mobile home.

Building. Any structure having a roof and intended for shelter, housing, or enclosure of persons, animals, or chattels, and having enclosing walls for fifty percent (50%) or more of its perimeter.

Building. A structure constructed or erected on the ground, with a roof supported by columns or walls. The term "building" shall be construed as if followed by the words "or part thereof."

Building. Any temporary or permanent structure, fence, wall or enclosure, built either above or below the ground.

Building. Any structure having a roof, including, but not limited to, tents, awnings, carports and other such enclosures.

Building. Combination of any materials, whether portable or fixed, having a roof, to form a structure affording shelter for persons, animals or property, but not a trailer.

Building. Structure forming a shelter for persons, animals or property and having a roof.

Building. A structure for support, shelter or enclosure of persons or property.

Building. Any structure for the shelter, support or enclosure of persons, animals, chattels, or property of any kind.

Building. A structure having walls and a roof designed and used for the housing, shelter, enclosure, or support of persons, animals or property.

Building. A building is a structure, built, erected, and framed of component structural parts designed for the housing, shelter, enclosure, or support of persons, animals, or property of any kind.

Building. Any structure used or intended for supporting or sheltering any use or occupancy.

Building. Any structure which is permanently affixed to the land having one or more floors and a roof, being bounded by either open space or lot lines, and used as a shelter or enclosure for persons, animals or enclosure for property. The term shall be used synonymously with "structure" unless otherwise noted in the code, and shall be construed as if followed by the words "parts or parts thereof."

Building. Any structure designed or intended for the support, enclosure, shelter or protection of persons, animals, or property, but not including signs or billboards.

Building. Any structure designed or built or used for the support, enclosure, shelter or protection of persons, animals, chattels, or property of any kind.

Building. A structure designed, built, or occupied as a shelter or roofed enclosure for persons, animals, or property. The term "building" shall be construed under the Code as if followed by the words "or parts thereof" and shall include tents, lunch wagons, dining cars, camp cars, trailers, and other roofed structures on wheels or other supports used for residential, business mercantile, storage, commercial, industrial, institutional, assembly, educational, or recreational purposes. For the purposes of this definition "roof" shall include an awning or other similar covering, whether or not permanent in nature.

Building. Shall mean a structure having a roof.

Building. A structure having a roof supported by columns or walls.

Building. Means an edifice, or portion thereof, constructed and designed for permanent or continuous occupancy for group habitation, assembly, business, educational, high hazard, industrial, institutional or storage purposes and the like. For the purpose of the Code, each portion of a building separated from other portions by a fire wall shall be considered a separate building.

Building. Any structure having a roof supported by columns or walls, for the housing or enclosure of persons, animals, chattels, or property of any kind.

Building. Any structure used, designed or intended for the protection, shelter, enclosure or support of persons, animals or property, when a building is divided into separate parts by unpierced walls extending from the ground up, each part shall be deemed a separate building.

Building. Means a structure having a roof supported by columns or walls for the shelter, support, or enclosure of persons, animals, and chattels; that is safe and stable, and adapted to permanent or continued occupancy. The term "Building" shall be construed as if followed by the words "or portions thereof."

Building. Is any structure built for the support, shelter or enclosure of persons, animals, chattels or property of any kind.

Building. A combination of any materials, whether portable or fixed, having a roof, to form a structure for shelter of persons, animals or property. For the purpose of this definition "roof" shall include an awning or any similar covering, whether or not permanent in nature. The word "building" shall be construed where the context requires as though followed by the words "or part or parts thereof."

Building. A structure wholly or partially enclosed within exterior walls, or within exterior and party walls, and a roof, affording shelter to persons or property. The word "building" for the purpose of the Code shall include the word "structure."

Building. Any structure enclosed and isolated by exterior walls constructed or used for residence, business, industry or other public or private purposes, or accessory thereto and also tents, lunch wagons, dining cars, trailers, freestanding billboards, and signs, fences, and similar structures whether stationary or movable.

Building. Any structure, either temporary or permanent, having a roof, and used or built for the shelter or enclosure of persons, animals, chattels, or property of any kind. This definition shall include tents, awnings or vehicles which function as a building.

Building. Any structure having enclosing walls and roof, permanently located on the land.

Building. Any structure built or used for the shelter, occupancy, enclosure, or support of persons, animals, or chattels and is construed as if followed by "other structure, or parts thereof" provided, however, that the term "building" or "structure" shall not include structures or installations ordinarily known as industrial, production, or manufacturing equipment, and the parts thereof. "Building" also includes the equipment therein which is subject to regulation under the Building Code.

Building. Any structure enclosed within exterior walls or fire walls, built, erected and framed of component structural parts, designed for the housing, shelter, enclosure and support of individuals, animals or property of any kind. It does not include a portable structure mounted on wheels such as a trailer.

Building. A combination of materials forming construction that is safe and stable and adapted to permanent or continued occupancy.

Building. A structure enclosed within exterior walls or fire walls, built, erected and framed of component structural parts designed for the housing, shelter, enclosure and support of individuals, animals, or property of any kind.

Building. Any structure built for the shelter or enclosure of persons, animals or chattels or any part of such structure when subdivided by division walls or party walls extending to or above the roof and without openings in such separate walls.

Building. Any structure built for the support, shelter or enclosure of persons, animals, chattels, or property of any kind which has enclosing walls for 50% of its perimeter. The term "building" shall be construed as if followed by the words "or part thereof." (For the purpose of the Code each portion of a building separated from other portions by a fire wall shall be considered as a separate building.)

Building. Means a combination of materials to form a construction that is safe and stable, and adapted to permanent or continuous occupancy for public, institutional, residence, business or storage purposes; the term "building" shall be construed as though followed by the words "or part thereof." For the purposes of the Code, each portion of a building, separated from other portions by a fire wall, shall be considered as a separate building.

Building. Any construction for the support, shelter or enclosure of persons, chattels or property of any kind. When a building is separated by party or division walls without openings.

Building. Any structure used or intended for supporting or sheltering any use or occupancy.

Building. A combination of materials to form a construction that is safe and stable, and adapted to permanent or continuous occupancy for assembly, business, educational, high hazard, industrial, institutional, mercantile, residential or storage purposes; the term "building" shall be construed as if followed by the words "or portion thereof." For the purposes of the code, each portion of a building separated from other portions by a fire wall shall be considered as a separate building.

Building. A structure that is completely enclosed by a roof and by solid exterior walls along whose outside faces can be traced, an unbroken line for the complete circumference of the structure, which is permanently affixed to a lot or lots, and used or intended for the shelter, support, or enclosure of persons, animals or property of any kind. The connection of two buildings by means of an open porch, breezeway, passageway, carport, or other such open structures, with or without a roof, shall not be deemed to make them one building.

Building. Any structure having a roof supported by columns or by walls and intended for the shelter, housing or enclosure of any person, animal or chattel.

Building. Means an edifice, or portion thereof, constructed and designed for permanent or continuous occupancy for group habitation, assembly, business, educational, high hazard, industrial, institutional or storage purposes and the like. For the purpose of the Code, each portion of a building separated from other portions by a fire wall shall be considered a separate building. Nothing herein contained shall be construed to obviate the necessity of submitting plans as provided in the Code.

Building. Any structure, either temporary or permanent, having a roof, and used or built for the shelter or enclosure of persons, animals, chattels, or property of any kind. This shall include tents, awnings or vehicles situated on private property and used for purposes of a building.

Building. Any structure having a roof supported by columns or walls used or intended to be used for the shelter or enclosure of persons, animals, or property. When such a structure is divided into separate parts by one or more unpierced walls extending from the ground up, each part is deemed a separate building, except as regards minimum side yard requirements as herein provided.

Building. Any structure having a roof supported by columns or walls, for the housing or enclosure of persons, animals, chattels, or property of any kind as described in the Code.

Building. An enclosed structure including service equipment therein. The term shall be construed as if followed by the phrase "structure, premises, or part thereof" unless otherwise indicated by the Code.

Building. Any structure used or intended for supporting or sheltering any use or occupancy, as listed in the Ordinance.

Building. The word "building" includes all structures of every kind regardless of similarity to buildings.

Building. Any structure designed or built for the support, enclosure, shelter, or protection of persons, animals, chattels, or property of any kind.

Building. A structure, wholly or partly enclosed within exterior walls, or within exterior walls and party walls, and a roof, affording shelter to persons and property.

Building. Any structure built for the support, shelter, or enclosure of persons, animals, chattels, or property of any kind, in accordance with the Building Code.

Building. Any structure used or intended for supporting or sheltering any use or occupancy. The term building shall be construed as if followed by the words "or portion thereof."

Building. Building shall be a structure, including utilities, enclosed with a roof and within exterior walls built, erected and framed of component structural part, designed for the housing, shelter, enclosure and support of individuals, animals or property of any kind.

Building. A structure wholly or partially enclosed within exterior walls, or within exterior and party walls, and a roof, affording shelter to persons, animals, or property, as required by Code.

Building. Shall mean any structure erected for the support, shelter or enclosure of persons, animals, chattels, or property of any kind.

Building. A structure having a roof and intended for the shelter, housing or enclosure of persons, animals, or chattels.

Building. Any structure, or part thereof, having a roof supported by columns or walls, for the housing or enclosure of persons, animals, chattels, or property.

Building. Anything constructed for the shelter or enclosure of persons, animals, chattels or movable property of any kind and which is permanently affixed to the land.

Building. A structure which encloses space; a structure which gives protection or shelter for any occupancy. The term "building" shall be construed as if followed by the phrase "or part thereof." When separated by fire walls, each portion so separated shall be deemed a separate building.

Building. Any structure built for the support, shelter or enclosure of persons, animals, chattels or movable property of any kind, and which is affixed to the land or attached to a structure which is affixed to the land. When two (2) building structures are separated by division walls from the ground up, and without openings, each portion of such building shall be deemed a separate building.

Building. Any structure designed, intended, or used for the support, enclosure, shelter or protection of persons, animals, chattels, or property, including tents, awnings, vehicles, movable structures and other similar facilities situated on a parcel and used for purposes of a building. When a structure is divided into separate parts by unpierced walls extended from the ground up, each part is deemed a separate building.

Building. Any structure with substantial walls and roof securely affixed to the land and entirely separated on all sides from any other structure by space or by walls in which there are no communicating doors, windows or openings; and which is designed or intended for the shelter, enclosure or protection of persons, animals or chattels. Any structure with interior areas not normally accessible for human use, such as gas holders, water tanks, coal bunkers, oil cracking towers and other similar structures, are not considered as buildings.

Building. A structure having a roof, supported by columns or walls, for the shelter, support, or enclosure of persons, animals or chattels.

Building (existing). A building erected prior to adoption of the Code or one for which a legal building permit has been issued.

Bulkhead. A retaining wall. Also the wall beneath a store display window.

Bulkhead. An enclosed structure on or above the roof of any part of a building enclosing a shaft, stairway, tank, or service equipment or other space not designed or used for human occupancy.

Bulkhead. The term "bulkhead" shall mean any structure above the roof of a structure enclosing stairways, shafts, tanks, elevator machinery, ventilating apparatus and other accessories to the structure, except where otherwise specifically provided.

Bulkhead. A structure above the roof of any building, enclosing a stairway, tank, elevator machinery or ventilating apparatus, or such part of a shaft as extends above a roof.

Bulkhead. That portion of the exterior walls of a building which is located immediately under the show-windows.

Bulkhead. The raised portion of a floor or roof, raised for the passage of persons, materials, equipment, and light and air, through the side of such raised portion, or for other purposes.

Buttress. A bonded column of masonry built as an integral part of the wall, projecting from either or both surfaces, and decreasing in area from base to top.

Buttress. A projecting part of a masonry wall built integrally therewith to furnish lateral stability which is supported on proper foundations.

Cellar. That portion of a story, the ceiling of which is less than four feet above the adjacent finished grade.

Cellar. That portion of a building, the ceiling of which is entirely below grade or less than four (4) feet six (6) inches above grade.

Cellar. That portion of the building wholly below, or with less than half of its ceiling height above the average finished grade of the ground adjoining the building.

Cellar. The occupiable portion of the building partly underground, having half or more than half of its clear height below the average grade of the adjoining ground. A cellar which is used as a habitable dwelling shall be considered a story.

Cellar. A story having more than one-half of its height below the average finished grade level. A cellar shall not be counted as a story for purposes of height regulations.

Cellar. A story having more than one-half (1/2) of its height below grade. A cellar is not included in computing the number of stories for purpose of height measurement.

Cellar. That portion of a building between floor and ceiling which is wholly or partly below grade and so located that the vertical distance from grade to the floor below is equal to or greater than the vertical distance from grade to ceiling.

Cellar. That space of a building that is partly or entirely below grade, which has more than half of its height, measured from floor to ceiling, below the average established curb level or finished grade of the ground adjoining the building.

Cellar. That portion of a building between floor and ceiling which is 50% or more below grade.

Cellar. The portion of the building partially underground, having half or more than half, of its clear height below the grade plane.

Cellar. Story having more than 1/2 of its height below the average level of the adjoining ground at the building line. A cellar shall not be counted as a story for purposes of height measurement.

Cellar. That portion of a building between floor and ceiling which is partly below and partly above grade, the ceiling which is less than 4 feet above grade.

Cellar. A story of which one-half or more is below grade.

Cellar. A story having more than one-half of its height, measured from the finished floor to finished ceiling, below the curb level at the center of the street front.

Cellar. An area below the first story having more than one-half (1/2) of its height below grade and used for utilities, storage, garage for occupants of the building, or janitor or watchman quarters. A cellar so used shall not be considered as a story.

Cellar. A portion of the building with more than one-half (1/2) its height measured from finished floor to finished ceiling below the average grade of the adjoining ground; and not considered a story.

Cellar. That portion of a building, between floor and ceiling which is more than 50 percent underground.

Cellar. That portion of a building below the first floor level having more than one-half of its height below the curb level at the center of the street in front of the building. Where the walls of a building do not adjoin a street or a building line, than a cellar is a story having more than one-half of its height below the average level of the ground on which the building stands.

Cellar. Portion of the building partly underground having half or more than half of its clear height below the average grade of the adjoining ground.

Cellar. A story having more than one-half of its height below the level of a street grade or ground nearest the building. A cellar shall be included in computing the height or number of stories of buildings referred to in the various sections of the code.

Cellar. Term used as the same definition for a basement.

Cellar. The portion of the building between floor and ceiling which is 50% or more below finished grade at the building line.

Cellar. Floor space in a building having more than one-half (1/2) of its ceiling height below grade. A cellar is not counted as a story.

Cellar. A story having more than one-half (1/2) of its height below grade.

Cellar. The portion of a building located partly or wholly underground and having half or more than half of its clear floor-to-joint height below the average grade of the adjoining ground.

Cellar. A construction partly underground, but having one-half or more of its clear height (measured from finished floor to finished ceiling below the curb level; except that where the curb level has not been legally established, or where every part of the building is set back more than 25 feet from a street line, the height shall be measured from the adjoining grade elevation. A cellar shall not be deemed a story unless otherwise specifically ordered by the Bureau. Occupiable rooms or areas are not allowed in cellars.

Center or complex. Building or group of buildings which are designed to use common facilities such as parking, walks, swimming pools, and recreational areas.

Closed building. Structure completely enclosed by a roof and walls of approved construction.

Completely enclosed building. Building having no outside openings, other than ordinary doors, windows, and ventilators.

Completely enclosed building. Building separated on all sides from the adjacent open area, or from other buildings or other structures, by a permanent roof and by exterior walls or party walls, pierced only by windows or normal entrances or exit doors.

Conforming building. An existing building conforming to all requirements of the Code.

Conforming structure. One which complies with all of the regulations of the Ordinance.

Covered mall. Covered or roofed area in which the least horizontal dimension is 30 feet or more and which is used as a pedestrian thoroughfare to connect 2 or more buildings.

Covered mall. A covered interior area open from lowest grade level to roof used as a pedestrian public way and connecting groups of buildings.

Covered patio. Attached or detached structure not exceeding 14′ feet in height, and enclosed on not more than 3 sides except for posts necessary for roof support.

Covered shaft. Any shaft used for any purpose and covered at the top.

Covered walkway. Walkway that has more than 50% of its perimeter open to the outdoors.

Detached building. A building surrounded by open space on the same zoning lot as the principal building but separated from the principal building by not less than ten (10) feet.

Detached building. A principal building surrounded by open space on the same lot.

Detached building. A building having no party wall in common with another building.

Detached structure. Any structure having no party wall or common wall with another structure. Bridges, tunnels, breezeways, and other similar means of connecting one structure to another shall not, for the purposes of the ordinance, be considered to constitute a party or a common wall.

Enclosed arcade. Arcade with less than 25% of its perimeter abutting a street or plaza.

Enclosed court. Court bounded on all sides by the exterior walls of a building or exterior walls and lot lines on which walls are allowable.

Enclosed court. Open unoccupied space surrounded on all sides by walls, or by walls and an interior lot line.

Enclosed or inner court. Court surrounded on all sides by the exterior walls of a building or structure or by such walls and an interior lot line.

Enclosed plaza. Plaza with less than 25% of its perimeter abutting a street or plaza.

Enclosed porch. Porch having at least 50% of the horizontal section of the exterior walls in glass.

Enclosed walkway. Walkway that has 50% or less of its perimeter open to the outdoors.

Enlargement. Increase in floor area of an existing building or structure, or an increase in the area of land use for an existing open use.

Enlargement. Addition to the floor area of an existing building, an increase in the size of another structure, or an increase in the portion of a tract of land occupied by an existing use of the district.

Entrance to arcade. Opening in a building wall, or between building walls, side lot lines, or rear lot lines, at ground level. Unobstructed by any solid matter.

Entrance to plaza. Opening in a building wall, or between building walls, side lot lines, rear lot lines, at ground level. Unobstructed by any solid matter, but this shall not be construed to prohibit air curtains.

Erected. Includes built, constructed, reconstructed, moved upon, or any physical operations on the premises required for the building or structure; excavations, fill, drainage, and the like, shall be considered a part of erection.

Exterior balcony. Narrow, open platform on the outside of a building wall and surrounded by a rail or balustrade.

Exterior balcony. Elevated platform attached to a building and enclosed on its lower portion, on one or more sides by railings.

Exterior balcony. An elevated floor space projecting beyond the outside faces of the exterior walls of a building. For the purposes of the definition, at least 25 percent of the perimeter of such a balcony shall remain open.

Fall-out shelter. An accessory structure which incorporates the fundamentals for fall-out protection, shielding mass, ventilation and space to live, and which is constructed of such materials and in such a manner as to afford the occupants substantial protection from radio-active fall-out.

Fall-out shelter. Any building, or any addition or structural alteration to an existing building, designed, intended or constructed for use as a shelter against radiation from radio-active fall-out; provided that a fall-out shelter which is wholly below ground level or wholly within the lines of any existing building, and which complies with the provisions of the Ordinance entitled "Buildings and Construction" and shall not be subject to the provisions of the Zoning Ordinance.

Floor. Any structure which divides a building horizontally and shall include the horizontal members, floor coverings and ceiling.

Gallery. That portion of the seating space of an assembly room having a seating capacity of more than ten (10) located above the balcony.

Gallery. Railing-enclosed platform projecting from and supported by an outer wall of a building and supported from the ground by columns.

Lobby. Enclosed space into which aisles, corridors, stairways, elevators or foyer may exit and provides access to exits.

Lobby. Has its commonly accepted meaning with respect to the occupancy of the building under consideration.

Lobby. Public lounge or space for waiting which is connected with an entrance or exit way which serves as a principal entrance or exit.

Lobby. The enclosed vestibule between the principal entrance to the building and the doors to the main floor of the auditorium or assembly room of a theatre or place of assembly, or to the main floor corridor of a business building.

Lobby. To a place of assembly. An enclosed vestibule between the principal entrance to the building and the door or doors leading to the main floor of the place of assembly.

Metal frame building. Building is of metal frame unprotected construction if the enclosing walls are of unprotected metal or unprotected metal in combination with other noncombustible materials and the other building elements are as set forth in the code unless otherwise exempted.

Mezzanine. A floor having an area not in excess of one-third (1/3) of the area of the structure at the level at which the mezzanine floor is located.

Mezzanine. An intermediate level between the floor and ceiling of any story, and covering less than thirty-three and one-third (33-1/3) percent of the floor area immediately beneath.

Mezzanine. An intermediate level between the floor and ceiling of any story and covering not more than one-third of the floor area of the room in which it is located.

Mezzanine. An intermediate floor between the floor and ceiling of any space. When the total gross floor area of all mezzanines occurring in any story exceeds 33-1/3 percent of all the gross floor area of that story such mezzanine shall be considered as a separate story.

Mezzanine. Intermediate floor between the floor and ceiling of any story, covering less than the floor area immediately below.

Mezzanine. Floor having an area not in excess of 1/3rd of the area of the floor area in which the mezzanine is located. Clear height above or below the mezzanine shall be not less than 7 feet.

Mezzanine. An intermediate floor between the floor and ceiling of any space that is completely open or provides adequate visibility.

Mezzanine floor. A floor within a story between the floor and ceiling thereof, having an area not over forty per cent of the area of the building at the level at which the mezzanine floor occurs. A floor of larger area separates two stories.

Mezzanine floor. A partial floor within a room.

Mezzanine or mezzanine floor. An intermediate floor, either open or enclosed.

Mezzanine or mezzanine floor. An intermediate floor placed in any story or room. When the total area of any such "Mezzanine Floor" exceeds 33-1/3 per cent of the total floor area in that room, it shall be considered as constituting an additional "Story". The clear height above or below a "Mezzanine Floor" construction shall be not less than seven feet (7').

Mezzanine or mezzanine floor. An intermediate floor placed in any story or room. When the total area of any such "mezzanine floor" exceeds 33-1/3 per cent of the total area in that room, it shall be considered as constituting an additional "Story."

Mezzanines. A mezzanine shall be counted as a story when the total floor area of all mezzanines opening on a common room or similar open space exceeds one-third (1/3) of the total open space area at the first mezzanine level. In any case the construction of a mezzanine shall comply with the type of materials and fire-resistance ratings as required for the building.

Miscellaneous buildings and structures. Buildings and structures of a permanent and temporary character shall be constructed and equipped to meet the requirements of the code commensurate with the fire and life hazard incidental to their use; and where not specifically provided for or classified in the code, the adopted classification shall be based on the character of construction, relation to the fire limits and the proximity to adjacent structures.

Moving buildings. No person shall be permitted to move any building. Which has been damaged to an extent greater than 50% of its appraised value by fire, weather, decay or otherwise; nor shall it be permissible to move any frame or unprotected noncombustible building of such character as is prohibited to be constructed within the fire limits to any point within the fire limits; nor shall it be permissible to move any building to a location at which the uses for which such building is designed are prohibited by the code.

Occupied space. Any room or space in which any person normally does or is required to live, work or remain for any period of time.

Occupied space. The following shall be included as occupied space: outside stairways, fire escapes, open porches, platforms, balconies, chimneys containing boiler flues or gas vents and all such projections beyond the face of the building walls. This provision shall not apply to unenclosed outside porches not exceeding 1 story in height, which do not extend into the required front or rear yard setback lines further than 10 feet, nor to one such unenclosed porch which does not extend into the side yard nearer than 5 feet to the side lot line.

Occupied space. Includes balconies, fire escapes, fire towers, chimneys, platforms, porches, and outside stairways and other projections and all other substantial additions which project from the face of the exterior wall to which they may be attached and shall be considered as part of the building or structure and not part of the yards or courts or unoccupied spaces. This provision shall not apply to one fireplace chimney projecting not more than 12 inches into the side yard space and not more than 8 feet in length, nor to unenclosed outside porches not exceeding one story in height which do not extend into front or rear yard a greater distance than 12 feet from the front or rear walls of the building, nor to one such porch which does not extend into the side yard a greater distance than 6 feet from the side wall of the building nor exceed 12 feet in its other horizontal dimension, or to cornices not exceeding 16 inches in width including gutters.

Occupied space. Total area of all buildings, or structures on any lot or parcel of ground projected on a plane excluding roof projection as allowed by the code.

Occupied space. Room or space in a building or structure in which any person normally does or is required to live, work or remain for any period of time.

Open porch. A roofed, open structure projecting from the front, side or rear wall of a building and having no enclosed features of glass, wood, or other material more than thirty (30) inches above the floor thereof, except the necessary columns to support the roof.

Open porch. A roofed piazza, porch, or porte-cochere not more than one story in height which may project beyond the main wall with columns supporting its roof presenting the minimum of obstruction to the view, and with no sash or other enclosure except screens placed between the columns.

Open porch. A roof partially supported by columns or wall sections. Any portion of such open porch which extends into a front or side yard shall be so constructed that the area of supporting columns or wall sections on any one side shall not exceed fifty percent (50%) of that side; and, further, that screen, lattice, grill work or other material which has a net ten percent (10%) closed area may be permitted, provided the maximum area of fifty percent (50%) is not exceeded. Such open porch shall be used solely for ingress and egress, and not for occupancy as a sleeping porch or washroom.

Open porch. Porch, deck or balcony, with or without a roof, which projects beyond the main wall of a building, but is roofed, shall be so constructed that the view sideways through the porch shall not be obstructed except by columns if essential for the support of the porch roof.

Open porch. Porch which has no enclosing features, except screens, and a protective rail not less than 42 inches above the floor, and the roof supports.

Open porch. Porch that has no walls or windows other than those of the main wall to which the porch is attached.

Patio. An open area or an accessory outdoor living structure not exceeding 14 feet in height and open on at least one side.

Penthouse. Structure built above a roof to enclose wholly or in part a stairway, tank or other equipment, or space that affords passage for air or light. If the aggregate area of all penthouses and other roof structures exceeds fifty percent (50%) of the area of the roof, they shall be considered as an additional story.

Penthouse. Enclosed structure other than a "roof structure," located on the roof of a building, extending not more than 12 feet above the roof of the building, and used primarily for living or recreational accommodations.

Penthouse. Structure located on the roof of the main building for purposes of living accommodations or mechanical equipment. When the area of the penthouse exeeeds 20 percent of the area of the roof or when the penthouse is to be occupied by persons, the penthouse shall be considered as another story.

Penthouse. Enclosed structure on or above a roof, except towers, spires, cupolas, and lanterns used primarily for architectural embellishment or as belfries, and shall include enclosures for stairways, elevators, elevator machinery, tanks, ventilating, or other equipment, storage, or for human habitation, use, or occupancy.

Penthouse. An enclosed structure above the roof of a building, extending not more than twelve (12) feet above the roof and occupying not more than thirty (30) per cent of the roof area.

Penthouse. An enclosed room or structure other than a "Roof Structure" located on the roof.

Penthouse. An enclosed structure other than a roof structure, located on the roof, extending not more than twelve (12) feet above a roof.

Penthouse. Any closed roof structure, other than a bulkhead which extends twelve feet or less above the roof of a structure and occupies thirty percent or less of the roof area.

Penthouse. Structure built above a roof, limited in use to the housing of machinery, mechanical equipment and stairways. A penthouse shall not be construed as a story.

Penthouse. An enclosed one-story structure extending above the roof of a building not exceeding twenty-five (25) percent of the area of the roof at the level on which such penthouse or penthouses are located.

Penthouse. Enclosed or partially enclosed structure extending above the main roof of a building or structure and/or enclosing a stairway, tank, elevator, machinery, mechanical equipment or other apparatus and not used for human occupancy.

Penthouse. A portion of a building, other than a roof structure, located on the roof structure, and used primarily for living quarters, recreation areas, or other use involving the more or less continuous presence of persons within. If a penthouse covers more than 1,000 square feet or 1/3 of the roof area of the supporting building, it shall be considered as an additional story.

Penthouse. An enclosed structure above the roof of a building, other than a roof structure, extending not more than twelve (12) feet above the roof, and occupying not more than thirty-three and one-third (33-1/3) percent of the roof area.

Porch. That portion of a building floor not entirely surrounded by the building walls and which is intended to serve as an entrance or exit to, or as a space auxillary to a building.

Porch. Roofed structure projecting from a building and separated from the building by the walls thereof.

Porch. Roofed structure with one or more open sides, erected against and projecting from, an exterior wall of a building.

Porch. Unenclosed exterior structure at or near grade attached or adjacent to the exterior wall of any building, and having a roof and floor.

Porches. Covered porches, whether enclosed or unenclosed, shall be considered as part of the principal building and shall not project into a required yard.

Remodeling. To remodel and/or alter means to change any building or structure which affects the structural strength, fire hazard, internal circulation or exits of the existing building or structure. This definition does not apply to maintenance, reroofing, or alterations to the heating and ventilating or electrical systems.

Remodeling. Any change, addition or modification in construction which involves a change in structure or in grade of occupancy.

Roof structure. A structure above the roof of any part of a building enclosing a stairway, tank, elevator machinery or service equipment, or such part of a shaft as extends above the roof; and not housing living or recreational accommodations.

Roof structure. A structure above a roof or any part of a building enclosing a stairway, tank, elevator machinery or ventilating apparatus, or such part of a shaft as extends above the roof.

Roof structure. An enclosed structure on or above the roof of any part of a building.

Roof structure. An unenclosed structure on or above the roof of any part of a building.

Roof structures. Any structure extending from and above the roof including a shelter for the protection of equipment necessary only to the operation of the building, but excluding a penthouse, roof sign or space for the purpose of providing additional floor area.

Structural alteration. A change in any of the supporting members of a building.

Structural alteration. Any change in or addition to the supporting or structural members of a building, such as the bearing walls, partitions, columns,beams, or girders, or any change which would convert an existing building into a different structure, or adapt it to a different use, or which in the case of a non-conforming use, would prolong the life of such use.

Structural alteration. Any change or rearrangement in the exterior or interior structural parts of any building.

Structural alteration. Any change in either the supporting members of a building, such as bearing walls, columns, beams or girders, or in the roof and exterior walls.

Structural alteration. The erection, strengthening, removal, or other change of the supporting elements of a building or structure. Such elements shall include, but shall not be limited to, footings, bearing walls, columns, beams, girders, joists, and decking.

Structural alteration. Any change in the supporting members of a building, such as bearing walls or partitions, columns, beams or girders, or any change in the roof or in the exterior walls, excepting such repair or replacement as may be required for the safety of the building.

Structural alteration. Any change or replacement of the supporting members of a building, such as bearing walls or partitions, columns, beams, or girders, or any substantial change in the roof or in the exterior walls.

Structural alterations. Any change in the supporting members of a building, such as bearing walls or partitions, columns, beams, or girders, or any complete rebuilding of the roof or the exterior walls.

Structural alterations. Any change in the supporting members of a building or in any substantial change in the roof structure or in the exterior walls.

Structural alterations. Any changes in the supporting members of a building, such as footings, bearing walls or partitions, columns, beams, or girders, or any substantial change in the roof or in the exterior walls, excepting such repair as may be required by an official governmental agency for the safety of the building.

Structural alterations. Any change, addition or modification in construction in the supporting members of a building, such as exterior walls, bearing walls, beams, columns, foundations, girders, floor joists, roof joists, rafters or trusses.

Structural alterations. Any change, except for repair or replacement, in the supporting members of a building, such as bearing walls, columns, beams or girders.

Structural alterations. Any change in the supporting members of a building or structure such as walls, columns, beams, girders, floor joists, roof joists, or in the exterior walls.

Structural alterations. Any change in the supporting members of a building or other structure, such as bearing walls, columns, beams, or girders.

Structural alterations. Any change in the physical structure of a building or structure.

Structural alterations. Any replacement or changes in the type of construction or in the supporting members of a building, such as bearing walls or partitions, columns, beams or girders, beyond ordinary repairs and maintenance.

Structural alterations. Any change or rearrangement in the bearing walls, partitions, columns, beams, girders, exit facilities, exterior walls, or roof of a building excepting such repair as may be required for the safety of the building, or an enlargement, whether by extending on a side or by increasing in height, or movement of the building from one location or position to another.

Structural alterations. Any change in the supporting members of a building, such as bearing walls, columns, beams or girders, or any substantial change in the roof or exterior walls.

Structural alterations. Any change, other than incidental repairs, in the supporting members of a building or structure, such as bearing walls or partitions, columns, beams or girders or any substantial change in the roof or exterior walls.

Structural alterations. Any change which would prolong the life of the supporting members of a building or structure, such as bearing walls, columns, beams or girders.

Structural alterations. Any change in the supporting members of a building, such as bearing walls, columns, beams or girders.

Structural repairs. The removal of any part of an existing building or structure not in violation of law for the purpose of its maintenance in its present class of construction and grade of occupancy.

Structure. An improvement upon land, other than the land itself, the use of which requires more or less permanent location on the ground or attachment to something having a permanent location on the ground, including, but not limited to, buildings, signs and pavements.

Structure. Anything constructed or erected, the use of which requires more or less permanent location on the ground or which is attached to something having a permanent location on the ground.

Structure. That which is built or constructed, or any piece of work artifically built up or composed of parts joined together in some definite manner, the use of which requires more or less permanent location on the ground, or which is attached to something having a permanent location on the ground.

Structure. Anything constructed or erected, the use of which requires permanent or temporary location on or in the ground, including, but without limiting the generality of the foregoing, advertising signs, billboards, backstops for tennis courts, pergolas, radio and television antennae, including supporting towers, and swimming pools, provided, however, that the definition shall not include underground tanks for the storage of any type of fuel.

Structure. Anything constructed or erected which requires a location on the ground, including a building or a swimming pool, but not including a fence or a wall used as a fence if the height does not exceed six (6) feet, or access drives or walks.

Structure. Anything constructed or erected the use of which requires location on the ground or attached to something having location on the ground.

Structure. Anything constructed or erected having location on or under ground or attached to something having location on or under the ground.

Structure. Anything constructed or erected, except fences, the use of which requires permanent location on the ground or attached to something having a permanent location on the ground.

Structure. Any construction or any production or piece of work artificially built up or composed of parts joined together in some definite manner.

Structure. Anything constructed or erected on, or under the ground or upon another structure or building.

Structure. Anything constructed or erected, whether portable, pre-fabricated, sectional or otherwise, which is permanent or temporary, located on and/or under the ground or attached to something so located.

Structure. The arrangement of parts or members securely joined together to form a unit for housing, amusement, edification or service of the people.

Structure. Anything constructed or erected which requires permanent location on the ground, or attached to something having permanent location on the ground, including signs and billboards, but not including fences or walls used as fences, six (6) feet in height or less.

Structure. A building or construction of any kind.

Structure. That which is built or constructed, an edifice or building of any kind, or any piece of work artificially built up or composed of parts joined together in some definite manner. The term "structure" shall be construed as if followed by the words "or part thereof."

Structure. A combination of materials to form a construction that is safe and stable, including among others, buildings, stadiums, reviewing stands, platforms, stagings, observation towers, radio towers, water tanks and towers, trestles, piers, wharves, sheds, coal bins, shelters, fences and display signs.

Structure. A combination of materials assembled at a fixed location to give support or shelter, such as a building, bridge, trestle, tower, framework, tank, tunnel, tent, stadium, reviewing stand, platform, bin, sign, flagpole and the like. A fence or wall over six (6) feet high is considered to be a structure.

Structure. That which is built or constructed, an edifice or building of any kind, or any piece of work artificially built up or composed of parts joined together in some definite manner. The term "structure" shall include "building."

Structure. Anything constructed or erected, the use of which requires more or less permanent location on the ground or which is attached to something having more or less permanent location on the ground.

Structure. Anything constructed or erected, which requires permanent location on the ground or attachment to something having such location, but not including a trailer.

Structure. Anything constructed or erected which requires permanent location on the ground, or attached to something having permanent location on the ground.

Structure. Anything constructed or erected, the use of which requires location on the land, or attachment to something having a permanent location on the land.

Structure. Anything constructed or erected with a fixed location on the ground, or attached to something having a fixed location on the ground.

Structure. An assembly of materials forming a construction for occupancy or use; including among others: buildings, stadiums, gospel and circus tents, reviewing stands, platforms, stagings, observation towers, radio towers, water tanks, trestles, piers, wharves, open sheds, coal bins, shelters, fences, and display signs. The word structure is construed as if followed by, "or parts thereof."

Structure. Anything constructed or erected, the use of which requires more or less permanent location on the ground or attached to something having a permanent location on the ground.

Structure. Anything constructed or erected, including a building which has permanent foundations on the ground, or anything attached to something having a permanent location on the ground.

Structure. That which is constructed on or under the ground or attached or connected thereto, including but not limited to: buildings, barriers, bridges, bulkheads, chimneys, fences, garages, outdoor seating facilities, parking areas, platforms, pools, poles, streets, tanks, tents, towers, sheds, signs, walls and walks; and excluding trailers and other vehicles whether on wheels or other supports.

Structure. Anything constructed or erected, the use of which requires a location on the ground, or attached to something having a location on the ground, including but without the generality of the foregoing, advertising signs, billboards, back stops for tennis courts, fences and pergolas.

Structure. Anything constructed or erected, the use of which required more or less permanent location on the soil, or attached to something having a permanent location on the soil. A sign, billboard, or other advertising medium detached or projecting and having a gross area of sixty (60) square feet or more shall be construed to be a structure in the ordinance.

Structure. Any combination of materials forming any construction the use of which requires location on the ground or attachment to something having location on the ground and including, among other things, fallout shelters, display stands, gasoline pumps, mobile dwellings (whether mobile or stationary at the time), outdoor signs, platforms, pools, porches, reviewing stands, sales stands, signs, stadiums, stagings, standpipes, tanks of any kind, towers of any kind, including radio and television towers and antennae, trellises and vending machines. The word "structure" shall be construed as though followed by the words "or part thereof."

Structure. An assembly of materials forming a construction for occupancy or use including among others, buildings, stadiums, public assembly tents, reviewing stands, platforms, stagings, observation towers, radio towers, water tanks, trestles, piers, wharves, open sheds, coal bins, shelters, fences, and display signs. The term structure shall be construed as if followed by the words or part thereof.

Structure. Any thing constructed or erected, the use of which requires more or less permanent location on the ground or attached to something having a permanent location on the ground. When a structure is divided into separate parts by an unpierced wall, each part shall be deemed a separate structure.

Structure. Any constructed or erected material or combination of materials, the use of which requires location on the ground, including, but not limited to, buildings, stadia, radio towers, sheds, storage bins, fences and signs.

Structure. That which is built or constructed. The term structure shall be construed as if followed by the words "or portion thereof."

Structure. The assembly of materials, forming a construction framed of component structural parts for occupancy or use, including buildings.

Structure. Anything constructed or erected, the use of which requires a permanent location on the ground or attachment to something having a permanent location on the ground.

Structure. That which is built or constructed, an edifice or building of any kind or any piece of work artificially built or composed of parts joined together in some definite manner and attached to the ground.

Structure. The assembly of materials forming a construction for occupancy or use, normally attached to the ground or to something located on or in the ground.

Structure. Anything which is constructed or erected and located on or under the ground, or attached to something fixed to the ground.

Structure. Any deliberate form or arrangement of building materials permanently located on the land.

Structure. A combination of materials assembled to give support or shelter, and affixed to real property. The term "structure" shall also be construed to mean a part of the structure.

Structure. Anything constructed or erected with a fixed location on the ground, or attached to something having a fixed location on the ground. Among other things, structures include buildings, mobile homes, walls, fences, billboards, and poster panels.

Structure. Anything constructed or erected which requires fixed location on the ground or attachment to something having fixed location on the ground.

Structure. An assembly of materials forming a construction framed of component structural parts for occupancy or use, including buildings.

Structure. Anything constructed or erected either upon or below the surface of the earth and which is supported directly or indirectly by the earth, but not including any vehicle which conforms to the State Vehicle Act.

Structure. Any building, fence, tower, edifice or building of any kind, or any piece of work artificially built up or composed of parts joined together in some definite manner which requires location on the ground or is attached to something having a location on the ground.

Structure. Any object constructed or installed by man, including, but without limitation, buildings, towers, smoke-stacks, and overhead lines.

Structure. Anything constructed or erected which is supported directly or indirectly on the earth, but not including any vehicle which conforms to the State Vehicle Act.

Structure. That which is built or constructed, an edifice or building of any kind, or any piece of work artificially built up or composed of parts joined together in some definite manner.

Structure. The term "structure" shall mean a building or construction of any kind.

Structure. That which is built or constructed.

Structure. An assembly of materials forming a construction for occupancy or use, including among others: buildings, stadia, tents, reviewing stands, platforms, stagings, observation towers, radio towers, tanks, trestles, open sheds, coal pockets, shelters, fences, and display signs.

Structure. Any combination of materials, including buildings, constructed or erected, the use of which requires location on the ground or attachment to anything having location on the ground, including among other things, signs, billboards, but not including utility poles and overhead wires.

Structure. A combination of materials assembled at a fixed location to give support or shelter, such as a building, framework, retaining wall, tent, reviewing stand, platform, bin, fence, sign, flagpole, mast for radio antenna or the like. The word "structure" shall be construed, where the context allows, as though followed by the words "or part or parts thereof."

Structure. Anything constructed or erected, the use of which requires more or less permanent location on the ground, or attached to something having a permanent location on the ground, including but without limiting the generality of the foregoing, advertising signs, backstops for tennis courts and pergolas.

Structure. Anything which is constructed or erected and the use of which requires more or less permanent location on ground or attachment to something having permanent location on ground, not however, including wheels; an edifice or a building of any kind; any production or piece of work, artificially built up or composed of parts and joined together in some definite manner. (Entrances not more than two feet above grade and vents not more than three feet above grade, which are features of bomb or fallout shelters, shall not be considered a structure or structural feature for purposes of setback computation.)

Structure. Anything constructed or erected, the use of which requires more or less permanent location on the ground or attached to something having a permanent location on the ground and shall include fences, tents, lunch wagons, dining cars, campcars or other structures on wheels or other supports and used for business or living purposes.

Structure. Anything built or erected, including among other things, buildings, stadiums, reviewing stands, bandstands, bleachers, booths, swimming pools, platforms, towers, bridges, trestles, bins, fences, barriers, poles, tanks above or below ground and signs; and shall also mean the supporting framework or supporting parts of a building. The term "structure" shall be construed as if followed by "or parts thereof."

Structure. Assembly of material forming a construction for occupancy or use (including, among others, buildings, stadiums, gospel or circus tents, reviewing stands, platforms, stagings, observation towers, radio and television towers, water tanks, trestles, piers, wharves, open sheds, coal bins, shelters, fences, and display signs).

Structure. That which is built of fabricated or manufactured building materials and placed either above or below the ground.

Structures to have access. Every building hereafter erected or moved shall be on a lot adjacent to a public street, with access to an approved private street, and all structures shall be so located on lots as to provide safe and convenient access for servicing, fire protection, and required off-street parking.

Substandard building. Any building or portion thereof for the premises on which the same is located in which there exists any of the conditions listed in the Code to an extent that it endangers the life, limb, health, property, safety or welfare of the public or the occupants thereof.

Terrace. Unenclosed exterior structure at or near grade having a paved, floored, or planted platform area adjacent to an entrance or to the exterior walls for a building or structure and having no roof.

Unenclosed arcade. Arcade with 25 percent or more of its perimeter abutting a street or plaza.

Unenclosed plaza. Plaza with 25 percent or more of its perimeter abutting a street or plaza.

Unenclosed porch. Roofed projection which has no more than fifty (50) percent of each outside wall area enclosed by a building or siding material other than meshed screens.

Unenclosed porch. Open porch with or without screened sides.

Walkway. Covered or roofed pedestrian thoroughfare used to connect two (2) or more buildings in which the least horizontal dimension of the thoroughfare is less than thirty (30) feet.

C-5 CHILD CARE FACILITIES

Child care center. Any place, home or institution which receives three or more children under the age of sixteen (16) years, for care apart from their natural parents, legal guardians or custodians, and received for regular periods of time for compensation; provided, however, this definition shall not include public and private schools, organized, operated or approved under the laws of this state, custody of children fixed by a court, children related by blood or marriage within the third degree to the custodial person, or to churches or other religious or public institutions caring for children within their institutional building while their parents or legal guardians are attending services or meeting or classes and other church activities.

Child day-care. Administering to the needs of infants, pre-school children, and school-age children outside of school hours by persons other than their parents for any part of the 24 hour day, for part or all of at least 2 consecutive weeks, but does not include the care of children in places of worship during religious services.

Child day-care center. Any place in which child day-care is provided for 5 or more infants, pre-school children, or school-age children outside of school hours in average daily attendance, other than the children of the Owner or Administrator of the center, with or without compensation.

Day camp. Organized recreational or educational enterprise for day use only.

Day camp. An organized recreational or educational enterprise for day use only and not overnight lodging except for owner or caretaker.

Day care center. Facility wherein a) the operator is provided with compensation and is licensed by the State, or by a State Agency, and wherein one (1) or more individuals is provided with care for less than 24 hours; includes without limitation by reason of enumeration, day nurseries, nursery schools, family day care homes, and other supplemental care facilities certified or licensed by the State.

Day care center. An agency other than a nursery for infants and toddlers two (2) years of age or younger, which is required to be licensed as a child care agency under the State, or is designed, arranged or intended to provide supplemental parental care during the day or night for a group or four (4) or more children not related to the proprietor under the age of sixteen (16) years for periods of more than one (1) hour, with or without compensation, but shall not include a church operated nursery for the convenience of persons attending church functions or worship services.

Day care center-nursery. Any detached single-family dwelling, detached child-care facility, apartment, duplex, condominium, or townhouse where more than five children under 12 years of age who are not members of the family of the superior or custodian in charge.

Day care centers. Place operated by any person who receives therein for pay children for daytime control, care and food, including all types of day care programs, such as day nurseries for the children of working mothers, nursery schools and kindergartens for children under the minimum age for admission to public schools, parent-cooperative nursery schools, kindergartens, play groups for school children or other similar units operated under any name whatsoever.

Day care homes. Dwelling where not more than seven (7) children under thirteen (13) years of age, including the children of the operator under thirteen (13) years of age, are housed and cared for between the hours of six (6) A.M. to twelve (12) o'clock midnight, while the natural parents are employed or otherwise engaged.

Day care homes. Dwellings where children up to 13 years of age are housed and cared for a period of 12 hours while the natural parents are employed or otherwise engaged.

Day nursery. Building or portion thereof appropriated to the care of 4 or more children, 16 years of age or less, apart from their parents, between the hours of 6 A.M. and 7 P.M.

Day nursery. Building whose primary use involves the daytime care of more than four (4) children under school age or children of older age with limiting physical and mental handicaps.

Day nursery. Place in a building where 5 or more children can be left by the parents for housing and care for a period of 8 hours.

Day nursery or nursery school. Any private agency, institution, establishment or place which provides supplemental parental care and/or educational work, other than lodging overnight, for six (6) or more unrelated children of pre-school age, for compensation.

Family day care home. Home operated by a person residing on the premises who receives pay for the day time care of one (1) to ten (10) children under 7 years of age; however, after school hours, school holidays, and school vacations, children under 17 years of age are permitted. The operating hours shall be from 7:00 a.m. to 6:00 p.m., provided further that all such homes be licensed by the Department of Public Welfare, and to conform to any other applicable laws.

Family day care home. Home operated by any person who receives therein, for pay, children, not to exceed six (6) in number, who are not related to such person and whose parents or guardians are not residents in the same house, for daytime supervision and care.

Infant. Child under 18 months of age including any child who has not yet reached the steady walking state or who requires the use of diapers.

Nursery. A home or institution where two (2) or more children are cared for, for hire by the day, week, month or year.

Nursery day school. Home or institution where two (2) or more children are cared for during the day. This includes a kindergarten.

Nursery school. School or day-care facility for five or more pre-elementary-school age children, other than those residents on the site.

Nursery school. Any building used routinely for the daytime care or education of pre-school age children, and including all accessory buildings and play areas.

Nursery school. Any structure, lot or premise where a commercial or institutional establishment is maintained or operated temporarily or permanently for the training and/or care (other than medical care) of pre-school age children.

Nursery school. Any place, however designated, operated for the purpose of providing training, guidance, education, or care for four (4) or more children under 6 years of age, separated from their parents or guardians, during any part of the day, including kindergartens, day nurseries, and day care centers.

Nursery school. Institution for the care of children, of preschool age, the activity of which shall be conducted between the hours of 7 a.m. and 7 p.m. Even though some instruction may be offered in connection with such care, the institution shall not be considered a "school" within the meaning of the ordinance.

Nursery school. Public or private school or kindergarten wherein day care or day care and education is provided for five (5) or more children aged five (5) years and under.

Nursery school. School for five (5) or more pre-elementary school age children, or use of a site or portion of a site for a group day-care program for five (5) or more children other than those resident on the site, including a day nursery, play group, or after-school group.

Nursery schools. Day care centers which receive children between the ages of two (2) and six (6) years and which are established and professionally operated primarily for educational purposes to meet the developmental needs of the children served.

C-6 COMMERCIAL, BUSINESS, MERCANTILE AND OFFICE BUILDINGS

Activity in a building. Includes but is not limited to the use or storage in any business of any fixed or movable equipment, or the use, storage, display, sale, delivery, offering for sale, production, or consumption in any business, or by any business invitee on the premises of the business, of any goods, wares, merchandise, products, or food, or the performance in any business of any work or services.

Adult book store. Any business establishment in which all or any portion of said premises is devoted to the sale or display of any book, magazine, newspaper, or other printed or written material or any picture, drawing, photograph, motion picture, or other pictorial representation or any statue or other figure, or any recording, transcription, or mechanical, chemical, or electrical reproduction, or any other articles, equipment, machines, or materials, which aforementioned articles or materials, by any means or in any manner are either sold or displayed.

Antique shop. The sale of fine art objects or household furnishings produced prior to the year 1830 and are characteristic of a specific period in a specific country.

Antique shop. Place of business where a product is sold or exchanged because of value derived due to the age as respects to modern or late date replicas.

Art gallery. Room or portion of a building used for the exhibition of works of art.

Art gallery. Room or building used for the exhibition of works of art.

Artist. One who practices the fine art which shall be limited to painting, drawing, sculpture, poetry, music, dancing and dramatic art.

Artist. One who practices the fine arts which shall be limited to dancing and dramatic art, drawing, music, painting, poetry, and sculpture. This is not deemed to include the business of teaching one or more of the arts.

Art, music and hobby stores. Stores selling materials and finished products related to arts, crafts and hobbies and shall include: art and art supply dealers, camera dealers, crafts supply dealers, stamp and coin shops, music stores, and any other store of a similar nature.

Automated service station. A place of business having pumps and/or storage tanks from which liquid gasoline and diesel fuel, and/or liquid lubricants are dispensed at retail directly into motor vehicles, and the primary portion of whose business is the repair, servicing or washing of motor vehicles. Sales and installation of auto accessories, washing, polishing, inspections and cleaning, but not steam cleaning may be carried on incidental to the sale of such fuel and liquid lubricants.

Bank. Building used primarily as an establishment for the custody, loan, exchange, or issue of money, for the extension of credit, and for facilitating the transmission of funds by drafts or bills of exchange.

Barber shop and beauty parlor. When established in a residential district a barber shop or a beauty parlor shall be incidental to the residential occupation, not more than 25 percent of the floor area of only one-story of a dwelling unit shall be occupied by such office, and only 2 barber chairs in a barber shop, and two (2) operators in a beauty parlor shall be permitted, and only one un-lighted name plate not exceeding 1 square foot in area containing the name and business of the occupant of the premises shall be exhibited.

Beauty parlor. Any premises, building, or part of a building, in which any branch of cosmetology or the occupation of a cosmetologist, as defined in the Code.

Business. Buildings in which the primary or intended occupancy or use is the transaction of administrative, business, civic, or professional service, and where the handling of goods, wares, or merchandise, in limited quantities, is incidental to the primary occupancy or use. Newsstands, lunch counters, barber shop, beauty parlors, and similar service facilities are considered as incidental occupancies or uses.

Business. Business occupancies are those used for the transaction of business for the keeping of accounts and records, and similar purposes.

Business. The word "business" or the word "commerce" when used in the Ordinance means the engaging in the purchase, sale, barter or exchange of goods, wares or merchandise; and the maintenance or operation of offices or recreational or amusement enterprises.

Business. Any occupation, employment or enterprise whereon merchandise is exhibited or sold or which occupies time, attention, labor and materials or where services are offered for compensation.

Business. An occupation, employment or enterprise which occupies time, attention, labor and/or materials for compensation; whether or not merchandise is exhibited or sold, or services are offered.

Business. An occupation, employment or enterprise which occupies time, attention, labor and materials, or wherein merchandise is exhibited, bought or sold, or where services are offered for compensation.

Business administration offices. Shall mean establishments engaged in the management of business enterprises.

Business building. A building occupied for the transaction of business, for the rendering of professional services, for the display or sale of goods, wares or merchandise, or for the performance of work or labor.

Business establishment. A place of business, the ownership, proprietorship, or management of which are separate and distinct from those of any other place of business.

Business establishment. A place of business wherein the ownership or management or type of such business is separate and distinct from that of any other business located on the same or adjacent lot.

Business occupancy. The occupancy or use of a building or structure or any portion thereof for the transaction of business or the rendering or receiving of professional services.

Business offices. Real estate, insurance and other similar offices; and the offices of the architectural, clerical, engineering, legal, dental, medical, or other established recognized professions, but excluding morticians, undertakers and funeral directors; in which only such personnel are employed as are customarily required for the practice of such business or profession and not exceeding a total of 5 persons at any one time.

Business or commerce. Purchase, sale or other transaction involving the frequent and regular handling or disposition of any article, substance, service or commodity for profit or livelihood, or the ownership or management of office buildings, offices, recreational or amusement enterprises, or the maintenance and use of offices by members of professions and trades rendering services.

Business or commerce. Shall mean the purchase, sale or other transaction involving the handling or disposition of any article, service, substance or commodity for livelihood or profit; or the management of office buildings, offices, recreational or amusement enterprises; or the maintenance and use of offices, structures and premises by professions and trades rendering services.

Business or commercial buildings. Buildings or parts of buildings, other than public or residential buildings, including among others, office buildings, stores, markets, restaurants, warehouses, freight depots, garages, factories, laboratories and the like.

Business service. Enterprise which provides other persons with planning, advice, technical aid and installs and maintains office equipment and data processing equipment and leases or sells such equipment.

Business service. Any commercial activity primarily conducted in an office, not involving the sale of goods or commodities available in the office, and not dispensing personal services, and including such businesses as real estate broker, insurance, accountants, financial institutions, studios, or any similar use.

Business service establishments. Establishments providing supplies and services to business and professions and shall include copy service, blue printing service, typing service, telephone answering service, office supply stores, delivery, and messenger service, advertising agencies, direct mail service, dental and medical laboratories and any other establishmet offering goods or services of a similar nature but not including establishments of a light industrial nature, such as printers, linen supply laundries, etc.

Business service establishments. Shall mean establishments providing supplies and services to business and professions and shall include copy service, blue printing service, typing service, telephone answering service, office supply stores, delivery and messenger service, advertising agencies, direct mail service, detective agencies, employment agencies, collection agencies, dental and medical laboratories and any other establishment offering goods or services of a similar nature, but not including establishments of a light industrial nature such as printers, linen supply laundries, etc.

Cabaret. The term "cabaret" shall mean any room, place or space in the city in which any musical entertainment, singing, dancing or other similar amusement is permitted in connection with the restaurant business or the business of directly or indirectly selling to the public food or drink.

Canvasser/solicitor. Any person who goes from house to house or from place to place, selling or taking orders for any product or service.

Capacity in persons. The establishment or use for which the maximum amount of persons can avail themselves of the services (or goods) in such an establishment, at any one time, with reasonable safety and comfort, as determined by the Fire Prevention Code of the City.

Casino. Any place where gaming is operated or maintained, except that "Casino" shall not be construed to include any place devoted to slot machines only.

Cigar store. A store or shop, or part thereof, place, building, structure or vehicle, where cigars, cigarettes or tobacco or all or any of them, are exposed for sale, or sold, and inclusive of sale of any cigars, cigarettes or tobacco, or all or any of them, from vending machines, or other apparatus or mechanical device.

Club. Association of persons for some common purpose, but not including a group organized primarily or which is actualy engaged to render a service which is customarily carried on as a business. A road house or tavern shall not be construed as a club.

Club. Buildings and facilities owned and operated by a corporation or association of persons for social or recreational purposes, but not operated primarily for profit or to render a service which is customarily conducted as a business.

Club. Building owned, leased, or hired by nonprofit association of purposes, who are bona fide members paying dues, the use of which is restricted to said members and their guests.

Club. Association of persons for some common nonprofit purpose, but not including groups organized primarily to render a service which is customarily carried on as a business.

Club. Organization or persons for special purposes or for the promulgation of sports, arts, sciences, literature, politics, or the like, but not operated for profit.

Club. Buildings and facilities owned and operated by a corporation, association, person or persons for a social, educational or recreational purpose, but not primarily for profit or to render a service which is normally carried on as a business.

Club. Membership organization formed for a specific purpose that may be fraternal, social or recreational and may or may not be profit-motivated.

Club—fraternity—lodge. Club, fraternity or lodge is an association of persons for some nonprofit purpose, but not including groups organized primarily to render a service which is carried on as a business.

Club house. House or building occupied in whole or in part as a residence by members of a club or organization, and whose members and guests have access to all parts or any part of the house or building. The term "club house" shall be understood to mean and include a fraternity house.

Club house. Building to house a club or social organization not conducted for profit and which is not an adjunct to or operated by or in connection with or as a public tavern, cafe or other like public place.

Club house. A club or social organization housed in a building and not conducted for profit and which is not an adjunct to or operated by or in connection with or as a public tavern.

Club or lodge. Association of persons for the promotion of some non-profit common object, as literature, science, politics, good fellowship, etc., meeting periodically, limited to members, with no more than one-third of the gross floor area occupied by the use used for residential occupancy.

Club or lodge. A private nonprofit association, corporation or other entity consisting of persons who are bona fide dues paying members which owns, leases or uses a building or portion thereof, the use of such premises being restricted primarily to members and their guests.

Club or lodge buildings. Building used by a fraternal, social, or similar organization for the private assembly of persons and not used by the general public as a place of assembly.

Commerce. The purchase, rental, sale or other transaction involving the handling of or disposition of any article, substance or commodity for profit or livelihood, or the ownership or management of office buildings, offices, recreational or amusement enterprises, automobile courts, garages, hotels, outdoor advertising and outdoor advertising structure, or shops conducted for the sale of personal services and other similar enterprises of the same class.

Commerical. Any building or structure that is occupied or used by any individual, group or business for puposes other than dwelling or industrial.

Commercial. Any use of structure or premises in which commodities are raised, produced, manufactured, rented, sold, or stored for profit, or in which plants or animals are propagated and/or raised, in excess of the consumption requirements of the family or families resident in said structure or on said premises and in contradistinction to a hobby or avocation.

Commercial building. Any building used in connection with direct trade with or service for the public.

Commercial buildings. Any structures or enclosures wherein goods or services are sold or offered or stored or wherein business is transacted whether public or private or mixed or wherein persons are housed in multiple occupancy or in which industry is conducted.

Commerical enterprise. A business establishment or development and the buildings, structures, and lands relevant thereto. A booster, meter or valve house for utilities or pipe lines, car sales lot, drive-in theater, implement sales lot, junk yard, lumber yard, road equipment storage yard, service station, shopping center, stadium, trailer camp or wrecking yard, etc., may be classified as a commercial enterprise. Clay, earth, gravel, sand and other pits, quarries and mining operations are also classified as commercial enterprises.

Commercial fisher. Commercial establishment for the receiving, processing, packaging, storage and wholesale or retail distribution and sale of products of the sea. Such an establishment may include facilities for the docking, loading, unloading, fueling, icing and provisioning of vessels and for the drying, maintenance and storage of equipment.

Commercial garage. A place of storage of one or more automobiles, trucks or other commercial vehicles, owned and used by not more than one (1) person.

Commercial hotel. Transient hotel which caters to the general public, whose primary market is persons traveling for business purposes.

Commercial purpose. An occupation, employment or enterprise which is carried on for profit by the owner, lessee or licensee, except for activities carried on by a not-for-profit organization which utilizes the proceeds of such activities solely for the purposes for which it is organized.

Commercial stable. A stable where horses are let, hired, used, boarded or sold on a commercial basis for remuneration.

Commercial use. A use operated for profit or compensation through buying or sale of commodities.

Commercial use. Any use operated for profit or compensation.

Commissary. Any establishment or concession preparing, serving or selling food products to occupants of buildings or employees or commercial or industrial enterprises for consumption upon the premises.

Community shopping center. Commercial establishment designed to provide the basic facilities found in a neighborhood center, with a wider range of commercial establishments. A community shopping center ranges from 30,000 to 300,000 square feet.

Convenience establishment. Small establishment designed and intended to serve the daily or frequent shopping or service needs of the immediate surrounding population. Filling stations and repair garages are specifically excluded from this definition.

Custom shop. Work, service or assembly made to order for individual customers for their own use or convenience.

Department store. Entire space occupied by one tenant or more than one tenant in an individual store where more than 100 persons commonly assemble on other than the street-level floor for the purpose of buying personable wearables and other merchandise.

Department store. Building which is used for the housing and sale at retail of a variety of goods which are located in departments according to a system.

Designed shopping center. Retail business or office development, planned as a unit, characterized by groups of retail, office, or combined uses having the common use of specifically designed off-street areas for access, parking, or service.

Discount stores. Retails stores selling goods and services across the counter under the concept of discounting usually well-known brands of merchandise at substantial discounts from customary or list prices.

Drive-in banks or other establishments. A bank or other establishment in which the facilities for dispensing services or consuming commodities, or both, are provided and intended to occur primarily in patrons' automobiles parked on or moving through the premises.

Drive-in business. Any place of business of premise which serves, sells or otherwise makes available its services to patrons situated in automobiles.

Drive-in facility. A facility or establishment which is designed, intended or used for transacting business with customers in autombiles.

Drug store or drug and chemistry supply house. A store or buildings used for receiving, compounding, storing, handling or dispensing medicinal preparations, drugs, chemicals, oils, volatile solvents and other substances which alone or in combination with other substances are of highly combustible, flammable or explosive nature.

Dry cleaning. Process of cleaning wearing apparel, textiles, leather goods or feathers by application of a volatile or flammable liquid.

Dry cleaning. The business of cleaning cloth, feathers or any type of fabric by the use of gasoline, naphtha, benzine or other petroleum or coal tar products; or cleaning by any methods which include the use of flammable, volatile or highly combustible or combustible material.

Dry dyeing. Process of changing the color of wearing apparel textiles, leather goods or feathers by the application of a dyeing solution in combination with volatile or flammable liquids.

Dry goods stores. Establishments selling clothing, blankets, linens, yardage, notions, small hardware and home appliances and shall include: dry goods stores, variety stores, gift stores, notions stores, sewing machine stores, yardage stores, shoe stores, clothing stores, and any other store of a similar nature.

Financial institutions. Establishments and offices offering financial services or counsel and shall include banks, savings and loan institutions, stock brokers offices, bonding companies, finance company offices, and any other institution of a similar nature.

Fine arts shops. Individual art pieces not mass-produced consisting of one or more of the following: paintings, drawings, etchings, sculptures, ceramics, inlays, needlework, knitting, weaving, and/or craftswork: leather, wood, metal or glass.

Fine arts studio. Limited to painting, drawing, sculpturing, poetry, music, dancing and dramatic art.

Food cooperative. An association of no more than one hundred (100) families that come together not more than once a week for distribution of food to the members, and which does not employ any paid staff.

Food stores. Establishments selling food and drink products for consumption off the premises and shall include: grocery stores, fruit and vegetable stores, bakeries, dairy stores, delicatessens, ice cream stores, butcher shops, liquor stores, candy stores, and any other store of a similar nature.

Furniture and appliance stores. Stores selling new or used furniture or appliances and providing incidental service and maintenance and shall include new and used furniture stores, appliance stores, antique dealers, carpet and linoleum dealers, and any other establishment of a similar nature.

Gambling or gaming establishments. Any place where gaming is operated and maintained. "Gaming" means and includes all games of chance or devices and any slot machines played for money or for checks or tokens redeemable in money, except, for the purpose of this title only "Gaming" shall not be construed to include slot machines when such slot machines are operated incidental or accessory to the conduct of a business permitted under the provision of the Code.

General membership club. A club or lodge established by and for membership organizations of a more inclusive or more general nature, with or without dining facilities and cocktail lounge, for the benefit of the membership and not for gain. This type of club (except country club) is not indigenous to any particular district or described residential area and is prohibited in all family residence districts, but admissible as a special exception to commercial districts.

General retail business. Enterprise for profit for the convenience and service of, and dealing directly with, and accessible to, the ultimate consumer: neither injurious to adjacent premises or to the occupants thereof by reason of the omission of cinders, dust, fumes, noise, odors, refuse matter, smoke vapor or vibration; nor dangerous to life or property. It shall include buildings or spaces necessary to a permitted use for making or storing articles to be sold at retail on the premises.

Goods. Any goods, wares, works of art, commodity, compound or thing, chattel, merchandise or personal property which may be lawfully kept or offered for sale.

Greenhouse. Building or structure constructed chiefly of glass, glass-like or translucent material, cloth or lath, which is devoted to the protection or cultivation of flowers or other tender plants for sale to the public.

Greenhouse. Building constructed mostly of glass, or similar material and which is used for the protection or cultivation of plants.

Gross leasable area. Total floor area designed for tenant occupancy and exclusive use, including basements, mezzanines, upper floors and storage area, and excluding areas devoted to housing of mechanical equipment, electrical substations and mail areas, and also excluding general office floor area having no connection to any of the retail establishments in the planned shopping center, up to twenty percent (20%) of the gross leasable floor area of the center.

Gross leasable area. Is the total floor area designed for tenant occupancy and exclusive use. The area of tenant occupancy is measured from the center lines of joint partitions to the outside of the tenant walls. All tenant areas, including areas used for storage, shall be included in calculating gross leasable area.

Gross leasable area. The gross leasable area is the total floor area designed for tenant occupancy and exclusive use. The area of tenant occupancy is measured from the center lines of joint partitions to the outside of the tenant walls.

Hand laundry. Laundry which occupies not more than 1200 square feet of floor space and within which not more than five (5) persons are engaged in laundry activities.

Hand laundry. Establishment where clothes are received for washing and ironing, and where the work is done by hand using not more than 2 each of washing and ironing machines of not more than 2 horsepower in the aggregate, and to or from which establishment the clothes are carried by the customers.

Hobby shop. Accessory use which is not conducted for remuneration, but soley as a hobby, pastime or means of education or entertainment. Such activity shall be carried on entirely within a building and only between the hours of 7 a.m. and 10 p.m.

Hobby shop. An accessory building in which is conducted an activity, which activity is not for remuneration, but solely as the hobby, pastime, means of education or entertainment of the occupants of the main building on which such accessory building is situated, and which activity is carried on entirely within a building and only between the hours established by the Code.

Homes association. Incorporated, nonprofit organization operating under recorded land agreements through which each lot owner of a development area is a member, and each lot is subject to charges of a proportionate share of the expenses for the organization's activities such as maintaining the common property.

Household and family service establishment. Shall mean a store or shop providing for the maintenance or repair of articles of normal home or family use and shall include: shoe repair shops, cleaning and dyeing establishments, coin operated laundries, lawn mower and saw sharpening, fix-it-shops, smaller appliance repair shops, tailor shops, photographers, and any other shop of a similar nature provided that establishments for the maintenance and repair of automobiles or tires shall not be deemed to be household and family service establishments.

Land sales presentation office. Means any building or room therein or portion thereof wherein land sales presentations, as defined in the Code, are presented, held or conducted.

Land sales presentation unit broker office. Any building or room therein or portion thereof wherein prospective customers are solicited to attend land sales presentations as defined in the Code.

Launderette. A business establishment which provides coin operated self-service type washing, drying, dry-cleaning and ironing facilities, provided that: (a) not more than four (4) persons, including owners, are employed on the premises at one time: and (b) no pick-up or delivery service is provided.

Launderette. (Which includes an automatic wet wash business). An establishment only for the washing and drying of clothing brought in by the customer and in which such washing and drying is performed with the use of mechanical equipment, and for which a fee is charged.

Laundromat. Business premises equipped with individual clothes washing and drying machines for the use of retail customers, exclusive of laundry facilities provided as an accessory use in an apartment house or an apartment house or an apartment hotel.

Laundromat. Business that provides washing, drying, and/or ironing machines or dry cleaning machines for hire to be used by customers on the premises.

Laundromat. Business providing for the hire and use on the premises of home type washing, drying and/or ironing machines.

Laundromat (self-service laundry). Business rendering a retail service by renting to the individual customer the temporary use of stationary equipment for the self-service of washing, drying, and otherwise processing of laundry and dry-cleaning with such equipment to be serviced and its use and operation supervised by the management.

Livery stable. Any place of business where horse-drawn wagons and carts are rented to the general public.

Market. Building or structure which is used for the sale and purchase of provisions.

Mercantile. Buildings in which the primary or intended occupancy or use is the display and sale to the public of goods, wares, or merchandise.

Mercantile building. Any building or a portion of which, used for the display and sale of merchandise.

Mercantile buildings. Include department stores, markets, dining rooms and restaurants seating or accommodating less than 100 persons, spaces used for retail or wholesale sales, and buildings of similar use.

Mineral or natural resources development. Commercial or industrial operations involving removal of timber, native vegetation peat, muck, topsoil, fill, sand, gravel, rock, or any mineral and other operations having similar characteristics.

Miscellaneous retail stores. Includes the following stores offering goods at retail: book stores, florists, stationery stores, drug stores, pet stores, but not pet hospitals, tobacco stores, travel agencies, jewelry stores, hardware stores, and paint stores.

Motor freight depot or trucking terminal. Place, building, or part thereof where merchandise, property, or freight transported by motor vehicles, including trailers, to or from outside the limits of the County is received, stored, transferred, loaded, unloaded, delivered, or dispatched, and shall include any parking service or gasoline filling station, service or repair shop or other accessory service operated in conjunction therewith.

Motorized cart track. Any area where motorized carts are operated for spectator viewing and/or participation.

Neighborhood shopping center. Group of commercial establishments providing for the sale of convenience goods or personal services. Its size in square footage must be between 5,000 and 30,000 square feet.

Neighborhood shopping districts. Designed for the convenience shopping and servicing of persons residing in adjacent resident areas, and to permit primarily such uses as are necessary to satisfy those basic shopping and service needs which occur frequently and so require retail and service facilities in relative proximity to places of residence.

Night club. Any room or space in which a business, primarily for the sale of food or drink, is operated, and in which entertainment is provided or dancing permitted.

Non-commercial recreation. Clubs or recreation facilities, operated by a non-profit organization and open only to bona-fide members of such non-profit organization.

Nursery. Land or greenhouses used to raise trees, shrubs, flowers and other plants for sale or for transplanting.

Nursery (plant material). Space, building or structure, or combination thereof, for the storage of live trees, shrubs, or plants offered for retail sale on the premises including products used for gardening or landscaping. The definition of nursery within the meaning of the Code does not include any space, building or structure used for the sale of fruits, vegetables or Christmas trees.

Occupancy. The purpose or activity for which a building or space is used or is designed or intended to be used.

Office. Business facility in which business administration, business service, and business transactions occur, but in which there is no display or sale of on-premises merchandise.

Office building. Building comprised of more than fifty percent offices, as compared with "offices" which are permitted in Commercial Districts and as compared with home occupations where offices are considered as a secondary or incidental use.

Office building. Any building consisting of one (1) or more offices used for the conducting of business.

Office building. Include banks, barber shops, beauty parlors, clinics, places of assembly for less than fifty (50) persons and which are not an accessory for another occupancy, mortuaries, buildings used primarily for offices, transportation stations accommodating less than 50 persons, studios, telegraph offices, telephone exchanges, and buildings of similar use.

Office building. Building designed for or used as the offices of professional, commercial, industrial, religious, institutional, public, or semipublic persons or organizations, provided no goods, wares or merchandise shall be prepared or sold on any premises except that a portion of an office building may be occupied and used as a drug-store, barbershop, cosmetologists shop, cigar stand or newstand, when such uses are located entirely within the building with no entrance from the street nor visible from any sidewalk and having no sign or display visible from the outside of the building indicating the existence of such use.

Office building. Any building consisting of one or more offices used for the conducting of business as listed in the Ordinance.

Offices. Space or rooms used for professional, administrative, clerical and similar uses.

Office space. Which is used primarily for administrative, stenographic, bookkeeping, or similar work.

Office space. Which is used primarily for the sale or trade of chattels located elsewhere, or for the sale or trade of tickets, certificates, or real estate, including the preparation of correspondence, records, and documents in connection therewith.

Office space. A place where a particular kind of business is transacted or a service is supplied, excluding retailing of a product.

Office space. A place in which functions, such as consulting record keeping, or clerical work of any type is performed.

Office space. A place in which a professional person, such as an architect, engineer, lawyer, or physician conducts his professional services.

Office space. Used primarily for the reception and transmission of messages for others, and operated as a business.

Office space. Used by professional persons for the rendering of services for others.

Open front store. Business establishment so developed that service to the patron be extended beyond the walls of the structure, not requiring the patron to enter the structure. The term "open front store" shall not include automobile repair stations or automobile service stations.

Outdoor advertising business. Provision of outdoor display space on a lease or rental basis only.

Owner. Any person firm or corporation owning or controlling property, including a duly authorized agent or attorney. Guardians or trustees shall also be regarded as owners.

Paint and oil store. Store or mercantile establishment in which paints, varnishes, oils, or other flammable liquids, used in painting and finishing are stored, handled and sold at retail.

Paint or lacquer spraying shop. A building or part thereof used for the application of combustible finishes by means of spraying or dipping, or by a brush.

Pawnbroker. A merchant who engages in the business of buying and selling and/or lending money for which he holds chattels as security for used personal property.

Personal service. Business which is neither conducted as the practice of a learned profession nor dealing primarily in wholesale or retail activities involving inventories of stock in trade on premises; e.g., barbers, beauticians, cobblers, dry-cleaning, realtors and tailors.

Personal service establishments. Shops or offices offering services for the health or welfare of the individual and shall include: barber shops, beauty shops, health and reducing studios, and any other establishment of a similar nature.

Personal services. Beauty parlor, shop or salon, barber-shop, massage, reducing, or slenderizing studio, steam or Turkish Baths, or any similar use.

Personal service shops. Business establishments, such as chiropody, massage, or similar personal service shops.

Pet shop. An establishment containing a store, room or group of rooms, cages or exhibition pens, not part of a kennel, wherein birds or animals commonly accepted or regarded as pets are kept for sale, excluding, however, such places wherein dogs are kept for sale.

Pharmacy. Place where medicines only are compounded or dispensed.

Place of business. Maintaining a bona fide address or location where all types of business may be transacted.

Place of employment. Includes every place, whether indoors or out or underground and the premises appurtenant thereto where either temporarily or permanently any industry, trade or business is carried on, or where any process or operation, directly or indirectly related to any industry, trade or business, is carried on, and where any person is, directly or indirectly, employed by another for direct or indirect gain or profit, but does not include any place where persons are employed in private domestic service which does not involve the use of mechanical power for farming.

Place of public accommodation. All public places of entertainment, amusement or recreation, all public places where food or beverages are sold for consumption on the premises, all public places which are conducted for the lodging of transients or for the benefit, use or accommodation of those seeking health or recreation and all establishments which cater or offer their services, facilities or goods to or solicit patronage from the members of the general public. Any residential house, or residence in which less than five (5) rooms are rented, or any private club, or any place which is in its nature distinctly private is not a place of public accommodation.

Planned shopping center. Group of separate commercial retail or professional or personal service establishments developed as a unit on a site having a minimum horizontal dimension of 400 feet in any direction.

Plant. (as distinguished from a shop). A business or manufacturing establishment such as a dry cleaning plant or a factory where the product worked upon may be received from or distributed to wholesale customers as well as retail customers.

Private club. Association for civic, social, cultural, religious, literary, political, recreational, or like activities, operated for the benefit of its members and not open to the general public.

Private club. Building and facilities or premises used or operated by an organization or association for some common purpose, such as, but not limited to, a fraternal, social, educational or recreational purpose, but not including clubs organized primarily for profit or to render a service which is customarily carried on as a business. Such organizations and associations shall be incorporated under the Laws of the State as nonprofit corporation or registered with the Secretary of State.

Private club. Property owned or leased and operated by a group or an association of persons and maintained and operated solely by and for the members of such a group or association and their guests and which is not available for unrestricted public access or use.

Private club. Any building or facility owned or operated by a corporation, association, person or persons, for a social, educational or recreational purpose no part of the income of which is distributed directly or indirectly to its members, owners, directors or officers.

Private club. An institution, used or intended to be used for an association of persons, whether incorporated or unincorporated, for some common purpose, but not including a group organized solely or primarily to render a service customarily carried on as a commercial enterprise.

Private non-commercial. Clubs or recreation facilities, operated by a non-profit organization and open only to members of the organization.

Profession. Occupation or calling requiring the practice of a learned art through specialized knowledge based on a degree issued by an institution of higher learning, e.g., Doctor of Medicine, Engineer, Lawyer.

Professional occupations. The office, of a member of a recognized profession, who is living on the premises.

Professional office. When conducted in a residential district a professional office shall be incidental to the residential occupation; shall be conducted by a member of the resident family entirely within a residential building, and shall include only the offices of doctors or practitioners, ministers, architects, landscape architects, professional engineers, lawyers, authors, musicians and other recognized professional occupations occasionally conducted within residences.

Professional office. Office of a doctor, practitioner, dentist, minister, architect, landscaper architect, professional engineer, lawyer, author, musician, or the other recognized professions. When established in a residential district, a professional office shall be incidental to the residential occupation, not more than 25 percent of the floor area of only 1 story of a dwelling unit shall be occupied by such office and only 1 unlighted name plate, not exceeding 1 square foot in area, containing the name and profession of the occupant of the premises shall be exhibited.

Professional office. A building more or less of residential, but not commercial character and containing one (1) or more offices in which there is no display of stock or wares in trade, commodity sold, nor any commercial use conducted other than the professional offices of a doctor, dentist, lawyer, architect, landscape architect, engineer, minister of religion, insurance agent, realtor or other similar professional services; but shall not include barbershops, beauty parlors nor similar services, nor general business offices.

Professional office. Office of a person engaged in any occupation, vocation or calling, not purely commercial, mechanical or agricultural in which a professed knowledge or skill in some department of science or learning is used by its practical application to the affairs of others, either advising or guiding them in serving their interest or welfare through the practice of an act founded thereon.

Professional services. Conduct of business in any of the following or related categories: law, architecture, engineering, medicine, dentistry, osteopaths, chiropractors, opticians, or consultants in these or related fields, studios of dancing, music and art.

Professional services. Shall include medical, dental, legal, architectural, engineering and similar services.

Professional use. The rendering of service of a professional nature by architects, engineers, and surveyors, doctors of medicine, osteopathy, dentistry and optometry, lawyers, accountants, consultants and practitioners who are recognized by the appropriate licensed professions, chiropractors, chiropodists, and naturopaths.

Professional use. The rendering of a lawful service of a recognized professional nature.

Proprietary club. Membership organization that is operated to generate income to the owner of the facilities.

Public market. Means any place or building wherein food stuffs are sold, or offered for sale, either wholesale or retail by 3 or more persons, acting independently of each other and conducting separate business of their own. Public market shall include farmers' markets where 3 or more farmers shall conduct business individually for the purpose of selling or offering for sale, either wholesale or retail.

Quarry-sand, pit-top soil stripping. Lot or land or part thereof used for the purpose of extracting stone, sand, gravel, or top soil for sale, exclusive of the process of grading a lot preparatory to the construction of a building for which application for a building permit has been made.

Real estate office. Any building or room therein or portion thereof, maintained by a real estate broker licensed pursuant to State Statutes.

Recreation and sport stores. Stores selling or renting sporting goods, camping equipment, bicycles, and other sports or recreation equipment and shall include: sporting goods stores, ski sales and rental stores, gun shops, camping equipment sales and rental stores, bicycle shops including bicycle repair, toy stores, and any other store selling goods or providing services of a similar nature.

Regional shopping center. Commercial establishment designed to provide a full scope of retail sales and services. It is designed to attract customers from an area of greater population than the County. Its size must be from 300,000 square feet up.

Repair shop. A shop for the adjustment, repair or reconditioning of personal, household or garden articles or devices.

Residential clubs. Club or lodge established by and for the residents of a particular residential district, or described family residence area for social, recreational, or cultural purposes, not for gain and entirely indigenous to a restricted, defined residential area as to membership, is admissible as a special exception to any residential district.

Restaurant. Food service operation as defined in the Code.

Retail. Sale to the ultimate consumer for direct consumption and not for resale.

Retail bakery. Combination bakery and baked goods sales shop at which more than 50 percent of the baked goods produced are sold at retail from the baked goods sales shop.

Retail florists. A business of selling at retail, cut flowers, growing flowers or plants, potteries, stands or holders containing plants or flowers.

Retail kosher butcher shop. A shop, store or establishment for the retail sale of meats or poultry to the public, which meats or poultry have been slaughtered, prepared, inspected and passed as "kosher" in accordance with the requirements set forth by traditional orthodox Judaism.

Retail nursery. Premises, whether or not enclosed, use for the retail sale of nursery stock and materials and, incidental thereto, the retail sale of gardening implements, packaged fertilizer, packaged soil conditioners and additives, pesticides and other materials and supplies customarily used in the cultivation and maintenance of a home garden or landscape.

Retail sales. Sales to the consumer.

Retail sales and service. Retail sales and service includes those business activities customarily providing retail convenience goods. Such uses includes department stores, variety stores, drug and sundry stores, restaurants, delicatessens, cafeterias, grocery and markets, gift shops, wearing apparel, home and auto supply, furniture and appliances, hardware, news-stands, book and stationery stores, shoe repair shops, luggage shops, bakeries, and candy shops (provided that products made on the premises are sold on the premises).

Retail sales area. Area in square feet within any enclosed structure or permanent outside sales or display facility devoted exclusively for the sale or display of goods or commodities.

Retail store. Establishment where goods are sold directly to the consumer for personal or household use, with or without processing on the premises for such retail sale, but excluding the processing, repair or renovating of furniture, bedding or fixtures.

Roadside stand. Structure not permanently fixed to the ground that is readily removable in its entirety, covered or uncovered and not wholly enclosed, and used solely for the sale of farm products produced on the premises. No such roadside stand shall be more than 300 square feet in ground area and there shall be not more than one roadside stand on any one premise.

Self-help laundry. Located in a building in which domestic type washing machines and/or driers are provided on a rental basis for use by individuals doing their own laundry.

Self-service laundry. Business rendering a retail service by renting to the individual customer at a fixed location, equipment for washing, drying and otherwise processing laundry, with such equipment to be serviced and its use and operation supervised by the management, and does not include processing the laundry by the management on behalf of the customer.

Service station. A place of retail business at which outdoor automotive refueling is carried on using fixed dispensing equipment connected to underground storage tanks by a closed system of piping, and at which goods and services generally required in the operation and maintenance of motor vehicles and fulfilling of motorist needs may also be available. The building consists of a sales office where automotive accessories and packaged automotive supplies may be kept or displayed. It may also include one or more service bays in which vehicle washing, lubrication and minor replacement, adjustment and repair services are rendered. An automotive service station building shall have no cellar or basement, but may have open pits if such pits are continually ventilated.

Shoe shine parlor. A place or establishment where 3 or more chairs are placed in use, or to be used for the purpose of shining or polishing shoes or boots with polish or other material, composition or mixture of any color whatever.

Shop. Use devoted primarily to the sale of a service or a product or products, but the service is performed or the product to be sold is prepared in its finished form on the premises. (Packaging is not considered to be preparation.)

Shop (as distinguished from a plant. Business establishment such as a barber shop, radio shop or dry cleaning shop for retail or service activities or both where the product worked on is received from or distributed to retail customers only.

Shopping center. Any combination of three (3) or more separately owned and operated retail businesses having common parking facilities.

Shopping center. Group of commercial establishments planned, developed, and owned and managed as a unit, with on-site parking and of similar architectural characteristics.

Shopping center. Group of retail stores, planned and designed for the site upon which they are built.

Shopping center. Lot or parcel of land having an area in excess of five (5) acres and upon which is located three (3) or more businesses.

Shopping center district. Retail business area within or adjacent to a residence district, characterized by a concentrated grouping of stores, shops and other uses herein permitted, ordinarily planned as a unit and built according to such plan.

Show window. Fixed window in a shop used exclusively for the display of merchandise or advertising matter.

Slaughterhouse. A place where cattle, sheep, hogs or other animals are killed or butchered for market or for sale; provided, however, that this shall not be taken to mean or include the killing of poultry or rabbits as permitted by this title.

Slaughter house. Any building, place or establishment in which is conducted the slaughtering of livestock or poultry for commercial purposes.

Snack bar. Retail food service establishment serving such items as coffee, tea, milk, soft drinks, sandwiches and pastry. All customer service shall be confined to the physical limits of the snack bar.

Store. Enclosed space used for the display and sale of merchandise, or sale of service, to the general public. Space used for cigar or newspaper stand and similar uses in a public lobby or similar location, is not deemed to be a store.

Store. Use devoted exclusively to the retail sale of a commodity or commodities.

Studio. Work room of a professional artist.

Studio. Room or series of rooms used by a broadcasting or television station primarily for the transmission of radio or television programs; except where such room is a "place of assembly" and is regulated by the Code.

Supermarket. Large retail enterprise offering for sale foodstuffs and related merchandise on a volume basis and providing or requiring considerable off-street parking on the site or in the immediate proximity to it.

Tenancy. Area of a building occupied by a person.

Tenant. Any tenant, licensee, occupant, person in charge or in possession of any dwelling or part thereof.

Tenant. Person, firm, or corporation possessed with the legal right to occupy premises.

Tenant. Person, persons, co-partnership, firm, or corporation occupying or using a structure.

Tenant. Person occupying a building or portion thereof.

Wholesale. The business of selling goods or merchandise to retailers or jobbers for resale to the ultimate consumer.

Wholesale sales. Sales to retailers or jobbers of articles in which they deal.

Workshop. Room in the building in which repair work incorporating carpentry, painting, welding, electrical work or similar activity is conducted, incidental to the uses of the Occupancy.

C-7 CONDOMINIUMS

Blanket encumberance. A trust deed, mortgage, judgment or other lien on a condominium including any lien or other encumberance arising as a result of the imposition of any tax assessment by a public authority.

Building. A building, or a group of buildings, each building containing one or more units, and comprising a part of the property provided that the property shall contain not less than two units.

Closing of the sale. The operation of transferring ownership of a condominium unit to the purchaser from the developer.

Common area. All areas accessible to and utilized by occupants of a building, including but not limited to vestibules, hallways, stairways, landings and common space as defined in the Declaration.

Common areas and facilities. Unless otherwise provided in the declaration or lawful amendments thereto include:

(a) The land on which the building stands and such other land and improvements thereon as may be specifically included in the declaration, except any portion thereof included in a unit.

(b) The foundations, columns, girders, beams, supports, main walls, roofs, halls, corridors, lobbies, stairs, stairways, fire escapes, and entrances and exits of the building.

(c) The basements, yards, gardens, parking areas and storage spaces.

(d) The premises for the lodging of janitors' or persons in charge of property.

(e) Installations of central services such as power, light, gas, hot and cold water, heating, refrigeration, air conditioning and incinerating.

(f) The elevators, tanks, pumps, motors, fans, compressors, ducts, and in general, all apparatus and installations existing for common use.

(g) Such community and commercial facility as may be provided for in the declaration and all other parts of the property necessary or convenient to its existence, maintenance and safety, or normally in common use.

Common elements. All of the condominium except the condominium units. "Common Elements" also includes limited common elements.

Common expenses. All sums lawfully assessed against the unit owners by the association of unit owners. Expenses of administration, maintenance, repair or replacement of the common areas and facilities. Expenses agreed upon as common expenses by the association of unit owners. Expenses declared common expenses by the provisions of the association, or by the declaration or the bylaws.

Common profits. The balance of all income, rents, profits, and revenues from the common areas and facilities remaining after the deductions of the common expenses.

Condominium. A form of property established pursuant to the Condominium Property Act of the State.

Condominium. A structure of two (2) or more units, the interior space of which are individually owned: the balance of the property (both land and building) is owned in common by the owners of the individual units. The size of each unit is measured from the interior surfaces of the exterior walls, floors, and ceiling. The balance of the property is called the common area.

Condominium. An apartment building in which the apartments are individually owned, like houses, and for which taxes and interest are deductible on income tax returns.

Condominium. A form of complete ownership by which the buyer has the entire undivided interest in an apartment or similar type dwelling, as opposed to ownership of the land upon which the dwelling is located.

Condominium. A system of separate ownership of individual units in multi-unit projects. In addition to the interest acquired in a particular unit, each unit owner is also a tenant in common in the underlying fee and in the spaces and building parts used in common by all unit owners.

Condominium. Ownership in fee of a specific area in a multiple dwelling. An apartment in an apartment building, with tenancy in common ownership of the common areas, such as corridors, parking lots, etc.

Condominium. A system of property ownership as defined and regulated by the provisions of the Condominium Property Act of the State.

Condominium. The ownership of single units in a multi-unit structure with common areas and facilities.

Condominium. Form of ownership of residential housing in which the individual owns his unit and the common areas are owned jointly.

Condominium. Unit in a multi-unit building in which the units are individually owned and the structure is owned jointly; regulated under the Zoning ordinance for use and occupancy on the basis of the jointly owned structure, with individually owned units treated as tenant spaces.

Condominium. An estate in real property consisting of an undivided interest in common in a portion of a parcel of real property together with a separate interest in air space.

Condominium. An estate in real property consisting of an undivided interest in common in a portion of a parcel of real property together with a separate interest in air space in a residential, industrial or commercial building on such real property, such as apartment, office or store.

Condominium. An estate in real property consisting of an undivided interest in common in portions of real property together with a separate interest in space in a residential, industrial or commercial building on such real property.

Condominium. (Association of unit owners.) All of the unit owners acting as a group in accordance with the bylaws and declaration.

Condominium dwelling. Multiple dwelling whereby the fee title to each dwelling unit is held independently of the others.

Condominium project. The sale of or plan by a developer to sell or the offering for sale of residential condominium units in an existing building or building to be constructed or under construction.

Condominium unit or unit. A separate three-dimensional area within the condominium identified as such in the declaration and on the condominium plat and shall include all improvements contained within such area except those excluded in the declaration.

Conversion or convert. The offering for sale by a developer or his agent of a condominium unit occupied or rented for any purpose by any person before commencement of a condominium project which includes such unit.

Declaration. A declaration referred to in the Condominium Property Act of the State.

Declaration. The instrument, duly recorded, by which the property is submitted to the provisions of the association, as hereinafter provided, and such declaration as from time to time may be lawfully amended.

Developer. Any person who submits property legally or equitably owned by him to the provisions of the Condominium Property Act of the State, including any successor to such developer's entire interest in the property; or any person who offers units legally or equitably owned by him for sale in the ordinary course of his business. "Developer" does not include a corporation owning and operating a co-operative apartment building unless more than six (6) units are to be sold to persons other than current stockholders of the corporation.

Limited common areas and facilities. Includes those common areas and facilities which are agreed upon by all the unit owners to be reserved for the use of a certain number of units to the exclusion of the other units, such as special corridors, stairways and elevators, sanitary services common to the units of a particular floor and the like.

Majority (or majority of unit owners). The owners of more than fifty per cent (50%) of the aggregate interest in the common areas and facilities as established by the declaration assembled at a duly called meeting of the unit owners.

Offering. Any inducement, solicitation, advertisement, publication or announcement by a developer to any person or the general public to encourage a person to purchase a condominium unit in a condominium or prospective condominium.

Person. A natural individual, corporation, partnership, trustee or other legal entity capable of holding title to real property.

Person. Individual, corporation, partnership, association, trustee, or other legal entity.

Property. Includes the land, the building, all improvements and structures thereon and all easements, rights and appurtenances belong thereto, and all articles of personal property intended for use in connection therewith, which have been or are intended to be submitted to the provisions of the Code.

Property report. The property report required in accordance with the provisions of the Condominium Property Act of the State and as amended by the Building Code.

Prospective purchaser. A person who visits the condominium project site for the purpose of inspection for possible purchase or who requests the property report.

Recordation. To file or record in the office of the county register of deeds in the county where the land is situated, in the manner provided by law for recordation of instruments affecting real estate.

Unit designation. The number, letter, or combination thereof designating the unit in the declaration.

Unit (or condominium unit). An enclosed space consisting of one or more rooms occupying all or a part of a floor or floors in a building of one or more floors or stories regardless of whether it is designed for residence, for office, for the operation of any industry or business, or for any other type of independent use and shall include such accessory spaces and areas as may be described in the declaration, such as garage space, storage space, balcony, terrace or patio, provided it has a direct exit to a thoroughfare or to a given common space leading to a thoroughfare.

Unit owner. A person, corporation, partnership, association, trust or other legal entity, or any combination thereof, who owns a unit with the building.

C-8 COVERED AND ENCLOSED MALL SHOPPING CENTERS

Anchor store. Is an exterior perimeter department store or major merchandising center having direct access to a mall, but having all required exits independent of a mall.

Anchor tenant. A tenant occupying a space more than three (3) floors in height, located usually at the mid-point or end of the mall building.

Covered mall. A covered interior area open from lowest grade level to roof used as a pedestrian public way and connecting groups of buildings.

Covered mall. A covered or roofed interior area used as a pedestrian public way and connecting tenant spaces and/or groups of tenant spaces housing individual or multiple tenants.

Covered mall building. Is a single building enclosing a number of tenants and occupancies such as retail stores, drinking and dining establishments, entertainment and amusement facilities, offices and other similar uses wherein two or more tenants have a main entrance into one or more malls. For the purpose of the Code anchor stores shall not be considered as part of the covered mall building.

Covered mall building. Wherein two (2) or more tenants have a main entrance into one (1) or more malls which are roofed interior areas providing common pedestrian facilities for the public wherein the distance of travel of one (1) of the exits from any point within a tenant space is measured to the mall.

Covered mall buildings. A covered mall building is a single building enclosing a number of tenants and occupancies such as retail stores, restaurants, places of assemblage, recreation facilities, motion picture theaters, offices, banks, specialty shops and anchor stores.

Covered mall center. A building enclosing a number of stores or shops and other occupancies leased by tenants.

Enclosed mall shopping center. A single structure composed of buildings or groups of buildings connected by a covered mall not exceeding eighty (80) feet in height; designed primmarily for mercantile use with multiple tenants.

Mall. A pedestrian area which is common to the various stores and shops and other occupancies by tenants located in a covered mall center.

Mall. A roofed or covered common pedestrian area within a covered mall building which serves as access for two or more tenants.

Mall. An area composed of sidewalks and landscaping which serves as a pedestrian thoroughfare between buildings, but is not dedicated to public use.

Mall. A mall is a roofed over common pedestrian area serving more than one (1) tenant located within a covered mall building.

Mall. Pedestrian way of passage within or between closed buildings Only areas at least fifty (50) feet within any open end mall shall be considered as being within the mall.

Tenant. An individual shop, store or other place of business occupying a given and well defined area or space within the covered mall center.

C-9 DWELLINGS AND DWELLING UNITS

Accessory living quarters. An accessory building used solely as the temporary dwelling of guests of the occupants of the premises; such dwellings have no kitchen facilities and not rented or otherwise used as a separate dwelling unit.

Apartment. Same as dwelling unit.

Apartment. A suite or rooms, with cooking facilities, used for living purposes. Each apartment shall be considered a dwelling unit.

Apartment. Any room or suite of two or more rooms occupied as a home for one or more persons and arranged or equipped for cooking meals.

Apartment. Shall mean a dwelling unit as defined in the Code.

Apartment. A room or suite of rooms which is occupied or which is intended or designed to be occupied by one family for living and sleeping purposes.

Apartment. A room or a suite of rooms occupied, or which is intended or designed to be occupied, as the home or residence of one individual, family or household, for housekeeping purposes.

Apartment hotel. A residential building designed or used for both two or more dwelling units and six or more guest rooms or suites of rooms.

Apartment hotel. Any building which contains dwelling units and also satisfies the definition of a hotel, as defined in the Code.

Apartment house. A residential building designed or used for three or more dwelling units or a combination of three or more dwelling units and not more than five guest rooms or suites of rooms.

Apartment house. A building containing more than two (2) dwelling units.

Apartment house. Multiple family dwelling and group dwelling.

Apartment house. Any building, or portion thereof, which is designed, built, rented, leased, let, or hired out to be occupied, or which is occupied as the home or residence of three or more families living independently of each other and doing their own cooking in the said building, and shall include flats and apartments.

Area of dwelling unit. Sum of the gross floor areas above the basement level, including those rooms, and closets, having the minimum ceiling height, light, ventilation and other features as required by the Building Code of the City.

Attached dwelling. A dwelling which is joined to another dwelling at one or more sides by a party wall or walls.

Attached dwelling. A dwelling joined to two (2) other dwellings by party walls.

Attached dwelling. A single-family dwelling joined to another single-family dwelling by a wall or walls or roof.

Attached dwelling. A dwelling which is joined to another dwelling at one or more sides by a party wall or walls or roof.

Bachelor apartments. One (1) room and bath with or without cooking facilities, in a multiple dwelling.

Bachelor dwelling unit. Dwelling unit for 1 or 2 adults with or without 1 bedroom.

Bathroom. Enclosed space containing one or more bathtubs or showers, and which may also contain water closets, lavatories or fixtures serving similar purposes.

Bathroom. Enclosed space containing one or more bathtubs or showers, or both, and which may also contain water closets, lavatories, or fixtures serving similar purposes.

Bedroom. A room in a dwelling unit containing not less than seventy (70) square feet of floor area, except kitchens, bathrooms, dining, living and family rooms.

Bedroom. As used in the Code "bedroom" shall mean any room within a residential dwelling unit which is designed to be used for sleeping purposes and contains a closet of sufficient size to hold clothing. One living room with entry closet shall not be considered a "bedroom" in each residential dwelling unit other than a studio or efficiency apartment.

Bedroom. A room used or designed, intended or furnished to be used for sleeping.

Bungalow court. Multiple dwelling.

Bungalow court. Two or more buildings used or intended for use as dwellings occupying a single lot or parcel of land, any portion of which is used or intended for use in common by the inhabitants of such dwellings including such dwellings as are usually in opposing rows separated by a walk-way, drive or court.

Bungalow court. Group of three (3) or more one-story detached buildings used or intended for use as dwellings, grouped about a plot or parcel of land any portion of which is used or intended to be used in common by the inhabitants of such dwellings, each building thereof having its own separate toilet and bath.

Bungalow court or group houses. A group of 3 or more detached one-story, one-family or two-family dwellings located upon a single lot together with all yard and open spaces as required by the ordinance.

Bungalow courts. A group of three (3) or more detached, one-story, single-family dwellings, arranged with common utilities and accessories under a common ownership.

Combined rooms. Two or more adjacent habitable spaces which by their relationship, planning and openness permit their common use.

Condominium dwelling. A privately owned dwelling unit in a multiple dwelling structure. Consisting of several rooms to be used primarily by a single family.

Condominium dwelling. A multiple dwelling, whereby the fee title to each dwelling is held independently of the others.

Corridor. A path of egress connecting more than one room or occupied space on any one floor; a hallway.

Density. The number of dwelling units per unit of land measure.

Detached dwelling. A dwelling which is entirely surrounded by open space on the same lot.

Detached dwelling. A dwelling unit in a detached building.

Detached dwelling. Dwelling having no party wall in common with an adjacent building.

Detached dwelling. A dwelling which is completely surrounded by permanent open spaces.

Detached dwelling. A dwelling which is entirely surrounded by open space.

Detached single family dwelling. Single-family dwelling surrounded by yards or other open spaces.

Duplex. Building designed and/or used exclusively for residential purposes and containing 2 dwelling units separated by a common party wall or otherwise structurally attached.

Duplex dwelling. Single dwelling, providing housekeeping units for not more than two (2) families with no interconnection between the two (2) units, except that it may have a single entrance; all other exterior characteristics shall be that of a one (1) family dwelling. Two (2) single housekeeping units connected by a breezeway or corridor shall not be classified as a duplex.

Dwelling. Any building or portion thereof which is designed or used exclusively for residential purposes, but not including a tent, cabin, trailer, or mobile home.

Dwelling. Building designed or used exclusively as the living quarters for one (1) or more familes.

Dwelling. Any house or building or portion thereof which is occupied in whole or it part as the home, residence or sleeping place of one or more persons either permanently or transiently.

Dwelling. Any building or portion thereof, which is designed for or occupied exclusively for residential purposes for not more than four (4) families.

Dwelling. Building containing one (1) or more dwelling units.

Dwelling. Structure occupied exclusively for residential purpose by not more than two (2) families.

Dwelling. House, apartment, or building, including dormitories, fraternities and sororities, used primarily for human habitation. The word dwelling shall not include hotels, motels, tourist courts, or other buildings for transients.

Dwelling. Any building or portion thereof designed or used exclusively as the residence of one (1) or more persons, but not including a tent, or recreation vehicle.

Dwelling. Building occupied exclusively for residence purposes and having not more than two (2) apartments, or as a boarding or rooming house serving not more than five (5) persons with meals or sleeping accommodations or both.

Dwelling. Building or portion thereof used exclusively for residential purposes, including one family, two-family and multiple dwellings, but shall not include hotels, boarding and lodging houses.

Dwelling. Building occupied exclusively for residence purposes and having not more than two (2) dwelling units or as a boarding or rooming house serving not more than 15 persons with meals or sleeping accommodations or both.

Dwelling. Structure occupied exclusively for residential purpose by not more than two (2) families as outlined in the Ordinance.

Dwelling. Any building, or portion thereof, which is designed or used exclusively for occupancy by a family.

Dwelling. Any building or portion thereof designed or used exclusively as a residence or sleeping place of one or more persons, but not including a tent, cabin, trailer or trailer coach, boarding or rooming house, hotel or motel.

Dwelling. Building containing one (1) or more dwelling units, but in the case of a building having two (2) or more portions divided by one (1) or more party walls forming a complete separation, each such portion shall be considered to be a separate dwelling.

Dwelling. Any building or any portion hereof, which is not an "Apartment House" or a "Hotel" as defined in the Rules and Regulations, which contains one (1) or more "Apartments" or "Guest Rooms", used, intended, or designed to be built, used, rented, leased, let or hired out to be occupied, or which are occupied for living purposes.

Dwelling. Building designed or occupied exclusively for residental use and permitted accessory uses.

Dwelling. Any building, or portion therof, which is designed or used exclusively for occupancy by a family in accordance with the Code.

Dwelling. Any building which is designed for or used exclusively for residential purposes. For the purpose of the ordinance, such building shall have a minimum area of 400 square feet.

Dwelling. House or other building used primarily as an abode for one or two families except that the word "dwelling" shall not include boarding or rooming houses, tents, tourist camps, hotels, trailers, trailer camps, or other structures designed or used primarily for transient residents.

Dwelling. One-family or multi-family dwelling other than mobile homes, automobile trailers, hotels, motels, labor camps, camp cars, tents, railroad cars, and temporary structures.

Dwelling. Building used as the living quarters for one (1) or more families, but not including a building of mixed occupancy.

Dwelling. Building or portion thereof designed or used as the residence or sleeping place of one (1) or more persons, including one-family, two-family and multiple dwellings, apartment-hotels, boarding and lodging houses, but not including motels, tourist cabins, trailers or trailer courts.

Dwelling. Any building designed or used exclusively for residential purposes.

Dwelling. A dwelling is a building, or portion thereof, containing one (1) or more dwelling units.

Dwelling. Any building occupied in whole or in part as the temporary or permanent home or residence of one or more families.

Dwelling. Any building or any portion thereof, which is not an "Apartment House," "Lodging House" or a "Hotel" as defined in the Code, which contains one or two "Dwelling Units" or "Guest Rooms," used, intended, or designed to be built, used, rented, leased, let, or hired out to be occupied, or which are occupied for living purposes.

Dwelling. Building designed or used as the living quarters for one (1) or more families.

Dwelling. Every building or shelter used or intended for human habitation or periodic activity.

Dwelling. Building occupied exclusively for residence purposes and having (1) one dwelling unit, (2) two dwelling purposes and having (1) one dwelling unit, (2) two dwelling units, or (3) one or two dwelling units with a total of not more than fifteen (15) boarders or roomers in these units served with meals or sleeping accommodations, or both.

Dwelling. Building or structure, or part thereof, used or occupied for human habitation or intended to be so used, and includes any yard, garden, out-houses, and appurtenances belonging thereto or usually enjoyed therewith.

Dwelling. Building or portion thereof designed, constructed, or altered as a residence for not more than four (4) families living independently of each other, each unit of which is designed for cooking upon the premises.

Dwelling. Building or portion thereof, designed or used exclusively for residential occupancy, but not including hotels, motels or mobile homes.

Dwelling. One (1) family unit with or without accessory buildings.

Dwelling. Every building or shelter used or intended for human habitation or periodic activity.

Dwelling. Any building or any portion thereof, which is not an "Apartment House," "Lodging House" or a "Hotel" which contains one (1) or two (2) "Dwelling Units" or "Guest Rooms", used, intended, or designed to be built, used, rented, leased, let, or hired out to be occupied, or which is occupied for living purposes.

Dwelling. Building used as the living quarters for one or more families, but not including a building of mixed occupancy as defined in the Code.

Dwelling. Any building or any portion thereof which is not an "Apartment House" or a "Hotel" as defined in the Building Code, or which contains one or more "Apartments" or "Guest Rooms" used, intended or designed to be built, used, rented, leased, let or hired out to be occupied, or which are occupied, or which are occupied for living purposes.

Dwelling. Any residential building, other than an Apartment House, Hotel or Apartment Hotel.

Dwelling. A residential building designed and used for human habitation only; not including the use of building or premises for the conduct of any business, commercial, or manufacturing purposes, except as provided by the Ordinance.

Dwelling. Any building or portion thereof, but not a trailer, which is designed and used exclusively for residential purposes.

Dwelling. A residential building or portion thereof; but not including hotels, motels, boarding houses, rooming houses, dormitories, nursing homes, institutions, tourist homes or mobile homes.

Dwelling. A building, or portion thereof, designed or used as living quarters including single-family dwellings, two-family dwellings, and muiltple-family dwellings, but not including hotels or motels or rooming and boarding houses.

Dwelling. A building or portion thereof, designed or used exclusively for residential occupancy (except for "Home Occupations"), including one-family dwelling units, two-family dwelling units, and multiple-family dwelling units, but not including house trailers or mobile homes, hotels, motels, boarding or lodging houses.

Dwelling. A building occupied exclusively for residential purposes and serving not more than two (2) housekeeping units used for cooking, living, or sleeping purposes.

Dwelling. A building or portion thereof designed or used exclusively for residential occupancy including one-family, two-family and multiple-family dwellings, not including, however, any other building wherein human beings may be housed.

Dwelling. Any building which is wholly or partly used or intended to be used for living or sleeping by human occupants.

Dwelling. Any building or portion thereof, which is not an apartment house, lodging house, or hotel as defined in the code which contains one or 2 dwellings or guest rooms used, intended, or designed to be built, used, rented, leased, let, or hired-out to be occupied, or are occupied for living purposes.

Dwelling. Any house or building or portion thereof which is occupied in whole or in part as a home, residential or sleeping place of one or more human beings either permanently or transiently.

Dwelling. Building designed and used in whole or in part for human habitation, and the premises thereof, including dormitories, fraternity houses, and lodging houses.

Dwelling. Building or portion thereof, used exclusively for residential occupancy, including one (1) one-family, two (2) family and multiple dwellings, but not including hotels, motels, lodginghouses, boardinghouses or tourist homes.

Dwelling. A building, or portion thereof, but not a house trailer, designed or used exclusively for residential occupancy, including single family dwellings, two (2) family dwellings, and multiple family dwellings, but not including hotels, or lodging houses.

Dwelling. Any building, or portion thereof, which is designed or used exclusively for occupancy by a family as defined in the Code.

Dwelling. House or building or portion thereof which is occupied in whole or in part as the home, residence or sleeping place of one or more familes.

Dwelling. Structure which is wholly or partly used or intended to be used for living, sleeping, cooking and eating.

Dwelling bathroom. A bathroom shall have an area of not less than 48 square feet and be planned to accommodate three (3) fixtures which include, a bathtub or shower, lavatory or wash basin and a water closet.

Dwelling building or structure. A building or structure used or designed or intended to be used, all or in part, for residential purposes.

Dwelling group. Group of two (2) or more detached single dwellings located on a parcel of land in one (1) ownership having a yard or court in common.

Dwelling group. Single family dwelling and one (1) or more single family or duplex dwellings located on a lot.

Dwelling group. Group or row of dwellings each containing one (1) or more dwelling units and all occupying one (1) lot or site, as defined herein, and having a court in common; including a bungalow court or apartment court, but not including an automobile court or automobile camp.

Dwelling group. Group of two (2) or more dwellings located on the same lot and having any yard or open space in common.

Dwelling house. Building used exclusively for residential purposes which may be occupied by no more than 3 families.

Dwelling house. A detached house designed for and occupied exclusively as the residence of not more than two (2) families, each living as an independent housekeeping unit.

Dwelling house. Detached house designed for and occupied exclusively as the residence of not more than two (2) families each living as an independent housekeeping unit.

Dwelling living room. Room shall be designed for living use. When it is used only as a living room and space is provided elsewhere for cooking and eating, this room shall have not less than 150 square feet of floor area. Where this room provides the only eating space in the unit space it shall be increased by not less than 40 square feet of floor area. Where this room provides the only space for living, cooking and eating, it shall have a floor area of not less than 220 square feet.

Dwelling room requirements. Every dwelling structure shall have not less than 2 rooms and one bathroom.

Dwellings. Dwellings shall be taken to mean and include all buildings which are used or designed and intended to be used as the home or residence if not more than two (2) separate and distinct families.

Dwellings. Consists of one (1) or more rooms, including a bathroom and complete kitchen facilities, which are arranged, designed, or used as living quarters for one (1) family or household.

Dwellings. Private residences intended for permanent occupancy by one (1) or more families, but does not include apartment houses, lodging houses, hotels, or motels.

Dwelling unit. Consists of one or more rooms which are arranged, designed, or used as living quarters for one family only. Individual bathrooms and complete kitchen facilities, permanently installed.

Dwelling unit. One (1) or more rooms with principal kitchen facilities designed as a unit for occupancy by only one (1) family for cooking, living, and sleeping purposes.

Dwelling unit. Suite of rooms providing complete living facilities for One (1) family, including permanent provisions for living, dining, cooking, sleeping, and sanitation.

Dwelling unit. Room or group of rooms located within a dwelling and forming a single habitable unit with facilities which are used or intended to be used for living, sleeping, cooking and eating.

Dwelling unit. Room or suite of rooms designed or used for living purposes including a family unit or a room for sleeping accommodations, including hotel and dormitory rooms.

Dwelling unit. Any building with two (2) or more habitable rooms which are occupied by one (1) family with facilities for living, sleeping, cooking and dining.

Dwelling unit. Any building or portion thereof which is designed or used exclusively for residential purposes by one (1) family, as defined by the ordinance.

Dwelling unit. One (1) room or rooms connected together, consisting of a separate, independent housekeeping establishment for one (1) family occupancy, and physically separated from any other rooms or dwelling units which may be in the same structure, and containing independent cooking, bathroom and sleeping facilities.

Dwelling unit. Living Unit.

Dwelling unit. One (1) or more rooms arranged for the use of one (1) or more individuals living together as a single housekeeping unit with cooking, living, sanitary and sleeping facilities.

Dwelling unit. One (1) or more rooms with provision for living, sanitary, and sleeping facilities arranged for the use of one (1) family.

Dwelling unit. Room or group of rooms in any building or portion thereof, designed to provide living, sleeping and culinary accommodations for the exclusive use of one (1) family.

Dwelling unit. Dwelling or portion thereof providing complete living facilities for one (1) family and in which not more than two (2) persons are lodged for hire with such family at any one time.

Dwelling unit. One (1) or more rooms in a dwelling or multiple dwelling or apartment hotel used for occupancy by one (1) family (including necessary servants and employees of such family) for living or sleeping purposes, and having only one (1) kitchen.

Dwelling unit. A room or suite of two (2) or more rooms that is designed for, or is occupied by one (1) family doing its own cooking therein and having only one (1) kitchen.

Dwelling unit. Residential accommodation including complete kitchen facilities permanently installed which are arranged, designed, used or intended for use exclusively as living quarters for one family and not more than an aggregate of two roomers or boarders. Where a private garage is structurally attached, it shall be considered as part of the building in which the dwelling unit is located.

Dwelling unit. One or more rooms in a dwelling, hotel or apartment hotel designed for occupancy by one family for living purposes and having its own permanent installed cooking and sanitary facilities.

Dwelling unit. A dwelling which consists of one or more rooms which are arranged, designed, or used as living quarters for one family only. Individual bathrooms and complete kitchen facilities, permanently installed, shall be included for each "dwelling unit".

Dwelling unit. A room or group of contiguous rooms which include facilities which are used or intended to be used for living, sleeping, cooking and eating, and which are arranged, designed or intended for use exclusively as living quarters for one (1) family maintaining a single and separate housekeeping unit.

Dwelling unit. A single unit providing independent living facilities for one (1) or more persons including permanent provisions for living, sleeping, eating, cooking, and sanitation.

Dwelling unit. One or more rooms, including permanent bathroom and kitchen facilities, which are arranged, designed, or used as living quarters for one (1) family only.

Dwelling unit. One (1) or more rooms, designed for or used by one (1) family.

Dwelling unit. One or more rooms in a dwelling or building that are arranged, designed, used or intended for use by one or more families.

Dwelling unit. One (1) or more rooms providing complete living facilities for one (1) family, including equipment for cooking or provisions for the same, and including room or rooms for living, sleeping and eating.

Dwelling unit. Room or group of rooms forming a habitable unit for one (1) family with facilities used or intended to be used for living, sleeping, cooking and eating.

Dwelling unit. One (1) or more rooms designed and equipped for one (1) family, or persons living together as one (1) family, to occupy as a residence, but shall not include tourist homes or cabins, lodging houses, hotels, motels, or other similar places offering overnight accommodations for transients.

Dwelling unit. Living accommodations for a single family whether a single family residence, a residence in a multi-family residential building or a single-family residence, a residence in a multi-family residential building or a single-family living unit in a transient lodging facility.

Dwelling unit. Building or entirely self contained portion thereof, containing a single complete housekeeping facility not in common with any other dwelling unit except for vestibules, entrance halls, porches or hallways.

Dwelling unit. Any structure, or part thereof, designed to be occupied as living quarters as a single housekeeping unit.

Dwelling unit. One or more rooms arranged for the use of one (1) or more individuals living together as a single housekeeping unit, with related cooking, living, sanitary and sleeping facilities as defined in the Ordinance.

Dwelling unit. Room or group of habitable rooms occupied by one (1) family for more or less permanent occupancy.

Dwelling unit. Building or portion thereof arranged or designed for occupancy by not more than one (1) family for living purposes and having cooking facilities.

Dwelling unit. A room or group of rooms within a building containing cooking accommodations and occupied exclusively by one (1) family. An apartment shall be considered a dwelling unit.

Dwelling unit. One or more rooms and a single kitchen in a dwelling or apartment hotel, designed as a unit for occupancy by not more than one family for living or sleeping purposes, and not having more than one kitchen or set of fixed cooking facilities, whether or not designed for use of occupants such as janitors, caretakers, servants or guests.

Dwelling unit. Consists of one (1) or more rooms for living purposes together with separate cooking and sanitary facilities and is accessible from the outdoors either directly or through an entrance hall shared with other dwelling units and is used or intended to be used by one (1) or more persons living together and maintaining a common household.

Dwelling unit. One (1) or more rooms connected together, constituting a separate, independent housekeeping establishment for owner occupancy, or rental, or lease on a weekly or a monthly basis, or with reference to longer periods, and physically separated from any other rooms or dwelling units which may be in the same structure, and containing independent cooking and sleeping facilities.

Dwelling unit. Room, or group of rooms, occupied or intended to be occupied as a separate living quarters for a single family or other group of persons living together as a household or by a person living alone.

Dwelling unit. Room or a group of rooms including cooking accommodations, occupied by one (1) family, and in which not more than two (2) persons, other than members of the family, are lodged or boarded for compensation at any one time.

Dwelling unit. One or more rooms connected together, but structurally divided from all other rooms in the same structure and constituting a separate, independent housekeeping unit for permanent residential occupancy by humans, with facilities for such humans to sleep, cook and eat.

Dwelling unit. Room or group of rooms which are arranged, designed or used as living quarters for the occupancy of one family containing bathroom and/or kitchen facilites.

Dwelling unit. Any portion of a building used, intended or designed as a separate abode for a family.

Dwelling unit. Structure or portion thereof providing independent complete cooking, living, sleeping and toilet facilities for one family.

Dwelling unit. Building or portion thereof, complete and permanent living facilities for one (1) family.

Dwelling unit. Space, within a dwelling comprising living, dining, sleeping room or rooms, storage closets, as well as space and equipment for cooking, bathing and toilet facilities, all used by only one family.

Dwelling unit. One (1) or more rooms within a building arranged, designed or used for residential purposes for one (1) family and containing independent sanitary and cooking facilities. The presence of cooking facilities conclusively establishes the intent to use for residential purposes.

Dwelling unit. A group of two or more rooms, one of which is a kitchen, designed for occupancy by one family for living quarters.

Dwelling unit. A group of two or more rooms, one of which is a kitchen, designed for occupancy by one family for living and sleeping purposes.

Dwelling unit. A single unit providing complete, independent living facilities for one (1) or more persons including permanent provisions for living, sleeping, eating, cooking and sanitation.

Dwelling unit. Any room or group of rooms located within a building or structure and forming a single habitable unit with facilities which are used or intended to be used for living, sleeping, cooking, and eating.

Efficiency dwelling unit. A room located within an apartment house or apartment hotel used or intended to be used for residential purposes which combines a kitchen and living and sleeping quarters therein, and which complies with the requirements of the Code.

Efficiency dwelling unit. A room located within an apartment house or apartment hotel used or intended to be used for residential purposes which has a kitchen and living quarters, sleeping quarters combined therein.

Efficiency dwelling unit. A dwelling unit consisting of one (1) room, exclusive of bathroom, kitchen, hallway, closets or dining alcove directly off the principal room, provided such dining alcove does not exceed one hundred (100) square feet in area.

Efficiency unit. A dwelling unit consisting of one principal room exclusive of bathroom, kitchen, hallway, closets, or dining alcove directly off the principal room.

Efficiency unit. A dwelling unit consisting of one principal room together with bathroom, kitchen, hallway, closets and/or dining room alcove directly off the principal room, provided such dining alcove does not exceed one hundred twenty-five (125) square feet in area. An efficiency unit shall contain a minimum of three hundred (300) square feet.

Efficiency unit. A dwelling unit consisting of one principal room exclusive of bathroom, kitchen, hallway, closets, or dining alcove directly off the principal room, providing the dining alcove does not exceed one hundred twenty-five (125) square feet in area.

End-row dwelling. Semi-detached dwelling.

Family. An individual or two or more persons related by blood or marriage or a group of not more than five persons (excluding servants) who need not be related by blood or marriage living together in a dwelling unit.

Family. An individual living alone, or two (2) or more persons related by blood or marriage, or an unrelated group of not more than five (5) persons, other than domestic employees, living together as a single housekeeping unit in a dwelling unit.

Family. One (1) or more persons occupying the premises and living therein as a single and separate housekeeping unit. A group occupying a boarding house, fraternity, club, hotel or motel shall not be deemed to be a family as the term is used in defining a dwelling unit.

Four family dwelling. Building designed for or occupied exclusively by four (4) families.

Group dwelling. One or more buildings containing dwelling units arranged around two or more sides of a court.

Group dwelling. Two or more one-family, two-family or multiple dwellings, apartment houses or boarding or rooming houses, located on the same lot.

Group house. A group of attached single-family dwellings, separated from each other by vertical walls, but designed as a single structure not more than two (2) rooms deep.

Group houses. Group of detached or semi-detached dwellings facing upon a open place, street or courtyard.

Group houses. Group of dwellings not more than two (2) rooms deep facing, a court, street, or yard.

Group housing. Two (2) or more buildings for dwelling purposes, erected or placed on the the same lot.

Guest. In connection with multiple-family occupancies means a person hiring a room for living and/or sleeping purposes.

Guest. In connection with single-family and two-family occupancies means a person sharing single-family accommodations without profit on these accommodations.

Guest house. As a part of multiple-family occupancies means a detached single-family dwelling occupied or intended to be occupied for hire.

Guest house or cottage. A separate dwelling structure located on a lot with one (1) or more main dwelling structures and used for the housing of guests or servants of the occupant of the premises; and such building shall not have a kitchen and shall not be rented, leased or sold separately from the rental, lease or sale of the main dwelling.

Guest or servant's house. A detached accessory building located on the same zoning lot as the principal building and containing living quarters for temporary guests or servants.

Guest room. In connection with multiple-family occupancies means a room in a building, occupied or intended to be occupied for hire.

Guest room. Any habitable room except a kitchen, designed or used for occupancy by one or more persons and not in a dwelling unit.

Guest room. A room occupied by one (1) or more persons not members of the family in which no cooking facilities are provided. Where such room is detached from the main building, it shall be considered a dwelling unit.

Habitable room. A residential room or space, having an area exceeding 59 square feet in which the ordinary functions of domestic life are carried on, and which includes bedrooms, living rooms, studies, recreation rooms, kitchens, dining rooms, and other similar spaces, but does not include closets, halls, stairs, laundry rooms, or bathrooms.

Habitable room. Any room meeting the requirements of regulations for sleeping, living, cooking, or dining purposes, excluding such enclosed spaces as closets, pantries, bath or toilet rooms, service rooms, connecting corridors, laundries, foyers, storage spaces, utility rooms and similar spaces.

Habitable room. A room or enclosed floor space arranged for living, eating, or sleeping purposes, not including bath or toilet rooms, laundries, pantries, foyers, or communicating corridors.

Kitchen. Any room used or intended or designed to be used for cooking or the preparation of food, including any room having a sink and either a 3/4 inch gas opening or provision for an electric stove.

Kitchen. Any room used, or intended or designed to be used for cooking and preparing food, except a light housekeeping room.

Kitchen. Space having a floor area of sixty square feet or more and which is used for the cooking of food.

Kitchen. Any room or part of a room used for cooking or the preparation of food.

Let. To give another person the right to occupy any portion of a dwelling, family unit, or rooming unit. The act of "letting" shall be deemed to be a continuing act for so long as the persons given the right to occupy premises, continues to do so. A further "letting" by any occupant of a dwelling, family unit, or rooming unit is for the purpose of the code, also a "letting" by the owner or operator of the dwelling.

Living room. Principal room designed for general living purposes in a dwelling unit. Each dwelling unit shall have a living room.

Multi-family apartment house. Building containing more than 2 dwelling units.

Multi-family apartment house. A building or portion thereof containing more than two (2) dwelling units and not classified as a one or two-family dwelling.

Multi-family apartment house. Any building or portion thereof used as a multiple dwelling for the purpose of providing three (3) or more separate (dwelling) units with shared means of egrees and other essential facilities.

Multi-family dwelling. Building containing 3 or more dwelling units arranged either side by side or one above the other.

Multi-family house. A building or portion thereof containing three or more dwelling units, including tenement house, apartment house or flat.

Multi-family dwelling. A building or portion thereof designed for or used by three (3) or more families or house-keeping units.

Multi-family dwelling. Building or portion thereof designed for occupancy by three (3) or more families.

Multi-family dwelling. Building consisting of three (3) or more dwelling units with varying arrangements of entrances and party walls.

Multi-family dwelling. Building containing two (2) or more dwelling units, not a hotel, apartment hotel, or motel.

Multi-family dwelling. Dwelling or portion thereof occupied or intended or designed to be occupied by two (2) or more families.

Multi-family dwelling. Residential building designed for or occupied by three (3) or more families with the number of families not exceeding the number of dwelling units.

Multi-family dwelling. Building arranged, intended or designed to be occupied by three (3) or more families, or by three (3) or more individuals or groups of individuals, living in separate apartments, including apartment houses and apartment hotels.

Multifamily house. Building or portion thereof containing 3 or more dwelling units; including apartment house, flat, and tenement house.

Multi-family residence. Residence containing three (3) dwelling units.

Multiple dwelling. Building or portion thereof, used or designed as a residence for 3 or more families living independently of each other.

Multiple dwelling. A building designed for or occupied by more than one (1) family, each family living independently of the other in separate dwelling units.

Multiple dwelling. A building or portion thereof, containing three (3) or more dwelling units; but not including motel, hotel or boarding house, as defined in the Code.

Multiple dwelling. A dwelling containing two dwelling units and not more than five guest rooms.

Multiple dwelling. Building or portion thereof used or designed as a residence for three (3) or more families, but not including automobile courts or motels.

Multiple dwelling. Building having accommodations for and occupied by more than two (2) families.

Multiple dwelling. Building or portion thereof used or designed as a residence for three (3) or more families living independently of each other.

Multiple dwelling. Building designed and/or used exclusively for family residential purposes containing more than one (1) but not more than four (4) dwelling units.

Multiple dwelling. Multiple dwelling occupied, as a rule transiently, as the more or less temporary abode of individuals or families with or without meals, and in which, as a rule, the rooms are occupied singly and not as dwelling units.

Multiple dwelling. Building or portion thereof designed for or used exclusively as a residence for three (3) or more families living independently of one another.

Multiple dwelling. Building or portion thereof designed for, or occupied by more than four (4) families, living independently of each other.

Multiple dwelling. Building or portion thereof occupied by more than two (2) families or housekeeping units.

Multiple dwelling. The same meaning as APARTMENT HOUSE.

Multiple dwelling. A building containing three or more dwelling units. Multiple dwelling shall not be deemed to include a hotel, hospital, school, convent, monastery, asylum or other public institution. This definition shall include buildings that are part hotel and part multiple dwelling and buildings that would be hotels save for the size and accommodation requirements.

Multiple-family dwelling. A building containing three (3) or more dwelling units.

Multiple-family dwelling. A building or portion thereof, designed or altered for occupancy by three (3) or more families living in separate dwelling units..

Multiple-family dwelling. A building, or portion thereof, containing three (3) or more dwelling units.

Multiple-family dwelling. A detached residential building containing three (3) or more dwelling units, including what is commonly known as an apartment building.

Multiple family dwelling. Dwelling designed for, constructed or under construction or alteration for, or occupied by three (3) or more families.

Multiple family dwelling. Any building under a single roof, with or without fire walls or partitions, designed for occupancy by 3 or more families.

Multiple family dwelling. Building, or portion thereof, other than a commercial or residential hotel, which contains 3 or more dwelling units regularly used or available for permanent occupancy only, by households or families. A row of town houses is construed as a multiple family dwelling.

Multiple family dwelling. Any building under a single roof, with or without fire wall partitions, designed for occupancy by three (3) or more families.

Multiple family dwelling. Building or portions thereof, other than a commercial or residential hotel, which contains three (3) or more dwelling units regularly used or available for permanent occupancy only, by households or families. A row of town houses (row dwellings) is construed as a multiple family dwelling.

Multiple family dwelling. Residential building designed for or occupied by three (3) or more families, with the number of families permitted in residence not exceeding the number of dwelling units provided.

Multiple family dwelling. Residence designed for or occupied by more families, with separate housekeeping and cooking facilities for each family.

Multiple family dwelling. Building designed for or occupied exclusively by five (5) or more families.

Multiple unit dwelling. Building so constructed, altered or used as to provide accommodation for more than one (1) family to dwell in separately.

Multiple unit dwelling. Single detached structure structurally divided into two (2) or more separate, independent dwelling units, each dwelling unit having but one (1) kitchen and housing only one of the following groups of persons living together as a single non-profit housekeeping unit: (a) any number of persons immediately related by blood, marriage or adoption or (b) five or fewer persons not necessarily related by blood, marriage or adoption; provided, however, that each dwelling unit having an individual outside entrance at ground level shall contain at least six hundred square feet of gross floor area.

Nonhabitable space. Space used as kitchenettes, pantries, bath, toilet, laundry, rest, dressing, locker, storage, utility, heater, and boiler rooms, closets, and other spaces for service and maintenance of the building, and those spaces used for access and vertical travel between stories.

Occupant. Any person or persons, including guests, in actual physical possession or occupancy of a unit of dwelling space on a regular basis. For purposes of assigning specific duties or responsibilities, the term "occupant", unless the text indicates otherwise, shall mean the tenant, lessee, head of the family or household, or other adult person or emancipated minor assuming basic responsibility for the continued renting or occupancy of the dwelling space.

Occupiable room. A room or space, other than a habitable room designed for human occupancy or use, in which persons may remain for a period of time for rest, amusement, treatment, education, dining, shopping, or similar purposes or in which occupants are engaged at work.

One-family detached dwelling. Is a one-family dwelling having a side yard on each side thereof.

One family dwelling. Building consisting of but one (1) dwelling unit with not more than six (6) lodgers or boarders.

One family dwelling. Detached building arranged, intended or designed to be occupied, or which is occupied, by not more than one (1) family and which has not more than one (1) kitchen.

One family dwelling. Detached building of permanent character, placed in a permanent location, which is designed for occupancy, for residential purposes only, by one family exclusively, and which, in addition, is not occupied or used for any purpose other than as a residence of one family exclusively or as a residence of one (1) family and not more than three (3) guests of each family if the rental of rooms by such family to its guests is clearly secondary and incidental to the occupancy of said building by said one (1) family as its own residence.

One family dwelling. Detached building used exclusively for occupancy by one (1) (including necessary servants and employees of such family) and containing one (1) dwelling unit.

One family dwelling. Building designed for or occupied by one (1) family only.

One family dwelling. Building containing one (1) dwelling unit with not more than five (5) lodgers or boarders.

One family dwelling. Detached building containing only one dwelling unit.

One family dwelling. Building containing only 1 kitchen, designed for the use of or to house not more than 1 family, including all necessary employees of such family.

One family dwelling. A structure containing not more than one (1) dwelling unit.

One family dwelling. A detached building designed for or occupied exclusively by one family.

One-family dwelling. A building containing one (1) dwelling unit with not more than five (5) lodgers or boarders.

One-family dwelling. A building containing exclusively a single dwelling unit.

One-family dwelling. Any detached building containing only one (1) dwelling unit.

One-family dwelling. A detached dwelling containing only one dwelling unit.

One family dwelling. Building consisting of a single dwelling unit only, separated from other dwelling units by open spaces. It may be attached to another one-family dwelling by a common fireproof wall.

One-family dwelling. Detached building used for residential occupancy by one (1) family.

One-family dwelling. Detached building containing but one (1) kitchen, designed or used to house not more than one-family, including all necessary household employees of such family.

Private dwelling. A dwelling occupied by one (1) family and so designed and arranged as to provide cooking and kitchen accommodations for one (1) family only.

Private dwelling. Includes every building intended or designed for or used as the home or residence of not more than 2 separate and distinct families or households, and in which not more than 15 rooms shall be used for the accommodation of boarders, and no part of which structure is used as a store or for any business purpose.

Private dwelling. Structure occupied exclusively for residence purposes by not more than two (2) families.

Private residence. Separate dwelling or a separate apartment in a multiple dwelling which is occupied only by the members of a single family unit.

Residence. Detached part or whole of a building occupied exclusively as a family dwelling place and the usual accessory occupancies.

Residence. Single family dwelling or a dwelling unit in a multiple family dwelling, which contains sleeping, bathroom, food refrigeration, cooking and dining facilities.

Room. A room is a space or area bounded by any obstruction to egress which at any time enclose more than 80 percent of the perimeter of the space or area. Openings of less than 3 feet (91.44 cm) clear width and less than 6 feet 8 inches (203.2cm) high shall not be considered in computing the unobstructed perimeter.

Room capacity. For determining the exits required, the minimum number of persons or the occupant content of any floor area shall in no case be taken less than specified in the Code.

Row dwelling. Any one of three (3) or more attached dwellings in a continuous row.

Row dwelling. Dwelling, the walls on two sides of which are party or lot-line walls.

Row house. Attached house in a row or group, each house containing not more than two (2) dwelling units and each house separated from adjoining houses in the same row or group by fire walls or fire separations.

Row house. One of a group of row of not less than four (4) nor more than twelve (12) attached, single-family dwellings designed and built as a single structure facing upon a street or place and in which the individual row houses may or may not be owned separately. For the purpose of the side yard regulations, the structure containing the row or group of row houses is to be considered as one (1) building occupying a single lot.

Semi-detached dwelling. A dwelling joined to one (1) other dwelling by a party wall.

Semi-detached dwelling. Dwelling, one side wall of which is a party or lot-line wall.

Semi-detached dwelling. Either of two (2) dwellings each of which has a single party wall in common with the other.

Semi-detached dwelling. Dwelling, one side wall of which is a party or side lot-line wall.

Single-family attached dwelling. Building containing a single dwelling unit and having one or more exterior walls in common with, or contiguous to, another such dwelling. An attached single-family dwelling is capable of being located on a separate lot.

Single family detached dwelling. Building designed for and occupied exclusively as a residence of only one family and having no party wall in common with an adjacent building.

Single-family dwelling. A building designed for or used exclusively for residence purposes by one (1) family or housekeeping unit.

Single-family dwelling. A detached residence for or occupied by one family only.

Single-family dwelling. A detached building having accommodations for and normally occupied by one (1) family only, with one (1) cooking arrangement only.

Single-family dwelling. A detached building having accommodations for and occupied by one (1) family only, plus full time domestic employees, statutory foster children, guests who are both gratuitous and non-permanent guests and one (1) other person.

Single-family dwelling. A building containing one (1) dwelling unit only.

Single-family dwelling. A building designed for or occupied exclusively by one (1) family.

Single-family dwelling. A building containing one (1) dwelling unit designed or altered for occupancy by one (1) family.

Single-family dwelling. A residential building containing one dwelling unit including detached, semi-detached and attached dwellings.

Single-family dwelling. Building, on a lot, designed and occupied exclusively as a residence for one (1) family.

Single-family dwelling. Dwelling designed for, constructed or under construction or alteration for, or occupied by not more than one (1) family.

Single-family dwelling. Dwelling having one dwelling unit from ground to floor and having independent and outside access.

Single-family dwelling. Detached residence designed for or occupied by one (1) family only.

Single-family dwelling. Building containing one (1) dwelling unit and designed or used exclusively for occupancy by one (1) family.

Single-family dwelling. Building regularly used or available solely for permanent occupancy by one family or household.

Single-family dwelling. Detached building arranged, intended or designed to be occupied by not more than one-family.

Single-family dwelling. Single detached dwelling for not more than one (1) family or one (1) housekeeping unit; provided, however, that no dwelling shall be erected in a "Residence District" having less than two (2) bedrooms and area of less than nine hundred twenty square feet (920 sq. ft.), and provided, further, that a three (3) bedroom dwelling erected in a "Residence District" shall have an area of not less than one thousand three hundred square feet (1,300 sq. ft.).

Single-family dwelling. Detached residential building, other than a mobile home, designed for and occupied by one (1) family only.

Single-family dwelling. Building regularly used or available solely for permanent occupancy by one family or household, as defined in the Code.

Single-family dwelling. A building designed for occupancy by one (1) family.

Single-family dwelling. A detached dwelling for use by a single family.

Single-family dwelling. A detached building designed for or used exclusively by one family.

Single family semi-detached dwelling. Building designed for and occupied exclusively as a residence for only one (1) family and having a party wall in common with an adjacent one (1) family dwelling.

Single unit dwelling. Single detached structure having but one (1) dwelling unit with a single kitchen and housing any number of persons immediately related by blood, marriage or adoption, living together as a single non-profit housekeeping unit, plus domestic servants employed for services on the premises.

Split-level dwelling. A one or two family dwelling having part floors at staggered levels, so arranged to have a habitable room or rooms both above and below the first floor level.

Terrace family dwelling. Building containing three (3) or more dwelling units arranged side by side, separated from each other by a fire wall and having separate means of egress and ingress from the outside.

Terraces. A group of two (2) or more, but not exceeding five (5) single-family dwellings separated by walls without openings, not more than two (2) rooms deep.

Three family dwelling. Building in which three (3) living units exist.

Three or more bedroom unit. Is a dwelling unit wherein for each room in addition to the three (3) rooms permitted in a two (2) bedroom unit, there shall be provided an additional area of two hundred (200) square feet to the minimum floor area of seven hundred (700) square feet. For the purpose of computing density, said three (3) bedroom unit shall be considered a four (4) room unit and each increase in a bedroom over three (3) shall be an increase in the room count by one (1) over the four (4).

Town house. Type of single-family dwelling unit constructed in a series, including three (3) or more units with some common walls.

Town house. Single-family dwelling forming one of a group or series of three (3) or more single-family dwellings separated from one another by party walls without doors, windows, or other provisions for human passage or visibility through such walls from basement or cellar to roof, and having roofs which may extend from one of the dwelling units to another.

Townhouse. Multi-family building comprised of single dwelling units attached by common fireproof walls, each unit having at least two separate exterior entrances.

Townhouse dwelling. Single family dwelling forming one of a group of two (2) or more attached single family dwellings separated from one another by party walls without doors, windows or other provisions for human passage or visibility through such walls from basement or cellar.

Two bedroom unit. Dwelling unit containing a minimum floor area of at least seven hundred (700) square feet per unit, consisting of not more than three (3) rooms in addition to kitchen, dining, and necessary sanitary facilities, and for the purposes of computing density shall be considered as a three (3) room unit.

Two family detached dwelling. Building designed for and occupied exclusively as a residence for two (2) families with one (1) family living wholly or partly over the other and having no party wall in common with an adjacent building.

Two-family dwelling. Building designed for or occupied exclusively by two (2) families.

Two-family dwelling. Building designed for occupancy by two (2) families.

Two-family dwelling. Detached residential building containing two (2) dwelling units, designed for occupancy by not more than two (2) families.

Two-family dwelling. A building designed for or occupied exclusively by two families, as defined in the Code.

Two-family dwelling. A dwelling containing two dwelling units.

Two-family dwelling. Any building containing only two (2) dwelling units.

Two-family dwelling. A residential building containing two dwelling units including detached, semi-detached and attached dwellings.

Two-family dwelling. A building containing exclusively two (2) dwelling units.

Two-family dwelling. A residential building containing not more than two (2) dwelling units entirely surrounded by open space on the same lot.

Two-family dwelling. A building containing two (2) dwelling units only.

Two-family dwelling. Building containing two separate dwelling units, and/or used exclusively by not more than two (2) families.

Two-family dwelling. Building containing not more than two (2) kitchens designed or used to house two (2) families living independently of each other, including all necessary employees of such families.

Two-family dwelling. A building designed for or used exclusively by two (2) families or housekeeping units.

Two-family dwelling. Dwelling occupied by two (2) families living independently of each other.

Two-family dwelling. Dwelling designed for, constructed or under construction or alteration for, or occupied by two (2) families.

Two-family dwelling. Building regularly used or available solely for permanent occupancy by two (2) families or households.

Two-family dwelling. Building occupied or constructed to be occupied exclusively by not more than two (2) families or housekeeping units.

Two-family dwelling. Building consisting of two (2) dwelling units which may be either attached side by side or one above the other, and each unit having either a separate or combined entrance or entrances.

Two-family dwelling. Residence designed for or occupied by two (2) families only, with separate housekeeping and cooking facilities for each.

Two-family dwelling. Dwelling having two (2) dwelling units and also known as a duplex.

Two-family dwelling. Building containing two (2) dwelling units and designed or used exclusively for occupancy by two (2) families living independent of each other; or two (2) one (1) family dwelling having a party wall in common.

Two-family dwelling. Building designed for or occupied exclusively by two (2) families, with separate housekeeping or cooking facilities for each family.

Two-family dwelling. Building, on a lot, designed and occupied exclusively as a residence for two (2) families with one (1) family living wholly or partly over the other.

Two-family dwelling. Building used exclusively for occupancy by two (2) families including necessary servants and employees of each such families living independently of each other and containing two (2) dwelling units.

Two-family dwelling. Building containing two (2) dwelling units with not more than five (5) lodgers or boarders per family, but not more than twenty (20) individuals.

Two-family dwelling. Building containing two (2) dwelling units with not more than four (4) lodgers or boarders per dwelling.

Two-family dwelling. Building designed for or occupied exclusively by two (2) families living independently of each other with separate entrance, and kitchens and baths; constitutes two (2) dwelling units; includes duplex.

Two-family dwelling. Detached or semi-detached building used for residential occupancy by two (2) families living independently of each other.

Two-family dwelling. Building containing not more than two (2) kitchens, designed or used to house not more than two (2) families, living independently of each other including all necessary household employees of each such family.

Two-family dwelling. Separate, detached building designed for or occupied exclusively as a residence by two (2) families.

Two-family dwelling. Dwelling occupied by two (2) families living independently of each other.

Two-family dwelling. Or duplex dwelling on any one (1) lot or two (2) single family dwellings on one (1) lot; provided, that such dwelling or dwellings conform in all respects to the setback, side line, area and coverage requirements as provided in the Ordinance.

Two-family dwelling. Building of a permanent character, placed in a permanent location, which is occupied or designed, intended or arranged for occupancy, for residential purposes, by two (2) families living independently of each other, and which, in addition, is not occupied or used for any purpose other than as a residence for not more than two (2) families living therein independently of each other or as a residence for said two (2) families, living therein as aforesaid, and not more than two (2) guests of each such family if the rental of rooms by each such family to its guests is clearly secondary and incidental to the occupancy of said building by such family as its residence.

Two-family dwelling. Detached building arranged, intended or designed to be occupied, or which is occupied, by two (2) families living independently of each other and which has not more than two (2) kitchens.

Two-family dwelling. Building having accommodations for and occupied exclusively by two (2) families.

Two family dwelling duplex. Detached building designed for or occupied exclusively by two (2) families independently of each other.

Two-family dwelling or duplex. Detached building under one (1) roof, arranged, intended or designed to be occupied by not more than two (2) families, each apartment extending from basement to roof of the structure with a separate entrance to each dwelling though there may be but one (1) front door and vestibule or hall.

C-10 FOOD SERVICES, DINING AND RESTAURANTS

Banquet hall. Any building or portion of building, which is used as a place wherein the public may assemble for the purpose of attending a banquet.

Cabaret. Any room, place or space in a building in which musical entertainment, singing, dancing or other similar amusement is permitted in connection with the restaurant business or the business of directly or indirectly selling food or drink.

Carhop service. Delivery service between the restaurant or other place where cooked food or beverages, or both, are served and delivered by the owner or operator, its agents or employees, to the occupants of automobiles parked on, in or about the restaurant and the like, premises, or adjacent premises or thoroughfare, for consumption there.

Carry-out restaurant. A restaurant where the majority of the food prepared and served is "carried out" for consumption off the premises.

City club. Membership organization and facility located in an urban center that frequently contains guest rooms, restaurant, function rooms and bar, as well as recreation facilities.

Dining room. Any building or part thereof, or any room or part thereof, in which food is dispensed or served.

Drive-in. Establishment designed or operated to serve a patron while seated in an automobile parked in an off-street parking space.

Drive-in. Establishment selling food or beverages to customers, some or all of whom customarily consume their purchases outdoors in or near their cars.

Drive-in. Business establishment so developed that its retail or service character is dependent on providing a driveway approach or parking spaces for motor vehicles as to serve patrons while in the motor vehicle rather than within a building or structure.

Drive-in establishment. A place of business operated for the sale, dispensing, or service of food, beverages, other commodities or service, or entertainment, designed and equipped to allow its patrons to be served or accommodated while remaining in their automobiles. It may also allow patrons to serve themselves and consume food, refreshments, or beverages in automobiles on the premises or elsewhere on the premises, but outside any completely enclosed structure. A place of business shall be considered a drive-in if, in addition to the consumption of food, refreshments, or beverages in automobiles or elsewhere on the premises outside any completely enclosed structure, it allows the consumption of such products within a completely enclosed structure.

Drive-in or carry-out restaurant. A business establishment where food or beverages prepared for immediate consumption are sold to the consumer for consumption away from the premises or where such food or beverages may be consumed in automobiles parked in the parking area or at a stand-up counter on the premises.

Drive-in restaurant. Any business where food (including frozen desserts) and beverages are sold, where an automobile parking area is provided, and where such food or beverages are consumed in automobiles parked in the parking area.

Drive-in restaurant. A restaurant where food (including frozen desserts) may be ordered from a motor vehicle, or where specific motor vehicle parking area is provided on the premises for the consumption of food.

Drive-in restaurant. Any establishment where food, frozen dessert or beverage is sold to the consumer and where motor vehicle parking space is provided and where such food, frozen dessert or beverage is intended to be consumed in the motor vehicle parked upon the premises or anywhere on the premises outside of the building.

Drive-in restaurant. Any establishment where food or beverages are dispensed and where such food or beverages are consumed on the premises as well as within a building.

Drive-in restaurant. Any building, structure and/or premises where food and/or beverages are sold to customers for consumption.

Drive-in restaurant or refreshment stand. Any place or premises used for sale, dispensing or serving of food, refreshments or beverages in automobiles, including those establishments where customers may serve themselves and may eat or drink the food, refreshments, or beverages on the premises.

Drive-in restaurants. Any establishment where food or beverages are dispensed and where such food or beverages are consumed on the premises but not within a building.

Eating and drinking establishment. Restaurant, food or drink dispensing stand, coffee shop, cafeteria, short order cafe, luncheonette, tavern, sandwich stand, soda fountain, inplant feeding establishment, hospital, institution, school lunchroom, private and semi-private club, camp, camp kitchen, church kitchen, caterer or any other place in which food or drink is prepared or stored for public consumption, distribution or sale and shall include vehicles used in connection therewith and all places where water for drinking or culinary purposes is available for public use.

Eating place. Retail establishment primarily engaged in the sale of prepared food and drinks for consumption on the premises.

Fast food restaurant. A restaurant operation which prepares food for a limited advertised menu, which can be consumed by its customers inside an enclosed place or outdoors in areas on the premises, using dishes and utensils which are disposable.

Food. All edible substances used in the preparation or serving of meals or lunches.

Food cover. Unit of food service provided to a customer. This term is not synonymous with "meal" because a food cover may comprise only a cup of coffee or a bowl of soup.

Food service establishment. Any fixed or mobile restaurant; coffee shop; cafeteria; short-order cafe; luncheonette; grille; tearoom; soda fountain; sandwich shop; hotel kitchen; smorgasbord; tavern; bar; cocktail lounge; night club; roadside stand; industrial feeding establishment; public or non-profit; hospital kitchen; nursing home kitchen; private, public or non-profit organization or institution routinely serving the public; catering kitchen; commissary or similar place in which the food or drink is prepared for sale or for service on the premises or elsewhere; and any other eating and drinking establishment where food is served or provided for the public with or without charge.

Food service operation. Two or more food serving areas supplied primarily by a common kitchen all of which are located in one building or structure and conducted under one operation shall be considered as a single food service operation. This definition shall not apply to operations serving only ice cream, frozen dessert novelties, milk served in the original container or through vending machines, soft drinks, nuts, candy, and similar confections.

Industrial catering truck. An industrial catering truck is a motor vehicle used for the purpose of dispensing and selling liquids from sanitary dispensers and/or ready-to-eat food and beverages which have been prepared and sealed, or packaged on premises having a valid health permit authorizing the preparation of food, other than the vehicle from which said drink, food, or beverages are sold.

Itinerant eating and drinking establishment. Any pushcart, wagon, truck or other food and drink vending vehicle from which food or drink is prepared, dispensed or sold.

Luncheon club. Membership organization whose primary purpose is to provide facilities for private dining during the luncheon period.

Lunch wagon. Any prefabricated structure brought in complete form to, or assembled on the site designed to be used for the purposes of a restaurant, whether standing on its own wheels or on a fixed foundation, whether or not connected with sewer or water mains.

Major kitchen. Kitchen with an area of 500 square feet, or more, in any building except a single-family or a two-family dwelling.

Mobile food service operation. One which may be moved without significant alteration of the structure or equipment after the structure and equipment has been moved from one location to another.

Mobile restaurant. A restaurant facility operating from a movable vehicle, trailer or boat which periodically or continuously changes location.

Mobile restaurant. One operating from a movable vehicle, trailer or boat which periodically or continuously changes location and wherein meals or lunches are prepared or served or sold to transients or the general public, excepting those vehicles used in delivery of pre-ordered meals or lunches prepared in a licensed restaurant. The term "mobile restaurant" does not include a common carrier regulated by the state or federal government.

Night club. Place of assembly used for the consumption of food and alcoholic beverages, and which normally affords entertainment through permitting dancing or through organized entertainment other than music.

Off-own premises consumption. For consumption on the premises of others than that of the caterer.

Off-own premises food caterer. One who provides prepared bulk food, including beverages, if any, for off-own premises consumption, with or without other service incidental thereto.

Restaurant. Includes any building, room or place wherein meals or lunches are prepared or served or sold to transients or the general public, and all places used in connection therewith. "Meals or lunches" shall not include soft drinks, ice cream, milk, milk drinks, ices and confections. The term "restaurant" does not apply to churches, religious, fraternal, youth or patriotic organizations, service clubs or civic organizations which occasionally prepare or serve or sell meals or lunches to transients or the general public, nor shall it include any private individual selling foods from a movable or temporary stand at public farm sales.

Restaurant. Building in which food is prepared and sold for consumption within the building, as opposed to a drive-in restaurant establishment where food may be taken outside of the building for consumption either on or off the premises.

Restaurant. Establishment other than a boarding house where meals which are prepared therein may be secured by the public.

Restaurant. Any restaurant except a drive-in restaurant or a take-out restaurant as defined in the Ordinance, coffee shop, cafeteria, short-order cafe, luncheonette, tavern, sandwich stand, drugstore and soda fountain serving food and all other eating or drinking establishments.

Restaurant. Includes any building, room, space or place wherein meals are prepared or served or sold to transients or the general public, and all accessory places used in connection therewith.

Restaurant. Any place in which food or nonalcoholic beverages are sold to the public for consumption on the premises, including, but not by way of limitation, coffee houses, lunchrooms and lunch wagons.

Restaurant. A business establishment where food or drink is prepared, served and consumed within a building. Serving or consuming food or drink outside of the building on the premises is prohibited.

Restaurant. A public eating place where all food is essentially served and consumed at a table or counter within a completely enclosed building.

Restaurant. Any building or part thereof, other than a boarding house, where food ready for consumption is sold at retail to the public.

Restaurant. An establishment where food is prepared, served and consumed.

Restaurant. Every building or part thereof and all outbuildings used in connection therewith, or any place or location, kept, used, maintained as advertised as, or held out to the public to be a place where meals, lunches or sandwiches are prepared and/or served, either gratutiously or for pay.

Restaurant operator or licensee. The person or firm legally responsible for the operation of the restaurant.

Room (dining). A room in which food or beverage is served for immediate consumption.

Specialty restaurant. Food service operation that has a theme that is carried out in the menu, service and decor. The menu is also frequently limited.

Standard restaurant. Any establishment where food, frozen desserts or beverages are available in a ready to consume state and where customers are normally provided with an individual menu are served their food or beverages from a kitchen fully equipped for the preparation of complete meals by a restaurant employee at the same table or counter at which said items are consumed and typically the food, frozen dessert, or beverage are served with reusable dishes and utensils. The fact that a "standard restaurant" shall incidently prepare food for off premise consumption in disposable containers shall not prevent it from being classified as a "standard restaurant."

Storage. After washing and cleaning of equipment and utensils, all food contact surfaces shall be handled and properly stored to protect these surfaces from contamination.

Table-service restaurant. Food service operation in which patrons are served only at tables.

Take-out restaurant. Any establishment which has as its principal business the preparation of food, frozen dessert, or beverage for sale to be consumed away from the premises of the establishment. This does not apply to restaurants which occasionally sell such items for consumption away from the premises and does not apply to drugstores or grocery stores.

Take-out service. A food establishment which serves or dispenses food through a window or door directly to the public for consumption on or off the premises is offering a take-out service; provided, any food establishment which serves or dispenses food inside its premises, but for consumption of such food off the premises only is offering a take-out service.

Temporary food service establishment. Any food service establishment which operates at a fixed location for a temporary period of time not to exceed two weeks in connection with a fair, carnival, circus, public exhibition, promotional activity, or similar transitory gathering.

Temporary restaurant. One operating for special occasions such as a fair, carnival, circus, public exhibition or similar gathering.

Temporary restaurant. A food facility operating for special occasions such as a fair, carnival, circus, public exhibition, or similar public gathering.

Vending machine. Any self-service device which, upon insertion of coin or token, or by other means, dispenses unit servings of food or beverage, either in bulk or in package, without the necessity of replenishing the device between each operation.

Washing aids. Includes but is not limited to brushes, dish mops, dish cloths, metal scrappers, and other hand aids used in pot and dishwashing. All washing aids shall be kept in clean and sanitary condition.

C-11 HOSPITALS, HOMES, HEALTH CARE AND INSTITUTIONS

Administrator of a home. Person who is responsible for the supervision of the home on a day-to-day basis. The same person may be both the operator and the Administrator.

Alarm annunciator. Remote annunciator, storage battery powered, provided to operate outside of the generating room in a location readily observed by operating personnel at a regular work station. Where a regular work station may be unattended periodically, an appropriately labeled derangement signal is located at the telephone switchboard. Annunciator indicates alarm conditions of the emergency or auxiliary power source.

Alcoholic nursing home. Nursing home for persons suffering from acute or chronic alcoholism.

Alternate power source. One or more generator sets intended to provide power during the interruption of the normal electrical service or the public utility electrical service intended to provide power during interruption of service normally provided by the generating facilities on the premises.

Ambulatory health care centers. A building or part thereof used to provide services or treatment to four (4) or more patients at the same time and meeting the requirements of the Code.

Ambulatory person. Person able to walk and physically able to leave the premises without assistance.

Anesthesia storage room. Room or place where explosive or flammable gases or liquids or powders which are used as anesthesia agents are stored or kept. The term shall include the room or place where machines used in the administering of these agents are stored or kept.

Anesthesiologist. Means a physician who is certified by the American Board of Anesthesiology or who has training and experience in the field of anesthesiology, substantially equivalent to that required for such certification.

Anesthetizing location. Areas in hospitals in which flammable anesthetics are or may be administered to patients. Such locations will include operating rooms, delivery rooms and anesthesia rooms, and will also include any corridors, utility rooms or other areas which are or may be used for administering flammable anesthetics to patients. Recovery rooms are not classed as anesthetizing locations unless used for administering flammable anesthetics.

Anesthetizing location. Any area in which any combustible anesthetic agent is administered in the course of examination or treatment, and shall include operating rooms, obstetrical rooms, anesthesia storage rooms, corridors, utility rooms and other areas if used for induction of anesthesia with combustible anesthetic agents. In an anesthetizing location the hazardous area is to be considered to extend five (5) feet above the floor.

Anesthetizing location receptacle. Receptacle designed for the Standard Use of Inhalation Anesthetics.

Anesthetizing room. Room where potentially explosive anesthetizing agents are used.

Asylum. Any building which, or a portion of which, is used for the lodging or care of physically or mentally disabled persons.

Attendant. Responsible person who is trained to provide personal care and services to patients and residents of homes.

Automatic. Self-acting, operating by its own mechanism when actuated by some impersonal influence as for example, a change in current strength, pressure, temperature, or mechanical configuration.

Autopsy table. Table or unit used for postmortem medical examinations.

Auxiliary power plants. Used for lighting or power purposes, may be located in boiler or heating equipment rooms if the fuel to be used is natural gas or diesel fuel. If the fuel to be used is gasoline, such equipment shall be located on the exterior of the building and shall be housed by at least eight (8) inch masonry walls. The use of liquified petroleum gas for fuel is prohibited.

Bedfast patient. One who requires complete bedside care and is helpless.

Boarding home for the aged. Any building, however named, which is operated for the express or implied purpose of providing service or domiciliary care for three or more elderly people who are not ill or in need of nursing care, and in which there is no agreement that such service shall include personal care or special attention.

Care home. Includes rest and nursing homes, convalescent homes, and boarding homes for the aged established to render domiciliary care for chronic or convalescent patients, but, excludes child care homes or facilities predominently for the care of mentally retarded, mentally disturbed, epileptic, alcoholic, and drug addicted patients.

Charitable institution. A nonprofit institution devoted to the housing, training, or care of children, or of aged, indigent, handicapped, or underpriviledged persons, but not including lodging houses, or dormitories providing temporary quarters for transient persons, organizations devoted to collecting or salvaging new or used materials, or organizations devoted principally to distributing food, clothing, or supplies on a charitable basis.

Child. Means a young person in the age range of 2 through 12 years.

Child. Child, adopted child, stepchild, foster child or grandchild, of whatever age.

Child boarding home. Building or portion thereof appropriated to the care and lodging, apart from their parents, on a 24 hour a day basis, of four (4) or more children sixteen (16) years of age or less.

Child care center. Any establishment which provides shelter, care, activity and supervision (with or without academic instruction) for five (5) or more children (between birth and six (6) years of age) between the hours of 7 A.M., and 7 P.M.

Child care center. Child care center means a place, home, building, or location where care is provided for four (4) or more children under the age of seventeen (17) years, not related to the operator, for compensation. Such term specifically includes nursery schools, kindergartens, or any other facility caring for children during either the day or night; but shall not include such facilities operated by the State, or any agency or unit of the State government, or governmental unit, county or federal.

Child care centers. Enterprise involving the care of three (3) or more children at one and the same time, either by day or night, which children are not foster children or related by blood or marriage to the operator.

Child care home. Private, public or semi-public facility licensed for the housing and care of children.

Child care institution. A child care facility where more than eight (8) children are received and maintained for the purpose of providing them with care or training or both. The term "child care institution" includes residential schools, primarily serving ambulatory handicapped children, and those operating a full calendar year, but does not include other such institutions listed in the Ordinance.

Child day care home. A family home which receives not more than eight (8) children, under nine (9) years of age, for care during the day. The maximum of eight (8) children includes the family's natural or adopted children under the age of eighteen (18) years and those children who are in the home under full-time care, as defined in the State Child Care Act.

Child night time home. A family home which receives no more than eight (8) children under nine (9) years of age, including the family's own children under eighteen (18) years of age, and those children who are in the home under full-time care, as described in the State Child Care Act for night time care.

Children's home. Any child-caring facility providing 24 hours a day care to eleven (11) or more children under twenty-one (21) years of age.

Clinic. A building or portion thereof designed for, constructed or under construction or alteration for, or used by two (2) or more physicians, surgeons, dentists, psychiatrists, physiotherapists, or practitioners in related specialties, or a combination of persons in these professions, but not including lodging of patients overnight, as specified in the Code.

Clinic. Clinic means any place, building or agency which provides diagnosis, treatment or advice to persons not residing in the facility.

Clinic. Building used as a medical institution for the diagnosis and treatment of patients or injured persons and not having overnight sleeping accommodations.

Clinic. Place for the examination and treatment of persons for medical, x-ray, psychiatric, and dental care by properly licensed persons, such as, but not limited to, physicians, dentists and psychiatrists.

Clinic. Establishment where patients, who are not lodged overnight, are admitted for examination and treatment by one person or group of persons practicing any form of healing or health building services to individuals, whether such persons be medical doctors, chiropractors, osteopaths, chiropodists, naturopaths, optometrists, dentists or any such profession, the practice of which is lawful in the State and in the City.

Clinic. Building used by a group of doctors for the medical examination or treatment of persons on an outpatient, or nonboarding basis only.

Clinic. Establishment used by physicians, surgeons, dentists, physiotherapists, psychiatrists, or practitioners, in related specialities, where patients who are not lodged overnight, are admitted for examination and treatment.

Clinic. Building or portion thereof designed for construction or under construction or alteration for, or used by two (2) or more physicians, surgeons, dentists, psychiatrists, physiotherapists, or practicioners in related specialties, or a combination of persons in these professions, but not including lodging of patients overnight, as specified in the Code.

Clinic. Institution in which a group of physicians and surgeons, with their allied professional assistants, offer therapeutic, diagnostic or preventive medical or dental care to ambulatory patients.

Clinic. Building designed for, constructed or under construction or alteration for, or used by one or more physicians, surgeons, dentists, psychiatrists, physiotherapists, or practitioners in related specialties, or a combination or persons in these professions.

Clinic. Any structure or premise used as an establishment for medical or surgical examination and/or treatment of persons classified as outpatients.

Clinic. A medical or dental office building, excepting however, that there shall be no facilities for overnight housing or care of, or boarding of, patients.

Clinic. A place for the medical or similar examination and treatment of persons as out patients.

Clinic. A place used for the care, diagnosis and treatment of sick, ailing, infirm and injured persons and those in need of surgical or medical attention but who are not provided with board and room or kept overnight on the premises.

Clinic: medical or dental. An organization of physicians or dentists licensed in the State, who have their offices in a common building. A clinic shall not include in-patient care.

Clinic, medical or dental. A building or portion thereof, the principal use of which is for offices of two (2) or more physicians or dentists, or both, for the examination and treatment of persons on an outpatient basis.

Communicable disease. Any contagious or infectious disease, and includes smallpox, chickenpox, diphtheria, scarlet fever, typhoid fever, measles, German measles, glanders, cholera, erysipelas, tuberculosis, mumps, anthrax, bubonic plague, rabies, poliomyelitis and cerebrospinal meningitis, and any other disease declared by the regulations to be a communicable disease.

Continuous power supply. Electrical system, independent of the alternate source, which supplies power with minimum interruption (twelve cycles or less) to one or more isolated power centers.

Convalescent home. Building used for the accommodation and care of persons recuperating from illness.

Convalescent home. Same as a nursing home.

Convalescent home. Building and premise in which two or more sick, injured or infirm persons are housed or intended to be housed for compensation.

Convalescent home. Home in which nursing, dieting or other personal services are furnished to convalescents, invalids and aged persons, but in which are kept no persons suffering from a mental sickness, disease, disorder or ailment, or from a contagious disease, and in which is performed no surgery, maternity or other primary treatment such as is customarily provided in sanitariums or hospitals, or in which no persons are kept or served who normally would be admitted to a mental hospital.

Convalescent home. Privately operated establishment devoted to the care, for pay, of three (3) or more persons during the recuperative stage of an illness, or in which persons may be cared for during an illness.

Convalescent home. Home for persons recovering from illnesses and after they have passed the stage where medical care and hospitalization are required.

Convalescent home. A convalescent home, a nursing home, or a rest home is a home for the aged, recuperating, chronically ill, or incurable persons, in which two (2) or more persons not of the immediate family are received, kept, or provided with food and shelter or care for compensation, but not including hospitals, clinics, or similar institutions devoted primarily to the diagnosis and treatment of disease or injury.

Convalescent home. A building where regular nursing care is provided for more than one (1) person not a member of the family which resides on the premises.

Convalescent home or nursing home. Shall mean any place or institution which makes provisions for bed care, or for chronic or convalescent care for one or more persons exclusive of relatives, who by reason of illness or physical infirmity are unable to properly care for themselves. Alcoholics, drug addicts, persons with mental diseases and persons with communicable diseases, including contagious tuberculosis, shall not be admitted or cared for in these homes licensed under the State as a convalescent and nursing home.

Convalescent homes. Includes homes for the aged, nursing homes, and such other institutions designed to take care of the aged or persons unable to care for themselves without supervision or assistance.

Critical branch. System of feeders and branch circuits connected to the alternate power source by one or more transfer switches and supplying energy to task illumination and selected receptacles serving areas and functions related to patient care.

Day care centers. Any licensed child care facility receiving more than eight (8) children for daytime care during all or part of a day. The term "day care centers" includes facilities commonly called "child care centers," "day nurseries," "nursery schools," "private kindergartens," "play groups," and centers and workshops for mentally or physically handicapped with or without stated educational purposes.

Day care home. A dwelling unit which receives not more than five (5) client children for care during the day and licensed as such by the State.

Day care homes. Family homes licensed to receive not more than eight (8) children for care during the day. The maximum of eight (8) children includes the family's natural or adopted children under age eighteen (18) and those children who are in the home under full time care.

Day treatment center. Day treatment center means any institution, place, building or agency which maintains and operates organized treatment services, activities and supervision under medical direction, for persons with mental illness or behavioral or emotional disorders for part of any 24-hour period but not including overnight accommodations.

Dental or medical clinic. Building in which a group of physicians, dentists or physicians and dentists and allied professional assistants are associated for the purpose of carrying on their profession. The clinic may include a dental or medical laboratory. It shall not include in-patient care or operating rooms for major surgery.

Dietitian. A person who is eligible for membership in the American Dietetic Association or who has training and experience substantially equivalent to that required for membership in said association.

Dispensary. Building, or part thereof, which is used primarily for medical care of sick or injured persons, incidental to an industrial, business, educational, athletic, or similar activity to which such building, or part thereof, is an accessory, and in which sleeping accommodations, if any, are intended for periods of time during regular working hours for the person affected.

Dormitory. As applied to a space in a home, means a room equipped with more than 2 beds.

Electrically susceptible patient. Patient being treated with an exposed electric conductor, such as a probe, catheter or electrode connected to the heart. These patients are likely to be concentrated in special treatment areas.

Emergency room. Any area in the hospital set up for the reception and treatment of persons in need of emergency medical care.

Emergency system. System of feeders and branch circuits meeting the requirements of NFPA or local Codes, connected to the alternate power source by transfer switch and supplying energy to an extremely limited number of prescribed functions vital to the protection of life and patient safety, with automatic restoration of electrical power within ten seconds after power interruption. Only circuits serving those areas or functions to prevent power interruption. Only circuits serving those areas or functions specifically required by Code or NFPA.

Equipment system. System of feeders and branch circuits arranged for delayed, automatic or manual connection to the alternate power source and which serves primarily three-phase power equipment.

Essential electrical systems for hospitals. Systems comprised of alternate sources of power, transfer switches, overcurrent protection devices, distribution cabinets, feeders, branch circuits, motor controls, and all connected electrical equipment, designed to assure designated areas and functions of a hospital continuity of electrical service during disruption of normal power sources, and also designed to minimize the interruptive effects of disruption within the internal wiring system.

Establishment for handicapped persons. Establishment for handicapped persons means any institution, place, building or agency which maintains and operates services for handicapped persons.

Facility for the mentally retarded. Facility for the mentally retarded means any institution, place, building or agency which maintains and operates facilities and care, treatment or schooling for mentally retarded persons.

Foster child care. Care and education of not more than five children unrelated to the residents by blood or adoption.

Foster family care. Care and education of not more than four children unrelated to the residents by blood or adoption.

Foster home. Shall be a home maintained by an individual or individuals having care or control of one (1), but not more than six (6) minor children, other than those related by blood, marriage or adoption, or those who are legal wards of such individuals, which is licensed by the Department of Economic Security.

Foster home. Home where three or more foster children reside and payment is received therefor.

Fracture room. Room where surgery or manipulation is performed for the setting of bones.

Full-time public health services. Public Health Services provided by medical officers of health. Public health nurses or public health inspectors who are employed full-time by the Department, a municipality or the Board of Health of a health unit, and includes such other full-time public health services as the regulations prescribe.

Group foster home. A licensed regular or special foster home suitable for placement of more than five minor children but not more than ten minor children.

Group foster home. Any individual, partnership, firm, corporation or association which provides full-time care for six (6) to fifteen (15) children not related by blood, adoption, or marriage to the person or persons maintaining or operating the home, who are received, cared for, and maintained for compensation or otherwise, including the provision of permanent free care.

Group foster home. Shall be a foster home suitable for the placement of more than six (6) minor children, but not more than ten (10) minor children, which is licensed by the State Department of Economic Security.

Group home. A residential building or portion thereof designed as a child care facility that provides supervised maintenance or personal care in a quasi-family setting for not more than ten (10) minor children, ages six (6) to under eighteen (18) years, not of common parentage, for part or all of the day or night.

Health care. Health care occupancies are those used for purposes such as medical or other treatment or care of persons suffering from physical or mental illness, disease or infirmity, and for the care of infants, convalescents, or infirm aged persons. Health care occupancies provide sleeping facilities for the occupants or are occupied by persons who are mostly incapable of self-preservation because of age, physical or mental disability, or because of security measures not under the occupants control.

Health facility. Health facility shall include buildings all or part of which accommodate patients or house services for patients such as laundry, power plant, laboratory or kitchen. Health facility shall not include separate buildings which are used exclusively to house personnel or provide activities not related to health facility patients.

Helpless person. Person who in an emergency, by reason of age or mental or physical condition is unable to evacuate the premises easily and readily without help and without impeding the progress of others.

Home. Institution, residence, facility, or other place, part or unit thereof, however named, in the state which is advertised, offered, maintained, or operated by the ownership, or management, for a period of more than 24 hours, whether for a consideration or not, for the express or implied purpose of providing accommodations and personal assistance, to three or more persons, who are dependent upon the services of others, or which provides skilled nursing and dietary care for persons who are ill or otherwise incapacitated, or which provides services for the rehabilitation of persons who are convalescing from illnesses.

Home for the aged. Facilities providing domiciliary care for ambulatory aged, including the service of meals and the provision of incidental nursing as required from time to time.

Home for the aged. Facilities providing domiciliary care for ambulatory aged, including the service of meals and the provision of incidental nursing as required from time to time, provided that the words "ambulatory aged" as used above shall not include any inmate of any prison or other correctional institution.

Home for the aged. Facilities in buildings providing domiciliary care for ambulatory aged, including the service of meals and the provision of incidental nursing as required from time to time, provided that the words "ambulatory aged" as used above shall not include any inmate of any penal or correctional institution.

Home for the aged, nursing home, convalescent home and rest home. Home for the aged or infirm in which one or more persons not of the immediate family are received, kept or provided with food, shelter and care for compensation; but not including hospitals, clinics or similar institutions devoted primarily to the diagnosis and treatment of the sick or injured.

Home for the aging. Home which provides personal assistance for a consideration or not, for three or more persons who are dependent on the services of others by reason of age and physical or mental impairment, but do not require skilled nursing care, for a consideration or not, for three or more persons who by reason of age and either illness, physical, or mental impairment require skilled nursing care.

Home-nursing-convalescent or old age. A building or buildings used for lodging, boarding, or nursing care on a 24 hour a day basis, of 3 or more convalescent or aged persons not able to care for themselves, but shall not include hospitable and mental or correctional institutions.

Hospital. Institution in which patients are given medical or surgical care and which is licensed by the State to use the title hospital without qualifying descriptive word.

Hospital. Any institution offering services and facilities for use beyond a 24 hour period by 2 or more unrelated persons who require diagnosis, treatment or care for illness, injury, deformity, infirmity, abnormality, or pregnancy.

Hospital. Institution in which sick or injured persons are given medical or surgical care, or for the care of contagious or infectious diseases or incurable patients.

Hospital. Includes "sanitarium," "sanatorium," "preventorium," "clinic," provided such institution is operated by, or treatment given under direct supervision of, a physician licensed to practice.

Hospital. Building or part thereof used for the medical, obstetrical or surgical care, on a 24-hour basis, of 4 or more patients. Hospital includes general hospitals, mental hospitals, tuberculosis hospitals, children's hospitals, and any such facilities providing inpatient care.

Hospital. Building, or parts thereof, wherein hospital treatment and care, of one or more of the types identified below, are provided to three (3) or more persons for a period of more than 24 hours during any one week: Medical, surgical, and/or obstetrical treatment and care, and specialized treatment and care of persons afflicted with communicable diseases, or specialized treatment and care of persons afflicted with mental illnesses or disabilities, including alcoholism and drug addiction.

Hospital. Any building or a portion of which, used for the confinement or treatment of sick or injured persons.

Hospital. Institution providing health services, primarily for in-patients, and medical or surgical care of the sick or injured, including as an integral part of the institution, such related facilities as laboratories, out-patient departments, training facilities, central service facilities and staff offices.

Hospital. Building or part thereof used for in-patient and out-patient medical or surgical care, on a twenty-four hour basis, of three or more patients. Hospitals includes general and specialized hospitals, mental hospitals, tuberculosis hospitals, children's hospitals, extended care facilities and other related facilities including clinics.

Hospital. Any building used primarily for the care of sick or injured persons or any day nursery or asylum, where such building shall have accommodation for bedridden or decrepit persons.

Hospital. Building or group of buildings, having room facilities for overnight patients, used for providing services for the inpatient medical or surgical care of sick or injured humans, and which may include related facilities, such as central service facilities, and staff offices; provided, however, that such related facilities must be incidental and subordinate to the main use and must be an integral part of the hospital operations.

Hospital. Any institution receiving inpatients and rendering medical care, including those wherein mentally retarded, mentally disturbed, epileptic, alcoholic, drug addicted chronically ill and physically handicapped patients are treated or cared for.

Hospital. Building or portion thereof designed or used for the diagnosis, therapeutic treatment, or other care of ailments, of bed patients who are physically or mentally ill.

Hospital. Institution for the diagnosis, care of treatment of two or more unrelated persons suffering from illness, injury or deformity, or for the rendering of obstetrical or other professional medical care other than in an emergency. The term "hospital" shall not be construed to include the office of a physician or practitioner. (For purposes of this definition "nursing care shall not be construed to be professional medical care.")

Hospital. Institution, other than a psychiatric facility or a mentally retarded facility, specializing in giving clinical, temporary and emergency services of a medical or surgical nature to injured persons and patients.

Hospital. A place for the treatment or care of human ailments, and unless otherwise specified, the term shall include sanitarium, preventorium, clinic, maternity home, rest home, homes for the aged, and convalescent homes.

Hospital. Means any institution offering services and facilities for use beyond a 24 hour period by 2 or more unrelated persons who require diagnosis of, treatment or care for illness, injury, deformity, infirmity, abnormality, or pregnancy.

Hospital. Hospital means any institution, place, building or agency with an organized medical or medical-dental staff which maintains and operates 24-hour inpatient services for the diagnosis, care and treatment of patients.

Hospital. A building or portion thereof used for the accommodation of sick, injured or infirmed persons.

Hospital. Any institution, place, building or agency which maintains and operates organized facilities for the diagnosis, care and treatment of human illnesses, including convalescence and including care during and after pregnancy or which maintains and operates organized facilities for any such purpose and to which persons may be admitted for overnight stay or longer.

Hospital. Any building or portion thereof used for the accommodation and medical care of sick, injured or infirmed persons. This would include sanitariums, institutions for the cure of chronic drug addicts and mental patients, rest homes, homes for the aged and alcoholic sanitariums.

Hospital ward. Room equipped with more than 2 beds.

Housing for the elderly. Housing project designed for the elderly persons which provides living unit accommodations and spaces for common use by the occupants in social and recreational activities and, when needed, incidental facilities and space for health and nursing services for the project residents.

Housing for the elderly. Project specially designed for elderly persons (62 years of age or over) which provides living unit accommodations and spaces for common use by the occupants in social and recreational activities and, when needed, incidental facilities and space for health and nursing services for the project residents.

Immediate restoration of service. Automatic restoration of operation with an interruption of not more than ten seconds as applied to those areas and functions served by the Emergency System, except for areas and functions served by the Emergency System, except for areas and functions for which NFPA or local Codes make specific provisions.

Infirmary. Building, or part thereof, which is used primarily for medical care of sick or injured persons incidental to an industrial, business, educational, athletic, or similar activity to which such building, or part thereof, is an accessory and in which there are accommodations intended for sleeping for extended periods of time.

Infirm persons. Institutionalized persons whose age or health is such that they require institutional care or treatment.

Institution. Establishment providing care, or, care and residence for children or adults.

Institution. Institution means any hospital, clinic, sanitarium, nursing home, school, nursery, day center or other place which maintains and operates services for the mentally ill, mentally retarded, alcoholic, drug addict or other incompetent person.

Institution. A building occupied by a non-profit association, corporation, or other entity primarily for public or semi-public use.

Institution. Building or group of buildings designed or used for non-profit, charitable, or public service purposes of providing board, lodging, health care for aged persons, indigent, or infirm, or for the purpose of performing educational or religious services and offering board and lodging to persons in residence.

Institutional building. Building in which persons are harbored to receive medical, charitable or other care or treatment, or in which persons are held or detained by reason of public or civic duty, or for correctional purposes.

Institutional building. A building owned, occupied and used by institutions of a civic, religious, charitable, educational, eleemosynary or philanthropic nature.

Institutional building. The occupancy or use of which or any portion thereof is by persons harboured or detained to receive medical, charitable or other care or treatment, or by persons involuntarily detained.

Institutional home. Building or part thereof designed for the housing and care of 3 or more aged or infirmed persons or children, and in providing hospital facilities.

Institutional occupancy. Applies to that portion of a building in which persons are confined to receive medical, charitable, educational, or other custodial care or treatment, or in which persons are held or detained by reason of public or civic duty, including among others, hospitals, asylums, sanitariums, police stations, jails, court houses with cells, and similar occupancies.

Institutional occupancy. Occupancy or use of a building or part thereof by persons harboured or detained to receive medical care or treatment or by persons involuntarily detained.

Institutional occupancy classification. Building classification based on occupancy which includes all buildings in which people suffering from physical limitations because of health or age are harbored for medical, charitable, or other care or treatment.

Institutional use. A structure occupied or an activity carried on by a non-profit entity for public benefit.

Institutional uses. Non-profit institutions such as churches, and schools teaching academic subjects at hospitals and libraries.

Institution for the aged. A retirement home which provides not less than twenty-five (25) nursing beds and fifty (50) sheltered care beds, and which is located on a lot with not less than five (5) acres in area.

Institutions of an educational, philanthropic and charitable nature. Institutional establishments maintained and operated on a non-profit basis.

Intensive nursing care units. Groups of beds, rooms, or wards, specifically designated by the hospital to provide intensive nursing care for critically ill patients and intended to be specifically staffed and organized for such service, distinct from a surgical, or obstetrical recovery unit forming a part of a surgical or obstetrical suite.

Intermediate care facility. Intermediate care facility means a facility, or an identifiable unit or distinct part of a facility, which provides supportive, restorative, and preventive health services in conjunction with a socially oriented program to its residents, and which maintains and operates 24-hour services including board, room, personal care and intermittent nursing care.

Isolated power system. Isolated power system is an assembly of electrical devices that provides local isolated power, a single grounding point, and distinctive receptacles for treatment of an electrically susceptible patient. Purpose is to assure that an intermingling of electrical apparatus located near or applied to the patient will not result in excessive leakage current in the patient's body, particularly the heart.

Laboratory. Medical or Dental: A laboratory which provides bacteriological, biological, medical, x-ray, pathological and similar analytical or diagnostic services to doctors or dentists. No fabricating is conducted on the premises, except the custom fabrication of dentures.

Life safety branch. System of feeders and branch circuits meeting the requirements of NFPA or local codes connected to alternate power source by a transfer switch, and functioning as a component of the emergency system.

Life support branch. System of feeders and branch circuits, connected to the alternate power source by a transfer switch and functioning as a component of the emergency system. Operation of this branch may be necessary for patient survival, and therefore requires maximum protection against failure. Only isolated power centers may be connected to this branch.

Limited care nursing home. Building, or parts thereof, wherein shelter, board, and limited nursing care are provided to three (3) or more persons, not related to the operator, for a duration of not less than 72 hours during any one week, such care being required because of physical or mental infirmities associated with old age or with those illnesses and disabilities, or phases thereof, not requiring hospital treatment and care.

Limited nursing care. Consists of those services and procedures employed in caring for the infirm and/or ill who require practical nursing care under the supervision of a physician.

Maternity care. Medical or nursing care of women for pregnancy, delivery, or the puerperal state, or for any condition intimately related to such pregnancy, delivery, or puerperal state.

Maternity department. That physical area in a hospital used directly for maternity and newborn care, including the infant formula room.

Maternity home. Lying-in hospital or an institution for the care of women in connection with childbirth and the puerperium.

Medical or dental clinic. Means an organization of specializing physicians or dentists, or both, who have their offices in a common building. A clinic shall not include in-patient care.

Medical or dental clinic. Building or buildings in which physicians, dentists, or physicians and dentists, and allied professional assistants are associated for the purpose of carrying on their professions.

Mental hospital. Hospital for the treatment and protection of persons suffering from a mental disease.

Mental infirmities. Those characteristics of ordinary senility or mental decline which do not cause the patient to be disturbing to others or to require restraint.

Mentally retarded facility. Consist of the following: "resident school," "resident care facility," "family home (mentally retarded)," "nursery" or "day care center."

Mentally retarded persons. Shall mean any retarded person who is unable, or likely to be unable, to physically or mentally respond to an oral instruction relating to fire danger and unassisted take appropriate action relating to such danger.

Mental nursing home. Nursing home for persons who are mentally incapacitated from causes other than simple senility, or who regularly require restraint.

Non-profit institution. Non-profit establishment maintained and operated by a society, corporation, individual, foundation or public agency for the purpose of providing charitable, social, educational, medical or similar services to the public, groups or individuals.

Nursery. That area in a newborn unit which constitutes a patient room for infants.

Nurses stations. Areas intended to provide a center of nursing activity for a group of nurses working under a nurse supervisor and serving bed patients, where the patient calls are received, nurses dispatched, nurses notes written, inpatient charts prepared and medications prepared for distribution to patients. Where such activities are carried out in more than one location within a nursing unit, all such separate areas are considered a part of the nurses station.

Nursing care. Nurseries for the full-time care of children, hospitals, homes for the aged, nursing homes, orphanages, sanitariums, and similar buildings (each accommodating more than five (5) persons).

Nursing care homes. Limited care nursing home is a building, or parts thereof, wherein shelter, board, and limited nursing care are provided to three or more persons, not related to the operator, for a duration of not less than 72 hours during any one week, such care being required because of physical or mental infirmities associated with old age or with those illnesses and disabilities, or phases thereof, not requiring hospital treatment and care.

Nursing, convalescent or old age home. Building or buildings used for lodging, boarding, or nursing care on a 24 hour a day basis, of 3 or more convalescent or aged persons not able to care for themselves, but shall not include hospitable and mental or correctional institutions.

Nursing home. A proprietary facility, licensed by a governmental authority, for the accommodation of persons who require skilled nursing care and related medical services, but are not in need of hospital care. A housing for the elderly project is distinguished from a nursing home in that an elderly project is primarily of a residential character with only incidental nursing facilities while a nursing home is primarily designed and used for the care of convalescent or ill persons.

Nursing home. Facility designed and intended to provide nursing service on a continuing basis to persons, the majority of whom require such service under trained professional nurses or physicians, and for whom medical records are maintained. The term "Nursing Home" shall include Post-Operative Centers, but shall not include any facility used for providing service to any inmate of any prison or other correctional institution.

Nursing home. Building used for the accommodation and care of persons with, or recuperating from, illness or incapacity, where nursing services are furnished.

Nursing home. Home used for the reception and care, for a consideration or not, of three or more persons who by reason of illness, physical, or mental impairment require skilled nursing care.

Nursing home. Facility other than a hospital in which continuing nursing care on a twenty-four hour basis by a graduate nurse, meeting standards adopted by the State Board of Health, is provided and in which medical services are prescribed or performed under the direction of a physician or surgeon licensed to practice for two (2) or more unrelated persons who are not acutely ill and not in need of hospital care.

Nursing home. Building or part thereof, used for the lodging, boarding and nursing care, on a 24 hour basis, of 4 or more persons who, because of mental or physical incapacity, may be unable to provide for their own needs and safety without the assistance of another person. Nursing Home, includes nursing and convalescent homes and infirmaries of homes for the aged.

Nursing home. An establishment licensed to care for infirm persons, regardless of age. Such a home shall not contain equipment for or provide care for those cases for which hospitalization is generally required.

Nursing home. A home for aged, chronically ill, infirm or incurable persons, or a place of rest for those suffering bodily disorders in which three (3) or more persons, not members of the family residing on the premises, are received, kept or provided with food and shelter or other care, but not including hospitals, clinics or similar institutions devoted primarily to the diagnosis or treatment of disease, injury, maternity cases or mental illness.

Nursing home. Any building which admits persons for care, treatment, or nursing by reason of sickness, injury, infirmity or other disability.

Nursing home. Institution caring for more than two (2) unrelated persons.

Nursing home. Nursing home means any institution, place, building or agency which maintains and operates 24-hour skilled nursing services for the care and treatment of chronically ill or convalescent patients, including mental, emotional or behavioral problems, mental retardation or alcoholism.

Nursing home. A facility other than a hospital in which continuing nursing care on a twenty-four hour basis by a graduate nurse licensed in the state or other nursing personnel meeting standards adopted by the Board of Health, is provided and in which medical services are prescribed or performed under the direction of a physician or surgeion licensed to practice in the state for two (2) or more unrelated persons who are not acutely ill and not in need of hospital care.

Nursing home. Structure operated as a lodging house in which nursing, dietary, and other personal services are rendered to convalescents, invalids, or aged persons, not including persons suffering from contagious or mental diseases, alcoholism, or drug addiction, and in which surgery is not performed and primary treatment, such as customarily is given in hospitals or sanitariums, is not provided. Convalescent home or a rest home shall be deemed a nursing home.

Nursing or convalescent home. Building designed or used in whole or in part to provide, for compensation, the care of the ill, senile, or otherwise infirm persons resident on the premises, not in need of hospital care.

Nursing or convalescent home. Building or structure having accommodations and where care is provided for invalid, infirm, aged, convalescent, or physically disabled or injured persons, not including insane and other mental cases, inebriate, or contagious cases.

Nursing or convalescent home. Home for the aged, chronically ill or incurable persons in which three or more persons not of the immediate family are received, kept or provided for with food and shelter or care for compensation; but not including hospitals, clinics or similar institutions devoted primarily to the diagnosis and treatment of the sick or injured.

Nursing or convalescent home. Any building where persons are housed or lodged and furnished with meals and nursing care for hire.

Nursing or convalescent home. Home, institution, building or residence, public or private, whether operated for profit or not, which provides maintenance, personal care or nursing for a period exceeding 24 hours to three or more ill, physically infirm or aged persons, who are not related by blood or marriage to the operator.

Nursing or convalescent home. Building designed or used in whole or in part to provide, for compensation, the care of the ill, senile, or otherwise infirm persons resident on the premises, not in need of hospital care.

Nursing or rest home, or convalescent home. Institution for the care of children, or the aged or infirmed, or a place of rest for those suffering bodily ailments, which does not include facilities for the treatment of a sickness or injuries or for surgical care.

Nursing unit. That physical area in a hospital containing patient rooms and adjacent service rooms, in a distinct physical section, for the purpose of providing nursing care to patients in such area.

Obstetrical room. Room for the delivery of babies.

Occupants. Patients, residents, and personnel housed in a home.

Old age home. A building used for the accommodation and care of persons of advanced age.

Operating room. Room where surgery is performed.

Operator of a home. Person, firm, partnership, association, or corporation which is required by code to obtain a license in order to open, manage, maintain, or operate a home.

Outpatient clinic. Any institution receiving outpatients only, and rendering medical service.

Patient. Person who needs skilled nursing or dietary care or rehabilitation service because of illness or incapacitation.

Patient. Shall mean a person housed in a special care home, who is not able to leave the premises unassisted.

Patient unit. That physical area in a patient room designated for an individual patient.

Personal care nursing home. Building, or parts thereof, wherein shelter, board, and personal nursing care are provided to three or more persons, not related to the operator, for a duration of not less than 72 hours during any one week, such care being required because of physical or mental infirmities associated with old age or with those illnesses and disabilities, or phases thereof, not requiring hospital treatment and care.

Personal nursing care. Consists of those services and procedures employed in caring for those individuals (the infirm) who, in whole or in part, require assistance in dressing, eating, bathing, and ambulation.

Philanthropic institution. A nonprofit, charitable institution devoted to the housing, training or care of children, or of aged, indigent, handicapped or underprivileged persons, but not including the following: office buildings, except as an accessory to and located on the same lot with an institutional activity, as listed above; hospitals, clinics or sanitariums, correctional institutions; institutions or homes for the insane or those of unsound mind; lodging houses or dormitories providing temporary quarters for transient unemployed persons; organizations devoted to collecting and salvaging new or used materials; or organizations devoted principally to distributing food, clothing or supplies on a charitable basis.

Psychiatric bed areas. Areas specifically set aside for active inpatient care of psychiatric patients, excluding bed areas not specifically assigned to psychiatric care and general care unit beds that may occasionally be assigned to psychiatric patients.

Psychiatric clinic. Psychiatric clinic means any clinic providing a direct service of referral, evaluation, diagnosis, care or treatment on a nonresidential basis to persons suffering from mental, emotional or behavioral disorders.

Psychiatric facility. Institution in which care of treatment is given to persons suffering from mental illness, disease, disorder or ailment. Such facilities include, but are not limited to "psychiatric hospitals," "long term care facilities," "resident treatment centers," "family homes (mentally ill)," "alcoholism hospital," and "facility for admission of drug addicts."

Radio-active material. One which emits ionizing radiation, such as alpha rays, beta rays or gamma rays.

Recovery room. Room set aside for the use of patients immediately following surgery.

Rehabilitation center for handicapped persons. Building or group of buildings, and accessory buildings and uses, devoted to providing services for handicapped persons. Such services can include physical and/or occupational therapy, related medical services, vocational training, placement service, and social and recreational activities suitable for disabled adults and children. Retail sales are permitted, but limited to disposal of articles which are the products of the therapy.

Residences. Including multiple housing wherein persons reside and are domiciled and including health care facilities.

Resident. Any individual who in general needs no service or care other than room and board, but who has a disability arising from a physical or mental disability or from prior institutionalization such that he or she requires assistance or supervision in such matters as dressing, bathing, diet or minor matters of health maintenance, and who is domiciled in a special occupancy boarding home. In no event shall a resident include any individual who is mostly incapable of self-preservation in the event of an emergency, nor an individual in need of skilled or intermediate nursing care.

Resident. Person who, by reason of age or infirmity, is dependent upon the services of others.

Residential care (halfway home). A private boarding home, institution, building, residence or other place, whether occupied for profit or not, which through its ownership or management provides residential care to three (3) or more persons who are not related to the applicant or owner by blood or marriage.

Residential care institutions or halfway houses. Residential care institution means, without limitation because of enumeration, any building, structure, institution, boarding home or other place for the reception and care of three or more unrelated individuals for not less than 72 hours in any week, who by reason of physical or mental disability, including mental retardation and mental illness are in the opinion of a licensed physician, in need of care, but not the care given in a nursing home as defined in previous definitions.

Residential-custodial care facility. Building or part thereof used for the lodging or boarding of three or more persons who are incapable of self care or have physical or mental limitations. This definition includes facilities in Nurseries (custodial care for children under six years of age), Mentally Retarded Care Centers, Mental Health, Drug, Alcoholism, Physical Rehabilitation Centers and the like.

Residential-custodial care facility. Building, or part thereof, used for the lodging or boarding of 4 or more persons who are incapable of self-preservation because of age, or physical or mental limitation. This definition includes facilities such as Homes for the Aged, Nurseries (custodial care for children under 6 years of age), and Mentally Retarded Care Institutions.

Rest home. Any place or institution which makes provision for bed care or for chronic or convalescent care for one or more persons exclusive of relatives, who by reason of illness or physical infirmity are unable to care for themselves; but in which no alcoholics, drug addicts, persons suffering from mental sickness, disease, disorder or ailment or from surgical or other primary treatments such as are customarily provided in sanitariums or hospitals are performed. "Rest Home" shall also mean any home operated as a boarding house for more than 15 persons.

Rest home. Home which provides personal assistance, for a consideration or not, for three or more persons who are dependent on the services of others by reason of age or physical or mental impairment but, who do not require skilled nursing care.

Rest home. Any building used or maintained to provide nursing, dietary and other personal services to convalescents, invalids, aged or infirm persons, but excluding cases of contagious or communicable diseases, and excluding surgery or primary treatments such as those customarily provided in sanitariums and hospitals.

Rest home. Premises used for the housing of and caring for the ambulatory, aged or infirm. There shall be only incidental convalescent care not involving either trained nurse or physician residing on the premises. There shall be no surgery, physical therapy or other similar activities, such as is customarily provided in sanitariums and hospitals.

Resthome-nursing home. A private home for the care of children or the aged or infirm or a place of rest for those suffering bodily disorders. Such home does not contain equipment for surgical care or for the treatment of disease or injury nor does it include maternity care or care of mental illnesses. Its minimum accommodation is six persons.

Restrained. As applied to persons in a building, means that such persons are locked in the building in such a manner that their freedom for normal egress to the outside of the building is dependent upon the unlocking or unbolting by others, of one or more doors or other barriers.

Restraint. The physical retention of a person within a room, cell or cell block by any means; or within the exterior walls of a building by means of locked doors unoperable by the person restrained. Restraint shall also mean the physical binding, strapping or similar restriction of any person in a chair, walker, bed or other contrivance for the purpose of deliberately restricting the free movement of ambulatory persons.

Retirement home. An establishment providing living accommodations and services principally to non-transient guests sixty (60) years of age or over. Accommodations provided may consist of dwelling units of any type, rooming units or nursing or sheltered care beds.

Sanitarium. Health station or retreat or other place where resident patients are kept, and where medical or surgical treatment is given to persons suffering from a sickness, disease, disorder or ailment other than a mental sickness, disease, disorder or ailment, but which does not specialize in giving clinical, temporary or emergency service.

Sanitarium. Building used as an institution for the recuperation and treatment of persons having physical or mental disorders.

Selected acute nursing areas. Areas within a nursing unit with beds grouped or located to facilitate concentrated nursing care or observation.

Selected receptacles. Minimal electrical receptacles to accommodate appliances ordinarily required for local tasks or likely to be used in patient care emergencies.

Self-contained battery powered life support equipment. Is portable, but entails constant responsibility for supervision and maintenance.

Sheltered care home. Facility other than a hospital or nursing home for two or more unrelated persons who are not acutely ill which renders personal care and assistance with meals, dressing, medications prescribed by a physician or surgeon licensed in the state, and other personal assistance of a similar type and includes homes for the aged and infirm who do not need skilled nursing care.

Sheltered care home. An establishment licensed to provide assistance, supervision or oversight to residents. A "sheltered care home" may not provide skilled or intermediate nursing services nor care for those cases for which hospitalization is generally required.

Skilled care nursing home. Building, or parts thereof, wherein shelter, board, skilled nursing care, and related medical and/or therapeutic services are provided to three or more persons, not related to the operator, for a duration of not less than 72 hours during any one week, such care and services being required because of physical or mental infirmities associated with old age or with those illnesses and disabilities, or phases thereof, not requiring hospital treatment and care.

Skilled nursing care. Consists of those services and procedures employed in caring for the infirm and/or ill who require professional nursing care under the supervision of one or more physicians.

Small Rest home. Rest home for 3, 4 or 5 persons.

Special care home. Any building, structure or portion thereof, (other than hospitals equipped or used for surgical or obstetrical care) used for the reception, housing or care, with or without compensation, of two and not exceeding a total of twelve patients and guests, not related to the operator, who, for any cause, require care or attention and are kept for a period of more than twenty-four hours.

Special placement residence. Any building or premises occupied by three or more persons not related to the owner, lessee or operator by blood, marriage or adoption, who upon their release as patients from any recognized mental institution, treatment ward for alcoholism, treatment center for narcotic addicts or as an inmate of any correctional penal institution, use such building or premises as living facilities in order to secure non-institutionalized care in their attempt to re-enter society as healthy, happy and useful human beings. The operator must reside at such premises.

Static systems. Consist of batteries, chargers and static inverters.

Suite of rooms. Two or more rooms which are arranged to be used as a unit and each room of which is accessible from within the unit without the use of a public exit way.

Supervisory care facility. A building or part thereof used for the lodging or boarding of four (4) or more mental health patients who are capable of self-preservation and who require supervision and who are receiving therapy, training or other health related care and who may have imposed upon them security measures not under their control.

Task illumination. Provision for the minimum illumination required to carry out necessary tasks in the described areas, including safe access to supplies and equipment, and access to exits.

Toilet room. Room or rooms including not less than one of each of the following plumbing fixtures; water closet, shower or bathtub and a lavatory which is located in or adjacent to the room in which the water closet is located.

Ward. Separate room in a hospital or a dormitory in which three or more persons are lodged or confined for treatment.

Welfare agency. Organization, public or private, offering professional social work services to individuals or groups.

Written record. Written record of inspection, performance, exercising period, and repairs, regularly maintained and available for inspection by the authority having jurisdiction.

X-ray room. Any room in which x-ray equipment is permanently installed.

C-12 HOTELS, MOTELS AND LODGING FACILITIES

Accessory sleeping quarters for transients. Means a space or building which is an accessory to another occupancy and which is used as sleeping quarters for 5 or more transients.

Apartment. A "Dwelling Unit".

Apartment. A space or suite of rooms which is intended for separate and independent living and under the control of the tenant.

Apartment. A room or suite of rooms occupied by one (1) person or one (1) family for living and sleeping purposes.

Apartment and lodging or rooming house combined. A building designed, constructed or altered in such manner that the sum of the number of rooms in the apartment units (excluding bathrooms and kitchens) plus the number of lodging rooms shall not exceed 15.

Apartment hotel. A building or group of buildings containing both apartments and rooming units in some combination and containing those services customary and appropriate to a hotel. Such services to be available for the convenience of the occupants thereof only.

Apartment hotel. A residential building designed or used for both, two or more dwelling units and six or more guest rooms or suites of rooms.

Apartment hotel. A building primarily for persons who have their residence therein, containing four (4) or more apartments which do not have kitchens.

Apartment hotel. A building or group of buildings containing a number of independent living quarters and in which a public dining room is provided.

Apartment hotel. A building designed for or containing both apartments and individual guest rooms or rental units, under resident supervision and which maintains an inner lobby through which all tenants must pass to gain access to apartments, rooms, or units.

Apartment hotel. A hotel with dwelling units in which all accommodations are provided in dwelling units and in which at least twenty-five (25) percent of the guest rooms are for occupancy by transient guests. An apartment hotel may have a dining room open to the public which is accessible only from an inner lobby or corridor.

Apartment hotel. A hotel in which a substantial portion of the accommodations include kitchen facilities, and in which at least eighty (80) percent of the units are occupied by permanent guests securing such accommodations by prearrangement for a continuous period of thirty (30) days or more.

Apartment hotel. A building, usually under resident supervision, made up of three (3) or more apartment units, arranged with common corridors and exits and maintaining an inner lobby or foyer, through which persons pass for access to the apartments.

Apartment hotel. Building containing four or more apartments without kitchens, primarily for persons who have their residences therein.

Apartment hotel. Hotel in which apartments are rented or designed to be rented in suites for terms of not less than one month, and in which there are no housekeeping facilities within the apartments.

Apartment hotel. Any apartment house or any hotel that is occupied as an apartment house, or as a combination apartment house and hotel with accommodations for transient guests with rooms or suites.

Apartment hotel. Building containing 4 or more apartments, any of which is occupied by transients.

Apartment hotel. Multiple-family dwelling designed for or containing both apartments and dwelling units without culinary facilities and which has an inner lobby through which all tenants must pass to gain access to apartments or dwelling units. The number of apartments with culinary facilities must be at least 10 and the number of dwelling units without culinary facilities shall not exceed 30 or shall not exceed the number of apartments, whichever is less.

Apartment hotel. Building designed for or containing both apartments and individual guest rooms or suites of which all tenants must pass to gain access to the apartments, and catering to permanent and not transient tenants, and which may furnish services ordinarily furnished by hotels, such as drug store, barber shop, cigar and news stands, dining rooms, when such uses are located entirely within the building with no entrance from the street nor visible from any sidewalk, and having no sign display visible from the outside of the building, indicating the existence of such use.

Apartment hotel. Any apartment house or any hotel that is occupied as an apartment house, or as a combination apartment house and hotel with accommodations for transient guests.

Apartment hotel. Building containing both dwelling units and rooming units, used primarily for permanent guests.

Apartment hotel. A multiple dwelling in which dwelling units are leased to permanent and, or transient tenants.

Apartment hotel. Apartment house or hotel that is occupied as an apartment house, or as a combination apartment house and hotel with accommodations for transient guests.

Apartment hotel. Shall include a building or part of a building used for permanent occupancy containing living, sleeping or eating facilities, provided not more than ten per cent (10%) of the dwelling units shall be designed for transient accommodations. Such eating facilities may include a dining room and private bar room accessible only from the interior of the building, and such eating facilities shall not have any exterior advertising.

Apartment hotel. Shall mean a building or portion thereof designed for or containing both individual guest rooms or suites of rooms and dwelling units.

Apartment hotel. A building designed for or containing both apartments and individual guest rooms or suites of rooms and apartments, wherein is maintained an inner lobby through which all tenants must pass to gain access to the apartments, and catering to permanent and not-transient tenants, and which may furnish services ordinarily furnished by hotels, such as drug store, barber shop, cigar and news stands, dining rooms, when such uses are located entirely within the building with no entrance from the street nor visible from any sidewalk, and having no sign display visible from the outside of the building, indicating the existence of such use.

Apartment house or apartment hotel. A building designed, constructed or altered to provide accommodations for more than 4 families living independently of each other, each unit of which is designed for cooking upon the premises.

Automobile court (tourist cabins). Group of two (2) or more detached or semi-detached buildings or cabins having less than three hundred (300) square feet of floor area for each unit and containing guest rooms and/or apartments, with automobile storage space provided in connection therewith, use and/or designed for use primarily by automobile transients.

Boarder (roomer, lodger). An individual not within the second degree of kindred to the person conducting the boarding house living within a household who pays a consideration for such residence and does not occupy such space as an incident of employment therein.

Boardinghouse. A dwelling in which three, four, or five rooms are occupied as guest rooms and in which food may be served to the occupants thereof. Any dwelling in which more than five rooms are occupied as guest rooms shall be deemed as a hotel. A boardinghouse shall not include institutions for persons requiring physical or mental care by reason of age, infirmity or disease.

Boardinghouse or lodginghouse. A building where, for definite periods, lodging with or without meals is provided for three or more persons but not exceeding 20 persons.

Bungalow court. A bungalow court is a group of three or more family units on one or more adjoining lots, having separate outside entrances on the ground floor level for each single family dwelling, including all open spaces required by the Code, and all maintained under one ownership.

Bungalow villas. A bungalow villa is a group of ten or more one-story dwelling units containing not more than two single family units per structure, located on one or more adjoining lots and having separate outside entrances on the ground floor level for each single family dwelling, designed to provide accommodations for transient or overnight guests. Structures may be designed for full residential use including cooking and similar facilities, and must be maintained under one ownership.

Commercial hotel. Establishment which provides sleeping rooms and/or dwelling units for pay, in which at least five (5) rooms or units or at least thirty per-cent (30%) of the accommodations, whichever is greater, are regularly available for transient occupancy, and which provides customary hotel services, including maid service, the furnishing and upkeep of linen, telephone or desk service, and the use and upkeep of furniture, to all transient occupants.

Commercial hotel. Transient hotel catering to a business clientele.

Community kitchen. A kitchen in a hotel used individually or collectively by the occupants of the building, but not used commercially to serve a dining room or the public.

Guest. Any person who is in the building under consideration for the purpose of sleeping, and any person who rents therein a room or suite of rooms which has sleeping facilities; except that the owner of the building and his regular employees therein are not guests.

Guest. Any person hiring or occupying a room for living or sleeping purposes.

Guest. Shall mean each occupant of any unit or any hotel, motel, tourist home or related establishment included in the definition of a hotel, motel or tourist home

Guest car ratio. For purposes of site planning, the number of parking spaces required for each living unit for use of guests.

Guest house. A structure for human habitation, containing one or more rooms with bath and toilet facilities, but not including a kitchen or facility which would provide a complete housekeeping unit.

Guesthouse. A detached accessory building located on the same premises with the main building, containing no more than one (1) dwelling unit for use by temporary guests of the occupants of the premises. Such quarters may have kitchen facilities or separate utility meters and may be rented but may not be sold or otherwise used as a separate dwelling.

Guest house (tourist home). Any dwelling in which rooms are rented for guests and for lodging of transients and travelers for compensation.

Guest room. Rooming unit of only one (1) room, if a guest room contains sleeping facilities for more than two (2) persons, each unit of sleeping facilities for two (2) persons or any portion thereof shall be considered as a separate guest room for density purposes only.

Guest room. Room in a hotel, motel or tourist home offered to the public for compensation in which room no provision is made for cooking and which room is used only for transient occupancy.

Guest room. Room occupied or intended, arranged, or designed for occupation, by one (1) or more guests. Every one hundred square feet (100 sq. ft.) of floor area in rooms occupied by more than two (2) guests shall be considered as an additional guest room.

Guest room. Any room or rooms used, or intended to be used by a guest for sleeping purposes. Every one hundred square feet (100 Sq. Ft.) of superficial floor area in a dormitory shall be considered to be a guest room.

Guest room. Room occupied by one or more guest for compensation and in which no provision is made for cooking, but not including rooms in a dormitory for sleeping purposes primarily.

Guest room. Any room or rooms used, or intended to be used, by guests for sleeping purposes. Every 100 square feet of superficial floor area in a dormitory is a guest room.

Guest room. A room occupied by one or more persons not members of the family, in which no cooking facilities are provided.

Guest room. A rooming unit of only one room. If a guest room contains sleeping accommodations for more than two persons, each unit of accommodations for two persons shall be considered as a separate guest room for purposes of calculating density.

Guest room. Any habitable room except a kitchen, designed or used for occupancy by one or more persons and not in a dwelling unit.

Guest room. A room occupied, or intended, arranged, or designed for occupation, by 1 or more guests. Every 100 square feet of superficial floor area in a dormitory is a guest room. A guest is any person paying, in money or services, for the use of a sleeping facility

Guest room or sleeping room. A room which is designed or intended for occupancy by not more than two (2) persons, but in which no provision is made for cooking and not including dormitories for sleeping purposes. All points of ingress or egress to such room shall be located through the main residence of which such room is an integral part.

Guest room or sleeping room. Room which is designed or intended for occupancy by, or which is occupied by, not more than two (2) persons, for compensation, not members of the family, but in which no provision is made for cooking and not including dormitories for sleeping purposes.

Guest room or unit. Any room for occupancy, meetings, display, conference or dining and shall include all other rooms integral to the hotel, motel or tourist home unit. In the case of a central toilet and/or shower room, the term "guest room" or "unit" shall also apply.

High-rise hotel. Hotel six (6) or more stories in height.

Hospitality suite (room). Facility used for entertaining, usually at conventions, trade shows, and similar meetings.

Hotel. Building or portion thereof in which ten (10) or more guest rooms are provided for occupancy, for compensation, by transient guests.

Hotel. Building in which board and lodging are provided and offered to the transient public for compensation.

Hotel. Building occupied as the more or less temporary abiding place of individuals who are lodged with or without meals and in which there are more than twelve (12) sleeping rooms usually occupied singly and no provision made for cooking in any individual room or apartment.

Hotel. Dwelling not consisting of living units and occupied by more than twenty (20) persons, all of whom may reach their living accommodations by passing through one central lobby.

Hotel. Building occupied as the more or less temporary abiding place of individuals who are lodged with or without meals; and in which there are not fewer than twenty (20) sleeping rooms, or apartments, and no provision made for cooking in any individual room.

Hotel. Building occupied as the more-or-less temporary abiding place of individuals who are lodged with or without meals in which there are more than ten (10) sleeping rooms usually occupied singly, and no provision made for cooking in any individual room or apartment.

Hotel. Building or group of buildings in which there are six (6) or more guest rooms where lodging with or without meals is provided for compensation, and where no provision is made for cooking in any individual room or suite, but shall not include jails, hospitals, asylums, sanitariums, orphanages, prisons, detention homes and similar buildings, where human beings are housed and detained under legal restraint.

Hotel. Building occupied as the more or less temporary abiding place of individuals who are lodged with or without meals; and in which there are not fewer than twenty (20) sleeping rooms, or apartments, and no provision made for cooking in any individual room or suite.

Hotel. Building containing sleeping rooms in which lodging is provided primarily for transient guests for compensation and which may include public dining facilities.

Hotel. Building containing twenty (20) or more individual sleeping rooms or suites, having each a private bathroom attached thereto, for the purpose of providing overnight lodging facilities to the general public for compensation with or without meals, excluding accommodations for employees, and in which ingress and egress to and from all rooms is made through an inside lobby or office supervised by a person in charge at all hours. Where a hotel is permitted as a principal use, all uses customarily and historically accessory thereto for the comfort, accommodation and entertainment of the patrons including the service of alcoholic beverages, shall be permitted.

Hotel. Building or portion thereof used as the more or less temporary abiding place of individuals who are lodged with or without meals and in which there are more than twelve (12) sleeping rooms usually occupied singly and in which provision for cooking is made preponderantly in a central kitchen.

Hotel. Any building containing six (6) or more guest rooms intended or designed to be used, or which are used, rented or hired out to be occupied, or which are occupied for sleeping purposes by guests.

Hotel. Building or other structure kept, used maintained, advertised as or held out to the public to be a place where sleeping accommodations are supplied for pay to transient or permanent guests or tenants, in which twenty (20) or more rooms are furnished for the accommodation of such guests; and having or not having one (1) or more dining rooms, restaurants or cafes where meals of lunches are served to such transients or permanent guests, such sleeping accommodations and dining rooms, restaurants or cafes, if existing, being conducted in the same building or buildings in connection therewith.

Hotel. Building occupied as a more or less temporary place of residence for individuals or groups of individuals who are lodged, without meals, in guest rooms without kitchens whether designated as a hotel, inn, club, or by an other name.

Hotel. Building other than a boarding house as defined herein, which building contains more than five (5) guest rooms.

Hotel. A building in which lodging is provided and offered to the public for compensation, and which is open to transient guests, in contradistinction to a boarding house or lodging house as defined in the Ordinance.

Hotel. A building which provides a common entrance, lobby, halls and stairways, and in which lodging is offered with or without meals principally to transient guests.

Hotel. A building in which lodging or boarding and lodging are provided and offered to the public and in which access to all rooms is normally made through a supervised inside lobby or office. As such, it is open to the public in contradistinction to a boarding house, a lodging house, or multiple-family dwelling and motel which are separately defined in the Ordinance.

Hotel. Any building containing ten (10) or more rooms, intended or designed to be used or which are used, rented or hired out to be occupied, or which are occupied by persons for sleeping purposes by paying guests.

Hotel. A building in which lodging or board and lodging are provided and offered to the public for compensation and in which ingress and egress to and from all rooms is made through an inside lobby or office supervised by a person in charge at all hours. As such it is open to the public in contradistinction to a boarding house, a lodging house, or an apartment hotel, which is separately defined in the Ordinance.

Hotel. A building or premises where lodging is provided, with or without meals, for more than twenty (20) persons.

Hotel. Building, or part thereof, designed or used primarily for sleeping accommodations for more than 10 guests or lodgers.

Hotel. Building other than a boarding house as defined in the ordinance, which building contains more than 5 guest rooms or suites.

Hotel. Building or that portion thereof containing more than five (5) rooming units or sleeping facilities for more than twenty (20) persons. Such rooming units or dwelling units being for residential purposes or transient purposes. A hotel shall provide customary hotel services such as bell boy service, room service, maid service, telephone, secretarial or desk service.

Hotel. Any building or portion thereof containing six (6) or more guest rooms used, designed or intended to be used, let or hired out to be occupied or which are occupied by six (6) or more individuals for compensation whether the compensation be paid directly or indirectly; this includes auto courts and motels.

Hotel. Any building containing six (6) or more rooms intended or designed to be used, or which are used, rented or hired out to be occupied, or which are occupied for sleeping purposes by guests.

Hotel. Building used as an abiding place of more than twenty (20) persons who for compensations are lodged, meals provided, no provision is made for cooking in individual rooms or suites and in which ingress and egress to and from all rooms is through an inside lobby or office supervised by a person in charge at all hours. As such it is open to the public in contradisdinction to a boarding house, or an apartment which are herein separately defined.

Hotel. Building occupied as the more or less temporary abiding place of individuals who are lodged with or without meals, and in which there are more than fifteen (15) sleeping rooms usually occupied singly and with no provision made for cooking in any individual room or apartment.

Hotel. Building in which lodging, with or without meals, is offered to transient guests for compensation and in which there are more than five (5) sleeping rooms with no cooking facilities in any individual room or apartment.

Hotel. Building containing five (5) or more rooms or suites of rooms which are used primarily by transient guests for the purpose of sleeping.

Hotel. Building occupied as a more or less temporary abiding place of individuals, who are lodged with or without meals, and in which there are more than fifteen (15) sleeping rooms, but no provisions for cooking in any individual suite.

Hotel. Any building containing six (6) or more rooms intended to be used, or which are used, rented or hired out to be occupied, or which are occupied for sleeping purposes of guests.

Hotel. Building arranged or used for sheltering, sleeping or feeding for compensation, of more than twenty (20) individuals.

Hotel. Building containing ten (10) or more sleeping rooms in which transient guests are lodged with or without meals, with no provisions for cooking in any individual room or suite.

Hotel. Building providing overnight accommodations in which access to each rental unit is provided by an entrance and central service core connected to interior halls.

Hotel. Building in which lodging is provided and offered to the public for compensation, and which is open to transient guests, in contradistinction to a boarding house or rooming house.

Hotel. Building containing rooms intended or designed to be used or which are used, rented, or hired out to be occupied or which are occupied for sleeping purposes by guests and transients and where only a general kitchen and dining room are provided within the building or in an accessory building.

Hotel. Building, or part thereof, designed or used primarily for sleeping accommodations for more than ten (10) guests or lodgers in rooms or suites.

Hotel. Structure that provides housing and frequently food and other services for persons away from home.

Hotel. Any building other than a boarding house where sleeping accommodations are provided for compensation.

Hotel. Any building occupied as the abiding place of persons, who are lodged with or without meals; in which, as a rule, the rooms are occupied singly for hire; and in which there are more than fifty (50) sleeping rooms.

Hotel. Group of rooming units for transient guests. A hotel may also contain a small number of dwelling units.

Hotel. Any building, including motels, having fifteen (15) or more sleeping rooms or where sleeping accommodations for more than thirty (30) persons are provided for hire, with or without meals.

Hotel. Building containing either or both guest rooms and apartments.

Hotel. Building designed to provide accommodation for short-time residence, with or without meals, providing for ten (10) or more sleeping rooms, and including customary accessory uses in connection with the principal use.

Hotel-motel-motor hotel. Any building or combination of buildings containing six or more rooms used for sleeping purposes by guests on a transient basis.

Hotel or motel. Includes a building or structure kept, used, or maintained as or advertised as, or held out to the public to be a hotel, motel, inn, motor court, tourist court, public lodging house or place where sleeping accommodations are furnished for a fee to transient guests with or without meals.

Hotel or motor hotel. Building or other structure kept, used, or maintained and advertised as or held out to the public to be a place where sleeping accommodations are supplied for pay, catering primarily to transient guests, in which ten (10) or more rooms are furnished for the accommodation of such guests, and/or cafes where meals are served to transient or other guests; such sleeping accommodations and dining rooms, restaurants and cafes, if existing, being conducted in the same building or accessory buildings in connection therewith.

Hotel or tourist park. Building or group of buildings providing sleeping accommodations intended primarily for transients travelling by automobile, arranged with a parking space to each lodging.

Hotels. Buildings or groups of buildings under the same management in which there are more than fifteen (15) sleeping accommodations for hire, primarily used by transients who are lodged with or without meals, whether designated as a hotel, inn, club, motel, or by any other name. So-called apartment hotels shall be classified as hotels because they are potentially subject to transient occupancy like that of hotels.

Kitchen. Any room used, intended or designed to be used for cooking or the preparation of food.

Lobby. Public lounge or waiting place adjacent to and connected with other spaces and a passageway which serves as a principal entrance or exit.

Lobby. Means a public lounge or space for waiting which is connected with an entrance or exit way which serves as a principal entrance or exit.

Lobby. Waiting room, or large hallway serving as a waiting room.

Lodging house. Any building containing less than ten rooms, intended or designed to be used or which are used, rented or hired out, or which are occupied for sleeping purposes by two or more paying guests.

Lodging house. The term lodging house shall be taken to mean a building or portion thereof used, or designed to be used, for the shelter of guests for a remuneration and containing not less than five, nor more than fifteen sleeping rooms above the first floor.

Men's cubicle hotels. Includes all lodging houses exclusively maintained for men, containing sleeping stalls, the separating partitions of which do not reach the ceiling.

Motel. Building or group of buildings containing apartments, and/or rooming units, and/or guest rooms each of which maintains a separate outside entrance. Such building or group of buildings being designed, intended, or used primarily for the accommodation of automobile travelers, and providing automobile parking conveniently located on the premises.

Motel. Includes the term motor hotel, tourist courts, and transient accommodations, primarily for those persons traveling by automotive vehicles and consisting of two (2) or more units or buildings designed to provide sleeping accommodations, and with customary accessory uses.

Motel. Building or buildings providing overnight accommodations principally for automobile travelers in which access to each rental unit is provided directly through an exterior door or by an entrance connected to a common interior hall leading to the exterior.

Motel. Series of attached, semi-detached or detached rental units containing bedroom, bathroom, and closet space. Units shall provide for overnight lodging and are offered to the public for compensation, and shall cater primarily to the public traveling by motor vehicle.

Motel. Building or group of buildings, whether detached or in connected units, used as individual sleeping or dwelling units designed primarily for transient automobile travelers, and providing for accessory off-street parking facilities and having individual unit entrances, opening to the outside. The term "Motel" includes buildings designated as auto courts, tourist courts, motor lodges and similar appellations.

Motel. Any building or portion thereof containing sleeping accommodations in ten (10) or more rooms for persons who are not members of a family as defined in the Code whether such establishment is designated as a hotel, inn, motel, motor lodge, motor hotel, or otherwise.

Motel. Permanent building or group of buildings containing rooms used, rented or hired out for the more or less temporary occupancy of overnight guests.

Motel. One or more buildings, each building containing one or more rooms used for transient occupancy.

Motel. Any building or group of buildings including either separate units or a row of units which have individual entrances and contain living and sleeping accommodations primarily for transient motorist occupancy but which do not contain cooking facilities in any individual unit. The term Motel includes any buildings or building groups designated as auto courts, motor lodges, tourist courts, or by any other title or sign intended to identify them as catering to motorists.

Motel. Multiple dwelling, intended primarily for motorists not over two (2) stories in height, in which the exit from each dwelling unit or sleeping room is directly to the exterior (includes, but is not limited to the terms motor court, motor hotel, tourist court).

Motel. Building or group of buildings containing apartments and/or rooming units and/or guest rooms, each of which maintains a separate outside entrance. Such building or group of buildings being designed, intended or used primarily for the accommodations of automobile travelers and providing automobile parking conveniently located on the premises. Occupancy is normally made up of overnight rentals. A motel shall be deemed a commercial occupancy.

Motel. Combination or group of two (2) or more detached, semi-detached, or connected permanent buildings occupying a building site integrally owned, and used as a unit to furnish living accommodations for transients only.

Motel. One (1) or more buildings, not more than two (2) stories in height, with an aggregate of five (5) or more rooms or suites of rooms which are used primarily by transient guests for the purpose of sleeping and which have an exit directly to the exterior of the building from each room or suite of rooms.

Motel. Series of attached, semi-attached or detached dwellings containing bedroom, bathroom and closet space, where each unit has convenient access to a parking space for the use of the unit's occupants. The units, with the exception of the apartment of the manager or caretaker, are devoted to the use of transients. No cooking facilities are offered. The definition includes tourist courts, auto courts and motor lodges.

Motel. Establishment that provides housing for persons away from home who are traveling by automobile.

Motel. Group of rooming units for transient guests one (1) or two (2) stories high with convenient parking space. A motel may also contain a small number of dwelling units.

Motel. Dwelling not consisting of living units and occupied by more than twenty (20) persons, in which there is no central lobby to reach individual living accommodations.

Motel. Automobile or tourist court.

Motel. A building or a group of two (2) or more buildings designed to provide sleeping accommodations for transient or overnight guests. Each building shall contain a minimum of ten (10) residential units or rooms which shall generally have direct, private openings to a street, drive, court, patio, or the like.

Motel. A building or group of buildings containing apartments, and/or rooming units, and/or guest rooms each of which maintains a separate outside entrance. Such building or group of buildings being designed, intended, or used primarily for the accommodation of automobile travelers, and providing automobile parking conveniently located to the space to be occupied.

Motel. Any building or group of buildings containing guest rooms or dwelling units, some or all of which have a separate entrance leading directly from the outside of the building with garage or parking space located on the lot and designed, used, or intended wholly or in part for the accommodation of automobile transients. Motel includes motor court, motor lodge and tourist court, but not a mobile home park.

Motel. An establishment consisting of a number of living or sleeping accommodations, designed primarily for lodging of transient automobile travelers, the physical layout of which affords ready access to guests' automobiles through outside doorway entrances other than through a common entrance, lobby and hallways.

Motel. A business establishment consisting of a group of attached or detached furnished lodging rooms, with bathrooms, designed for use by transient automobile passengers. A motel furnishes customary hotel services such as maid service, laundering, telephone and secretarial and desk service. Less than fifty (50) percent of the living and sleeping accommodations in a motel are occupied or designed for occupancy by persons other than transient automobile passengers. Cooking facilities are not permitted in any lodging rooms.

Motel. A structure designed or intended to provide temporary sleeping accommodations for automobile transients and having off-street parking spaces immediately adjacent to the sleeping rooms.

Motel. An establishment consisting of a group of attached or detached living or sleeping accommodations with bathroom and closet space, located on a single zoning lot and designed primarily for use by tourists. A "motel" furnishes customary hotel services.

Motel. Building or group of buildings, whether detached or in connected units, used as individual sleeping or dwelling units designed primarily for transient automobile travelers and providing accessory off-street parking facilities. Motel includes buildings designated as motor lodges, auto courts and similar appellations.

Motel. Building or group of buildings having units containing sleeping accommodations which are available for temporary occupancy by transients and providing off-street parking facilities adjacent or convenient thereto.

Motel. Combination or group of 2 or more detached, semi-detached or connected permanent buildings occupying an integrally owned site, and used as a unit to provide living accommodations for transients only.

Motel and motel court. Building or a group of buildings in which overnight lodging is provided and offered to the public for compensation and catering primarily to the public traveling by motor vehicle.

Motel, motor court, tourist court, or motor lodge. Building in which lodging, or boarding and lodging are provided and offered to the public for compensation. As such, it is open to the public in contradistinction to a boarding or lodging house, or multiple dwelling; same as a hotel, except that the buildings are usually designed to serve tourists traveling by automobile, ingress and egress to rooms need not be through a lobby or office, and parking usually is adjacent to the dwelling unit.

Motel or hotel. Structure or portion thereof or a group of structures containing guest rooms where lodging is provided to transients for compensation.

Motel or hotel. Structure or portion thereof or a group of attached or detached structures containing completely furnished individual guest rooms or suites, occupied on a transient basis for compensation, and in which more than 60 per cent (60%) of the individual guest rooms and suites are without kitchens or cooking facilities.

Motel or hotel units. Each motel or hotel unit shall have a minimum of 200 square feet of living space excluding all other spaces, except lavatory and closet space.

Motel or motor court. Building or a group of two (2) or more detached or semi-detached buildings containing rooms or apartments having separate ground floor entrances provided directly or closely in connection with automobile parking or storage space serving such rooms or apartments, which building or group of buildings is designed, intended, or used principally for the providing of sleeping accommodations for automobile travelers and is suitable for occupancy at all seasons of the year.

Motel or motor court. Land used or intended to be used or occupied by a group of two (2) or more detached or semi-detached buildings, except trailers, or by a multiple building containing guest rooms, with automobile parking space and incidental utility structures which are provided in connection therewith, all of which is used or designed for use primarily by automobile transients.

Motel or motor lodge. Building or group of buildings where lodging primarily for the temporary occupancy of automobile travelers, food, drink, and other incidental services are provided for compensation. The terms motel or motor lodge shall include auto court and tourist court.

Motel or tourist court. Group of attached or detached buildings containing individual sleeping or living units where a majority of such units open individually and directly to the outside, and where a garage is attached or a parking space is conveniently located to each unit, all for the temporary use by automobile tourists or transients, and such words shall include auto courts and motor lodges.

Motel—tourist cabins. Dwelling designed and occupied on a non-housekeeping basis.

Motor court. Building or group of buildings containing one or more guest rooms having separate outside entrances for each such room or suite of rooms and for each of which rooms or suites of rooms automobile parking space is provided.

Motor hotel. Full-service hotel for motorists.

Murphy bed. Standard bed that folds or swings into a wall or cabinet in a closet like fashion.

Number of exits. Every place of assembly and every floor area intended for occupancy in a hotel shall be provided with not less than two (2) independent well-separated exits when its area is greater than 1,000 square feet or when its occupant load is greater than sixty persons.

Occupant load. Total number of occupants or persons that may occupy a hotel or portion thereof at any one time, based on the allotment of a certain portion of the floor area for each person.

Outside room. Room on the perimeter of the building facing outward with an exposure more desirable than that of an inside room.

Passenger elevator. An elevator designed and used for carrying persons.

Public space. Any area in the hotel which generally is made accessible to the general public, including dining rooms, bars, lobby, and function rooms.

Residential hotel. Building or group of buildings used for hotel purposes having a gross coverage and exceeding forty percent (40%) of the building site and providing no outside entrances for business purposes and using no street frontage for business or business displays.

Residential hotel. Provides sleeping rooms and/or dwelling units for pay at least seventy per cent (70%) of the accommodations for regularly used or available for permanent occupancy, and provides customary hotel services, including maid service, the furnishing and upkeep of linen, telephone or desk service, and the use and upkeep of furniture to all transient occupants.

Residential hotel. Hotel catering to long-term guests who have made the property their home and residence.

Residential use. Includes single and multiple dwellings, hotels, motels, dormitories, and mobile homes.

Residential use. Shall be deemed to include single and multiple dwellings, hotels, motels, dormitories and trailer parks.

Resort. Shall mean a group or groups of buildings containing more than five (5) dwelling units and/or guest rooms and providing outdoor recreational activities which may include golf, horseback riding, swimming, shuffleboard, tennis and similar activities. A resort may furnish services customarily furnished by a hotel, including a restaurant, cocktail lounge, and convention facilities.

Resort hotel. A building or group of buildings used for hotel purposes on a building site which contains at least five (5) acres, having a gross coverage not exceeding twenty per cent (20%) of the building site, providing no outside entrances for business purposes, providing no outside business displays, and having not more than one (1) identification sign which does not exceed twenty (20) square feet in area.

Resort hotel. A building or group of buildings used for hotel purposes on a building site which contains at least five (5) acres, having a gross coverage not exceeding twenty per cent (20%) of the building site, providing no outside entrances for business purposes, providing no outside business displays, and having not more than one (1) identification sign which does not exceed twenty (20) square feet in area.

Resort hotel. Caters to vacationing guests providing recreational and entertainment facilities, often a destination hotel.

Resort hotel. Hotel which caters mainly to vacationers or tourists, usually offering more recreational facilities than other hotels.

Resort hotel. Building or group of buildings used as a hotel, and in which there are one or more dining rooms available for meals to be served and where entertainment may be provided transient guests. Additionally that in which there are facilities used, designed or intended to be used in order to serve or sell intoxicating beverages. Also, those which permit the operation of gambling or games of chance.

Retirement hotel. An establishment where meals are provided as part of the price of the accommodations, which caters primarily to nontransient guests and which either holds itself out to the public as a retirement facility or selectively caters to or solicits older persons or has admission standards based on age. A retirement hotel may not offer any type of long-term care, including nursing or sheltered care services.

Rollaway bed. Portable utility bed approximately 30 inches by 72 inches; also called a cot.

Room. A room is a space or area bounded by any obstructions to exit passage which at any time enclose more than 80 percent of the perimeter of the area. In computing the unobstructed perimeter, openings less than 3 feet clear width and less than 6 feet 8 inches high shall not be considered.

Room. Every compartment in a hotel, apartment house, rooming house, motor court or trailer court, including parlors, dining rooms, sleeping porches, kitchens, offices, sample rooms, living rooms, sleeping room; but not including halls, bathrooms, toilet rooms, closets, pantries or store rooms. And in an apartment having no designated dining room, any dining space wherever located shall be classed as a dining room.

Room. Every compartment in any building, including parlors, dining-rooms, sleeping rooms and porches, kitchens, offices, stores, samplerooms, living rooms, but not including halls, bathrooms, closets, pantries or storage or equipment spaces.

Rooming unit. A suite of rooms forming a single habitable unit used or intended to be used for residential or transient purposes but not including cooking facilities of any kind. Wherein a rooming unit is designed to be divided into separate guest rooms then each separate sleeping room shall be counted as a guest room for density purposes.

Rooming unit. A room or group of rooms forming a single habitable unit, other than a dwelling unit, which is rented or available for rent for sleeping purposes with or without cooking facilities.

Service elevators. Elevators for use by employees (room service, housekeeping and maintenance, etc.) on hotel business and are not readily visible to the guests.

Sleeping room. A room, other than a guest room, in which no cooking facilities are provided.

Sleeping room. Room used for sleeping primarily for single tenant occupancy.

Sleeping room. Means a room used primarily for sleeping by one or more persons.

Sleeping rooms. Occupied individually and not as suites, and suites in a hotel, or in an addition made to a hotel shall be separated from adjacent rooms, suites and corridors or other interior exits by a 3/4 hour fire separation. (Some areas and localities require 1 hour separation).

Smoke-proof barriers. Every hotel that is not of noncombustible construction shall have smoke-proof barriers erected in such manner and in such locations as the building department orders.

Sof-a-bed. Sofa with fixed back and arms that unfolds into a standard single or double bed; also called a hideabed.

Standpipe and hose system. Every hotel four or more stories in height and every addition four or more stories in height made to a hotel, shall have a standpipe and hose system comprised of the components and materials, and designed, installed and maintained in the manner prescribed by the regulations.

Suite of rooms. Two (2) or more rooms which are arranged to be used as a unit and each room of which is accessible from within the unit without the use of a public exit way.

Tourist camp. Tract of land, with or without buildings, or other equipment, on which one (1) or more camp cabins are located, or where temporary accommodations are provided for two (2) or more automobile trailers or house cars, open to the public free or for a fee.

Tourist camp. Group of three (3) or more detached one-story buildings used, or primarily intended for use, as dwellings for tourists coming and going by automobiles and grouped about a plot or parcel of land, any portion of which is used or intended to be used in common by the inhabitants of such dwellings and having toilet and bath facilities common to the use of all occupants.

Tourist camp. Group of attached or detached cottages, cabins or similar buildings, containing individual sleeping or living units, but not including kitchen or cooking spaces for the accommodation of transient guests; the parking or supporting by a foundation of two or more trailers should be termed a tourist camp.

Tourist camp. Group of two (2) or more detached or semi-detached buildings containing guest rooms or apartments or both, with automobile storage space provided in connection therewith, used or designed for use primarily by automobile transients. Land used or intended to be used for camping purposes by automobile transients.

Tourist court. Group of attached or detached buildings designed, constructed or under construction or alteration for guest rooms or dwelling units intended primarily for automobile transients, each unit having a separate entrance opening out-of-doors or into a foyer, with parking space appropriately located on the lot for use by guests of the court, operation of such court to be supervised by a person in charge at all hours. Tourist courts include auto courts, motels, motor courts, motor hotel, and motor inns.

Tourist court. Group of attached or detached buildings containing individual sleeping or living units with attached garage or parking space conveniently located to each unit, all for the temporary use by automobile tourists or transients; includes auto courts, motels, or motor lodges.

Tourist court. Group of attached or detached buildings containing individual sleeping or living units without cooking or kitchen facilities with at least one parking space for each unit located on the same premises and convenient to each unit, all for the temporary use by automobile tourists or transients; includes auto courts, motels, or motor lodges.

Tourist court. Building or group of buildings containing one (1) or more guest rooms having separate outside entrances for each such room or suite of rooms and for each of which rooms or suites of rooms automobile parking space is provided.

Tourist court. Building or group of buildings designed and used to provide guest rooms primarily for automobile transients, each room or unit having a separate entrance opening out-of-doors or into a foyer, with parking space provided on the lot for use of guests of the court, operation of such court to be supervised by a person in charge at all hours. Tourist courts include auto courts, motels, motor courts, motor hotels, and motor inns.

Transient. Means a person who resides in the same building for a period of less than 30 days.

Transient purposes. The intent to use and/or the use of a room or a group of rooms for the living, sleeping and housekeeping activities of persons on a temporary basis of an intended tenure of less than one (1) month.

Twin-double. Two (2) double beds; a room with two (2) such beds capable of accommodating four persons.

Unit. The word "unit" shall mean and refer to houses, apartments, group of rooms, or a single room occupied or intended for permanent or transient occupancy as separate living quarters.

Vestibule. An enclosed space, with doors or opening protectives, to provide protected passage between the exterior and interior of a building, or between spaces within a building.

C-13 HOUSING AND MINIMUM REQUIREMENTS

Accessory living quarters. Shall mean living quarters within an accessory building for the sole use of persons employed on the premises or for temporary use by guests of the occupants of premises, which living quarters have no kitchen facilities and are not rented or otherwise used as a separate dwelling

Accessory room. Any room or enclosed floor space used for eating, cooking, bathrooms, water closet compartments, laundries, pantries, foyers, hallways, and similar floor spaces. Rooms designated as recreation, study, den, family room, office, etc., in addition to habitable rooms are considered accessory rooms.

Accessory room. Any room in a dwelling unit other than a bedroom, bathroom, kitchen or one (1) living room or living-dining room combination, including, but not necessarily limited to, a den, library, family room, game room, patio room.

Apartment. Means a suite of two or more rooms wherein occupants live independently of other occupants of the same building.

Apartment. One (1) or more rooms comprising a dwelling unit or serving as the home or residence of an individual, or a family or a household.

Apartment. A room or a suite of rooms within an apartment house, arranged, intended or designed to be used as a home or residence of one family with kitchen facilities for the exclusive use of the one family.

Apartment building. A building which is used or intended to be used as a home or residence for three (3) or more families living in separate apartments, in which the yard areas, hallways, stairways, balconies and other common areas and facilities are shared by families living in the apartment units.

Apartment buildings. Buildings containing three or more living units with independent cooking and bathroom facilities, whether designated as apartment house, tenement, garden apartment, or by any other name.

Apartment house. Any building or portion thereof used as a multiple dwelling for the purpose of providing three (3) or more separate dwelling units which may share means of egress and other essential facilities.

Apartment house. A building made up of three or more apartment units so arranged that each unit has direct access, without common corridors, to a means of egress from the building, and which may or may not maintain an inner lobby for its tenants.

Apartments for the elderly. An apartment building specifically designed for housing elderly individuals who are capable of self preservation.

Basement. The portion of any building partly or wholly below grade.

Bathroom. A room containing a bathtub, tub or shower compartment; water closet and lavatory or other similar facilities provided immediately adjacent thereto.

Building. A structure which encloses space; a structure which gives protection or shelter for any occupancy. The term "building" shall be construed as if followed by the phrase "or part thereof." When separated by fire walls, each portion so separated shall be deemed a separate building.

Building. Any structure having a roof supported by columns or walls for the housing or enclosure of persons or property of any kind.

Cluster housing. Shall mean housing which is perceived as a complex of closely related structures.

Coach house. A structure separated from the principal building on a lot and containing a dwelling unit. Constructed in conjunction with, though not necessarily above, storage space which could be utilized by automobiles.

Communal kitchen. A kitchen within a dwelling building used by the occupants of more than one (1) dwelling unit, or shared by any persons residing in a rooming occupancy.

Condominium. An estate in real property consisting of an undivided interest in common in a portion of a parcel of real property together with a separate interest in air space.

Corridor. A passageway, hallway, or similar designation for horizontal travel from or through subdivided areas.

Detached house. One which has yard areas on all four sides.

Discriminating and discriminate. Means to segregate, separate, exclude or treat any person unequally only because of race, color, religion, ancestry or national origin.

Dormitory. Any dwelling (other than a fraternity or sorority house) occupied primarily as a place of temporary abode by persons attending educational institutions.

Dormitory. A space in a unit where group sleeping accommodations are provided, with or without meals, for persons not members of the same family group, in one room, or in a series of closely associated rooms under joint occupancy and single management, as in college dormitories, fraternity houses, military barracks and ski lodges.

Dwelling. Any building which is wholly or partly used or intended to be used for living or sleeping by human occupants.

Dwelling building. A building used or designed or intended to be used, all or in part, for residential purposes.

Dwelling unit. A room or group of rooms located within a dwelling building and forming a single habitable unit with living, sleeping, cooking, eating and sanitary facilities used or intended to be used by one (1) family.

Dwelling unit. One or more rooms arranged for the use of one or more individuals living together as a single family unit, with cooking, living, sanitary, and sleeping facilities.

Efficiency apartment. A dwelling unit of not more than one (1) room in addition to kitchen and bath.

Efficiency apartment. A dwelling unit which has only one (1) combined living and sleeping room, said dwelling unit, however, may also have a separate room containing only kitchen facilities and also a separate room containing only sanitary facilities.

Efficiency living unit. Any room having cooking facilities and used for combined living, dining, and sleeping purposes.

Efficiency unit. A dwelling unit consisting of one principal room exclusive of bathroom, kitchen, hallway, closets, or dining alcove directly off the principal room, providing the dining alcove does not exceed one hundred twenty-five square feet in area.

Elderly housing. Planned multiple shelter of not less than 20 dwelling units under one (1) or more roofs, but upon one premises, designed and erected to be used exclusively for the housing of persons 60 years or more of age except as provided herein, and in which the normal operations of housekeeping and preparation of meals is customarily performed by the tenants or occupants and provision is made therefore in each dwelling unit, excepting laundry and heating equipment; provided further that one (1) dwelling unit may be occupied without restriction of age by a custodian or manager.

Factory built housing. Housing that is partly or totally built in a factory and then transported in sections or as a complete unit to a site where it is erected or stationed and provided with the necessary services to make it a habitable unit that, when occupied, is void of transport features such as wheels, tires, axles, brakes or lamps.

Family. A single individual living on the premises as a separate housekeeping unit; or a collective body of persons living together upon the premises as a single housekeeping unit in a domestic relationship based on birth, marriage, or other domestic bond.

Family unit. A room or group of rooms used or intended to be used as a housekeeping unit for living, sleeping, cooking and eating.

Fraternity or sorority house. A building containing no more than one (1) dwelling unit and more than two (2) rooming units or guest rooms. Such rooming units or guest rooms shall be for residential purposes only.

Guest house. Detached living quarters of permanent construction, clearly subordinate and incidental to the main building on the same lot, and intended for use without compensation by occasional guests of the occupants of the main building.

Guest house. Living quarters within an accessory building for the sole use of persons employed on the premises or for temporary use by guests of the occupants of premises, which living quarters having no kitchen facilities and are not rented or otherwise used as a separate dwelling.

Habitable attic. Attic which has a stairway as a means of access and egress and in which the ceiling area at a height of seven and one-third (7-1/3) feet above the attic floor is not more than one-third (1/3) the area of the floor next below.

Habitable room. Space used for living, sleeping, eating or cooking, or combinations thereof, but not including bathrooms, toilet compartments, closets, halls, storage rooms, laundry and utility rooms, basement recreation rooms and similar spaces.

Habitable room. Room occupied by one or more persons for living, eating, sleeping, or working purposes. It does not include toilets, laundries, serving and storage pantries, corridors, cellars, and spaces that are not used frequently or during extended periods.

Habitable room. Any room meeting the requirements of the Code for sleeping, living, or dining purposes excluding such enclosed spaces as closets, pantries, bath or toilet rooms, service rooms, connecting corridors, laundries, unfinished attics, foyers, storage spaces, cellars, utility rooms and similar places.

Habitable room. Any room or enclosed floor space arranged for living and/or sleeping purposes.

Habitable room. A room designed and used for living, sleeping, eating or cooking. Bathrooms, toilet compartments, closets, halls, storage or utility spaces and similar areas are not considered habitable rooms.

Habitable room. A room in a Residential Unit used for living, sleeping, eating or cooking purposes, but excluding baths, toilet rooms, storage spaces and corridors.

Habitable room. A room or enclosed floor space arranged for living, eating, or sleeping purposes, but does not include a room used as a bathroom, water closet compartment, laundry, pantry, foyer, hallway, or other accessory floor space.

Habitable room. A room used or intended to be used for living, sleeping, eating or cooking purposes but does not include bathrooms, toilet rooms, laundries, pantries, foyers, corridors, storage spaces, stairways or closets.

Habitable room. Room occupied by one (1) or more persons for living, eating, sleeping, or working purposes. It does not include toilets, laundries, serving and storage pantries, corridors, cellars, and spaces that are not used frequently or during extended periods.

Habitable room. Room or enclosed floor space arranged for living, eating, or sleeping purposes, not including bath or toilet rooms, laundries, pantries, foyers, or communicating corridors.

Habitable room. Room designed and used for living, sleeping, eating or cooking. Bathrooms, toilet compartments, closets, halls, storage or utility spaces and similar areas are not considered habitable rooms, as defined in the Code.

Habitable room. Room which is designed or may be used for living, sleeping, eating or cooking. Storerooms, bath rooms, toilets, closets, halls or spaces in attics are not habitable rooms.

Habitable room. Any room occupied in a place of habitation, refuge or detention, as a kitchen, dining room, living room, parlor, bedroom, or library, in distinction from a closet, bathroom, water-closet room, corridor, laundry, furnace or boiler room, storage room or other utility room.

Habitable room. Any room meeting the requirements of the Code for sleeping, living, or dining purposes excluding such enclosed spaces as closets, pantries, bath or toilet rooms, service rooms, connecting corridors, laundries, unfinished attics, foyers, storage spaces, cellars, utility rooms and similar places as defined in the Code.

Habitable room. Room in which persons sleep, eat or carry on their usual domestic or social vocations or avocations, but not including a private laundry, bathroom, toilet room, dressing room, pantry, storeroom, corridor and similar places not used by persons frequently or for extended periods.

Habitable room. Room occupied by or designed for occupancy by one or more persons for living, sleeping, eating or cooking including kitchens serving a dwelling unit, but not including bathrooms, water closet compartments, laundries, serving and storage pantries, corridors, cellars, attics, recreation rooms, and spaces that are not used frequently or during extended periods.

Habitable room. A room used or intended to be used for living, sleeping, eating, or cooking purposes but does not include bathrooms, toilet rooms, laundries, pantries, foyers, corridors, storage spaces, stairways or closets as defined in the Code.

Habitable room. Space used for living, sleeping, eating or cooking, or combinations thereof, but not including bathrooms, toilet compartments, closets, halls, storage rooms, laundry and utility rooms, basement recreation rooms and similar spaces as defined in the Building Code.

Habitable room. Any room meeting the requirements of the Code for sleeping, living, cooking or dining purposes excluding such enclosed spaces as closets, pantries, bath or toilet rooms, service rooms, connecting corridors, laundries, unfinished attics, foyers, storage spaces, cellars, utility rooms and similar spaces.

Habitable room. A room in a residential unit used for living, sleeping, eating or cooking, but excluding baths, toilets, storage spaces or corridors.

Habitable space. Space which may be used for living, sleeping, dining, or cooking, except that kitchenettes are not considered habitable spaces.

Habitable space. Space occupied by one or more persons for living, sleeping, eating, or cooking, except as provided for under the definition of "assembly space".

Habitable story. Story which has at least four (4) feet between the ground level and the ceiling joists and has enough area to provide a habitable room with net floor-to-ceiling distance of 7 feet 6 inches over half the floor area of the room, constitutes a habitable room as defined in the Building Code.

Half story. Story finished as living accommodations located wholly or partly within the roof frame and having a floor area at least half as large as the story below. Space with less than 5 feet clear headroom shall not be considered as floor area.

Half story. The uppermost story containing usable space, usually wholly or partly under a sloping roof, wherein the area of usable floor space having a clear height of 7 feet or more is 60 percent or less of the floor area of the story directly below. Whenever such usable floor area of an uppermost story exceeds 60 percent of the floor area of the story directly below, it shall be deemed to be a full story.

Half story. A space under a sloping roof which has the line of intersection of roof decking and wall face not more than three (3) feet above the top floor level, and in which space not more than two-thirds (2/3) of the floor area is finished off for use. A half story containing independent apartment of living quarters shall be counted as a full story.

Half story. The uppermost story lying under a sloping roof, the usable floor area of which, at a height of four feet above the floor does not exceed two-thirds (2/3) of the floor area in the story directly below, and the height above at least two hundred (200) square feet of floor space is seven feet six inches (7′-6″).

Hotel. A building or part thereof in which rooms are available for hire as lodging to the general public, which rooms are occupied transiently and singly, rather than as dwelling units.

House. Building used for sleeping quarters and usual accessories to the occupants.

House or household. Includes a dwelling house, lodging house, hotel, and also a students' residence, fraternity house or other buildings in which any persons in attendance as students, pupils, or teachers or employees in any capacity in or about a University, College, school or other institution of learning resides or is lodged.

Housing. Means any improved property which is used or occupied or is intended to be used or occupied as a home or residence.

Housing for the elderly. The term shall mean new construction of multiple-family units or remodeling of existing multiple-family dwelling units, the occupation of which shall be limited to persons 62 years of age or more; provided that if two (2) or more persons occupy a unit, at least one (1) shall be 62 years of age or more, with a maximum annual income of such amount as may be determined by the Board by resolution.

Human occupancy. Use of a space or spaces in a building or structure in which any human does or is required to live, work or remain for continuous periods of time exceeding 2 or more hours.

Kitchen. Space having a floor area of 60 square feet or more and which is used for the cooking of food.

Kitchen. Any room or space used or intended or designed to be used for cooking or the preparation of food.

Kitchen. Room used or adapted for cooking and containing a stove, range, hot-plate or other cooking apparatus, which burns coal, oil, gas or other fuel or is heated by electricity, except electric appliances consuming less than eighteen hundred (1800) watts.

Kitchen. Space, 60 square feet or more in floor area, used for cooking or preparation of food.

Kitchenette. Space, less than 60 square feet in floor area, used for cooking or preparation of food.

Kitchenette. Space having a floor area of less than 60 square feet and which is used for the cooking of food.

Let. To give another person the right to occupy any portion of a dwelling family unit, or rooming unit. The act of "letting" shall be deemed to be continuing act for so long as the persons given the right to occupy premises, continues to do so, in accordance with lease agreement.

Livability space ratio. For the purpose of site planning the minimum square foot of non-vehicular outdoor area which shall be provided for each square foot of total floor area. The LSR is found by dividing the open space minus the open car space minus 1/2 the ground area of any open carports by the total floor area.

Living unit. Residential unit providing complete, independent living facilities for one family including permanent provisions for living, sleeping, eating, cooking and sanitation.

Living unit. Any enclosed floor space consisting of one or more habitable rooms, with or without accessory rooms used by a person(s) or family.

Lot coverage. Shall mean the area of land which is covered by a building on a particular site. Lot coverage shall be the percentage of net lot area which is covered by the gross floor area of the first floor.

Member of a household. Person residing, boarding or lodging in a house.

Migrant labor housing. One (1) or more buildings or structures, trailers, or vehicles, together with the land appertaining thereto, established, operated or used as living quarters for seasonal, temporary or migrant workers, provided they are public lodging establishments, are not located within the farm, and the laborer pays rent. This definition shall not apply to forestry or tobacco farm operation.

Minimum habitable room height. Clear height from finished floor to finished ceiling of not less than seven and one-half (7-1/2) feet, except that in attics and top half-stories, the height shall be not less than seven and one-third (7-1/3) feet over not less than one-third (1/3) the floor area when used for sleeping, study, or similar activity.

Minimum habitable room size. A room with a minimum dimension of seven (7) feet and a minimum area of seventy (70) square feet, between enclosing walls or partitions, exclusive of closet and storage space.

Minimum habitable room size. No room shall have a dimension less than 8 feet and a minimum total area of 80 square feet, between enclosing walls or partitions, exclusive of closet and storage spaces.

Minimum use requirement. In no case shall a room, suite or group of rooms comprising a family dwelling unit, in any dwelling, be so occupied as to provide less than 800 cubic feet of air space per occupant, exclusive of cubic air space of bathroom, toilet rooms, closets, stairways, attics, utility rooms, and basements. No bedroom or room used as a bedroom, in any dwelling, shall be so occupied as to provide less than 300 cubic feet of air space per occupant, exclusive of the cubic air space of bathrooms, toilet rooms, and closets.

Multi-family house. Building occupied as the home or residence of individuals, families or households living independently of each other, of which three (3) or more are doing cooking within their apartments; including among others tenement house, apartment house, and flat.

Multi-family house. Building or portion thereof containing three (3) or more dwelling units, including tenement house, apartment house, flat.

Multi-family residence. Building designed and/or used exclusively for residential purposes and containing more than one (1) dwelling unit.

Multiple dwelling. A dwelling building occupied for residence purposes by more than two (2) families, or by individuals living in a rooming occupancy.

Net floor area. Net floor area shall be the actual occupied area, not including accessory unoccupied areas or thickness of walls.

Nonhabitable space. Space used as a kitchenette, bath, toilet, closet, storage, utility or heater room, recreation room below grade, garage, and similar spaces as well as halls, corridors, and stairways within a dwelling house.

Non-habitable space. Space used as kitchenettes, pantries, bath, toilet, laundry, rest, dressing, locker, storage, utility, heater, and boiler rooms, closets, and other spaces for service and maintenance of the building, and those spaces used for access and vertical travel between stories.

Non-habitable space. Space used as a kitchenette, pantry, laundry, closet, bath, toilet, rest, dressing, locker, storage, utility, heater, or boiler room; and other spaces used only for service and maintenance of the building; and those spaces used only for access and vertical travel between stories.

Non-residential usable floor area. Measurement of usable floor area for non-residential uses shall be the sum of the area of the first floor, as measured to the exterior face of the exterior walls, plus that area, similarly measured, of all other stories that are accessible by a fixed stairway, ramp, escalator or elevator, which may be made fit for occupancy, the measurement shall include the floor area of all accessory buildings measured similarly.

Occupant. A person over one (1) year of age, living, sleeping, cooking or eating in, or having actual possession of, a dwelling unit or a room used for rooming occupancy.

Occupied area. Any place or area in which any person is, shall be or may be required to inhabit, abide or sojourn.

One family dwelling. A building containing one (1) dwelling unit with not more than five (5) lodgers or boarders.

One family house. Building which is rented, leased, let or hired out to be occupied or is occupied or is intended, arranged or designed to be occupied as the home or residence of not more than one (1) family. All such buildings whether built singly, or in conjunction with others as double houses or terraces or attached or semi-detached rows, shall be deemed one (1) family houses when each such house complies with the definition of "Building."

Parties in interest. All individuals, associations and corporations who have interests of record in a multiple dwelling, and who are in actual possession thereof; and any person authorized to receive rents payable for housing space in a multiple dwelling.

Patio. An open area or an accessory outdoor living structure not exceeding 14 feet in height and open on at least one side.

Patio home. An attached or detached single family dwelling constructed with no side yard on one or more sides of the lot.

Place of abode. Building or part of a building, such as apartment building, row house, rooming house, hotel or dormitory.

Premises. Land, improvements thereon, or any part thereof.

Public hallway. A public corridor or space separately enclosed which provides common access to all the exitways of the building in any story.

Public space. A legal open space on the premises, accessible to a public way or street, such as yards, courts, or open spaces permanently devoted to public use which abuts the premises.

Residence building. A building in which sleeping accommodations are provided, except when classed as an institutional building.

Residential building. A principal building arranged, designed, used or intended to be used for residential occupancy by one (1) or more families. Residential buildings include, but are not limited to the following types: (1) single-family dwellings, (2) two-family dwellings, (3) multiple-family dwellings, and (4) a row of one or two-family attached dwellings developed initially under single ownership or control. Buildings used in whole or in part as hotels, rooming houses, boarding houses, nursing homes, retirement homes, dormitories, fraternities and sororities and other institutional housing shall be considered as non-residential buildings.

Residential building. A building, or parts thereof, designed or used for one or more family units or designed or used for sleeping accommodations.

Residential floor areas. Sum of the gross horizontal areas of the several floors of the dwelling, exclusive of garages, cellars, and open or roofed porches measured from the exterior faces of the exterior walls of a dwelling.

Residential purposes. Any building used for residential purposes, shall be taken to mean one or two-family residences, apartment houses, and multiple dwellings.

Residential unit. Room or suite of rooms, with or without cooking facilities, which are occupied as the residence of a single individual, family or group individuals.

Residential usable floor area. Measurement of "usable floor area" for residential uses shall be in the sum of the area of the first floor, as measured to the exterior face of the exterior walls, plus that area,—similarly measured, of all other stories having more than ninety (90) inches of headroom that are accessible by affixed stairway and which may be made usable for human habitation; but excluding the floor area of uninhabitable basements, cellars, garages, accessory buildings, attics, breezeways, and unenclosed porches.

Room. Every compartment in any building, including living rooms, dining rooms, sleeping rooms, porches, kitchens, offices, stores, sample rooms, parlors, but not including halls, bathrooms, closets, pantries or storage or equipment spaces.

Room. Space within a building completely enclosed with walls, partitions, floor and ceiling, except, when used for habitation to be provided with the necessary openings for light, ventilation.

Room capacity. For determining the exits required, the minimum number of persons or the occupant content of any floor area shall in no case be taken less than the square foot area of the floor area divided by occupancy allowance per person in square feet for the use of the area which will give the allowable occupant content.

Roomer. A nontransient paying guest in the home of another person.

Room-habitable. An enclosing subdivision in a residential building commonly used for living purposes, but not including any lobby, hall, closet, storage space, water closet, bath, toilet, slop sink or general utility room or service porch.

Room height. Minimum clear height of a habitable room shall be 6 feet 10 inches, including clear height under beams or other projections from the ceiling. In rooms with sloping ceilings, the minimum clear height of 6 feet 10 inches shall be required for only 40 per cent of the room area, and where side wall height is less than 3 feet 6 inches, the area beyond that height shall not be made part of a habitable room.

Rooming house. Any dwelling or that part thereof in which space is let by the owner, operator or occupant of one (1) or more rooming units to three (3) or more persons who are not husband or wife, son or daughter, mother or father, grandparents, grandchildren, sister or brother or niece or nephew of the owner or operator or tenant or his spouse of any of these, but any child lawfully under the care of any of the above members of a family shall not be deemed a roomer.

Rooming house. A building or portion thereof containing lodging rooms which accommodate three (3) or more persons who are not members of the keeper's family, and where lodging is provided for compensation, whether direct or indirect.

Rooming occupancy. The term describes a use wherein a room or group of rooms in a dwelling building are used or intended to be used for sleeping, but not for cooking or eating purposes, provided however, that where one (1) or two (2) roomers are so accommodated within a dwelling unit occupied by a family, the accommodations so provided shall not be deemed to be a rooming occupancy.

Rooming unit. Any room or group of rooms forming a single habitable unit, used or intended to be used, for living and sleeping but not for cooking or eating purposes. A room occupied by a person who is permitted to prepare meals anywhere in a dwelling is a family unit and not a rooming unit unless it is an institutional unit defined in the Code or in a convent, monastery, a bona fide not-for-profit club, or in a dormitory, fraternity or sorority affiliated with an educational or (other) charitable institution.

Rooming unit. Any room, or group of rooms, forming a single habitable unit used for living and sleeping but which does not contain cooking or eating facilities.

Rooming unit. Any "habitable room" or group of "habitable rooms" forming a single habitable unit not used or intended to be used for cooking or eating purposes.

Rooming unit. Suite of rooms forming a single rental unit used or intended to be used for residential or transient purposes, but not including cooking facilities of any kind; wherein a rooming unit is designed to be divided into separate guest rooms then each separate sleeping room shall be counted as a guest room for density purposes.

Rooming unit. Any room or group of rooms forming a single habitable unit, used or intended to be used, for living and sleeping but not for cooking or eating purposes. A room occupied by a person who is permitted to prepare meals anywhere in a dwelling is a family unit and not a rooming unit.

Row house. A row house is a place of abode not more than 2 stories in height, arranged to accommodate 3 or more attached row living units in which each living unit is separated from the adjoining unit by an unpierced vertical occupancy separation of not less than one-hour fire-resistive construction, extending from the basement or lowest floor to the under side of the roof boards.

Row houses. Attached one-family and two-family residential structures, in groups of not less than 3 and not more than 12, erected in a row as a single building, on adjoining lots, each being separated from the adjoining unit or dwelling by a masonry party wall extending from the basement or cellar through the roof, with separate entrances for each unit or dwelling, which entrances shall face the same street.

Sanitation facilities. Water closet, lavatory or sink, and shower or bathtub.

Semi-detached building. Building which has only one (1) party wall in common with an adjacent building.

Semi-detached house. Surrounded on three (3) sides by yard area and so constructed that one wall is on a side lot line, and abuts the neighboring house.

Senior citizens housing. Dwellings, including multiple dwellings, owned and operated by an educational, religious or philanthropic organization, no part of the earnings of which inures to the benefit of any private shareholder, contributor or individual, having accommodations for and occupied exclusively by persons who are sixty-two (62) years of age or over.

Servants quarters. Accessory building located on the same lot or grounds with the main building, and used as living quarters for servants employed on the premises and not rented or otherwise used as a separate domicile.

Servant's quarters. As accessory to multiple-family occupancies means accommodations for such number of servants and other employees as are required by the main occupancy and which accommodations may be detached and may or may not include separate cooking facilities.

Sleeping accommodations. Room, space, or portion thereof, used primarily for sleeping purposes.

Sleeping room. Room rented for sleeping or living quarters, but without cooking facilities and with or without an individual bathroom. In a suite of rooms not part of a dwelling unit, each room which provides sleeping accommodations is construed as one sleeping room.

Sleeping room. A room designed for sleeping use and shall have a floor area of not less than 120 square feet. Where more than one bedroom is provided, two bedrooms shall have not less than 120 square feet each, the third shall have not less than 100 square feet, and the other bedrooms shall have not less than 80 square feet of floor area.

Student cooperative housing. Facility used for housing students who therein largely perform their own household maintenance and meal preparation and who have a vote in the management of their household affairs. Such housing is associated with and recognized by the University and supervised by it in relation to membership, resident supervision, and operations, in the same fashion as fraternities and sororities are supervised.

Tenement house. Any house or building or portion thereof which is rented, leased, let or hired out to be occupied or is occupied as the home or residence of three (3) or more families, living independently of each other and doing their cooking upon the premises.

Toilet room. Room containing one (1) or more waterclosets; and may also contain one (1) or more lavatories, urinals and other plumbing fixtures.

Townhouse. A single-family dwelling unit constructed in a series or group of attached units with property lines separating such units.

Townhouse. A single-family dwelling forming one (1) of a group or series of three (3) or more attached single-family dwellings separated from one another by party walls without doors, windows or other provisions for human passage or visibility through such walls from basement to roof, and having roofs which may extend from one of the dwelling units to another.

Two family dwelling. A building containing two (2) dwelling units with not more than five (5) lodgers or boarders per family, but not more than twenty (20) individuals.

Two family house. Two (2) family house is a building which is rented, leased, let or hired out to be occupied or is occupied or is intended, arranged or designed to be occupied as the home or residence of not more than two (2) families. All such buildings whether built singly, or in conjunction with others as double houses or terraces or attached or semi-detached rows, shall be deemed two (2) family houses when each such house complies with the definition of "Building" of this definition.

Unit. One or more rooms arranged for the use of one or more individuals living together as a single housekeeping unit, with cooking, living, sanitary and sleeping facilities.

C-14 INDOOR ASSEMBLY FOR RECREATION AND ENTERTAINMENT

Adult live entertainment establishment. An establishment which features dancers, go-go dancers, exotic dancers, strippers, or other similar entertainers, any of whom perform topless or bottomless.

Assembly hall. Auditorium in which the public assembles for the purpose of amusement, entertainment, instruction, worship, transportation, sports, military drills, or similar purposes, with admission either public or restricted.

Ballroom. Room in a hotel or other building ordinarily used for gatherings of people usually with common social or fraternal or business interests, or for other related functions such as dancing, dining, conferences, seminars or similar affairs or events.

Banquet hall. Means any building, or portion of building, which is used as a place wherein the public may assemble for the purpose of attending a banquet.

Bingo parlor. Any room or space in which the game of chance known as "bingo," or any similar game of chance or competition, is conducted and in which the general public is permitted to be present and may participate.

Bingo parlor. Any room or space in which the game of chance known as "bingo," or any similar game of chance or competition, in which the general public participates or at which the general public is permitted to be present, is conducted or permitted.

Cabaret. The term shall mean a place of business other than a "night club" located in a hotel or motel having fifty or more guest rooms, where liquor, beer or wine is sold, given away or consumed on the premises, and where music or other entertainment is permitted or provided for the guest of said hotel or motel only.

Commercial recreation facilities. Operated as a business and open to the general public for a fee.

Dance hall. Any room or space in which the greater part of the floor area is reserved for general dancing.

Dance hall. The term "dance hall" shall mean any room, place or space in which dancing is carried on and to which the public may gain admission, either with or without the payment of a fee.

Dance hall. Any building, or a portion of which, used as a place for dancing, whether or not the public is admitted; the term shall include ballrooms but this shall not be interpreted to mean that dance halls or ballrooms within a residence shall be within the scope of the Code.

Dance hall. Any room, place or space in a building in which the requirements of the code have been complied with, can be used for public dancing, either with or without the payment for admission.

Dance hall. A building or part thereof the primary purpose of which is for dancing by a gathering of people.

Dancing. Shall not apply to exhibitions or performances where persons paying admission do not participate.

Entertainment. An engaging or diverting presentation of, or participation in, including but not limited to, live singing, dancing, musical instrumentation, dramatic, prosaic, or poetic activities.

Family game center. A supervised indoor amusement and recreation facility, the main use of which consists of mechanical, electronic, and electrical devices for entertainment and amusement, and which may include but not be limited to such facilities as card games, ping pong tables, and billiard tables.

Function room. Special room which is used primarily for private parties, banquets, and meetings; also called Banquet Room.

Gymnasium. Any building or a portion of which, is used for athletic exercises or games.

Lodge hall. Any building or a portion of which, used as an assembling place for conducting the business of a social, fraternal, or secret organization.

Pool or billiard parlor. An establishment the primary purpose of which is to provide a place of public assemblage in which the game commonly known as pool or billiards or games of a similar nature are played.

Public dance. Any dance to which admission can be had either with or without payment of a fee, wardrobe charge or otherwise or any other dance in which the public generally may participate.

Public dance hall. Any room, space or place in which dancing is carried on and to which admission can be had either with or without the payment of a fee, wardrobe charge or otherwise.

Public hall. Any building or a part of which, where the public may assemble for the purpose of entertainment, instruction, to witness athletic contests, or for other similar purposes.

Public recreation. Publicly owned or operated recreation facilities.

Public recreation facilities. Operated as a nonprofit enterprise by the Township or any other governmental entity or any non-profit organization and open to the general public.

C-15 INDUSTRIAL BUILDINGS AND STRUCTURES

Assembly. Joining together of completely fabricated parts to create a finished product.

Bakery. Include establishments which manufacture quantities of goods for retail elsewhere than on the premises.

Bakery. Building or part of a building wherein is carried on the production, packing, or storing of bread, cake, pies, or other bakery products including any separate room used for the convenience or accommodation of the workers. No retail sales are conducted.

Basic manufacture. First operation or operations which transform a material from its raw state to a form suitable for fabrication.

Brewery. Building where brewing is carried on in the preparation of an alcoholic beverage.

Cosmetic plant or establishment. Any place, premises, building, part of building, cellar or basement, apartment or room occupied or used therein for the having, holding, collection, handling, production, processing, mixing, compounding, manufacture, packing, storage, distribution or sale of articles used (for or as cosmetics) for or intended to be rubbed, poured, sprinkled or sprayed on, introduced into, or otherwise applied to the human body or any part thereof for cleansing, beautifying, promoting attractiveness or altering the appearance.

Creamery. Building where butter or cheese is made, or where milk or cream is processed.

Distillery. Building where distilling of alcoholic liquors is carried on.

Drug plant or establishment. Any place, premises, building, part of building, cellar or basement, apartment or room occupied or used therein for the having, holding, collection, handling, production, processing, mixing, compounding, manufacture, packing, storage, distribution or sale of medical drugs.

Factory. A building, the primary use of which is for the processing of materials or manufacture of products.

Food plant or establishment. Any place, premises, building, part of a building, cellar or basement, apartment or room occupied or used therein for the having, holding, collection, handling, production, preparation, processing, manufacture, packing, storage, distribution or sale of articles used for food and drink intended for consumption by man or other animal and shall include any bakery, confectionery, cannery, packing house, slaughterhouse, dairy, creamery, milk plant, cheese factory, restaurant, hotel, grocery, meat market, delicatessen, bottling or nonalcoholic beverage plant, pickle or pickling plant, candy manufacturing plant, ice cream plant, oyster shucking establishment or egg breaking establishment.

Hazardous manufacturing processes or storage. Includes but is not limited to such processed products as matches, artificial flowers, cork, cotton, carpet linings, cotton rag sorting rooms, brooms, mattresses, rubber, feathers, paper, pasteboard, shoddy mills, cereal, flour, grist and starch mills, sugar refineries, oil refineries, distilleries, rendering plants, drying rooms and other occupancies of equal fire and life hazard.

Hazardous processes. Processes which cause and produce dust, lint, or other airborne particles or matter liable to ignition and explosion.

Hazardous-use buildings. Building, or parts thereof, used for the manufacturing, processing, storage or use of materials which are, highly combustible, inflammable or explosive, highly corrosive, toxic or noxious alkalies, acids, fume hazardous, poisonous, irritant or corrosive gases, dust and fine particles and chemicals as listed and in the code.

Hazardous use general units. Any hazardous use unit other than a hazardous use storage unit and other than a hazardous use industrial unit.

Hazardous use industrial units. Any hazardous use unit designed, intended or used for industrial purposes, including any operation or process incident to the producing, fabricating, assembling, developing, moulding, pressing, preparing or adapting for use, repairing or refinishing of any high hazard material, high hazard product, article or substance or high hazard parts or appliances of any product or article.

High hazard buildings. All buildings and structures or parts thereof shall be classified in the high hazard use group which are used for the storage, manufacture or processing of highly combustible or explosive products or materials which are likely to burn with extreme rapidity or which may produce poisonous fumes or explosions; for storage or manufacturing which involves highly corrosive, toxic or noxious alkalies, acids or other liquids or chemicals producing flame, fume, explosive, poisonous, irritant or corrosive gases; and for the storage or processing of any materials producing explosive mixtures of dust or which result in the division of matter into fine particles subject to spontaneous ignition.

High hazard group. All buildings, or portions of buildings which are designed or used for storage, manufacture, processing or sale of highly combustible, flammable or explosive products which may burn with extreme rapidity or from which poisonous fumes or toxic gases may be emitted, or for the manufacture, storage or sale of highly corrosive, toxic or noxious alkalies, acids or other liquids or chemicals shall be classified in the high hazard use group. Motor vehicle service stations shall not be included in the high hazard classification.

High-hazard industrial buildings. Include industrial buildings whose use or occupancy involves substances or products which are volatile, flammable, or explosive, or which are hazardous when exposed to moisture, or which burn with extreme rapidity, or which when burning or subjected to heat, produce toxic fumes or gases in quantities and under conditions dangerous to the safety or health of any person. Such substances and products include those having the fire-hazard characteristics of any of those listed in the Code.

Industrial. Any building or structure that is used for the primary purposes of manufacturing or any related uses.

Industrial. Buildings in which the primary or intended occupancy or use is the manufacture or processing of products of all kinds, including operations such as making altering assembling, bottling, canning, finishing, handling, mixing, packaging, repairing, cleaning, laundering and similar operations.

Industrial building. A building, or parts thereof, used primarily for manufacturing or in which five (5) or more persons are engaged in fabricating, assembling, or processing of products or materials, except hazardous uses.

Industrial building. Any structure used for manufacturing purposes.

Industrial buildings. Buildings in which work is performed in the fabricating, assembly, manufacture, processing, repair, or rebuilding of materials, assemblies, or products. Industrial buildings do not include buildings or parts of buildings used for trade or custom work, or service customarily performed for the residents of the locality, provided that not more than 5 skilled craftsmen are engaged on the premises.

Industrial buildings. Include bakeries, breweries, creameries, distilleries, electric substation buildings, factories, laundries, mills, plants, pumping stations, slaughterhouses, smokehouses, workshops, and buildings of similar use.

Industrial buildings. Buildings used in connection with production or process work or with storage or warehousing.

Industrial classification. Includes buildings or parts thereof, in which 5 or more persons are engaged in the fabrication, assembly, manufacturing, or processing of products or materials, except that when such products, materials, or those involved in their manufacture are highly combustible or explosive, the occupancy shall comply with the requirements for hazardous occupancies when the latter are more restrictive than the corresponding requirements for the industrial classification. Where less than 5 persons are engaged in work as described herein, the occupancy shall be in the business classification.

Industrial establishment. Building or part of a building (other than office or exhibit space) where persons are employed in manufacturing processes or in the handling of material, as distinguished from dwellings, offices, and like occupancies.

Industrialized unit. Assembly of materials or products comprising all or part of a total structure which, when constructed, is self-sufficient, or substantially self-sufficient and when installed constitutes the structure or a part of the structure, except for preparation for its placement.

Industrial occupancy. Applies to an entire building or to that portion of a building used for manufacturing, processing, or storage of materials or products, including among others, chemical, food, candy, and ice cream factories, ice making plants, meat packing plants, refineries, perishable food warehouses and similar occupancies, provided the entire building is occupied by a single tenant.

Industrial occupancy. Occupancy or use of a building or structure or any portion thereof for assembling, fabricating, finishing, manufacturing packaging or processing operations; except when classed as a high hazard occupancy.

Industrial occupancy. Occupancy or use of a building or part thereof for assembling fabricating, manufacturing, processing, repairing or storing of goods and materials.

Industrial occupancy. Buildings in which work or labor is performed in connection with the fabrication, assembly, processing, etc., of products or materials shall be called Industrial Occupancy, this shall include, among others, the occupancies lited in the Code, but does not include buildings used for any purpose involving highly combustible, inflammable, or explosive products or materials.

Industrial occupancy. Includes all buildings and parts of buildings and in which work or labor is performed in connection with the fabrication, assembly, manufacturing, repairing, cleaning, assorting, packaging, or processing of products or materials not included under High Hazard Occupancy and which are neither highly combustible or flammable, nor explosive.

Industrial Park. Tract of land subdivided and developed according to a comprehensive development plan, in a manner which provides a parklike setting for industrial establishments.

Industry. Storage repair, manufacture, preparation or treatment of any article, substance or any commodity for commercial use.

Industry. Any activity involving the manufacturing or treatment of any commodity including the assembly, packaging, canning, bottling, or processing of any item. To change any commodity in composition, form, size, shape, texture, or appearance is deemed to be an industrial process.

Industry. Manufacturing activity.

Industry. The manufacture, fabrication, processing, reduction or destruction of any article, substance or commodity, or any other treatment thereof, in such a manner as to change the form, character or appearance thereof, including, but not limited to the following: Animal hospitals, bakeries, bottling plants, building or contractors' yards, cleaning and dyeing establishments, creameries, dog pounds, junkyards, laundries, lumberyards, milk bottling and distributing stations, stockyards, storage elevators, truck storage yards, warehouses, wholesale storage and other similar types of enterprise.

Laundry. Building which is used for the washing of clothes or other articles of cloth, or for ironing in connection therewith, except where such work is performed as a home occupation and there are no employees to do laundering of articles for those who are not members of the household.

Light manufacturing. Manufacturing or processing of materials employing electricity or other unobjectionable motive power, utilizing hand labor or unobjectionable machinery or processes, and free from any objectionable odors, fumes, dirt, vibration, or noise.

Light service industries. Include bakeries, creameries, soft drink bottling plants, cleaning and dying plants, laundries, printing and engraving plants, and any other plant of a similar nature producing a product or providing a service.

Low-hazard industrial buildings. Include Industrial Buildings in which the use and occupancy involve materials or substances which create not more than a "low fire hazard" as defined in the Code.

Machinery. Any mechanical apparatus or contrivance or parts thereof used in production, processing, preparation, compounding, mixing, bottling, manufacture, packing, storage or distribution of food, drugs, device or cosmetics.

Major occupancy. Principal occupancy for which a building or part thereof is used, and shall be deemed to include the subsidiary occupancies which are an integral part of the principal occupancy.

Manufacture. All operations required to produce a product.

Manufacturer and/or manufacturing. Processing and/or converting of raw, unfinished or finished materials, or products or any, or either of them, into an article or articles or substance of different character, or for use for a different purpose; industries furnishing labor in the case of manufacturing or the refinishing of manufactured articles.

Manufacturing. The assembly, fabrication, treatment, processing, rebuilding, blending, or molding of materials into a finished product.

Mill. Building which is used to house machinery for the production of a product by continuous repetition of substantially the same action.

Mixed occupancy. Term applied to a building occupied or used for different purposes in different parts of the building.

Mixed occupancy. When a building is used for 2 or more occupancies, classified within different occupancy groups, it shall be considered a mixed occupancy.

Mixed occupancy. Occupancy of a building or land for more than one use.

Mixed occupancy. Occupancy of a building in part for residential use and in part for some other use not accessory thereto.

Mixed occupancy. Occupancy of a building in part as the living quarters for one (1) or more families and in part for some other use not accessory thereto. For the purpose of determining off-street parking requirements, that part of such buildings used as living quarters shall be considered as dwellings.

Mixed occupancy. Shall apply to a building occupied or used for different purposes in different parts. When the occupancies are cut off from the rest of the building by tight partitions, floors, and ceilings and protected by self-closing doors, the requirements for each type of occupancy shall apply for its portion of the building. For example, the cold storage spaces in retail frozen food lockers, hotels and department stores, in buildings occupied by a single tenant might be classified under Industrial occupancy, whereas other portions of the building would be classified under other occupancies. When the occupancies are not so separated, the occupancy carrying the more stringent requirements shall govern.

Moderate-hazard industrial buildings. Include Industrial Buildings which are not low-hazard buildings, but in which the use and occupancy involve uses, materials, or substances which create not more than a "moderate fire hazard" as defined in the Code.

Non-relocatable structure. Factory built unit for use on permanent foundations.

Plant. As applied to an industrial enterprise, means the buildings and grounds contiguously located with respect to each other comprising an industrial enterprise under one ownership.

Plant. As applied to a building, means a building used primarily to house equipment for the production, conversion, or distribution of energy or for the manufacture or production of a particular product.

Retail manufacturing. Baking, confectioner, dressmaking, dyeing, laundry, dry cleaning, printing, tailoring, upholstering, and similar establishments.

Slaughterhouse. Building used for the slaughtering of animals and the scalding, dressing, butchering and storage of carcasses for human consumption, but not including the rendering, smoking, curing or other processing of meat, fat, bones, offal, blood, or other by-products of the permitted operations.

Slaughterhouse. Any establishment in which animals or fowls are killed, or parts thereof are dressed or prepared as meats for wholesale distribution for human consumption.

Smokehouse. Building used to cure meat or fish with smoke.

Smokehouse. Building, or part of a building, designed, intended or used for no purpose other than that of smoking meats or fish.

Special industrial buildings. Include low or moderate-hazard industrial buildings which are designed and suitable only for types of operations characterized by a relatively low density of employee population. Special industrial buildings include, but are not limited to the following: Electric Substation Buildings, Power Plants, Pumping Stations, Steel Mills, and Portland Cement Plants.

Wholesale. The sale of tangible personal property for resale by a licensed retailer and not the sale of tangible personal property for consumption by the purchaser.

Wholesaling. The selling of goods or merchandise to retailers or jobbers for resale to the ultimate consumer.

Workshop. Building used primarily for the work of one trade, or for work of a specific kind, except as regulated by the Code.

C-16 MOBILE HOMES AND TRAILERS

Accessory structure. Any structural addition to the mobile home which includes awnings, cabanas, carports, Florida rooms, porches, ramadas, storage cabinets and similar appurtenant structures.

Access to park streets. Each mobile home lot within a mobile home park should have direct access to a park street. The access should be an unobstructed area.

Air-conditioning or comfort-cooling equipment. All of that equipment intended or installed for the purpose of processing the treatment of air so as to control simultaneously its temperature, humidity, cleanliness and distribution to meet the requirements of the conditioned space.

Area of house trailer spaces. Based upon the gross area of the park, the number of individual unit spaces shall be not more than ten (10) per gross acre. The minimum area of any space for a house trailer shall be not less than three thousand (3,000) square feet with no dimension less than forty (40) feet. No such space shall be located less than twenty-five (25) feet from street lot lines or interior lot lines. House trailers shall be located on each space so that there will not be less than fifteen (15) feet to any other house trailer or building within the park.

Automobile camp. Land or premises used or occupied, for compensation, by campers traveling by automobile or otherwise, or occupied by trailer coaches or movable dwellings, rooms or sleeping quarters of any kind.

Automobile trailer. A vehicle without motive power, designed to be drawn by a motor vehicle and to be used for human habitation or for carrying persons and property, including a trailer coach or house trailer.

Awning. Shade structure supported by posts or columns and partially supported by a mobile home installed, erected, or used on a mobile home lot.

Awning window. Shade structure supported wholly by the mobile home or vehicle to which it is attached.

Boat trailer. A vehicular structure, without its own motive power designed to transport a recreational vessel for recreational and vacation use, and which is eligible to be licensed or registered and insured for highway use.

Branch-circuit panelboard. Necessary equipment, consisting of circuit-breakers or switch and fuses and their accessories, intended to constitute the main control and means of cutoff for the supply to the mobile home.

Building. Roofed structure erected for permanent use.

Building. Building is any permanent structure built for the support, shelter, or enclosure of persons, animals, chattels, or property of any kind.

Cabana. Any portable demountable, or permanent cabin, small house, room, enclosure, or other building or structure erected, constructed or placed on any trailer site within six feet (6′) of any trailer coach on the same site in a trailer park and used for human habitation. Cabana does not include awning or a private toilet and bath constructed in accordance with the provisions of the ordinance and other County ordinances.

Cabana. Cabana is any portable, demountable or permanent cabin, small house, room, enclosure, or other building erected, constructed or place to be used in conjunction with any automobile house trailer or camp car.

Camp car. Includes "MOTOR HOME" and includes motor vehicles designed for temporary human habitation for travel or recreational purposes, and includes trucks on which a camping facility has been installed either permanently or temporarily.

Camp car. Vehicle with or without motive power, which is designed or used for human habitation.

Camper trailer. A folding or collapsible vehicular structure, without its own motive power, designed as temporary living quarters for travel, camping, recreation and vacation use, and eligible to be licensed or registered and insured for highway use.

Camp ground. Any area or tract of land used to accommodate two (2) or more camping parties, including cabins, tents, house trailers, or other camping outfits.

Camping area. Any area or tract of land where one or more lots are rented or leased or held out for rent or lease to accommodate camping parties.

Camping party. Person or group of not more than 10 persons occupying a campsite or lot for not more than 30 days annually.

Camping party. One or more persons occupying a camp site or lot.

Camping trailer. A folding unit of canvas or other materials, mounted on wheels, which opens over the top of the trailer base.

Camping trailer. Vehicular portable unit mounted on wheels and constructed with collapsible partial side walls which fold for towing by another vehicle and unfold at the campsite to provide temporary living quarters for recreational, camping, or travel use.

Campsite. Area within an incidental camping area occupied by a camping party.

Camp site. Parcel of land occupied or intended for occupancy by only one of the following: tent, tent trailer, pickup camper, or camping trailer.

Carport. Awning or shade structure for a vehicle or vehicles which may be free-standing or partially supported by a mobile home.

Chassis-mount camper. A portable structure designed to be affixed to a truck chassis, and constructed to provide temporary living quarters for recreational, travel or camping use.

Chassis mounted trailer. An accommodation body that attaches onto a truck chassis and is not intended for removal.

Circulating air supply. Circulating air supply taken from outside the recreational vehicle, or from the conditioned area inside the mobilehome, or from both sources. Every air-conditioning or comfort-cooling system designed to replace required natural ventilation, designed to take not less than one-fifth (1/5) of the air supply from outside the recreational vehicle.

Combination compartment. Shower stall or recess that provides for a toilet and is of such size and proportions that it may not be occupied by more than one (1) person at one time.

Commercial coach. Vehicle, with or without motive power, designed and equipped for human occupancy for industrial, professional, or commercial purposes, and includes a trailer coach.

Common area. Any area or space designed for joint use of tenants occupying mobile home developments.

Community management. Person who owns or has charge, care or control of the Mobilehome Development.

Community system (water or sewerage). Central system which serves all living units and is not publicly owned.

Converted house trailer. Trailer rendered immobile and placed on either a temporary or permanent foundation for use as sleeping or living quarters (but still considered as a mobile home).

Density. Number of mobile homes or mobile home stands per gross acre.

Density. Density of mobile homes shall be regulated by mobile home lot requirements and separation requirements established by local authorities.

Dependent mobilehome. One not equipped with a toilet for sewage disposal.

Dependent trailer coach. A trailer coach which does not have a flush toilet and bath and shower.

Dependent unit. A recreational vehicle which requires connection or access to external sources of electricity, water and sewage systems.

Diagonal tie. Any tiedown designed to resist horizontal or shear forces and which deviates not less than 30 degrees from a vertical direction.

Drainage lateral. That portion of the mobile homes park drainage system extending to an individual mobile home lot drain inlet.

Drainage system. All the piping within or attached to the vehicle structure that conveys sewage or other liquid wastes to the drain outlet.

Drain connector. Removable extension, consisting of all pipes, fittings and appurtenances, from the drain outlet to the drain inlet serving the mobilchome.

Drain inlet. Terminal of the park drainage lateral at a mobilehome lot to which a mobilehome drain connector is attached.

Drain outlet. Discharge end of the vehicle main drain to which a drain connector may be attached.

Driveway. Minor private way used by vehicles and pedestrians on a Mobilehome Lot or for common access to a small group of lots or common facilities.

Dry camp. Camping area where a public supply of potable water is unavailable within the camping area.

Easement. Vested or acquired right to use land, other than as a tenant, for a specific purpose; such rights being held by someone other than the owner who owns title to the land.

Electrical distribution system. Electrical wiring system consisting of wiring, fixtures, equipment, and appurtences installed and maintained in accordance with applicable State and City electrical codes and regulations governing such systems.

Expandable vehicle. Enclosed room, semi-enclosed room, or roofed porch which expands outward from the basic vehicle by means of rollers, hinges, or other devices or arrangements, but is designed as a structural portion of the vehicle, and is carried on or within the vehicle while traveling on the highway.

Exterior wall. Wall or element of a wall which defines the exterior boundaries of a mobilehome.

Feeder. Circuit conductors between the mobile park service, or the generator switchboard of an isolated plant, and the branch circuit overcurrent protective devices.

Feeder. That part of the electric distribution system from the service entrance equipment to the mobile home.

Feeder assembly. Overhead or undervehicle feeder conductors, including the grounding conductor together with the necessary fittings and equipment or a cord set listed for mobile homes designed for the purpose of delivering energy from the source of electrical supply to the branch-circuit panelboard within the mobile home.

Fence. Vertical structure designed and erected as a free-standing unit, the surface of which is more than 50 percent open.

Fire extinguishers in auto and trailer camps. Installed and maintained in good repair, 3/4-inch standpipes connected to sufficiently large water mains, to which shall be permanently affixed good grade garden hose, not greater than 100 feet in length with a Biston type garden hose nozzle or its equivalent attached. Such hose shall be enclosed in a cabinet approved by the Fire Department and plainly marked "FOR FIRE ONLY." A sufficient number of these stand-pipe units shall be installed in each auto and trailer camp so that at least one nozzle will reach any building, trailer, or trailer site.

Fire protection. Portable fire extinguishers of a type approved by the Fire Department shall be provided in service buildings and at all other locations designated by the Fire Chief and maintained in good operating condition. Locations of fire extinguishers and fire safety rules and regulations shall be posted in conspicuous places.

Fire protection. Every park equipped at all times with fire extinguishing equipment in good working order, of such type, size and number and so located within the park as to satisfy applicable reasonable regulations of the Fire Department. No open fires permitted at any place which may endanger life or property. No fires left unattended at any time.

Floor area. Area included within the surrounding exterior walls of a mobilehome or portion thereof.

Folding tent trailer. Canvas folding structure, mounted on wheels and designed for travel and vacation use.

Free standing awning. Shade structure supported entirely by columns or posts and not attached to or supported by mobile home or vehicle.

Garbage receptacles. Metal garbage cans with tight-fitting covers shall be provided in quantities adequate to permit disposal of all garbage and rubbish. Garbage cans shall be located on the mobile home space. The cans shall be kept in sanitary condition at all times. Garbage and rubbish shall be collected and disposed of as frequently as may be necessary to insure cleanliness in the area.

Gas piping. All piping and tubing installed in or attached to a vehicle to convey fuel gas to fuel-burning, heat-producing appliances.

Gas supply connector. Detachable flexible connector designed to convey fuel gas from the gas supply to the vehicle gas supply connection.

Ground anchors. Installed at each mobile home stand when a mobile home is located thereon to permit tiedowns of mobile homes to the ground.

Group model. Two or more manufacturer designed vehicles which constitute one model.

Health-safety and general welfare. There shall be provided in each mobile home court such other improvements as the Township may require whereby such requirements shall at all times be in the best interests of the public's health, safety and general welfare and may include, but shall not be limited to, garbage and trash collection and disposal facilities as approved by the State Department of Health, and adequate park lighting system, as required by the State Department of Safety.

Horse trailer. A vehicular structure, without its own motive power, not exceeding twenty (20) feet in length, eight (8) feet in width, and ten and one-half (10-1/2) feet in height, designed and/or used primarily for the transportation of horses and which, in combination with the carrying vehicle, is eligible to be licensed or registered and insured for highway use.

House trailer. Includes any structure intended for or capable of human habitation, mounted upon wheels and capable of being driven, propelled or towed from place to place without change in structure or design, by whatsoever name or title it is colloquially or commercially known; provided that this definition shall not include transport trucks or vans equipped with sleeping space for a driver or drivers. Removal of wheels after arrival of such structure at its destination shall not disqualify it from this definition. It shall also include the terms for mobile homes as applied in the ordinance.

House trailer. Nonself-propelled vehicle containing living or sleeping accommodations, which is designed and used for highway travel.

House trailer. A living unit less than forty (40) feet in length or less than ten (10) feet in width.

House trailer. A vehicle without motive power, designed to be drawn by a motor vehicle and used for human habitation and/or the carrying of property.

Housing. Living units, dwellings and/or other structures that shelter or cover.

Hurricane zones. Not more than 12 feet on centers beginning from the front line of the mobile home stand (congruent with the front wall of the mobile home). Not more than 6 feet open end spacing provided at the rear line of the mobile home stand unless additional tiedowns are installed.

Incidental camping area. Any area or tract of land where camping is incidental to the primary use of the land for agriculture, timber management, or water or power development purposes, and where one or more campsites used for camping are rented or leased or held out for rent or lease. The density of usage shall not exceed 25 camping parties within a radius of 265 feet from any campsite within the incidental camping area.

Independent mobile home. One equipped with a toilet for sewage disposal.

Independent trailer coach. A trailer coach which has a flush toilet and a bath or shower.

Independent unit. A recreational vehicle which may operate independently of connection or access to external sources of electricity, water and sewage systems for a period of from one to seven days. Such a unit has its own battery and/or gas system to operate lights, refrigerator, stove and heater, a large water tank with a pressure system, and a holding tank with a toilet.

Interior finish. Surface material of walls, fixed or movable partitions, ceilings and other exposed interior surfaces affixed to the mobile home structure including any material such as paint or wallpaper. Interior finish does not include decorations or furnishings which are not affixed to the mobile home structure.

Length of vehicles. Distance measured from the extreme front to the extreme rear of a vehicle (including expandable rooms, hitch, coupling, tongue, and other attachments).

License. Written document issued by the enforcing agency allowing a person to operate and maintain a mobile home development under the provisions of this regulation.

Listed. Equipment or materials included in a list published by a nationally recognized testing laboratory that maintains periodic inspection of production of listed equipment or materials and whose listing states either that the equipment or material meets nationally recognized standards or has been tested and found suitable for use in a specified manner.

Live load. Weight superimposed by the use and occupancy of the mobile home including wind load and snow load, but not including dead load.

Living unit. Residential unit providing complete, independent living facilities for one family including permanent provisions for living, sleeping, eating, cooking and sanitation.

Longitudinal center. Midpoint betwen the right and left side of the vehicle unit containing the utility connections.

Lot. Lot is any area or tract of land or portion of a mobile home park, travel trailer park, recreational trailer park, temporary trailer park, or tent camp designated or used for the occupancy of one mobile home, travel trailer, camp car, or camping party, and shall include a site.

Lot area. Total area reserved for exclusive use of the occupants of a mobile home.

Lot line. Line bounding the Lot as shown on the accepted plot plan.

Low-voltage circuits. Furnished and installed by the recreational vehicle manufacturer, other than those related to braking, are subject to the Code. Circuits supplying lights subject to Federal or State regulations in accordance with applicable government regulations, but not lower than provided by the Code.

Manufacturer. Person engaged in the business of manufacturing or constructing mobile home accessory buildings or structures.

Mobile home. Transportable, single-family dwelling unit suitable for year-round occupancy and containing the same type of water supply, waste disposal, and electrical conveniences as immobile housing.

Mobile home. Manufactured relocatable residential unit providing complete, independent living facilities for one family including permanent provisions for living, sleeping, eating, cooking and sanitation.

Mobile home. Detached residential dwelling unit designed for transportation, after fabrication, on streets or highways on its own wheels or on flatbed or other trailer, and arriving at the site where it is to be occupied as a dwelling complete and ready for occupancy, except for assembly operations, location on jacks or other temporary or permanent foundations, connections to utilities, etc. A travel trailer is not to be considered as a mobile home.

Mobile home. Factory-assembled structure or structures equipped with the necessary service connections and made so as to be readily movable as a unit or units on its (their) own running gear and designed to be used as a dwelling unit(s) without a permanent foundation.

Mobile home. Vehicle designed and equipped for human habitation and for being drawn by a motor vehicle.

Mobile home. Any vehicle or similar portable structure having been constructed with wheels whether or not such wheels have been removed and having no foundation other than wheels, jacks or skirtings and so designed or constructed as to permit occupancy for dwelling or sleeping purposes.

Mobile home. Vehicular, portable structure built on a chassis and designed to be used without a permanent foundation as a dwelling when connected to indicated utilities.

Mobile home. One-family dwelling unit of vehicular portable design, built on a chassis and designed to be moved from one site to another and to be used without a permanent foundation.

Mobile home. Structure that is permanently equipped with the necessary axles, wheels, tires, brakes and lamps or reflectors to permit it to be moved about on a highway and is used, or is intended to be used, as a year-round shelter or dwelling place for persons when it is furnished and connected to appropriate sewer, water and electrical services.

Mobile home. Vehicle with or without its own motive power, equipped for or used for living purposes and mounted on wheels or designed to be so mounted and transported. Mobile homes shall include vehicles designed for temporary dwellings primarily for, travel, recreational and vacation uses and such temporary dwelling vehicles shall be known as travel coaches.

Mobile home. Single family dwelling designed for transportation after fabrication on streets and highways on its own wheels or a flatbed or other trailer, and arriving at the site where it is to be occupied as a dwelling complete and ready for occupancy, except for minor and incidental unpacking and assembly operations, location on jacks or permanent foundations, connection to utilities and the like.

Mobile home. Any vehicle which at any time was used or maintained for use as a conveyance upon highways or public streets, or waterways; so designed and so constructed as to permit occupancy thereof as a dwelling unit or sleeping place for one or more persons whether attached or unattached to a permanent foundation. Nothing in the ordinance shall be construed as permitting a mobile home in other than an approved mobile home park.

Mobile home. Detached single-family dwelling unit with all the following characteristics: (a) designed for full time occupancy and containing sleeping accommodations, flush toilet, tub or shower bath, and kitchen facilities, with plumbing and electrical connections provided for attachment to outside systems; (b) designed to be transported after fabrication on its own wheels or on flatbed or other trailers of detachable wheels; or by other means; (c) arriving at the site where it is to be occupied as a dwelling complete and ready for occupancy except for minor and incidental unpacking and assembly operations, location on and connection to foundation supports, connection to utilities and the like. Anchorage must comply with design load requirements of the building code.

Mobile home. Vehicular portable home designed and built for long-term residential occupancy. A mobile home shall be a dwelling unit for density and parking requirement purposes, but shall not be considered a single or multiple dwelling, apartment, efficiency apartment, rooming unit or guest room. A mobile home shall be larger than eight (8) feet in width and thirty-two (32) feet in length, and shall be classified as a mobile home whether or not its wheels, rollers, skids or other rolling equipment have been removed, or whether or not any addition thereto has been built on the ground and contains sleeping accommodations, a flush toilet, a tub, or shower, and kitchen facilities with plumbing and electrical connections provided for attachment to outside systems.

Mobile home. Manufactured relocatable living unit.

Mobile home. Vehicle, portable structure, or part thereof with or without wheels, designed for over the road movement, and used for permanent living or sleeping quarters.

Mobile home. Movable living unit or similar portable structure having no foundation other than wheels, jacks, or blocks, sometimes referred to as trailers or trailer homes.

Mobile home. Vehicle designed for human habitation and built on a chassis for being drawn upon the highway by a motor vehicle. Must be over eight feet in width or over forty feet in length.

Mobile home. A manufactured single family dwelling not less than forty (40) feet in length nor less than ten (10) feet in width, designed for transportation on streets or highways, on its own wheels or transported by any truck or rail type transportation, to any location ready to be occupied as a living quarter.

Mobile home. Any vehicle used, or so constructed as to permit its use, as a conveyance upon the public streets or highways and duly licensed as such, which shall include self-propelled and non-self propelled vehicles, and so designed, constructed, reconstructed or added to by means of an enclosed addition or room, in such manner as will permit occupancy thereof as a dwelling unit or rooming unit for 1 or more persons and having no foundation other than wheels. The term shall be synonymous with house trailer, automobile trailer, trailer coach and trailer home.

Mobile home accessory building or structure. Building or structure which is an addition to or supplements the facilities provided a mobile home. It is not a self-contained, separate, habitable building or structure. Examples are awnings, cabanas, ramadas, storage structures, carports, fences, windbreaks, or porches.

Mobile home accessory building or structure. Any awning, portable, demountable or permanent cabana, ramada, storage cabinet, carport, fence, windbreak or porch established for the use of the occupant of the mobile home.

Mobile home and commercial coach model. Specific design of vehicle as designated by the manufacturer which is based on width and/or type of construction.

Mobilehome community. Mobile home Development and related qualities and facilities, including the Mobilhomes and all of the people living within the development.

Mobile home condominium ownership. That situation where the lot on which the mobile home stands is owned by the occupant (or his legal authorized tenant) who pays any levied taxes directly to the county, municipality or state taxing body, irrespective of the ownership of the total subdivision in which the "plot" is situated. A monthly maintenance charge is applicable when central recreational facilities are provided. These facilities are ordinarily operated through a management contract with the subdivision developer.

Mobile home court. Any lot or portion thereof housing or intended to house two (2) or more mobile homes.

Mobile home development. Any lot, tract or parcel of land used or offered for use in whole or in part, with or without charge, for the parking of occupied mobile homes.

Mobile home development. Contiguous parcel of land which has been planned and improved for the placement of mobile homes.

Mobile home drainage system. All the piping within or attached to the vehicle structure that conveys sewage or other liquid wastes to the drain outlet.

Mobile home drain connector. Removable extension connecting the mobilehome drain outlet to the drain inlet on the mobilehome lot.

Mobilehome drain outlet. Discharge end of the vehicle drainage system to which the drain connector is attached.

Mobile home dwelling. Moveable or portable dwelling over thirty-two feet in length or over eight (8) feet wide, constructed to be towed on its own chassis and designed to be installed, with or without a permanent foundation for occupancy as a residence. A mobile home may include one or more components that can be retracted for towing and subsequently expanded for additional capacity, or two or more units separately towable, but designed to be joined into one integral unit, as well as a portable dwelling composed of a single unit. A mobile home does not include a recreational vehicle as defined in the Code.

Mobilehome lot. Any area designated, designed, or usable for the occupancy of one mobilehome on a temporary, semi-permanent or permanent basis.

Mobile home lot. That parcel of a mobile home park that provides facilities for long term occupancy of a mobile home and has a minimum size of 6,000 square feet.

Mobile home lot. Designated portion of a mobile home park designed for the accommodation of one mobile home and its accessory buildings or structures for the exclusive use of the occupants.

Mobile home lot. Plot of ground within a mobile home park designed for the accommodation of one (1) mobile home.

Mobile home or house trailer. Vehicular, portable structure built on a chassis whether or not it is designed to be placed upon a permanent foundation.

Mobile home park. Any area or tract of land established by permit, containing one or more mobile home lots, does not include a "recreational trailer park," "temporary trailer park," or "travel trailer park."

Mobile home park. Any lot, tract or parcel of land licenses and used or offered for use in whole or in part, with or without charge, for the parking of occupied mobile homes and travel trailers subject to the Code and used solely for living and/or sleeping purposes. Travel trailers shall not occupy more than fifteen (15) percent of the spaces of the total mobile home park.

Mobile home park. Parcel of land under single ownership which has been planned and improved for the placement of mobile homes for residential use having a density pattern of 8 or 9 mobile home spaces to the acre. Each site is rented to individual owners of mobile homes.

Mobile home park. Parcel of contiguous property that has a minimum size of 12 acres and is platted for the development of a minimum of 50 mobile home lots.

Mobile home park. Parcel (or contiguous parcels) of land which has been so designated and improved that it contains two or more mobile home lots available to the general public for the placement thereon of mobile homes for occupancy.

Mobile home park. Any area or tract of land where one (1) or more mobilehome lots are rented or leased or held out for rent or lease to accommodate mobile homes used for human habitation. The rental paid for any such mobile home shall be deemed to include rental for the lot it occupies.

Mobile home park. Any lot, tract or parcel of land licensed and used or offered for use in whole or in part, with or without charge, for the parking of occupied mobile homes and travel trailers and used solely for living and/or sleeping purposes. Travel trailers shall not occupy more than fifteen (15) percent of the spaces of the total mobile home park at approved locations.

Mobile home park. Tract of land of a minimum of six acres, located on a collector street, designed and developed for the placement of mobile home residential units thereon, and to be used for dwelling purposes only, except as provided herein. Mobile home spaces are to be for rent or lease only, and each mobile home space shall be located on an access easement of a minimum width of thirty (30) feet. No sale or display of mobile homes is permitted in a mobile home park.

Mobile home park. Any plot of ground upon which two or more mobile homes, occupied for dwelling or sleeping purposes, are located, regardless of whether or not a charge is made for such accommodation.

Mobile home park. Tract of land used or intended to be used for the parking of mobile homes together with the necessary improvements and facilities upon the land.

Mobile home park. Lot, parcel, or tract of land used as the site of occupied mobile homes, including any building, structure, or vehicle or enclosure used as a part of the equipment of such mobile home park, and licensed to operate under the provisions of the code.

Mobile home park electrical wiring system. All of the electrical wiring, fixtures, equipment and appurtenances related to electrical installations within a mobile home park, including the mobile home service equipment.

Mobile home park or trailer park. Any lot or portion of a lot upon which two or more mobile homes or trailers occupied for dwelling or sleeping purposes are located regardless of whether or not a charge is made for such accommodations.

Mobile home park service. Conductors and equipment for delivering electrical energy from the electrical supply system, or the generator of an isolated plant, to the wiring system of the mobilehome park.

Mobile home park sewer. That part of the park drainage system beginning from a point two (2) feet downstream from lowest drainage lateral or building drain connection and terminating at the public sewer or private sewage disposal system.

Mobile home park street. Private way which affords principal means of access to abutting individual mobile home lots and auxiliary buildings.

Mobile home park water service outlet. That portion of the park water distributing system, including fittings, devices and appurtenances, provided with a fitting for connecting a mobile home water connector.

Mobile homes. Vehicular portable structures built on a chassis without permanent foundations. They are intended for use as dwelling units, lodging, offices, stores or similar occupancies.

Mobile home service equipment. That equipment containing the disconnecting means, overcurrent protective devices, and receptacles or other means for connecting a mobile home feeder assembly.

Mobile home space. Unit of land, conforming with the requirements of the code within a mobile home development intended to accommodate one mobile home.

Mobile home stand. Part of an individual lot which has been reserved for placement of the mobile home, appurtenant structures, or additions.

Mobile home stand. A plot or section of ground within a mobile home park designed and situated so as to provide for the parking of the mobile home, the necessary open space around the mobile home and the placement of its accessory structures.

Mobile home stand. An adequate foundation for the placement of the mobile home, appurtenant structures, or additions.

Mobile home subdivision. Parcel of land subdivided into lots, each lot individually owned and utilized as the site for placement of a single mobile home and its facilities.

Mobile home subdivision. Subdivision for residential use by mobile homes exclusively.

Mobile home water connector. Removable extension connecting the vehicle water distributing system to the park water service outlet.

Mobile home water distributing system. All the water piping within or attached to the vehicle, including the water supply connection.

Mobile home water supply connection. Fitting or point of connection of the vehicle water distributing system designed for connection to a water connector.

Mobile housing development. Area of land, subdivided into separate plots, each furnished with on-site outlets offering utility services directly to a mobile home intended for placement in accordance with a pre-planned layout and proximity arrangement.

Mobile industrial or commercial structure. Structure not intended as a dwelling unit, towable on its own chassis and for use without a permanent foundation. Such structures are built specifically for commercial or industrial use such as construction offices, bunk houses, wash houses, kitchen and dining units, libraries, TV units, industrial display units, laboratory units and medical clinics.

Mobile office. Structure that contains one or more office units and that is designed to be transportable, either by traction or under its own power. This definition shall apply regardless of whether or not the wheels or other devices for mobility are actually in place, and regardless of alterations or additions to the original structure.

Mobile residence. Structure that contains one or more dwelling units and that is so designed as to be transportable, either by traction or under its own power. This definition shall apply regardless of whether or not the wheels or other devices for mobility are actually in place, and regardless of alterations or additions to the original structure.

Mobile residence park. Any space, however designated, that is occupied or designated for occupancy by four (4) or more mobile residences.

Model. Specific design or style of a mobile home accessory building or structure, as designated by the manufacturer, designed as a specific assembly of component structural parts.

Modern campground. A tract of land where recreational units are accommodated and water flush toilets and water under pressure are available at a service building or a water outlet and sewer connection are available at each site.

Motor home. Self-propelled motor vehicle designed to provide temporary living quarters for recreation or travel.

Motor home. Vehicular unit designed to provide temporary living quarters for recreational, camping or travel use built on or permanently attached to a self-propelled motor vehicle chassis or on a chassis cab or van which is an integral part of the completed vehicle.

Motorized home. Portable dwelling designed and constructed as an integral part of a self-propelled vehicle, and limited to eight (8) feet in width, 4500 pounds in weight and twenty-nine (29) feet in length.

Multiple vehicle. Two (2) or more separately licensed vehicles each of which is designed, constructed or reconstructed, to be attached directly to the other without intervening structures.

Nonhurricane zones. Not more than twenty-four (24) feet on centers beginning from the front line of the mobile home stand (congruent with the front wall of the mobile home). Not more than six (6) feet open-end spacing provided at the rear line of the mobile home stand unless additional tiedowns are installed.

Occupancy. Purpose for which a mobile home, is designed to be used.

Occupied area. That area of an individual mobile home lot which has been covered by a mobile home and its accessory structures.

Occupied area. Total of all of the lot area covered by a mobile home and roofed mobile home accessory buildings and structures on a mobile home lot.

Occupied area. Total of all the space covered by a mobile home and mobile home accessory buildings and structures on a mobile home lot.

Occupied area. The space covered by a mobile home and mobile home accessory buildings and structures on a mobile home lot, allowed by the Code.

Open space. All mobile home courts shall provide not less than ten percent (10%) of the total land area for public open space purposes and such lands shall be improved whereby the same will be accessible to all families residing within said tract and whereby such open space may be used for recreational purposes.

Operator. Person to whom a permit to operate is issued by the enforcement agency, or person operating a mobile home park.

Permanent building. Any building except a mobile home or a mobile home accessory building or structure.

Permanent mobile home space. Location of an individual trailer that is not part of a mobile home park.

Permit. Written document issued by the enforcing agency permitting the construction, alteration or expansion of a Mobilehome Development.

Pick-up camper. Structure designed primarily to be mounted on a pickup or truck chassis and with sufficient equipment to render it suitable for use as a temporary dwelling for travel, recreational or vacation uses.

Pickup coach. Structure designed primarily to be mounted on a pickup or truck chassis and with sufficient equipment to render it suitable for use as a temporary dwelling for travel, recreational and vacation uses, and limited to eight (8) feet in width, 4500 pounds in weight and twenty-nine (29) feet in length.

Plot. Parcel of land consisting of one or more lots or portions thereof which is described by reference to a recorded plot or metes and bounds.

Porch. Outside walking area having the floor elevated more than eight (8) inches above grade.

Power supply assembly. Conductors, including the grounding conductors, insulated from one another, the connectors, attachment plug caps, and all other fittings, grommets, or devices installed for the purpose of the delivering panel within the Mobile Home.

Power supply cord. Flexible cord assembly of conductors, including a grounding conductor, connectors, attachment plug cap, and all other fittings, grommets or devices, designed for the purpose of delivering energy from the source of electrical supply to the branch circuit distribution panelboard within the vehicle.

Prefabricated house. Is not to be considered a mobile home.

Primary system. That part of the electrical wiring system of the mobile home park distributing electrical energy in excess of 240 volts.

Primitive campground. A tract of land where recreational units are accommodated and water is furnished from a hand pump well and sewage is disposed of by means of a sanitary privy.

Private street. Private way which affords principal means or access to abutting individual mobilehome Lots and auxiliary buildings.

Property. Plot with any buildings or other improvements located thereon.

Property line. Recorded boundary of a plot.

Public camp. Any land used or intended to be used in occupancy by persons using automobile house trailers, tents, cabins or other temporary quarters of any kind.

Public street. Public way which affords principal means of access to abutting properties.

Public system (water or sewerage). System which is owned and operated by a local governmental authority or by an established public utility company which is adequately controlled by a governmental authority. Such systems are usually existing systems serving a municipality, a township, an urban county, or a water or sewer district established and directly controlled under the laws of a state.

Ramada. Any free-standing roof, or shade structure, installed or erected above an occupied mobile home or any portion thereof.

Recreational trailer park. Any area or tract of land, within an area zoned for recreational use, where one or more lots are rented or leased or held out for rent, or lease to owners or users of recreational vehicles or tents and which is occupied for temporary purposes.

Recreational vehicle. Vehicular type unit thirty-two (32) feet or less in length and eight (8) feet or less in width, primarily designed as temporary living quarters for recreational, camping, or travel use, which either has its own motive power or is mounted on or drawn by another vehicle.

Recreational vehicle. Camp car, motor home, travel trailer or tent trailer, with or without motive power, designed for temporary human habitation for travel, recreation, or emergency purposes, with a living area less than 220 square feet, excluding built-in equipment such as wardrobes, closets, cabinets, kitchen units or fixtures, bath and toilet rooms, and is identified as recreational vehicle by the manufacturer.

Recreational vehicle. Portable structure intended as a temporary accommodation for travel, vacation or recreational use.

Recreational vehicle. Vehicular type unit primarily designed as temporary living quarters for recreational, camping, or travel use, which either has its own motive power or is mounted on or drawn by another vehicle. The basic entities are: travel trailer, camping trailer, truck camper and motor home.

Recreational vehicle. Provides temporary living quarters which has its own motive power or may be drawn by a vehicle.

Recreational vehicle. Any building or structure designed and/or used for living or sleeping purposes and equipped with wheels to facilitate movement from place to place, and automobiles when used for living or sleeping purposes and including pick-up coaches (campers), motorized homes, travel trailers and camping trailers.

Recreational vehicle model. Specific design of vehicle as designated by the manufacturer which is based on type or size of plumbing, heat-producing or electrical equipment or installations.

Recreational vehicle park. Any lot or parcel of land used or intended to be used for the accommodation of two (2) or more recreational vehicles for transient dwelling purposes.

Recreational vehicle park driveway. A thoroughfare provided and maintained in a recreational vehicle park as the principal means of access to the abutting recreational vehicle sites.

Recreational vehicle site or recreational vehicle space. A plot of land in a recreational vehicle park, used or intended to be used for the accommodation of not more than one (1) recreational vehicle and one (1) tow motor vehicle which is not in itself a recreational vehicle.

Rights-of-way. Area, either public or private over which the right of passage exists.

Roof jack. That portion of a heater flue or vent assembly, including the cap, insulating means, flashing and ceiling plate, located in and above the roof of a mobile home.

Screening and landscaping. All mobile home parks shall be effectively screened on each side by a landscaped planting strip. Such planting strips shall be properly maintained at all times and may consist of grass, hardy shrubs, and trees. The width of the planting strip shall be of sufficient size to effectively screen the use of the property.

Secondary system. That part of the electrical wiring system of the mobile home park distributing electrical energy at nominal 115/230 volts, single phase, or other approved voltage.

Self-contained mobile home. One equipped with a toilet, water storage tank for potable water, and sewage holding tank.

Self-contained system. Complete factory-made and factory-tested system in a suitable frame or enclosure which is fabricated and shipped in one or more sections and in which no refrigerant-containing parts are connected in the field other than by companion or block valves.

Self-contained vehicle. Vehicle equipped with a toilet and waste-holding tank for the temporary retention of liquid borne waste. Such tank may be an integral part of the toilet.

Service building. Structure housing toilet, lavatory, and such other facilities as may be required in a mobile housing development.

Service building. Building housing toilet, lavatory and such other facilities as may be required by the Code.

Service building. Housing sanitation facilities be permanent structures complying with all applicable ordinances and statutes regulating buildings, electrical installations and plumbing and sanitation systems.

Service equipment. Necessary equipment, usually consisting of circuit breaker or switch and fuses, and their accessories, which consitututes the overcurrent protection and disconnecting means for the supply to a primary system, a secondary system, a mobilehome or a building.

Setbacks. Mobile home should be safely located from any park property boundary line abutting upon a public street or highway.

Sewage and refuse disposal. Waste from showers, bath tubs, flush toilets, urinals, lavatories and slop sinks in service and other buildings within the park be discharged into a public sewer system in compliance with applicable ordinances or into a private sewer and disposal plant or septic tank system of such construction and in such manner as will present no health hazard.

Sewage disposal lines. Located in trenches of sufficient depth to be free of breakage from traffic or other movement and separated from the park water supply system at a safe distance. Sewers, at a grade which will insure a velocity of two (2) feet per second when flowing full. All sewer lines constructed of materials according to the Code, adequately vented, and watertight joints.

Sewer connection. Sewer connection consist of all pipes, fittings, and appurtenances from the drain outlet of the mobile home to the inlet of the corresponding sewer riser pipe of the sewerage system serving the Mobile Housing Development.

Sewer riser pipe. That portion of the sewer lateral which extends vertically to the ground elevation and terminates at each mobile home lot.

Site. Parcel of land consisting of one or more lots or portions thereof which is described by reference to a recorded plat or by metes and bounds.

Slide-in camper. An accommodation body that fits into a standard vehicle and is designed to be easily removable.

Stabilizing devices or piers. May be used for mobile homes not permanently attached to foundations.

Storage cabinet. Structure located on a mobile home lot which is designed and used solely for the storage and use of personal equipment and possessions of the mobile home occupants.

Storage structure. Structure located on a mobile home lot which is designed and used solely for the storage of personal equipment and possessions of the occupants.

Streets. Street improvements within a mobile home court shall be in accordance with the locational criteria and related standards of and shall be constructed to the standards and required improvements of the ordinance. All streets shall be properly drained and shall be kept free of debris or other obstructions so as to provide access for fire, police or other emergency vehicles.

Structural requirements. Mobile homes designed and constructed as a completely integrated structure capable of sustaining the design load requirements of the Code, and capable of transmitting these loads to running gear or stablizing devices without causing an unsafe deformation or abnormal internal movement of the structure or its structural parts. When shipping supports are required, they should be installed in accordance with the manufacturer's instructions.

Structure. That which is built or constructed, an edifice or building of any kind, or any piece of work artificially built up or composed of parts joined together in some definite manner.

Temporary trailer park. Any area or tract of land where one or more lots are rented or leased or held out for rent or lease to owners or users of recreational vehicles and which is established for one operation not to exceed 11 consecutive days, and is then removed.

Tent. Any enclosed structure or housing fabricated entirely or in major part of cloth, canvas or similar flexible material.

Tent camp. Area or tract of land where one or more lots or sites are rented or leased or held out for rent or lease for the exclusive use of tent campers.

Tent trailer. A vehicular, portable structure built on its own chassis, having a rigid or canvas top and side walls which may be folded or otherwise condensed for transit.

Tiedown. Any device designed for the purpose to anchor a mobile home to ground anchors.

Tourist or trailer camp. An area containing one (1) or more structures, designed or intended to be used as temporary living facilities for two (2) or more families and intended primarily for automobile transients, or providing spaces where two (2) or more tents, auto trailers, or camping vehicles can be or are intended to be parked.

Trailer. Any structure used for sleeping, living, business, or storage purposes, having no foundation other than wheels, blocks, skids, jacks, horses, or skirting, and which is, has been, or reasonably may be, equipped with wheels or other devices for transporting the structure from place to place, whether by motive power or other means. The term "trailer" shall include camp car and house car. For the purpose of the Ordinance a trailer is a single-family dwelling when located outside of a licensed trailer camp and shall conform to all regulations thereof.

Trailer. Any vehicle, house car, camp car, or any portable or mobile vehicle on wheels, skids, rollers, or blocks, either self propelled or propelled by any other means, which is used or designed to be used for residential, living, sleeping, or commercial purposes and herein referred to as a trailer.

Trailer. Includes trailer coach, house trailer, mobile home, automobile, trailer, camp car or any self-propelled or non self-propelled vehicle constructed, re-constructed or added to be means of accessories in such a manner as will permit the use and occupancy thereof for human habitation, storage or conveyance of machinery, tools, or equipment, whether resting on wheels, jacks, or other foundations and used or so constructed that is or may be mounted on wheels or any similar transportation devices and used as a conveyance of highways and streets, but not including those vehicles that are attached to an automobile or truck for the sole purpose of transporting goods or farm animals. Trailers shall not be considered buildings, dwellings or structures for the purposes of the ordinance.

Trailer. Any vehicle used for living or sleeping purposes.

Trailer. Any vehicle, covered or uncovered, used for living, sleeping, business or storage purposes, having no foundation other than wheels, blocks, skids, jacks, horses, or skirtings and which is, has been, or reasonably may be equipped with wheels or other devices for transporting the vehicle from place to place, whether by motive power or other means. The term "trailer" shall include camp car and house car.

Trailer. Mobile home or similar portable structure having foundation other than wheels, jacks, or skirtings, and so designed or constructed as to permit occupancy for dwelling or sleeping purposes.

Trailer. A vehicle or portable structure constructed so as to permit occupancy thereof for lodging or dwelling purposes or for use as an accessory structure in the conduct of a business, trade, or occupation, and which may be used as a conveyance on streets and highways, by its own or other motive power.

Trailer. Any vehicle, house car, camp car, or any portable or mobile vehicle on wheels, skids, rollers or blocks, either self propelled or propelled by any other means, which is used or designed to be used primarily for residential, living, sleeping or commercial purposes or for the transportation of boats or other recreational equipment, or for other similar purposes.

Trailer. A structure used for living, sleeping, business or storage purposes, having no foundation other than wheels, blocks, skids, jacks, horses, or skirting and which is, has been, or reasonably can be, equipped with wheels or other devices for transporting the structure from place to place, whether by motor power or other means. The term "trailer" shall include camp car and house car.

Trailer. Vehicle, without motive power, designed for living quarters and for being drawn by a motor vehicle.

Trailer. Any vehicle or structure not structurally anchored to a foundation, propelled by an attached vehicle or other propelling apparatus, for residential, commercial, hauling or storage purposes.

Trailer. Any portable or movable structure or vehicle including trailers designed for living quarters, offices, storage, or for moving or hauling freight, equipment, or merchandise of any kind, and including boats and boat trailers.

Trailer. Any vehicle used, or so constructed as to permit its use, as a conveyance upon the public streets or highways and duly licensed as such, which shall include self-propelled and non-self-propelled vehicles, and so designed, constructed, reconstructed or added to by means of an enclosed addition or room, in such manner as will permit the occupancy thereof as a dwelling unit and having no permanent foundation other than wheels, blocks, jacks or skirtings, and shall include without limiting the general definition thereof, house trailer, mobile home, or other enclosure or vehicle which is so designed, constructed or reconstructed.

Trailer. Any structure used for living, sleeping, business or storage purposes, having no foundation other than wheels, blocks, skids, jacks, horses or skirtings, and which is, has been, or reasonably may be equipped with wheels or other devices for transporting the structure from place to place.

Trailer camp. Any lot on which are parked two (2) or more trailers, or one or more trailers for a longer period of time than forty-eight (48) hours.

Trailer camp. Lot or premises used for occupany by two or more trailers for any length of time, or by one trailer for an aggregate of more than thirty (30) days in any one calendar year.

Trailer camp. Any site, lot, field, or tract of land privately or publicly owned or operated, upon which two or more house trailers used for living, eating or sleeping quarters are, or are intended to be, located, such establishment being open and designated to the public as places where temporary residential accommodations are available whether operated for or without compensation.

Trailer camp. Any lot, tract or site used, or intended to be used, for parking of two or more trailers, and shall include any building, structure, vehicle or enclosure used or intended for use as a part of the equipment of such trailer camp.

Trailer camp. Any lot, piece or parcel of land set aside and offered by any person to the transient public for the parking and accommodation of two or more automobile trailers which are to be occupied for sleeping or eating.

Trailer camp or court. Any premises where one or more trailer coaches are parked for living and sleeping purposes, or any premises used or set apart for the purpose of supplying to the public parking space for one or more trailer coaches for living and sleeping purposes, and which include any buildings, structures, vehicles or enclosure used or intended for use as a part of the equipment of such trailer camp or court.

Trailer camp-trailer park-trailer lot. Any area or premises on which space available for two or more trailers is rented or held out for rent, or on which free occupancy or camping for such number is permitted to trailer owners or users for the purpose of securing their trade, herein referred to as a trailer camp, but not including automobile or trailer sales lots on which unoccupied trailers are parked for purposes of inspection and sale.

Trailer coach. Readily movable vehicle with wheels designed or used for sleeping or living quarters, and propelled either by its own power or by any other power-driven vehicle to which it may be attached. A trailer coach may be referred to as a "Trailer," "Mobile Home" and "Camper."

Trailer coach. Any vehicle designed; used, or so constructed as to permit its being used as a conveyance upon the public streets or highways and duly licensable as such, and constructed in such a manner as will permit occupany thereof as a dwelling or sleeping place for one or more persons.

Trailer coach. Any vehicle or similar portable structure having no foundation other than wheels, jacks or skirtings and so designed or constructed as to permit occupancy for dwelling or sleeping purposes.

Trailer coach or mobile home. Trailer coach or mobile home refers to any vehicle used or so constructed as to permit its being used as a conveyance upon the public streets or highways and duly licensable as such and shall include self-propelled and non-self-propelled vehicles so designed, constructed and reconstructed in such a manner as will permit the occupancy thereof as a dwelling or sleeping place for one (1) or more persons and having no foundation other than wheels and jacks.

Trailer coach or mobile home park. Trailer coach or mobile home park refers to any site, lot, field or tract of land upon which one (1) or more occupied mobile homes or trailer coaches are harbored, either free of charge or for revenue purposes, and shall include any building, structure, tent, vehicle or enclosure used or intended for use as a part of the equipment of such mobile home or trailer coach park.

Trailer coach park. Any plot of ground upon which two or more trailer coaches, occupied for dwelling or sleeping purposes, are located, regardless of whether or not a charge is made for such accommodation.

Trailer coach space. A plot of ground within a trailer coach park designed for the accommodation of any trailer coach.

Trailer court. Any area of land used to accommodate two or more trailers, trailer coaches, or vehicles used for human habitation or for carrying persons and property.

Trailer court — trailer camp. Plot of ground designed and laid out for occupancy by house trailers, either on a permanent or temporary basis, to be used as living quarters.

Trailer dwelling. Trailer used as a temporary dwelling for recreational, vacation or travel purposes and which is not more than eight feet in width, and is not more than twenty-nine feet in length provided that the length limitation shall not apply if the gross weight does not exceed four thousand five hundred pounds.

Trailer house or house trailer. Any vehicle used for living or sleeping purposes or any conventional trailer built as a living or sleeping unit, whether or not it is used for that purpose.

Trailer (mobile home). Vehicle used for living or sleeping purposes, and standing on wheels or on rigid supports.

Trailer or mobile home. Any vehicle, covered or uncovered, used for living, sleeping, business or storage purposes, having no foundation other than wheels, blocks, skids, jacks, horses, or skirtings and which is, has been, or reasonably may be equipped with wheels or other devices for transporting the vehicle from place to place, whether by motive power or other means. The term "trailer" shall include camp car and house car.

Trailer or tourist camp. Area containing one or more structures designed or intended to be used as temporary living facilities of two (2) or more families and intended primarily for automobile transients or providing spaces where two (2) or more tents or auto trailers can be parked provided that no person or family occupies the parking or living facilities of such a camp for more than thirty (30) consecutive days.

Trailer park. Site or portion of a site which is used or intended to be used by persons living in trailers or mobile homes on a permanent or transient basis.

Trailer park. Any lot, tract or parcel of land licensed and used or offered for use in whole or in part, with or without charge, for the parking of occupied trailer coaches used solely for living and/or sleeping purposes.

Trailer park. Area providing spaces where one or more auto trailers can be or are intended to be parked, with flush toilet and bathing facilities provided on the site.

Trailer park. Any lot, tract or parcel of land, used or offered for use in whole or in part with or without charge for the parking of three (3) or more trailer coaches or mobile homes, used solely for living and/or sleeping purposes.

Trailer park or camp. Any place, area, or tract of land upon which is located any trailer or camp car, but not including a location where a camp car or trailer is not inhabited.

Trailer park or trailer court. Area providing a minimum of three (3) spaces where trailers or mobile homes can be or are intended to be parked.

Trailer sanitation station. Plumbing receptor desgned to receive the discharge of sewage holding tanks of self-contained vehicles, and a water hose connection for washing the receptor.

Trailer site. That portion of a trailer park set aside and designated for the occupancy of a trailer coach and including the area set aside or used for parking and buildings or structures such as awnings, cabanas or ramadas which are accessory to the trailer coach.

Travel/Recreational trailer. A portable structure built on a chassis designed to be used as a temporary dwelling for travel recreational and vacation uses. When factory equipped for the road, it shall have a body width of not more than eight (8) feet and a body length of not more than thirty-two (32) feet.

Travel trailer. Vehicular, portable structure built on a chassis and designed to be used for temporary occupancy for travel, recreational or vacation use; with the manufacturer's permanent identification "Travel Trailer", thereon; and when factory equipped for the road, being of any length provided its gross weight does not exceed forty-five hundred (4500) pounds, or being of any weight provided its overall length does not exceed twenty-eight (28) feet.

Travel trailer. Vehicular portable home designed as a temporary dwelling for travel, recreational and vacation uses. Such home shall not exceed eight (8) feet in width and thirty-two (32) feet in length, and shall be classified as a travel trailer whether or not its wheels, rollers, skids or other rolling equipment have been removed, or whether or not any addition thereto has been built on the ground.

Travel trailer. Any vehicle designed, built and maintained so that it may be drawn on the highway and primarily built, furnished and used, or intended to be used, for overnight or short term shelter.

Travel trailer. Vehicular, portable structure built on a chassis, designed to be used as a temporary dwelling for travel, recreational and vacation uses, premanently identified "travel trailer" by the manufacturer.

Travel trailer. Vehicular, portable structure built on a chassis, designed to be used as a temporary dwelling for travel and recreational purposes, having a body width not exceeding eight (8) feet and a body length not exceeding thirty (30) feet.

Travel trailer. Vehicular unit mounted on wheels, designed to provide temporary living quarters for recreational, camping, or travel use, of such size or weight as not to require special highway movement permits when drawn by a motorized vehicle, and with a living area of less than 220 square feet, excluding built-in equipment (such as wardrobes, closets, cabinets, kitchen units or fixtures) and bath and toilet rooms.

Travel trailer park. Any area or tract of land or a separate designated section within a mobilehome park, when one (1) or more lots are rented or leased or held out for rent or lease to owners or users of recreational vehicles used for travel or recreational purposes.

Travel trailer park. Park designed to accommodate travel trailers for temporary occupancy of not over thirty (30) days duration.

Travel trailer park. Any lot, tract or pacel of land licensed and used or offered for use in whole or in part, with or without charge, for the parking of occupied mobile homes, travel trailers, pick-up campers, converted buses, tent trailers, tents or similar devices used for temporary portable housing and used solely for living and/or sleeping purposes.

Truck camper. Portable unit which may be loaded into or affixed to the bed or chassis of a truck for living quarters.

Truck camper. Portable unit constructed to provide temporary living quarters for recreational, travel, or camping use, consisting of a roof, floor, and sides, designed to be loaded onto and unloaded from the bed of a pick-up truck.

Unit system. Self-contained system which has been assembled and tested prior to its installation and which is installed without connecting any refrigerant-containing parts. A unit system may include factory-assembled companion or block valves.

Utility connections. Utility connections serving the mobile home shall be located to properly service the mobile home placed on the stand.

Utility improvements. All mobile home courts shall provide to each lot line a continuing supply of safe and potable water as approved by the State Department of Health as well as a sanitary sewage disposal system in accordance with, and as approved by, the State Department of Health, all such systems being provided to the lot lines of all lots in any such Mobile Home Court.

Utility Trailer. A vehicular structure, without its own motive power, not exceeding twenty (20) feet in length, eight (8) feet in width, and ten and one-half (10 1/2) feet in height, designed and, or used primarily for the transportation of all manner and types of motor vehicles, goods or materials and eligible to be licensed or registered and insured for highway use.

Vehicle. Mobilehome, recreational vehicle or commercial coach.

Vehicle parking. Mobile home parks should be designed to include at least two (2) automobile parking spaces for each mobile home lot.

Water connection. Water connection consists of all pipes, fittings and appurtenances from the water riser pipe to the distribution system within the mobile home.

Water connector. Removable extension connecting the mobilehome water distribution system to the water supply.

Water distribution system. All the water piping within or attached to the mobile home.

Water riser pipe. That portion of the water supply system serving the development which extends vertically to the ground elevation and terminates at a designated point at each mobile home lot.

Water storage tank. Tank installed in a vehicle for the purpose of storing potable water.

Water supply. Accessible, adequate, safe, and potable supply of water provided for each mobile home.

Water supply. Adequate supply of pure water for drinking and domestic purposes supplied by pipes to all buildings, if provided, and mobile home lots within the park, to meet the requirements of the park.

Water supply connection. Fitting or point of connection of the mobile home water distribution system designed for connection of a water connector.

Windbreak. Vertical wall structure designed and erected as a free-standing unit, the vertical surface of which is not more than 50 percent open.

Wind load. Lateral or vertical pressure or uplift on the mobile home due to wind blowing in any direction.

Window. Glazed opening on the exterior of the mobile home, including glazed doors.

Window awning. Shade structure supported wholly by the mobilehome or building to which it is attached.

C-17 MOTOR VEHICLE—HOUSING, SERVICE AND SALES

Accessory garage. Accessory building or part of a main building used only for the storage of motor vehicles as an accessory unit.

Accessory parking area. Open or enclosed private area, other than a street, used for the free parking of passenger automobiles for occupants, their guests or customers, of a main building.

Attached garage. Private garage which is structurally attached to a principal building and which has livable floor area adjoining one or more walls thereof.

Auto laundry. A building, or portion thereof containing facilities for washing more than two (2) automobiles, using production line methods with a chain conveyor, blower, steam cleaning service, or other mechanical devices.

Auto laundry. A manned facility, enclosed in a building, for washing automobiles or other vehicles and utilizing such devices as a vehicle conveyor, blowers, steam cleaners, waxers, or any other such mechanical devices.

Automobile laundry. A building or portion thereof where automobiles are washed, using a conveyor, blower, steam-cleaning equipment or other mechanical device of production-line nature.

Automobile laundry. Structure, or area, used for the purpose of cleaning or reconditioning the exterior and interior surfaces of automotive vehicles, but not including an incidental one-bay washing facility in a gasoline service station.

Automobile laundry. A building or portion thereof, where automobiles are washed using production line methods with a chain conveyor, blower, steam cleaning or other mechanical devices.

Automobile filling and service station. Any place of business having pumps and storage tanks at which fuels or oils for the use of motor vehicles are dispensed, sold or offered for sale at retail and where minor repairs, services, and inspections may be carried on and rendered incidental to the sale of such fuels and oils.

Automobile filling station. A building, structure, premises, enclosure or other place used for the sale or offering for sale at retail of automobile fuels or oils, except hardware stores, painting and decorating shops, dyeing and cleaning shops, tailor shops, or drug stores, where such fuels or oils are not regularly dispensed to automobiles.

Automobile or trailer sales area. Any space used for display, sale, or rental of motor vehicles or trailers, in new or used and operable condition.

Automobile or trailer sales area. Open area, other than a street, used for the display, sale or rental of new or used motor vehicles or trailers in operable condition and where only minor repair work is done.

Automobile repair station. Place where, along with the sale of engine fuels, the following services may be carried out: general repair, engine rebuilding, rebuilding of reconditioning of motor vehicles, collision service, such as body, frame, or fender straightening and repair, painting and undercoating of automobiles.

Automobile sales and storage lot. Open, off-street area where two (2) or more operable motor vehicles are stored or offered or displayed for sale or advertising purposes.

Automobile sales lot. Any place outside a building where two (2) or more vehicles are offered for sale or are displayed for sale or advertising purposes.

Automobile service station. Place of business having pumps and/or storage tanks from which liquid fuel and/or lubricants are dispensed at retail directly into the motor vehicle. Sales and installation of auto accessories, washing, polishing, inspections and cleaning, but not steam cleaning may be carried on incidental to the sale of such fuel and lubricants.

Automobile service station. Is a building or place of business where gasoline, oil and grease, batteries, tires and automobile accessories are supplied and dispensed directly to the motor vehicle trade at retail, and where the following services may be rendered.

Automobile service station. Place where gasoline or any other automobile engine fuel (stored only in underground tanks) kerosene or motor oil and lubricants or grease (for operation of automobiles) are retailed directly to public on premises; including sale of minor accessories and services for automobiles.

Automobile service station. Any building or premises used for dispensing, sale or offering for sale at retail to the public, gasoline stored only in underground tanks, kerosene, lubricating oil or grease for the operation of automobiles and including the sale and installation of tires, batteries and other minor accessories and services for automobiles, but not including major automobile repairs; and including washing of automobiles where no production line methods are employed. When dispensing, sale or offering for sale of motor fuels or oil is incidental to the conduct of a public garage, the premises shall be classified as a public garage.

Automobile service station (gas station). A building or portion thereof or premises used for dispensing or offering for sale at retail any automotive fuels or oils; having pumps and storage tanks thereon, or where battery, tire and other similar services are rendered, but only if rendered wholly within lot lines. Automobile service stations do not include open sales lots as defined in the ordinance, and the open storage of motor vehicles and the repair, rebuilding or reconstruction of motor vehicles is not permitted.

Automotive laundry. Building or other structure or premises, or part thereof, used for the washing of motor vehicles.

Automotive service garage. A garage where no repair work is done except exchange of parts and maintenance requiring no open flame, cutting, welding, or the use of highly flammable liquids.

Automotive service garage. Garage where no repair work is done except exchange of parts and maintenance requiring no open flame, cutting, welding, or the use of highly flammable liquids, only allowed by the Code.

Automotive services. Shall mean establishments for sale or rental of new and used cars, trucks, and trailers, gasoline service stations; automobile and truck repair garages, body fender and paint shops; tire shops and tire recapping plants; stores for the sale of new auto parts supplies and accessories; and any other establishment of a similar nature, but not including auto wrecking or the storage or sale of used parts.

Automotive services. Public garages, provided they conform to the applicable provisions of the Code, include; service stations, provided the washing or lubricating of motor vehicles is performed inside the building, and parking lots.

Automotive services. Establishments for the sale or rental of new and used cars, trucks and trailers, gasoline service stations, automobile and truck repair garages, body, fender and paint shops, tire shops and tire recapping plants, stores for the sale of new auto parts supplies and accessories, and any other establishment of a similar nature, but not including auto wrecking or the storage or sale of used parts.

Automotive services. Public garages, provided they conform to the applicable provisions of the Code and includes service stations, provided the washing or lubricating of motor vehicles is performed inside the building.

Automotive service station. Place of retail business at which outdoor automotive refueling is carried on using fixed dispensing equipment connected to underground storage tanks by a closed system of piping, and at which goods and services generally required in the operation and maintenance of motor vehicles and fulfilling of motorist needs may also be available. The building consists of a sales office where automotive accessories and packaged automotive supplies may be kept or displayed. It may also include one or more service bays in which vehicle washing, lubrication and minor replacement, adjustment and repair services are rendered. An automotive service station building shall have no cellar or basement.

Automotive service station. Place of business having pumps and/or storage tanks from which liquid fuel and/or lubricants are dispensed at retail directly into the motor vehicle. Sales and installation of auto accessories, washing, polishing, inspections and cleaning, but not steam cleaning, may be carried on incidental to the sale of such fuel and lubricants.

Automotive service station. Place of retail business at which outdoor automotive refueling is carried on using fixed dispensing equipment connected to underground storage tanks by a closed system of piping, and at which goods and services generally are required in the operation and maintenance of motor vehicles and fulfilling of motorist needs may also be available, including a sales office where automotive accessories and packaged automotive supplies may be kept or displayed. It may also include one or more service bays in which vehicle washing, lubrication and minor replacement, adjustment and repair services are rendered. An automotive service station building shall have no cellar or basement.

Auto park or market. Open land area used for the storage and sale of complete and operative automobiles, new or used in accordance with the Zoning Regulations.

Auto repair shop. A shop intended and used for the repair for renumeration of one or more vehicles.

Auto park or market. Open land area used for the storage and sale of complete and operative vehicles, new or used.

Auto sales lot. Lot used primarily for the sale of new or used motor vehicles fit for transportation and complying with requirements of ordinances of the City and statutes of the State regarding such new or used vehicles.

Auto sales lot. Any land used primarily for the sale of new or used motor vehicles fit for transportation and complying with requirements of ordinances of the City and statutes of the State regarding such vehicles.

Basement garage. Garage having its floor level more than one (1) foot below grade shall be designated as a basement garage.

Basement parking garage. Enclosed parking garage located in a basement, and includes an underground parking garage.

Basement parking garage. Portion of the basement area of a structure that is used for the storage of private vehicles.

Basement parking garage. Enclosed parking garage located in a basement or cellar.

Boathouse. Building which is used for the housing, storage, maintenance, or repair of one (1) or more boats.

Boats and boat trailers. Includes boats, floats and rafts, plus the normal equipment to transport the same on the highway.

Boat yards and ways. Commercial establishment which provides such facilities as are customary and necessary to the construction or reconstruction or repair or maintenance or sale of boats or marine engines or marine equipment, and marine services of all kinds, including, but not limited to, rental of covered or uncovered boat slips or dock space or enclosed dry storage space or marine railways or lifting or launching services.

Carport. Garage attached to a dwelling and having one or more open sides.

Carport. Canopy or shed, attached to the main building, open on two or more sides, for the purpose of providing shelter for one or more vehicles.

Carport. Structure having a roof supported by columns or walls.

Carport. Covered area open on one or more sides for the storage of passenger vehicles, boats, or trailers.

Carport. Is a permanent roofed structure with not more than two enclosed sides used or intended to be used for automobile shelter.

Carport. Roofed structure providing space for the parking of motor vehicles and enclosed on not more than two sides. For the purposes of the ordinance a carport attached to a principal building shall be considered as part of the principal building and subject to all yard requirements herein.

Carport. Space for the housing or storage of motor vehicles and enclosed on not more than two sides by walls.

Carport. Roofed area having at least one side open to the weather, provided it was designed to be used for motor vehicles.

Carport. Accessory structure permanently attached to a dwelling having a roof supported by columns, but not otherwise enclosed.

Carport. An accessory building or portion of a main building with two (2) or more open sides designated or used for the parking of motor vehicles. Enclosed storage facilities may be provided as part of a carport.

Carport. Building used solely for the storage of motor vehicles and containing no enclosing walls, screens, lattice or other material other than the wall or walls of the building to which it attaches, or other than a storage room as permitted in the Code. The area of supporting columns or wall sections of any one side other than a storage room shall not exceed fifty percent of that side. Screen, lattice, grille work or other material which has a net of ten percent (10%) closed area may be permitted provided the maximum area of fifty percent (50%) is not exceeded.

Carport. Permanently roofed structure, open on three sides and attached to the rear of the main structure, or attached to a rear accessory structure, designed and used for the shelter in storage of automobiles owned or operated by the occupants of the main building.

Carport. Roofed area attached to a dwelling and open on three sides.

Carport. Roofed space having at least one side open to the weather, primarily designed or used for motor vehicles. This term is usually related to small one and two family dwellings. In multifamily properties, a garage may have one or more sides open to the weather.

Carport. Structure not more than one story in height, without walls, doors or other enclosure, on at least two sides, the floor of which rests upon the ground, used exclusively for the storage or parking of not more than two motor vehicles, and which is accessory to private dwelling

Carport. An open-sided roofed shelter for motor vehicles, boats or trailers, usually formed by extension of the roof from the side of a building.

Carporte. A covered area for sheltering a motor vehicle and which is not more than 75 percent enclosed by walls.

Carwash. Building or structure erected for the primary purpose of washing automobiles.

Carwash. Building which is used exclusively for the cleaning and washing of motor vehicles, and in which specialized equipment is used for this purpose.

Commercial storage of vehicles. Assembling or standing of vehicles offered or to be offered for sale or to be repaired or being repaired.

Commercial vehicle. Any vehicle designed, intended or used for transportation of people, goods or things, other than private passenger vehicles and private trailers, used for non-profit transportation of goods and boats.

Community garage. Building or part thereof for the storage of automobiles of residents of the vicinity and in which ordinary maintenance service on such vehicles may be provided.

Community garage. Group of private garages, detached or under one roof, arranged in a row or around a common means of access, and erected for use of residents in the immediate vicinity.

Community garage. Series of private garages located jointly on a common lot, having no public shop or service in connection therewith.

Community garage. Accessory building, having no public shop or service in connection therewith, for the storage of non-commercial vehicles.

Enclosed parking garage. Garage having exterior enclosure walls and used for the parking of motor vehicles.

Enclosed parking garage. Garage having exterior enclosure walls and having less than 50 percent of the area of each of any 2 sides of the garage permanently open to the outside air at each story.

Filling station. Any building, structure, premises, enclosure or other place used for the dispensing, sale, or offering for sale of any motor vehicle fuels or oils having pumps and storage tanks, located wholly within the lot lines. When such dispensing, sale or offering for sale of any fuels or oils is incidental to the conduct of a public garage, the premises shall be classified as a public garage.

Filling station. Building or premises, or portions thereof arranged or designed to be used for the retail sale of oil, gasoline, or other fuel for the propulsion or lubrication of motor vehicles, including facilities for changing tires, tube repairing, polishing, greasing, washing or minor servicing of such motor vehicles, but excluding high speed automotive washing, steam cleaning, body repairing, major transmission or chassis or motor repairing, bumping and painting of any metal parts of such motor vehicles.

Filling station. Any building, structure or land used for the sale at retail of motor vehicle fuels, lubricants or accessories, or for the servicing of automobiles or repairing of minor parts and accessories, but not including major repair work such as motor replacement, body and fender repair or spray painting.

Filling station. Any building or premises used for the dispensing, sale, or offering for sale at retail of any automobile fuels or oils.

Filling station. Any building, structure, or land used for the dispensing, sale or offering for sale at retail of any automobile fuels, lubricants, tires, or accessories except that indoor car washing, minor motor adjustment, and flat tire repair may be performed when incidental to the conduct of a filling station.

Filling station. Buildings on lots used for the purpose of supplying motor fuel to tanks of motor vehicles for immediate use.

Filling station. Any building or premises used for the dispensing, sale, or offering for sale at retail or any motor fuels, oils, or lubricants. When the dispensing, sale, or offering for sale is incidental to the conduct of a public garage, the premises are classified as a public garage.

Filling station. Buildings and premises where gasoline, oil, grease, batteries, tires and automobile accessories may be sold and where in addition routine automotive servicing and parts replacement may be done. Tire recapping and regrooving and major automotive mechanical and body work, painting, welding, storage, auto wrecking and motor over haul are activities specifically excluded from this definition.

Filling stations (buildings and structures). By filling station is meant one or more pumps, tanks, and other pieces of equipment used in the storage and dispensing of liquid fuels and arranged for the sale of such liquid fuels to the public.

Garage. Building or portion thereof in which a motor vehicle containing gasoline, distillate or other volatile, flammable liquid in its tank, is stored, repaired, or kept.

Garage. Building or portion of a building designed or used for the storage or housing of self-propelled vehicles.

Garage. Any building which or portion of which, is used for the storage or repairs of automobile equipment, or as stalls for placing cars other than the Owner's car.

Garage. Building, or part of a building, which accomodates or houses self-propelled vehicles. For the purposes of the code the term vehicle includes land, air and water vehicles.

Garage. Building or portion of a building in which one or more self-propelled vehicles carrying volatile flammable liquid for fuel or power are kept for use, sale, storage, rental, repair, exhibition, or demonstrating purposes, and all that portion of a building which is on or below the floor or floors in which such vehicles are kept and which is not separated therefrom by suitable cutoffs.

Garage. A building, shed or enclosure, or part thereof, in which a motor vehicle containing a flammable liquid in its fuel tank is housed or stored or repaired.

Garage. Building, shed or enclosure or any portion thereof in which a motor vehicle, other than one in which the fuel storage tank is empty, is stored, housed or kept.

Garage. Building or part of a building designed or used for the display, shelter, storage or servicing of motor vehicles containing flammable fuel and having a floor area exceeding 800 square feet.

Garage. Building which is used for the housing, storage, maintenance, or repair of one or more motor vehicles containing a flammable liquid or gas, except that a "service station building", as defined in the Code, is not a garage.

Garage. Building or portion thereof in which is housed or stored one or more motor vehicles containing or using a volatile flammable fluid for fuel or power, or in which such motor vehicles are painted, repaired or serviced.

Garage or carport. Accessory structure or a portion of a main structure, having a permanent roof, and designed for the storage of motor vehicles.

Gasoline filling station. Structure, building or premise, or a portion thereof, where volatile flammable oil for retail supply to motor vehicles is stored or sold.

Gasoline pump. A fixed unit of equipment installed to dispense a liquid produced, prepared or compounded for the purpose of generating power by means of internal combustion or that may be used for that purpose.

Gasoline service station. Any building, structure, or land used or combination thereof for the sale at retail, of motor vehicle fuels, lubricants, accessories, incidental or for the servicing of automobiles or repairing of minor parts and accessories, but not including major repair work, such as motor replacement, body and fender repair, or spray painting.

Gasoline service station. Any area of land, including structures thereon, that is used primarily for the sale of gasoline or other motor vehicle fuel; accessory uses may include the sale of oil, other lubricating substances or motor vehicle accessories, or facilities for lubricating, washing, the incidental replacement of parts, or for motor service to passenger automobiles or trucks not exceeding one and one-half tons rated capacity; but shall not include general repair, rebuilding, or reconditioning of engines, motor vehicles or trailers, or collision service, body repair, frame straightening, painting, undercoating, vehicle steam cleaning or upholstering.

Gasoline service station. Business establishment operated at a fixed location at which gasoline is offered for sale at retail, and, when sold, is dispensed from fixed tanks by pump, or otherwise, directly into the fuel system storage tanks of automobiles or other motor vehicles.

Gasoline service station. Building or structure designed or used for the retail sale or supply of fuels, lubricants, air, water, and other operating commodities for motor vehicles, and including the customary space and facilities for the installation of such commodities on or in such vehicles, but not including any operation included specifically outlined in the Code.

Gasoline service station. Any building or premises used for the dispensing, sale, or offering for sale at retail of any motor fuels, oils or lubricants. When such activities are incidental to the conduct of a public garage, the use or premises is classified as a public garage.

Gasoline service station. Place of business at which the principal service is the retail sale of gasoline.

Gasoline service station. Any area of land, including structures thereon, or any building or part thereof, that is used for the sale of gasoline or other motor vehicle fuel or accessories, and which may or may not include facilities for lubricating, washing, or otherwise servicing motor vehicles, but which shall not include painting or body and fender repairs.

Gasoline service station. Any building, structure, or land used or combination thereof for the sale at retail, of motor vehicle fuels, lubricants, accessories, incidental or for the servicing of automobiles or repairing of minor parts and accessories, but not including major repair work, such as motor replacement, body and fender repair, or spray painting unless permitted by Code.

Gasoline service station. Any area of land, including structures thereon, that is used for the sale of any motor vehicle fuel or lubricating substance or which is used for the sale of motor vehicle accessories, or which has facilities for lubricating, washing or servicing motor vehicles other than the painting thereof by any means.

Gasoline service station. Commercial establishment that sells petroleum products for automobiles and trucks as well as some vehicular parts and accessories and deals in automobile repair of a light nature excluding major motor work and body repair and repainting.

Gas station. Any building or premises used for the retail sale of liquified petroleum products for the propulsion of motor vehicles, and including such products as kerosene, fuel oil, packaged naptha, lubricants, tires, batteries, antifreeze, motor vehicle accessories, and other items customarily associated with the sale of such products; for the rendering of services and making of adjustments and replacements to motor vehicles, and the washing, waxing, and polishing of motor vehicles, as incidental to other services rendered and the making of repairs to motor vehicles except those of a major type. Repairs of a major type are defined to be spray painting, body, fender, clutch, transmission, differential, axle, spring, and frame repairs; major overhauling of engines requiring the removal of engine cylinder head or crankcase pan; repairs to radiators requiring the removal thereof; or complete recapping or retreading of tires.

Marina. Boating establishment which provides covered or uncovered boat slips or dock space, charter and sight-seeing boat dockage, dry boat storage, marine fuel and lubricants, marine supplies and accessories, restaurants or refreshment facilities, boat and boat motor sales or rentals. Pleasure boat and boat motor outfitting, maintenance and repair is permitted as an accessory use, however, no dredge, barge or other work boat dockage or service is permitted, and no boat manufacturing or major reconstruction is permitted.

Marina. Place for docking or storage of pleasure boats or providing services to pleasure boats and the occupants thereof, including minor servicing and repair to boats while in the water, sale of fuel and supplies, or provision of lodging, food, beverages, and entertainment as accessory uses. A yacht club shall be considered a marina, but a hotel, motel, or similar use, where docking of boats and provision of services thereto is incidental to the other activities shall not be considered a marina, nor shall boatdocks accessory to a multiple dwelling where no boat related services are rendered.

Marine service station. Those portions of properties where flammable or combustible liquids and liquefied petroleum gases used as fuel for floating craft are stored and dispensed from fixed equipment on shore, piers, wharves, floats, or barges into the fuel tanks of floating craft; and includes all facilities used in connection therewith, and is considered as intended for servicing small craft.

Mechanical garage. Any building or premises where automotive vehicles are repaired, rebuilt, reconstructed, or stored, for compensation.

Mechanized parking garage. Parking garage in which motor vehicles are raised or lowered to tier levels by mechanical devices installed in the garage, and in which such vehicles are conveyed to stalls or areas by such mechanical devices or by employees, and in which the public is not admitted except in the drive-in and drive-away areas.

Motor fuel service station. Structure, building, or premise or any portion thereof where a flammable fluid is stored, housed or sold for supply to motor vehicles.

Motor vehicle accessories. Sale of kerosene, package naptha, Stoddard solvent, lubricants, tires, batteries, anti-freeze, and items of motor vehicles accessories customarily associated with such retail business.

Motor vehicle care. Washing, polishing, and application of wax and similar finish protectives.

Motor vehicle repair shop. Building, structure or enclosure in which the general business of repairing motor vehicles is conducted including a public garage.

Motor vehicle repair shop. A building or portion of a building arranged, intended and designed to be used for making repairs to motor vehicles.

Motor vehicle service station. Structure located on premises where flammable liquid is stored for supply to motor vehicles and which is used for general servicing of motor vehicles such as greasing or oiling, but not for the repair or storage of motor vehicles.

Non-storage garage. Garage in which no volatile inflammable oil other than that contained in the fuel storage tanks of motor vehicles, is handled, stored or kept.

Open sales and rental lot. Any land used or occupied for the purpose of buying, selling, or renting new or used motor vehicles, boats, trailers, aircraft, recreational or camping equipment, or other commodities, and for the storage thereof prior to sale or rental.

Open sales lot. Land used or occupied for the purpose of buying, selling or renting merchandise stored or displayed out-of-doors prior to sale. Such merchandise includes, but is not limited to, automobiles, trucks, motor scooters, motorcycles and boats.

Open sales lot. Any premises used or occupied for the purpose of buying and selling merchandise, passenger cars, trucks, commercial trailers, mobile homes, motor scooters, motorcycles, boats and monuments, or for the storing of same prior to sale.

Parking facility. Area or structure, located off the right-of-way of any street, that is used for the parking of motor vehicles, including driveways and aisle areas.

Parking garage. Building, land, or portion thereof designed or used for the temporary storage of motor-driven vehicles, with or without the retail dispensing, sale, or offering for sale of motor fuels, lubricants, and tires, or indoor car washing, minor motor adjustment, and flat tire repair when such operations are incidental to the storage of motor-driven vehicles.

Parking garage. Building used for the storage of passenger vehicles either for a fee, or as a condition of doing business or being employed by a firm providing such parking.

Parking garage Structure or series of structures for the temporary storage or parking of motor vehicles, having no public shop or service in connection therewith, other than for the supplying of motor fuels and lubricants, air, water, and other operating commodities to the patrons of the garage only.

Parking garage. Structure or part thereof used for the storage, parking or servicing of motor vehicles, but not for the repair thereof.

Parking garage. Building, or part thereof, other than an accessory or repair garage, used for the storage of passenger vehicles and which may include servicing of said vehicles as an incidental use, but not the repair thereof, and keeping any such vehicles for hire.

Parking level. Floor in a structure on which vehicles are parked.

Parking or storage garage. Any building, except one herein defined as a private garage, used exclusively for parking of self-propelled vehicles, and with not more than two (2) pumps for the incidental sale of gasoline.

Parking space. Impervious, hard surfaced area, enclosed in the main building or in an accessory building or unenclosed, having a rectangular area of not less than one hundred and sixty (160) square feet, with a minimum width of eight and a half (8 1/2) feet when unenclosed, or one hundred and eighty (180) square feet with a minimum width of nine feet (9′) when individually enclosed on two (2) or more sides, exclusive of driveways, permanently reserved for the storage of one automobile.

Parking structure. Unenclosed or partially enclosed structure for the parking of motor vehicles, with no provisions for the repairing or servicing of such vehicles.

Private garage. Accessory building or a portion of a principal building designed or used solely for the storage of motor vehicles owned and used by the occupants of the building to which it is accessory.

Private garage. Accessory building, having not more than one-thousand (1000) square feet of usable floor area, to be used for the storage of non-commercial motor vehicles and not more than one (1) commercial vehicle of not more than one (1) ton capacity; there shall be no public shop or services in connection therewith.

Private garage. Accessory building or portion of a main building, enclosed on three (3) or more sides and designed or used for the shelter or storage of vehicles owned or operated by the occupant of the main building

Private garage. Accessory building designed or used for the storage of not more than three (3) motor driven vehicles owned and used by the occupants of the building to which it is accessory, where no servicing for profit, is conducted.

Private garage. Garage intended for private use, but in which space may be rented for storage only to the occupants of the building to which such garage is accessory. Provided, however, where a building contains less than four (4) families, private garages may be erected for the storage only for four (4) vehicles, and in which space may be rented to other than the occupants of the building to which such garage is accessory, but in no case shall space be used for more than one (1) commercial vehicle.

Private garage. Garage for the storage of motor vehicles not conducted as a business nor used for the storage of more than one (1) commercial vehicle per family, which commercial vehicle or vehicles shall be owned by a person residing on the premises. Trucks or trailer trucks in excess of 2 ton capacity are not considered to be within the meaning of commercial vehicle as used above, but a tractor designed for hauling a trailer truck is considered to be within such meaning.

Private garage. Enclosed space accessory to a dwelling for the storage of automobiles provided that no business, occupation or service for profit is conducted therein or therefrom and that space therein for more than one (1) car is not leased to non-residents of the premises.

Private garage. Garage with a capacity of not more than four (4) power-driven vehicles for storage only, and which is erected as an accessory to a dwelling. A private garage may exceed a four (4) vehicle capacity provided the area of the lot whereon such a private garage is to be located shall contain not less than one thousand (1000) square feet for each vehicle stored.

Private Garage. Garage for the storage of not more than three (3) motor vehicles used exclusively by the owner or the tenant of the premises and which is not equipped for repairing or servicing motor vehicles.

Private garage. Building designed and used only for the storage of non-commercial motor vehicles as an accessory use.

Private garage. Detached, fully enclosed accessory building or a portion of the principal building used for the storage of passenger vehicles, boats or trailers.

Private garage. Garages which are provided for the storage of motor vehicles owned by tenants of buildings on the premises, and with maximum undivided space used for storage of not more than four automobiles, or trucks of one ton or less capacity, but not exceeding 850 square feet, shall be considered private garages. All other garages shall be considered private garages.

Private garage. Building, or a portion of a building, not more than one thousand square feet in area, in which only motor vehicles used by the tenants of the building or buildings on the premises are stored or kept.

Private garage. Garage used for storage only, with a capacity for not more than 4 self-propelled vehicles, trailers, or trucks of one ton or less capacity, and in which garage space for not more than 2 vehicles or trailers may be rented to persons not occupants of the premises.

Private garage. Building, or part of a building, not more than one story high and having an area not exceeding 800 square feet and designed or used for the storage of passenger motor vehicles containing flammable fuel.

Private garage. Accessory building or portion of a main building used for the storage of self-propelled vehicles used by the occupants of the premises, and which may include space for not more than one passenger vehicle used by others.

Private garage. Assessory building or a portion of the principal building, used for storage of automobiles of the occupants of the principal building.

Private garage. Acessory building occupied by the passenger motor vehicles of the families residing on the same lot. This may include one commercial vehicle under five ton capacity. Non-commercial vehicles of persons not resident on the lot may occupy up to one-half the capacity of such garage.

Private garage. Enclosed space for the storage of not more than three (3) motor vehicles, provided that no business, therein for more than one (1) car is leased to a non-resident of the premises.

Private garage. Building or space used as an accessory to or a part of a main building permitted in any residence district, and providing for the storage of motor vehicles and in which no business, occupation or service for profit is in any way conducted.

Private garage. Garage intended for the shelter or storage of vehicles owned or operated by the members of families resident upon the premises.

Private garage. Any building or part thereof used or occupied as a private garage for the storage of not more than four (4) passenger automobiles, or three trucks not exceeding 1-1/2 ton capacity, with gasoline or other volatile flammable fuel in their storage tanks.

Private garage. Accessory building not exceeding 900 square feet in floor area designed or used for storage of vehicles only, not to exceed 4 storage spaces, which shall not be occupied by any vehicles other than motor-driven vehicles of one ton capacity or less or uncovered trailers that are not used for living or sleeping purposes; and provided further that not more than 2 said storage spaces may be rented to persons not occupants of the main building of the lot upon which the private garage is located.

Private garage. Building or part thereof accessory to a main building and providing for the storage of automobiles and in which no occupation or business for profit is carried on.

Private garage. Detached accessory building or portion of a main building, used only for the storage of passenger vehicles or a commercial vehicle as permitted by code.

Private garage. Accessory building or portion of a main building on the same lot and used for the storage only of private, passenger motor vehicles, not more than 2 of which are owned by others than the occupants of the main building.

Private garage. Building, or a portion of a building, not more than 1,000 square feet in area, in which only motor vehicles used by the tenants of the building or buildings on the premises are stored or kept. The garage structure shall be enclosed on three sides and provided with doors if said garage is constructed to the front of the main building and/or buildings, or if said garage fronts onto a side street or an alley. If the garage is constructed at the rear of the main building and not visible from the street, the garage shall be enclosed on three sides with the provision of doors optional.

Private garage. Building devoted partially or wholly to the storage of a motor vehicle or motor vehicles in connection with which no public service is rendered and no business conducted.

Private garage. Detached accessory building or portion of a main building, used for the storage of self-propelled vehicles where the capacity does not exceed three vehicles, or not more than one per family housed in the building to which such garage is accessory, whichever is the greater, and not more than one-third the total number of vehicles stored in such garage shall be commercial vehicles. Storage space for not more than three (3) vehicles may be rented for vehicles of other than occupants of the building to which such garage is accessory.

Private garage. Accessory building or portion of a principal building used for vehicular storage only, and having a capacity adequate to accomodate the automobiles or light trucks owned and registered in the name of the occupants of the principal building. The term includes carport and, when related to the context, shall relate to the storage of one or more vehicles.

Private garage. Accessory building or space for the storage only of not more than two (2) motor-driven vehicles.

Private garage. Accessory building used for the storage of motor vehicles and farm equipment, which may include one commercial vehicle, owned and used by the owner or tenant of the premises, and for the storage of not more than two (2) private non-commercial vehicles owned and used by persons other than the owner or tenant of the premises.

Private garage. Buildings or parts of buildings enclosed on three (3) or more sides used for the storage of motor vehicles owned by tenants of a building located on the same property, which have an undivided floor area not exceeding 750 square feet.

Private garage. Building, or a portion of a building, in which only motor vehicles used by the tenants of the building or buildings on the premises are stored or kept, and with space for not more than ten (10) automobiles.

Private garage. Garage with capacity for not more than four (4) self-propelled vehicles for storage only; provided, that a private garage may exceed a four (4) vehicle capacity if such garage is to serve an apartment house of multi-family dwelling located on the same lot.

Private garage. Accessory to a principal building, either attached to it or separate, and is used only for storage purposes.

Private garage. Garage for housing only, with a capacity for not more than three (3) vehicles. A garage exceeding a three (3) vehicle capacity, intended primarily for housing of cars belonging to occupants of the premises, shall be considered a private garage if the lot whereupon such garage is located contains not less than fifteen hundred (1500) square feet for each vehicle capacity.

Private garage. Any accessory building or part of a principal building used for the storage of motor vehicles owned and used by the owner or tenant of the premises, and for the storage of not more than two (2) motor vehicles owned and used by persons other than the owner or tenant of the premises. Not more than one (1) commercial vehicle or truck may be stored in a private garage.

Private garage. Accessory building for parking or storage of not more than that number of vehicles as may be required in connection with the permitted use of the principal building. In residential areas the storage of not more than one (1) commercial vehicle of a rated capacity not exceeding three-fourths (3/4) ton is permitted.

Private garage. Building or an accessory portion of the main building intended for an used to store the private motor vehicles of the families resident upon the premises, and in which no business, service, or industry connected directly or indirectly with motor vehicles is carried on, and provided that not more than one-half (1/2) of the space may be rented for the private vehicles of persons not resident on the premises, except that all of the space in a garage of one (1) or two (2) car capacity may be so rented. Such a garage shall not be used for more than one (1) commercial vehicle per family resident upon the premises, but no such commercial vehicle shall exceed five (5) tons capacity.

Private garage. Building or part thereof used or intended for the storage of not more than four (4) passenger motor vehicles and in which there are no facilities for repairing or servicing such vehicles.

Private garage. Building, or a portion of a building, not more than one thousand (1,000) square feet in area, in which only motor vehicles used by the tenants of the building or buildings on the premises are stored or kept, but not serviced.

Private garage. Building, accessory to a one or two-family dwelling used exclusively for the parking or temporary storage of passenger automobiles.

Private garage. An accessory building enclosed by walls on all sides and having a permanent roof, intended for or used for the housing of motor vehicles, boats or trailers, constructed either as a part of the main building or detached therefrom, and with a capacity for housing not more than four (4) motor vehicles.

Private garage. An accessory building or an accessory portion of the principle building, including a carport, which is intended for and used for storing the privately owned motor vehicles, boats and trailers of the family or families resident upon the premises, and in which no business, service or industry connected directly or indirectly with motor vehicles, boats and trailers is carried on.

Private garage. An accessory building designed and used primarily for the storage of motor vehicles owned and used by the occupants of the building to which it is accessory and in which no occupation or business for profit is carried on. Not more than one (1) motor vehicle in excess of one and one-half (1-1/2) ton capacity may be stored in a private garage.

Private garage. A garage intended for and used by the private motor vehicles of the families resident upon the premises, except that garage space may be rented for the private motor vehicles of persons not resident on the premises.

Private garage. Private garage which is not an integral structural part of a main building may be located in the required side and/or rear yards, but not less than three (3) feet from any property line, provided it is situated not less than ten (10) feet farther back from the street line than the rear-most portion of the main building. Nothing in the Code shall be construed to prohibit the erection of a common or joint garage which is not an integral structural part of a main building on adjoining lots. A garage which is connected to a main building by a breezeway or similar structure shall be considered an integral structural part of the main building.

Private garage. Accessory building or portion of a main building designed or used solely for the storage of motor-driven vehicles, boats, house trailers, and similar vehicles owned and used by the occupants of the building to which it is necessary.

Private storage garage. Any garage, except those defined as a private or public garage.

Public garage. Any garage, other than a private garage or community garage, available to the public and which is used for the storage, repair, greasing, washing, rental, sales, servicing, adjusting, or equipping automobiles or other motor vehicles.

Public garage. A garage for the storage of motor vehicles conducted as a business.

Public garage. Any premises used for storage, repair, rental, greasing, servicing, adjusting or equipping of motor vehicles.

Public garage. Any premises used for the storage or care of motor driven vehicles, or place where any such vehicles are equipped for operation, repaired or kept for remuneration, hire or sale, including general repair; rebuilding or reconditioning of motors, motor vehicles or trailers; collision service, including body frame or fender straightening or repair, body painting and refinishing; and steam cleaning of motor vehicles.

Public garage. Building or portion of a building, except that herein defined as a private garage or as a repair garage, used for the storage of motor vehicles, or where any such vehicles are kept for remuneration or hire; in which any sale of gasoline, oil and accessories is only incidental to the principle use.

Public garage. Garage other than a private garage where motor-driven vehicles are stored, equipped for operation, repaired or kept for remuneration, hire or sale.

Public garage. Building, other than a private or storage garage, one or more stories in height, used solely for the commercial storage, service or repair of motor vehicles.

Public garage. Any other than a private garage or a public storage garage.

Public garage. Garage building designed for the storage of five (5) or more motor vehicles and is for the use of the residents of the multifamily project or their visitors.

Public garage. Building or part thereof, other than a private garage or a community garage, for the storage of motor vehicles and in which service station activities may be carried on.

Public garage. Building other than a private or storage garage used for the care, repair or storage of self-propelled vehicles or where such vehicles are kept for remuneration, hire or sale. This includes premises commonly known as filling stations or service stations.

Public garage. Building, not a private garage, used for the repair or servicing of motor vehicles owned and used by persons other than the owner-tenant of the premises and/or for the storage of more than two (2) motor vehicles owned used by persons other than the owner or tenant of the premises.

Public garage. Public garage for the purpose of the Code shall be any building or part thereof exceeding 750 square feet in area wherein is kept or stored a motor vehicle having any gasoline or other volatile inflammable oil in its fuel storage tank, or where the painting, repairing, or greasing of motor vehicles is performed. A garage used as a repair shop for motor vehicles is performed. A garage used as a repair shop for motor vehicles shall be considered a public garage for the purpose of the Code.

Public garage. Any building or premises used for the storage of one or more non-commercial self-propelled vehicles or motorcycles, including storage for repair, demonstration, sale, rental, spot painting or adjustment of equipment. Repairs shall not include body and fender work and paint spraying. A salesroom conducted exclusively for the exhibition of not more than ten (10) vehicles as defined above shall not be classed as a public garage.

Public garage. Any building, other than that herein defined as a private garage, used for the storage or care of motor vehicles, or where any such vehicles are equipped for operation, repaired, or kept for remuneration, hire or sale.

Public garage. Garage in which parking fees are charged or in which motor vehicles are repaired or any garage which is not included in the term "Private Garage."

Public garage. Any premises, except those described as a private or storage garage, used for the storage or care of self-propelled vehicles, or where any such vehicles are equipped for operation, repaired or kept for remuneration, hire or sale.

Public garage. Building or portion thereof, other than a private or storage garage, designed or used for equipment, repairing, hiring, servicing, selling or storing vehicles.

Public garage. Garage building designed for the storage of five (5) or more motor vehicles and is for the use of the residents of the multifamily project or their visitors, where spaces are identified by the residents.

Public garage. Garage other than a private garage, available to the public, operated for gain, and which is used for storage, repair, rental, greasing, washing, servicing, adjusting or equipping of automobiles or other vehicles.

Public garage. Any building or premises, except those described as a private or storage garage, used for the storage or care of motor vehicles, or where any such vehicles are equipped for operation, repaired or kept for remuneration, hire, or sale.

Public garage. Any garage other than a private garage available to the public, operated for gain, and used for storage, repair, rental, greasing, washing, sales, servicing, adjusting or equipping of automobiles or other motor vehicles.

Public garage. A building or premises used for housing only of motor vehicles and where no equipment or parts are sold and vehicles rebuilt, serviced, repaired, hired or sold, except that fuel, grease or oil may be dispensed within the building to vehicles stored therein.

Public garage. A building or portion thereof, other than a private or storage garage, designed and used for equipping, servicing or repairing motor vehicles, leasing, selling and storing of motor vehicles may be included.

Public garage. Any premises used for the housing or care of more than four (4) motor vehicles (excluding community garages) or where any such vehicles are equipped for operation, repaired, or kept for renumeration, hire or sale, not including exhibition or show rooms for model cars. A building or premises used primarily as an automobile body and fender shop, an auto laundry, an automotive machine shop, a welding shop, an auto repainting shop or a shop engaged in the repair or testing of airplane engines is not considered to be a public garage.

Public garage. Building or portion thereof, other than private or parking garage, designed and used for equipping, servicing, repairing, hiring, selling or storing motor driven vehicles.

Public garage. Any garage, except a private garage, used for the storage or care of self-propelled vehicles.

Public garage. Garage other than a private garage, used for housing or care of automobiles or other self-propelled vehicles, or where any such vehicles are equipped for operation, repairing or kept for remuneration, hire or sale.

Public garage. Building, land or portion thereof, other than a private or storage garage, designed or used for equipping, servicing, repairing, hiring, selling, or storing motor-driven vehicles.

Public garage. Building structure or any portion thereof used for housing or repairing motor vehicles, including rooms for storing, exhibiting or showing cars for sale.

Public garage. Building, other than a private or storage garage, one or more stories in height, used solely for the commercial storage, service or repair of motor vehicles when permitted by the Fire Code.

Public garage. Building other than a private or storage garage, used for storage, equipping, repairing, hiring or selling of motor driven vehicles.

Public garage. Any enclosed space for the storage of one or more vehicles other than a private garage.

Public garage or repair garage. Any garage, other than a private garage or gasoline service station, which is used for storage, repair, rental, greasing, washing, servicing, adjusting or equipping of automobiles or other motor vehicles.

Public garage or service station. Any building, premises or land in or upon which there is an establishment devoted primarily to the business or supplying and selling motor fuel or other petroleum products, supplies or equipment for use in and upon motor vehicles and repairing, servicing, washing, storing or reconditioning of motor vehicles.

Public or repair garage. Any premises, except those described as a private or parking garage, used for the storage or care of self-propelled vehicles, or where any such vehicles are equipped for operation, repaired or kept for remuneration, hire or sale.

Public or storage garage. Building or part thereof other than a private garage or a community garage for the storage of motor vehicles and in which service station activities may be carried on.

Public storage garage. One for storage and service of motor vehicles for the public in which no machine or mechanical work is done.

Ramp type. Parking structure provided with inclined driveway for transporting vehicles to various parking levels.

Repair garage. Structure or portion thereof, other than a private or parking garage, used for the storage, sale, care, repair or refinishing of motor vehicles.

Repair garage. Building, land, or portion thereof other than a private or storage garage, designed or used for equipping, servicing, repairing, hiring, selling, or storing motor-driven vehicles.

Repair garage. Garage wherein major repairs may be made to more than two motor vehicles at a time.

Repair garage. Building or part thereof in which general repair of any kind is performed on motor vehicles.

Repair garage. Structure or part thereof where motor vehicles or parts thereof are repaired or painted.

Repair garage. Building or space for the repair or maintenance of motor vehicles, but not including factory assembly of such vehicles, auto wrecking establishment or junk yards.

Repair garage. Any building or premises which may be designed and used for the purposes of a "filling station" and also for major automotive mechanical repairs and body work and other customary and incidentally related activities.

Repair garage. Main or accessory building used or designed for repairing motor vehicles; a service garage if accessory to an automobile salesroom.

Repair garage. A building or portion thereof, other than a private garage, designed or used for equipping, servicing, or storing motor driven vehicles, including doing general repair and body work.

Sales garage. Garage or part thereof which is used exclusively for the display of motor vehicles for sale, and the offices and sales areas which are an accessory thereto.

Service garage. Repair garage accessory to an automobile sales room and primarily for the repair and servicing of automobiles of the make sold in the salesroom.

Service garage. "Major repairs" as applied to motor vehicles means body, fender, clutch, transmission, differential, axle, spring, or frame repairs; repair of an engine or motor requiring the removal of the cylinder head or crankcase pan; repairs to a radiator requiring the removal thereof, spray-painting; recapping or retreading of tires; and repairs of similar extent or hazard.

Service garage. Garage or part thereof which is used for the repairing or servicing of motor vehicles, except that a "service station building," as defined in the Code, is not a service garage.

Service station. Building, structure or land used for dispensing, sale or offering for sale at retail any automobile fuels, lubricants, or accessories and in connection with which is performed general automotive servicing as distinguished from automotive repairs.

Service station. Place where gasoline or any other motor fuel, lubricating oil, or grease for the operation of motor vehicles is offered for sale to the public and deliveries are made directly into the vehicle, including sale of accessories, performance of minor repairs and lubrication, and the washing of automobiles where no chain conveyor or blower is used.

Service station. Building and land including pumps, tanks and grease racks used for the retail sales of gasoline, lubricants, batteries, tires and other automobile accessories, and performing minor services and repairs.

Service station. Any business engaged primarily in the servicing of automotive vehicles, including the sale and delivery of fuel, lubricant's and other products necessary to the operation of automobile vehicles including the sale and installation of accessories, tires, batteries, seat covers, tire repair, cleaning facilities, minor engine tune-up and wheel balancing and aligning, brake service, but not including mechanical or body repair facilities, the sale or rental of vehicles or trailers.

Service station. Building or lot where gasoline, oil, greases and accessories are supplied and dispensed to the motor vehicle trade, also where battery, tire and other similar services are rendered.

Service station. Filling station which supplies motor fuel and oil to motor vehicles including grease racks or elevators and providing minor tire and battery servicing and sale of motor vehicle accessories.

Service station. Any commercial establishment conducted primarily for the purpose of retailing lubrication oils, gasoline or other motor fuel for internal combustion engines for motor vehicles from storage tanks above or below the ground with or without other incidental service.

Service station. Premises, including the buildings and equipment, used for selling at retail of products, other than liquefied petroleum gases, for the propulsion of motor vehicles.

Service station. Any building, structure or land used for dispensing, sale or offering for sale at retail of any automobile fuels, oils or accessories and in connection with which is performed general automotive servicing as distinguished from automotive repairs.

Service station. Any building, place or location designed to supply motor vehicles with gasoline, oil, grease and supplies, and for the inspection, testing and examination and the repair thereof, and equipped with gasoline pumps and oil pumps maintained for the purpose of selling gasoline and oil.

Service station. Any building, structure, premises, or other place used for dispensing, sale, or offering for sale of any motor fuel or oils, having pumps and underground storage tanks of a total capacity of not more than thirty thousand (30,000) gallons; also where battery, tire, and other similar services are rendered, located wholly within lot lines. When such dispensing, sale, or offering for sale of any fuels, or oils is incidental to the conduct of a mechanical garage, the premises shall be classified as a mechanical garage.

Service station. Building, structure or land used for dispensing, sale or offering for sale at retail any automobile fuels, lubricants, or accessories and in connection with which is performed general automotive servicing as distinguished from automotive repairs, and replacement of parts.

Service station. Includes buildings on lots used for the purpose of supplying motor fuel to tanks of vehicles for immediate use. They shall have no cellars or basements but may have open pits if such pits are continually ventilated. Where the exterior wall of a service station is located 7′ or less from a property line or adjacent structure, no openings shall be placed in such walls; and all walls shall be of one hour fire-resistive construction.

Service station. Retail place of business engaged primarily in the sale of motor fuels, and which may supply goods and services generally required in the operation and maintenance of motor vehicles. These may include sale of petroleum products; sale and servicing of tires, batteries, automotive accessories and replacement items, lubrication services; washing of cars where no chain conveyor or blower is used. Operations outside shall be limited to the dispensing of gasoline, oil and water; changing tires, replacement and adjustment of automotive accessories such as windshield wipers, lights and batteries and similar minor customer needs.

Service station. Use of premises primarily for the retail sale of products other than liquefied petroleum gases, for the propulsion of motor vehicles, but including such products as kerosene, fuel oil, packaged naptha, Stoddard Solvent lubricants, tires, batteries, anti-freeze, motor vehicle accessories and other items customarily associated with the sale of such products; and for the rendering of services and making of adjustments and replacements to motor vehicles, and the washing, waxing, and polishing of motor vehicles, as incidental to other services rendered; and the making of repairs to motor vehicles, except those of a major type. Repairs of a major type are defined to be spray painting, body, fender, clutch, transmission, differential axle, spring, and frame repairs; major overhauling of engines requiring the removal of engine cylinder head or crankcase pan; repairs to radiators requiring the removal thereof; or complete recapping or retreading of tires.

Service station. Place of retail business at which outdoor automotive refueling is carried on using fixed dispensing equipment connected to underground storage tanks by a closed system of piping, and at which goods and services generally required in the operation and maintenance of motor vehicles and fulfilling of motorist needs may also be available. The building consists of a sales office where automotive accessories and packaged automotive supplies may be kept or displayed. It may also include one or more service bays in which vehicle washing, lubrication and minor replacement, adjustment and repair services are rendered. An automotive service station building shall have no cellar or basement, but may have open pits if such pits are continually ventilated.

Service station. Service station to supply motor fuel and oil to motor vehicles and including grease rack or elevators and providing minor tire and battery servicing and sales of motor vehicle accessories.

Service station building. Building which is an accessory to a service station and which is used as a shelter for the operators of the service station, or for any of the purposes for which a "service station" may be used, under the definition in the Code.

Service stations. A building or portion thereof where gasoline, oil and greases are supplied and dispensed to the motor vehicle trade, also where tire, battery, washing, polishing and lubrication services are rendered and minor adjustments are made.

Storage garage. Building or portion thereof, except if defined as a private garage, providing storage for motor vehicles which may have facilities for washing and cleaning only.

Storage garage. Enclosed space for the storage of three (3) or more motor vehicles pursuant to previous arrangement, and not to transients, and at which automobile fuels and oils are not sold and motor vehicles are not equipped, repaired, hired or sold.

Storage garage. Building or portion thereof designed or used exclusively for term storage or motor driven vehicles, and at which motor fuels and oils are not sold and motor-driven vehicles are not equipped, repaired, hired or sold.

Storage garage. Building, not a private or public garage, one story in height, used solely for the storage of motor vehicles (other than trucks), but not for the service or repair thereof, nor for the sale of fuel, accessories, or supplies.

Storage garage. Any building or premises, other than a private or public garage, used exclusively for the parking or storage of motor vehicles, which shall contain space for a minimum of twenty-five automobiles or trucks.

Storage garage. Building, not a private or public garage, one story in height, used solely for the storage of motor vehicles (other than trucks), but not for the service or repair thereof, nor for the sale of fuel, accessories, or supplies, or washing, cleaning and polishing.

Storage garage. Building or operation thereof designed or used exclusively for the storage or parking of automobiles. Services other than storage at such storage garage shall be limited to refueling, washing, waxing and polishing.

Storage garage. Building, land, or portion thereof designed or used for storage only of five (5) or more motor-driven vehicles pursuant to previous arrangement and not to transients, and at which automobile fuels and oils are sold and motor-driven vehicles are not equipped, repaired, hired, or sold.

Storage garage. Any building or premises or part thereof, used only for housing of motor driven vehicles pursuant to previous arrangement and not transients, and at which automobile fuels and oils are not sold, and motor driven vehicles are not equipped, repaired, hired or sold.

Storage garage. Building or part thereof intended for the storage or parking of motor vehicles and which contains no provision for the repair or servicing of such vehicles.

Storage garage. Garage in which volatile inflammable oil, other than that contained in the fuel storage tanks of motor vehicles, is handled, stored or kept.

Storage garage. Building or portion thereof designed or used exclusively for term storage by prearrangement of motor-driven vehicles, as distinguished from daily storage furnished transients, and at which motor fuels and oils are not sold, and motor-driven vehicles are not equipped, repaired, hired, or sold.

Storage garage. Building or portion thereof designed and used exclusively for housing more than four (4) vehicles.

Storage garage. Main or accessory building, other than a private garage, used for the parking or temporary storage of passenger automobiles, and in which no service shall be provided for renumeration.

Storage garage. A building or premises used for housing only of motor vehicles and where no equipment or parts are sold and vehicles are not rebuilt, serviced, repaired, hired or sold, except that suel, grease or oil may be dispensed within the building to vehicles stored therein.

Storage of nonoperating motor vehicles. Shall not include automobile wrecking. The presence on any lot or parcel of land of five (5) or more motor vehicles which for a period exceeding thirty (30) days have not been capable of operating under their own power, and from which no parts have been or are to be removed for re-use or sale shall constitute prima facie evidence of the storage of nonoperating motor vehicles.

Super service station. A filling station which supplies motor fuel and oil to motor vehicles, and includes, as a part of its service or equipment, grease racks or elevators, wash racks or pits, tire repairs, battery servicing and repairing, ignition service, sales of motor vehicle accessories and other customary services for automobiles, including tire recapping where the equipment for such recapping does not exceed four (4) moulds, but excluding painting, body work and repairs and steam cleaning.

Underground parking garage. Enclosed parking garage located below grade with no building above the garage.

Used car lot. Any land used or occupied for the purpose of buying and selling second-hand motor vehicles and storing said motor vehicles prior to sale, provided that such vehicles are in running condition as evidenced by a inspection sticker, affixed thereto.

Used car lot. Area for the display and sale of usable used cars.

Used car lot. Lot or group or contiguous lots, used for the storage, display and sales of used automobiles and where no repair work is done except the necessary reconditioning of the cars to be displayed and sold on the premises.

Used car sales lot. Lot or parcel of land used for the display and sale of used automobiles.

Vehicle. Any self-propelled conveyance designed and used for the purpose of transporting or moving persons, animals, freight, merchandise, or any substance, and shall include passenger cars, trucks, buses, motorcycles, scooters, but shall not include tractors, construction equipment, or machinery, or any device used in performing a job as stated above.

Vehicle major repair. Any general repair, rebuilding or reconditioning, collision service including body, frame, or fender straightening or repair; painting or paint shop; mechanical car wash establishments; but not including any operations which require the heating or burning of rubber.

Wharf. Structure built on the shore of a body or stream of water to which vessels may tie up for the loading and discharge of passengers or cargo. Not a bulkhead.

C-18 PARKING STRUCTURES

Automobile parking structure. A structure used for the parking or storage of automobiles.

Automotive lift. A fixed mechanical device for raising an entire motor vehicle above the floor level, but not through successive floors of the building or structure.

Commercial garage. Any building or part thereof wherein is kept or stored any truck exceeding 1-1/2 ton capacity, or more than four passenger automobiles; or more than three trucks of 1-1/2 ton capacity or less, with gasoline or other volatile flammable fuel in their storage tanks; or which is used for the business of repairing or general servicing of motor vehicles.

Commercial garage. Any building or premises used for the storage of one or more buses, trucks, tractors, trailers, bulldozers and other heavy motor-driven equipment, including storage for manufacture, repair, demonstration, sale, rental, painting, adjustment or inspection of the foregoing.

Commercial garage. Garage building designed for the storage of five or more motor vehicles, but is operated for the public generally.

Garage. Any building or portion of which, used either for the storage or repairs of automobile equipment, or as stalls for cars other than the car of the owner.

Garage. A building, or parts thereof, designed or used for the shelter, storage or servicing of motor vehicles containing flammable fuel.

Garage. A building or part of building or premises in which one or more motor vehicles, excluding motorcycles, are kept for storage, manufacture, repair, demonstration, sale, rental, painting, oiling, greasing, adjustment of equipment, or washing, including also a place of storage of motor vehicles for such work.

Group garage. A place of storage of five (5) or more motor vehicles, whether such place of storage be in one (1) or more buildings.

Hoist type. A parking structure in which mechanical lift equipment provides a combination vertical and horizontal operation for semi-automatic parking of vehicles in a multi-level structure.

Manlift. A continuous belt driven lifting device used for transporting attendants between various floor levels.

Manlifts. A power-operated belt device with steps and hand-holds for transporting persons in a vertical position through successive floors or levels of the building or structure.

Mechanized parking garage equipment. Special devices in mechanical parking garages that operate in either stationary or horizontal moving hoistways, that are exclusively for the conveying of automobiles, and in which no persons are normally stationed on any level other than the receiving level and in which each automobile during the parking process is moved by means of a power driven transfer device, on and off the elevator directly into parking spaces or cubicles.

Motor vehicle. A self-propelled vehicle powered by an internal combustion motor.

Off-street parking facility. A structure, lot or portion of either which is used exclusively for the temporary storage of passenger motor vehicles.

Open-air parking garage. A structure having not less than fifty (50) percent of the area of the exterior walls open at each story and is used for the parking of motor vehicles.

Open air parking garage. Garage having not less than 50 percent of two sides of the garage open to the air at each story and used for the parking of motor vehicles.

Open deck parking garage. Structure with exposed structural members which has at least two sides without enclosing walls and which is used only for parking passenger automobiles having a capacity of not more than nine (9) persons each.

Open parking decks and garages. Enclosed structures, or parts of structures, for the parking or storage of automobiles under and on the deck level or levels.

Open parking garage. Parking garage having 50 percent or more of the area of each of 2 or more sides of the garage permanently open to the outside air in each story of the building.

Open parking structure. Unenclosed or partially enclosed structure for the parking or motor vehicles.

Open parking structure. Structure for the parking of passenger cars wherein two or more sides of such structures are not less than fifty percent open on each floor or level for fifty percent of the distance from the floor to the ceiling, and wherein no provision for the repairing of such vehicles is made.

Open parking structure. Structure for the storage of more than three (3) passenger automobiles in which at least two exterior walls are open.

Open parking structure. A structure for the parking of motor vehicles having at least 75 per cent of two exterior sides of each story permanently open.

Parking area. One or more parking spaces and may also include access drives, aisles, ramps and maneuvering area.

Parking building. A public garage designed and used primarily for the parking of automobiles.

Parking facility. A multi-level structure in which each level is used primarily for the purpose of storing passenger motor vehicles, and which does not necessarily have enclosing walls.

Parking garage. Building or structure wholly devoted to the parking of motor vehicles, and where a charge is made for storage or parking vehicles.

Parking garage. A garage for passenger motor vehicles involving only the parking or storing of automobiles and not including automobile repair or service work or the sale of gasoline or oil.

Parking garage. Any building or portion thereof used for the storage of four or more automobiles in which any servicing may be provided is incidental to the primary use for storage purposes, and where repair facilities are not provided.

Parking level. A floor in a structure on which vehicles are parked.

Parking lift (automobile) Mechanical device for parking automobiles by movement in any direction.

Parking space. Space within a building, lot or parking lot for the parking or storage of one (1) automobile. The space shall not be less than a nine (9) foot by twenty (20) foot area, exclusive of drives, streets, alleys or aisles giving ingress and egress thereto.

Parking structure. An unenclosed or partially enclosed structure for the parking of motor vehicles, with no provisions for the repairing or servicing of such vehicles.

Parking structure (open). A structure for the parking of passenger cars wherein two (2) or more sides of each structure are not less than fifty (50) percent open on each floor or level for fity (50) percent of the distance from the floor to the ceiling and wherein provisions for the repairing of such vehicles is not made. Such open parking structures are not classified as public garages, but shall comply with the requirements of the Code.

Private garage. A building or portion of a building not more than 1000 square feet in area, in which only motor vehicles used by the tenants of the building or buildings on the premises are stored or kept. A double garage is a garage 400 square feet or more in area with a width and length each not less than 18 feet.

Private garage. A detached accessory building or a portion of the principal building used or intended for use by the occupants of the premises for storage of passenger vehicles or trailers.

Private garage. A detached accessory building or a portion of a main building designed or used for the parking of temporary storage of automobiles owned and used by the occupants of the premises.

Private parking area. Any land area, being part of the same lot or tract on which is erected a building or structure, or being adjacent and contiguous to said lot or tract, designated by the owner, operator or occupant of said building or structure for the parking, without charge, of motor vehicles of occupants, customers or employees in said building or structure, where said land area is not in a more restricted zone than said building or structure, and where no cusomter parking shall be permitted after posted time on any portion of said area.

Public garage. The structure or portion thereof, other than a private garage, used for the storage, sale, hire, care, repairing or refinishing of any vehicles.

Public garage. Any building or portion thereof, other than a private garage, designed or used for servicing, repairing, equipping, hiring, selling or storing of motor-driven vehicles.

Public garages. A place in which one (1) or more motor vehicles or parts thereof, excluding motorcycles, are kept for storage, manufacture, repair, demonstration, sale, rental, painting or servicing.

Public parking decks. Special structure limited in use only to the temporary parking of motor vehicles.

Ramp type. A parking structure provided with inclined driveway for transporting vehicles to various parking levels.

C-19 PENAL, DETENTION AND CORRECTIONAL FACILITIES

Administrative offices. The office space required for the operation of the facility based on the size of the facility and the type of inmates to be detained. Spaces should be provided for the following functions: booking, fingerprinting, photographing, show-up room, interrogation, and should be located inside the security area of the building.

Adult camp, ranch or farm. A corrections facility, usually located in a rural area, for persons sentenced to confinement for a short period of time.

Arsenal. The arsenal so located as to be secure from any possible access by prisoners or outside persons and at the same time be quickly accessible to jail officers.

Auditorium-gymnasium. A multi-purpose room provided for the assembly of large number of inmates, and for the use of physical exercise and the playing of team games.

Cell block. A group or cluster of single and/or multiple occupancy cells or detention rooms immediately and directly accessible to a day or activity room.

Cell design. Cells designed for the use of one prisoner. The minimum clear size of an interior cell should be approximately fifty (50) square feet, with a ceiling height of not less than eight (8) feet. Each cell should contain a rigidly constructed bed frame fastened to the wall or floor, a prison-type toilet and washbowl, a small shelf-type table with wall-bracket seat, a small shelf easily observable from the door, and a few sturdy hooks for towels and clothing. The total housing facility should be designed to provide approximately seventy-five (75) square feet per inmate, including the cells and day room areas.

Chapel. A room designed and furnished to provide religious comfort for inmates and visitors.

Community daycare housing. Housing in structures without physically restricting the construction, security fences and security hardware. The housing is usually converted apartment buildings or private homes. The housing is not intended to be detention facilities.

Consolidated jails. Jails used for the detention of prisoners awaiting court action and those few short-sentence prisoners who require maximum security. An institution serving several jurisdictions.

Control center. Every jail provided with a control center, manned around the clock, through which telephone and other communications are channeled, emergency alarm systems operated, official count records cleared, key control established, and all day-to-day operations centralized and controlled. The control room should be centrally located, with good visibility, and should be protected by bullet-resistant glass.

Detention. Temporary care of persons alleged to be delinquent who require secure custody in a physically restricting facility pending court disposition or execution of a court order.

Detention and correctional occupancies. Used for purposes such as jails, detention centers, correctional institutions, reformatories, houses of correction, pre-release centers, and other residential restrained care facilities where occupants are confined or housed under some degree of restraint or security, and provide sleeping facilities for four (4) or more residents and are occupied by persons who are generally prevented from taking self-preservation action because of security measures not under the occupants' control.

Detention and correctional occupancies. Are those used to house occupants under some degree of restraint or security. Detention and correctional occupancies are occupied by persons who are mostly incapable of self-preservation because of security measures not under the occupants' control.

Detention facility. A local confinement institution for which the custodial authority is usually 48 hours or more; some persons can be confined in such facilities pending adjudication and for short term sentences.

Disciplinary and restraint rooms. Every jail provided with facilities for the isolation and control of problem prisoners, including special restraint rooms for the housing of violent cases. It is recommended that restraint rooms be finished in ceramic tile and equipped with recessed light fixture, and unbreakable combination water closet and lavatory with control valve located outside the room, fixed-sash windows with laminated safety glass and forced-air ventilation. The room should contain no structural projections or articles of furniture. For cleaning purposes the floor should slope to a drain outside the room.

Drains. For ease of cleaning, all kitchens, dining rooms, corridors, dormitories, and cell blocks should contain floor drains.

Facility. The actual physical setting in which a detention program or agency functions.

Facility. A place, an institution, a building or part thereof, set of buildings or an area whether or not enclosing a building or set of buildings which is used for the lawful custody of persons.

Feeding. Every jail should maintain a properly equipped kitchen. In large jails a central dining room equipped for cafeteria-style service is recommended. In small jails where this is not feasible, tables located in the day rooms are used. In either case, a minimum of 24 inches of table space per man should be provided.

Foot-candle. A unit for measuring the intensity of illumination. It is the amount of light thrown on a surface one foot away. For recommended levels of illumination see the IES Lighting Handbook.

Guard facilities. Facilities provided for guards which should include, rest rooms, locker rooms, dining room and reading areas.

Guard houses. Include the construction of guard houses containing communication equipment, alarm system, arsenal rack, and lockers for clothing for inclement weather. All glass, laminated safety glass.

Halfway house. A residential facility located in the community which provides early release opportunities for inmates and similar services to pre-trial and pre-sentence clients, probationers, parolees, ex-offenders, and out-clients.

High hazard area. The kitchen and furnace or boiler room where a fire or explosion could occur; cell sections located in the basement; and any storage rooms which contain flammable equipment and supplies.

Holding facility or lockup. A temporary confinement facility for which custodial authority is usually less than 48 hours where arrested persons are held pending release, adjudication or transfer to another facility.

House of correction. Building used as an institution to which offenders are committed for punishment, discipline, or reformation.

Infirmary. Space provided where a jail physician may conduct sick call, examine inmates in privacy, and render routine medical treatment. In addition, facilities should be available where inmates suffering from illnesses which do not require hospital care be housed and properly cared for. Space for secure storage of medical supplies should be provided adjacent to the infirmary area.

Inmate. Any person, whether sentenced or unsentenced, who is confined in a detention or holding facility.

Inside security cells. Space provided for prisoners requiring maximum security. The calculated national average in any jail rarely exceeds 20 percent of the population.

Institutional use group. All buildings, or portions of buildings, which are designed or which are used for harboring persons for penal, or correctional confinement.

Jail. Building used for the confinement of persons held in lawful custody for minor offenses or pending judicial proceeding.

Jail. A confinement facility, usually operated by a local law enforcement agency, which holds persons detained pending adjudication and/or persons committed after adjudication for short term sentences, while intended for the confinement of adults, sometimes hold juveniles as well.

Jail. This institution houses both offenders awaiting court action and those serving short sentences.

Juvenile camp, ranch or farm. A correctional facility, usually located in a rural area, which has confinement authority over juveniles committed after adjudication.

Juvenile hall. Facility designed for the reception and temporary care of minors detained in accordance with the provisions of the juvenile law.

Laundry. A laundry is a necessity in every jail and space and facilities should be provided near service area locations. In small jails, home-type washing and drying equipment may be sufficient; in large jails, commercial type equipment would be required.

Library. A room used for the shelving of books, publications, and manuscripts and all other types of reading matter, for use by the inmates with sufficient space for reading.

Life safety code (NFPA No. 101). Manual published by the National Fire Protection Association, and revised when required, specifying minimum standards for fire safety necessary in the public interest; chapter 14—New Detentions and Correctional Occupancies, is devoted to corrections facilities, and chapter 31—Operating Features, Section 31-5 (1981 edition).

Location. Wherever located, the jail building should be set far enough back from the property line to prevent improper contacts between inmates and persons on the outside.

Lockup. A security facility, usually operated by the police department, for temporary detention of persons held for investigation or awaiting a preliminary hearing.

Lockups. Facilities used for short periods of detention immediately following arrest or during court sessions may be located in or near the courthouse or local government center. Facilities should never be located in basement rooms, or attics, or other space unfit for prisoner living quarters.

Maintenance shops. Shops provided and space and facilities for work activities.

Multiple-occupancy rooms. Minimum and medium custody prisoners can be housed in squad rooms or dormitories which should have a capacity of not less than four (4) nor more than fifty (50) inmates. The ceiling height should be at least 10 feet and at least 75 square feet of floor space should be available for each inmate. Multiple-occupancy rooms should contain at least one toilet and one washbowl for each eight (8) inmates or fraction thereof, and a minimum of one shower head for each fifteen (15) inmates or fraction thereof.

Nonsecure facility. A facility not characterized by the use of physically restricting construction, hardware and procedures which provides its residents access to the surrounding community with minimal supervision.

Place of detention. Building or part of a building used as a place of abode and wherein persons are forcibly confined, such as asylums, mental hospitals and jails.

Planning the new jail. The new facility planned as a joint effort of the jail administrator and the governing board, working with the architect and a staff of consultants from the supervisory state agency or qualified consultants from other jurisdictions.

Prison. Building for the safe custody or confinement of criminals or others committed by lawful authority.

Rated cell or room capacity. The officially stated number of inmates which a cell or room is designed to house.

Rated facility capacity. The officially stated number of inmates which a detention/corrections facility is designed to house.

Receiving and discharge. Space provided where prisoners are received, searched, showered, and issued clothing prior to entrance into living quarters. Discharge of prisoners can be accomplished in the same space.

Reformatory. Building used as an institution to which young offenders are committed for training or reformation.

Restrained. Persons that are locked in a building in such a manner that their freedom for normal egress to the outside of the building is dependent upon the unlocking or unbolting by others, of one or more doors or other barriers.

Restrained. As applied to persons in a building, means that such persons are confined or corporally restrained in the building in such a manner that the freedom for normal egress from the building is dependent upon the unlocking or unbolting by others, of one or more doors or barriers, or the removal of restraints.

Restricted detention. Includes such facilities as mental hospitals, mental/sanitoriums, jails, prisons, reformatories, and buildings where personal liberties of inmates are similarly restrained.

Safety equipment. Includes such equipment as chemical extinguishers, hoses, nozzles, water supplies, alarm systems, sprinkler systems, emergency exits and fire escapes, first aid kits, stretchers, and emergency alarms.

Safety vestibule. A grille cage at least six (6) feet square, located at the entry/exits, that divides the inmate areas from the remainder of the facility. The safety vestibule should have two (2) doors or gates, only one (1) of which opens at a time, permitting entry to or exit from inmate areas in a safe and controlled manner.

Safety vestibules. All entrances to security sections of the jail and to the cell blocks and dormitories provided with a double gate arrangement which will permit the locking of one (1) gate before the other is opened.

Sally port. A square or rectangular enclosure situated in the perimeter wall or fence of the facility, containing gates or doors at both ends, only one (1) of which opens at a time.

Sanitation and safety. All facilities of confinement requires provisions for high standards of sanitation and safety for the inmates and the administrative personnel.

Secure facility. A facility which is designed and operated so as to ensure that all entrances and exits are under the exclusive control of the staff of such facility, whether or not persons being detained have freedom of movement within the perimeters of the facility or are controlled through the use of locked rooms and buildings, fences, or physical restraint.

Security custody. The degree of restriction of inmate movement within a detention/corrections facility, usually divided into maximum, medium and minimum risk levels.

Security devices. Includes but is not limited to locks, gates, doors, bars, fences, screens, ceilings, floors, walls and barriers used to confine and control inmates. Also included are electronic monitoring equipment, security alarm systems, security light units, auxiliary power supply, and other equipment used to maintain facility security.

Security perimeter. The outer portions of a facility which actually provide for the secure confinement of inmates. The perimeter may vary for individual inmates, depending upon their security classification.

Shelter facility. Any public or private facility, other than a juvenile detention or correctional facility, designated to provide either temporary placement for alleged or adjudicated status offenders prior to the issuance of a disposition order, or a longer term care under a juvenile court disposition order.

Stairs. Stairs should be avoided for the vertical transportation of prisoners awaiting booking, dressing in, and the locking up of prisoners who may be intoxicated and incapable of being ambulatory. If the facilities are located on several floors, an elevator is recommended.

Storage space. Adequate space and facilities provided for storage of food supplies, including proper coolers and refrigerated areas, mattresses and bedding, inmate uniforms, maintenance equipment, cleaning supplies, and rooms for the sterilization and storage of prisoners' personally owned clothing.

Type of building. The building designed to provide custodial security of inmates who are to be confined in a building that is well lighted, ventilated and fire safe as possible using the requirements of the local, state or Life safety codes.

Visiting rooms. Visiting room arrangements which provide the maximum security in the separation of visitors from prisoners by a reinforced-glass partition which prevents the passage of contraband. The exchange of conversation is conducted over a voice-powered or self contained transistorized handset telephone and the conversation can also be monitored if considered necessary.

Windows. The building, so designed that inmates have no unsupervised access to windows. Windows in security areas should be equipped with steel bars or steel detention sash. Windows in prisoners' living quarters should have a minimum total space which is equal to at least one-eighth (1/8) of the floor space which the windows serve.

Workhouse, jail farm, or camp. Institutions which house minimum custody offenders serving short sentences.

Worker's quarters. Facilities provided for prisoners who are engaged in outside work-release programs, and housed in separate quarters which permit no contact with other inmates.

C-20 PLACES OF ASSEMBLY

Aisle. Passageway between rows of seats, or between rows of seats and a wall in a place of assembly, or between desks, tables, counters or other equipment or construction.

Arena. Place of assembly with a performance or exhibition area which is usually surrounded by fixed seats arranged in tiers.

Arena stage. A stage or platform open on at least three sides to audience seating. It may be with or without overhead scene handling facilities.

Armory. Building used by a military organization and containing a drill hall or similar area of assembly.

Armory hall. Any building which, or a portion of which, is used as a drill hall for members of any military organization.

Assembly. Buildings in which the primary or intended occupancy or use is the assembly for amusement, athletic, civic, dining, educational, entertainment, patriotic, political, recreational, religious, social, sports, or similar purposes.

Assembly building. Building used, in whole or in part, for the gathering together of persons for such purposes as deliberation, worship, entertainment, amusement or awaiting transportation.

Assembly building. Building or portion of a building used for the gathering together of 50 or more persons for such purposes as deliberation, worship, entertainment, amusement or awaiting transportation or of 100 or more persons in drinking and dining establishments.

Assembly building. A building used, in whole or in part, for the gathering together of persons.

Assembly hall. Building or part of a building seating or accommodating 100 or more persons and used for the assembly of persons for education, instruction, entertainment, or amusement, including places where persons congregate to hear speakers or lecturers or to listen to operas, concerts, or musical entertainments presented without the use of portable scenery; and banquet halls, dance halls, skating rinks and gymnasiums.

Assembly hall. Buildings used as an auditorium in which the public assembles, and whose primary and intended use is for the assembly of persons for the purpose of amusement, entertainment, instruction, worship, transportation, sports, military drills or similar purposes, with admission either public or restricted.

Assembly halls. In the assembly hall classification are included all buildings, or parts of buildings, other than theaters, which will accommodate more than 100 persons for entertainment, recreation, worship or dining purposes.

Assembly halls. In the assembly hall classification are included all buildings, or parts of buildings, other than theaters, which will accommodate more than 100 persons for entertainment, instruction, recreation, worship or dining purposes. Every assembly hall (commercial use) which will accommodate not more than 100 persons shall conform to the requirements of the Code, covering factories, office and mercantile buildings.

Assembly halls and roof gardens above first story. Where assembly halls are provided above the first story, limitation of occupancy, type of construction and exit facilities shall apply.

Assembly halls (without a stage). Include, but are not limited to, amusement park buildings; arenas; armories; art galleries; auditoriums; ballrooms; bath houses; bowling alleys; courtrooms; dance halls; dining rooms; exhibition halls; gymnasiums; lecture halls; libraries; motion picture theaters; natatoriums; music halls; museums; recreation centers; restaurants; shelters; skating rinks; and transportation stations, but are without a stage.

Assembly room. A room appropriated to the gathering together of persons for such purposes as deliberation, instruction, worship, entertainment, amusement, dining, or awaiting transportation.

Assembly space. Room or space used by a gathering of people for amusement, cultural, entertainment, exercise, governmental activities, social, sporting events, or for worship, or similar activities, where the size of such gathering of people is as set forth in the provisions of the building code, for a specific type of assembly, and comes within the classification of occupied space.

Assembly space. A room or space where more than ninety-nine persons congregate or gather for amusement, athletic, civic, dining, educational, entertainment, patriotic, political, recreational, religious, social, sports, or similar purposes.

Assembly space. Any part of a place of assembly, exclusive of a stage, that is occupied by numbers of persons during the major period of occupancy. Every tier of seating shall be considered a separate assembly space.

Assembly use group. Buildings, or portions of buildings, which are designed or which are primarily used for public assembly for the purpose of amusement, education, entertainment, recreation, religion, sports events, and civic and similar purposes, shall be classified in the assembly use group.

Assembly use group. All buildings, or portions of buildings, which are designed or which are primarily used for public assembly for the purpose of amusement, entertainment, recreation, education, religion, travel, civic or similar purposes, shall be classified in the assembly use group, based on allowable occupancy.

Auditorium. Assembly hall in which persons may assemble to hear or see concerts, plays, lectures, athletic events or similar performances.

Auditorium. That part of a church, theater, or other place of assembly, assigned to the audience as provided under the requirements of the code.

Auditorium. Means that portion of a building used as a place of assemblage.

Auditorium. Shall mean a building or portion thereof used or designed for use by an audience, including but not limited to, public assembly, religious services, meetings, lectures, dances, entertainment, and similar uses. The largest room of a church, school or college used for the assembly of persons shall be deemed to be the auditorium.

Auditorium. That portion of a building used as a place of assemblage.

Auditorium. Building or portion thereof used or designed for use by an audience, including but not limited to, public assembly, religious services, meetings, lectures, dances, entertainiment, and similar uses. The largest room of a church, school or college used for the assembly of persons shall be deemed to be the auditorium when meeting the requirements of the Code.

Auditorium. That part of a church, theater, or other place of assembly, assigned to the audience.

Auditorium. That portion of an assembly room with fixed seats, movable seats, or terraced floor.

Auditorium area. Separate area used exclusively for the actual exhibition of motion pictures to patrons seated in fixed seats arranged for this type of entertainment.

Balcony. Gallery or mezzanine floor above the main floor of an auditorium, theatre or church. Where such a balcony extends over 33 per cent or more of the horizontal area within the outer walls of the building it shall be considered a story.

Balcony. Within an auditorium, is a floor, inclined, stepped, or level, above the main floor, the open side or sides of which shall be protected by a rail or railings. Where a balcony of an auditorium has means of egress at two (2) or more levels opening into separate foyers, one above another, each portion thereof served by such a foyer shall be considered a separate balcony for the purpose of the code.

Balcony. Shall mean that portion of a building which projects into and above another floor in the same room or space.

Clubhouse. Any house or building occupied, in whole or in part, as a residence by members of a club or organization, and whose members and guests have access to all parts or any part of the house or building. The term shall also be understood to mean and include a fraternity house.

Club or lodge building. Building used by a fraternal, social, or similar organization for the private assembly of persons and not used by the general public as a place of assembly.

Collecting safe area. A safe area that receives occupants from the assembly space it serves, as well as from other safe areas.

Continental seating. Seating in rows across the auditorium with access from aisles along each sidewall and from the safe area outside and adjacent to the auditorium walls.

Convention hall. An assembly or meeting place for delegates for action on particular matters such as political, fraternal, veterans affairs and the like.

Cross over aisle. An aisle in a place of assembly usually parallel to rows of seats, connecting other aisles or an aisle and an exitway.

Dais. Small and low "platform" or "rostrum," may be placed on a "stage." May be permanent, temporary, or portable.

Dining room. Place of assembly used by persons for consumption of food and beverages.

Dressing room. Room used or intended to be used by a performer or performers for dressing or changing of clothing.

Enclosed platform. Partially enclosed portion of an assembly room the ceiling of which is not more than five feet (5') above the proscenium opening, and which is designed or used for the presentation of plays, demonstrations, or other entertainment wherein scenery, props, decorations, or other effects are to be installed or used.

Fly gallery. Long narrow walkway at an elevation above the top of the proscenium opening of the space above a stage used (principally) for storing or manipulating stage equipment or scenery.

Foldable grandstand. Assembly of prefabricated units used to produce a grandstand which can readily be folded, rolled or telescoped in a front-to-back direction into a comparatively small space when not in use and can readily be extended for use as a support for audiences.

Foyer. The enclosed space surrounding or in the rear of the auditorium of a theatre or other place of assembly which is completely shut off from the auditorium and is used as an assembly or waiting space for the occupants.

Foyer. Is an area or space within a building and located between a lobby and main entrance and the main floor.

Foyer. Any space in a place of assembly which is used for ingress, egress, the distribution of people to the aisles, and as a space for waiting.

Gallery. Any seating tier above a balcony within a place of assembly.

Gallery. Is that portion of the seating capacity of a theatre or assembly room having a seating capacity of more than ten persons and located above a balcony.

General stage. Stage is a partially enclosed portion of an Assembly Building, cut off from the audience section by a proscenium wall, which is designed or used for the presentation of plays, demonstrations, or other entertainment "Stages" shall be classified as "working stage" and "non-working stage."

Gymnasium. Building or part thereof used for athletic exercises or for athletic games.

Large assembly. Shall include theaters and places of assembly having a capacity of one thousand (1,000) or more persons. Also, Large Assembly shall include theaters and places of assembly having a working stage and having a capacity of seven hundred (700) or more persons.

Licensed place of public assembly. The term "Licensed Place Of Public Assembly" as used in the Code shall mean any room or space which is used or occupied as a "Place Of Assembly" as defined in the Code, when the lawful use, occupancy or operation of such place is contingent upon the issuance of a license by the fire department, the police department or the department of licenses.

Lobby. The enclosed vestibule between the principal entrance to the building and the doors to the main floor of the auditorium or assembly room of a theatre or place of assembly or the main floor corridor of a business building.

Longitudinal aisle. Passageway perpendicular to the direction in which seat rows are placed.

Minor places of assembly. Places of assembly designed or used for a gathering of fewer than 100 persons, but more than 49 persons.

Music hall. Building or part thereof, the primary purpose of which is for musical entertainment of a gathering of people.

Night club. Place of assembly used for the consumption of food and alcoholic beverages, and which normally affords entertainment through permitting dancing or through organized entertainment other than music.

Non-working stage. Partially enclosed portion of an Assembly Building, cut off from the audience section by a proscenium wall of not less than one-hour fire-resistive construction, without the equipment common to the Working Stage (such as fly gallery and Gridiron) and of such dimensions that such equipment cannot be installed (but flat scenery may be used on such stage). A fireproof curtain is not required for a non-working stage, but if there is a fabric or other curtain it shall be of incombustible materials or treated with an approved fire retardant. The depth of the stage may be more or less than fifteen (15) feet.

Outdoor places of assembly. Include grandstands, bleachers, coliseums, stadiums, drive-in theaters, tents, and similar structures, and enclosed areas which are designed or used for outdoor gatherings of an aggregate of 200 or more persons. Outdoor places of assembly are regulated by the Code.

Place of assembly. Room or space used for assembly or educational occupancy for 100 or more occupants.

Place of assembly. Floor area that has an occupant load based on fifteen (15) square feet or less per person.

Place of assembly. A room or space accommodating fifty (50) or more individuals for religious, recreational, educational, political, social or amusement purposes, or for the consumption of food and drink, including all connected rooms or space with a common means of egress and entrance.

Place of assembly. Room or space in which provision is made for the seating of one hundred (100) or more persons for religious, recreational, educational, political, social or amusement purposes or for the consumption of food or drink. Such room or space shall include any occupied connecting room or space in the same story, or in a story or stories above or below, where entrance is common to the rooms or spaces.

Place of assembly. Room or space accommodating one hundred (100) or more individuals for religious, recreational, educational, political, social and amusement purposes or for the consumption of food and drink, including all connected rooms or space with a common means of entrance and exit.

Place of assembly. Room or space which is occupied by seventy-five (75) or more persons and which is used for educational, recreational or amusement purposes and shall include assembly halls in school structures; dance halls; cabarets; night clubs; restaurants; any room or space used for public or private banquets, feasts, socials, card parties or weddings; lodge and meeting halls or rooms; skating rinks; gymnasiums; swimming pools; billiard, bowling, and table tennis rooms; halls or rooms used for public or private catering purposes; funeral parlors; markets; recreational rooms; concert halls; broadcasting studios; school and college auditoriums; and all other places of similar type of occupancy.

Place of assembly. Any building designed, constructed, reconstructed, remodeled, altered, used, or intended to be used, for fifty (50) or more persons to assemble therein for any of the following: Dance halls; cabarets; restaurants; including the type of restaurant commonly known as a night club; all places in which alcoholic beverages are sold or for sale to be consumed on the premises; any room or space used for public or private banquets, feasts, dances, socials, card parties, or weddings or religious services except in the case of funerals in private homes; lodge and meeting halls or rooms; skating rinks; gymnasiums; swimming pools; billiard; pool, bowling and table tennis rooms; halls or rooms used for public or private catering purposes; funeral parlors; recreation rooms; broadcasting studios; school and college auditoriums; and all other places of similar occupancy. Nothing in this paragraph shall apply to a single family or two family dwelling, or to a place of incarceration or detention, a convent, a monastery, a church, a synagogue, a theatre, a special hall, a public hall, or a schoolhouse.

Place of assembly. Room or space in which seventy-five (75) or more persons are congregated for religious or recreational, educational, political, social or amusement or for the consumption of food or drink. Such room or space shall include any occupied appurtenant rooms or space.

Places of assembly. Other than school assembly halls, and outdoor places of assembly.

Places of assembly. For fewer than fifty (50) persons and used as an accessory to another occupancy, should conform to the provisions of the Code which are applicable to a building of such occupancy; otherwise such places of assembly shall conform to the provisions of the Code which are applicable to business buildings.

Places of assembly. Include but are not limited to all buildings or portions of buildings used for gathering together fifty (50) or more persons for such purpose as deliberation, worship, entertainment, dining, amusement or awaiting transportation.

Platform. Platform is a raised section of floor within the assembly hall or auditorium area, and setting on the floor thereof, not enclosed above the platform floor level, and usually of relatively small area as compared to the auditorium seating area. A platform may be of permanent, temporary, or portable construction; it may have "flat" movable scenery and draw curtains.

Platform. A portion of an assembly room which may be raised above the level of the assembly floor and which may be separated from the assembly space by a wall and proscenium opening provided the ceiling above the platform shall be not more than five (5) feet above the proscenium opening.

Platform (enclosed). A partially enclosed portion of an assembly room the ceiling of which is not more than five feet (5′) above the proscenium opening and which is designed or used for the presentation of plays, demonstrations, or other entertainment wherein scenery, drops, decorations, or other effects may be installed or used.

Podium. Small "dais" of size sufficient to accommodate one or two persons, such as a band or orchestra conductor or a soloist. A "Podium" may be located on a Stage, Platform, Rostrum, or Dais, or the floor of the audience section of a place of assembly. A Podium is almost always a portable construction.

Podium stage. A small "dais" of size sufficient to accommodate one or two persons, such as a band or orchestra conductor or a soloist. A "podium" may be located on a Stage, Platform, Rostrum. A podium is almost always a portable construction.

Projection block. That portion of a theater or assembly room containing a projection room alone or in combination with other rooms appurtenant to the operation thereof.

Property room. Room for the storage of any materials of a theatrical or similar performance, except scenery, commonly known and described as stage properties.

Proscenium. Vertical plane of separation between an auditorium and a stage.

Proscenium. Wall which divides the stage, scenery and dressing rooms from the space designated for the audience.

Proscenium wall. Wall in a place of assembly which divides the stage and scenerey space from the seating space.

Proscenium wall. A wall between an assembly room and a stage.

Public space. Space within a building for public use where more than ninety-nine persons congregate, such as lobbies, lounges, reception, ball, meeting, lecture and recreation rooms, banquet and dining rooms and their kitchens, and swimming pools.

Raised platform. A raised portion of floor to be used for simple stage purposes that involves a minimum of fire hazard, so located that it extends not more than eighteen (18) feet behind the probable curtain line of the proscenium opening and of an area limited to seventeen and one-half (17.5) percent of the assembly room floor area of 1,550 square feet, whichever is less.

Rostrum. Usually used for single or small group of persons such as lecturers. Scenery or curtains are not included. May be permanent, temporary, or portable.

Row of seats. Group of adjoining seats arranged side by side.

Safe area. Area which receives occupants from normal entry exits discharged into it from the place of assembly it serves as well as from other safe areas.

Safe dispersal area. Area which will accommodate a number of persons equal to the safe capacity of the stand and building it serves, in such a manner that no person within the area need be closer than 50 feet from the stand or building. Dispersal areas normally based upon an area of not less than 3 square feet per person.

Seating capacity. Number of seats within an auditorium or other hall when fastened to the floor; the number of persons who may be seated within an auditorium or hall allowing six square feet of floor area per person unless fixed seats are provided.

Seating capacity. Number of seats permitted in any area or occupancy of assembly as based upon floor areas, as stated in the Code.

Seating capacity. The number of seats permitted in any area or occupancy of assembly as based upon floor areas and use group.

Seating capacity. Seating capacity of a theater, auditorium or any room or place of public assemblage in which seats are not fixed, shall be determined on the basis of 7 square feet of floor, balcony and gallery area per person; in the case of fixed seats, such as pews or benches, the seating capacity shall be based on one (1) person to each 18 inches of pew or bench length.

Seating capacity. Signs stating the maximum seating capacity shall be conspicuously posted by the Owner of the building in each assembly room, auditorium or room used for a similar purpose where fixed seats are not installed. It shall be unlawful to remove or deface such notice or to permit more than this legal number of persons within such space.

Seating capacities—moveable seating. Capacity of dance floors or the playing areas of gymnasiums, board rooms, conference rooms, dining rooms, and assembly areas of less concentrated use, when such areas or floors are not to be used for general assembly purposes, shall be determined on the basis of 15 square feet of floor area per person.

Seating section. A area of seating bounded on all sides by aisles, cross over aisles, walls or partitions.

Stage. Space in a theater or assembly room separated from the auditorium equipped for theatrical or similar performances that provide for the use of curtains, portable or fixed scenery, lights, or mechanical appliances. Recesses at the front of an auditorium used or designed soley for the mounting of a motion picture screen and its required sound equipment containing no fixed or movable scenery other than curtains of flame-resistive material shall not be deemed to be a stage.

Stage. Raised platform with its scenery and theatrical accessories on which the performance in a theatre, concert hall, auditorium or place of entertainment, takes place. This definition shall not include an unenclosed raised platform placed on an open floor to elevate the performers, musicians or speakers, provided no curtain, scenery or other theatrical accessories associated with the stage not provided. A back drop of incombustible materials, or materials treated so as not to ignite or support combustion, may be provided.

Stage. A space within a place of assembly which is designed or used for the presentation of plays, demonstrations, entertainment, or other such purpose wherein scenerey, drops, or other effects may be installed or used, and where the distance between the top of the proscenium opening and the ceiling of the stage is more than 10 feet, with safety provisions required by the Fire Code.

Stage. A partially enclosed portion of an assembly building which is designed or used for the presentation of plays, demonstrations, or other entertainment wherein scenery, drops, or other effects may be installed or used, and where the distance between the top of the proscenium opening and the ceiling above the stage is more than five (5) feet.

Stage. A partially enclosed, unenclosed, or raised platform portion of an assembly building which is designed or used for the presentation of plays, demonstrations or other entertainment wherein scenery, drops, or other effects may be installed or used.

Stage. A partially enclosed portion of a building which is designed or used for the presentation of plays, demonstrations, or other entertainment wherein scenery, drops or other effects may ;be installed or used.

Stage. The term "stage" shall mean the raised platform with its scenery and theatrical accessories on which the performance in a theatre, concert hall, auditorium, or place of entertainment, takes place. This definition shall not include an unenclosed raised platform placed on an open floor to elevate the performers, musicians or speakers, provided no curtain, scenery or other theatrical accessories associated with the stage not provided. A back drop of incombustible materials, or materials treated so as not to ignite or support combustion, may be provided.

Stage. Space within a place of assembly which is designed or used for the presentation of plays, demonstrations, entertainment, or other such purpose wherein scenery, drops, or other effects may be installed or used, and where the distance between the top of the proscenium opening and the ceiling of the stage is more than 10 feet.

Stage. Partially enclosed portion of an assembly building which is designed or used for the presentation of plays, demonstrations, or other entertainment wherein scenery, drops, or other effects may be installed or used, and where the distance between the top of the proscenium opening and the ceiling above the stage is more than five feet (5′), and in accordance with the Fire Code.

Stage block. That portion of a theater or assembly room containing only the stage or the stage in combination with dressing rooms, storage and property rooms, workshops and other rooms appurtenant to the operation thereof.

Stage platform. Platform is a raised section of floor within the assembly hall or auditorium area, and setting on the floor thereof, not enclosed above the platform floor level, and usually a relatively small area as compared to the auditorium seating area. A relatively small areas as compared to the auditorium seating area. A platform may be of permanent-temporary, or portable construction; it may have "flat" movable scenery and draw curtains.

Stage podium. Small "dais" of size sufficient to accommodate one or two persons, such as a band or orchestra conductor or a soloist. A "podium" may be located on a Stage, Platform, Rostrum, or Dais, or the floor of the audience section of a place of assembly. A podium is almost always of portable construction.

Stage rostrum. Usually used for single or small group of persons such as lecturers, no scenery or curtains. May be permanent, temporary, or portable.

Stage–working. Working stage is a partially enclosed portion of an Assembly Building, cut off from the audience section by a proscenium of masonry of not less than 4 hour fire-resistance construction, and which is equipped with a fireproof and smoke-proof curtain, and the depth from the proscenium curtain to the back wall shall be not less than fifteen (15) feet.

Stage workshop. Any shop or room in which carpentry, electrical work incidental to the preparation, operation or maintenance of any stage is done.

Thrust stage. That portion of a stage that projects into the audience on the audience side of a proscenium wall or opening.

Tier. Main floor, mezzanine, loge, balcony, gallery or other similar level, on which seats are provided.

Tier. Level of seating separated from other levels by floors, walls or partitions in structures of Assembly Occupancies.

Tier. Term "tier," as used in connection with exits or seats in special occupancy structures, shall mean an orchestra floor, mezzanine, lodge, balcony, gallery, or other similar level in the auditorium of such special occupancy structure in which seats are provided for the audience.

Tier of seating. A general level of seating, such as an orchestra (usually the main tier), a balcony or gallery.

Working stage. Portion of an assembly building which is cut off from the audience section by a proscenium wall, provided with an opening, so arranged that curtain or drops may be lifted more than ten feet (10′) or so that their lower edges are higher than one-half of the height of the proscenium opening.

Working stage. Partially enclosed portion of an Assembly Building, cut off from the audience section by a proscenium wall of masonry of not less than 4 hour fire-resistance construction, and which is equipped with scenery loft, gridiron, fly-gallery, and lighting equipment, and the proscenium opening shall be equipped with a fireproof and smoke-proof curtain, and the depth from the proscenium curtain to the back wall shall be not less than fifteen (15) feet.

C-21 PUBLIC SERVICE STRUCTURES AND SYSTEMS

Carbarn. A structure for the storing and repairing of electric street cars and buses, and other electrical conveyances.

Company. Any person, firm or corporation now or hereafter engaged as a public utility in the business of furnishing gas, electric, water or telephone services to domestic, commercial or industrial consumers within the city limits.

Electric distribution center. A terminal at which electric energy is received from the transmission system and is delivered to the distribution system only.

Electric substation. A terminal at which electric energy is received from the transmission system and is delivered to other elements of the transmission system and, generally, to the local distribution system.

Electric substation. An assemblage of equipment for purposes other than generation or utilization, through which electric energy in bulk is passed for the purpose of switching or modifying its characteristics to meet the needs of the general public, provided that in Agricultural or Residence Districts an electric substation shall not include rotating equipment, storage of materials, trucks or repair or housing of repair crews.

Essential services. The erection, construction, alteration, or maintenance of public utilities or municipal or other governmental agencies of underground or overhead gas, electrical, steam, or water transmission or distribution systems or collection, communication, supply, or disposal systems, including poles, wires, mains, drains, sewers, pipes, conduits, cables, fire alarm boxes, police call boxes, traffic signals, hydrants, and other similar equipment and accessories in connection therewith reasonably necessary for furnishing adequate services by such public utilities or municipal or other governmental agencies or for the public health, safety, or general welfare.

Essential services. The erection, construction, alteration or maintenance by a public utility, for the purpose of furnishing adequate service by such public utilities for the public health, safety or general welfare, of gas, electrical, steam, or water transmission or distribution systems, including elevated water towers or tanks; of collection, communication, supply or disposal systems; of poles, wires, mains, drains, sewers, pipes, cables, fire-alarm boxes, police call boxes, hydrants, and other similar equipment and accessories in connection therewith, other than the following; buildings, electrical sub-stations and receiving or transmission towers.

Essential services. Includes the erection, construction, alteration or maintenance, by public utilities or municipal or other governmental agencies, of underground or overhead gas, electrical, steam or water transmission or distribution system, collection, communication, telephone exchange, supply or disposal systems, fire alarm boxes, police call boxes, traffic signals, hydrants and all similar services which are necessary for the general health, safety, or welfare.

Essential services. Facilities and structures owned or maintained by a government, a public agency or a public utility company for the purpose of and directly necessary for rendering or providing communication, electric, gas, sewer, water or comparable service of a public utility nature, and in fact, used in the rendition of such service.

Essential services. Essential services shall be permitted as authorized and regulated by law and other Ordinances, it being the intention hereof to exempt such essential services from the application of the Ordinance.

Essential services. The phrase "Essential Services" means the erection, construction, alteration, or maintenance by public utilities or by governmental departments or commissions of such underground or overhead gas, electrical, steam, or water transmission or distribution systems and structures, collection, communication, supply or disposal systems and structures, including towers, poles, wires, mains, drains, sewers, pipes, conduits, cables, fire alarm boxes, police call boxes, street lights, traffic signals, hydrants, and other similar equipment, and accessories in connection therewith, but not including buildings or microwave radio relay structures, as are reasonably necessary for the furnishing of adequate service by such public utilities or governmental departments or commissions or as are required for protection of the public health, safety or general welfare. For the purpose of this definition, the word "building" does not include "structures" for essential services.

Fire station. Building or part of a building designed or used as a place for the housing of one or more pieces of fire fighting or salvaging equipment, together with sleeping quarters, locker rooms, toilet rooms, heating plant and such other rooms or spaces as required by the fireman or the equipment.

Police station. Building or part of a building used by the police department for administrative offices and detention facilities and may include sleeping quarters, courtrooms and such other rooms or spaces as may be required.

Power plant. Building used primarily to house equipment for the production, conversion, or distribution of energy.

Public building. Building in which persons congregate for civic, political, educational, religious, social or recreational purposes including, but not limited to courthouses, schools, colleges, libraries, museums, exhibition buildings, lecture halls, churches, assembly halls, lodge rooms, dance halls, theaters, taverns, bath houses, armories, recreation piers, stadiums, passenger stations, bowling alleys, skating rinks, gymnasiums, city halls, clubs, grandstands, motion picture theaters, auditoriums, restaurants, etc.

Public building. Premises, buildings or parts thereof used for civic political, educational or recreational purposes; or in which persons are harbored to receive medical, charitable or other care or treatment; or in which persons are held or detained for public or civic duty or for correctional purposes. Public buildings include, among others, hotels, restaurants, court houses, theaters, schools, colleges and churches.

Public building. A building owned and operated, or owned and intended to be operated, by a public agency of the United States of America, or the State, or any of their subdivisions.

Public building. The term "public building" means and includes any structure; including exterior parts of such building, such as a porch, exterior platform or steps providing means of ingress or egress, used in whole or in part as a place of resort, assemblage, lodging, trade, traffic, occupancy, or use by the public or by 3 or more tenants.

Public conveyance. Any railway car, street car, cab, bus, airplane or other vehicle transporting passengers for hire.

Public land uses. Any land use operated by or through a unit or level of government, either through lease or ownership, such as municipal administration and operation, county buildings and activities, state highway offices and similar land uses; and federal uses such as post offices, bureau of public roads, and internal revenue offices, and military installations, etc.

Public off-site sanitary sewage disposal. Sanitary sewage collection system in which sewage is carried from individual lots by a system of pipes to a central treatment and disposal plant.

Public open space. Any publicly-owned open area, including, but not limited to, the following: parks, playgrounds, beaches, waterways, parkways, and streets.

Public service agencies. The provisions of the Code shall not apply to installations for electric supply or communication agencies in the generation, transmission or distribution of electricity, or the operation of signals, or the transmission of intelligence, or located within or on buildings or premises used exclusively by such agency, or on public thoroughfares.

Public sewer. Plans for new plumbing systems or alterations to existing plumbing systems shall be accompanied by a diagram showing the relative elevation of the floor at the lowest fixture in the structure and the top of the street sewer lateral referred to in the established datum, when such public or private sewer is available. The plans shall show the size, number and location of all new sewer connections, subject to the approval of the Water Control Board.

Public sewer. A sewer entirely controlled by public authority.

Public space. Plot or area of land outside of the building other than a street or highway dedicated or devoted to public use by legal mapping or any other lawful procedure, or a space within a building devoted to public use.

Public space. A plot or area of land outside of a building or structure dedicated or devoted to public use by a legal mapping or any other lawful procedure, or a space within a building devoted to public use.

Public space. Space within a building for public use, such as lobbies and lounges, reception, ballroom, meeting, lecture, banquet and dining rooms and their kitchens, and swimming pools.

Public space. A legal open space on the premises, accessible to a public way or street, such as yards, courts or open spaces permanently devoted to public use which abuts the premises.

Public system. Water or sewerage system which is owned and operated by a local government authority or by a local utility company adequately controlled by a governmental authority.

Public system. A water or sewerage system which is owned by a local unit of government.

Public use. Use of any land, water, or buildings by a public body for a public service or purpose.

Public utility. One which is granted a franchise by the City to perform such services as are necessary to fulfill the obligations as indicated by such authorization or franchise, or any City, State or Federal owned utility.

Public utility. Any person, firm, corporation, municipal department or board, duly authorized to furnish and furnishing under governmental regulation to the public, electricity, gas, steam, telephone or telegraph, water, communication or transportation.

Public utility. Any person, firm, corporation or governmental agency authorized to furnish to the public under regulation electricity, gas, steam, communications services, transit, drainage, waste disposal, and/or water.

Public utility. Any person, firm, corporation, municipal department or board, duly authorized to furnish and furnishing general community services to the public, under federal, state or municipal regulations, including, but not restricted to electrical, gas, steam, communication, transportation, sewer or water service.

Public utility. Any facility for rendering electrical, gas, communications, transportation, water supply, sewage disposal, drainage, garbage or refuse disposal and fire protection for the general welfare of the public.

Public utility. Any facility for rendering electrical, gas, communications, transportation, water supply, sewage disposal, drainage, garbage or refuse disposal and fire protection to the general public.

Public utility. Any person, firm, corporation, municipal department or board duly authorized to furnish and furnishing under public regulation, to the public, electricity, gas, heat, power, steam, telephone, telegraph, transportation or water.

Public utility. Any person, firm, corporation, municipal department or board, duly authorized to furnish and furnishing under state or municipal regulations to the public: electricity, gas, steam, communications, telegraph, transportation or water.

Public utility. A public utility is one which is granted a franchise by the City to perform such service as are necessary to fulfill the obligations as indicated by such authorization or franchise, or any City, County, State or Federal owned utility.

Public utility. Any person, firm, or corporation, municipal department, board or commission duly authorized to furnish and furnishing under Federal, State, or Municipal regulations to the public: gas, steam, electricity, sewage disposal, communication, telegraph, transportation, or water.

Public utility building. Includes telephone exchange buildings, transformer stations and substations, gas regulator stations and similar structures.

Quasi-public use. Use serving a community or public purpose, and operated by a non-commercial entity, or by a public utility.

Radio transmitting stations. The provisions of the Code shall not apply to electrical equipment used for radio transmission, except the equipment and wiring for power supply and the installation of radio towers and antennae, whether erected on buildings or on the ground.

Railway utilities. The provisions of the Code shall not apply to the installations or equipment employed by a railway utility in the exercise of the functions as a public carrier and located outdoors or in buildings used exclusively for that purpose.

Roundhouse. A structure for the storing and repairing of locomotives using any fuel other than a volatile flammable liquid.

Special uses. Includes public utilities, transportation terminals and facilities not owned and operated by a public body.

Telephone central office. Building or part of a building used for the transmission and exchange of telephone or radio telephone messages; provided that, in Residence Districts, such use shall not include the transaction of business with the public, storage of materials, trucks, or repair facilities, or housing of repair crews.

Telephone central office. Building and its equipment used for the transmission and exchange of telephone or radio telephone messages between subscribers and other business of a telephone company, providing that in the Residential Districts a telephone central office shall not include public business facilities, storage of materials, trucks, or repair facilities, or housing of repair crews.

Telephone exchange building. Building with its equipment, used or to be used for the purpose of facilitating transmission and exchange of telephone messages between subscribers and other business of the telephone company; but in a Residence District, as established by the ordinance, not to include public business facilities, repair facilities, storage of plant materials or spare parts (other than those carried for the particular building), or storage of equipment, automobiles or trucks or housing or quarters for installation, repair or trouble crew.

Transmission lines. Electric power line bringing power to receiving substation or a distribution substation.

Transmission system. Wires or conductors and associated apparatus and supporting structures, whose exclusive function is the transmission of electrical energy between generating stations, substations and transmission lines or other utility systems.

Utilities. Gas service and equipment therefor; electric service and equipment therefor; water supply, including hot water, and equipment therefor; heat and equipment therefor; refrigeration service and equipment therefor; and housebell system and equipment therefor.

Utility service installation. Any structure or installation by utility company deemed to be necessary for the safe or efficient operation of that utility.

Water facility. Any water works, water supply works, water distribution system or part thereof, designed, intended or constructed to provide or distribute potable water.

Water (street) main. A water-supply pipe for public or community use controlled by public authority.

C-22 PUBLIC AND RELIGIOUS USE BUILDINGS

Armory hall. Any building which, or a portion of which, is used as a drill hall for members of any military organization.

Church. A legally approved structure, used and approved on a permanent basis, primarily for the public worship of God.

Church. Any building which, or a portion of which, is used as an assembling place by members of a religious body for the conducting of services or worship.

Church. Building, together with its accessory buildings and uses, where persons regularly assemble for religious worship, and which building, together with its accessory buildings and uses, is maintained and controlled by a religious body organized to sustain public worship.

Church. Any building, structure, or part thereof, the primary purpose of which is for religious worship; or for religious instruction in connection with a church program, such as a Sunday School.

Church. A permanently located building wherein persons regularly assemble for religious worship and which is maintained and controlled by a religious body to sustain public worship, and church related uses.

Church. Any building, or portion of a building erected and used for the purpose of religious worship and where religious services are held at regularly stated intervals.

Church building. Building wherein persons regularly assemble for religious worship and which is maintained and controlled by a religious body organized to sustain public worship, together with all accessory buildings and uses customarily associated with such primary purpose.

Church, synagogue or temple. Shall mean a permanently located building commonly used for religious worship. Churches, synagogues or temples shall conform to the Building Code and are subject to Development Review Board.

Community use. Facility or land used for public recreation, safety, cultural or educational activities, which is owned and managed by a public semi-public or non-profit organization.

Convent. Area containing one or more buildings accommodating persons-usually nuns devoting their activities to a religious life.

Convent. A dormitory.

Institution. Building or group of buildings designed or used for the non-profit, charitable, or public-service purposes of providing board, lodging, and health care for persons aged, indigent, or infirm; or a building or group of buildings for the purpose of performing educational or religious services and offering board and lodging to persons enrolled for training.

Institutional. Buildings in which the primary or intended occupancy or use is for persons domiciled or detained under supervision.

Institutional occupancy. Buildings in which more than six people are detained for penal or correctional purposes; or in which the liberty of the inmates is restricted, or places of involuntary detention.

Institutional uses. Those uses organized, established, used or intended to be used for the promotion of a public, religious, educational, charitable, cultural, social, philanthropic activities normally operated on a nonprofit basis.

Library. Building or part thereof devoted to reading and to a collection of books, manuscripts, and similar reading matter, which are for use, but not generally for sale.

Library. Place in which books, manuscripts, musical scores or other literary and artistic materials are kept for use and only incidentally for sale.

Library research. Place in which books, manuscripts, musical scores or other literary and artistic materials are kept for use, and only incidentally for sale, as reference in the inquiry and investigation of facts aimed at interpretation and discovery.

Monastery. Parcel of land containing one or more buildings used for religious retirement or of seclusion for persons under religious vows, especially monks.

Monastery. Area containing one or more buildings used for religious retirement or of seclusion for persons under religious vows, especially monks, representing any order.

Monastery. A dormitory.

Museum. Institution, including historical museums, devoted to the procurement, care and display of objects of lasting interest or value.

Museum. Non-profit, non-commercial establishment operated as a repository or a collection of nature, scientific or literary curiosities or objects of interest or works of art, not including the regular sale or distribution of the objects collected.

Museum. Building or part thereof in which are preserved and exhibited objects of permanent interest in one or more of the arts and sciences.

Parish house. Residence for a minister, priest or rabbi in connection with the operation of a church, synagogue or temple.

Place of assembly. An enclosed room or space in which 75 or more persons gather for religious, recreational, educational, political or social purposes, or for the consumption of food or drink, or for similar group activities, but excluding such spaces in dwelling units; or an outdoor space in which 200 or more persons gather for any of the above reasons.

Public building. Any building, other than a single-family, two-family, or three-family dwelling house, and their accessory buildings. The Building Code does not apply to buildings owned by and used for a function of the United States Government.

Public building. Building in which persons congregate for civic, political, educational, religious, social or recreational purposes.

Public building. Building in which persons congregate for civic, political, educational, religious, social, or recreational purposes; including, among others, court houses, schools, colleges, libraries, museums, exhibition buildings, lecture halls, churches, assembly halls, lodge rooms, dance halls, theatres, bath houses, armories, recreation piers, stadiums, passenger stations, bowling alleys, skating rinks, gymnasiums, city halls, grandstands, motion picture theatres, auditoriums, clubs and restaurants.

Public building. Includes any structure, including any building, including exterior parts of such building, such as a porch, exterior platform or steps providing means of ingress or egress, used in whole or in part as a place of resort, assemblage, lodging, trade, traffic, occupancy, or use by the public or by 3 or more tenants.

Public buildings. Any structures or enclosures in which persons assemble for entertainment, convention, education, instruction, worship, or coming together for any purpose, whether such use is public or private or mixed.

Public monument. Building, pillar or similar structure erected in memory of the dead or a person or event either by a public agency or controlled by a public agency.

Public museum. Structure owned by the city, and operated by an institution, no part of the net earnings of which inures to the benefit of any private shareholder or individual, which maintains a supervised public education program, and which operates a structure or structures in which are preserved and exhibited objects of permanent interest in one or more of the arts and sciences, available to school children and to the general public.

Public space. Means a plot or area of land outside of a building or structure dedicated or devoted to public use by a legal mapping or any other lawful procedure, or a space within a building devoted to public use.

Semi-public building. Building used for semi-congregation of persons including, but not limited to: office buildings, banks, retail or wholesale stores, super-stores, etc.

Semi-public buildings. Office buildings, stores, and similar buildings shall be considered as semi-public buildings.

Semi-public land uses. Philanthropic and charitable land uses including: Y.M.C.A.'s, Y.W.C.A.'s, Salvation Army, churches and church related institutions, orphanages, humane societies, private welfare organizations, nonprofit lodges and fraternal orders, hospitals, Red Cross, and other general charitable institutions.

Semi-public use. Use of any land, water, or buildings; a semi-public body.

Transportation station. Railroad station, bus station, air line terminal, or other public transportation building, used for the shelter of persons waiting for transportation facilities for which such building is an accessory.

C-23 RADIO AND TELEVISION ANTENNAS AND TOWERS

Antenna–community antenna television system. Any facility which, in whole or in part receives directly or indirectly from the air and amplifies or otherwise modifies electronic or microwave signals transmitting programs broadcast by one or more television stations and/or originates or purchases programs or electronic or microwave signals and distributes such signals or any of them by wire or cable to subscribing members of the public who pay for such service, but such term shall not include any such facility which serves only the residents of an apartment dwelling complex under common ownership control or management consisting of not more than two buildings and commercial establishments located on the premises of such an apartment complex.

Antenna elements. That portion or portions of the outside antenna system for television and radio receiving apparatus or equipment, which is connected to a television or radio receiver.

Antenna system. The combination of any components comprising an inter-related system for radio or television, as defined in "antenna elements."

Control line. The lead, wires, or cable which serves to control or put "in" or "out" of operation, from a remote location, the rotator, antenna top booster, or other accessory part of the antenna system.

Guy line. A wire or cable secured at one end to the vertical structure whose opposite end is secured in such manner as to maintain or aid to maintain the structure upright and immobile.

Height of antenna system. The overall vertical length of the antenna system above the ground, or if any such system be located on a building, then above that part of the level of such building upon which the system rests.

Lightning arrestors. Lightning arrestors approved by the Underwriters Laboratories Inc., shall be used on all lines not directly connected.

Mast. That type of structure composed of cylindrical section or sections arranged and secured end to end so as to compose a "flag-pole" style support.

Rotator. A rotating mechanism which will orient the antenna elements in a horizontal plane to the most favorable reception from various transmitting stations.

Self-supported. As regards vertical structures, one which depends only on its base mount to support it vertically, using no guy lines or attachment of any kind for support.

Sub-tower. A structural arrangement of tower legs used to support a mast and antenna system.

Tower. That type of structure composed of uprights secured together and strengthened by means of crossmembers spaced vertically throughout its height.

Transmission line. The lead, wires, or cable which serves to electrically convey the television or radio signal from the antenna elements to the receiving equipment.

Vertical structure. A fabricated metal or treated wood supporting structure to elevate the antenna elements and any associated equipment to a height deemed necessary for adequate operation. The vertical structure may be a tower, mast or a combination of tower and mast erected from the ground or located on a building.

C-24 RESIDENTIAL BUILDINGS

Accessory living quarters. Living quarters within an accessory building, for the sole use of persons employed on the premises; such quarters having no kitchen facilities and not rented or otherwise used as a separate dwelling.

Accessory sleeping quarters for nontransients. Space or building which is an accessory to another occupancy and which is used for sleeping quarters for five or more nontransients.

Apartment. Portion of a residential building used as a separate housekeeping unit.

Apartment. Room or suite of two (2) or more rooms in a multiple dwelling, occupied or suitable for occupancy as a residence for one family.

Apartment. Suite of rooms or portion of a building which is designed, intended or used for living purposes including cooking facilities.

Apartment. Room or suite of rooms, including bath and culinary accommodations, in a two-family dwelling intended or designed for use as a residence by a single family.

Apartment. Dwelling unit occupied by an individual, family, or household for living and sleeping purposes.

Apartment. Room or a suite of rooms occupied, or which is intended or designed to be occupied, as the home or residence of one individual, family or household, for housekeeping purposes.

Apartment. Space or suite of rooms which is intended for separate and independent living and under the control of the tenant.

Apartment. Any room or suite of two (2) or more rooms occupied as a home for one (1) or more persons and arranged or equipped for cooking meals.

Apartment. One (1) or more rooms in an apartment house or dwelling occupied or intended or designed for occupancy by one (1) family for sleeping or living purposes and containing one (1) kitchen.

Apartment. Room or suite of rooms with sanitation facilities and with or without cooking facilities, and occupied as the home or residence of a single family, individual, or group of individuals.

Apartment. Same as Living Unit. Generally used in connection with multi-family buildings.

Apartment. A room or suite of rooms with culinary facilities designed for or used as living quarters for a single family.

Apartment. A room or suite of rooms which is occupied or which is intended or designed to be occupied by one family for living and sleeping purposes.

Apartment. Means a room, or a suite of two (2) or more rooms, in a building occupied, or which is intended or designed to be occupied, as the home or residence of an individual, family or household.

Apartment. A space or suite of rooms which is intended for separate and independent living and under the control of the tenant or occupant.

Apartment. Any room or suite of two or more rooms occupied as a home for one or more persons and arranged or equipped with cooking facilities.

Apartment. A portion of a residential building used as a separate housekeeping unit.

Apartment. Shall mean a room or suite of two or more rooms in a multiple dwelling, occupied or suitable for occupancy as a residence for one family, with culinary facilities.

Apartment. Shall mean a suite of rooms or portion of a building which is designed, intended or used for living purposes including cooking equipment.

Apartment. A room or suite of rooms, including bath and culinary accommodations, in a two-family dwelling intended or designed for use as a residence by a single family of 4 persons.

Apartment. A dwelling unit occupied by an individual, family, or household for living and sleeping purposes, including cooking facilities.

Apartment. A dwelling unit in a multi-family building.

Apartment. A room or a suite of rooms occupied, or which is intended or designed to be occupied, as the home or residence of one individual, family or household, for housekeeping purposes, defined in the Building Code.

Apartment. A dwelling unit.

Apartment. Room or suite of two (2) or more rooms, which is designed or intended for occupancy by, or is occupied by, one family doing its own cooking therein, or by one person doing his or her own cooking therein.

Apartment. Room or a suite of rooms occupied, or which is intended or designed to be occupied, as the home or residential purposes for one individual, family or household.

Apartment. Single building containing four (4) or more living units.

Apartment. Multi-family building comprised of three (3) or more dwelling units arranged one above the other and side by side, each unit having at least one entrance connected to a common interior hall leading to the exterior.

Apartment. Room or suites of rooms in a multiple dwelling, or where more than one living unit is established above non-residential uses, intended or designed for use as a residence by a single family including culinary accommodations.

Apartment. Suite of rooms within a building arranged, designed or used for residential purposes for one family, and containing independent sanitary and cooking facilities. The presence of cooking facilities conclusively establishes the intent to use for residential purposes. Each apartment shall be considered a dwelling unit.

Apartment. Room or suite of rooms in a hotel offered to the public for compensation and arranged or designed for permanent occupancy, with facilities for sleeping, cooking and eating.

Apartment. Part of a building consisting of a room or rooms intended, designed, or used as a residence by an individual or single family; also known as multi-family residence.

Apartment. Any building, or portion thereof, which is designed, built, rented, leased, let or hired out to be occupied, or which contains units for three or more families living independently from each other.

Apartment. One or more rooms in a multiple apartment building arranged, intended or designed or occupied as the residence of a single family, individual, or group of individuals, with one food preparation unit.

Apartment. One or more rooms occupied as a home or residence for an individual or a family or a household. The existence of, or the installation of, sink accommodations and/or cooking facilities within a room or suite of rooms shall be deemed sufficient to classify such rooms or suite of rooms as an apartment. The floor area for an apartment shall be not less than required by applicable zoning regulations.

Apartment. A household unit suitable for occupancy by one or more persons in an apartment house or building, a part of which may be used for commercial purposes.

Apartment building. Multiple dwelling or part thereof designed or used primarily for family unit occupancy and containing three (3) or more family units.

Apartment building. Structure containing three (3) or more dwelling units each with direct access to a common passageway, court, or street, and arranged, intended or designed for multiple occupancy by three (3) or more families living independently of each other.

Apartment building. Single residential structure designed and constructed to contain four (4) or more separate dwelling units regardless of the internal arrangement of such units or the ownership thereof.

Apartment house. A building made up of three (3) or more apartment units so arranged that each unit has direct access, without common corridors, to a means of egress from the building, and which may or may not maintain an inner lobby for its tenants.

Apartment house. A building containing two (2) or more apartments.

Apartment house. Building containing four (4) or more apartments.

Apartment house. Building containing three (3) or more apartments.

Apartment house. Any building or portion thereof which is designed, built, rented or leased, let or hired out to be occupied, or which is occupied as the home or residence of three (3) or more families living independently of each other and doing their own cooking in the building, and shall include flats and apartments.

Apartment house. Any house or building which or a portion of which, is used as two (2) complete apartments above the first story, or one (1) complete apartment and one (1) or more business establishments above the first story or any apartment or apartments above the second story; provided, occupants of the apartments and business establishments have a common right in the halls, stairways, yard, cellar or other parts of the building.

Apartment house. Building designed for and occupied exclusively as a residence for three (3) or more families living independently of one another.

Apartment house. Same as a multiple dwelling.

Apartment house. Any building used and/or arranged for occupancy as dwellings for three (3) or more families as separate housekeeping units.

Apartment house. Any building, or portion thereof, which is designed, built, rented, leased, let or hired out to be occupied, or which is occupied as the home or residence of more than two (2) families living independently of each other and doing their own cooking in the said building, and shall include flats and apartments.

Apartment house. Building other than a hotel or motel, designed for or used to house three (3) or more families, living independently of each other, including all necessary employees of such families.

Apartment house. Any building, or portion thereof, which is designed, built, rented, leased, let, or hired out to be occupied, or which is occupied as the home or residence of three or more families living independently of each other and doing their own cooking in the said building.

Apartment house. Apartment house is any building or portion thereof, which is designed, built, rented, leased, let, or hired-out to be occupied, or which is occupied as the home or residence of three or more families living independently of each other in dwelling units as defined in this Code.

Apartment house. A multiple dwelling in which dwelling units are leased to permanent tenants.

Apartment house. Any house or building or a portion of which, used as two complete apartments above the first story, or one complete apartment and one or more business establishments above the first story, or any apartment or apartments above the second story, the occupants of such apartments and business establishments having a common right in the halls, stairways, yard, cellar, or other common areas of the building, as defined in each lease.

Apartment house. A multiple dwelling containing 3 or more dwelling units.

Apartment house. Any building, or portion thereof, which is designed, built, rented, leased, let or hired out to be occupied, or which is occupied as the home or residence of more than two (2) families living independently of each other and doing their own cooking in the said building, and shall include flats and apartments, as defined in the Ordinance.

Apartment house. Shall mean a multiple dwelling, of 2 or more units.

Apartment house. Any building, or portion thereof, which is designed, built, rented, leased, let or hired out to be occupied as a residence of three or more families living independently of each other and doing their cooking in each dwelling unit.

Apartment house. Apartment house is any building or portion thereof, which is designed, built, rented, leased, let or hired-out to be occupied, or which is occupied as the home or residence of three (3) or more families living independently of each other in dwelling units as defined in the Zoning Regulations.

Apartment house. Multiple dwelling in which dwelling units are leased to individual permanent tenants.

Apartment house. Any structure or building or a portion of which, used as two (2) complete apartments above the first story, or one complete apartment and one or more business establishments above the first story, or any apartment or apartments above the second story, the occupants of such apartments and business establishments having a common right in the halls, stairways, yard, cellar, attic, storage areas, or other common areas of the building.

Apartment house. Multiple dwelling building containing three (3) or more dwelling units.

Apartment house. Building housing more than four (4) dwelling units, whether designated as apartment house, apartment hotel, tenement, garden apartment, or by any other name.

Apartment house. Building or that portion thereof containing more than four (4) dwelling units or efficiency apartments.

Apartment house or apartment hotel. A building designed, constructed or altered to provide accommodations for more than 4 families living independently of each other, each unit of which is designed for cooking upon the premises.

Bachelor apartment. One (1) or more rooms in an apartment house or dwelling occupied or intended or designed for occupancy by one (1) family for sleeping or living purposes and containing not more than one (1) kitchen and utility room, one (1) sleeping room, one (1) bathroom and incidental closet space.

Bachelor apartment. Same as an Efficiency Dwelling Unit.

Basement. That portion of a building between the floor and ceiling which is wholly or partly below grade and having more than one-half (1/2) of its height below grade.

Bathroom. Enclosed space containing one or more bathtubs or showers, and which may also contain water closets, lavatories or fixtures serving similar purposes.

Bedroom. Any room within a residential dwelling unit which is designed to be used for sleeping purposes and contains a closet of sufficient size to hold clothing. One living room with entry closet shall not be considered a "bedroom" in each residential dwelling unit other than a studio or efficiency apartment.

Boarders. Incidental keeping of nontransient boarders or lodgers for compensation by a resident family; provided, that no more than one-half (1/2) of the bedrooms be so used.

Combined rooms. Two (2) or more adjacent habitable spaces which by their relationship, planning and openness permit their common use.

Corridor. A hallway, passageway or other compartmented space providing the occupants with access to the required exitways of the building or floor area.

Covered shaft. An interior enclosed space extending through one (1) or more stories of a building, connecting openings in successive floors, or floors and roof, and covered at the top.

Court. An open unoccupied area, other than a yard, on the same lot with a building and bounded on two or more sides by such a building. "Court Apartment" means any multiple dwelling arranged around two or three sides of a court which opens into a street.

Culinary or cooking facilities. Space in a dwelling arranged, intended, designed, or used for the preparation of food for a family. Facilities may include a sink, stove, cabinets, and refrigerator or any combination of these arranged in such space. A refrigerator alone shall not constitute culinary or cooking facilities under this definition.

Dormitory. Building whose principal use is "accessory sleeping quarters" for nontransients.

Duplex. A building designed and/or used exclusively for residential purposes and containing two dwelling units separated by a common party wall or otherwise structurally attached.

Dwelling. Any building or portion thereof designed or used exclusively as the residence of one or more persons, but not including a tent, or recreation vehicle.

Dwelling duplex. A residence building designed for, or used as the separate homes or residences of two separate and distinct families, but having the appearance of a single family dwelling house. Each individual unit in the duplex shall comply with the definition for a one-family dwelling.

Dwelling unit. Any building or portion thereof which contains living facilities, including provisions for sleeping, eating, cooking and sanitation, as required by the code, for not more than one (1) family.

Dwelling units. Means a suite of rooms providing complete living facilities for one family, including permanent provisions for living, dining, cooking, sleeping, and sanitation.

Efficiency. Accommodations that include kitchen facilities.

Efficiency apartment. A dwelling unit of not more than 1 room in addition to kitchen facilities, and a bathroom.

Efficiency apartment. One story building with all living units on the ground floor level and each living unit having its own outside entrance.

Efficiency apartment. An apartment consisting of not more than one (1) habitable room together with kitchen or kitchenette and sanitary facililties.

Efficiency apartment. Apartment consisting of not more than one (1) habitable room, together with cooking and sanitary facilities.

Efficiency apartment. Dwelling unit containing not over three hundred and fifty (350) square feet of floor area, and consisting of not more than one (1) room in addition to kitchen, dining and necessary sanitary facilities, and for the purposes of computing density shall be considered as one (1) room.

Efficiency apartment. Dwelling unit of not more than one (1) room in addition to kitchen and bath.

Efficiency dwelling unit. A dwelling unit containing only one habitable room.

Efficiency living unit. Any room having cooking facilities used for combined living, dining, and sleeping purposes, and meeting the other requirements of the Building code.

Efficiency living unit. Single family dwelling unit which does not contain a separate bedroom, but which may have two (2) rooms, exclusive of bathroom, if one such room shall be a separate kitchen.

Efficiency living unit. Any room having cooking facilities used for combined living, dining, and sleeping purposes.

Efficiency unit. Single family dwelling unit which does not contain a separate bedroom, but which may have two (2) rooms exclusive of bathroom, if one such room shall be a separate kitchen.

Family. Individual or any number of individuals related by blood or marriage, or a group of not more than five (5) individuals not so related, living together.

Family. One person living alone or a group of 2 or more persons living together under one management.

Family. One or more persons living together, whether related to each other by birth or not, and having common housekeeping facilities.

Family. A single person, or two or more persons related by blood, or marriage, and living together and maintaining a common household, with not more than four boarders, roomers or lodgers; or a group of not more than 4 persons, not necessarily related by blood or marriage, and maintaining a common household.

Family. Consists of one person living individually or a group of persons living as a single household unit, using common housekeeping facilities, not to include, however, more than three persons unrelated by blood, marriage or adoption.

Family. Single individual living upon the premises as a separate housekeeping unit; or a collective body of persons living together upon the premises as a single housekeeping unit in a domestic relationship based upon birth, marriage, or other domestic bond.

Family unit. Room or group of rooms used or intended to be used as a housekeeping unit for living, sleeping, cooking and eating.

Family unit. Room or group of rooms designed or used as a housekeeping unit for a family or for a group of not more than ten (10) persons other than a family and providing living, sleeping, eating, cooking, and sanitation facilities complying with the requirements of the Code.

Family unit. One-family dwelling or that part of a multi-family dwelling arranged for the use of one (1) or more persons living and cooking together as a single housekeeping unit, with cooking, living, sanitary and sleeping facilities.

Floor area. (a) The term "floor area" shall mean any floor space within a story of a structure enclosed on all sides by either exterior walls, fire walls, or fire partitions. Adjoining rooms having openings in dividing partitions in excess of one-quarter of the length of sulch partitions, whether or not separated by rolling, folding, sliding or other forms of movable enclosures, shall be considered as one area. (b) The term "net area" for any floor shall mean the gross area within the exterior walls less the area occupied by enclosed stair, elevator and other permanent shafts completely enclosed in fire partitions.

Floor area. The area included within the surrounding exterior walls of a building or portion thereof, exclusive of vent shafts and courts. The floor area of a building, or portion thereof, not provided with surrounding exterior walls shall be the usable area under the horizontal projection of the roof or floor above.

Fraternity or sorority building. Building, rented, occupied or owned by a general or local chapter of some regularly organized college fraternity or sorority, or by or on its behalf by a building corporation or association composed of members of the local chapter of such fraternity or sorority, as a place of residence.

Fraternity or sorority house. Building containing no more than one (1) dwelling unit and more than two (2) rooming units or guest rooms. Such rooming units or guest rooms shall be for residential purposes only.

Garage apartment. Accessory building, not a part of or attached to the main building, a portion of which contains living quarters and space for at least one (1) automobile.

Garage apartment. Accessory or subordinate building, not a part of or attached to the principal building containing living facilities for not more than one (1) family.

Garage apartment. A dwelling for one (1) family erected as a part of a private garage.

Garden apartment. Group of two (2) or more multiple dwellings not over two (2) stories in height, located on the same lot, that offer each dwelling unit direct access to an open yard area.

Garden apartments. Building or group of buildings containing four (4) or more dwelling units not exceeding two and one-half (2-1/2) stories or thirty-five (35) feet in height whichever is the lesser, and surrounded by open grassy areas or containing open courts for use of the tenants.

Garden apartments. Group of buildings not more than 2-1/2 stories in height, each building to contain not more than 12 dwelling units, with a minimum distance between buildings of 20 feet and with no building having a frontage of more than 150 feet between side yards, and no portion of the building below the first floor occupied as a dwelling unit except an apartment below grade, if any, provided for a building superintendent.

Garden apartments. Garden type apartment is one which is generally located in a structure containing not less than four (4) apartments and up to eighteen (18) units; usually not exceeding three (3) stories in height; sometimes designed around courts or common green spaces; often having private balconies or patios; and, frequently exhibiting different facades and design features between structures in a garden apartment complex.

Garden type apartments. Apartment building not exceeding three (3) stories in height, excluding the basement.

Garden type apartments. Group of dwelling units on one (1) lot in one (1) or more buildings, connected or detached from one another, no one of which exceeds two (2) stories and an attic in height, each unit having separate front and rear exterior entrances opening directly into the yard area.

Guest house. Shall mean an attached or detached accessory building used to house guests of the occupants of the principal building, and which is never rented or offered for rent. Any guest house providing cooking facilities shall be considered a dwelling unit.

Guest house. A detached living quarters of permanent construction, clearly subordinate and incidental to the main building on the same lot, and intended for use without compensation by occasional guests of the occupants of the main building.

Guesthouse. A single family building in the rear yard area of a residence which is not occupied year around but which is used as temporary residence, only. Such a building shall conform to the requirements for accessory buildings, except that a sink, bathtub and cooking facilities may be provided. Only nonpaying and personal guests of the occupant of the principal residence shall occupy a guesthouse. Year around occupancies shall not be permitted by the same guest, nor shall the owner occupy the guesthouse and rent the principal residence.

Habitable attic. A habitable attic is an attic which has a stairway as a means of access and egress and in which the ceiling area at a height of seven and one-third (7-1/3) feet above the attic floor is not more than one-third (1/3) the area of the floor next below.

Habitable room. A room or enclosed floor space arranged for living, eating, and sleeping purposes (not including bathrooms, water closet compartments, laundries, pantries, foyers, hallways and other accessory floor spaces).

Habitable room. Room or enclosed floor space in a basement, first or upper story arranged for living, eating or sleeping purposes, not including bath or toilet rooms, laundries, pantries, foyers or communicating corridors or cellar recreation rooms, with means of light and ventilation complying with the Code.

Habitable space. Space occupied by one (1) or more persons for living, sleeping, eating, or cooking. Kitchenettes are not classified as habitable spaces.

Half-bath toilet room. Enclosed space, containing one (1) or more water-closets and a lavatory.

High rise apartment. Multiple family dwelling of six (6) or more stories above the ground level of the principal entrance.

High rise apartment building. Apartment building exceeding six (6) stories in height.

High rise apartments. Residential structure four (4) or more stories in height containing four (4) or more separate dwelling units each with its own access to a common stairwell or elevator well.

Home. Place of human habitation.

Household unit. An apartment.

Household unit. Room or suite of rooms equipped with, or containing one (1) or more facilities for cooking food and which room or suite of rooms is occupied as a residence of a single family, individual, or group of individuals.

Housekeeping residential use. All residential uses except those used for transient purposes.

Housekeeping unit. Room or combination of rooms containing living, sleeping and kitchen facilities for one (1) family.

Housing space. That portion of a multiple dwelling rented or offered for rent for living or dwelling purposes in which cooking equipment is supplied; and includes all privileges, services, furnishings, furniture, equipment, facilities and improvements connected with the use or occupancy of such portion of the property. The term shall not mean nor include public housing or dwelling space in any hotel, motel or established guest house, commonly regarded as a hotel, motel or established guest house, as the case may be in the city ordinance.

Kitchen. Any room or part of a room which is designed, built, used or intended to be used for food preparation and dishwashing; but not including a bar, butler's pantry or similar room adjacent to or connected with a kitchen.

Kitchen. Space, 60 sq. ft. or more in area, used for cooking and preparation of food.

Kitchen. A room used or adapted for cooking and containing a stove, range, hot-plate or other cooking apparatus, which burns coal, oil, gas or other fuel or is heated by electricity.

Kitchen. Means a space having a floor area of 60 square feet or more and which is used for the cooking of food.

Kitchenette. Means a space having a floor area of less than 60 square feet and which is used for the cooking of food.

Landscaping. The provision of plantings and related improvments for the purpose of beautifying and enhancing a property and for the control of erosion and the reduction of glare, dust and noise.

Livability space. That portion of open space not devoted to motor vehicle parking or circulation and which is landscaped, or improved as outdoor living space or recreation space for occupants of the premises. Public street and alley rights-of-way included in the computation of land area shall not be included as livability space. Beneficial open space included in the computation of land area may be included as livability space subject to the above limitations and provided that such space is directly accessible to occupants of the adjacent premises and is available for their leisure time use.

Livable room. Any room used for normal living purposes in a residence structure and shall not include kitchens, laundry rooms, bathrooms or storerooms.

Living unit. Residential unit providing complete, independent living facilities for one family including permanent provisions for living, sleeping, eating, cooking and sanitation.

Living unit. Room or rooms comprising the essential elements of a single housekeeping unit. Facililties for the preparation, storage and keeping of food for consumption within the premises shall cause a unit to be construed as a living unit, but shall be conveniently accessible to the living unit.

Main building. Any building having the predominant land use which is not an accessory building.

Model home. A one-family dwelling having all of the following characteristics: (a) Said dwelling is constructed upon a proposed lot previously designated as a model home site by the Advisory Agency in a subdivision for which the Advisory Agency has approved or conditionally approved a tentative map but for which a final map has not yet been recorded. (b) The proposed lot upon which the model home is constructed is recognized as a legal building site for the duration of the model home permit.

Multi-family apartment house. Building containing more than two (2) dwelling units.

Multi-family apartment house. Any building or portion thereof used as a multiple dwelling for the purpose of providing three (3) or more separate dwelling units with shared means of egress and other essential facilities.

Multi-family apartment house. Any building or portion thereof used as a multiple dwelling for the purpose of providing three (3) or more separate dwelling units with shared means of egress.

Multi-family dwelling. Building or structure designed, erected, occupied or intended to be occupied as living quarters for more than one (1) family, and having cooking facilities for each occupancy.

Multifamily house. Means a building occupied as the home or residence of individuals, families or households living independently of each other, of which 4 or more are doing cooking within their apartments; including tenement house, apartment house, or flat. A row of 4 or more single family houses not separated by Fire Walls is considered to be a multifamily house.

Multiple dwelling. Building containing three (3) or more dwelling units; or building containing living, sanitary and sleeping facilities occupied by one (1) or two (2) families and more than four (4) lodgers residing with either one of such families; or building with one (1) or more sleeping rooms, other than a one- or two-family dwelling, used or occupied by permanent or transient paying guests or tenants; or building with sleeping accommodations for more than five (5) persons used or occupied as a club, dormitory, fraternity or sorority house, or for similar uses; or building used or occupied as a convalescent, old-age or nursing home, but not including private or public hospitals or public institutions.

Multiple dwelling. Building arranged, intended or designed to be occupied, or which is occupied, by three or more families or groups of individuals living independently of each other in separate housekeeping units or apartments. The term "Multiple Dwelling" includes the term "Apartment House," "Bungalow Court," "Apartment Hotel," "Row House," "Terrace," and "Tenement House."

Multiple dwelling. Building not a Single Family Dwelling nor a Two-Family Dwelling. Designed for and occupied exclusively for dwelling purposes by three (3) or more families living independently of one another, not a row house, but customarily called an "Apartment House."

Multiple dwelling. Building or portion thereof designed for and occupied by more than two (2) families including tenement houses, row houses, apartment houses and apartment hotels.

Multiple dwelling. Building, portion thereof, or buildings used for occupancy by three (3) or more families (including necessary servants and employees of each such families) living independently of each other, and containing three (3) or more dwelling units. "Multiple dwelling" shall include apartment houses, bungalow courts, and group houses.

Multiple dwelling. Building or portion thereof used or designated as a residence for three (3) or more families as separate housekeeping units, including apartments and apartment hotels.

Multiple dwelling. Building other than a "Dwelling House" "Row House" or "Institution" occupied in whole or in part as a residence. It shall include apartment houses, rooming houses, and other buildings defined by the Code.

Multiple dwelling. Multiple dwelling occupied more or less permanently for residence purposes by more than two (2) families and in which the rooms are occupied in apartments, suites, or groups, each comprising a dwelling unit.

Multiple dwelling. A building containing two (2) or more dwelling units.

Multiple dwelling. A single building providing separate living quarters for three (3) or more families.

Multiple dwelling. A building containing three or more dwelling units. Multiple dwelling shall not be deemed to include a hospital, school, convent, monastery, asylum or other public institution.

Multiple family dwelling. Building used or designed as a residence for three (3) or more families living independently of each other and doing their own cooking therein, including apartment houses, apartment hotels, flats, and group houses.

Multiple family dwelling. Building in which four (4) or more living units exist. Multiple family units shall be the same as apartment.

Multiple family dwelling. Building or group of buildings, designed for and occupied by three (3) or more families with separate housekeeping and cooking facililties for each family to include apartments, townhouses, condominiums, apartment houses and apartment hotels.

Multiple family housing development or project. Three or more single family buildings, or more than one two-family building or more than one multiple family building on a building site, or any combination thereof.

Net floor area. For the purpose of determining the number of persons for whom exitways are to be provided, net floor area shall be the actual occupied area, not including accessory unoccupied areas or thickness of walls.

Nontransient. Person resident in the same building for a period of 30 days or more.

Occupant. Person residing in a dwelling unit who is the responsible head of a family; or, if there be no apparent head of a family, then the person actually occupying such dwelling unit. The owner and occupant may be the same person.

Occupant. Any person or persons in actual possession of, and living in a dwelling, dwelling unit, rooming house, or rooming unit, including the owner and operator.

One bedroom unit. Is a dwelling unit containing a minimum floor area of at least five hundred (500) square feet per unit, consisting of not more than two (2) rooms in addition to kitchen, dining, and necessary sanitary facilities, and for the purposes of computing density shall be considered a two (2) room unit.

One-room apartment. Apartment containing one (1) livable room.

Owner. Any person who has legal title to any dwelling, with or without accompanying actual possession thereof; or, who has equitable title and is either in actual possession or collects rents therefrom. Or, who as executor, executrix, trustee, guardian, or receiver of the estate of the owner, or as mortgagee or as vendee in possession either by virtue of a court order or by agreement or voluntary surrender of the premises by the person holding the legal title, or as collector of rent, has charge, care or control of any dwelling or rooming house, or any person who is a lessee or assignee subletting or assigning any part or all of any dwelling shall have joint responsibility over the portion of the premises sublet or assigned.

Permanent guest. A person who occupies or has the right to occupy a residence accommodation for a period of thirty (30) days or more.

Permanent guest. A person who occupies or has the right to occupy a hotel accommodation, boarding house or lodging house as his or her domicile and place of permanent residence.

Permanent guests. Persons who occupy or have the right to occupy a lodging house or hotel accommodation as their place of permanent residence.

Permanent occupancy. Rental of housing accommodations or rooms on a week to week, month to month, or year to year basis with a fixed rent for each period of occupancy.

Permanent occupancy. Occupation of a sleeping room or dwelling unit as a domicile or place of permanent residence. The permanency of residency may range from less than a month to a lifetime.

Plumbing fixtures. Within each living unit there shall be provided the following plumbing fixtures: A kitchen sink properly located to facilitate food preparation and dishwashing, a water closet located in the bathroom, a bathtub or shower located in a bathroom, and a lavatory located in a bathroom. Each of the plumbing fixtures shall be permanently installed and connected to a plumbing system.

Private. As applied to an exit way, toilet room, or other part of a building, means that such exit way or other part of the building is an adjunct to not more than one (1) room or one (1) suite of rooms.

Public. As applied to an exit way, toilet room, or other part of a building, means that such exitway or other part of the building is subject to common use by those who occupy or enter the building and includes those parts of the building which are not included within the means of "private."

Public view. Any premises, or any part thereof, or any building or any part thereof, which may be lawfully viewed by the public, or any member thereof, from a sidewalk, street, alleyway, licensed open air parking lot, or from any adjoining or neighboring premises.

Rental room. Room rented for permanent, not transient residence, as evidenced by a rental charge on a weekly or monthly basis, but not permitting cooking.

Residence. Building containing only dwelling units. The term "residence" or any combination thereof shall not be deemed to include hotel, boarding house, rooming house, motel, hospital, or other accommodations used for transient occupancy or mobile residences.

Residence. A detached part or whole of a building occupied exclusively as a family dwelling place and the usual accessory occupancies.

Residence building. Building in which sleeping accommodations are provided.

Residence hall. Dormitory as applied to a building.

Residential accommodations. Any building or part of a building used or intended to be used for sleeping accommodations by a person or group of persons. Other housekeeping facilities may be provided.

Residential building. Building or portion thereof designed or used for human habitation.

Residential buildings. Includes all buildings or parts of buildings in which families or household units are housed, or in which sleeping accommodations are provided, other than buildings classified as "Institutional buildings".

Residential dwelling unit. Building or portion of a building planned, designed or usable as a residence for one (1) family only, living independently of other families included. In said unit (for example, a one-family dwelling, and four-family dwelling, each apartment four-family dwelling, each apartment in an apartment house and each unit of a condominium or townhouse.)

Residential occupancy. Any building or part of a building in which a person or group of persons are provided with sleeping accommodations. Other housekeeping accommodations may also be provided.

Residential occupancy. Occupancy or use of a building or part thereof by persons for whom sleeping accommodation is provided by who are not harboured or detained to receive medical care or treatment or are not involuntarily detained.

Residential occupancy. Buildings in which families or households live or in which sleeping accommodations are provided, and all dormitories, shall be classified as Residential Occupancy.

Residential occupancy. Applies to that portion of a building in which sleeping accommodations are provided for two (2) or more families, or more than twelve (12) persons. Residential occupancy shall include club houses, convents, dormitories, hotels, lodging houses, multiple story apartments, studios, tenements, and similar occupancies.

Residential purposes. Intent to use and/or the use of a room or group of rooms for the living, sleeping and housekeeping activities of persons on a permanent or semi-permanent basis of an intended tenure of one month or more.

Residential structure. Any building or part of a building constructed with or as sleeping accommodations for a person or group of persons. Other housekeeping accommodations also may be provided.

Residential use. Shall be deemed to include single and multiple dwellings, hotels, motels, dormitories, and mobile homes.

Room. Any enclosed division of a building containing over seventy (70) square feet of floor space and commonly used for living purposes, not including lobbies, halls, closets, storage space, bathrooms, utility rooms and unfinished attics, cellars or basements. An enclosed division is an area in a structure bounded along more than 75% of its perimeter by vertical walls or partitions, or by other types of dividers which serve to define the boundaries of the division.

Room. Space in an enclosed building, or space set apart by a partition or partitions, and any space in a building used or intended to be used for non-residential purposes.

Room. An unsubdivided portion of the interior of a building, but not including an enclosed show window in a building.

Roomer. Nontransient paying guest in the home of another person.

Service space. Space provided in a building to facilitate or conceal the installation of building service facililties, such as chutes, ducts, pipes, shafts or wires.

Shaft. An enclosed space within a building, whether for air, light, elevator-dumb-waiter or any other purpose connecting a series of 2 or more openings in successive floors and covered either by a skylight or by the roof. A vent shaft shall be used solely to ventilate or light water-closets, compartments or bathrooms.

Sleeping room. Room used primarily for sleeping by one or more persons.

Sleeping room space. Every sleeping room shall be of sufficient size to afford at least 400 cubic feet of air space for each occupant over 12 years of age and 200 cubic feet for each occupant under 12 years. For cabins and cottages having less than 3 sleeping rooms, each room shall be of sufficient size to afford at least 50 square feet of floor area, having a minimum ceiling dept of 7 feet and at least 400 cubic feet of air space for each occupant. No greater number of sleeping occupants than the number thus established shall be permitted in any such rooms.

Spacing of buildings. The required minimum horizontal distance between any wall of 2 or more buildings facing or overlapping each other in any manner either parallel or oblique. Such distance is measured at any given point and any given level by projecting or prolonging vertically and horizontally the perimeter lines of each wall from the lowest habitable floor to the ceiling of the highest habitable floor.

Suite. A group of habitable rooms designed as a unit, and occupied by only one family; but not including a kitchen or other facilities for the preparation of food, with entrances and exits which are common to all rooms comprising the suite.

Suite of rooms. Two (2) or more rooms which are arranged to be used as a unit and each room of which is accessible from within the unit without the use of a public exit way.

C-25 SCHOOLS AND SCHOOL FACILITIES

Advanced school buildings. Buildings and parts of buildings in which students receive instruction or training more advanced than that afforded in secondary schools and includes, but is not limited to, the following: University Buildings—College Buildings—Business and Trade School Buildings—Seminary Buildings.

Assembly (assembly space). Any part of a place of assembly, exclusive of a stage, that is occupied by numbers of persons during the major period of occupancy. Every balcony tier of seating shall be considered a separate assembly space.

Assembly hall. A hall or room, including the balconies thereof, if any, in which persons may assemble in a manner as permitted in the code.

Business school. Privately owned schools offering instruction in accounting, secretarial work, business administration, the fine or illustrative arts, trades, dancing, music, and similar subjects.

Campus. A contiguous area of land constituting and making up the grounds of a college or university containing the main buildings or within the main enclosure; provided, however, that for the purpose of this definition the contiguity of any land area involved shall not be deemed to be destroyed by the presence of public rights of way.

Classroom. A room with desks or equivalent used for group instruction purposes for ten (10) or more students.

Classroom. Room in a school building used primarily for the instruction of students in a group.

Classroom instruction. Such may consist of a class or group of persons assembled either in a classroom auditorium or assembly room where visual education is a part of the curriculum or course of training and the subjects are of a vocational or educational nature.

College. Building which, or a portion of which, is used for advanced educational purposes, or may be a particular building for a subject matter or course as part of a university.

College. Building which, or a portion of which, is used for advanced educational purposes.

College. Includes junior college or university supported by public funds, or a private college, junior college or university which gives comparable general academic instruction and degrees.

Common exit way. Exit way used by pupils from more than one schoolroom or from a school assembly hall.

Educable mentally handicapped children. Means children between the ages of 5 and 21 years who because of retarded intellectual development as determined by individual psychological examination are incapable of being educated profitably and efficiently through ordinary classroom instruction but who may be expected to benefit from special educational facilities designed to make them economically useful and socially adjusted.

Educational. Educational occupancies include all buildings used for the gathering of groups of six (6) or more persons for purposes of instruction.

Educational institution. Any school, college or university supported wholly or in part by public funds and any other college or university giving general academic instruction as prescribed by the State Board of Education.

Educational institution. Public, parochial or private pre-primary, primary, grammar or high school; a private preparatory school or academy providing courses of instruction substantially equivalent to the courses offered by public high schools for preparation for admission to college or universities which award B.A. or B.S. degrees; a junior college, college or university, either public or parochial or founded or conducted by or under the sponsorship of a religious or charitable organization, or private when such junior college, college or university is not conducted as a commercial enterprise for profit. Nothing in this definition shall be deemed to include trade or business schools or colleges.

Educational institutions. Colleges or universities supported wholly or in part by public funds and other colleges or universities giving general academic instruction, as prescribed by the State Board of Education.

Educational occupancy. Occupancy or use of a building or structure or any portion thereof by persons assembled for the purpose of learning or of receiving educational instruction.

Educational occupancy. Includes all buildings and parts of buildings in which people come together in classrooms for educational or instructional purposes; except that all parts of such buildings equipped with a stage for the use of portable scenery or used for viewing motion pictures projected from nitrocellulose film and accommodating an assembly or gathering of 50 or more persons, and all other rooms or spaces in such buildings used or intended to be used for the assembly or gathering of 100 or more persons, shall be classified as Assembly Occupancy; irrespective of the gathering; also, all parts of such buildings used or intended to be used as nurseries for children under five years of age shall be classified as Institutional Occupancy, when accommodating six (6) or more children.

Educational occupancy. The occupancy or use of a building or structure or any portion thereof by persons assembled for the purpose of learning or of receiving educational instruction; including among others, academies, colleges, libraries, schools, and universities.

Educational uses. Schools, colleges, public libraries, public museums and art galleries.

Education occupancy. Occupancy or use of a building or structure or any portion thereof by persons assembled for the purpose of learning or of receiving educational instruction.

Elementary and high schools. Any institution of learning which offers instruction in the several branches of learning and study required to be taught in the public schools by the Education Code of the State.

Elementary and high schools. Institutions of learning which offer instructions in the several branches of learning and study required to be taught in the public schools by the Education Code of the State. High schools include Junior and Senior classifications.

Elementary school buildings. Buildings and parts of buildings in which pupils receive instruction or training as afforded in a kindergarten, or in one or more of the first six (6) grades and may include grades 7 and 8 or child day-care as afforded in a day-care center.

Gifted children. Means children between the ages of 5 and 21 years whose mental development, as determined by individual examination, is accelerated beyond the average to the extent that they need and can profit from specially planned educational services.

Maladjusted children. Means children between the ages of 5 and 21 years who because of social or emotional problems are unable to make constructive use of their school experience and require the provisions of special services designed to promote their educational growth and development. Maladjusted children to receive special educational facilities shall be designated under regulations adopted by the school board, by a court of competent jurisdiction, or in districts exceeding 500,000 inhabitants by the superintendent of schools.

Nursery school. A school or day-care facility for five (5) or more pre-elementary-school age children, other than those resident on the site.

Open plan school. A building intended to be used as a specific established Code type school use, which consists principally of individual teaching areas that are separated from each other only by informal dividers, such as cabinets, bookcases, partitions of less than floor to ceiling height.

Open plan school. School building in which the school consists principally of individual teaching areas that are separated from each other only by informal dividers such as cabinets, bookcases, and partitions of less than floor to ceiling height.

Physically handicapped children. Means children, other than those with a speech defect, between the ages of 3 and 21 years who suffer from any physical disability making it impracticable or impossible for them to benefit from or participate in the normal classroom program of the public schools in the school districts in which they reside and whose intellectual development is such that they are capable of being educated through a modified classroom program.

Playground. Accessory to a public or parochial school. When operated in conjunction with a public school or a church and parochial school, the playground area may be used for temporary open parking of passenger automobiles only during the time such playground area is not being used for recreational purposes.

Pre-school. Day care and education of five (5) or more children aged five (5) years and under.

Pre-school. Facility engaged in educational work with four (4) or more pre-school children not related to the proprietor. Children enrolled are not necessarily in need of supplemental parental care.

Pre-school child. Excludes infants.

Private business school or business school or school. Means an educational institution privately owned or operated by an owner, partnership, or corporation, offering business courses for which tuition is charged, in such subjects as typewriting, shorthand (manual or machine), filing and indexing, receptionist's duties, keypunch, teletype, penmanship, bookkeeping, accounting, office machines, business arithmetic,

English, business letter writing, salesmanship, personality development, leadership training, public speaking, real estate, insurance, traffic management, business psychology, economics, business management, and other related subjects of a similar character or subjects of general education when they contribute values to the objective of the course of study; but shall not include non-profit schools conducted by bona fide eleemosynary, religious, or public institutions exempt from property taxation under the laws of the State. Classes in any of the subjects herein referred to which are taught or coached in homes or elsewhere are included in the term "school"

Private school. Privately owned schools having a curriculum essentially the same as ordinarily given in a public elementary or high school. The term includes day nurseries and kindergartens.

Private school. Building used for the purpose of elementary or secondary education and having a curriculum essentially the same as ordinarily given in a public school.

Private school room. Room occupied by not more than 15 persons for instruction by a private tutor.

Public school fraternity, sorority or secret society. Means any organization, composed wholly or in part of public school pupils, which seeks to perpetuate itself by taking in additional members from the pupils enrolled in such school on the basis of the decision of its membership rather than upon the free choice of any pupil in the school who is qualified by the rules of the school to fill the special aims of the organization.

School. Any building or a portion of which, used as a place of instruction. The school may be either public or private, or it may be operated in conjunction with a church or other organization.

School. A place of general instruction including college but not including business colleges, nursery schools, dancing schools, riding academies or specialized trade or vocational schools.

School. Buildings used for the purpose of elementary or secondary education, which meets all requirements of the compulsory education laws of the State and not providing residential accommodations.

School. Building or group of buildings involving assemblage for instruction, education or recreation; including primary and elementary schools, secondary schools, academies, colleges, universities, trade schools, business schools and colleges, and the like, whether supported by private or public funds or any combination thereof.

School. Elementary school or a high school or a college where regular supervised fire drills are held in which pupils are trained in rapid dismissal from the building. Such fire drills shall be held several times each semester, including summer classes.

School. Place of general instruction including public and private colleges, but not including business colleges, nursery schools, dancing schools, riding academies or specialized trade or vocational schools.

School. Place for instruction in the arts and sciences, but excluding institutions whose primary purpose is the teaching of dancing, physical culture, music (unless as a home occupation), trades or industries, riding academies, or the combination of any two (2) or more of these.

School. Unless otherwise specifically described in the Ordinance of the City or other applicable laws or regulations, the term school shall mean any building used for educational purposes by five or more persons at one time.

School. An elementary school, high school, or college, either public or private.

School. Any building or premises in which a regular course of public or private instruction is afforded to not less than ten (10) pupils at one time, exclusive of rooms in buildings separate from or attached to churches used for the primary purpose of religious instruction.

School assembly hall. Assembly Hall used in connection with schools.

School assembly hall. Any room which is used for a gathering of 150 or more persons and which is located in a school building, or in a building which is primarily an accessory to a school building, and includes, among others, a gymnasium or an auditorium used for such a gathering. A school assembly hall in connection with a school building does not constitute a mixed occupancy building. In school buildings a room or area devoted solely to library use, does not constitute a school assembly hall.

School board. Means school principal, directors, board of education and board of school inspectors of public and private schools.

School bus. Means every motor vehicle operated by or for a public or governmental agency or by a private or religious organization solely for the transportation of pupils in connection with any school activity.

School classifications. Includes buildings, or parts thereof, in which people come together in classroom assemblies for educational or instructional purposes, including, among others, schools, colleges, universities, and academies. EXCEPTION: Parts of school buildings used for the congregation or gathering of 100 or more persons in one room shall comply with the requirements applicable to an assembly occupancy regardless of whether such gathering is of an educational or instructional nature or not.

School lunch program. Means the program whereby certain types of lunches called balanced, nutritious lunches adopted as standard types and designated by the Superintendent of Public Instruction, are furnished to students.

School nursery. Public or private agency engaged in educational activity with preschool children who are not necessarily in need of supplemental parental care.

School occupancy. Building in which people come together for education or instructional purposes shall be called School Occupancy.

School of general instruction. Public, parochial, or private school or college giving regular instruction at least five (5) days each week (except for holidays) for a normal school year of not less than seven (7) months; but not including: (a) a school of special instruction as defined herein, or (b) a nursery school unless conducted as a part of a school of general instruction, or (c) a riding school, however designated, or (d) a school for mental defectives.

School of special instruction. School with three (3) or more employees primarily devoted to giving instruction in vocational, professional, commercial, musical, dramatic, artistic, linguistic, scientific, religious, or other special subjects, but not including: (a) nursery school, or (b) riding school, however designated, or (c) school for mental defectives.

School or college. Shall mean, unless otherwise specified, private or public places of general instruction but shall not include day nursery schools, dancing schools, riding academies, or trade or specialized vocational schools.

Schoolroom. Room used for the education or training of persons and includes among others, classrooms, study, recitation, play, domestic, science, manual training, and drafting rooms, and does not include school assembly hall.

Schools. A place of general instruction in the arts and sciences including college, but excluding institutions such as business colleges or vocational schools, or whose primary purpose is the teaching of physical culture, music or dancing, unless a home occupation, trades, or industries, riding academies, or the combination of any two (2) or more of these.

Schools—elementary—junior high and high schools. Institution of learning which offers instruction in the several branches of learning and study required to be taught in the public schools by the Education Code of the State; and in which no pupil is physically restrained.

School structure. Structure devoted entirely to school purposes and activities incidental to school use.

Secondary school buildings. Buildings and parts of buildings, in which pupils receive instruction or training as afforded in one or more of grades 7 to 12, inclusive, except that the inclusion of grades 7 and 8 shall not result in the classification of a building or a part of a building as a secondary school where such grade is a part of an elementary school as defined in the Education Code of the State. Secondary schools also include buildings and parts of buildings in which pupils in the ninth through the twelfth grade age-group receive systematic vocational training and may include pupils in the seventh and eighth grade age-group receiving such training.

Special educational facilities. Includes special schools, special classes, special instruction, special reader service for visually handicapped children, transportation, maintenance, equipment, therapy, psychological examination, salaries of teachers, including teachers in home and hospital instruction for children whose physical handicaps are such that attendance in regular or special class is not possible and other professional personnel, and other special educational services required by the child because of his disability if such services are approved by the Superintendent of Public Instruction and the child is eligible.

Special schools. Schools or studios offering instruction in arts, trades, and professions in a classroom or studio setting, and not including any school teaching industrial skills or building trades. Special schools shall include music and dance studios, secretarial schools, language schools, photography schools, cooking schools, dressmaking and millinery schools and, any other class or school of a similar nature.

Trade or business school. Secretarial school or college, or business school or college, when not public and not owned or conducted by or under the sponsorship of a religious or charitable organization; school conducted as commercial enterprise for teaching instrumental music, dancing, barbering or hairdressing or for teaching industrial skills in which machinery is employed as a means of instruction. This definition shall not be deemed to include "educational institution" as defined in the Education Code.

Trade or industrial school. Establishment, public or private, for the purpose of training students in skills required for the practice of trades or industries.

Trade or industrial school. School, public or private, offering training to students, in skills required for the practice of building trades and in industry.

C-26 SIGNS AND OUTDOOR DISPLAY STRUCTURES

Accessory sign. Sign which is accessory to the principal use of the premises.

Accessory sign. Computed and based on the area of the principal front of the building only and may be subdivided into not exceeding four units and not exceeding the permissible signs area. Accessory signs as permitted, shall be used only to advertise the principal use of the building.

Advertising device. A device other than a recognized or standard type of sign that is placed, or affixed, to advertise or to attract attention, or to promote publicity for an individual, firm, organization, product or event and includes devices of a decorative nature.

Advertising or billboard sign. Advertising sign or billboard is a sign which directs attention to a business, commodity, service, or entertainment conducted, sold, or offered elsewhere than upon the same zoning lot.

Advertising sign. Shall include (A) A non-illuminated professional or identification sign not exceeding two (2) square feet in area and attached flat against the structural surface of a building. (B) A sign not in excess of twelve (12) square feet in area pertaining only to the rent, lease or sale of premises upon which it is displayed. (C) A sign or bulletin board not in excess of twelve (12) square feet in area for the purpose of displaying the name and activities of the church, institution, club or organization on whose premises it is displayed. (D) Directional or informational signs of a public or quasi-public nature not exceeding eight (8) square feet in area and which indicates the name or location of a community, public or private institution, church or the name and place of meeting of an official or civic club or body.

Advertising sign. A sign which includes any copy and/or graphics relating to any service, product, person, businss, place, activity or organization in addition to simple identification, excluding directional information.

Advertising sign. A sign which directs attention to a profession, business, commodity, service or entertainment other than one conducted, sold or offered upon the premises where such sign is located, or on the building to which the sign is affixed.

Advertising sign. A sign on which is portrayed information which directs attention to a business commodity, service or entertainment or other activity not related to the use on the lot upon which such sign is located. Advertising signs include "billboards."

Advertising sign. Sign which directs attention to a business, commodity, service or entertainment conducted, sold or offered elsewhere than on the lot and only incidentally on the lot, if at all.

Advertising sign. Sign that is not an identification or directional sign.

Advertising structure. Any structure erected for advertising purposes, with or without any advertisement display thereon, situated upon or attached to real property, upon which any poster, bill, printing, painting, device, or other advertisement of any kind whatsoever may be placed, posted, painted, tacked, nailed or otherwise fastened, affixed or displayed.

Advertising structure. Any sign, billboard, surface, object or structure used for advertising purposes.

Animated sign. Any sign having a conspicuous and intermittent variation in the illumination or physical position of any part of the sign; provided, however, that a slow rotation of the sign not be considered animation.

Approved combustible plastic. Plastic material more than one-twentieth (1/20) inch in thickness which burns at a rate of not more than two and one-half (2-1/2) inches per minute when subjected to standard tests for flammability of plastics in sheets of six-hundredths (0.06) inch thickness.

Approved combustible plastic. Plastic material which, when tested in accordance with Standard Method of Test for Flammability of Plastics over 0.05 inch in thickness, burns at a rate not exceeding 2-1/2 inches per minute in sheets of 0.06 inch thickness.

Area of a sign. The number of square feet of surface including the border or frame.

Area of a sign. Consists of one or more letters, symbols, or other parts, affixed to or mounted upon a building or affixed to any other approved mounting, which sign does not have a border or frame, is all of the area of the surface to which the sign is attached lying within the extremities of the sign.

Artisans' signs. Signs of mechanics, painters, and other artisans may be erected and maintained during the period such persons are performing work on the premises on which such signs are erected.

Awning, canopy, roller curtain or umbrella sign. Any sign painted, stamped, perforated, or stitched on the surface area of an awning, canopy, roller curtain or umbrella.

Awning sign. An awning shall include any structure made of cloth or metal with a metal frame attached to a building and projecting over a public area, having sufficient surfaces for sign work, which can be raised to a position flat against the wall of the building when not in use, but having the sign work visible to the public.

Awning sign. A sign on or attached to a temporary retractable shelter that is supported entirely from the exterior wall of a building.

Banjo sign. Sign having a total area of not more than fifty (50) square feet, the advertising content of which is not closer than ten (10) feet to the surface of the ground.

Banner. Any sign having the characters, letters, illustrations or ornamentations applied to cloth, paper, balloons or fabric of any kind with only such material for a foundation.

Banners, balloons, posters, etc. Includes signs which contain or consist of banners, balloons, posters, pennants, ribbons, streamers, spinners or other similarly moving devices.

Billboard. Any sign attached to the land or attached to any building which does not advertise the business conducted on the premises to which such sign is attached.

Billboard. Any notice, sign or advertisement, pictorial or otherwise, with an area of 300 or more square feet, and also all signs used as an outdoor display for the purpose of making anything known, the origin or place of sale of which is not on the lot with such display, except that governmental notices shall not be considered to be billboards.

Billboard. Any sign or advertisement used as an outdoor display by painting, posting or affixing on any surface, of a picture, emblem, word, figures, numerals or lettering for the purpose of making anything known.

Billboard. Includes all structures, regardless of the material used in the construction of the same, that are erected, maintained or used for public display of posters, painted signs, wall signs, whether the structure be placed on the wall or painted on the wall itself, pictures or other pictorial reading matter which advertise a business or attraction which is not carried on or manufactured in or upon the premises upon which said signs or billboards are located.

Billboard (poster panel). Board, panel or tablet used for the display of printed or painted advertising matter.

Billboards. Any framework for signs advertising merchandise, services or entertainment, sold, produced, manufactured or furnished at a place other than the location of such structure.

Building master identification signs. Building Master identification signs are signs which identify the name of a multiple-tenant commercial building.

Building or wall sign. Sign, other than a roof sign, which is supported by a building or wall.

Business identification sign. A sign that is affixed to or placed or mounted upon a commercial establishment or property solely to name and identify the business conducted on that property. The sign must be affixed to the establishment or to be placed upon the same property as the establishment. A business identification sign may display the registered or commonly used name or the registered trademark, or both, of a commercial establishment. No other message, symbol or device shall appear upon such sign.

Business sign. A sign which directs attention to a profession or businss conducted, or to a commodity, service or entertainment sold or offered, upon the premises where such sign is located, or on the building to which such sign is affixed.

Business sign. A sign which directs attention to a business, commodity, service, entertainment or other activity conducted upon the lot on which such sign is located.

Business sign. Sign which directs attention to a business or profession conducted upon the lot.

Business sign. One which directs attention to a business or profession conducted or to a commodity, service, or entertainment sold or offered on the premises where such sign is located.

Change-panel signs. A sign designed to permit an immediate change of copy which may be other than the name of the business.

Closed sign. Sign in which the projected area of letters, figures, strips and structural framing exposed to wind consists of at least seventy (70) per cent of the area within the perimeter of the framing.

Closed sign. Display sign in which the entire area is solid or rightly enclosed or covered.

Combination sign. Any sign incorporating any combination of the features of pole, projecting, and roof signs.

Combustible ground sign. Any sign constructed in whole or in part of combustible material erected or maintained upon the ground and not attached to any building.

Commercial sign. Any sign belonging to or controlled by the Owner or occupant of a building or premises which is used to identify the building or premises or the products or services sold therein or thereon.

Construction sign. Sign identifying the contractor, developer, architect and financing agency which is erected on the lot in which the premises under construction are located.

Detached sign. Any sign not attached to or painted on a building, but which is affixed to the ground.

Direction sign. Any sign permanently or temporarily erected to denote the route to any city, town, village, historic place, shrine or hospital; signs directing and regulating traffic; notices of any railroad, bridge, ferry or other transportation or transmission company necessary for the direction or safety of the public; signs, notices or symbols for the information of aviators as to locations, directions and landings; and conditions effecting safety in aviation; signs, notices or symbols as to the time and place of civic meetings, and signs or notices erected or maintained upon private property giving the name of the Owner, Lessee, or Occupant of the premises, or the street number.

Directional sign. Sign for the purpose of traffic control which is located on private property.

Directory sign. Sign containing the name of a building, complex or center.

Directory signs. Directory signs are used to guide pedestrians to individual businesses within a multiple-tenant commercial building.

Display sign. Structure that is arranged, intended, designed or used as an advertisement, announcement, or direction, and includes a sign, sign screen, billboard, and advertising devices of every kind.

Display surface. Surface made available either by the structure or the sign facing for the mounting of letters and decorations.

Display surface. The area made available by the sign structure for the purpose of displaying the advertising message.

Display surface. Means the surface made available by the structure, either for the direct mounting of letters and decoration, or for the mounting of facing material intended to carry the entire advertising message.

Electric sign. Fixed or portable, self-contained electrically-illuminated appliance with words or symbols designed to convey information or attract attention.

Electric sign. Any model, character, illustration, device or representation illuminated by means of electricity.

Electric sign. Any sign containing electrical wiring, but not including signs illuminated by an exterior light source.

Entry-way sign. A freestanding sign used to identify the entrance to a project or facility.

Erect. To build, construct, attach, hang, place, suspend, or affix and shall also include the painting of wall signs.

Facia sign. Single-faced building or wall sign which is parallel to its supporting wall.

Facing. Means the surface of the sign upon, against, or through which the message of the sign is exhibited.

Facing surface. The face or surface of the sign upon, against or through which the message is displayed or illustrated on the sign.

Fence signs. Signs painted on the surface of enclosure or division fences, or on picket or other ornamental fences.

Field advertising sign. A sign that advertises a business not conducted, or a product or service not available, upon the property on which the sign is located.

Fin sign. A sign which is supported wholly by a one-story building of an open-air business or by poles placed in the ground or partly by such pole or poles and partly by a building or structure.

Fixed projecting sign. Any sign projecting at an angle from the outside wall or walls of any building and rigidly affixed thereto.

Flashing sign. An illuminated sign on which the artificial light is not maintained constant or stationary in intensity or color at all times when such sign is in use. A revolving sign, or any advertising device which attracts attention by moving parts operated by mechanical equipment, or movement is caused by natural sources, whether or not illuminated with artificial lighting, shall be considered a flashing sign.

Flashing sign. Sign, the illumination of which is not kept stationary or constant in intensity at all times when in use except that illuminated signs which indicate the temperature, date, or similar public service information shall not be considered flashing signs.

Flashing sign. Moving or animated sign or any illuminated sign on which the artifical or reflected light is not maintained stationary and constant in intensity or color at all times when in use. Any revolving illuminated sign shall be considered a flashing sign.

Flashing sign. Any illuminated sign on which the artificial light is not maintained stationary or constant in intensity and color at all times when the sign is in use.

Flashing sign. Any sign having a conspicuous and intermittent variation in the illumination of the sign.

Flat or wall sign. Any sign erected parallel to the face or outside wall of any building and supported throughout its length by the wall of the building.

Flat sign. Any sign attached to and erected parallel to the face of, or erected or painted on the outside wall of a building and supported throughout its length by such wall or building; or any sign in any way applied flat against a wall.

For sale or lease sign. Sign on a lot denoting that same lot or the premises thereon are for sale, lease or rent.

Free-standing or attached signs. The area includes all lettering, wording, and accompanying design and symbols, together with the background, whether open or enclosed, on which they are displayed.

Freestanding sign. A permanent or portable sign which is neither attached to a wall of a building nor within a building, including a sign which is installed upon the roof of a building.

Free standing sign. Sign which is supported by one or more upright or braces in or upon the ground and not attached to any building or wall.

Freestanding wall sign. A sign consisting of individual letters on a wall which is integrated architecturally with the building.

Government building signs. Signs erected on a municipal, state or federal building which announce the name, nature of the occupancy and information as to use of or admission to the premises.

Grand opening sign. A sign advertising the introduction, promotion, announcement of a new business, store, shopping center, office or the announcement, introduction, promotion of an established business changing ownership or management.

Gross surface area of sign. Entire area within a single continuous perimeter enclosing the extreme limits of such and in no case passing through or between any adjacent elements of same. However, such perimeter does not include any structural or framing elements lying outside the limits of the sign and not forming an integral part of the display.

Gross surface area of sign. The entire area within a single continuous perimeter enclosing the extreme limits of a sign.

Ground sign. Freestanding sign supported by one or more uprights, braces, or pylons located in or upon the ground or to something requiring location on the ground including "billboards" or "poster panels" so called.

Ground sign. Sign which is supported by one or more uprights, poles or braces in or upon the ground.

Ground sign. Display sign supported by one or more uprights or braces in or upon the ground surface, and includes ground-supported pole signs and billboards.

Ground sign. A billboard or similar type sign which is supported by one or more uprights, poles, or braces in or upon the ground other than a combination sign, fin sign or pole sign.

Ground sign. Any sign supported by uprights or braces placed in or upon the ground and not attached to any building.

Ground sign. Outdoor sign supported by one or more uprights or braces in or upon the ground and shall include ground supported pole signs and billboards.

Ground sign. A free-standing sign which is supported by one (1) or more uprights or braces. A sign or an accessory structure shall be considered a ground sign.

Ground sign. A sign which does not extend or project into or over a public way and is supported by one or more uprights or braces that are in or upon the ground.

Height of a sign. The height of a sign with a border or frame is the vertical distance from the ground on which it stands to the highest extremity of the sign.

Height of a sign. The height of a sign, without a border or frame, affixed to or mounted upon any building or other approved mounting, is the vertical distance from the ground to the top of the letter, symbol, or other part of the sign which is the highest.

Holiday sign. Sign in the nature of a decoration, clearly incidental and customary and commonly associated with any national, local, or bona fide religious holidays, provided that there be on the sign, no names of firms or products.

Home occupation sign. Sign identifying or advertising a home occupation, provided such sign is no larger than one hundred forty-four (144) square inches and erected on the premises housing such home occupation.

Home occupation signs. Signs advertising home occupations, bearing the name and occupation of the practitioner.

Horizontal projecting sign. Any projecting sign which is greater in width than in height.

Identification sign. A sign indicating the name and address of a building, or the name of an occupant thereof, or the type of permitted occupation conducted on the premises. Identification signs include "nameplates."

Identification sign. A sign, design or symbol which clearly defines or illustrates the name and/or primary nature of the business establishment.

Illuminated sign. Illuminated sign is any sign designed to give forth any artificial light, or designed to reflect light from one or more sources, natural or artificial.

Illuminated sign. Display sign whose exterior surface is made legible through the use of external artificial light or a reflected light and in which the illuminating units are not contained within or directly upon the exterior surfaces.

Illuminated sign. Sign designed to give forth artificial light or through transparent or translucent material from a source of light within such sign, including but not limited to neon and exposed lamp signs.

Illuminated sign. Any sign which has characters, letters, figures, designs or outline illuminated by electric lights or luminous tubes as a part of the sign proper.

Incombustible material. Any material which will not ignite at or below a temperature of twelve hundred (1200) degrees Fahrenheit and will not continue to burn or glow at that temperature.

Incombustible sign material. Incombustible material is any material which will not ignite at or below a temperature of 1,200° Fahrenheit, during an exposure of five minutes, and which will not continue to burn or glow at that temperature.

Indirect lighting. A source of external illumination located a distance away from the sign, which lights the sign, but which is itself not visible to persons viewing the sign from any normal position of view.

Indirectly-illuminated sign. Illuminated, non-flashing sign whose illumination is derived from an external artificial source so arranged that no direct rays of light are projected from such source into any residential district or public street.

Indirectly lighted sign. Sign illuminated by artificial light reflecting from the sign face, the light source not visible from any street right-of-way.

Individual letter sign. Letters or figures individually fashioned from metal or other approved materials and attached to the wall of a building, but not including a sign painted on a wall or other surface.

Information sign. A sign that conveys a message other than advertising, which sign is placed solely for the guidance of individuals or groups.

Insignias and flags. Includes such insignias, flags and emblems of the United States, the state, and municipal and other bodies of established government, or flags which display the recognized symbol of a non-profit and/or non-commercial organization.

Institutional bulletin board. On-premises sign containing a surface area upon which is displayed the name of a religious institution, school, library, community center or similar institution and the announcement of its services or activities.

Internal–indirect lighting. A source of illumination entirely within the sign (generally a free standing letter) which makes the sign visible at night by means of lighting the background upon which the free standing character is mounted. The character itself shall be opaque, and thus will be silhouetted against the background. The source of illumination shall not be visible.

Internal lighting. A source of illumination entirely within the sign which makes the contents of the sign visible at night by means of the light being transmitted through a translucent material but wherein the source of the illumination is not visible.

Internally illuminated sign. Sign illuminated by an artificial light source which is not visible, but which reaches the eye through a diffusing medium.

Length of a sign. The length of a sign with border or frame is the distance between the extremities of the sign measured horizontally.

Length of a sign. The length of a sign, without border or frame, affixed to or mounted upon a building or other approved mounting, is the horizontal distance between the first and last extremities of the lettering, symbols, or other parts of the sign.

Letters and decorations. Include the letters, illustrations, symbols, figures, insignia and other devices employed to express and illustrate the message of the sign.

Location. Means a lot, premises, building, wall, or any place whatsoever upon which a sign is erected, constructed and maintained.

Maintenance. The replacing or repairing of a part or portion of a sign made unusable by ordinary wear, tear or damage beyond the control of the owner or the reprinting of existing copy without changing the wording, composition or color of said copy.

Mansard and parapet signs. A sign permanently affixed to a wall or surface designed to protect the edge of a roof, constructed no more than 20° from vertical.

Marquee sign. Sign attached to or supported by, or hung from, a marquee.

Marquee sign. Any sign projecting from, attached to or hung from a marquee and said marquee shall be known to mean a canopy or covered structure projecting from and supported by a building, when such canopy or covered structure extends beyond the building, building line or property line.

Marquee sign. Sign attached to or hung from a marquee or canopy.

Marquee sign. Display sign attached to or hung from marquee or fixed canopy projecting from and supported by the building and extending beyond the building wall, building line or street lot line.

Menu board. A permanently mounted sign displaying the bill of fare of a drive-in or drive thru restaurant.

Monumental sign. Free standing sign affixed to a sign monument.

Monument sign. Structure, built on grade, that forms an integral part of the sign or its background is in conformance with the zoning requirements of the district in which it is located.

Nameplate. A sign indicating the name and/or address of a building or the name of an occupant thereof and the nature of a permitted occupation therein.

Name plate sign. Sign which states the name or address or both of the profession or business on the lot where the sign is located.

Neon strip lighting. Neon strip lighting shall be prohibited above the roof level of any building except on approved pylons and parapets.

Non-accessory sign. Sign which is not accessory to the principal use of the premises.

Noncombustible ground sign. Any sign constructed of noncombustible materials, including the foundation, supporting framework, poles or posts erected or maintained upon the ground and not attached to, and independent of, any part of the building.

Non-conforming sign. Sign, outdoor advertising structure, or display of any character, which was lawfully erected or displayed, but which does not conform with standards for location, size, or illumination for the district in which it is located by reason of adoption or amendment of this ordinance, or by reason of annexation of territory to the City.

Non-flashing sign. An illuminated sign on which the artificial light is maintained stationary and constant in intensity and color at all times when such sign is illuminated. For the purpose of the Ordinance, any moving, illuminated sign shall not be considered a non-flashing sign.

Nonstructural trim. Nonstructural trim is the molding battens, caps, nailing strips, latticing, and walkways which are attached to the sign structure.

Number of signs. Sign shall be considered to be a single display surface or display device containing elements organized, related, and composed to form a unit. Where matter is displayed in a random manner without organized relationship of elements, or where there is reasonable doubt about the relationship of elements; each element shall be considered to be a single sign.

Obsolete sign. A sign that advertises an activity, business, product or service that no longer is conducted on the premises on which the sign is located.

Off-premise sign. A structure which bears a sign which is not appurtenant to the use of the property where the sign is located, or a product sold or a service offered upon the property where the sign is located, and which does not identify the place of business where the sign is located as a purveyor of the merchandise or services advertised upon the sign. Permanent off-premise signs are prohibited absolutely.

On-premises sign. Sign the primary purpose of which is to identify and/or direct attention to a profession, business, service, activity; product, campaign or attraction manufactured, sold, or offered upon the premises where such sign is located.

Open sign. Sign in which the projected area exposed to wind does not conform to the definition of "Closed Sign."

Open sign. Display sign in which at least 50 per cent of the enclosed area is uncovered, or open to the transmission of wind.

Outdoor advertising display sign. Any fabricated sign, including its structure, consisting of any letter, figure, character, mark, point, plane, marquee sign, design, poster, pictorial picture, stroke, stripe line, trademark, reading matter or illuminating device, constructed, attached, erected, fastened, or manufactured in any manner whatsoever so that the same shall be used for the attraction of the public to any place, subject, person, firm, corporation, public performance, article, machine or merchandise whatsoever, and displayed in any manner whatsoever out doors for recognized advertising purposes.

Outdoor advertising sign. Attached or free standing structure constructed and maintained for the purpose of conveying to the public, information, knowledge or ideas. Such structure may be double faced or V-type, but shall contain no more than four (4) signs in any one unit and not more than two (2) signs side by side. The structure shall have a total length of not more than sixty (60) ft.

Outdoor advertising signs. An attached or free standing structure constructed and maintained for the purpose of conveying to the public, information, knowledge or ideas.

Outdoor advertising structure. Anything constructed or erected, either free standing or attached to the outside of a building, for the purpose of conveying information, knowledge or ideas to the public about as subject either related or unrelated to the premises upon which located.

Outdoor advertising structure. Structure of any kind or character erected or maintained for outdoor advertising purposes, upon which any advertising sign may be placed.

Outdoor advertising structure. Includes without limitation by reason of enumeration, any fabricated sign including its structure, consisting of any letter, figure, character, mark, point, plane, marquee sign, design, poster, pictorial picture, stroke, stripe, line, trademark, reading matter, or illuminating devices, constructed, attached, erected, fastened, or manufactured in any manner whatsoever so that the same shall be used for advertising purposes only.

Outdoor advertising structure. Any poster panel, billboard, painted bulletin or other structure, device, surface, or display used for advertising purposes which is not located on the premises of the business advertised.

Outdoor commercial advertising device. Visible, immobile contrivance or structure in any shape or form, the purpose of which is to advertise any product or service, campaign, event, etc.

Outdoor display structure. Any structure erected or attached outdoors used for advertising or display, or for the affixment, attachment, or support of a sign or signs, or for any similar purposes, and shall include billboards, and displays for special occasions such as Christmas displays.

Outdoor sign. Any arrangement of letters, figures, symbols, or other devices, used for advertising, announcement, direction, or declaration, intended to attract or inform the public, affixed or attached to the exterior walls of a building or other structure, or upon constructed surfaces erected, attached, or supported outdoors. The word "sign" shall be interpreted to include the structure of such signs.

Outdoor signs. Includes all fabricated signs and their supporting structures erected on the ground or attached to or supported by a building or structure.

Painted wall sign. Any sign painted on the outside wall of any building.

Permanent sign. A sign which is firmly attached to a structure supported by a foundation.

Point of purchase sign. Any structure, device, display board, screen, surface or wall with characters, letters or illustrations placed thereto, thereon or thereunder by any method or means whatsoever where the matter displayed is used for advertising on the premises a product actually or actively offered for sale or rent thereon or therein or services rendered.

Point of sale sign. Shall mean any sign advertising or designating the use, occupant of the premises, or merchandise or products sold on the premises.

Point of sale sign. A sign that advertises a product or service available, or a business conducted, upon the property on which the sign is located.

Pole or ground sign. Any sign erected upon a pole or poles and which is wholly or be totally independent of any building for support.

Pole sign. A sign wholly supported by a sign structure in the ground.

Pole sign. Sign attached to or hung from a pole.

Pole sign. Free standing sign other than a portable sign or a monumental sign.

Political sign. Any advertising structure or banner used in connection with a local, state or national election campaign.

Political sign. Sign advertising any candidate or issue that is to be voted upon in any primary or other election that is specifically authorized or established by the laws of any political body corporate or political subdivision existing by virtue of the laws of the State.

Portable display sign. A display surface temporarily fixed to a standardized advertising structure which is regularly moved from structure to structure at periodic intervals.

Portable sign. Free standing sign not permanently anchored or secured.

Portable sign. A sign which is not permanently installed.

Posted sign. Tablet, card, or plate which defines the use, occupancy, fire grading and floor loads of each story, floor or parts thereof for which the bulding or part thereof has been approved.

Price sign. A permanently mounted sign displaying the retailing cost of a gallon of gasoline on the premises of a service station.

Private identification sign. A sign that is affixed to or placed or mounted upon a private or residential property placed or mounted upon a private or residential property solely to name or identify the property, the occupant, or the owner. A sign placed by a private club or association to name or identify the club, club premises, etc., is also classified as a private identification sign.

Projecting sign. Any projecting sign or marquee sign extending beyond the building wall or affixed to any hood, canopy or marquee.

Projecting sign. Sign attached to a structure and which projects from the exterior wall of the structure so that both sides are visible.

Projecting sign. Sign, other than wall sign, which projects from and is supported by a wall of a building or structure.

Projecting sign. Display sign which is attached directly to a building wall or other structure and extends more than fifteen (15) inches beyond the building wall or parts thereof, or other structure, building line, or property line.

Projecting sign. A sign which is affixed to a structure and extends beyond the wall or parts thereof more than twelve (12) inches.

Projecting sign. Any sign which is attached to a building or other structure and extends beyond the line of such building or structure the surface of that portion of the building or structure to which it is attached.

Projecting sign. Outdoor sign which is affixed or attached to a building wall or other structure and extends more than twelve (12) inches beyond the face of the building wall or other structure, or more than twelve (12) inches beyond the building line, street line, or alley line.

Projection. The distance by which a sign extends over public property or beyond the building line.

Public sign. A sign of a non-commercial nature and in the public interest, erected by or upon the order of a public official in the performance of his public duty, such as safety signs, danger signs, trespassing signs, traffic signs, memorial plaques, signs of historical interest and all other similar signs, including signs designating hospitals, libraries, schools, airports and other institutions or places of public interest or concern.

Pylon sign. Refers to an advertising structure projecting from the wall or extending over the roof of any building, comprising a framework and display surface, the structural members of which are an integral part of the building upon which such sign is erected.

Raised or embossed sign. Includes any lettering, characters or numbers affixed to the building or structure.

Real estate sign. Any sign erected by the Owner, or his agent, advertising the real property upon which the sign is located for rent or for sale, but shall not include rooming house signs.

Roof sign. Outdoor advertising display sign erected, constructed, or maintained above the roof of any building.

Roof sign. Any sign erected, constructed and maintained wholly upon or over the roof of any building with the principal support on the roof structure.

Roof sign. A sign erected upon or above a roof or parapet of a building or structure.

Roof sign. Sign in which the display of letters and figures is above roof level.

Roof sign. Sign erected upon or above a roof of a building or other structure.

Roof sign. Sign which is erected, constructed or maintained upon, and projects above or beyond the roof or parapet.

Roof sign. A sign, any portion of which is erected, constructed and maintained above any portion of the roof of a building.

Roof sign. Any sign which is an independent structure, and which is fastened to or supported by the roof. No roof sign shall project outward beyond any wall of a building, or if erected on a flat roof, sign shall not extend beyond the edge of the roof. If a flat sign or cantilever sign is partially braced to the roof, same shall not be classified as a roof sign.

Roof sign. A sign which is erected, constructed, or maintained above the roof of a building and does not project more than twelve (12) inches beyond the wall line of the building.

Sale and rental signs. Ground signs or wall signs used exclusively for the sale or lease of property on which they are erected, having an area not exceeding twenty four (24) square feet may be constructed entirely of combustible materials; provided that such ground signs shall be located not less than ten (10) feet from any building or public way.

Sandwich sign. Any sign which is not permanently affixed to any structure on the site or permanently ground-mounted; any portable sign.

Service sign. Sign identifying rest rooms and other service facilities.

Sign. Any structure or part thereof, or device attached thereto, or painted thereon, located outside a building that is arranged intended, designed for, or used as (or which displays or includes any letter, word motto, banner, flag, pennant, insignia, device or representation which is in the nature of an advertisement, announcement, direction or attraction.)

Sign. "Sign" is a name, identification, description, display, or illustration which is affixed to, painted, or represented directly or indirectly upon a building, structure, or piece of land, and which directs attention thereto.

Sign. Any words, letters, parts of letters, figures, numerals, phrases, sentences, emblems, devices, designs, trade names, or trademarks by which anything is made known, such as are used to designate an individual, a firm, an association, a corporation, a profession, a business, or a commodity or products, which are visible from any public street or highway and used to attract attention.

Sign. Any object or structure on the surface of which are any letters or symbols, or which is intended to identify or to attract attention to any property or premises, and which is placed or maintained in, upon, or over, or is within view of any sidewalk, street, lane, alley or other public place or way within the corporate limits of the City.

Sign. Every name, identification, description, announcement, declaration, demonstration, display, flag, illustration or insignia, and the structure displaying or supporting any of the same, affixed directly or indirectly to or upon any building or outdoor structure, or erected or maintained upon a piece of land, which directs attention to an object, product, place, activity, person, institution, organization or business, except air navigational signs.

Sign. Use of any words, numerals, figures, devices, designs, or trademarks by which anything is made known such as are used to show an individual firm, profession or business, and are visible to the general public.

Sign. Any advertisement, announcement, direction or communication produced in whole or in part by the construction, erection, affixing or placing of a structure on any land or on any other structure, or produced by painting on or posting or placing any printed, lettered, pictured, figured or colored material on any building, structure, or surface. Signs placed or erected by governmental agencies or nonprofit civic associations for a public purpose in the public interest shall not be included herein, nor shall this include signs which are a part of the architectural design of a building.

Sign. Any medium including its structure and component parts, which is used or intended to be used to attract attention to the subject matter for advertising purposes.

Sign. Any lettering or symbol made of cloth, metal, paint, paper, wood, or other material of any kind whatsoever placed for advertising, identification or other purposes on the ground or on any bush, tree, rock, wall, post, fence, building, structure, vehicle, or on any place whatsoever. The term "placed" shall include constructing, erecting, posting, painting, printing, tacking, nailing, gluing, sticking, carving, or otherwise fastening, affixing, or making visible in any manner whatsoever beyond the boundaries of a site.

Sign. Any device or part thereof which is situated outdoors or indoors, but visible from the street or adjacent property which is used to advertise, identify, display, direct, or attract attention to an object, person, institution, organization, business, product, service, event or location by any means including words, letters, figures, designs, symbols, fixtures, colors, motion illumination or projected images.

Sign. Any lettered or pictorial device designed to inform or attract attention.

Sign. Any device on which lettered, figured or pictorial matter is displayed for the purpose of visually bringing the subjects to which it is appertaining to the attention of the public while viewing the same from outdoors.

Sign. Structure used for the display of letters and figures for advertising, announcement, instruction or direction.

Sign. Sign is any structure or part thereof, or any device attached to, painted on, or represented on a building or other structure, upon which is displayed or included any letter, work, model, banner, flag, pennant, insignia, decoration, device, or representation used as or which is in the nature of, an announcement, direction, advertisement, or other attention-directing device. A sign shall not include a similar structure or device located within a building except for illuminated sign within show windows. Sign includes any billboard, but does not include the flag, pennant, or insignia of any nation or association of nations, or of any state, city or other political unit, or of any political, charitable, educational, philanthropic, civic, professional, religious, or like campaign, drive, movement, or event.

Sign. Structure, building wall, including special lighting effects and lighting devices on buildings, or other outdoor surface, or any device used for the purpose of bringing the subject thereof to the attention of the public, or to display, identify and publicize the name and product or services of any person. It shall include a temporary sign, pylon or pole sign, marquee, awning, canopy or street clock, and shall include any announcement, declaration, demonstration, display illustration, insignia or sound used to advertise or promote the interest of any person when the same is placed indoors for view of the general public from out-of-doors. Roof sign is a sign erected upon or supported by the roof of a building.

Sign. Any device designed to inform or attract the attention of persons by the display of characters, letters, illustrations or any ornamentations.

Sign. Any outdoor advertising having a permanent location on the ground or attached to or painted on a building including bulletin boards, and poster boards.

Sign. Includes every sign, billboard, ground sign, wall sign, roof sign, illuminated sign, projecting sign, temporary sign, marquee, awning, canopy, and street clock, and shall include announcement, declaration, demonstration, display, illustration or insignia used to advertise or promote the interests of any person when same is placed out of doors in view of the general public.

Sign. Any sign, notice, advertising device, or any part thereof, whether it contains words or not and includes any device that is used solely to attract attention.

Sign. Any medium including its structure and component parts, which is used or intended to be used to attract attention to the subject matter for advertising purposes other than paint on the surface of a building.

Sign. Any display of characters, letters, illustrations or any ornamentations, or the complete structure on which any such characters, letters, illustrations or ornamentations are stated or applied (except buildings to which the same may be attached), used for identification, directional, advertising or promotional purposes; provided, however, that said sign shall not be construed so as to include self-contained fixtures approved by the National Board of Fire Underwriters, or nonelectrical display wholly contained within a store building.

Sign. Any outdoor thing or device employed to publicly display notice, advertisement or propaganda by the use of letters, words, characters, figures, pictures or other graphical means. A tablet, plaque, dedication or memorial inscription, name, date, street number, or like marking consisting entirely of incombustible materials, incorporated as component parts of a structure, or any painting on fences, building walls or like structures shall not be classed as a sign.

Sign. Any advertising device, notice or any part thereof whether it contains words or not and includes any device that is used solely to attract attention.

Sign. Includes a closed sign, display sign (but not including billboards and poster panels or outdoor advertising signs), ground signs, marquee sign, open sign, roof sign, wall sign and projecting sign, or as otherwise defined in the Building Code.

Sign. Structure, building wall, or other outdoor surface, or any device used for visual communication which is used for the purpose of bringing the subject thereof to the attention of the public, or to display, identify and publicize the name and product or services of any person.

Sign area. Entire area within a single continuous perimeter enclosing the extreme limits of the actual message or copy area. It does not include any structural elements outside the limits of such sign and not forming an integral part of the display. Only one side of a double-faced or V-type sign structure shall be used in computing allowable sign area.

Sign area. Area of a sign shall be computed as the entire area within a single continuous rectilinear perimeter of not more than eight straight lines enclosing the extreme limits, writing, representation, emblem, or design, together with a material or color forming an integral part of the display or used to differentiate the sign from the background against which it is placed. Sign supports shall not be included in determining sign area unless they are an integral part of the display. The area of a sign or the total area of all signs on a site shall be the total area that would be visible, whether legible or not, to off-site observer, having an unobstructed view of the site from any single point within a horizontal distance of one hundred (100) feet from the site boundary at an elevation not more than one hundred (100) feet above the site boundary.

Signboard. Any structure or part thereof on which lettered or pictorial matter is displayed for advertising or notice purposes.

Signfacing. The opaque or transparent surface or surfaces of the sign, upon, against, or through which the message of the sign is exhibited.

Sign frontage. The length in feet of the ground floor level of a building front or side facing a street (or facing a right-of-way accessible from a street) that is occupied by an individual business.

Sign-on-site. Sign relating in its subject matter to the premises on which it is located or to products, accommodations, services, or activities on the premises. On-site signs do not include signs erected by the outdoor advertising industry in the conduct of the outdoor advertising business.

Signs. Outdoor advertising displays, means any letter, figure, character, mark, plane, point, marquee sign, design, poster, pictorial, picture, stroke, stripe, line, trademark, reading matter or illuminated service, which shall be so constructed, placed, attached, painted, erected, fastened or manufactured in any manner whatsoever, so that the same shall be used for the attraction of the public to any place, subject, person, firm, corporation, public performance, article, machine or merchandise, whatsoever, which are displayed in any manner whatsoever out of doors.

Sign structure. Any structure which supports or is capable of supporting any sign as defined in the code. A sign structure may be a single pole and may or may not be an integral part of the building.

Sign surface area. Entire area within a parallelogram, triangle, circle, semi-circle or other geometric figure, including all of the elements of the matter displayed, but not including blank masking, frames, or structural elements outside the advertising elements of the sign and bearing no advertising matter.

Size of sign. The smallest rectangle in which the sign, including decorative borders or symbols, will fit.

Sky or roof sign. Deemed to be any letter, word, model, sign or device in the nature of an advertisement, announcement or direction supported wholly or in part over or above any wall, building or structure.

Small professional or announcement signs. Shall be limited to not over one (1) square foot in area, if fixed flat to the main wall of a residence building. Name and announcement signs firmly fixed to the main wall of public or semi-public buildings, and which are not over six (6) square feet in area; signs having an area of not exceeding six (6) square feet on horticultural or agricultural buildings, and real estate signs are displayed behind the prevailing front building line of that block and further, provided, that there shall be but one (1) such sign on each lot which shall be used only to advertise the premises upon which it is erected.

Snipe sign. Any sign under twenty (20) square feet in area made of any material, including paper, cardboard, wood and metal, when such sign is tacked, nailed, posted, pasted, glued, or otherwise attached to trees, poles, fences or other objects, and the advertising matter appearing thereon is not applicable to the premises upon which said sign is located.

Special displays. Special decorative displays used for holidays, public demonstrations or promotion of civic welfare or charitable purposes, when authorized by the municipal authorities, on which there is no commercial advertising.

Spectacular sign. Any sign of noncombustible materials, erected or maintained independently of any building, upon a structure of structural steel or concrete, and exceeding twenty-five (25) feet in height.

Stop or danger signs. Any sign which uses the word "stop" or "danger" or presents or implies the need or requirement of stopping, or the existence of danger, or which is a copy or imitation of official signs. Red, green or amber (or any color combination thereof) revolving or flashing light giving the impression of a police or caution light shall be considered a prohibited sign, whether on a sign or on an independent structure.

Street banners. Signs advertising a public event providing that specific approval is granted under regulations established by the City Council.

Street clock. Any timepiece erected upon a standard upon the sidewalk or on the exterior of the building or structure placed and maintained by some person for the convenience of the public and for the purpose of advertising their place of business.

Structural trim. Includes the molding, battens, cappings, nailing strips, latticing and platforms which are attached to the sign structure.

Structure. That which is built or constructed, an edifice or building of any kind or any piece of work artificially built up or composed of parts joined together in some definite manner.

Subdivision sign. Any sign located either on or off a subdivision tract, which indicates the direction to or advertises the location, existence, or sale of a subdivision or any part thereof.

Surface area of a sign. Computed as including the entire area within a regular geometric form or combination of regular geometric forms comprising all of the display area of the sign and including all of the elements of the matter displayed. Framed and structural members not bearing advertising matter shall not be included in computation of surface area.

Swinging projecting sign. Any sign projecting at an angle from the outside wall or walls of any building, which is supported by only one rigid support, irrespective of the number of guy wires used in connection therewith.

Swinging signs. Signs which swing or otherwise noticeably move as a result of wind pressure because of the manner of their suspension or attachment.

Temporary sign. Shall mean signs to be erected on a temporary basis, such as signs advertising the sale or rental of the premises on which located; signs advertising a subdivision of property;

signs advertising construction actually being done on premises on which the sign is located; signs advertising future construction to be done on the premises on which located and special events, such as public meetings, sporting events, political campaigns or events of a similar nature.

Temporary sign. A sign or cloth or other combustible material, with or without a frame, which is usually attached to the outside of a building on a wall or store front, intended for a limited period of display.

Temporary sign. Sign intended for a limited period of display.

Temporary sign. Any sign erected and maintained by a contractor, sub-contractor or material-man upon any property upon which or for which such contractor, sub-contractor or material-man is furnishing labor or materials.

Temporary sign. A display sign, banner, poster, or other semi-permanent advertising device constructed of cloth, canvas, fabric or other light temporary material, with a structural frame, intended for a limited period of display; including decorative displays for holidays or public demonstrations, but not including signs, banners or other devices wholly of cloth, paper or plastic.

Temporary sign. Free-standing sign not attached to the structure it serves and/or intended as a permanent fixture or designed to be such in the judgment of the Zoning Inspector.

Temporary window signs. Allowed only if they advertise special sales or events lasting no more than 15 days. They may cover no more than 30 percent of the area of the window in which they appear.

Traffic sign. A temporary or permanent sign designed solely for the purpose of regulating vehicular or pedestrian traffic, the warning of danger, or the providing of directions to public facilities or streets; provided that, where possible, such signs shall be of the same type as in use by the State.

Vehicle signs. Signs accessory to the use of any kind of vehicle, providing the sign is painted or attached directly to the body of the vehicle.

Vertical projecting sign. Any projecting sign which is greater in height than in width.

Wall area sign. The width multiplied by the height of that section of the building occupied by the advertiser, facing any one public way.

Wall sign. Includes all flat signs of solid face construction which are placed against a building or other structure and attached to the exterior, front, rear or side wall of any building or other structure.

Wall sign. Any sign attached to or erected against the wall of a building or structure, with the exposed face of the sign in a plane parallel to the plane of the wall.

Wall sign. Any sign on a readily detachable surface, or plane, that may be affixed against and parallel to the front, rear or side of any building.

Wall sign. Flat sign which is attached directly to the wall of a structure in order that the sign is approximately parallel to the wall surface and only one (1) side is visible.

Wall sign. Any sign attached to the wall of a building or structure, with the exposed face of the sign in a plane parallel to the plane of said wall.

Wall sign. Display sign which is affixed or attached to the wall of a building or other structure and which extends not more than fifteen (15) inches from the face of the wall.

Wall sign. Sign which is attached to a wall of a building and projects not more than fifteen (15) inches from such wall except that lighting reflectors may project from such wall a distance not exceeding eight (8) feet.

Wall sign. Outdoor sign affixed or attached to the wall of a building or other structure and projecting not more than twelve (12) inches from the face of the wall.

Wall sign. A sign which is supported wholly or partially by an exterior wall of a building and extends not more than twelve (12) inches therefrom.

Wind loads. For the purpose of design, and except for roof signs, wind pressure shall be taken upon the gross area of the vertical projection of all signs at not less than fifteen (15) pounds per square foot for those portions less than sixty (60) feet above the ground, and at not less than twenty (20) pounds per square foot for those portions more than sixty (60) feet above the ground. (Verify wind loads with Local Authorities).

Wind sign. Any sign in the nature of a series of two or more banners, flags, pennants or other objects or material which call attention to a product or service, fastened in such a manner as to move upon being subjected to pressure by wind or breeze.

Window area sign. The width multiplied by the height of the total window area facing any one public way.

C-27 STORAGE AND WAREHOUSE BUILDINGS

Bulk storage. Floor area devoted to the storage of merchandise, materials and other non-automobile traffic generating uses.

Coal bin. Structure used for the storage of coal for a commercial purpose.

Cold storage plant. Building used for the storage of food or other material and for the preservation thereof by ice or refrigeration for a commercial purpose.

Fabrication. Manufacturing, excluding the refining or other initial processing of basic raw materials such as metal ores, lumber or rubber. Fabrication relates to stamping, cutting or otherwise shaping the processed materials into useful objects.

Factory. Building, the primary use of which is for the processing of materials or manufacture of products.

Freight depot. Building used for storage incidental to transportation of freight or cargo.

Hazardous storage. Includes but is not limited to such products as nitro-cellulose, or products composed in whole or in part of nitro-cellulose or similar flammable materials such as films, combs, pens, pencils, toilet articles and the like.

Hazardous use storage units. Any hazardous use unit designed, intended or used for the storage of high hazard materials, high hazard products and all other high hazard storage uses.

High piled storage. Includes combustible materials in closely packed piles more than 15 feet high or materials or pallets or in racks more than 12 feet high. For highly combustible materials such as rubber goods and certain plastics, the critical height of piling may be as low as 8 feet.

Low-hazard storage buildings. Includes storage buildings in which the use and occupancy involve materials or substances which create not more than a "low fire hazard" as defined in the Code. Low fire-hazard materials and substances include those which have the fire-hazard characteristics of any of the following: aluminum, iron, steel, and copper, when not finely divided; asbestos; chalk and crayons; glass; ivory; porcelain and pottery; and talc and soapstone.

Lumber storage shed. Shed which is used for the orderly storage of lumber.

Lumber storage sheds of type V–wood frame construction. Located with respect to other buildings, other such sheds, and interior lot lines other than railroad right-of-way lines, so that their supporting posts or columns are not less than 30 feet (minimal) therefrom and so that their eaves and canopies are not less than 20 (minimal) feet therefrom, except that no fire separation is required for a side or back of a lumber storage shed which is fully protected by a fire wall having a fire-resistance rating of not less than 2 hours.

Open shed. Structure used for storage in which at least one of the long sides and at least 40% of the perimeter is not enclosed.

Petroleum bulk storage. Building or structure for the storage of lubricating oils with a flash point of three hundred (300) degrees F. or higher and storage space for not more than one motor vehicle.

Petroleum bulk storage buildings. Warehouses for the bulk storage of not more than fifty thousand (50,000) gallons of lubricating oils with a flash point of not less than three hundred (300) degrees F. in approved sealed containers may be erected outside the fire limits of masonry wall (type 3) construction not more than five thousand (5000) square feet in area and not more than one (1) story or twenty (20).

Separated storage. Storage in the same fire area but physically separated by as much space as practicable, using sills or curbs as safeguards, or by intervening storage of non-hazardous, compatible commodities.

Storage. Act of depositing goods, wares and merchandise in any structure, part of a structure or, warehouse, gratuitous or otherwise, shall be called storage.

Storage. Buildings in which the primary or intended occupancy or use is the storage of, or shelter for, goods, merchandise, products, vehicles, or animals.

Storage building. A building for the housing, except for purely display purposes, of airplanes, automobiles, railway cars or other vehicles of transportation, for the sheltering of horses, livestock, or other animals, or exclusively for the storage of goods, wares, or merchandise, not excluding in any case, offices incidental to such uses.

Storage buildings. Include buildings which are used for sheltering or confining fowls or animals, coal bins, cold storage plants, freight depots, lumber storage sheds, grain storage bins, warehouses, and buildings and structures of similar use.

Storage garage. The term "storage garage" shall mean a garage in which volatile inflammable oil, other than that contained in the fuel storage tanks of motor vehicles, is handled, stored or kept.

Storage occupancy. Occupancy or use of a building or structure or any portion thereof for the storage of goods, wares, merchandise, raw materials, agricultural or manufactured products, including parking garages, or the sheltering of live stock and other animals; except when classed as a high hazard occupancy.

Storage occupancy. Includes all buildings and other structures and parts thereof, and all premises, used for the storage of goods, wares, or merchandise not highly combustible, flammable or explosive.

Storage rooms. Every room exceeding 500 square feet in the area appropriated for the storage of combustible materials, other than those enumerated in the Code. Accessory to retail sales area by a one-hour fire-resistive partition and ceilings with openings protected by Class "C" fire doors.

Truck terminal. Commercial facility where truck freight is stored, handled, and dispatched between various locations by way of different major truck carriers and including facilities for the storage and repair of trucks and trailers while awaiting consignment.

Truck terminal. Temporary parking of motor freight vehicles or trucks of common carriers during loading and unloading and between trips, including necessary warehouse space for storage of transitory freight and office space.

Warehouse. A building used and occupied for the storage of merchandise, furniture or other combustible or flammable material, and iron, steel and similar material.

Warehouse. Storage building which is used for the storage of goods, wares, food, beverages, merchandise, or other chattels, and includes a building used for the storage of unused new automobiles or aircraft, but does not include a garage or hangar.

Warehouse. Structure or part of a structure, for storing goods, wares and merchandise, whether for the Owner or for others, and whether the same being public or private warehouse.

Warehouse. Any building used exclusively for the storage of merchandise or other products.

Warehousing and wholesaling. Structure for the purpose of storing and/or wholesaling goods and materials including freight storage and forwarding, general warehousing cold storage, and any other type of warehousing or wholesaling, provided that all operations except the loading and parking of trucks shall be enclosed within a building.

C-28 TEMPORARY BUILDINGS, STRUCTURES, AREAS AND USES

Exhibition areas. Any building or buildings, area or areas, within a building used for temporary exhibition of goods, wares, merchandise or equipment.

Minor recreational uses of property. Those recreational uses of property which do not have permanent structures.

Reservoir space. Temporary storage space, exclusive of street area or required parking spaces, for a vehicle waiting for service or admission.

Semi-public parking area. Open area other than a street, alley or place, used for temporary parking of more than four (4) self-propelled vehicles and available for public use whether free, for compensation, or as an accommodation for clients or customers.

Temporary building. A building not designed to be permanently located in the place where it is or where it is intended to be placed or affixed.

Temporary building. A building not designed to be permanently located at its present site, or where it is intended to be temporarily erected, placed or affixed.

Temporary food service establishment. Any food service establishment which operates at a fixed location for a temporary period of time in connection with a fair, carnival, circus, public exhibition or other transitory gathering.

Temporary restaurant. A food facility operating for special occasions such as a fair, carnival, circus, public exhibition, or similar public gathering.

Temporary seating facilities. Those which are intended for use at a location for not more than ninety (90) days.

Temporary sign. A sign which is constructed of materials which are subject to rapid deterioration by weather and/or extended use.

Temporary signs. A sign constructed of cloth, fabric or other light temporary material with or without a structural frame intended for a limited period of display; including decoration displays for holidays or public demonstrations.

Temporary structure. A movable structure not designed for human occupancy nor for the protection of goods or chattles and not forming an enclosure.

Temporary structures for use or protection of general public. Before any construction or demolition operation or project is commenced, plans for all temporary structures used for the protection of the general public, including sheds, scaffolds, trestles, foot bridges, guard fences and other similar structures required for such project or operation, shall be filed with the Building Commissioner and shall be approved by the Building Department before any work is commenced.

Temporary trailer park. Shall mean any area or tract of land where one or more spaces are rented or leased or held out for rent or lease, or for which free occupancy is permitted to accommodate recreational vehicles and which is established for one operation not to exceed 10 consecutive days, and is then removed.

Temporary use or building. Use or building permitted by the Zoning Board to exist during periods of construction of the main building or use, or for special events.

Temporary wood frame structures and tents. Include platforms, reviewing stands, gospel tents, circus tents, and other structures erected to serve their purpose for a limited time.

Tent. Shelter or structure which is not an appendage to a building, nor a roof structure, the covering of which is wholly or partly of canvas or other pliable material which is supported and made stable by standards, stakes, and ropes.

C-29 THEATERS

Adult theater. Shall mean (A) An enclosed building or open-air drive-in theater regularly used for presenting any film or plate negative, film or plate positive, film or tape designed to be projected on a screen for exhibition, or films, glass slides or transparencies, either in negative or positive form, designed for exhibition by projection on a screen depicting, describing or relating to 'specified sexual activities' or characterized by an emphasis on matter depicting, describing or relating to 'specified anatomical areas.' (B) An enclosed building or open-air drive-in theater regularly used for presenting any film or plate negative, film or plate positive, film or tape designed to be projected on a screen for exhibition, or films, glass slides or transparencies, either in negative or positive form, designed for exhibition by projection on a screen and which regularly excludes all minors.

Aisle. Means the clear width and length of an area which is provided for ingress or egress between rows of seats, or between rows of seats and a wall, or between desks, tables, counters, machines, or other equipment or materials, or between such articles or materials and a wall.

Aisle (longitudinal). Means an aisle approximately at right angles to the rows of seats served.

Aisle (transverse). Means an aisle approximately parallel to the rows of seats between which it passes.

Arena or round theater. Any building, or portion of building, which is used for the presentation of performances on the floor level without elevated stage, scenery, curtains, apparatus, or stage properties.

Arena or round theater. Any building, or portion of building, which is used for the presentation of performances on the floor level without elevated stage, scenery, curtains, apparatus or stage properties, except that a limited amount of furniture may be used and there may be bleachers, tiered seats, or floor level seats on one or more sides.

Balcony. That portion of the seating space of an assembly room, the lowest part of which is raised 4 feet or more above the level of the main floor.

Cross aisle. An aisle in a place of assembly usually parallel to rows of seats connecting other aisles or an aisle and an exit.

Cross aisle. Aisle connecting longitudinal aisles.

Drive-in-theater. An open-air theater designed for viewing by the audience from motor vehicles.

Foyer. Enclosed space surrounding or in the rear of the auditorium of a theater or other place of assembly which is completely shut off from the auditorium and is used as an assembly or waiting space for the occupants.

Foyer. A room adjoining an auditorium and serving as the principal entrance to any seating level.

Foyer. A lobby, corridor or passage, one or more in combination, adjacent to the auditorium of a theater or assembly hall at the level of the main floor or a balcony thereof and into which one or more exits therefrom open, in the path of normal egress from the building.

Foyer. Any space in a place of assembly which is used for ingress, egress, the distribution of people to the aisles, and as a space for waiting.

Foyer. Enclosed space surrounding or in the rear or front of an assembly building and used as an assembly or waiting space for the occupants.

Foyer. Enclosed space and passageway into which aisles, corridors, stairways, or elevators may exit and from which the public has access to exits.

Foyer wall. Wall separating the assembly area from the foyer, lobby or exit passageways.

Indoor theater. Building designed and/or used primarily for the commercial exhibition of motion pictures or live shows to the general public.

Longitudinal aisle. Aisle approximately at right angles to the rows of seats served.

Main aisle. Aisle which is not dependent upon travel along a subsequent aisle to provide access to an exit from a room.

Motion picture house. A building or portion of a building used for the showing of motion pictures with a recess at the front of the auditorium used and designed solely for the mounting of a motion picture screen and its relayed sound equipment and containing no fixed or movable scenery other than curtains of flame resistant material.

Motion picture projection room. A room or booth enclosing a motion picture projector and constructed in accordance with specifications issued by the Building Department.

Motion picture theater. Any building or part of a building regularly used for private or corporate profit as a place of assemblage for the witnessing of motion pictures, and not having a stage capable of being used for theatricals, and not using movable scenery.

Motion picture theater. Theater used primarily for the projection of motion pictures.

Projection room. Any room appropriated to the use of motion picture projection machine using film over 7/8″ wide or cellulose-nitrate base film.

Projection room. Room in a theater or assembly hall containing a projector for moving pictures.

Proscenium arch. The opening of the proscenium wall that permits a view of the stage.

Proscenium wall. The wall separating the stage from the auditorium in a theater.

Proscenium wall. A fire-resistive wall that separates a stage or enclosed platform from the public of spectator's area of an auditorium or theater.

Scenery and scenic elements. Any or all of those devices ordinarily used on a stage in the presentation of a theatrical performance, such as back drops, side tabs, teasers, borders or scrim, rigid flats, set pieces, and all properties, but not including costumes.

Seating capacity. Seating capacity for a theater, auditorium, or any room or place of public assemblage in which seats are not fixed, shall be determined on the basis of seven (7) square feet of floor, balcony, and/or gallery area per person, and in the case of fixed seats, such as pews or benches, the seating capacity shall be based on one person to eighteen (18) inches of pew or bench length.

Seating capacity. Shall mean, where seats are fixed, the number of persons for whom seats are provided, and where seats are not fixed, or provided, shall be calculated on the basis of the areas listed in the Building Code.

Seating section. An area of seating bounded on all sides by aisles, cross aisles, walls or partitions.

Seats. The seating capacity of a particular building as determined by the specifications and plans and filed with the Zoning Officer; in the event individual seats are not provided, each 20 inches of benches or similar seating accommodations shall be considered as one seat for the purpose of the Ordinance.

Stage. That portion of an assembly room or area adjacent thereto which is used for any exhibition.

Stage. Space designed primarily for theatrical performances with provision for quick change scenery and overhead lighting, including environmental control for a wide range of lighting and sound effects and which is traditionally, but not necessarily, separated from the audience by a proscenium wall and curtain opening.

Stage. Place used for theatrical presentations or other entertainments, whereon movable scenery, or other accessories are used.

Stage. A portion of a place of assembly or building which is used for presentations of plays and other entertainments, wherein scenery, drops, or other effects may be installed or used, and where the distance between the top of the proscenium opening and the ceiling of the stage is more than five (5) feet.

Stage. A partially enclosed portion of an assembly room wherein scenery drops or other effects may be installed and used, and which is cut off from the audience section by a proscenium wall, and where there is more than five feet of open space above and on the stage side of the proscenium opening.

Stage (dais). A small and low "platform" or "rostrum" may be placed on a "stage". May be permanent, temporary, or portable.

Stage (general). A partially enclosed portion of an Assembly Building, cut off from the audience section by a proscenium wall, which is designed or used for the presentation of plays, demonstrations, or other entertainment. "Stages" shall be classified as "working stage" or "non-working stage".

Stage lift. A movable section of a stage floor, designed to carry scenery between staging areas and the stage, and also used to be raised to and temporarily retained at elevations above or below the stage level.

Standing room. That space not required for legal aisles, floor passages, and exits, in auditoriums or assembly halls.

Summer motion picture theater. Any building, or part of a building, used as a place of assemblage not over 14 weeks per year for witnessing motion pictures; such does not have a stage capable of being used for theatricals nor movable scenery.

Summer motion picture theater. Any building or part of a building used for commercial purposes as a place of assemblage not over fourteen (14) weeks per year for the witnessing of motion pictures and not having a stage capable of being used for theatricals and not using movable scenery.

Summer theater. A building, or part of a building, used as a place of assemblage not over 14 weeks per year for witnessing amateur or professional theatricals, motion pictures, vaudeville or operatic performances in which scenery, apparatus, or stage properties are employed. The stage shall be separated from the auditorium by a proscenium wall.

Theater. Any building, or part of a building, with fixed seats, which is regularly used as a place of assemblage for witnessing amateur or professional theatricals, motion pictures, vaudeville, dramatic, operatic or pantomime performances, in which scenery, apparatus, or stage properties are employed. The stage shall be separated from the auditorium by proscenium wall.

Theater. Any building or part of a building regularly used for commercial purposes as a place of assemblage with fixed seats for the witnessing of amateur or professional theatricals, motion pictures, vaudeville, dramatic, operatic, or pantomime performances in which scenery, apparatus or stage properties are employed. The stage shall be separated from the auditorium by prosenium wall.

Theater. Building which has a stage, and to which the general public is admitted to witness theatrical, vaudeville, burlesque, dramatic, or operatic performances, or to witness motion or televised pictures.

Theater. Buildings or parts of buildings in which persons congregate to witness spectacular vaudeville, burlesque, dramatic or operatic performances when either transient scenery or more than one set of permanent scenery are used; or when such permanent scenery, including the drop curtain exceeds 1,000 square feet in area; or where either more than 3 entrances (scenery term) are used; or when any such entrances exceed 15 feet in depth or 25 feet in width; or when either a movable stage or traps are used.

Theater. Building, or part thereof which contains an assembly hall with or without stage which may be equipped with curtains and permanent stage scenery or mechanical equipment adaptable to the showing of plays, operas, motion pictures, performances, spectacles and similar forms of entertainment.

Theater. Building or part of a building to which the general public is admitted to witness regular theatrical, vaudeville, burlesque, dramatic, or operatic performances, entertainment, or exhibitions in which scenery is used, or to witness the projection of motion pictures.

Theater. Building, or any portion thereof, which is used for public dramatic, operatic, motion-picture, or other performances.

Theater. Place of public assembly intended for the production and viewing of the performing arts or the screening and viewing of motion-pictures and consisting of an auditorium with permanently fixed seats intended solely for a viewing audience.

Theater. Structure used for dramatic, operatic, motion pictures, or other performance, for admission to which entrance money is received and no audience participation or meal service allowed.

Theater. Building or part thereof which contains an assembly hall, having a stage which may be equipped with curtains and permanent stage scenery or mechanical equipment adaptable to the showing of plays, operas, motion pictures, performances, spectacles and similar forms of entertainment.

Theater. An assembly unit designed or used primarily for theatrical performances and containing a Type I stage.

Transverse aisle. Aisle approximately parallel to the rows of seats between which it passes.

Working stage. Permanent stage in connection with any occupancy, which is equipped, or adaptable for equipment, with a rigging loft or fly gallery, and which is used for theatrical, musical and like performances.

Working stage. A portion of an assembly building which is cut off from the audience section by a proscenium wall, provided with an opening, so arranged that curtain or drops may be lifted more than ten feet (10′) or so that their lower edges are higher than one-half of the height of the proscenium opening.

Working stage. Partially enclosed portion of an Assembly Building, cut off from the audience section by a proscenium wall of masonry of not less than four (4) hour fire-resistive construction, and which is equipped with scenery, loft, gridiron, fly-gallery, and lighting equipment, and the proscenium opening shall be equipped with a fire-proof and smoke-proof curtain, and the depth from the proscenium curtain to the back wall shall be not less than fifteen (15) feet.

C-30 TRANSIENT ACCOMMODATIONS

Apartment. A facility where at least fifty (50) percent of the rental units are equipped for housekeeping and is usually located in a vacation destination area. Units typically provide a fully equipped kitchen and generally have a living room and one or more bedrooms, although they also may be studio-type rooms with kitchen equipment in an alcove that can be closed off from living/sleeping areas.

Cottages. Individual buildings, usually containing one rental unit equipped for housekeeping. May have separate living room and bedrooms; occasionally have more than one rental unit per building as in "duplex cottages." Parking usually located at each unit.

Hotel. A multistory building which is most often located in or near the downtown business district, large business parks in highly populated neighborhoods, and also may be situated in a resort area. More extensive services usually include coffee shops, dining rooms, cocktail lounges, room service, convenience shops, gift shops, valet, laundry services, banquet and meeting spaces. Typically caters to public transportation travelers; parking is limited and on/off the premise parking facilities is provided.

Lodge. Typically a country inn or lodge having large reception areas, two or more stories with service and convenience facilities in the building. Usually has adequate parking facilities.

Motel. A building having one or more stories. Food service such as a coffee shop or snack bar is available. Often has some recreational facilities such as a swimming pool and playground. Some have tennis courts. Ample convenient parking is usually available at the unit door.

Motor hotel. Similar to a hotel in services provided, but often has recreational facilities such as swimming pool, gameroom, health club and saunas. Caters to motorists and provides ample parking.

Motor inn. Usually larger than a motel with two or more stories and offers recreational facilities, more extensive food service and beverages. Usually provides ample parking.

Place of public accommodation. Includes but not be limited to: any tavern, road house, hotel, motel, trailer camp, summer camp, day camp, or resort camp, whether for entertainment of transient guests or accommodation of those seeking health, recreation or rest.

Ranch. A combination of a lodge having guest rooms, and individual cottages or small motel buildings, providing food and beverage services, but featuring western-type outdoor recreational facilities.

Resort. The operation may apply to the use of any type or combination of types of facilities, but removed from urban and metropolitan areas. Offers extensive recreational facilities on the premises; may cater to specific interests such as golf, tennis, fishing, horseback riding etc.

Transient. Person who resides in the same building for a period of less than 30 days.

1. Category D

FUNCTIONAL REGULATIONS D-1 THROUGH D-19

D-1 ALCOHOLIC BEVERAGES AND DISPENSING OUTLETS

Alcohol. The product of distillation of any fermented liquid, whether rectified or diluted, whatever may be the origin thereof, and shall include synthetic ethyl alcohol. It shall not include denatured alcohol or wood alcohol.

Alcoholic beverage. Means beverages controlled by the Alcoholic Beverage Control Board.

Alcoholic beverages. The definitions contained in the State laws are the same definitions of terms used in the city code.

Alcohol liquor. Includes the four (4) varieties of liquor such as, alcohol, spirits, wine, and beer, and every liquid or solid, patented or not, containing alcohol, spirits, wine, or beer, and capable of being consumed as a beverage by a human being.

Bar. An establishment, including but not limited to a cocktail lounge, discotheque, nightclub, or tavern, the main use of which is to serve spiritous liquors for on-site consumption. Such facility may serve food, as well as, provide dancing and entertainment.

Bar. An establishment, the main use of which is to serve spiritous liquors for on-Site consumption. A bar usually has a counter and stools and may serve food.

Bar. An establishment whose primary business is the serving of alcoholic beverages to the public for consumption on the premises.

Bar. Establishment the main use of which is to serve spirituous liquors to be consumed on the premises. Food may or may not be served. Usually a counter and stools are present.

Beer. A beverage obtained by alcoholic fermentation of an infusion or concoction of barley, or other grain, malt, and hops in water, and shall include, among other things, beer, ale, stout, lager beer, near beer, porter beer and the like.

Bottle club. An establishment in which an operation whether formally organized as a club, having a membership list, dues, officers and regular meetings or not, maintaining premises where persons who have made their own purchases of alcoholic liquors congregate for the express purpose of consuming such alcoholic liquors.

Bottle sales. Sales of liquor by the full bottle. Also called Package Sales.

Cabaret. Shall mean a place of business other than a "night club" located in a hotel or a motel having fifty (50) or more guest rooms, where liquor, beer or wine is sold, given away or consumed on the premises and where music or other entertainment is permitted or provided for the guests of said hotel or motel only, which place of business is duly licensed as a "cabaret."

Club. A corporation organized under the state laws, not for a pecuniary profit, solely for the promotion of some common object other than the sale or consumption of alcoholic liquors.

Cocktail lounge. Place for service or consumption of alcoholic beverages.

Cocktail lounge. Establishment which is licensed to sell intoxicating liquors by the drink.

Cocktail lounge. Establishment the main use of which is to serve spirituous liquors to be consumed on the premises. Food may or may not be served. A lounging area is provided and a counter is not necessarily present.

Distributor. A person importing or causing to be imported into the state, or purchasing or causing to be purchased within the state, alcoholic liquors for sale or resale to retailers licensed in accordance with the provisions of the law.

Manufacture. Includes but is not limited to the methods or processes to distill, rectify, ferment, brew, make, mix, concoct, process, blend, bottle, or fill an original package with any alcoholic liquor or other preparation of drinks for serving by authorized persons, for consumption on the premises where sold.

Manufacturer. Includes every brewer, fermenter, distiller, rectifier, winemaker, blender, processor, bottler, or person who fills or refills an original package and others engaged in brewing, fermenting, distilling, rectifying, or bottling alcoholic liquors.

Night club. Any place of business located within any building or establishment under one roof and on one floor, established and operated for the purpose of supplying entertainment or music, or both, and providing meals and refreshments prepared on the premises, having a seating capacity of not less than forty people at tables; having an aggregate floor space of not less than 2,200 square feet, and providing a dance floor containing not less than 308 square feet; such floor space providing for dancing to be free from chairs, tables or other obstructions at all times.

Nonbeverage user. A product which contains alcoholic liquor, and is used in laboratories, hospitals, and sanatoria for nonbeverage purposes.

Nonintoxicating beverage. All beverages containing more than one-half of one percent (1/2 of 1%) alcohol by volume, and not more than three and two-tenths percent (3.2%) alcohol by weight.

Original package. Includes any bottle, flask, jug, can, cask, barrel, keg, hogshead, or other receptacle or container whatsoever, used corked, or capped, sealed, and labeled by the manufacturer of alcoholic liquor, to contain and to convey any alcoholic liquor.

Package store. A retail outlet for the sale of alcoholic beverages.

Person. Any natural person, corporation, partnership or association.

Private club. Any association, person, firm, or corporation, key club, bottle club, locker club, pool club, or any other kind of club or association, excluding the general public from its premises, place of meeting, congregating, or operating or exercising control over any other place where persons are permitted to drink alcoholic beverages other than a private home.

Restaurant. Any public place kept, used, maintained, advertised, and held out to the public as a place where meals are served, and alcoholic liquors may be served under proper license.

Restaurants. Vendors holding a license from the state beverage department for the sale of alcoholic beverages for consumption on the premises in restaurants, which are restricted by the zoning regulations to making such sales with the service of food only, shall make no sales of such alcoholic beverages on week days except between the hours of 8:00 a.m. and 1:00 a.m. on the following day, and shall make no sales of beer on Sundays except between the hours of 10:00 a.m. and 1:00 a.m. on the following Monday; and shall make no sales of other alcoholic beverages on Sundays except between the hours of 1:00 p.m. and 1:00 a.m. on the following Monday. Sales of alcoholic beverages for consumption off the premises shall not be permitted.

Retail dealer. Includes any and all persons, firms, corporations, associations, or concessionaires who sell, dispense any nonintoxicating beverages within the corporate limits of the City, without regard as to the place where such beverages may be consumed or used.

Sale. Any transfer, exchange, or barter in any manner or by any means whatsoever for a consideration, and shall include all sales made by any person, whether principal, proprietor, agent, servant, or employee.

Spirits. Any beverage which contains alcohol obtained by distillation, mixed with water or other substance in solution, and shall include brandy, rum, whiskey, gin, or other spirituous liquors, and such liquors when rectified, blended, or otherwise mixed with alcohol or other substances.

Tavern. A place where alcoholic liquors are sold at retail to the general public, and where no other kind of business is being maintained or conducted; provided, however, that in such taverns, cigars, cigarettes, tobacco, nuts, jerky, popcorn, potato chips and pretzels may be sold or given away. Otherwise, however, no lunches, foodstuffs or so-called "free lunches" shall be either sold or given away in such taverns, including groups designated as bars or cocktail lounges.

Wine. Any alcoholic beverage obtained by the fermentation of the natural contents of fruits or vegetables, containing sugar, including such beverages when fortified by the addition of alcohol or spirits.

D-2 CONSUMER SALES

Canvasser and solicitor. Shall be deemed to mean and include any person who goes from house to house or from place to place in the City selling or taking orders for or offering to sell or take orders for photographs, pictures, enlargements thereof, tinting or other services pertaining to photography, for future delivery or performance.

Canvassing. Includes door to door soliciting or soliciting by the use of circulars, visitations, or any other means where the canvasser, or his employer has not been invited or requested by the owner as defined in the Code.

Direct seller. Any individual who, for him/herself, or for another (including but not limited to partnership, association or corporation) sells goods, or takes sales orders for the later delivery of goods, at any location other than the permanent business place or residence of the individual, partnership, association or corporation, and includes, but is not limited to peddlers, solicitors and transient merchants. For purposes of the code, the acceptance of a "donation" in exchange for goods, or an order for goods, shall be deemed an act requiring compliance with all of the regulatory provisions of the code including registration and posting of a bond.

Goods. Include personal property of any kind, and goods provided incidental to services offered or sold.

Pawnbroker or secondhand dealer. Shall be defined as a merchant who engages in the business of buying and selling and/or lending money for which he holds chattels as security for any of the following classes of used personal property: Watches, jewelry, sporting equipment, guns, knives, electrical appliances, hand tools, silverware, cameras, typewriters, radios and televisions.

Peddler. Any person commonly referred to either as a peddler or hawker, who goes from place to place or from house to house by traveling on the streets and carries with him goods, wares and merchandise for the purpose of selling or delivering them to consumers or any person who has goods, wares and merchandise of any description sent from place to place or from house to house by traveling on the streets for the purpose of selling and delivering goods to consumers.

Permanent merchant. A direct seller who, for at least one year prior to the consideration of the application of the ordinance to the merchant: Has continuously operated an established place of business in the city.

D-3 CONTRACTING

Air conditioning. Shall include all work and operations ordinarily performed in connection with the construction, operation, maintenance and repair of air conditioning systems, whether such systems control all or part only of the following properties of the air in any building or structure; temperature, humidity, oxygen content, content of dust, dirt, carbon dioxide, carbon monoxide, or other noxious or poisonous gases; together with the adequate distribution of air throughout the building and the intake of fresh air and the discharge of used air. Where such systems provide only for heating and humidifying the air they may be classified as heating systems.

Apprentice. A person at least 16 years of age who is covered by a written agreement, with an employer, an association of employers, or an organization of employees acting as employer's agent, and approved by the Apprenticeship council; which apprentice agreement provides for hours of reasonably continuous employment for such person for his participation in an approved schedule of work experience and for at least 144 hours per year of related supplemental instruction. The required hours for apprenticeship agreements may vary in accordance with standards adopted by local or State joint apprenticeship committees, subject to approval of the State Apprenticeship council and Commissioner of Labor.

Apprentice electrician. Shall mean any employed person at least sixteen years of age who is engaged in learning a recognized skilled trade, through actual work experience under the supervision of craftsmen, which training should be supplemented by properly coordinated studies of related technical and supplementary subjects; who has entered into a written agreement (hereinafter called an apprentice agreement) with an employer, an association of employers, or a local joint apprenticeship committee, providing for not less than four thousand hours of reasonably continuous employment for such person.

Awarding authority. An owner, or a duly authorized agent of an owner, who awards an original building or construction contract.

Boiler, hot-water heating and steam fitting contractor. Contractor whose principal contracting business is the execution of contracts requiring the art, ability, experience, knowledge, science and skill to intelligently install fire-tube and water-tube steel power boilers and hot-water heating low pressure boilers, including all steam fitting and piping, hot-water piping, fittings, valves, gauges, pumps, radiators, convectors, fuel oil tanks, fuel oil lines, chimneys, flues, heat insulation and all other devices, apparatus and equipment appurtenant thereto, in such a manner that power boiler installations, hot-water heating systems and steam fitting can be executed, fabricated and installed, or to do any part, or any combination of any thereof.

Cabinet and millwork contractor. Contractor whose principal contracting business is the execution of contracts, usually subcontracts, requiring the art, ability, experience, knowledge, science, and skill to intelligently cut, surface, join, stick, glue and frame wood and wood products, in such a manner that, under an agreed speciication, acceptable cabinet, case, sash, door, trim, nonbearing partition, and such other mill products as are by custom and usage accepted in the building and construction industry, as cabinet and millwork, can be executed; including the placing, erecting, fabricating and finishing in buildings, structures and elsewhere of such cabinet and millwork, or to do any part or any combination of any thereof.

Cement and concrete contractor. Contractor whose principal contracting business is the execution of contracts, usually subcontracts, requiring the art, ability, experience, knowledge, science, and skill to intelligently proportion, batch and mix aggregates consisting of sand, gravel, crushed rock or other inert, materials having clean uncoated grains of strong and durable minerals, cement and water or to do any part, or any combination of any thereof, in such a manner that, under an agreed specification, acceptable mass, pavement, flat and other cement and concrete work can be poured, placed, finished and installed; including the placing and setting of screeds for pavements or flatwork; but shall not include those contractors whose sole contracting business is: The application of plastic coatings on; or the construction of forms and formwork for the casting, forming and shaping of; or the placing and erecting of steel or bars for the reinforcing of, mass, pavement, flat and other cement and concrete work.

Construction contract. Any contract for the construction, reconstruction, alteration or repair of any public building or other public work or public improvement, including highways.

Contractor. Any person, who undertakes to or offers to undertake to or purports to have the capacity to undertake to or submits a bid to, or does himself or by or through others, construct, alter, repair, add to, subtract from, improve, move, wreck or demolish any building, highway, road, parking facility, railroad, excavation or other structure, project, development or improvement, or to do any part thereof, including the erection of scaffolding or other structures or works in connection therewith. The term contractor includes subcontractor and specialty contractor.

Contractor. A person who contracts with an owner to improve real property.

Contractor. Any person who has entered into a construction contract with a contracting body.

Drywall contractor. Contractor whose principal contracting business is the execution of contracts, usually subcontracts, requiring the art, ability, experience, knowledge, science and skill to intelligently lay out, fabricate and install gypsum wall board assemblies and products, including the taping operation incidental thereto; but, shall not include the work of any other specialty-contractor classification defined in the Code.

Electrical construction. Shall be held to include and govern all work and materials used in installing, maintaining and/or extending a system of electrical wiring for the use of light, heat, or power, and all appurtenances, apparatus, or equipment used in connection therewith, inside of or attached to any building or structure, lot or premises.

Electrical contractor. Means any person, firm, copartnership, corporation, association, or combination thereof who undertakes or offers to undertake with another to plan for, lay out, supervise, and install or to make additions, alterations, and repairs in the installation of wiring apparatus and equipment for electric light, heat, and power for a fixed sum, price, fee, percentage, or other compensation.

Electrical contractor. Shall be held to mean a contractor doing work on any premises, or in any building or structure requiring the installation, repair, alteration, addition or changes to any system of electrical wiring, apparatus or equipment for light, heat or power and who is a master electrician.

Electrical contractor. Contractor whose principal contracting business is the execution of contracts, usually subcontracts, requiring a knowledge of the art and science of placing, installing, erecting or connecting of any electrical wires, fixtures, appliances, apparatus, raceways or conduits, or any part thereof, which generate, transmit, transform or utilize electrical energy in any form and in connection with which electrical energy is used for any purpose whatsoever.

Electrical sign contractor. Contractor whose principal contracting business is the execution of contracts requiring the art, ability, experience, knowledge, science and skill to intelligently fabricate, install and erect electrical signs, including the wiring of such electrical signs.

Electrician. Shall be held to mean a person who is engaged in the trade or business of electrical construction, and who is qualified under the terms and provisions in the Code.

Elevator installation contractor. Contractor whose principal contracting business is the execution of contracts, requiring the art, ability, experience, knowledge, science and skill to intelligently fabricate, erect and install sheave beams, motors, sheaves, cable and wire rope, guides, cab, counter-weights, doors, including sidewalk elevator doors, automatic and manual controls, signal systems, and all other devices, apparatus and equipment appurtenant to the safe and efficient installation and operation of electrical, hydraulic and manually operated elevators.

Excavating, grading, trenching, paving and surfacing contractor. Contractor whose principal contracting business is the execution of contracts requiring the art, ability, experience, knowledge, science, and skill to intelligently dig, move and place material forming the surface of the earth, other than air and water, in such a manner that a cut, fill, excavation, grading, trenching, backfilling, tunneling and any similar excavating, grading and trenching operation, can be executed with the use of those hand and power tools and machines that use and custom has established for, or which tools and machines are now used to dig, move and place that material forming the surface of the earth, other than air and water; including the use of explosives in connection therewith; including the mixing, fabricating and placing of paving and surfacing consisting of graded mineral aggregates and to do any part, or any combination of any thereof.

Fire protection engineering contractor. Contractor whose principal contracting business is the execution of contracts requiring the art, ability, experience, knowledge, science, and skill to intelligently lay out, fabricate and install all types and kinds of fire protection systems, including all apparatus, devices and equipment appurtenant thereto, in such a manner that acceptable fire protection work can be laid out, executed, fabricated and installed; or to do any part thereof.

General building contractor. Contractor whose principal contracting business is in connection with any structure built, being built, or to be built, for the support, shelter and enclosure of persons, animals, chattels or movable property of any kind, requiring in its construction the use of more than two unrelated building trades, or crafts, or to do or superintend the whole or any part thereof. This does not include anyone who merely furnishes materials or supplies as defined in the Code, without fabricating them into, or consuming them in the performance of the work of the general building contractor.

General engineering contractor. Contractor whose principal contracting business is in connection with fixed works requiring specialized engineering knowledge and skill, including the following divisions or subjects: irrigation, drainage, water power, water supply, flood control, inland waterways, harbors, docks and wharves, shipyards and ports, dams and hydroelectric projects, levees, river control and reclamation works, railroads, highways, streets and roads, tunnels, airports and airways, sewers and sewage disposal plants and systems, waste reduction plants, bridges, overpasses, underpasses and other similar works, pipelines and other systems for the transmission of petroleum and other liquid or gaseous substances, parks, playgrounds and other recreational works, refineries, chemical plants and similar industrial plants requiring specialized engineering knowledge and skill, powerhouses, power plants and other utilities plants and installations, mines and metallurgical plants, land leveling and earthmoving projects, excavating, grading, trenching, paving and surfacing work and cement and concrete works in connection with the above mentioned fixed works.

Glazing contractor. Contractor whose principal contracting business is the execution of contracts, usually subcontracts, requiring the art, ability, experience, knowledge, science, and skill to intelligently select, cut, assemble and install all makes and kinds of glass and glass work, and execute the glazing of frames, panels, sash and doors, in such a manner that under an agreed specification, acceptable glass work and glazing can be executed, fabricated and installed; but shall not include the manufacture, or fabrication, or installation in any building or structure of any frame, panel, sash or door upon or within which such frame, panel, sash or door, such glass work or glazing has been executed or installed.

Heating or Heating Work. Shall include all work and operations ordinarily performed in connection with the construction, operation, maintenance, and repair of all heating systems, whether coal, oil, gas, or electricity be the primary source of heat, and whether transmitted by hot air, hot water, steam, or directly radiated.

Home improvement. The repairing, remodeling, altering, converting, or modernizing of, or adding to, residential property.

House and building moving, wrecking contractor. Contractor whose principal contracting business is the execution of contracts requiring the art, ability, experience, knowledge, science, and skill to intelligently raise, crib, underpin, and move or remove buildings and structures so that alterations, additions, repairs and new substructures may be built, constructed and completed under the existing permanently retained portions of such building or structure as remains undisturbed after being raised, cribbed or underpinned, including the complete rehabilitation thereof and the alterations and additions thereto, either upon the present site, or upon another site to which such building or structure may be removed, including the wrecking and demolition of buildings and structures in such a manner that house and building moving, wrecking and demolition can be executed and completed.

Independent contractor. Any person who renders service for a specified recompense for a specified result, under the control of his principal as to the result of his work only and not as to the means by which such result is accomplished.

Insulation and acoustical contractor. Contractor whose principal contracting business is the execution of contracts requiring the art, ability, experience, knowledge, science and skill to intelligently install any insulating materials in buildings and structures for the sole purpose of temperature control and examine surfaces, prepare acoustical layouts, select and install preformed architectural acoustical materials in such a manner that building and construction insulation and acoustical work can be executed and installed in accordance with a layout.

Journeyman electrician. A person having the necessary qualifications, training, experience and technical knowledge to wire for, install, and repair electrical apparatus and equipment in accordance with the standard rules and regulations governing such work.

Journeyman electrician. Shall be held to mean a person who possesses the necessary qualifications, training and technical knowledge to install electrical wiring, apparatus, or equipment for light, heat, or power, and who is qualified under the terms and provisions herein, and he shall be capable of doing electrical work according to the plans and specifications furnished to him by a duly licensed electrical contractor or master electrician, and in accordance with the electrical code.

Labor. Includes labor, work, or service whether rendered or performed under contract, subcontract, partnership, salary plan, or other agreement if the labor to be paid for is performed personally by the person demanding payment.

Landscaping contractor. Contractor whose principal contracting business is the execution of contracts, usually subcontracts, requiring the art, ability, experience, knowledge, science and skill to intelligently grade and prepare plots and areas of land for architectural horticulture and the decorative treatment, arrangement, planting and maintenance of gardens, lawns, shrubs, vines, bushes, trees and other decorative vegetation; construct pools, tanks, fountains, pavilions, conservatories, hot and green houses, retaining walls, fences, walks, drainage and sprinkler systems; arrange, fabricate and place garden furniture, statuary and monuments, in connection therewith, or to do any part or any combination of any thereof in such a manner that, under an agreed specification, acceptable landscaping projects can be executed.

Lathing contractor. Contractor whose principal contracting business is the execution of contracts, usually subcontracts, requiring the art, ability, experience, knowledge, science, and skill to intelligently examine surfaces and specify, select, apply and affix wood and metal lath, or any other material or product prepared or manufactured, to provide key or suction bases for the support of plaster coatings, in such a manner that, under an agreed specification, acceptable lathing can be executed and installed; including the channel iron work for the support of metal or other fireproof lath, and the channel iron and metal, or other fireproof lath work for solid plaster partitions; but shall not include any plaster work, or the erection of any wall, ceiling or soffit to which such key or suction plaster bases are applied.

Lien. A charge, imposed upon specific property, by which it is made security for the performance of an act.

Maintenance-electrician. Shall be held to mean a person who is a journeyman electrician qualified as to his knowledge and of the electrical industry, pertaining to maintenance thereof. He shall not be employed by more than one person, firm or corporation at any one time. The qualification of journeyman electrician would be acceptable without further examination for a maintenance electrician's certificate. The work of the maintenance electrician shall be confined to the repair of existing branch circuits, fixtures, apparatus or equipment connected thereto, contained and used upon the premises or in building owned, occupied or controlled by the person, firm or corporation by whom the maintenance electrician is employed. His work shall not include the installation, alterations or replacement of service conductors, service equipment, or any feeder to any center or centers of distribution. All work shall comply with all rules and regulations governing this work.

Masonry contractor. Contractor whose principal contracting business is the execution of contracts, usually subcontracts, requiring the art, ability, experience, knowledge, science, and skill to intelligently select, cut and lay brick and other baked clay products, rough, cut and dressed stone, artificial stone and precast blocks, structural glass brick or block, laid at random or in courses, with or without mortar, or to do any part, or any combination of any thereof, in such a manner that, under an agreed specification, acceptable brick and other baked clay products, stone, and structural glass brick or block masonry can be executed, fabricated and erected, but shall not include those contractors whose sole contracting business is: The application of tile to existing surfaces; or the execution, fabrication and erection of poured cement and concrete masonry.

Master electrician. A person having the necessary qualifications, training, experience and technical knowledge to properly plan, lay out, and supervise the installation of wiring, apparatus, and equipment for electric light, heat, power, and other purposes in accordance with the standard rules and regulations governing such work.

Master electrician. Shall be held to mean a person who possesses the necessary qualifications, training and technical knowledge to plan, layout,and supervise the installation of electrical wiring, apparatus or equipment for light, heat, or power, and who is qualified under the provisions in the Code.

Original contractor. (general contractor). Any person who has an original building or construction contract with and from an awarding authority.

Ornamental metals contractor. Contractor whose principal contracting business is the execution of contracts, usually subcontracts, requiring the art, ability, experience, knowledge, science, and skill to intelligently assemble, cast, cut, shape, stamp, forge, fabricate and install, sheet, rolled and cast, brass, bronze, copper, cast iron, wrought iron, monel metal, stainless steel, steel, and any other metal, or any combination of any thereof, as have been, or now are, used in the building and construction industry for the architectural treatment and ornamental decoration of buildings and structures, in such a manner that, under an agreed specification, acceptable ornamental metal work can be executed, fabricated and installed; but, shall not include the work of a sheet metal contractor.

Owner. Is hereby defined to mean the owner who caused the building, improvement, or structure, to be constructed, altered, or repaired (or his successor in interest at the date a notice of completion or cessation from labor is filed for record) whether the interest or estate of such owner be in fee, as vendee under a contract of purchase, as lessee, or other interest or estate less than the fee; and where such interest or estate is held by two or more persons as joint tenants or tenants in common, any one or more of the co-tenants may be deemed to be the "owner" within the meaning of this definition; provided, that any notice of completion or cessation from labor signed by less than all of such co-owners shall recite the names and addresses of all such co-owners; and provided further, that any notice of completion signed by a successor in interest shall recite the names and addresses of his transferor or transferors.

Painting and decorating contractor. Contractor whose principal contracting business is the execution of contracts, usually subcontracts, requiring the art, science, knowledge, experience, skill and ability to intelligently examine surfaces and specify and execute the preliminary and preparatory work necessary to bring such surfaces to a condition where, under an agreed specification acceptable work can be executed thereon with the use of any, or all, of the following: Paints, pigments, oils, turpentines, japans, driers, thinners, varnishes, shellacs, stains, fillers, waxes, cement, water and any other vehicles, mediums and materials that may be mixed, used and applied to the surfaces of buildings, edifices, structures, monuments and the appurtenances thereto, of every kind, type and description in their natural state or condition, or constructed or fabricated of any material or materials whatsoever that can be painted and decorated.

Person. As used in the Code includes an individual, a firm, copartnership, corporation, association or other organization, or any combination of any thereof.

Pipeline contractor. Contractor whose principal contracting business is the execution of contracts requiring the art, ability, experience, knowledge, science and skill to fabricate and install pipelines for the conveyance of water, gas, petroleum and other fluid substances, including the application of protective coatings and the trenching, boring, shoring, backfilling, compacting, paving and surfacing necessary to complete the installation of such pipelines.

Plastering contractor. Contractor whose principal business is the execution of contracts, usually subcontracts, requiring a knowledge of the art and science of coating surfaces with a mixture of sand, gypsum plaster, quick-lime or hydrated lime and water, or sand and cement and water, or a combination of such other materials as create a permanent surface coating, and which coatings are usually applied with a plasterer's trowel over any surface which offers a mechanical key for the support of such coating, or to which such coating will adhere by suction, and/or the afixation of lath or any other material or product prepared or manufactured to provide a base for such coating.

Plumbing. Is hereby defined to be the system of pipes, fixtures, apparatus and appurtenances, installed upon the premises, or in a building, to supply water thereto and to convey sewage or other waste therefrom.

Plumbing contractor. Contractor whose principal contracting business is the execution of contracts, usually subcontracts, requiring a knowledge of the art and science of creating and maintaining sanitary conditions in buildings, structures, and works where people or animals live, work and assemble, by providing a permanent means for a supply of safe, pure and wholesome water, ample in volume and of suitable temperatures for drinking, cooking, bathing, washing and cleaning, and to cleanse all waste receptacles and like means for the reception and speedy and complete removal from the premises of all fluid and semi-fluid organic wastes and other impurities, including piping for a safe and adequate supply of gases and liquids for any purpose whatsoever in connection with the use and occupancy of such premises including piping for vacuum and air systems for medical, dental and industrial purposes.

Plumbing or plumbing work. Shall include all work and operations ordinarily performed by plumbers in, about or in connection with, the construction, operation, maintenance and repair of all water and sewer systems, drainage systems and gas lines.

Promise. Includes promise, undertaking, contract, or agreement, whether written or oral, express or implied.

Public works. Construction, alteration, demolition or repair work done under contract and paid for in whole or in part out of public funds, except work done directly by any public utility company pursuant to order of the Public Utilities Commission or other public authority.

Refrigeration contractor. Contractor whose principal contracting business is the execution of contracts, usually subcontracts, requiring the art, ability, experience, knowledge, science, and skill to intelligently construct, erect and install devices, machinery and units for the control of air temperatures below fifty degrees Fahrenheit (50° F.) in refrigerators, refrigerator rooms, and insulated refrigerator spaces and the construction, erection, fabrication and installation of such refrigerators, refrigerator rooms, and insulated refrigerator spaces, temperature insulation, air conditioning units, ducts, blowers, registers, humidity and thermostatic controls, or any part or any combination of any thereof, in such a manner that, under an agreed specification acceptable refrigeration plants and units can be executed, fabricated and installed, but, shall not include those contractors who install gas fuel or electric power services for such refrigerator plants or other units.

Reinforcing steel contractor. Contractor whose principal contracting business is the execution of contracts requiring the art, ability, experience, knowledge, science and skill to intelligently fabricate, place and tie steel reinforcing bars (rods), of any profile, perimeter, or cross-section, that are or may be used to reinforce concrete buildings and structures, in such a manner that steel reinforcing bar (rod) work for concrete buildings and structures can be fabricated, placed and tied.

Roofing contractor. Contractor whose principal contracting business is the execution of contracts, usually subcontracts, requiring the art, ability, experience, knowledge, science and skill to intelligently examine surfaces and specify the preliminary and preparatory work necessary to bring such surfaces to a condition where, under an agreed specification, acceptable work can be executed and fabricated thereon with such material or materials as do seal, waterproof and weatherproof such surfaces by such means and in such manner as to prevent, hold, keep and stop water, its derivatives, compounds, and solids, from penetrating and passing any such protective material, membrane, roof, surface or seal thereby gaining access to material or space beyond such weatherproof, waterproof or watertight material, membrane, roof, surface or seal with the use of any, or all, of the following: Asphaltum, pitch, tar, felt, flax, shakes, shingles, roof tile, slate and any other material or materials, or any combination of any thereof, that use and custom has established as usable for, or which material or materials are now used as, such waterproof, weatherproof or watertight seal for such membranes, roofs and surfaces; but shall not include a contractor whose sole contracting business is the installation of devices or stripping for the internal control of external weather conditons.

Sewer, sewage disposal, drain and cement pipe laying contractor. Contractor whose principal contracting business is the execution of contracts requiring the art, ability, experience, knowledge, science and skill to intelligently fabricate concrete and masonry sewer, sewage disposal and drain structures and lay cast-iron, steel, concrete, vitreous and nonvitreous pipe for sewers, sewage disposal, drains and irrigation, including the excavating, grading, trenching, backfilling, paving and surfacing only in connection therewith; the fabrication and erection of sewage disposal plants, cesspools and septic tanks and the appurtenances thereto, in such a manner, that sewers, sewage disposal, drains and cement pipe laying can be fabricated and installed.

Sheet metal contractor. Contractor whose principal contracting business is the execution of contracts requiring the art, ability, experience, knowledge, science, and skill to intelligently select, cut, shape, fabricate and install sheet metal such as cornices, flashings, gutters, leaders, rainwater downspouts, pans, kitchen equipment, duct work, patented chimneys, metal flues, etc., or to do any part, or any combination of any thereof, in such a manner that sheet metal work can be executed, fabricated and installed.

Sign-journeyman. Shall be held to mean a person who possesses the necessary qualifications, training and technical knowledge to install electrical sign apparatus or equipment for signs and who is qualified under the terms and provisions herein, and he shall be capable of doing electrical sign work according to the plans and specifications furnished to him by a duly licensed sign contractor or sign master and in accordance with the electrical code.

Sign-master. Shall be held to mean a person who possesses the necessary qualifications, training and technical knowledge to plan, lay out and supervise the installation of any electrical sign apparatus or equipment, in or on said sign and shall be permitted to connect to the existing sign outlet provided by others and who is qualified under the terms and provisions in the Code.

Specialty contractor. A contractor whose operations as such are the performance of construction work requiring spcial skill and whose principal contracting business involves the use of specialized building trades or crafts.

Structural steel contractor. Contractor whose principal contracting business is the execution of contracts requiring the art, ability, experience, knowledge, science and skill to intelligently fabricate and erect structural steel shapes and plates, of any profile, perimeter or cross-section, that are or may be used as structural members for buildings and structures, including riveting, welding and rigging only in connection therewith, in such a manner that structural steel work can be fabricated and erected.

Swimming pool contractor. Contractor whose contracting business is the execution of contracts requiring that specific art, ability, experience, knowledge, science and skill in connection with the use of those building or construction industry trades, crafts or skills as are necessary to the construction of a swimming pool. A licensee classified pursuant to the provisions of the Code shall not use or perform a building or construction industry trade, craft or skill except when such trade, craft or skill is required in connection with the construction of a swimming pool.

Subcontractor (specialty contractor) Any person who has a contract with and from an original contractor.

Tile contractor. Contractor whose principal contracting business is the execution of contracts requiring the art, ability, experience, knowledge, science, and skill to intelligently examine the surfaces and specify and execute the preliminary and preparatory work necessary to bring such surfaces to a condition where burned-clay tile work can be executed thereon by first preparing a base or sub-surface upon the original surface and then upon such prepared base or sub-surface execute ceramic, encaustic, faience, quarry, semi-vitreous, vitreous, and other tile work, and to which prepared base or sub-surface such tile will adhere by suction or is held upon such prepared base or sub-surface by adhesives, including all pseudo tile products, in such a manner that acceptable tile work may be executed, fabricated and installed; but shall not include hollow or structural partition tile.

Wages. Includes all amounts for labor performed by employees of every description, whether the amount is fixed or ascertained by the standard of time, task, piece, commission basis, or other method of calculation.

Warm-air heating, ventilating and air-conditioning contractor. Contractor whose principal contracting business is the execution of contracts requiring the art, ability, experience, knowledge, science and skill to intelligently fabricate and install warm-air heating systems complete with warm-air appliances, ducts, registers and flues with or without air filters, humidity and thermostatic controls; ventilating systems complete with blowers, ducts, plenum chambers, registers, with or without air filters, humidity and thermostatic controls; air-conditioning systems complete with air-conditioning unit, ducts, registers, air filters, humidity and thermostatic controls; in such a manner that warm-air heating, ventilating and air-conditioning systems can be executed, fabricated and installed, or to do any part, or any combination of any thereof.

Water Conditioning contractor. Contractor whose contracting business is the execution of contracts requiring the art, ability, experience, knowledge, science, and skill to install water conditioning equipment with the use of only such pipe and fittings as are necessary to connect the water conditioning equipment to the water supply system and to by-pass all those parts of the water supply system within the premises from which conditioned water is to be excluded. This classification excludes fixed works requiring specialized engineering knowledge and skill as provided in the Code.

Welding contractor. Contractor whose principal contracting business is the execution of contracts requiring the art, ability, experience, knowledge, science and skill to intelligently use gases and electrical energy to create temperatures of sufficient heat to cause metals to become permanently affixed, attached, joined and fabricated in such a manner that welding can be executed.

Well drilling contractor. Contractor, whose principal contracting business is the execution of contracts requiring some practical elementary knowledge of geology, hydrology, the occurrence of water in the ground, water-levels in wells, the prevention of surface and subsurface contamination and pollution of the ground-water supply and the art, ability, experience, knowledge, science and skill to intelligently bore, drill, excavate, case, cement, clean and repair water-wells, or to do any, or any combination of any, or all, such boring, drilling, excavating, casing, cementing, cleaning and repairing with hand or powered tools or rigs.

Wood flooring contractor. Contractor whose principal contracting business is the execution of contracts requiring the art, ability, experience, knowledge, science, and skill to intelligently select, cut, lay, finish, and repair wood floors and flooring, in buildings and structures previously built and presently under construction, including the scraping, sanding, filling, staining, shellacking and waxing of such wood floors and flooring, in such a manner that wood flooring can be laid, fabricated and installed, or to do any part, or any combination of any thereof.

D-4 EARTHQUAKE DESIGN CRITERIA

Base. The level at which the earthquake motions are considered to be imparted to the structure or the level at which the structure as a dynamic vibrator is supported.

Box system. A structural system without a complete vertical load-carrying space frame. In this system the required lateral forces are resisted by shear walls or braced frames as hereinafter defined in the Building Code.

Box system. A structural system where the vertical load is carried by bearing walls and structural framing and where the lateral stability and lateral force resisting system consists of shear walls or braced frames.

Braced frame. A truss system or its equivalent which is provided to resist lateral forces in the frame system and in which the members are subjected primarily to axial stresses.

Braced frame. A vertical truss or its equivalent which is provided to resist lateral forces in which the members are subjected primarily to axial stresses.

Dead load. The weight of all permanent construction including walls, floors, roofs, partitions, stairways and of fixed service equipment.

Dual bracing system. Consists of a moment resisting space frame and shear walls which meet the following design criteria: (a) the space frame and shear walls shall resist the total lateral force in accordance with their relative rigidities considering the interaction of the shear walls and space frame. (b) the shear walls acting independently of the resisting portions of the space frame shall resist the total lateral force. (c) the resisting space frame shall have the capacity to resist not less than twenty-five (25) percent of the total lateral force.

Ductile moment-resisting space frame. A moment-resisting space frame complying with the requirements for a ductile moment-resisting space frame as given in the Code.

Duration of load. The period of continuous application of a given load, or the aggregate of periods of intermittent application of the same load.

Earthquake load. The assumed lateral load acting in any horizontal direction on the structural frame due to the kinetic action of earthquakes.

Foundation level. The lowest of any of the following: (a) the bottom of any spread or combined footing or foundation mat; (b) the bottom of any pile cap; (c) the top of any pier or caisson.

Impact load. The load resulting from moving machinery, elevators, craneways, vehicles, and other similar forces and kinetic loads.

Lateral force resisting system. That part of the structural system to which the total lateral forces prescribed in the Code.

Lateral force-resisting system. That part of the structural system assigned to resist the lateral forces prescribed in the Code.

Lateral soil load. The lateral pressure in pounds per square foot due to the weight of the adjacent soil, including due allowance for hydrostatic pressure.

Liquefaction. A term used to describe a group of phenomena occuring in saturated conhesionless sandy and silty soils consisting of a large decrease in effective stress (total stress minus pore pressure) accompanied by large deformations under either static or cyclic loading. The term cyclic mobility should also be included within the scope of the definition of liquefaction.

Live load. The weight superimposed by the use and occupancy of the building, not including the wind load, earthquake load, or dead load.

Moment-resisting space frame. A space frame designed to carry all vertical loads and in which the members and joints are capable of resisting design lateral forces by bending moments.

Moment-resisting space frame. A vertical load-carrying space frame in which the members and joints are capable of resisting forces primarily by flexure.

Primary member. Any member of the structural frame of a building or structure used as a column; grillage beam; or to support masonry walls and partitions; including trusses, isolated lintels spanning an opening of eight (8) feet or more; and any other member required to brace a column of a truss.

Secondary member. Any member of the structural framework other than a primary member including filling-in beams of floor systems.

Shear wall. A wall designed to resist lateral forces parallel to the wall.

Space frame. A three-dimensional structural system composed of interconnected members, other than bearing walls, designed to function as a complete self-contained laterally stable unit with or without the aid of horizontal diaphragms or floor bracing systems.

Space frame. A three-dimensional structural system without bearing walls, composed of interconnected members laterally supported so as to function as a complete self-contained unit with or without the aid of horizontal diaphragms or floor-bracing systems.

Steel joist. Any secondary steel member of a building or structure made of hot or cold-formed solid or open-web sections, or riveted or welded bar, strip or sheet steel members or slotted and expanded or otherwise deformed rolled sections.

Structural steel member. Any primary or secondary member of a building or structure consisting of a rolled steel structural shape other than formed steel, light gage steel or steel joist members.

Vertical load-carrying space frame. A space frame designed to carry all vertical loads.

Wind load. the lateral pressure on the building or structure in pounds per square foot due to wind blowing in any direction.

D-5 ENERGY CONSERVATION

Accessible. Means having access thereto, but which first may require the removal or opening of an access panel, door or similar obstruction.

Air conditioner. Means one or more factory made asemblies which include an evaporator or cooling coil and an electrically driven compressor and condenser combination, and may include a heating function.

Automatic. Means self-acting, operating by its own mechanism when actuated by some impersonal influence, as for example, a change in current strength, pressure, temperature, or mechanical configuration.

Building envelope. Means the elements of a building which enclose conditioned spaces through which thermal energy may be transferred to or from the exterior.

Central air conditioner. Means an air conditioner which is not a room air conditioner.

Coefficient of performance (COP)—Cooling. Means the ratio of the rate of net heat removal to the rate of total energy input, expressed in consistent units and under designated operating conditions. British thermal units shall be converted to kilowatt hours at the rate of 3413 British thermal units per kilowatt-hour.

Coefficient of performance (COP)—Heat Pump, Heating. Means the ratio of the rate of useful heat output delivered by the complete heat pump unit (exclusive of supplementary heating) to the corresponding rate of energy input, in consistent units and under designated operating conditions. British thermal units shall be converted to kilowatt hours at the rate of 3413 British thermal units per kilowatt-hour.

Conditioned space. Means space, within a building, which is provided with a positive heat supply or a positive method of cooling.

Degree day, heating. Means a unit, based upon temperature difference and time, used in estimating fuel consumption and specifying nominal annual heating load of a building. For any one day, when the mean temperature is less than 65°F, there exist as many degree days as there are Fahrenheit degrees difference in temperature between the mean temperature for the day and 65°F. The number of degree days for specific geographical locations shall be those listed in the Code.

Energy efficiency ratio (EER). Means the ratio of net cooling capacity in Btu/hr to total rate of electric input in watts under designated operating conditions.

General lighting. Means lighting designed to uniformly illuminate an entire area.

Gross square feet of conditioned floor area. Means the sum of the enclosed areas of conditioned space on all floors of the building, including basements, mezzanines, and intermediate floor tiers and penthouses, measured from the exterior faces of exterior walls and the centerline of walls separating conditioned and unconditioned spaces of the building.

Heat pump. Means an air conditioner which is capable of heating by refrigeration, and which may or may not include a capability for cooling.

HVAC system. Means a system that provides either collectively or individually the processes of comfort heating, ventilating, and/or cooling within or associated with a building.

Infiltration. Means the uncontrolled inward air leakage through cracks and interstices in any building element and around windows and doors of a building, caused by the pressure effect of wind and/or the effect of differences in the indoor and outdoor air density.

New energy. Means electrical or chemical energy converted to thermal or mechanical energy expressly for the purpose of comfort heating or cooling.

Nondepletable energy source. Means energy source which cannot be exhausted by use, such as wind and solar energy.

Outside air. Means air taken from outdoors and not previously circulated through the system.

Packaged terminal air conditioner. Means a room air conditioner consisting of a factory-selected combination of heating and cooling components, assemblies or sections, intended to serve an individual room or zone and constructed in a manner which complies with the definition contained in the Standard for Packaged Terminal Air Conditioners approved by the Air-Conditioning and Refrigeration Institute in 1976, known as ARI 310.

Plenum. Means an air compartment connected to one or more air inlets or outlets.

Readily accessible. Means capable of being reached quickly for operation, renewal, or inspection, without requiring those to whom ready access is requisite to climb over or remove obstacles or to resort to the use of portable access equipment.

Recool. Means the application of cooling as a secondary process to either preconditioned primary air or recirculated room air.

Recovered energy. Means energy utilized which would otherwise be wasted from an energy utilization system.

Reheat. Means the application of heat as a secondary process to either preconditioned primary air or recirculated room air.

Room air conditioner. Means a factory encased air-conditioner designed as a unit for mounting in a window or through a wall, or as a console. It is designed for delivery of conditioned air to an enclosed space without ducts. "Room air conditioner" includes packaged terminal air conditioners.

Service systems. Means the HVAC, service water heating, electrical distribution, and illuminating systems provided in a building.

Service water heating. Means heating of water for domestic or commercial purposes other than comfort heating.

Shading coefficient. Means the ratio of the solar heat gain through a glazing system corrected for external and internal shading to the solar gain through an unshaded single light of double strength sheet glass under the same set of conditions.

Skylight. Means any opening in the roof surface which is glazed with a transparent or translucent material.

Task-oriented lighting. Means lighting designed specifically to illuminate one or more task locations, and generally confined to those locations.

U value (overall coefficient of thermal transmittance). Means the heat flow rate through a given construction assembly, air to air, expressed in Btu per hour per square foot of area, per °F temperature difference.

Ventilation air. Means that portion of supply air which comes from outside plus any recirculated air that has been treated to maintain the desired quality of air within a designated space.

Zone. Means a space or group of spaces within a building combined for common control of heating or cooling.

D-6 FIRE SAFETY

Accessibility. (a) Premises which are not readily accessible from public roads and which the fire department may be called upon to protect in case of fire shall be provided with access roads or fire lanes so that all buildings on the premises are accessible to the fire department apparatus. (b) Access roads and fire lanes shall not be obstructed.

Building. A structure wholly or partially enclosed within exterior walls, or within exterior and party walls, and a roof, affording shelter to persons, animals, or property.

Combustible. Material or combination of materials which is not noncombustible.

Corridor. Passageway or hallway which provides a common way of travel to an exit or to another passageway leading to an exit.

Exit. That portion of the way of departure from the interior of a building or structure to the exterior at street, or grade level accessible to a street, consisting of: (a) corridors, stairways and lobbies enclosed in construction having a fire-resistance rating, including the door opening thereto from a habitable, public or occupied space; or (b) an interior stairway; or (c) a horizontal exit; or (d) a door to the exterior at grade; or (e) an exterior stairway, or ramp.

Exit lighting and exit signs. (a) Exits shall be adequately lighted at all times when a building or structure is occupied. (b) Exit signs shall be maintained in a clean and legible condition, unobstructed by decorations, furnishings or equipment and illuminated at all times when the building or structure is occupied.

Fire alarm system. An installation of equipment for sounding a fire alarm.

Fire and smoke-detecting system. An installation of equipment which automatically actuates a fire alarm when the detecting element is exposed to fire, smoke or abnormal rise in temperature.

Fire load. The combustible contents within a building during normal use.

Fire protection equipment. Apparatus, assemblies or systems either portable or fixed, for use to prevent, detect, control or extinguish fire.

Flammable. Capable of igniting within 5 seconds when exposed to flame, and continuing to burn.

Flash point. Shall be the minimum temperature in degrees Fahrenheit at which a flammable liquid will give off flammable vapor as determined by appropriate test procedure and apparatus as specified in the Code.

Fuel oil. Shall mean kerosene or any hydrocarbon oil conforming to nationally recognized standards and having a flash point not less than 100°F.

Liquefied petroleum gas (LP gas). Shall mean any material which is composed predominantly of the following hydrocarbons, or mixtures of them: propane, propylene, butane (normal butane of isobutane) and butylenes.

Noncombustible. Material or combination of materials which will not ignite, support combustion, or liberate flammable gas when subjected to fire when tested in accordance with generally accepted standards.

Oxidizing materials. Shall mean and include substances that readily yield oxygen to stimulate combustion.

Shaft. A vertical opening or enclosed space extending through two or more floors of a building, or through a floor and roof.

Smoke-detectors. Devices which are activated by smoke or products of combustion.

Sprinkler system. A system of piping and appurtenances designed and installed so that heat from a fire will automatically cause water to be discharged over the fire area to extinguish it or prevent its further spread.

Standpipe system. An installation of piping and appurtenances, whereby all parts of a building can be quickly reached with an effective stream of water.

Structure. An assembly of materials forming a construction framed of component structural parts for occupancy or use, including buildings.

Volatile. Capable of emitting flammable vapors at a temperature below 75° Fahrenheit.

D-7 FOODS AND FOOD PREPARATION

Adulterated foods. The condition of a food if it bears or contains any poisonous or deleterious substance in a quantity which may render it injurious to health.

Approved. Acceptable to the Health Authority following his determination as to conformance with appropriate standards and good public health practice.

Approved. Food or drink, a source of food or drink, a method, a device or a piece of equipment meeting requirements of the division of Health and State Department of Health.

Beverage dispensing. Use of the common drinking cup is prohibited. The use of any fountain, cooler or dispenser for filling glasses or other drinking receptacle where the top rim of receptacle comes in contact with any part of the appliance is prohibited.

Commercial cooking and/or dishwashing equipment. The term "commercial cooking and/or dishwashing equipment" shall mean equipment used in a licensable food establishment for cooking food and/or washing utensils and which may produce steam, mist, particulate matter, vapors or smoke which are to be removed through a local exhaust ventilating system.

Commercial food processing establishment. Establishment in which food is processed or otherwise prepared and packaged for human consumption and which is subject to sanitary regulations and periodic sanitary inspection by a Federal, State, or local governmental inspection agency.

Commissary. Any establishment or concession preparing, serving or selling food products to occupants of buildings or employees or commercial or industrial enterprises for consumption upon the premises.

Commissary. Shall mean catering establishment, restaurant, or any other place in which food, containers or supplies are kept, handled, prepared, packaged, or stored, and directly from which vending machines are serviced.

Container. Means the package, wrapper, or other receptacle in which food may be placed and includes, but is not limited to, any cup, mug, glass, jar, can, bottle, box or bag.

Convenience food. Food item whose condition greatly simplifies preparation and portion control. It may be fresh (an egg, for example), frozen, dried or canned.

Dried meat. The product obtained by subjecting fresh meat or cured meat to a process of drying with or without the aid of artificial heat, until a substantial portion of the water has been removed.

Drink. Includes any and all liquid used or intended to be used as a beverage for human consumption, irrespective of any nourishing quality thereof.

Easily cleanable. Readily accessible and of such material and finish, and so fabricated that residue may be completely removed by normal cleaning methods.

Employees' handwashing facilities. Employees' hand-washing facilities shall be separate from utensil washing facilities and shall be located in or immediately adjacent to the food preparation area whenever possible in existing restaurants.

Fast food. Generic term applied to the limited menu, quick service type of restaurant.

Fluid milk products. Includes the following products: milk, cream, certified milk, skim milk, skimmed milk, non-fat milk, non-fat fortified milk, fortified skim milk, fortified skimmed milk, flavored milk, dairy drink, buttermilk, cultured buttermilk, cultured skim milk, cultured milk, cultured sour cream, cultured salad cream, yogurt, cultured or sour half-and-half, Vitamin D milk, Vitamin D fluid milk products, homogenized milk, modified milk, ice cream mix, ice milk mix and half-and-half.

Food. Any raw, cooked, or processed edible substance, beverage, or ingredient used or intended for use or for sale in whole or in part for human consumption.

Food. Includes candy, confectionary and ice cream, and any and all products, manufactured or natural and prepared, whether solid, semi-solid or liquid, which is used or intended to be used for human consumption and for nourishment of the human body.

Food. Any article used for food for man and articles used for components of any such article and includes poultry food products.

Food contact surfaces. Those surfaces of equipment and utensils with which food normally comes in contact, and those surfaces with which food may come in contact, or which drain back onto surfaces normally in contact with food.

Food establishment. Any place where food intended for human consumption is kept, stored, manufactured, prepared, processed, dressed, slaughtered, handled, sold or offered for sale, either at wholesale or retail.

Food in unsanitary containers. It shall be unlawful for any person to use any container which is in an unsanitary condition, due to the presence of food residue, decomposed food, offal or filth of any kind, for the storage, display, delivery or sale of any foodstuffs.

Food product containing poultry product. Any article of food for human consumption which is prepared in part from any edible portion of dressed poultry or from any product derived wholly from such edible portion, if such edible portion or product does not comprise a substantial portion of such article of food.

Food service operation. Means a food operation in which food is served, or prepared and served, for consumption in or about the food establishment; or in which food is prepared for service and consumption elsewhere.

Frozen desserts. Includes ice cream, frozen custard, French ice cream, French custard ice cream, sherbet, fruit sherbet, ice milk, ice, water ice, quiescently frozen confection, quiescently frozen dairy confection, whipped cream confection, bisque tortoni, artificially sweetened ice cream, or artificially sweetened ice milk.

Grease hood. Any commercial cooking hood which is at or over equipment that produces, or may produce, grease vapors.

Hamburger (chopped meat.) Shall consist of chopped fresh beef with or without the addition of beef fat as such and seasoning, and shall not contain more than 30 percent beef fat.

Leaf lard. Lard rendered at moderately high temperatures from the internal fat of the abdomen of the hog, excluding that adherent to the intestines, and has an iodine number not greater than 60.

Maintenance. All utensils and food contact surfaces of equipment, used in the preparation or serving of food or drink, including food-storage utensils shall be thoroughly cleaned before each use.

Meat. The properly dressed flesh derived from cattle, swine, sheep or goats, sufficiently mature and in good health at the time of slaughter, but is restricted to that part of the striated muscle which is skeletal or that which is found in the tongue, in the diaphragm, in the heart or in the esophagus, and does not include that found in the lips, in the snout or in the ears, with or without the accompanying and overlaying fat, and the portions of bone, skin, sinew, nerve and blood vessels, which normally accompany the flesh and which may not have been separated from it in the process of dressing it for sale.

Meat food products. Any articles of food or any articles that enter into the composition of food which are not prepared meats, but which are derived or prepared, in whole or in part, by a process of manufacturer from any portion of the carcasses of cattle, swine, sheep or goats, if such manufactured portion be all, or a considerable and definite portion of the articles, except such preparations as are for medicinal purposes only.

Meat loaf. The product consisting of a mixture of comminuted meat with spice or with cereals with or without milk or eggs pressed into the form of a loaf and cooked.

Milk products. Includes ice cream, ice milk, sherbets, butter, butter oil, the various types of cheeses, dried milk, dried skim milk, and any other food for human consumption made from milk or cream or both.

Multi-service items. Includes all utensils which are reused for food preparation, eating and drinking, and shall be of materials which are non-toxic in nature.

Mutton. Meat derived from sheep nearly one year of age or older.

Perishable food. Any food of such type or in such condition as may spoil.

Permanent ware. China, glassware, silver and linen that is made of materials that are intended to be reused and long-lasting.

Pork. Meat derived from swine.

Potentially hazardous food. Any perishable food which consists in whole or in part of milk or milk products, eggs, meat, poultry, fish, shellfish or other ingredients capable of supporting rapid and progressive growth of infectious or toxigenic micro-organisms.

Poultry. Any kind of domesticated bird, including but not being limited to chickens, turkeys, ducks, geese, pigeons, guineas or fowl.

Poultry food product. Any article of human food or any article intended for or capable of being so used which is prepared or derived in whole or in substantial part, from any edible portion of dressed poultry.

Prepared meats. The products obtained by subjecting meat to a process of comminuting, drying, curing, smoking, cooking or seasoning, or of flavoring, or to any combination of such processes.

Rabbit. Includes all species ordinarily considered as rabbits, such as wild and domestic rabbits, jack rabbits and hares.

Safe temperatures. As applied to potentially hazardous foods, shall mean temperatures of 45° F., or below, and 140° F. or above.

Sanitize. A method of effective bactericidal treatment of clean surfaces of equipment and utensils by a process which has been approved and acceptable to the licensing agency, as being the most effective method or process in effectively destroying micro-organisms, including pathogens.

Sausage meat. Fresh meat or prepared meat, and is sometimes comminuted. The term sausage meat is sometimes applied to bulk sausage containing no meat by-products.

Single-service articles. Cups, containers; lids, or closures; plates, knives, forks, spoons, stirrers, toothpicks; paddles; straws, place mats, napkins, doilies, wrapping materials; and all similar articles which are constructed wholly or in part from paper, paperboard, molded pulp, foil, wood, plastic, synthetic or other readily destructible materials, and which are intended by the manufacturers and generally recognized by the public as for one use only, then to be discarded.

Single-service items. Includes but is not limited to paper or plastic coated plates, cups, plastic knives, forks and spoons and plastic or waxed straws. Single-service utensils and containers shall be made of non-toxic materials but shall be disposed of after use.

Smoked meat. The product obtained by subjecting fresh meat, dried meat or cured meat to the direct action of the smoke either of burning wood or of similar burning material.

Special frozen dietary foods. Includes any frozen milk products which includes a combination of one or more of the ingredients used in the manufacture of ice cream no matter under what coined or trade name it may be sold or offered for sale, the process of which is similar to the process of manufacture of ice cream, and which contains added vitamins, minerals or edible carbohydrates.

Utensil. Receptacles, implements, cutlery, containers, kettles, cans, tanks, grinders, reels, scales, utensils, scrapers, agitators, splitters, saws, extractors, filters, sieves, sifters, trays, ladles, rollers, pestles, instruments, fillers, cutters, covers, kitchenware, tableware, glassware or other equipment with which food or drink, devices, drugs or cosmetics come in contact.

Veal. Meat derived from young cattle one year or less of age.

Venison. Flesh derived from deer.

Wholesome. In sound condition, clean, free from adulteration, and otherwise suitable for use as human food.

D-8 HAZARDOUS MATERIALS, ARTICLES, SUBSTANCES AND PRODUCTS

Acid. Means a corrosive and combustible material when used in conjunction with other ingredients.

Ammunition. Means a metal or other shell containing a fulminate or black or smokeless powder for the purpose of propelling projectiles or shot; or black or smokeless powder packed for use as a propelling charge or for saluting purposes.

Black powder (gunpowder). Means an explosive substance composed of sulphur, charcoal and either sodium or potassium nitrate.

Class I flammable liquid. Liquid having a flash point at or below 20°F.

Class II flammable liquid. Liquid having a flash point above 20°F and below 70°F.

Class III flammable liquid. Liquid having a flash point above 70°F.

Combustible dust. Fine particles of matter liable to spontaneous ignition or explosion or constituting a dust hazard, such as lint, shavings, sawdust, flour, starch, sulphur, metal powders, and powdered plastics, except when handled, stored, or confined to eliminate the hazard.

Combustible liquid. Any liquid having a flash point of one-hundred degrees Fahrenheit or more, but less than six hundred degrees Fahrenheit.

Corrosive liquids. Includes hydrochloric, nitric, sulphuric, perchloric, hydrofluoric, and other similar acids; alkaline caustic liquids; and all other liquids which react chemically with common materials such as wood or iron, or which may emit hazardous or irritant fumes, or generate heat and cause fire when in contact with organic matter or water, or cause severe damage when in contact with living tissue, or are otherwise similarly hazardous to life and property.

Corrosive liquids. Shall include those acids, alkaline caustic liquids, and other corrosive liquids which, when in contact with living tissue, will cause severe damage of such tissue by chemical action; or in the case of leakage will materially damage or destroy other containers of other hazardous commodities by chemical action and cause the release of their contents; or are liable to cause fire when in contact with organic matter with certain chemicals.

Corrosive liquids. Includes hydrochloric, nitric, sulphuric, hydrofluoric, perchchloric, and other corrosive acids; alkaline caustic liquids and other corrosive liquids.

Dangerous chemical. Any substance which is dangerous to life, limb, or property while being processed, stored or transported, when so designated in the Code.

Explosion hazard gases. Includes but is not limited to the following gases; acetylene, ether, ethyl chloride, ethylene, liquified petroleum gases, hydrogen, illuminating gas, methyl chloride gas, and similar gases susceptible to explosion.

Explosion hazard gases. Acetylene gas, ammonia gas, ether gas, ethyl chloride gas, ethyene gas, liquefied hydrocarbon gases, liquefied petroleum gases, hydrogen gas, illuminating gas, methyl chloride gas, and any other gas which is a poisonous, irritant or corrosive gas, and is also a gas susceptible to explosion under any condition or is not a poisonous, irritant or corrosive gas, but is a gas susceptible to explosion under any condition.

Explosion hazards. Every structure, room or space occupied for uses involving explosion hazards shall be equipped and vented with explosion relief systems and devices arranged for automatic release under predetermined increase in pressure as herein provided for specific uses or in accordance with accepted engineering standards and practices.

Explosive. Any chemical compound or mechanical mixture that is intended for the purpose of producing an explosion; that contains any oxidizing and combustible units, or other ingredients, in such proportions, quantities, or packing that an ignition by fire, by friction, by concussion, by percussion, or by detonator of any part of the compound or mixture may cause such a sudden generation of highly heated gasses that the resultant gaseous pressures are capable of producing destructive effects on contiguous objects, or of destroying life or limb.

Explosive gases. Acetylene, ether, ethyl chloride, ethylene, hydrogen illumination gas, petroleum, gases, methyl chloride gas, and oxygen.

Explosive material. Any dangerous chemical classified as an explosive material in the Code.

Fire hazard. Any building, device, appliance, apparatus, equipment, tank, vehicle, combustible waste, fence, or vegetation which, in the opinion of the Fire Department, is in such a condition as to cause a fire or explosion.

Fireworks. A combustible or explosive composition, or any substance or combination of substances or articles prepared for the purpose of producing a visible or an audible pyrotechnic effect by combustion, explosion, deflagration or detonation.

Flammable. Capable of igniting within 5 seconds when exposed to flame and continuing to burn.

Flammable. Combustible; subject to easy ignition and rapid flaming combustion.

Flammable aerosol. Aerosol which is required to be labeled "Flammable" under the Federal Hazardous Substances Labeling Act.

Flammable and combustible solids. Pyroxylin products, nitrocellulose, asphalt, coal tar, pitch, waste paper and rags, feathers, straw, hemp, excelsior, kapok, and greases and fats under 300 degrees "F" flash point.

Flammable anesthetic. Compressed gas which is flammable and administered as an anesthetic and includes among others, cyclopropane, divinyl ether, ethyl chloride, ethyl ether and ethylene.

Flammable cryogenic fluids. Those cryogenic fluids which are flammable in their vapor state.

Flammable dust. Any solid material sufficiently comminuted for suspension in still air which, when so suspended, is capable of self-sustained combustion.

Flammable fiber. Any free burning material in a fibrous or shredded form such as: cotton, sisal, rayon, henequin, ixtle, jute, hemp, tow, cocoa fiber, oakum, kapok, Spanish moss, excelsior, shredded paper and other materials of a similar nature.

Flammable gas. Any gas having a flammability range with air greater than 1% by volume.

Flammable liquid. A liquid having a flash point below two hundred (200) degrees Fahrenheit and having a vapor pressure of forty (40) pounds per square inch absolute or less at one hundred (100) degrees Fahrenheit.

Flammable liquid container. Any can, bucket, barrel, tank or other vessel in which flammable liquids are stored or kept for sale.

Flammable liquids. Include but not limited to, acetone, amyl acetate, amyl alcohol, alcoholic liquors, benzol, carbon bisulfide, ether, ethyl alcohol, ethyl acetate, collodion, gasoline, kerosene, methyl acetate, naptha, petroleum, paint, varnish, lacquer, whiskey, driers and similar liquids and solutions.

Flammable material. Any material that will readily ignite from common sources of heat; and any material that will ignite at a temperature of 600 degrees Fahrenheit or less.

Flammable materials. Include but are not limited to asphalt, tar, pitch, resin, and other similar products and materials. Also includes gas, liquid, or solids which are easily ignited and burns quickly and rapidly.

Flammable solid. Includes solid substances other than one classified as an explosive, which is liable to cause fires through friction, through absorption of moisture, through spontaneous chemical changes, or as a result of retained heat from manufacturing or processing.

Flash point. The lowest temperature for which a flash is observed in flammable liquids as determined by ASTM Designation D323: "Reid Method of Test for Vapor Pressure of Petroleum Products."

Flash point. The minimum temperature in degrees Fahrenheit at which a liquid will give off flammable vapor as determined by appropriate test procedure and apparatus. The flash point of liquids having a flash point below 175°F shall be determined in accordance with the Standard Method of Test for Flash Point by means of the Tag Closed Tester, ASTM D56. The flash point of liquids having a flash point of 175°F or higher shall be determined in accordance with the Standard Method of Test for Flash point by means of the Pensky-Martens Closed Tester ASTM D 93.

Flash point. Minimum temperature in degrees Fahrenheit at which a flammable liquid will give off flammable vapor as determined by appropriate test procedure and apparatus.

Flash point. Minimum temperature at which a liquid gives off vapor within a test vessel in sufficient concentration to form an ignitable mixture with air near the surface of the liquid.

Flash point. For the purpose of the Code the flash points of liquids shall be determined in the manner required by the relevant standards of the American Society for Testing Materials.

Flash point. Minimum temperature at which flammable liquid will give off vapor sufficient to form ignitable mixture in the air immediately above the liquid.

Fume hazard gases. Ammonia gas, chlorine gas, phosgene gas, sulphur dioxide gas and any other gas which has by testing, determined to be a poisonous irritant or corrosive gas and is not susceptible to fire or explosion.

Fume or explosion hazard building. Building, or part of a building, designed, intended or used for the purpose of manufacturing, compressing or storing any fume hazard gas or any explosion hazard gas, either at a pressure of more than fifteen pounds per square inch or in a quantity of more than twenty-five hundred cubic feet.

Guncotton. Nitrocellulose chemically known as hexanitrocellulose, and generally used alone or in combination with other substances as a blasting explosive or as a propelling charge, including also all cellulose nitrates of a higher degree of nitration.

Hazardous areas. Areas of structures, buildings or parts thereof, used for the purposes that involve highly combustible, highly flammable, or explosive products or materials which are likely to burn with extreme rapidity or which may produce poisonous fumes or gases, including highly toxic, or noxious alkalies, acids, or other liquids or chemicals, which involve flame, fume, explosive, poisonous or irritant hazards; also uses that cause division of material into fine particles of dust subject to explosion or spontaneous combustion, and uses that constitute a high fire hazard because of form, character, or volume of the material used.

Hazardous chemicals. Chemicals having serious flame or explosion hazards when coming into contact with water or moisture, such as metallic sodium, metallic potassium, sodium peroxide, calcium phosphide, yellow phosphorous, metallic magnesium powder, aluminum powder, calcium carbide, red phosphorous and similar chemicals and solutions.

Hazardous gases. Includes but not limited to such gases as ammonia, chlorine, phosgene, carbon bisulphide and other toxic irritant, corrosive or fume hazard gases such as acetylene, ether, ethyl chloride, ethylene, liquified hydo-carbons, and methyl chloride gas.

Hazardous material. Any material included under the definitions of Flammable Dust, Flammable Fiber, Combustible Liquid, Dangerous Chemical, Flammable Gas, Liquefied Flammable Gas, and Flammable Liquid.

Hazardous piping. Any service piping conveying oxygen, flammable liquids, flammable gases or toxic gases.

Hazardous roofing materials. Readily ignitible and hazardous roofing materials. Roofing materials approved for use usually carry classification markings from the Underwriters Laboratories Inc.

High fire hazard. Building use which involves the storage, sale, handling, manufacture, or processing of volatile, flammable, or explosive substances or products, or which burn with extreme rapidity or when burning or subjected to heat, produce large volumes of smoke, or produce poisonous fumes or gases in quantities and under conditions dangerous to the health or safety of any person.

High hazard. All uses which involve the storage, sale, manufacture or processing of highly combustible, volatile flammable or explosive products which are likely to burn with extreme rapidity and produce large volumes of smoke, poisonous fumes, gases or explosions in the event of fire.

Highly hazardous material-product-article or substance. Any material, product, article or substance which is liable to burn with rapidity, or while burning to emit poisonous or noxious fumes or while burning to cause explosions.

Highly inflammable liquids. Such as carbon bisulphide, naphtha, ether, benzol, styreme, butadiene, collodion, ethyl, acetate, amyl acetate, acetone, amyl alcohol, kerosene, turpentine, petroleum paint, varnish, dryer, gasoline, alcohol and oil in bulk quantities, and inflammable liquids used and stored in paint mixing and spraying rooms.

Highly toxic materials. Materials so toxic to man as to afford an unusual hazard to life and health during fire fighting operations. Such materials as parathion, TEEP (tetraethyl phosphate), HETP (hexaethyl tetra phosphate) and similar insecticides or pesticides. Any material classed as Poison "A" or Poison "B" by the Federal Department of Transportation, and is considered highly toxic.

Inflammable fibrous materials. Includes such materials as hay, straw, broomcorn, hemp, tow, jute, sisal, kapok, hair, excelsior, oakum and the like. (Inflammable and flammable are identical in meaning).

Liquefied flammable gas. Any liquid or gas which is a liquid while under pressure and having a vapor pressure in excess of 27 pounds per square inch absolute at a temperature of 100 degrees Fahrenheit, and a flammability range with air greater than 1% by volume.

Moisture hazard substance. Magnesium powder, calcium carbide, metallic sodium, and sodium peroxide.

Moisture hazard substances. Substances having serious flame or explosion hazard when coming in contact with water or moisture; such as aluminum powder, magnesium powder, barium peroxide, calcium carbide, metallic potassium, metallic sodium, metallic calcium, phosphorous pentassulphide, sodium peroxide, strontium peroxide, potassium peroxide, sulphuric acid, zinc powder, cyanides, and similar chemicals.

Ordinary hazard. Where the amount of combustibles or flammable liquids present is such that fires of moderate size may be expected. These may include mercantile storage and display, auto showrooms, parking garages, light manufacturing, warehouses not classified as extra hazard, school shop areas, etc.

Poisonous, corrosive, or fume-hazard substance. Hydrochloric, nitric, sulphuric, hydrofluoric, perchloric, and other corrosive acids, corrosive, toxic, or noxious alkalies, cyanides, ammonia, chlorine, phosgene, sulphur dioxide, and similar substances providing like hazards.

Poisonous-irritant. Corrosive, or Fume Hazard Gases: such as ammonia, chlorine, phosgene, sulphur dioxide, and similar poisonous, irritant, corrosive, or fume hazard gases.

Smokeless powder. A propellent for small arms or cannon, in the combustion of which smoke is largely eliminated having for its explosive base nitrocellulose in varying proportions.

Soluble cotton. Pyroxylin or nitrocellulose, including all cellulose nitrates below that chemically known as hexanitrocellulose, and soluble in a volatile flammable liquid.

D-9 HEALTH AND SANITATION

Ashes. The residue from the burning of wood, coal, coke or other combustible materials.

Baler. A machine used to compress and bind a quantity of solid waste or other material.

Bulky waste. The large items of solid waste such as appliances, furniture, large auto parts, trees and branches, stumps, flotage, and the like.

Carry container. A container used to transfer solid wastes from premises to a collection vehicle.

Catch basin. An enlarged and trapped inlet to a sewer designed to capture debris and heavy solids carried by storm or surface water.

Collector. Any person who is engaged in the collection or transportation of solid waste.

Combustible rubbish. Miscellaneous burnable materials.

Composting. A controlled microbial degradation of organic waste yielding a nuisance free product of potential value as a soil conditioner.

Construction and demolition wastes. The waste building materials and rubble resulting from construction, remodeling, repair, and demolition operation on houses, commercial buildings, pavements, and other structures.

Disposal area. Any site, location, tract of land, area, building, structure or premises used or intended to be used for partial or total solid waste disposal.

Domestic refuse. All those types of refuse which normally originate in a residential household or apartment house.

Food waste (garbage). Animal and vegetable waste resulting from the storage, handling, preparation, cooking or serving of foods.

Food waste (garbage) grinder. Device for pulverizing food waste (garbage) for discharge into the sanitary sewerage system.

Hazardous wastes. Those wastes that can cause serious injury or disease during the normal storage, collection and disposal process, including but not limited to explosives, pathological and infectious wastes, radioactive materials, and dangerous chemicals.

Incinerator. Any equipment, device or contrivance and all appurtenances thereof used for the destruction by burning of solid, semi-solid, liquid, or gaseous combustible wastes.

Incinerator residue. Solid materials remaining after reduction in an incinerator.

Industrial waste. Solid wastes which result from industrial processes and manufacturing operations such as factories, processing plants, repair and cleaning establishments, refineries and rendering plants.

Junk. A collection of sorted salvageable materials.

Non-combustible refuse. Miscellaneous refuse materials that are unburnable at ordinary incinerator temperatures of at least 1300°F.

Open dump. An area on which there is an accumulation of solid waste from one or more sources without proper cover materials.

Person. Any individual, firm, partnership, company, corporation, trustee, association, or any other private or public entity.

Premises. A building, together with any fences, walls, sheds, garages, or other accessory buildings appurtenant to such building, and the area of land surrounding the building and actually or by legal construction forming one enclosure in which such building is located.

Putrescible wastes. Wastes that are capable of being decomposed by microorganisms with sufficient rapidity as to cause nuisances from odors, gases, and similar objectionable conditions. Kitchen wastes, offal, and dead animals are examples of putrescible components of solid waste.

Residue. The solid materials remaining after burning, comprising ash, metal, glass, ceramics, and unburned organic substances.

Rubbish. Nonputrescible solid wastes, including ashes, consisting of both combustible and non-combustible wastes, such as paper, cardboard, tin cans, yard rubbish, wood, glass, bedding, crockery, or litter of any kind.

Solid waste refuse. Putrescible and nonputrescible solid wastes, except body wastes, and including abandoned vehicles, food waste (garbage), rubbish, ashes, incinerator residue, street cleanings, tree debris, and solid market and industrial wastes.

Solid waste storage. The temporary on-site storage of solid waste.

Street refuse. Material picked up by manual or mechanical sweeping of alleys, streets and sidewalks, litter from public litter receptacles, and dirt removed from catch basins.

Suspended solids. Solids that either float on the surface of, or are in suspension in water, sewage or other liquids, and which are removable by laboratory filtering.

Waste. Useless, unwanted, or discarded materials resulting from normal community activities. Wastes include solids, liquids, and gases. Solid wastes are classed as refuse.

Yard rubbish. Prunings, grass clippings, weeds, leaves, and general yard and garden wastes.

D-10 HIGHWAYS, ROADS AND BRIDGES

Base course. The layer or layers of specified or selected material of designed thickness placed on a subbase or a subgrade to support a surface course.

Bridge. A structure with a total clear span of more than 20 feet measured along the centerline of roadway, over a stream, watercourse, or opening. When used in a general sense the term bridge includes grade separations.

Culvert. A structure, not classified as a bridge, which provides an opening under the roadway.

Earth grade. The completely graded roadway before placing the pavement structure.

Grade separation. A structure which provides for highway traffic, pedestrian traffic, or utilities to pass over or under another highway or the tracks of a railway.

Highway. The entire width between the boundary lines of every way publicly maintained when any part thereof is open to the use of the public for purposes of vehicular travel.

Maximum unit weight. The maximum weight per cubic foot of compacted soil or surfacing or base course aggregate as determined by AASHO T 99 for cohesive soils and by the Michigan Cone Test for granular soils. The applicable method shall be as specified under Maximum Unit Weight, 2.08.01 for soils, and 3.01.08 for surfacing and base course aggregates.

Pavement structure. Any combination of subbase, base course, and surface course, including shoulders, placed on a subgrade.

Plan grade. Vertical control grade shown on plans.

Roadbed. That portion of the roadway between the outside edges of finished shoulders, or the outside edges of berms back of curbs or gutters, when constructed.

Right-of-way. A general term denoting land, property or interest therein acquired for or devoted to a highway, as shown on the plans.

Roadside. That portion of the right-of-way outside of the roadway.

Roadside development. Those items necessary to the complete highway which provide for the preservation of landscape materials and features; the rehabilitation and protection against erosion of all areas disturbed by construction through seeding, sodding, mulching and the placing of other ground covers; such suitable planting and other improvements as may increase the effectiveness and enhance the appearance of the highway.

Roadway. That portion of the right-of-way required for construction, limited by the outside edges of slopes and including ditches, channels, and all structures pertaining to the work.

Shoulder. The portion of the roadway contiguous with the traveled way for accommodation of stopped vehicles, for emergency use, and for lateral support of base and surface courses.

Sidewalk. That portion of the roadway primarily constructed for the use of pedestrians.

Subbase. The layer of specified material of designed thickness placed on the subgrade as a part of the pavement structure.

Subgrade. That portion of the earth grade upon which the pavement structure is to be placed.

Substructure. All of that part of a structure below the bridge seats or below the skewbacks of arches, including backwalls, wing walls, and wing protection railings except backwalls designed integrally with the superstructure.

Superstructure. All of that part of a structure not classified as substructure.

Surface course. One or more layers of a pavement structure designed to accommodate the traffic load, the top layer of which resists skidding, traffic abrasion, and the disintegrating effects of climate. The top layer is sometimes called "Wearing Course."

Temporary road. A temporary road includes all roadways, culverts, and structures necessary to facilitate the movement of highway and pedestrian traffic around a construction operation until such time as the traffic may use the designated route.

Temporary route. An existing road over which the traffic is temporarily diverted.

Temporary structure. A temporary bridge, culvert, or grade separation required to maintain traffic during the construction or reconstruction of a bridge, grade separation, or culvert.

Traveled way. The portion of the roadway for the movement of vehicles, exclusive of shoulders.

D-11 HOME OCCUPATIONAL USES

Home occupation. A home occupation is an occupation or a profession which is customarily carried on in a dwelling unit, or in an attached, closed building, and is carried on by a member or members of the family residing in the dwelling unit, and is clearly incidental and secondary to the use of the dwelling unit for residential purposes, and which conforms to all of the regulations and conditions listed in the Ordinance for Residential Districts.

Home occupation. An accessory use customarily conducted within a dwelling by the resident thereof, which is clearly secondary to the use of the dwelling for living purposes and does not change the character thereof or have any exterior evidence of such secondary use other than a small name plate of not more than 4 in. × 12 in; and provided that in no case shall more than 15% of the floor area of any dwelling unit exclusive of the secondary use be used.

Home occupation. Occupation carried on in a dwelling by the resident thereof as a secondary use in connection with which there is no noise, odor, or bright lights noticeable beyond said dwelling; no person employed, no advertising or advertising signs larger than 144 square inches, no displays and no storage, buying, and selling of a commodity shall be permitted on the premises.

Home occupation. Any use customarily conducted entirely within a dwelling and carried on solely by the inhabitant thereof, and which use is clearly incidental and secondary to the use of the dwelling for dwelling purposes, and does not change the character thereof, and in which 25% of the dwelling is used for home occupation, and in which any signs advertising said home occupation are limited to one unlighted sign, not over 2 square feet in area attached to the dwelling, and also which there is no public display of goods.

Home occupation. Occupation for gain or support conducted only by members of a family residing on the premises and conducted entirely within the dwelling, provided that no article is sold or offered for sale except such as may be produced by members of the immediate family residing on the premises.

Home occupation. A home occupation is an occupation or profession which: (1) Is customarily carried on in a dwelling unit, or in an attached building, provided no commodity is sold upon the premises. (2) Is carried on by a member of the family residing in the dwelling unit. (3) Is clearly incidental and secondary to the use of the dwelling unit for residential purposes.

Home occupation. Occupations which in general include personal services as the professions of a doctor, dentist, osteopath, chiropractor, chiropodist, optometrist, artist, engineer, architect, lawyer, accountant, and the occupations of a dressmaker, beautician and barber; such profession or occupation shall be carried on by one member of a family residing in the dwelling.

Home occupation. Any occupation or activity carried on by a member of the family residing on the premises, in connection with which; there is no sign other than a nonlighted and nonreflecting name plate not more than one square foot in area, which name plate may designate the home occupation carried on within, in letters not to exceed 2 inches in height, and which name plate must be clearly visible at the entrance to the premises where said home occupation is carried on and must be attached to the building wherein the home occupation is conducted, and where there is no commodity sold upon the premises, except that which is prepared on the premises in connection with such occupation or activity.

Home occupation. Profession or use customarily and historically conducted entirely within a dwelling unit by the permanent inhabitants thereof only, which use is clearly incidental and secondary to the use of the dwelling for dwelling purposes and does not change the character or appearance thereof.

Home occupation. Any occupation for commercial or financial gain carried on by a member of the family residing on the premises in connection with which there shall not be used any name plate or sign nor any artificial lighting or any display that will indicate from the exterior that the building is being utilized in whole or in part for any purpose other than that of a single family dwelling, and in connection with which occupation there shall not be kept upon the premises any stock in trade or any commodity for sale, nor shall there be employed any persons other than a member of the family residing on the premises, nor shall there be used any mechanical equipment except for particular domestic or household purposes.

Home occupation. Profession or use customarily conducted entirely within a dwelling by the permanent inhabitants thereof only, which use is clearly incidental and secondary to the use of the dwelling for dwelling purposes and does not change the character or appearance thereof.

Home occupation. Gainful occupation conducted by members of a resident family wholly within a dwelling or in a building accessory.

Home occupation. Occupation for gain and support conducted entirely and only by the members of a family within a residential building and provided that no article is sold or offered for sale except such as may be produced in the household by members of the family, and that no display of products shall be visible from the street, and that no accessory building shall be used in conjunction with such home occupation.

Home occupation. Conduct of an art or profession, the offering of a service, the conduct of a business, or the handcraft manufacture of products in a dwelling in accord with the regulations prescribed in the ordinance.

Home occupation. Any occupation or profession carried on by a member of the family residing on the premises, provided no commodity is sold thereon, no person is employed other than such a member of the family; and that no mechanical equipment is used for such as is not ordinarily used for domestic purposes.

Home occupation. A customary personal service occupation such as dressmaking, millinery, and food preparation; provided that such occupation shall be solely by members of the resident family and in the main building or in an accessory building only, that not more than the equivalent of 30% of the area of the first floor of the principal building shall be used for such purposes, that no display or advertising other than a small name plate, and no display of products made shall be visible from the street.

Home occupation. Occupation carried on within the walls of a dwelling unit and not visible or noticeable in any manner or form from outside the walls of the dwelling.

Home occupation. Accessory use in a residential area consisting of an occupation carried on entirely within a dwelling and only by members of the family permanently living therein, where products are not offered for sale from the premises, and where no evidence of the occupation is visible or audible from the exterior of the dwelling and that no commercial vehicles are kept on the premises or parked overnight on the premises unless otherwise permitted by the ordinance.

Home occupation. Any gainful activity engaged in, at or from a dwelling unit, its accessory buildings or on the lot, but not including a boarding house, rooming house, nursing home or tourist home.

Home occupation. Accessory use which is clearly incidental or secondary to the residential use of the dwelling unit, and is customarily carried on within a dwelling unit or accessory building by one or more occupants of such dwelling unit; does not permit selling articles produced elsewhere than on the premises, having exterior display of goods visible from the outside, or making external alterations which are not customary in residential buildings; and includes the following occupations: the professional practice of medicine, dentistry, architecture, engineering, law; and individuals as artists, beauticians, barbers and veterinarians, excluding any land use for stables or kennels; and does not permit the employment of more than 2 persons not living on the premises.

Home occupation. Accessory use of a service character customarily conducted within a dwelling by a resident thereof, which is clearly secondary to the use of the dwelling for living purposes and does not change the character thereof or have any exterior evidence of such secondary use other than a small name plate not more than 4 inches in width and 18 inches in length and provided that in no case shall more than 15% of the floor area of any apartment in such dwelling exclusive of any accessory building be used for such home occupation. The office of a physician, surgeon, dentist, or other professional person, who offers skilled services to clients and is not professionally engaged in the purchase of or sale of economic goods, including violin, piano, or other individual musical instrument instruction limited to a single pupil at a time, shall be deemed to be Home Occupations; and the occupations of a dressmaker, milliner, or seamstress, each with not more than one paid assistant shall also be deemed to be Home Occupations; dancing instruction, band instrument instruction in groups, tea rooms, tourist homes, beauty parlors, real estate offices, convalescent homes, mortuary establishments, and stores, trades, or business of any kind is not herein excepted.

Home occupation. Commercial or professional activity conducted in a dwelling and meeting the provisions of the ordinance for secondary uses.

Home occupation. Occupation or profession carried on as a subordinate use by a member of a family residing on the premises.

Home occupation. An occupation or profession conducted in a dwelling unit by a member of the family residing therein, which is clearly incidental and secondary to the use of the dwelling unit for residential purposes.

Home occupation. Any use conducted entirely within a dwelling and carried on by persons residing in that dwelling unit, of which the use is clearly incidental and secondary to the use of the dwelling for dwelling purposes, also, that it does not change the character thereof nor adversely affect the uses permitted in the district of which it is a part, and in connection with which there is no display nor stock in trade. The home occupation shall not include the sale of commodities except those which are produced on the premises and shall not involve the use of any accessory building or yard space, or activity outside of the main building not normally associated with residential use. Signs, merchandise or other articles shall not be displayed for advertising purposes and there shall not be any employees. Clinics, surgeries, hospitals, barbershops, beauty parlors, business offices and professional offices shall not be deemed home occupations.

Home occupation. Any occupation carried on solely by the inhabitants of a dwelling which is clearly incidental and secondary to the use of the dwelling for dwelling purposes, which does not change the character thereof, and which is conducted entirely within the main or accessory building; provided that no trading and merchandising is carried on and in connection with which there is no display of merchandise or advertising sign other than one nonilluminated name plate, not more than two (2) square feet in area attached to the main or accessory building, and no mechanical equipment is used except such as is customary for purely domestic or household purposes.

Home occupation. Includes the use of premises by any person for any profession or occupation which meets the requirements of the ordinance. Home occupation also includes the use of the premises by a cosmetologist when the dwelling is occupied by such cosmetologist as his or her home and no other person is employed to assist in the conduct of such occupation or business.

Home occupations. Customary home occupations such as dressmaking, millinery, hairdressing, manicuring, laundry, preserving and cooking; provided, that such occupations shall be conducted solely by home occupants in the main building and that no more than 1/2 of the area of one floor shall be used for the purpose and that no displays of products shall be visible from the street.

Home office. Shall include any office, studio, or room used for rendering of a service or advice or consultations for a fee when such use is conducted entirely within a dwelling which is bona fide residence of the principal practitioner; no other persons are engaged in the occupation in excess of one employee of the principal practitioner; the architectural style shall not be changed from its residential character; no outside storage or display shall be allowed; the use shall not generate continuous traffic to the site.

Home professional office. Includes such spaces as an office or studio of a resident physician, dentist, lawyer, architect, engineer, or teacher as herein restricted; provided that no more than 2 persons are employed who are not members of the family; and that such office shall be in the main building or an accessory building and shall not occupy more than the equivalent of 30% of the area of the first floor of the principal building; for the purpose of this definition, a teacher shall be restricted to a person giving individual instruction in a musical instrument, in singing, or in academic or scientific subjects to a single pupil at a time. A Home Professional Office shall not include the office of any person professionally engaged in the purchase or sale of economic goods. Dancing instruction, band instrument or voice instruction in groups, tea rooms, tourist homes, beauty parlors, barber shops, hairdressing and manicuring establishments, real estate offices, convalescent homes, mortuary chapels, and stores, trades or business of any kind or nature not herein excepted shall not be deemed to be Home Professional Offices.

Incidental home occupations. Any use conducted entirely within a dwelling and carried on by the occupants thereof, which is clearly incidental and secondary to the use of the dwelling for dwelling purposes and does not change the character thereof, and connection with which there is no display, no stock-in-trade, nor commodity sold which is not produced on the premises and no person, not a resident on the premises, is employed specifically in connection with the incidental home occupancy.

D-12 POLLUTION REGULATIONS

Air pollutant. Dust, fumes, gas, mist, smoke, vapor, odor, particulate matter, or any combination thereof, except that such term shall not include uncombined water in the atmosphere unless it presents a safety hazard.

Air pollution. The presence in the outdoor atmosphere of one or more air pollutants in sufficient quantities and of such characteristics and duration as are likely to be injurious to public welfare, to the health of human, plant or animal life, or to property, or which interferes with the reasonable enjoyment of life and property.

Control device. Any device which has as its primary function the control of emissions from fuel burning, refuse burning, or from a process, and thus reduces the creation of, or the emission of, air pollutants into the atmosphere or both.

Emission. The act of releasing or discharging air pollutants into the outdoor atmosphere from any source.

Episode stage. A level of air pollution in excess of the ambient air quality standard which may result in an imminent and substantial danger to public health or welfare. This term shall include alert, warning, and emergency stages.

Existing source. Equipment, machines, devices, articles, contrivances, or installations which are under construction or in operation on the effective date of this regulation, except that any existing equipment, machine, device, article, contrivance, or installation which is altered, replaced, or rebuilt after the effective date of the regulation shall be defined as a new source.

Fossil fuel. Natural gas, petroleum, coal, and any form of solid, liquid, or gaseous fuel derived from such materials.

Fossil fuel fired steam generating unit. A furnace or boiler, or combination of furnaces or boilers connected to a common stack, used in the process of burning fossil fuel for the primary purpose of producing steam by heat transfer.

Fuel burning equipment. Any furnace, boiler, apparatus, stack, and all appurtenances thereto, used in the process of burning fuel for the primary purpose of producing heat or power by indirect heat transfer.

Fugitive dust. Solid, airborne particulate matter emitted from any source other than through a stack.

Incinerator. Any furnace used in the process of burning solid waste for the primary purpose of reducing the volume of the waste by removing combustible matter.

Modification. Any physical change in, or change in the method of operation of, a stationary source which increases, or decreases the amount of any air pollutant emitted by such facility, or which results in the emission of any air pollutant not previously emitted, except that such term shall not include the following: (a) Routine maintenance, repair, replacement; (b) An increase in the production rate, if such increase does not exceed the operating design capacity of the affected facility; (c) An increase in hours of operation, if such increase does not exceed the operating design capacity of the facility; (d) Use of an alternative fuel or raw material if, prior to the date any standard under the Code becomes applicable to such facility, the affected facility is designed to accommodate such alternative use.

Multiple chamber incinerator. Any incinerator consisting of three or more refractory lined combustion chambers in series, physically separated by refractory walls, interconnected by gas passage ports or ducts and employing adequate design parameters necessary for maximum combustion of the material to be burned. The combustion chamber shall include as a minimum, one chamber principally for ignition, one chamber principally for mixing, and one chamber for combustion.

New source. Equipment, machines, devices, articles, contrivances, or installations built or installed on or after the effective date of the regulation, or existing at such time which are later altered, repaired, or rebuilt. Any such equipment, machines, devices, articles, contrivances, or installations, moved to a new address, or operated by a new owner, or a new lessee, after the effective date of the regulation, shall be considered a new source.

Odor. That property of an air pollutant which affects the sense of smell.

Opacity. A state which renders material partially or wholly impervious to rays of light and causes obstruction of an observer's view.

Organic solvents. Volatile organic compounds which are liquids at standard conditions, and which are used as dissolvers, viscosity reducers, or cleaning agents.

Particulate matter. Any finely divided material which exists as a liquid or solid under standard conditions, with the exception of uncombined water.

Person. Includes individuals, firms, partnerships, companies, corporations, trusts, associations, organizations, or any other private or public entities.

Process. Any action, operation, or treatment of materials, including handling and storage thereof, which may cause the discharge of an air pollutant, or pollutants, into the atmosphere, excluding fuel burning and refuse burning.

Process weight. The total weight in pounds of all materials introduced into any specific process.

Process weight per hour. The process weight divided by the number of hours in one complete operation, excluding any time during which equipment is idle.

Smoke. Small gas-borne particles resulting from incomplete combustion, consisting predominantly, but not exclusively, of carbon, ashes, or other combustible material.

Solid waste. Refuse, more than 50 percent of which is waste consisting of a mixture of paper, wood, yard wastes, food wastes, plastics, leather, rubber, and other combustibles, and noncombustible materials such as glass and rock.

Source. Any property, real or personal, which emits or may emit any air pollutant.

Stack. Any chimney, flue, conduit, or duct arranged to conduct emissions to the outdoor atmosphere.

Standard conditions. A dry gas temperature of 70° Fahrenheit and a gas pressure of 14.7 pounds per square inch absolute.

Stationary source. Any building, structure, facility, or installation which emits or may emit air pollutants.

Submerged fill pipe. Any fill pipe, the discharge opening of which is entirely submerged when the liquid level is 6 inches above the bottom of the tank. This term shall also include, when applied to a tank which is loaded from the side, a fill pipe adequately covered at all times during normal working of the tank.

Volatile organic compounds. Any compound containing carbon and hydrogen or containing carbon and hydrogen in combination with any other element which has a vapor pressure of 1.5 pounds per square inch absolute or greater under actual storage conditions.

D-13 PROFESSIONAL QUALIFICATIONS

Architect. Within the meaning of the Code, shall be deemed to be a duly registered and licensed architect.

Architect. A person licensed to practice the profession of architecture under the Education Law of the State.

Architect. An individual of good moral character, who after architectural education and after experience on architectural projects developed under the immediate supervision of a licensed architect with both education and experience acceptable to the State Board of Architects, and who by examination has satisfied the Board as to his proved competence in: (a) Architectural administration including the application of codes and laws related to a building, a structure or a group or groups of these units and their environment; (b) The theory, history, practice and aesthetics and their application to architecture; (c) The analysis, planning, design, usage and the inspection of construction of buildings and structures, their component parts, related spaces both internal and external and their environment. (d) The site development, structural, sanitary, mechanical, electrical, and other components pertaining thereto; (e) The execution and administration of these disciplines, the related design professions, and other related skills.

Architect. A person who, by reason of his knowledge of the mathematical and physical sciences, and the principles of architecture and architectural engineering acquired by professional education, practical experience, or both, is qualified to engage in the practice of architecture as attested by his registration as an architect.

Architectural practice. Any service or creative work requiring architectural education, training and experience, and the application of the mathematical and physical sciences and the principles of architecture and architectural engineering to such professional services or creative work as consultation, evaluation, design and review of construction for conformance with contract documents and design, in connection with any building, planning or site development. A person shall be deemed to practice or offer to practice architecture who in any manner represents himself to be an architect, or holds himself out as able to perform any architectural service or other services recognized by educational authorities as architecture.

Building. Any structure consisting of foundations, floors, walls and roof, having footings, columns, posts, girders, beams, joists, rafters, bearing partitions, or a combination of any number of these parts, with or without other parts or appurtenances thereto.

Designer and designer of engineering systems. Means the holder of a current designer's permit granted by the examining board. Design services which may be performed by designers, within the meaning and intent of these rules, includes and is limited to the preparation of plans and specifications, and consultation, investigation and evaluation in connection with such preparation of plans and specifications, in the specific fields.

Engineer. A duly registered and licensed engineer.

Engineer. Within the meaning of the Code, shall be deemed to be a duly registered and licensed structural engineer.

Engineer. Within the meaning of the Code, shall be deemed to be a duly registered and licensed engineer.

Engineer. The Registered Civil Engineer employed by the owner or by the Subdivider to prepare the Subdivision Map or Record of Survey Map and improvement plans.

Engineer. A professional engineer who, by reason of special knowledge of the mathematical and physical sciences and the principles and methods of engineering analysis and design, acquired by professional education or practical experience, is qualified to practice engineering as attested by his registration as a professional engineer.

Engineering practice. Any professional service or creative work requiring engineering education, training and experience and the application of special knowledge of the mathematical, physical and engineering sciences to such professional services or creative work as consultation, research investigation, evaluation, planning, surveying, design, location, development, and review of construction for conformance with contract documents and design, in connection with any public or private utility, structure, building, machine, equipment, process, work or project. Such services and work include plans and designs relating to the location, development, mining and treatment of ore and other minerals. A person shall be deemed to be practicing or offering to practice engineering if he practices any branch of the profession of engineering, or by verbal claim, sign, advertisement, letterhead, card or any other manner represents himself to be a professional engineer, or holds himself out as able to perform or does perform any engineering service or other service or recognized by educational authorities as engineering. A person employed on a full-time basis as an engineer by an employer engaged in the business of developing, mining and treating ores and other minerals shall not be deemed to be practicing engineering for the purposes of this definition if he engages in the practice of engineering exclusively for and as an employee of such employer and does not hold himself out and is not held out as available to perform any engineering services for persons other than his employer.

Geological practice. Any professional service or work requiring geological education, training, and experience, and the application of special knowledge of the earth sciences to such professional services as consultation, evaluation of mining properties, petroleum properties, and ground water resources, professional supervision of exploration for mineral natural resources including metallic and nonmetallic ores, petroleum, and ground water, and the geological phases of engineering investigations.

Geologist. A person, not of necessity an engineer, who by reason of his special knowledge of the earth sciences and the principles and methods of search for and appraisal of mineral or other natural resources acquired by professional education and practical experience is qualified to practice geology as attested by his registration as a professional geologist. A person employed on a full-time basis as a geologist by an employer engaged in the business of developing, mining or treating ores and other minerals shall not be deemed to be engaged in "geological practice" for the purposes of this definition if he engages in geological practice exclusively for and as an employee of such employer and does not hold himself out and is not held out as available to perform any geological services for persons other than his employer.

Landscape architect. A person who, by reason of his professional education, practical experience, or both, is qualified to engage in the practice of landscape architecture as attested by his registration as a landscape architect.

Landscape architectural practice. The performance of professional services such as consultations, investigation, reconnaissance, research, planning, design, or responsible supervision in connection with the development of land and incidental water areas where, and to the extent that the dominant purpose of such services is the preservation, enhancement or determination of proper land uses, natural land features, ground cover and planting, naturalistic and esthetic values, the settings and approaches to buildings, structures, facilities, or other improvements, natural drainage and the consideration and the determination of inherent problems of the land relating to erosion, wear and tear, light or other hazards. This practice shall include the location and arrangement of such tangible objects and features as are incidental and necessary to the purposes outlined in this paragraph, but shall not include the design of structures or facilities with separate and self-contained purposes for habitation or industry, such as are ordinarily included in the practice of engineering or architecture; and shall not include the making of cadastral surveys or final land plats for official recording or approval, nor manditorially include planning for governmental subdivisions.

Land surveying. Within the meaning and intent of this section means any service comprising the determination of the location of land boundaries and land boundary corners; the preparation of maps showing the shape and area of tracts of land and their subdivisions into smaller tracts; the preparation of maps showing the layout of roads, streets and rights of way of same to give access to smaller tracts; and the preparation of official plats, or maps, of said land in the state.

Land surveyor. Person who engages in the practice of surveying tracts of land for the determination of their correct locations, areas, boundaries and description, for the purpose of conveyancing and recording, or for establishment or reestablishment of boundaries and plotting of lands and subdivisions.

Practice of architecture. Any one or combination of the following practices by a person: The planning, designing or supervision of the erection, enlargement or alteration of any building or of any appurtenance thereto other than exempted buildings. "Practice of architecture" does not include any contractor or his duly appointed superintendent or foreman directing the work of erection, enlargement or alteration of any building or any appurtenance thereto, under the supervision of a registered architect or registered professional engineer.

Practice of architecture. Within the meaning and intent of this definition includes any professional service, such as consultation, investigation, evaluation, planning, architectural and structural design, or responsible supervision of construction, in connection with the construction of any private or public buildings, structures, projects, or the equipment thereof, or addition to or alterations thereof, wherein the public welfare or the safeguarding of life, health or property is concerned or involved.

Practice of professional engineering. Within the meaning and intent of the code includes any professional service, requiring the application of engineering principles and data, wherein the public welfare or the safeguarding of life, health or property is concerned.

Professional engineer. A person who by reason of his knowledge of mathematics, the physical sciences and the principles of engineering, acquired by professional education and practical experience, is qualified to engage in engineering practice as hereinafter defined in the Building Code.

Real estate broker. Includes a person, firm or corporation who, for a fee, commission or other valuable consideration, or by reason of promise or reasonable expectation thereof, lists for sale, sells, exchanges, buys or rents, or offers or attempts to offer a sale, exchange, purchase, or rental of real estate or an interest therein.

Real estate office. Any building or room therein or portion thereof, maintained by a real estate broker licensed.

Real estate salesman. Includes any person who, for compensation, valuable consideration or commission, or other thing of value, or by reason of a promise.

D-14 PUBLIC BEHAVIOR

Civil emergency. 1. A riot, disorderly picketing or demonstrating or unlawful assembly characterized by the use of actual force or violence or any threat to use force if accompanied by immediate power to execute by three (3) or more persons acting together without authority of law. 2. Any natural disaster or man-made calamity, including flood, conflagration, cyclone, tornado, earthquake, explosion or complete electrical blackout, within the corporate limits of the City, resulting in the death or injury of persons or the destruction of property to such an extent that extraordinary measures must be taken to protect the public health, safety and welfare.

Curfew. A prohibition against any person or persons walking, running, loitering, standing, remaining or motoring upon any alley, street, highway, public property or vacant premises within the corporate limits of the City, excepting persons officially designated for duty with reference to said civil emergency.

Improper dress or improperly dressed. The failure to wear any clothing or covering or the failure to wear opaque clothing or covering over the private (genital, pubic, buttocks) areas of the body or over that portion of the upper torso or upper body normally known as and generally described as breasts or chest.

Loiter. To stand around or remain, or to park or remain parked in a motor vehicle, at a public place or place open to the public, and to engage in any conduct prohibited under the Code. "Loiter" shall also mean to collect, gather or congregate in or be a member of a group or a crowd of people who are gathered together in any public place or place open to the public, and to engage in any conduct prohibited under the Code.

Panhandling. Any type of begging or accosting of others for money, services, food or any other objects; solicitation of money, services, food or any other objects for which any service or object is offered in return, if the solicitor is not licensed by or registered with the municipality and said license or registration identification tag is not publicly displayed at the time of said solicitation; the harassment of citizens by persons attempting to entice or procure service, money, food or any other objects by the use of promises, threats, fraud or artifice; any peddling, soliciting, distributing or hawking in violation of the provisions of the Code.

Place open to the public. Any place open to the public or any place to which the public is invited, and in, on or around any privately owned place of business, private parking lot or private institution, including places of worship, cemeteries or any place of amusement and entertainment whether or not a charge of admission or entry thereto is made. It shall include the elevators, lobbies, halls, corridors and areas open to the public of any store, office or apartment building.

Public nuisances, private nuisances. (a) A nuisance is unlawfully doing an act, or omitting to perform a duty, or is any thing or condition, which either (1) annoys, injures, or endangers the comfort, repose, health, or safety of others; (2) offends decency; (3) unlawfully interferes with, obstructs, or tends to obstruct, or renders dangerous for passage, any lake or navigable river, stream, canal, or basin, or any public park, square, street, or other public property; or (4) in any way renders other persons insecure in life or in the use of property. (b) A public nuisance is one which affects at the same time an entire community or neighborhood or any considerable number of persons, although the extent of the annoyance or damage inflicted upon the individuals may be unequal. (c) Every nuisance not included in subsection (b) above is a private nuisance.

Public place. Any public street, road or highway, alley, lane, sidewalk, crosswalk or other public way, or any public resort, place of amusement, park, playground, public building or grounds appurtenant thereto, school building or school grounds, public parking lot or any vacant lot.

D-15 PUBLIC PROPERTY PARKING

Operator. Includes every individual who operates a vehicle as the owner thereof, or as the agent, employee or permittee of the owner, or is in actual physical control of a vehicle.

Parking meter. Includes any mechanical device or meter not inconsistent with the Code placed or erected for the regulation of parking by authority of the Code. Each parking meter installed shall indicate, by proper legend, the legal parking time established by the city and, when operated, shall at all times indicate the balance of legal parking time and, at the expiration of such period, shall indicate illegal or overtime parking.

Parking meter space. Any space within a parking meter zone or parking meter lot adjacent to a parking meter and which is duly designated for the parking of a single vehicle by lines painted or otherwise durably marked on the curb or on the surface of the street or lot adjacent to or adjoining the parking meters.

Parking meter zone. Includes any street or lot upon which parking meters are installed and in operation, either at the present time or in the future.

Park or parking. The standing of a vehicle whether occupied or not, upon a street otherwise than temporarily for the purpose of, and while actually engaged in, receiving or discharging passengers or loading or unloading merchandise or in obedience to traffic regulations, signs or signals or an involuntary stopping of the vehicle by reason of causes beyond the control of the operator of the vehicle.

Public area. Includes all public ways, parks, and other lands owned or leased by the city.

Public way. Includes all public streets, roads, boulevards, alleys and sidewalks.

Street. Any public street, avenue, road, alley, highway, lane, path or other public place located in the city and established for the use of vehicles.

Vehicle. Any device in, upon, or by which any person or property is or may be transported upon a highway, except a device which is operated upon rails or tracks.

D-16 REDEVELOPMENT

Blighted area. Shall mean an area in which a majority of buildings have declined in productivity by reason of obsolescence, depreciation or other causes to an extent they no longer justify fundamental repairs and adequate maintenance.

Blighted area. An area in which there is a predominance of buildings or improvements (or which is predominantly residential in character), and which, by reason of dilapidation, deterioration, age or obsolescence, inadequate provision for ventilation, light, air, sanitation, or open spaces, high density of population and overcrowding, unsanitary or unsafe conditions, or the existence of conditions which endanger life or property by fire and other causes, or any combination of such factors, substantially impairs the sound growth of the community, is conducive to ill health, transmission of disease, infant mortality, juvenile delinquency and crime, and is detrimental to the public health, safety, morals or welfare; provided, no area shall be considered a blighted area nor subject to the power of eminent domain.

Bonds. Shall mean any bonds (including refunding bonds), interim certificates, certificates of indebtedness, debentures or other obligations.

Federal government. Shall include the United States of America or any agency or instrumentality, corporate or otherwise, of the United States of America.

Improve. To build, effect, alter, repair, or demolish any improvement upon, connected with, or on or beneath the surface of any real property, or to excavate, clear, grade, fill or landscape any real property, or to construct driveways and private roadways, or to furnish materials, including trees and shrubbery, for any of such purposes, or to perform any labor upon such improvements.

Improvement. All or any part of any building, structure, erection, alteration, demolition, excavation, clearing, grading, filling, or landscaping, including trees and shrubbery, driveways, and private roadways, on real property.

Owner. A person who owns an interest in the real property improved and for whom an improvement is made and who ordered the improvement to be made. "Owner" includes successors in interest of the owner and agents of the owner acting within their authority.

Real property. The real estate that is improved, including lands, leaseholds, tenements and hereditaments, and improvements placed thereon.

Real property. Lands, lands under water, structures and any and all easements, franchises and incorporeal hereditaments and every estate and right therein, legal and equitable, including terms for years and liens by way of judgment, mortgage or otherwise.

Redeveloper. Any individual, partnership or public or private corporation that shall enter or propose to enter into a contract with a commission for the redevelopment of an area under the provisions of the ordinance.

Redevelopment. The acquisition, replanning, clearance, rehabilitation or rebuilding of an area for residential, recreational, commercial, industrial or other purposes, including the provision of streets, utilities, parks, recreational areas and other open spaces.

Redevelopment. May include a program of repair and rehabilitation of buildings and other improvements, and may include the exercise of any powers under the Municipal Code with respect to the area for which such program is undertaken.

Slum area. Shall mean any area where dwellings predominate, which, by reason of depreciation, overcrowding, faulty arrangement or design, lack of ventilation, light or sanitary facilities, or any combination of these factors, are detrimental to the public safety, health or morals.

Urban renewal area. Shall mean a slum area or a blighted area or a combination thereof which the municipality designates as appropriate for an urban renewal project.

Urban renewal plan. Shall mean a plan, as it exists from time to time, for an urban renewal project, which plan shall be sufficiently complete to indicate such land acquisition, demolition and removal of structures, redevelopment, improvements, and rehabilitation as may be proposed to be carried out in the urban renewal area, zoning and planning changes, if any, land uses, maximum density and building requirements.

Urban renewal project. Shall mean undertakings and activities of a municipality in an urban renewal area for the elimination and for the prevention of the development or spread of slums and blight, and may involve slum clearance and redevelopment in an urban renewal area, or rehabilitation or conservation in an urban renewal area, or any combination or part thereof in accordance with an urban renewal plan. Such undertakings and activities may include:

D-17 SOLAR SYSTEMS

Absorptance. The ration of the amount of radiation absorbed by a surface to the amount of radiation incident upon it.

Absorptivity. The capacity of a material to absorb radiant energy.

Active solar system (flat plate or concentrating collector based). A system characterized by the use of powered mechanical equipment to move the heat transfer fluid (liquid or gas) through a collector and from a collector to load or storage.

Auxiliary energy subsystem. Equipment utilizing conventional energy sources both to supplement the output provided by the solar energy system and to provide full energy backup during periods when the solar H or DHW systems are inoperable.

Cathodic protection. Corrosion protection against electrolytic reactions.

Chemical compatibility. The ability of materials and components in contact with each other to resist mutual chemical degradation, such as the chemical degradation caused by electrolytic action or plasticizer migration.

Collector efficiency (instantaneous). The ratio of the amount of energy removed by the transfer fluid per unit of aperture (entrance window area) over a 15 minute period to the total incident solar radiation onto the same collector area for the same 15 minute period (as defined by NBSIR 74-635).

Collector subsystem. The assembly for absorbing solar radiation, converting it into thermal energy, and transferring the thermal energy to a heat transfer fluid.

Combined system (combined collectors and storage devices). A combined component system characterized by a system with integral construction and operation of the components such that the solar radiation collection and storage phenomena cannot be measured separately in terms of flow rate and temperature changes.

Control subsystem. An assembly of devices and its electrical, pneumatic or hydraulic auxiliaries used to regulate the processes of collecting, transporting, storing and utilizing energy.

Design life. The period of time during which a solar energy system or component is expected to perform without major maintenance or replacement.

Dielectric fitting. An insulating or nonconducting fitting used to isolate electrochemically dissimilar materials.

Emittance. The ratio of the radiant energy emitted by a body to the radiant energy emitted by a black body at the same temperature.

Facility. Means a bulding or structure including appliances, heating or cooling equipment, industrial or manufacturing processes to be served by the solar energy system.

Flow condition. The condition existing in the solar energy system when the heat transfer fluid is flowing through the collector under normal operating conditions.

Heat generated cooling. The use of thermal energy to operate an absorption refrigerating unit.

Heating degree days. The number of degrees that the daily mean temperature is below 18.3°C. (65°F.).

Maximum "flow" temperature. The maximum temperature obtained in a component when the heat transfer fluid is flowing through the system.

Maximum "no-flow" temperature. The maximum temperature obtained in a component when the heat transfer fluid is not flowing through the system.

Maximum service temperature. The maximum temperature to which a component will be exposed in actual service, either with or without the flow of heat transfer fluid.

Operating energy. The conventional energy required to operate the H, HC and HW systems, excluding any auxiliary energy which supplements the solar energy collected by the systems (e.g., the electrical energy required to operate the energy transport and control subsystems).

Outgassing. The emission of gases by component materials usually during exposure to elevated temperature or reduced pressure.

Passive solar system (integral collector, storage and building). A passive system characterized by collector and storage components which are an integral part of the building. Auxiliary energy may be used for control purposes but heating is achieved by natural heat transfer phenomena. Roof ponds, modified walls, roof sections with skylights, or similar applications where solar energy is used to supply a measurable fraction of the building heating requirements are examples of passive systems.

Solar collector. Any device or assembly of components which collects solar energy and uses such energy directly or transforms it to another usable form of energy.

Solar concentrator. A reflective surface or refracting lens for directing insolation onto the absorber surface.

Solar constant. The insolation on a surface in space at the earth's distance from the sun, 428 BTU/hr/ft^2 (1.94 Langleys/min).

Solar cover plate. A transparent plastic, glass plate or other material placed over the absorber plate of a solar collector to avoid heat losses and weathering of the absorber plate.

Solar energy. For the purpose of the code, the radiant energy of the sun in the forms of direct, diffuse or reflected radiation.

Solar energy system. An assembly of subsystems and components which is designed to convert solar energy into thermal energy.

Solar medium. The material in an assembly used for storing solar energy in its transformed state, be it thermal or electrical.

Transmittance. The ratio of the radiant flux transmitted through and emerging from a body to the total flux incident on it.

D-18 VEHICLES AND TRANSPORTATION MEANS

Authorized emergency vehicle. Vehicles of the fire department, police vehicles, ambulances or other emergency vehicles.

Automobile. A two axle private motor vehicle designed and used primarily for the conveyance of not more than nine (9) persons.

Bicycle. Every device propelled by human power upon which any person may ride, having two (2) tandem wheels either of which is more than twenty (20) inches in diameter.

Bicycle. A two-wheel vehicle, propelled by human power, having a tandem arrangement of wheels equipped with tires either of which is over twenty (20) inches in diameter.

Bus. Every motor vehicle designed for carrying more than ten (10) passengers and used for the transportation of persons; and every motor vehicle, other than a taxicab, designed and used for the transportation of persons for compensation.

Bus. A motor vehicle designed for carrying more than nine (9) passengers and used for the transportation of persons.

Coach. A self-propelled vehicle so designed and constructed to comfortably seat four (4) or more persons.

Commercial car. A motor vehicle having motor power designed and used for carrying merchandise or freight, or used as a commercial tractor.

Commercial tractor. A motor vehicle having motive power designed or used for drawing other vehicles and not so constructed as to carry any load thereon, or designed or used for drawing other vehicles while carrying a portion of such other vehicle or vehicles, or the load thereon or both.

Commercial unit. Includes any commercial tractor, truck, trailer, semi-trailer, pole trailer or commercial car; or any combination of these vehicles, when a commercial tractor, truck or commercial car is actually connected to one or more trailers, semi-trailers or pole trailers for the purpose of drawing such vehicles.

Cycle. Every device propelled by human power upon which any person may ride, having one or more wheels.

Driver. Every person who drives or is in actual physical control of a vehicle.

Emergency vehicle. Includes vehicles used by the fire department, police and state highway police, as vehicles for emergency purposes for municipal, county state, or public utilities, when identified as such as required by law. This also includes publicly or privately owned ambulances.

Mini-bike. Any self-propelled device primarily used for transportation or sport including, but not limited to, motorcycles, go-carts, motorized bicycles, off-the-road trail bikes and all-terrain vehicles.

Motorcycle. Every motor vehicle having a seat or saddle for the use of the rider and designed to travel on not more than three (3) wheels in contact with the ground, but excluding a tractor.

Motorcycle. A motor vehicle having a saddle for the use of the operator and designed to travel on not more than three (3) wheels in contact with the ground, but excluding a tractor.

Motor vehicle. Every vehicle which is self-propelled and every vehicle which is propelled by electric power obtained from overhead trolley wires, but not operated upon rails.

Motor Vehicle. A vehicle propelled or drawn by power other than muscular power or power collected from overhead electric wires, except road rollers, traction engines, power shovels, power cranes and other equipment used in construction work and not designed for or employed in general highway transportation, hole digging machinery, well drilling machinery, ditch digging machinery, farm machinery, threshing machinery, hay baling machinery, and agricultural tractors and machinery used in the production of horticultural, floricultural, agricultural and vegetable products.

Motor Vehicle. A passenger vehicle, motorcycle, motor truck, truck-trailer, trailer or semitrailer which is propelled or drawn by mechanical power.

Motor vehicle. A passenger vehicle, motor scooter, motorcycle, truck, truck-trailer, trailer, or semi-trailer propelled or drawn by mechanical or electrical power.

Motor vehicle. A self-propelled wheeled vehicle designed primarily for transportation of persons or goods along public streets.

Motor vehicle. Any conveyance propelled by an internal combustion engine.

Ordinary vehicle of commerce. Any vehicle that can move along the streets and highways with other vehicles and can turn corners with facility without impeding traffic.

Pole trailer. A trailer or semi-trailer attached to the towing vehicle by means of a reach or pole, or by being boomed or otherwise secured to the towing vehicle, and ordinarily used for transporting long or irregular shaped loads, such as poles, pipes or structural members capable, generally, of sustaining themselves as beams between the supporting connections.

Public hack or cab. A vehicle plying for hire, for which public patronage is solicited upon the streets; any motor vehicle carrying passengers for hire; any motor vehicle carrying passengers for hire, operating from or to a railroad station, airport, ship or boat landing dock.

Railroad. A carrier of persons or property upon cars, other than streetcars, operated upon stationary rails.

Railroad. A carrier of persons or property operating upon rails placed principally on a private right-of-way.

Railroad Train. A steam engine, electric or other motor, with or without cars coupled thereto, operated upon rails, except streetcars.

Railroad train. (locomotive). Includes the mechanically operated portion of a train (a combination of cars) which is motivated by steam, electric or deisel oil, and is coupled to other cars for carrying passengers, mail and freight.

School bus. Motor vehicle owned by a public or governmental agency, or privately owned and operated for the transportation of children to or from school or school-sponsored or authorized activities.

Taxicab. A motor vehicle, commonly called "taxi," which is: Constructed so as to comfortably seat not less than 4 passengers exclusive of the driver; and engaged in the business of carrying passengers for hire; and held out, announced or advertised to operate on and over the public streets of the city; and accepts persons who may offer themselves for transportation from a place within the city; and not operated over a fixed route.

Taxicab. A coach driven by mechanical power, capable of carrying from four (4) to six (6) persons, in which the charges are recorded on a meter.

Taxicab. Vehicles propelled by motor power engaged for the purpose of carrying persons for hire upon the streets, roads and highways located within the city limits.

Trailer. A vehicle designed or used for carrying persons or property wholly on its own structure and for being drawn by a motor vehicle, and includes any such vehicle when formed by or operated as a combination of a "semi-trailer" and a vehicle of the dolly type such as that commonly known as a "trailer dolly."

Truck. A motor vehicle except trailers and semi-trailers designed and used to carry property.

Vehicle. Any horse-drawn or self-propelled or motorized vehicle, including, but not by way of limitation, cart, wagon, dray, truck or other carriage or vehicle.

Vehicle. A device or equipment, upon, or by which any person or property is or may be transported or drawn upon a highway, except devices or equipment moved by power collected from overhead electric wires, or used exclusively upon stationary rails or tracks, and except devices or equipment other than bicycles moved by human power.

Vehicle. Any horse driven cart or wagon, truck or other self-propelled vehicle.

D-19 X-RAY SAFETY

Cineradiography. The making of a motion picture record of the successive images appearing on a fluorescent screen.

Contact therapy. Irradiation of accessible lesions usually employing a very short source-skin distance and potentials of 40-50 KV.

Dead-man switch. A switch so constructed that a circuit-closing contact can only be maintained by continuous pressure by the operator.

Diagnostic-type tube housing. An X-ray tube housing so constructed that the leakage radiation at a distance of 1 meter from the target cannot exceed 100 milliroentgens in 1 hour when the tube is operated at any of its specified ratings.

Filter. Material placed in the useful beam to absorb preferentially the less penetrating radiations.

Interlock. A device for precluding access to an area of radiation hazard either by preventing entry or by automatically removing the hazard.

Leakage radiation. All radiation coming from within the tube housing except the useful beam.

Primary protective barrier. A barrier sufficient to attenuate the useful beam to the required degree.

Protective barrier. A barrier of attenuating materials used to reduce radiation exposure.

Scattered radiation. Radiation that, during passage through matter, has been deviated in direction.

Secondary protective barrier. A barrier sufficient to attenuate stray radiation to the required degree.

Shutter. A device, generally of lead, fixed to an X-ray tube housing to intercept the useful beam.

Stray radiation. Radiation not serving any useful purpose. It includes leakage and secondary radiation.

Therapeutic-type tube housing. (a) For X-ray therapy equipment not capable of operating at 500 kVp or above, an X-ray tube housing so constructed that the leakage radiation at a distance of 1 meter from the source does not exceed 1 roentgen in an hour when the tube is operated at its maximum rated continuous current for the maximum rated tube potential. (b) For X-ray therapy equipment capable of operating at 500 kVp or above, an X-ray tube housing so constructed that the leakage radiation at a distance of 1 meter from the source does not exceed either 1 roentgen in an hour or 0.1 percent of the useful beam dose rate at 1 meter from the source, whichever is greater, when the machine is operated at its maximum rated continuous current for the maximum rated accelerating potential. (c) In either case, small areas of reduced protection are acceptable provided the average reading over any 100 square centimeters area at 1 meter distance from the source does not exceed the values given above.

Useful beam. That part of the radiation which passes through the window, aperture, cone, or other collimating device of the tube housing.

Part Two

BUILDING CODES

2. Article 1

GENERAL PROVISIONS 1-A THROUGH 1-K

1-A CODE AND ORDINANCE ENFORCEMENT

Abatement. Imposition of the penalties described in the ordinance shall not preclude the city from instituting an appropriate action or proceedings to prevent an unlawful erection, construction, reconstruction, addition, alteration, conversion, removal, demolition, maintenance or use, or to restrain, correct or abate a violation, or to prevent the occupancy of a building or structure or portion thereof, or of the premises, or to prevent an illegal act, conduct, business or use in or about any premises.

Administrative official. Appropriate officer or other authority charged with the administration and enforcement of the code or any other particular laws or regulations pertaining to fire, health, or public safety, or his duly authorized representative.

Administrator. Executive officer of the jurisdictional area.

Applicability. Any appropriate criterion which is acceptable to the building code or adopted regulations.

Applicable governing body. A city, county, state, state agency or other political government subdivision or entity authorized to administer and enforce the provisions of the Code, as adopted or amended.

Application of regulations. No land shall be used or occupied, and no building shall be erected, altered, used or occupied, except in conformity with the regulations established for the district in which such land, building or structure is located. In cases of mixed occupancy, the regulations for each use shall apply to the portion of the building or land so used.

Appropriate authority having jurisdiction. Departments of the government and agents thereof that have authority over the subject that is regulated.

Approval. Shall mean that a building or structure constructed under the provisions of the Code, or a material or product used therein, meets the requirements of the Code and has been approved by the Building Official.

Approvals by inspector. Any material or method of construction meeting the requirements of the code shall be approved by the building inspector within a reasonable time after completion of the tests. All such approvals and the conditions under which they are issued shall be reported and kept on file in the office of the building department, open for public inspection.

Approved. (1) Materials, devices or construction—accepted by the authority having jurisdiction under the provisions of the Code by reason of tests or investigations conducted by it or by an agency satisfactory to the authority or by reason of accepted principles or tests by national authorities, technical or scientific organizations. (2) Occupancy or use—accepted by the authority having jurisdiction under the provisions of the Code with the basic requirements of the Code.

Approved. As to materials and types of construction, refers to approval as the result of investigation or tests, national authorities, technical, or scientific organizations accepted as meeting the requirements of the Code.

Approved. As to materials and types of construction, refers to approval by the Building Council and the Building Commissioner as a result of investigation and tests conducted by him, or by reason of accepted principles or tests by national authorities, technical or scientific organizations.

Approved. The sanction and endorsement by the Commissioner of Buildings under the provisions of the code.

Approved. As to structures erected or moved into the City shall mean, consideration being given to the temporary nature of such structures in the application of standards which may be more appropriate for permanent installations.

Approved. Passed upon favorably by the person, board, or other authority authorized by state or city laws to give approval on the matter in question in the application or enforcement of the provisions of the Building Code.

Approved. Accepted or acceptable under an applicable specification stated or cited in the Code, or accepted as suitable for the proposed use by the appropriate administrative official.

Approved. Approval granted by the department under the regulations stated in the Code.

Approved. Approved by the Building Commissioner or other authority having jurisdiction.

Approved. Approved by the authority having jurisdiction or the appropriate authority having jurisdiction.

Approved. As applied to a material, device, or mode of construction, means approved by the building official in accordance with the provisions of the Code, or by other authority designated by law to give approval in the matter in question.

Approved. As applied to a material, device or method of construction, means approved by the Commissioner under the provisions of the Code, or by other authority designated by law in the matter in question, to give approval.

Approved. In accordance with the Code is determined by by the Building Official, or passed upon favorably by the Board of Building Code Appeals, or by any other authority designated by the Code to give approval in the matter in question.

Approved. Means acceptable to the department based on its determination as to conformance with appropriate standards and good public health practices.

Approved. As to materials and types of construction, refers to approval by the Building Official as the result of investigation and tests conducted by him, or by reason of accepted principles or tests by national authorities, technical or scientific organizations.

Approved. Accepted by the Chief of the Fire Department or Chief of the Bureau of Fire Prevention, as a result of their investigation and experience or by reason of test, listing or approval by Underwriters' Laboratories, Inc., the National Bureau of Standards, the American Gas Association Laboratories or other nationally recognized testing agencies.

Approved. As to materials and methods of construction or as to equipment and its installation, refers to approval by the Appeals Boards as provided in the Code, or by a nationally recognized authority, as listed for the particular material, method or equipment in the Rules and Regulations of the Code.

Approved. Approved by the building official, board or administrative authority designated by law with the administration of the building code and building regulations.

Approved. Approved by the Building Official or other authority having jurisdiction.

Approved. Means approval granted by the department of industrial, labor.

Approved. Refers to approval by the Building Official as a result of investigation and/or tests conducted by him or by reason of accepted principles or tests by national authorities, technical, or scientific organizations.

Approved. Accepted by the authority having jurisdiction under the provisions of the Code by reason of tests or investigations conducted by it or by an agency satisfactory to the authority, based upon nationally accepted test standards or principles.

Approved. As to materials and types of construction, refers to approval by the Director as the result of investigation and tests conducted by him, or by reason of accepted principles or tests by national authorities, technical or scientific organizations and meeting Code requirements.

Approved. Approved by the enforcement officer under the regulations of the Code or approved by an authority designated by law or the Code.

Approved. The term "approved," as applied to any material device or mode of construction, shall mean approved by the board or legally approved by the superintendent under the provisions of the Code or by any other authority legally designated to give approval of the matter in question.

Approved. Means approved by the Building Official.

Approved materials and methods. The use of all materials or methods of construction which meet the specified strength, durability, sanitary and fire resistive requirements of the code and when they are accepted engineering practices shall be permitted and approved.

Approved rules. Legally adopted rules of the Building Commissioner.

Approved rules. Legally adopted rules of the administrative official or of a recognized authorative agency.

Approved rules. The legally adopted rules of the building official or of a jurisdiction.

At-site construction. Nothing in the code shall be deemed to prohibit at-site construction and erection of buildings or structures when designed in compliance with the provisions of the code and the minimum requirements prescribed in the code.

Authority having jurisdiction. Municipal council and the authorized agent thereof having the authority over the subject that is regulated.

Authority having jurisdiction. Duly authorized representative or agency having legal enforcement responsibility in cases where the code is applied with the force of law. Where the code is applied on a contractural basis, the contract shall specify the individual or agency to act as the authority having jurisdiction.

Authority having jurisdiction. The duly authorized representative or agency having legal enforcement responsibility in cases where the LIFE SAFETY CODE is applied with the force of law.

Authority to enter upon private property. The building inspector, and any inspector under his direction, in the performance of their function and duties under the provisions of the code, may enter upon any land and make examinations and surveys as deemed necessary in the administration and enforcement of the code.

Basic classification of construction materials. All materials and methods used in the design and construction of buildings and structures shall be classified as controlled materials and ordinary materials as defined in the Code. The design and construction shall be based on the assumptions, limitations and methods of stress determination of recognized design procedures.

Basic code. The Code used in the application of its provisions and requirements, referred to as the "basic code" the "code" or the "basic building code."

Building. A structure built or used for shelter, occupancy, enclosure, or support of persons, animals, or chattels and is construed as if followed by "other structure, or parts thereof;" provided, however, that the term "building" or "structure" shall not include structures or installations ordinarily known as industrial, production, or manufacturing equipment, and the parts thereof. "Building" also includes the equipment therein which is subject to regulation under the Code.

Building commissioner. Usually holds the title of the Building Official.

Building commissioner. Is the Chief Building Commissioner or any regularly authorized deputy.

Building department. Department or division of the government of a municipal corporation or township, or county, which department or division has been created in conformity with the law for the purpose of enforcing a building code which is applicable to all buildings and structures within the boundaries of said municipal corporation, township or county and which building codes are equivalent to and not in conflict with the State building codes.

Building official. Shall mean the official of the jurisdiction charged with the administration of the Building Code.

Building official. The officer or other designated authority charged with the administration and enforcement of the Code, or his duly authorized representative.

Building official. The officer, or other person, charged with the administration and enforcement of the ordinance, or his duly authorized representative.

Building official. The term "Building Official" whenever employed in the ordinance shall mean the "Building Official" who shall be the person vested with executive authority to see that all laws, ordinances and Codes regulating building construction are observed and enforced. The "Public Works Director" may designate a "Deputy Building Official" who shall perform such duties as are prescribed by the "Building Official."

Building official. The officer charged with the administration and enforcement of the Code, or his regularly authorized deputy.

Chief enforcement official. Means the building inspector or commissioner of buildings in a municipal corporation or county having a building department, or the health commissioner or his authorized representative in health districts, whichever one has jurisdiction.

City. A political subdivision which has the authority to adopt a code for the regulation of building construction within its jurisdiction.

Code. The purpose of the code is to provide minimum standards to safeguard life or limb, health, property, and public welfare by regulating and controlling the design, construction, quality of materials, use and occupancy, location and maintenance of all buildings, structures and utilities and certain equipment specifically regulated by the Code.

Code enforcing authority. Under present building codes the enforcing authority is the Building Official, who may be the Building Commissioner, the Chief Building Inspector, or the Administrative Officer of the department or Division.

Conflict. Any building regulation that is incompatible, inconsistent, in opposition or at variance with another building regulation or law on the same building or structure.

Construction. Includes alterations and repairs and operations incidental to construction.

Construction. Includes all labor and materials used in the framing or assembling of component parts in the erection, installation, enlargement, alterations, repair, moving, conversion, razing, demolition or removal of any appliance, device, building, structure or equipment.

Due public notice. When the law requires "due public notice" for any purpose whatsoever the City shall make all of the arrangements to have the particular item published one time in a city newspaper 15 days before the pronounced action will take place, stating the date, place, time and the nature of the business with which the notice is concerned.

Duplication. Any building regulation or standard that duplicates, repeats or corresponds to another State building regulation or law on the same building or structure.

Electrical code. Code or the regulations enforced by the governmental agency to control the installation and maintenance of electric services and equipment in a building.

Enforcement officer. A person lawfully empowered to enforce the regulations of the Code.

Existing. That which is in existence as the date when the Zoning ordinance or Building code goes into effect, as existing buildings, structures, or exit facilities.

Existing. Building, structure, or equipment completed or in the course of construction or use or occupied prior to the effective date of applicable rules of the code.

Existing. Constructed, installed, or situated, or being constructed, installed or situated prior to the adoption of the Building code.

Existing building. A building erected prior to the adoption of the code, or one for which a legal building permit has been issued.

Existing building. A building erected prior to the adoption of the rules and regulations, or one for which a legal building permit has been issued.

Existing building. Building erected or partially erected prior to the effective date of the Building code or a completed building for which a legal building permit was issued under the provisions of the original regulations or previous code.

Existing building. A building, structure or part thereof which has been completed and ready for occupancy.

Existing installations. The provisions of the code shall apply to the addition of or replacement of any major apparatus in existing buildings.

Existing structure. Structure in existence and in use prior to the effective date of the new Building code.

Existing structure repairs. All repairs shall be made with materials meeting the standards of the new code. Where repairs are made in excess of 25% of the surface area of any wall or partition which exceeds 10 feet in length or floor, ceiling or roof which exceeds 200 square feet, the entire wall, partition, floor, ceiling or roof shall conform to the new code.

Fire code. Shall herein be defined as rules which are duly promulgated by the State Fire Prevention Commission and shall include such state and federal laws containing provisions therein for their enforcement by the State Fire Marshal.

Interpretation of standards. In their interpretation and application, the provisions of the ordinance shall be held to be minimum requirements. Where the ordinance imposes a greater restriction than is imposed or required by other provisions of law or by other rules or regulations or ordinances, the provisions of the ordinance shall govern.

Interpretation of the ordinance. The standards prescribed by the ordinance shall be held to be minimum requirements. When standards are greater than those contained in any other existing provisions of the law, the standards of the Ordinance shall apply. However, when private covenants or deed restrictions are greater than those imposed by the ordinance, the same shall have precedence.

Jurisdiction. Shall mean the legally-constituted authority which has adopted the Code as law or ordinance.

Materials and methods of construction. Nothing in the Code shall be construed to prevent the use of any material or method of construction whether or not specifically provided for in the

Code, if, upon presentation of plans, methods of analysis, test data or other necessary information to the Building Department by the interested person or persons, the Building Department is satisfied that the proposed material or method of construction complies with specific provisions of or conforms to the intent of the Code.

Model code. A building code developed by a regional federation of building officials. Codes are continually reviewed and updated by committees of building officials. Model codes in the United States are the Basic Building Code (BOCA), National Building Code (NBC), Standard Building Code (SBC), and the Uniform Building Code (UBC).

Municipality. Corporate governmental unit with the authority to adopt a code under due legislative statutes.

New building. Includes buildings, additions thereto, and alterations thereof, for which complete drawings have not been approved by the department of industrial labor, construction is not in progress, prior to listed date.

New building. Includes all new buildings and all remodeled buildings where forty (40) percent or more of the remodeled building is new work and when plans for such new or remodeled buildings are approved after the date of promulgation of the regulations.

New materials. All new materials not specifically provided for shall be tested and approved in accordance with the provisions of the Code for strength, durability and fire-resistance; or the building commissioner shall accept the reports of accredited testing authorities complying with the approved rules to assist him in his determination.

Nomenclature. A consistent system of wording, adopted or approved by the Commission, designating or defining a building standard in language and phrases commonly used.

Notice of unsafe buildings or structures. Upon determining that a building or structure or portion thereof is unsafe, the Building Department shall serve or cause to be served on the Owner, or some one of the Owners' Executors, Administrators, Agents, Lessees or other persons who may have a vested or contingent interest in the same, a written notice containing the description of the building or structure or portion thereof is unsafe, and an order requiring the same to be made safe and secure or removed, as may be deemed necessary by him. If the person to whom such notice and order is addressed cannot be found after diligent search, then such notice and order shall be sent by certified or registered mail to the last known address of such person; and a copy of such notice shall be posted in a conspicuous place on the premises to which it relates. Such mailing and posting shall be deemed adequate service.

Ordinance. A law or regulation passed by a municipality; a law enacted by a legislative body of a municipal corporation; a local law that applies to to persons and things subject to the local jurisdiction.

Overlap. Any building regulation or standard which relates to the same subject matter as another State building regulation or standard without wholly duplicating or coming into appreciable conflict with it.

Penalties. Any and all persons who shall violate any of the provisions of the Code or fail to comply therewith, or who shall violate or fail to comply with any order or regulation made thereunder, or who shall build in violation of any detailed statement or plans or specifications submitted and approved thereunder, or any permit issued thereunder, shall severally for each and every such violation and noncompliance respectively forfeit and pay a penalty as listed in the Code.

Performance standards. Standards developed from tests of materials, assemblies of materials, or equipment which will effectively and adequately provide the intended safety based upon established criteria.

Precautions during building operations. Provisions of the Code shall apply to all construction operations in connection with the erection, alteration, repair, removal or demolition of buildings and structures. Execution of the requirements shall be regulated by the approved rules and the safety Code for building construction listed in the Code.

Pre-code building. A building already erected on the effective date of the Code, or thereafter erected, as provided in the Code, under permit for its construction subject to the provisions of law in effect prior to such effective date. A pre-code building may be altered, repaired, enlarged, moved, or converted to other uses, only in conformity with the provisions of the Code and subject to the permit.

Public authority. Any authority or any officer who is in charge of any department or branch of the government of the city relating to health, fire, building, regulations, or to other Codes or Ordinances concerning non-residential premises in the city.

Purpose of code. For the purpose of protecting the health, safety, and welfare of the people of the city there is hereby enacted a Code known as the "Non-Residential Property maintenance Code of the City," which establishes minimum standards for non-residential premises, determines the respective responsibilities of owners, operators, and occupants of applicable structures and buildings now in existence or which may hereafter be constructed or established, provides for the enforcement of provisions pertaining to such standards and responsibilities and provides penalties for the violation of the Code.

Purpose of the code. To provide for safety, health and public welfare through structural strength and stability, means of egress, adequate light and ventilation and protection of life and property from fire and hazards incident to the design, construction, alteration, removal or demolition of buildings and structures. It is also the purpose of the Code to prescribe regulations consistent with nationally recognized good practice for the safeguarding to a reasonable degree of life and property from the hazards of fire and explosion arising from the storage, handling and use of hazardous substances, materials and devices, and from conditions hazardous to life or property in the use or occupancy of buildings or premises.

Quality of materials and methods of construction. Any material or method of construction failing to conform to the requirements of the code shall not be used. Whenever there is reason to doubt the quality of a material or method of construction to be used in a building or structure, the building department may require that tests be made to establish its suitability or to determine whether it conforms to the intent of the code. Such tests shall be made at the expense of the owner.

Regulations. The whole body of regulations, text, charts, tables, diagrams, maps, notations, references and symbols contained or referred to in the Code.

Regulations and statutes. Nothing shall be contained in the code which shall be construed to nullify any rules, regulations, statutes of agencies governing the protection of the public or workmen from health and other hazards involved in manufacturing, mining, and other processes and operations which generate toxic gases, dust or other elements dangerous to the respiratory system, eyesight or health.

Removal or made safe. When a building or structure or any portion thereof is found unsafe upon inspection by the Building Official, he shall order such building or structure or any portion thereof to be made safe or secured or taken down and removed.

Repair. Replacement of existing work with the same kind of material used in the existing work, not including additional work that would affect the structural safety of the building, or that would affect or change required exit, lighting or ventilating facilities, or that would affect a vital element of an elevator, plumbing, gas piping, wiring or heating installation, or that would be in violation of a provision of the Code.

Scope of the code. Provisions of the Code apply to the construction, alteration, equipment, use and occupancy, location and maintenance of and additions to buildings and structures and to appurtenances such as vaults, areaways and street encroachments, hereafter erected and, where expressly stated, existing on land or over water and buildings and structures and equipment for the operation thereof hereafter moved or demolished in the municipality. The provisions of the Code based on occupancy also apply to conversions of existing buildings and structures or portions thereof from one occupancy classification to another.

Scope of the code. The provisions of the Code shall govern the construction, reconstruction, alteration, repair, demolition, removal, use or occupancy, and the standards of materials, including materials used for finish and trim, to be used in such construction, reconstruction, alteration, repair, demolition, removal, use or occupancy of any building, portion of a building or room, specified in the Code. They shall apply to existing and proposed buildings as provided in the Code.

Scope of the code. The provisions of the Code shall control the classification of all buildings as to use group and type of construction; and in the interpretation of the Code the words and terms used shall be construed as defined in the Code.

Where a given word or term is defined in a section of the Code, the definition shall be used in construing such word or term wherever appearing in any following section of the Code.

Subsidewalk space. No permit shall be issued for the construction, erection, repair or alteration of any building or structure, if in one or more walls abutting a public way windows or other openings are placed below the level of such public way, the lighting or ventilation of which will require the use of subsidewalk space, until the applicant therefor has first obtained specific authority for such use as provided in the Code.

1-B CONSTRUCTION TYPE CLASSIFICATIONS

Air-supported structure. A structure constructed of a single diaphragm lightweight flexible fabric or film, or any combination thereof, which derives its primary support and stability from inflation pressure, together with anchorage attached to the base of the structure.

Cold-formed steel construction. That type of construction made up entirely, or in part, of steel structural members cold-formed to shape from sheet or strip steels such as roof deck, floor and wall panels, studs, floor joists, roof joists and other structural elements.

Combustible frame construction. Type of construction in which the structural elements, including enclosing walls, are entirely or in part of wood or other materials not more combustible than wood.

Construction classification. A classification of buildings into types of construction which is based on the fire resistance of the walls, floors, roof and other structural members.

Construction classification. Where a building is constructed of two or more types of construction, the construction classification of the entire building shall be the lowest of such types of construction.

Construction classification. Classification of buildings into types of construction which is based upon the fire properties of walls, floors, roofs, ceilings, and other elements.

Exterior protected construction. Type of construction in which the exterior walls are of noncombustible construction having a fire resistance rating as required by the code and which are structurally stable under fire conditions and in which the interior structural members and roof are wholly or partly of combustible construction.

Exterior-protected construction. Type of construction in which all exterior walls are of noncombustible materials providing fire resistance as required in the code.

Fireproof construction. That type of construction in which the walls, floors, roof and structural members are of approved masonry, reinforced concrete or other approved incombustible material meeting all the requirements of these regulations and having a minimum fire resistance as indicated in the Code. Fire-retardant treated wood complying with definitions in the Code may be used as specified.

Fireproof construction. When all of the exterior walls of the building or structure are of masonry or reinforced concrete, or of other approved materials or combination of materials, and in which all the structural members are of non-combustible materials, and provide fire-resistance, not less than stipulated in the building code.

Fireproof construction. As applied to buildings, in which walls are of approved masonry or reinforced concrete; and the structural members of which have fire resistance ratings sufficient to withstand the hazard involved in the occupancy, but not less than 4-hour rating for bearing walls, firewalls, isolated piers, columns and wall supporting griders; a two hour rating for walls and girders other than already specified, and for beams, floors, roofs and floor fillings; and a 2-hour rating for fire-partitions.

Fireproof construction. Fireproof-fire-resistive construction.

Fire-resistive. Type of buildings, where the structural frame shall be of fire-protected structural steel or iron or of reinforced concrete. The exterior walls, inner court walls, and walls enclosed vertical openings shall be of fire-resistive construction. The roof construction and floors shall be of fire-resistive materials. Exterior doors and windows shall be of fire-resistive construction only when specified in the Code.

Fire-resistive construction. Type of construction in which all structural elements, including walls, bearing partitions, floors, ceilings, roofs and their supports, are of non-combustible materials providing fire resistance as required in the code.

Fire-resistive construction. That type of construction in which the walls, partitions, columns, floors and roof are non-combustible with sufficient fire resistance to withstand the effects of a fire and prevent its spread from story to story.

Fire-resistive construction. Fabricated units or assemblies of units of construction which have a fire-resistive rating of not less than 3/4 hours.

Fire-resistive construction. All buildings or parts of buildings required to be fire-resistive construction shall be constructed of such non-inflammable material as stone, steel, concrete, brick, tile, expanded metal lath with plaster on steel studs when specifically permitted by the regulations, and such other materials as may be approved from time to time. Fire-retardant treated wood complying with definition in the Code may be used as specified.

Fire-resistive construction. That type of construction in which the walls, partitions, columns, floors, roof, ceilings and other structural members are noncombustible with sufficient fire resistance to withstand the effects of a fire and prevent its spread from one story to another.

Fire-resistive construction. That type of construction intended to provide a resistance to fire and the prevention of spread of fire in degrees varying with the manner and type of occupancy.

Fire-resistive construction. Construction in which all exterior walls are of masonry or reinforced concrete, or of other approved materials or combinations of materials and in which all the structural members are of non-combustible materials, and provide fire-resistance not less than stipulated in the code.

Fire-resistive construction type. All fire-resistive type buildings and structures shall be constructed with enclosure walls of masonry, reinforced concrete, or other approved incombustible materials with a fire-resistive rating as required by the code but in no case less than 1 hour; and with interior walls, floors, roofs, permanent partitions, exitways and structural elements designed and protected with incombustible materials to afford a fire-resistive rating of not less than 1 hour; except that the fire-resistive rating may be waived where unprotected steel is permitted under provisions of the code.

Fire-resistive type. Building or structure of fire-resistive construction of which the walls, partitions, piers, columns, floors, ceilings, roof and stairs are built of noncombustible material, with a fire-resistive rating as specified in the code.

Fire retardant construction. Fabricated units or assemblies of units of construction which have a fire-resistive rating of not less than 3/4 hours.

Formed steel construction. That type of construction used in floor and roof systems consisting of integrated units of sheet or strip steel plates which are shaped into parallel steel ribs or beams with a continuous connecting flange deck; generally attached to and supported on the primary or secondary members of a structural steel or reinforced concrete frame.

Frame construction. Type of construction in which the exterior walls, bearing walls, partitions, floor and roof construction are constructed wholly or partly of wood stud and joist assemblies.

Frame construction. As applied to buildings, means that in which exterior or party walls are wholly or partly of wood. Buildings of exterior masonry veneer or stucco on wood frame, constituting wholly or in part the structural supports of the building or its loads, are frame buildings within the meaning of this definition.

Frame construction. That type of construction in which the walls, partitions, floors and roof are wholly or partly of wood or other combustible material.

Heavy timber. Building code designation for a particular type of construction with good fire endurance.

Heavy timber. Exterior protected construction in which the interior structural members are of heavy wood timbers or in combination with noncombustible structural members.

Heavy timber construction. That type of combustible construction in which a degree of fire safety is attained by placing limitations on the sizes of wood structural members and on thickness and composition of wood floors and roofs, by avoidance of concealed spaces under floors and roofs, and by use of approved fastenings, construction details and adhesives for structural members.

Heavy timber construction. A type in which fire-resistance is attained by the sizes of heavy timber members (sawn or glued-laminated) being not less than indicated in the Code or by providing fire-resistance not less than one-hour where materials other than wood are used; by the avoidance of concealed spaces under floors and roofs; by the use of approved fastenings, construction details, and adhesives for structural members; and by providing the required degree of fire-resistance in exterior and interior walls.

Heavy timber construction. Approved type of wood construction in which a degree of fire endurance is attained by placing limitations on the minimum sizes of wood structural assemblies.

Heavy timber construction. Type of construction in which all interior structural elements shall be of heavy timber as described in the code.

Heavy timber construction. That type of construction in which the exterior walls are of masonry or other noncombustible materials having equivalent structural stability under fire conditions and a fire-resistance rating of not less than 2 hours; the interior structural members including columns, beams and girders, are of heavy timber, in heavy solid or laminated masses, but with no sharp corners or projections or concealed or inaccessible spaces; the floors and roofs are of heavy plank or laminated wood construction, or of any other material providing equivalent fire-resistance and structural properties, or construction is as set forth in the generally accepted standards.

Incombustible construction type. All incombustible type buildings shall be constructed entirely of steel, concrete or other approved incombustible materials without specified fire-resistive rating except that the enclosure walls shall be required to afford the fire resistance specified in the code when erected within the fire limits; and exitways shall be constructed of not less than 3/4 hours in other than one-family dwellings.

Light gage steel construction. That type of construction in which the structural frame consists of studs, floor joists, arch ribs, rafters, steel decks and other structural elements are composed and fabricated of cold-formed sheet or strip steel members less than three-sixteenths (3/16) inch thick.

Light incombustible. Type of buildings where the structural framework shall be of steel, iron, masonry, or reinforced concrete and exterior walls shall be of incombustible materials. Partitions, floors and roof construction shall be of incombustible materials except as set forth in the Code. Foundations shall be of masonry or reinforced concrete.

Masonry wall construction type. All masonry wall type buildings shall be constructed with exterior, fire and party walls of approved masonry complying with the provisions of the code or of other approved incombustible materials with a fire-resistive rating of not less than 2 hours, with roofs, floors, and interior framing wholly or partly of wood or other approved materials of similar combustible characteristics of approved unprotected incombustible materials; except that girders and their supports carrying walls of masonry shall in all cases be protected to afford the require fire-resistive rating of the walls supported thereon.

Mill construction. Building or structure in which the walls are of approved masonry or reinforced concrete, and an interior framing of wood, with plank or laminated wood floors and roofs, and in which the interior structural elements are arranged in heavy solid masses and smooth flat surfaces assembled to avoid thin sections, sharp projections, and concealed or inaccessible spaces. The interior framing may be partly or entirely of protected steel or concrete, and the floors and roofs may be constructed in whole or in part of noncombustible materials.

Mill construction. A type of building with heavy timber frame, masonry bearing walls, and laminated floors.

Mixed types of construction. When 2 or more types of construction occur in the same building and are not separated by the fire separation specified in the Building code for mixed occupancies, the entire building shall then be subject to the occupancy restrictions of the least fire resistive type of construction used in the building.

Moderate fire hazard. A building use which involves the storage, sale, handling, manufacture, or processing of materials which are likely to burn with moderate rapidity, or produce a considerable volume of smoke, but which do not produce either poisonous fumes in quantities dangerous to the health of any persons, or explosion in the event of fire.

Noncombustible. Construction, as applied to a building, means that in which all structural members including walls, floors, roofs and their supports, are of steel, iron, concrete, or of other incombustible materials, and in which the exterior walls have not less than a two-hour fire resistance rating as tested for an interior and an exterior fire.

Non-combustible construction. That type of construction in which a degree of fire safety is attained by the use of non-combustible materials for structural members and other building assemblies.

Non-combustible construction. Assembly such as a wall, floor or roof having components of noncombustible material.

Non-combustible construction. That type of construction in which the walls, partitions, columns, floors, roof, ceilings and other structural members are non-combustible, but which does not qualify as fire-resistive construction.

Non-combustible construction. Includes protected non-combustible construction and unprotected non-combustible construction.

Non-combustible construction. Type of construction, in which all structural elements, including walls, bearing partitions, floors, ceilings, roofs and their supports, are of non-combustible materials but which are generally not fire protected except as required by the code.

Noncombustible construction. That type of construction in which the walls, partitions, columns, floors and roof are noncombustible and have less fire resistance than required for fire-resistive construction in the Code.

Ordinary. Construction, as applied to a building, means that in which exterior walls and bearing walls are of approved masonry or reinforced concrete and in which the structural elements are wholly or partly of wood of smaller dimensions than required for Heavy Timber construction, or of other materials not protected as required for Heavy Timber construction.

Ordinary construction. That type of construction in which the exterior walls are of noncombustible or approved limited combustible materials, or are approved limited combustible assemblies, and have fire resistance ratings not less than set forth in the code. The interior structural members are wholly or partly of wood of smaller dimensions than required for heavy timber construction or are of noncombustible, approved limited combustible, or other combustible materials.

Ordinary construction. That type of construction in which the exterior walls are of masonry, reinforced concrete, or other approved incombustible materials meeting the requirements of the regulations and having a fire-resistance not less than the requirements indicated in the Code and in which the interior framing is of partly or wholly combustible materials.

Ordinary construction. Construction in which the exterior walls and bearing walls are of three (3) hour fire resistance construction and in which the structural members including columns, floors, and roof construction are wholly or partly of wood of smaller dimensions than required for heavy timber construction or of steel or iron not protected as required for semifireproof construction.

Ordinary construction. Construction in which the exterior walls are of masonry or reinforced concrete or of approved materials or assembly of materials that provide fire-resistance. Also which the interior framing is partially or wholly of unprotected wood, or of unprotected iron or steel.

Ordinary construction. Type of construction in which the construction other than heavy timber construction in which the structural elements of the interior framing are entirely or in part of wood or other materials not more combustible than wood.

Ordinary construction. That type of construction in which the exterior walls are of masonry or other noncombustible materials having equivalent structural stability under fire conditions and a fire-resistance rating of not less than 2 hours, the interior structural members being wholly or partly of wood of smaller dimensions than those required for heavy timber construction.

Ordinary masonry. Type of buildings where the interior load bearing construction may be masonry or reinforced concrete walls, a structural frame of steel reinforced concrete, or wood. Exterior walls shall be of fire-resistive materials. Partitions, floors and roof framing may be of wood.

Protected construction. That in which all structural members are constructed, chemically treated, covered or protected so that the individual unit or the combined assemblage of all such units has the required fire-resistance rating specified for its particular use or application in the Code and includes protected-frame, protected-ordinary and protected-noncombustible construction.

Protected exterior. Exterior enclosure walls which are veneered and insulated or otherwise constructed of materials which in themselves or in combination with collateral materials are classified as incombustible, to develop the specified fire-resistance to exterior fire exposure; and which are supported on adequate foundations or supports of the required fire-resistive rating.

Protected ordinary. Exterior protected construction in which all interior structural members are protected with at least a 1-hour fire resistance rating.

Rigid frame construction. Continuous or restrained frame type, for which the design is predicated on the assumption that the end connections and junctions of all main members within the frame have sufficient rigidity to hold virtually unchanged the original angles between such main members.

Semi-fireproof. Construction, as applied to a building, means that in which the structural members, including interior and exterior bearing walls and exterior non-bearing walls, are of approved incombustible construction having the necessary strength and stability and having a fire resistance rating of not less than shown in the Code. A combustible roof may be used when it is protected by an approved automatic sprinkler system and the ceiling of the top story is of Semi-fireproof construction.

Semi-rigid frame construction. Construction of the partially restrained type for which the design is predicated on the assumption that the connections of beams and girders possess a dependable and known moment capacity intermediate in degree between the complete rigidity of "Rigid

Frame Construction," and the complete flexibility of "Simple Frame Construction."

Simple frame construction. Construction of the unrestrained type for which the design is predicated on the assumption that the ends of beams and girders are connected for shear only and are free to rotate under load.

Special types of construction. Includes types of construction involving materials and methods of construction which require special and different consideration in design, construction, or equipment with respect to other types of construction as classified in the Code.

Steel joist construction. Steel joist construction consists of decks or top slabs defined in the Code, supported by separate steel members referred to as steel joists. Any steel member suitable for supporting floors and roofs between the main supporting girders, trusses, beams, or walls when used as hereinafter stipulated shall be known as a "steel joist." Such steel joists may be made of hot or cold formed sections, strip or sheet steel, riveted or welded together, or by expanding.

Wood frame construction. That type of construction in which the structural parts and materials are of wood or are dependent upon a wood frame for support, including a construction having a non-combustible exterior veneer.

Wood frame construction. That type of construction in which the exterior walls, partitions, floors, roof and other structural members are wholly or partly of wood or other combustible materials assembled to provide a specified fire resistance rating.

Wood frame construction. That in which the enclosing walls are of wood or other combustible materials, including construction having exterior masonry veneer, stucco, or metal which is dependent upon wood for support, stability or rigidity and in which interior framing is of wood or other combustible materials.

Wood frame (protected). Building is of wood frame protected construction if the structural parts and enclosing walls are of protected wood, or protected wood in combination with other materials, with fire-resistive ratings as set forth in the Code. If such enclosing walls are veneered, encased or faced with stone, brick, tile, concrete, plaster or metal, the building is also termed a wood frame protected building.

Wood frame (unprotected). Building is of wood frame unprotected constuction if the structural parts and enclosing walls are of unprotected wood, or unprotected wood in combination with other materials. If such enclosing walls are veneered, encased or faced with stone, brick, tile, concrete, plaster or metal, the building is also termed a wood frame unprotected building.

1-C DESIGN LOADS

Concentrated load. A conventionalized representation of an element of dead or live load whereby the entire load is assumed to act either at a point or within a limited area.

Concentrated loads. Live load concentrated upon a specified small area of a floor, roof, wall or other supported surface or structural member.

Concurrent loads. Two or more elements of dead or live load that, for purposes of design, are considered to act simultaneously.

Dead load. The weight of all permanent structural and nonstructural components of a building.

Dead load. The stationary, permanent loads; that is, the weight of all the material used in construction of the building (or section).

Dead load. The weight of walls, floors, roofs, partitions and other permanent portions of the structure.

Dead load. The weight of the fixed, permanent structural members and service equipment in a structure.

Dead load. The weight of all permanent construction in a building.

Dead load. The weight of materials built into the construction of a building including walls, permanent partitions, floors, roofs, framing and all other permanent stationary construction entering into and becoming part of a building.

Dead load. The weight of all permanent construction, including walls, framing, partitions, floors, stairways, roof and all fixed building service equipment.

Dead load. The "dead load" when applied to a structure shall include the weight of walls, permanent partitions, framing, floors, roofs, columns and their fireproofing, and all other permanent stationary construction entering into a structure.

Dead load. The weight of all permanent construction including walls, floors, roofs, partitions, stairways, and of fixed service equipment.

Dead load. The constant weight of a building or structure including all inherent equipment; does not include variable live loads such as furniture, merchandise or people.

Dead load. Is the weight of the walls, permanent partitions, framing, floors, roofs and all other permanent stationary construction forming a part of the building.

Dead load. The weight of all permanent structural and non-structural components of a building, such as walls, floors, roofs, and fixed service equipment.

Dead load. The weight of materials forming a permanent part of the building, including permanent partitions.

Dead load. Weight of all permanent construction, including walls, framing, floors, roof, partitions, stairways, and fixed building service equipment.

Dead load. Materials, equipment, constructions, or other elements of weight supported in, on, or by the building (including its own weight) that are intended to remain permanently in place.

Dead loads. All buildings and structures, and parts thereof, shall be designed and constructed to support in addition to the minimum superimposed live loads required by the code, the actual dead weight of all component members; and in addition thereto, an allowance for the weight of partitions, ceiling and floor finishes, and concentrated loads such as safes, mechanical apparatus and similar equipment.

Design bearing pressure. Maximum allowable net pressure on soil or rock.

Design capacity. Load that a foundation is designed to transfer to the supporting soil or rock.

Design load. Total load which a structure is designed to safely sustain.

Design load. Total load which a building or structure or member is designed to sustain safely with exceeded specified deformation.

Design load. Total load which a structure is designed to sustain safely, includes live and dead loads.

Design properties. Properties of soil or rock used in proportioning and determining the design capacity of a foundation.

Duration of load. The period of continuous application of a given load, or the aggregate of periods of intermittent applications of the same load.

Earthquake load. The assumed lateral load acting in any horizontal direction on the structural frame due to the kinetic action of earthquakes.

Eccentric loads. Walls supporting eccentrically applied loads including eccentric loads produced by the deflection of floor and roof members shall be analyzed for stability and strength. Maximum unit stresses shall not exceed those specified in the code.

Equivalent uniform load. A conventionalized representation of an element of dead or live load, used for the purposes of design in lieu of the actual dead or live load.

Horizontal force. A horizontal force caused by wind pressure or earthquake effect.

Impact load. A kinetic load of short duration such as that resulting from moving machinery, elevators, craneways, vehicles, etc.

Impact load. Load resulting from the sudden application of live loads.

Impact load. The load resulting from moving machinery, elevators, craneways, vehicles, and other similar forces and kinetic loads.

Imposed load. All loads, exclusive of dead load, that a structure is to sustain.

Lateral soil load. The lateral pressure in pounds per square foot due to the weight of the adjacent soil, including due allowance for hydrostatic pressure.

Lateral soil load. The lateral pressure in pounds per square foot due to the weight of the adjacent soil, including due allowance for hydrostatic pressure, used in the design of retaining and foundation walls.

Liveload. Variable loads imposed on a structure by people, furnishings, merchandise, and equipment not inherent to the structure.

Liveload. Load imposed by the occupancy; it does not include the wind load.

Live load. The weight of all moving and variable loads that may be placed on or in a building such as snow, wind, occupancy, etc.

Live load. All loads except dead loads.

Live load. All live loads assumed for the purposes of design shall be the greatest loads that probably will be produced by the intended uses and occupancies; provided that the minimum live loads to be considered are uniformly distributed.

Live load. The weight superimposed by the use and occupancy of the building or structure, not including the wind load, earthquake load or dead load.

Live load. Any load other than a "dead load" or a "horizontal force."

Live load. All occupants, materials, equipment, constructions or other elements of weight supported in, on or by a building that will or likely to be moved or relocated during the expected life of the building.

Live load. The weight superimposed by the use and occupancy of the building, not including the wind load, earthquake load, or dead load.

Live load. The load in pounds per square foot superimposed by the use and occupancy of the building or structure, not including the wind load, earthquake load or dead load.

Live loads. Loads other than dead loads to be assumed in the design of the structural members of a building. It includes loads resulting from snow, rain, earthquake and those due to occupancy, including movable partitions.

Live loads. Weight imposed solely by the occupancy.

Live loads. The weight imposed on a structure by loads other than dead, wind, or impact loads or other than loads resulting from soil or hydrostatic pressure.

Live loads. All loads except dead and lateral loads.

Live loads. All buildings and structures, and parts thereof, shall be designed and constructed to support the minimum superimposed live loads uniformly distributed in pounds per square foot of horizontal area in addition to the dead loads.

Live loads. The planned loads on the structure must carry under normal conditions, such as people or furniture and equipment, that would be moved on the surface. These loads are generally assigned by the building code for the type of structure; for example, a heavy-equipment storage warehouse, a house, or an office building. Live loads are generally considered to be uniform loads.

Live loads. All imposed, fixed, or transient loads other than "Dead Loads."

Live loads. The load or weight to be supported on floors or other portions of buildings incidental to their occupancy; the pressure of wind, the weight of snow, and all loads other than dead loads.

Nonconcurrent loads. Two or more elements of dead or live load which for purposes of design, are considered not to act simultaneously.

Occupant load. The number of occupants of a space, floor or building for whom exit facilities shall be provided.

Racking load. Load applied in the plane of an assembly in such manner as to lengthen one diagonal and shorten the other.

Rebound. Recovery of displacement due to release or reduction of applied load.

Required live load. The live loads to be assumed in the design of buildings and structures shall be the greatest load produced by the intended use and occupancy, but in no case less than the minimum uniformly distributed unit loads required in the Code for specific uses.

Uniform live load. Average live load applied uniformly over a floor, roof or wall or along a beam or girder. Except for roofs, uniform live loads do not occur in residential construction, but are used for simplicity of calculations.

Uniformly distributed load. A conventionalized representation of an element of dead or live load as a load of uniform intensity, distributed over an area.

Wind load. Lateral pressure on a structure due to action of wind.

Wind load. The lateral pressure on the building or structure in pounds per square foot due to wind blowing in any direction.

1-D ENGINEERING TERMINOLOGY AND ABBREVIATIONS

a = conductance of air space. Thermal conductance of an air space; the amount of heat, expressed in BTU transmitted in one hour across an air space of one square foot area for each degree F. temperature difference.

Allowable soil pressure. The maximum stress permitted in soil of a given type and under given conditions.

Allowable stress. The maximum stress permitted at a given point in a structural member under given conditions.

Ambient. Near or in contact with.

Amplitude decrement factor. Difference between the maximum and the mean temperatures of a heat wave passing through a wall, roof or floor; dependent upon the thickness, mass, specific heat and orientation of the wall or roof.

Axis. Term used by architects and engineers to describe a center line.

Bearing surface. Contact surface between a foundation unit and the soil or rock upon which it bears.

Bearing value. Maximum allowable unit stress which may be applied to a soil without causing settlements damaging to a supported structure.

British thermal unit (B.T.U.). The amount of heat required to raise the temperature of one pound of pure water one degree F. to 60 degrees F.

British thermal unit. (BTU). A unit of measurement of the quantity of heat required to raise the temperature of 1 pound of water 1 degree F. The mean Btu is usually used, which is 1/180 of the heat required to raise the temperature of 1 pound of water from 32° F. to 212° F. at a constant atmospheric pressure of 14.69 psi.

British thermal unit (B.T.U.) The amount of heat required to raise the temperature of one pound of pure water one degree Fahrenheit at 60° Fahrenheit.

BTU. British thermal unit; approximately the heat required to raise one pound of water from 59 to 60 degrees Fahrenheit.

B.T.U. (british thermal unit). A measurement of heat, i.e., the amount of heat required to raise one pound of water one degree Fahrenheit.

BTU Rating. The listed maximum capacity of any apparatus, expressed in British Termal Units input per hour.

Cantilever. Structural member supported at one end only, or a structural member that projects beyond its support.

Centerline. Line dividing a space into two equal spaces.

Center to center. Measurement between centers of two adjoining parallel structural elements.

Condensation. Beads or droplets of water, and frequently frosted in extremely cold weather, that accumulates on the inside of the exterior covering of a building when warm, moisture-laden air from the interior reaches a point where the temperature no longer permits the air to sustain the moisture it holds.

Control joint. Spaces between the abutting portions of a building or structure to relieve longitudinal stresses due to contraction of materials.

C = Thermal conductance. Is applied to specific material as used, either homegeneous or heterogeneous, solid or gaseous, for the thickness of construction stated, not per inch of thickness and the time rate of heat flow is expressed in BTU's (per hour) (square foot) (Fahrenheit degree average temperature difference between (2) surfaces).

Decibel. A unit of measurement of the loudness of sound. A division of a logarithmic scale for expressing the ratio of two amounts of power or energy. The number of decibels denoting such a ratio is 10 times the logarithm of the ratio.

Density. Mass of a substance in a unit volume. When expressed in the metric system, it is numerically equal to the specific gravity of the same substance.

Factor of safety. Quotient obtained by dividing the amount of the breaking load or ultimate strength of a material or device when loaded in a certain manner, by the load under consideration for the material or device when applied in a similar manner.

Failure. Condition of distress in a structural member which upon removal of a load does not react to recovery of the member to its original physical shape and state and causes a permanent and substantial reduction in the load carrying capacity of the member.

F = film or surface conductance. The time rate of heat exchange by radiation conducted and convection of a unit area of surface with its surroundings and is usually expressed in BTU's (per hour) (square foot of surface) (Fahrenheit degree temperature difference).

Fixture unit. Measure of the hydraulic load of a fixture on a drainage system.

Fuel contributed classification. Comparitive measure of fuel contribution of a material or an assembly in the flame-spread test of ASTM E-84.

Gage. The term to express the thickness of metal sheets and wire. Sheet steel and iron shall be measured with the United States Standard Gage as published in the U.S. Bureau of Standards Circular No. 391, latest edition. The diameter of wire shall be measured with Steel Wire Gage (Washburn & Moen) as published in the U.S. Bureau of Standards Circular No. 67 latest edition.

Gradient. Slope, or rate of increase or decrease in elevation, or a surface, road or pipe, expressed in inches of rise or fall per horizontal linear foot or percent.

Incline. Degree of inclination with respect to the horizontal.

K = thermal conductivity. The time rate of heat flow through an homegeneous material under steady state conditions through unit area per unit temperature gradient in the direction perpendicular to the isothermal surface and is expressed in BTU's (per hour) (square foot) (Fahrenheit degree per inch of thickness).

Lateral force resisting system. That part of the structural system to which the lateral forces prescribed in the Code are assigned.

Negative gage pressure. Amount by which atmospheric pressure within a space is less than the atmospheric pressure immediately outside the space.

Perm. Unit measure of permeance of a material, paticularly vapor barriers. One perm indicates the passage of one grain of moisture per square foot per hour under a vapor pressure differential of one inch of mercury.

PSIG. Pounds per square inch gage.

Racking load. Load applied in the plane of an assembly in such manner as to lengthen one diagonal and shorten the other.

Radioactive materials. Nuclear radiation shall not be emitted to exceed quantities established as safe by the U.S. Atomic Energy Commission.

Rated input. That maximum amount of fuel expressed in Btu per hour which a heating appliance is designed to burn completely, and which is marked on any appliance meeting the requirements and specifications of the American Gas Association or other similar recognized agency.

Relative humidity. Amount of water vapor expressed as a percentage of the maximum quantity that could be present in the atmosphere at a given temperature. (the actual amount of water vapor that can be held in space increases with the temperature).

Residual deflection. Deflection resulting from an applied load, remaining after removal of such load.

Resistant. Used as a suffix (such as absorption-resistant, moisture-resistant, etc.) means material constructed, protected, or treated so that it will not be injured readily when subjected to the specific material or condition.

Restrained support. Flexural member where the supports and/or the adjacent construction provides complete or partial restraint against rotation of the ends of the member and/or partial restraint against horizontal displacement when subject to a gravity load and/or temperature change.

R = thermal resistance. The reciprocal of a heat transfer coefficient as expressed by U, C or f. It is expressed in Fahrenheit degree per BTU (hour) (square foot).

Saturation coefficient. Ratio of absorption by 24-hour submersion in water at room temperature to that after 5 hour submersion in boiling water.

Secondary stress. Stress resulting from causes other than the initial effects of design loads including, but not limited to stresses due to dimensional charges tending to occur in a structure because of the application of design loads or because of temperature changes or foundation settlements.

Seismic design. Construction designed to withstand earthquake forces.

Shear. Tendency produced by loads to deform a structural member by sliding one plane over another.

Simple support. Flexural member where the supports and/or the adjacent construction allows free rotation of the ends of the member and horizontal displacement when subject to a gravity load and/or a temperature change.

Slope ratio. Relation of horizontal distance to vertical rise or fall such as 2 feet horizontal to 1 foot vertical would be designated as 2 to 1 (2:1).

Sound power. The rate at which sound energy is radiated by a source.

Sound power level. The ratio, expressed in decibels, of the sound power of a source to the reference power of 10^{-12} watts.

Sound pressure level. The square ratio, expressed in decibels of a sound pressure to a reference pressure of 0.0002 dynes per square centimeter.

Specific gravity. Ratio of the weight of a body to the weight of an equal volume of water at 4° C. or other specified temperature.

Square. Unit of measure-100 square feet-usually applied to roofing material. Side-wall coverings are often packed to cover 100 squre feet and are sold on that basis.

Stress. The internal distributed force that resists the change in shape and size of a body subjected to external forces.

Stressed skin. A design in which frame and skin or sheathing are joined so that the skin may aid in resisting strains.

Structural failure. Rupture; loss of sustaining capacity or stability; marked increase in strain without increase in load; deformation increasing more rapidly than the increase in imposed load.

Terms (abbreviations).

bhp	brake horsepower
Btu	British thermal unit
C	centigrade
cfm	cubic feet per minute
cps	cycles per second
cu. ft.	cubic feet
db	decibel
dia	diameter
F	Fahrenheit
fpm	feet per minute
fps	feet per second
fsp	fire standpipe
ft.	foot
gal.	gallon
gypm	gallons per minute
gyps	gallons per second
h.p.	horsepower
hr.	hour
in.	inch
INR	impact noise rating
I.P.S.	iron pipe size
lb.	pound
mph	miles per hour
oz.	ounce
P.C.E.	pyrometric cone equivalent
pcf.	pounds per cubic foot
psf	pounds per square foot
psi	pounds per square inch
psia	pounds per square inch absolute
psig	pounds per square inch guage
rpm	revolutions per minute
sec.	second
swp	steam working pressure
sq. ft.	square foot
sq. in.	square inch
sq. yd.	square yard
STC	sound transmission class
Tag	Tagliabue
wwp	water working pressure

Thermal conductance. Time rate of heat flow through a unit area of a body, of given size and shape, per unit temperature difference. Value is expressed in Btuh per s.f. per degree F.

Thermal conductivity. Time rate of heat flow through a unit area of a homogeneous material under the influence of a unit temperature gradient. Value is expressed in Btuh per s.f. per degree F. per inch.

Thermal inertia. Property which modifies the effect of the U value on the heat transmission of a building element by expanding the time scale or time lag.

Time lag. Delay caused by heat storage and its subsequent release by the structure; increases as the mass of the wall increases.

Ton of refrigeration. The unit of capacity of refrigeration equivalent to the removal of heat of the rate of twelve thousand (12,000) B.T.U. per hour.

U = overall coefficient of heat transmission. As applied to the usual combinations of materials and also to single materials such as window glass and includes the surface conductance on both sides and is expressed in BTU's (per hours) (Square foot) (Fahrenheit degree temperature difference between air on the inside and air on the outside of a wall, floor, roof or ceiling).

"U" the overall coefficient of heat transmission. Amount of heat (BTU) transmitted in one hour per square foot of the wall, floor, roof, or ceiling for a difference in temperature of one degree F. between the air on the inside and outside of the wall, floor, roof, or ceiling.

Unit stress. Intensity of force acting on a structural member.

Urea-formaldehyde. Thermal setting synthetic resin used in producing rigid foams.

Urethane. Synthetic resin similar to polyurethane and used in making rigid foam.

Urethane and isocyanurate foamed plastics. Can be purchased as systems, comprised of two liquids that, when mixed, can form a rigid mass 30 times their original volume, or as rigid preformed materials. Acts as a sound attenuation agent for acoustical insulation. Urethane's 60 to 62 lbs. per cubic foot buoyancy affords a flotation medium ideal for marine use.

U = u=factor. Overall heat transmission co-efficient; the amount of heat, expressed in BTU transmitted in one hour through one square foot of a building section (wall, floor or ceiling) for each degree F of temperature difference between air on the warm side and air on the cold side of the building section.

U—value (transmission). The overall co-efficient of heat transmission through composite materials. The reciprocal of the (the sum of the R-values for each element between the inside and outside air). (BTU/hr./sq. ft./°F.).

Vapor. The gaseous form of substances which are normally in a solid or liquid state and which can be changed to these states either by increasing the pressure or decreasing the temperature alone.

Volatile fluid. A fluid which vaporizes easily.

Yield strength. Stress at which a material exhibits a specified limiting permanent set.

Yield stress. Maximum unit stress for which the unit strain is proportional to unit stress.

1-E MACHINERY AND EQUIPMENT

Building service equipment. The mechanical, electrical and elevator equipment, including piping, wiring, fixtures and other accessories, which provide sanitation, lighting, heating, ventilation, fire-fighting and transportation facilities essential for the habitable occupancy of the building or structure for its designated use and occupancy.

Building service equipment. The mechanical, electrical and elevator equipment, including piping, wiring, fixtures and other accessories essential to the sanitation, lighting, heating, ventilation, fire-fighting and transportation facilities for the designated use and occupancy of the building or structure.

Building service equipment. The mechanical, electrical, plumbing and elevator equipment, including piping, wiring, fixtures, and other accessories which constitute the sanitation, lighting, heating, ventilating, air conditioning, fire extinguishing, and transportation facilities provided or require in a building.

Equipment. Self-contained systems and apparatus attached to or built into the building and used for mechanical or electrical processing, comfort, safety, sanitation, communication or transportation within a building.

Equipment. Consists of all mechanical, electrical or storage devices and fixtures requiring conformance with the code in regard to construction, installation, operation, and alteration.

Equipment. A general term including material, fittings, devices, appliances, fixtures, apparatus and the like, used as a part of, or in connection with, an electrical installation.

Independent pole scaffold. A scaffold supported by multiple rows of uprights, and not depending on the building for support.

Machinery. Stationary steam boilers, stationary engines of every type, both steam and electrical, air compressors; refrigerating (not domestic) ice plants; elevators; air conditioning for cold, warm or hot air, steam and water heating; laundry and cleaning, all processes; sugar and syrup refinery, distilleries; textile and woodworking machines, or any other machines that are co-incident with and supported by or in the building.

Machinery. Includes but is not limited to stationary steam boilers, stationary engines of every type, both steam and electrical, air compressors; refrigerating ice plants; elevators and escalators; air conditioning for cold, warm, or hot air, steam and water heating; laundry and cleaning, all processes; sugar and syrup refineries, distilleries; textile and woodworking machines, or any other machines that are co-incident with and supported by or in the building.

Power-operated scaffold. Any form of scaffold that is propelled vertically by the use of power machinery.

Service equipment. Equipment, including all components thereof, which provides sanitation, power, light, heat, cooling, ventilation, air-conditioning, refuse disposal, fire-fighting, transportation, or similar facility for a building which by design becomes a part of the building, and which is regulated by the provisions of the Code.

Single pole scaffold. A platform resting on putlogs or crossbeams, the outer ends of which are supported on ledgers secured to a single row of posts or uprights, and the inner ends of which are supported by a wall.

Unsafe equipment. Whenever any doubt arises as to the structural quality or strength of scaffolding plank or other construction equipment, such material shall be replaced; provided, however, the building commissioner may accept a strength test to two and one-half (2-1/2) times the superimposed live load to which the material or structural member is to be subjected. The member shall sustain the test load without failure.

1-F MAINTENANCE AND REPAIRS

Access doors. Doors installed in walls, ceilings and other concealed spaces for the purpose of entering the space for maintenance.

Electrical repairs. Connection or disconnection of wiring, fixtures, lampholders, switches and other devices or equipment which are part of the fixed or permanent wiring in the building.

Maintenance. Replacing of a part or parts of a building, mechanical equipment, mechanical and electrical functions, which need replacements or repair due to ordinary wear or tear, or by the weather.

Maintenance. Plumbing drainage and sewage disposal systems shall be maintained in a sanitary and serviceable condition.

Maintenance. The replacing of a part or parts of a building, which have been made unusuable by ordinary wear or tear, or by the weather.

Maintenance and use. It shall be unlawful to maintain, occupy or use a building or structure, or part thereof, that has been erected or altered in violation of the provisions of the code, and no building or structure shall be occupied unless it is in a safe and habitable condition as prescribed in the code.

Maintenance electrician. Shall be held to mean a person who is qualified as prescribed as a master electrician but who must be regularly employed to maintain and make minor repairs to the electric wiring, apparatus and equipment which is installed, contained and used upon the premises or in buildings owned, occupied or controlled by the person, firm or corporation by whom the maintenance electrician is employed.

Mechanical repairs. Repairs to mechanical equipment that falls within the scope of the regulations shall be as set forth in the code.

Minor repair. The repair of not more than ten (10) square feet of the surface of a secondary structural member.

Minor repair. Renewal or repair of any part of a building in keeping with the existing type of construction and arrangement of parts; the occupancy of the building, for maintenance purposes when the structural parts of the building are not affected.

Rehabilitation. Rehabilitation of one or more properties to a satisfactory improved physical condition overcoming existing deterioration and aiding in the improvement of its neighborhood. Rehabilitation may include additional new construction, buildings or additions.

Repair. The replacement of existing work with the same kind of material used in the existing work, not including additional work that would change the structural safety of the building, or that would affect or change required exit facilities, a vital element of an elevator, plumbing, gas piping, wiring or heating installation, or that would be in violation of a provision of law or ordinance. The term "Repair" or "Repairs" shall not apply to any change of construction.

Repair. Replacement or renewal, excluding additions, of any part of a building, structure, device or equipment, with like or similar materials or parts, for the purpose of maintenance of such building, structure, device or equipment.

Repair. The renewal, replacement, or reinforcement of any existing part of a building, in keeping with its existing type of construction, arrangement of parts and occupancy, for maintenance purposes, including replacements and reinforcement because of fire damage and damage caused by the force of objects and the elements against a building.

Repair. The replacement of existing work with equivalent materials for the purpose of its maintenance; but not including any addition, change, or modification in construction, exit facilities or building service equipment.

Repair. The replacement of existing work with equivalent materials for the purpose of its maintenance; but not including additional work that would affect safety, or affect required exit facilities, or a vital element of an elevator, plumbing, gas piping, wiring, ventilating or heating installation.

Repair. Reconstruction or renewal of any part of an existing building for the purpose of its maintenance. The word "Repair" or "Repairs" shall not apply to any change in construction.

Repair. The replacement of existing work with equivalent materials for the purpose of its maintenance, but not including any addition, change or modification in construction, exit facilities, or permanent fixtures or equipment.

Repair. The replacement of existing work with similar materials for the purpose of its maintenance.

Repair. The reconstruction of part of a structure.

Repairs and replacements. Anything that does not require the alteration of its former condition.

Structural faults. Where during the course of alterations or repairs of a portion of an existing building, the involved or uncovered portion is discovered to be structurally unsound, the remaining portions of the building of similar construction shall be thoroughly investigated by a registered structural engineer. All portions determined to be unsound shall be made to comply with the applicable structural requirements of the Code.

1-G PERMITS, INSPECTIONS AND OCCUPANCY COMPLIANCE

Additional tests. The building inspector may require tests to be repeated, if at any time there is reason to believe that a material no longer conforms to the requirements on which its approval was based.

Bond. Means a written obligation or undertaking, under seal, whereby an applicant for or holder of a permit engages and agrees to indemnify the city for any loss, damage or injury resulting from his acts under such permit.

Building. Means any permanent or temporary stationary structure or enclosure, built above or below the ground.

Building permit required. It shall be unlawful to commence the excavation or filling of any lot for any construction of any building, or to begin construction of any building, or to commence the moving or alteration of any building or to commence the development of land for a use not requiring a building, until the building inspector has issued a building permit for such work.

Building permits. No person shall erect, construct, alter, repair, demolish, remove or move any building or structure, nor shall any person commence any sandblasting, liquid washing compressed air cleaning or steam cleaning of exterior surfaces of any building unless he has obtained a permit therefor from the Building Department. A separate permit shall be obtained for each separate building or structure except that a group of temporary structures erected on one site for a limited period of time may be included on one permit.

Building permits. Before proceeding with the erection, alteration or removal of any building, a permit shall first be obtained by the owner or his agent from the building department. The application shall be made in writing and upon printed forms furnished by the building department. To determine satisfactory compliance with requirements the application shall be accompanied by two (2) complete sets of plans and specifications conforming to the requirements of the Code. When plans and specifications shall be found to conform with the provisions of the Code, the building department shall issue a permit, and when a private water supply or sewage disposal system is necessary, he shall notify in writing the local health official. One copy of the plans and specifications shall remain on file with the records of the building department and shall be returned to the owner upon completion and final inspection of the building. The other set is to be stamped and kept at the construction site for reference until the completion of the building.

Building permits for construction. Permits for construction and use in accord with the provisions of the zoning ordinance and the building code, each application of which shall be deemed to be an application for a certificate of use or occupancy and compliance.

Certificate of occupancy. Certificate or statement, signed by the administrative Officer or Building Official, setting forth either that a building or structure complies with the Zoning ordinance or that a building or structure or parcel of land may be lawfully occupied for specified uses, or both.

Certificate of occupancy. No vacant land shall be occupied or used and no building erected, altered or moved shall be occupied until a certificate of occupancy shall have been issued by the Zoning Commissioner. Such certificate shall indicate that the building or premises or part thereof, and the proposed use thereof, are in conformance with the provisions of the ordinance.

Certificate of occupancy. No building hereafter constructed, erected, or altered under a zoning permit shall be occupied or used in whole or in part for any use whatsoever and no change in use of any building or part of building shall hereafter be made, until the builder, owner or occupant has been issued a certificate of occupancy by the zoning officer indicating that the building or use complies with the provisions of the ordinance. A certificate of occupancy shall be granted or denied within 10 days of the date of written application therefore. No fee shall be charged for a certificate of occupancy.

Certificate of occupancy required. No building permit, or permit for excavation for any building shall be issued before application has been made for certificate of occupancy.

Certificate of occupancy required. No vacant land shall be occupied or used, except for agricultural purposes, and no building hereafter erected or structurally altered shall be occupied or used until a certificate of occupancy shall have been issued within 3 days after the application has been made, provided such use is in conformity with the provisions of the ordinance.

Certificate of use and occupancy. The certificate issued by the building official which permits the use of a building in accordance with the approved plans and specifications and which certifies compliance with the provisions of law for the use and occupancy of the building in its several parts together with any special stipulations or conditions of the building permit.

Certificate of use and occupancy. A Certificate issued by the administrative official which permits the use of a building in accordance with the approved plans and specifications and which certifies compliance with the provisions of law for the use and occupancy of the building in its several parts together with any special stipulations or conditions of the permit.

Certificate of use or occupancy and compliance. The certificate shall certify that the subject use of any land or structure, in whole or part conforms to the requirements of the ordinance and of the building code, and shall be issued to legal, nonconforming uses in accordance with the provisions of the ordinance and shall certify such nonconforming status.

Chief building inspector. The head of the Building department or his authorized deputy.

Chief enforcement official. The chief of the division, or building inspector or commissioner of buildings in municipal corporations or counties having building departments, whichever has the jurisdiction.

Code compliance certificate. The certificate provided by the manufacturer to the State Building Inspector which warrants that the manufactured building or building component complies with the Code.

Combustible occupancy permit. Permit issued by the Fire department under the provisions of the municipal code, except that such permit when issued for refrigerating systems containing more than 20 lbs. of refrigerant, or for the transfer of carbonic acid to a container of lower pressure for use only by the operator at his own retail pressure for use only by the operator at his own retail soda water stand, shall not be considered as a combustible occupancy permit for purposes of the municipal code.

Conduct of tests. Tests shall be conducted under the supervision of the building inspector except that duly authenticated tests or certification by a competent person or laboratory may be accepted by him in lieu of tests under his supervision.

Conforming. Complying with the requirements of the applicable zoning ordinances and building codes.

Construction contrary to permit. It shall be unlawful for any owner, agent, architect, structural engineer, contractor, or builder engaged in erecting, altering, or repairing any building to make any departure from the drawings or plans, as approved by the Building department, of a nature which involves any violation of the provisions of the code on which the permit was issued.

Construction without permit. No person shall begin work for which a building permit is required or any work of excavation in preparation therefor until the permit has been obtained.

Coordinated inspections. All departments involved in the issuing of permits shall be responsible for the scheduled inspections of all buildings or structures relating to the operation, equipment, fire protection, housekeeping, maintenance, safety and sanitary conditions of the use and occupancy of all buildings or structures and shall report all violations to the Administrative Official for further action.

Cost. The cost of new construction, alteration or repairs may be evidenced by a contract arrived at by receipt and acceptance of sealed bids, or in the absence of such evidence may be determined, computed or estimated by the Administrative Official in accordance with generally accepted methods of estimating prescribed by the approved rules, as for example valuation of listings of buildings from nationally accepted organizations such as Marshall and Stevens or published cost data by R.S. Means.

Cost estimate. Estimate used for the purpose of fixing the permit fee shall include the estimated value of the building including the value of painting, plumbing, cooling, heating, electrical work, and the estimated value of all materials and labor entering into the building or structure, but not paving outside the building, nor fixtures or specialized equipment.

Cost estimated. For the purpose of fixing permit fees as required by the code, means the estimated total value of a building including the value of painting, plumbing, cooling and heating, electrical work, and the estimated value of all materials and labor entering into the building or structure, but not paving outside the building, nor fixtures, nor specialized equipment. The estimated cost shall include the reasonable market value of all the components of such building or structure whether or not the material was obtained without cost, or at reduced cost, and whether or not the builder performs all or part of the labor himself without actual outlay of money.

Cost of alterations or repairs. Cost of alterations or repairs shall be construed as the total actual combined cost of such alterations or repairs made within a specific period of time, including the value of rebates, discounts, free services and other gratuities not normally granted or received in such operations.

Cubic content. Cube or cubage of a building is the actual cubic space enclosed within the outer surfaces of the outside or enclosing walls and contained between the outer surfaces of the roof and 6 inches below the finished surfaces of the lowest floors. It includes, but is not limited to, the volume contained in the basement or cellar, dormers, penthouses, vaults, roofed porches, and enclosed appendages, but does not include the volume contained in courts and light shafts which have no roof, or the volume contained in outside steps, cornices, and parapet walls.

Cubic content. The area within the outside limits of the exterior walls of a building, multiplied by the average height of the portion considered. Average height shall be the height from the top of the lowest floor to the mid-height of sloping roofs or the top of flat roofs. Open porches, decks and balconies may be ignored, but enclosed porches, sun porches, and similar permanent finished rooms shall be computed, at full volume as for the main building.

Cubic content. Is the cube or cubage of a building and is the actual cubic space enclosed within the outer surfaces of the outside or enclosing walls and contained between the outer surfaces of the roof and 6 inches below the finished surfaces of the lowest floors and includes, but is not limited to, the volume containing in dormers, penthouses, vaults, enclosed appendages, and one-half the volume contained in porches, but does not include the volume contained in courts and light shafts which have no roof, or the volume contained in outside steps, areaways, cornices, and parapet walls.

Department of safety and permits. Receives, processes and issue permits in accord with the provisions of the ordinance and any other applicable regulations.

Disregard of unsafe notice. If a person served with a notice or order to repair or remove an unsafe building or structure or portion thereof and should fail, within a reasonable time, to comply with the notice or order, the building department shall advise the proper department of the facts in the case and the City shall institute an appropriate action in the courts to compel compliance.

Disregard of violation notices. In case a violation notice or order is not properly complied with, the Building department shall notify the proper department of such noncompliance and the City upon such notice shall institute an appropriate action or proceeding at law or in equity, to restrain, correct or remove such violation, or the execution of work thereon, or to restrain or correct or the erection or alteration of, or require the removal of, or to prevent the occupancy or use of the building or structure erected, constructed, added to or altered, in violation of, or not in compliance with the provisions of the code.

Driveways. No permit shall be issued for the construction, erection, repair, or alteration of any building or structure designed or intended for use as a garage or any other business, the operation of which will require a driveway across a public walk, until the applicant therefor has first obtained a permit as prescribed by the code.

Enforcement officer. Person lawfully empowered to enforce the requirements and regulations of the code.

Exceptions. A permit shall not be required for any minor repairs, as may be necessary to maintain existing parts of buildings, but such work or operations shall not involve the replacement or repair of any structural load-bearing members, nor reduce the means of exit, affect the light or ventilation, room size requirements, sanitary or fire-resistive requirements, use of materials not permitted by the building provisions of the code, nor increase the height, area, or capacity of the building.

Expiration of permit. A permit for work for which no work is commenced within 6 months after issuance shall expire by limitation and a new permit secured before work is started. A permit may be revalidated depending on the cause of the delay.

Foundation inspection. A foundation inspection shall be made after the foundation is poured or set in place, prior to back-filling and framing.

Framing or rough-in inspection. The inspection shall be made after all framing, masonry walls, or fireplace vents and chimneys are completed, including the roof structure, fire blocking, wall bracing, sheathing, heating and cooling ductwork, or other appurtenances and accessories which may be concealed, and after plumbing, electrical, and fire rough-in inspections have been posted by those inspection departments. No mechanical, electrical or plumbing systems which are to be concealed shall be covered before this inspection has been made and approved by the Building Official.

Inspecting engineer. Registered professional engineer, qualified by training and experience, who is required under the code to be employed by the owner for a specific purpose subject to approval by the building department and by the registered professional engineer or registered architect whose seal appears on the drawings or computations submitted with the building permit application.

Inspection. When inspection of any construction operation reveals that any unsafe or illegal conditions exist, the building commissioner shall notify the owner and direct him to take the necessary remedial measures to remove the hazard or violation.

Inspectors. All officials, officers, or employees of the City entrusted with the enforcement of the Code.

Insurance inspectors or special inspectors. Inspectors of mechanical equipment, such as air-conditioning and cooling towers, boilers, unfired pressure vessels, hot water heaters, elevators and escalators, shall apply for a certificate of competency issued by the Mechanical Inspection Department, Public Works Division.

Issuance of a permit. The building department shall notify the applicant that his plans have been examined and are or are not approved for permit, if approved the applicant shall within a given period of time secure the plans and pay the permit fees, if not approved the applicant can review the plans with the reviewers and if necessary make all of the revisions to the plans as directed in accordance with the provisions of the code and then secure the permit by paying the permit fees.

Notices of violations. Whenever the Building Department is satisfied that a building or structure, or any work in connection therewith, the erection, construction, addition or alteration, execution of which is regulated, permitted or forbidden by the Code, is being erected, constructed, added to or altered, in violation of the provisions or requirements of the code or in violation of a detailed statement or plan submitted and approved thereunder, or of a permit or certificate issued thereunder, he shall serve a written notice or order upon the person responsible therefor directing discontinuance of such illegal action and the remedying of the condition that is in violation of the provisions or requirements of the code.

Obstruction of streets. Before any permit shall be granted to any person, for the obstruction of any street or sidewalk, an estimate of the cost of restoring said street and sidewalk, to a condition equally as good as before it was obstructed with a fair additional margin for contingent damages, shall be made by the Department of Streets. The deposit shall be made at the time the permit is issued and any balance left on deposit after the street and sidewalk has been repaired will be returned to the owner.

Occupancy certificate. Official statement certifying that a building, other structure or parcel of land is in compliance with the provisions of all applicable codes, or is a lawfully existing noncomforming building or use and hence may be occupied and used lawfully for the purpose designated thereon.

Occupancy certificate. It shall be unlawful to use or permit the use of any building or premises or part thereof hereafter created, erected, altered, changed, or converted wholly or partly in its use until a certificate of occupancy is issued by the Building Department, to the effect that the building or premises or the part thereof so created, erected, altered, changed or converted and the proposed use thereof conforms to the applicable provisions of the code.

Permit. A written authorization by the Building Official to proceed with construction, alteration, repair, installation or demolition.

Permit. Authorization by the Commissioner of Buildings to proceed with construction, alteration, installation or demolition.

Permit. Permission or authorization in writing by the authority having jurisdiction to perform work regulated by the Code and, in the case of an occupancy permit, to occupy any building or part thereof.

Permit. Authorization by the Building Department to proceed with any excavation, construction or demolition regulated by the Code.

Permit application. Applications for building permits shall be in such forms as shall be prescribed by the Building Department. Every such application for a permit shall be accompanied by a copy of every recorded easement on the lot on which the building is to be erected, and on the immediately adjoining lots, showing the use or benefit resulting from such easement. All such applications shall be accompanied by drawings, plans, and specifications in conformity with the provisions of the Code. Where alterations or repairs in buildings are made necessary by reason of damage by fire, the permit shall not be issued until the Department shall cause a thorough inspection to be made of the damaged premises with the view of testing by the Building Department for the construction, erection, addition to or alteration of any building or structure unless the applicant therefor shall furnish to the said commissioner a certificate or other written evidence of the proper federal officer or agency that the proposed construction is not prohibited by any order, rule or directive of an agency of the United States Government.

Permit fees. No permit as required by the Building Code shall be issued until the fee prescribed in the ordinance shall have been paid. Nor shall an amendment to a permit be approved until the additional fee, if any, due to an increase in the estimated cost of the building or structure, shall have been paid.

Permit fees. Fees for the issuance of permits for new buildings, alterations, and other structures and other permits shall be payable to the Director of Revenue when such permits are issued, as required by the Code.

Permit required. It shall be unlawful to proceed with the erection, enlargement, alteration, repair, removal, or demolition of any building, structure, or structural part thereof within the city unless a permit therefor shall have first been obtained from the Building Department. Such permit shall be posted in a conspicuous place upon the exterior of the premises for which it is issued, and shall remain so posted at all times until the work is completed and approved.

Permit to move building. No building or structure shall be moved until a permit has been obtained from the Building Inspector, and no permit shall be issued for moving a building or structure over a public highway or other public open area controlled by the city to a new location unless the application for the permit is accompanied by formal approval and permission of the Public Works Department, and over a state controlled highway from the State Department of Public Works and Highways.

Pier inspection. Where special foundations are required such as drilled and poured in place concrete piers, driven piling of all types, caissons, and other extraordinary types, the Building Official shall make at least one inspection and more if the size of the job warrants it. Reinforcing required in the above cases shall be set to allow adequate inspections.

Posting of building permit. Copy of the permit shall be posted in a conspicuous place on the premises during the prosecution of the work and until the completion of the same.

Prohibition. No person shall commence or continue any of the work concerned with the construction, alteration, or addition thereto, unless the owner of the building has obtained a building permit.

Records. The books containing the records of the department of Building Inspection relating to building permits shall be open to public examination at all reasonable times. Plans, specifications, statements or other papers relating to any application shall be open to the Building Official and his subordinates but are not public records. The Building Official shall not allow anyone except the owner of the property or his agent to copy any plan or specification on file in the department of Building Inspection, nor shall be allow the removal of any plan or specification from the office except for the purpose of another City department or pursuant to court order.

Records. It shall be the duty of the Building Department Inspector to keep a record of all building permits and certificates of occupancy issued, and copies shall be furnished on request to any person having a proprietary or tenancy interest in the building or land involved.

Reproduction cost. The reproduction cost of a building or structure shall be the estimated cost of constructing a new building of like size, design and materials at the site of the original structure, assuming such site to be clear.

Restoration of unsafe building or structure. Building or structure or part thereof declared unsafe by the Building Inspector may be restored to safe condition, provided that if the damage or cost of reconstruction or restoration is in excess of 50 percent of the value of the building or structure, exclusive of foundations, such buildings or structures, if reconstructed or restored, shall be made to conform to the requirements, of the Code with respect to materials and type of construction, but no change of use or occupancy shall be compelled by reason of such reconstruction or restoration.

Restoration of use. When an accident involves the failure or destruction of any part of the system, operating mechanism, or of the structure housing the equipment, the re-use of the installation shall be unlawful until it has been made safe, and approval shall be secured for any installation requiring a permit under the provisions of the Code.

Revocation. If the work in, upon, or about any building or structure shall be conducted in violation of any of the building provisions of the code, it shall be the duty of the Building Official to revoke the permit for the building or wrecking operations in connection with which such violation shall have taken place. It shall be unlawful, after revocation of such permit, to proceed with such building or wrecking operations unless such permit shall first have been reinstated or reissued by the Building Official. Before a permit so revoked may be lawfully re-issued or reinstated, the entire building and building site shall first be put into condition corresponding with the requirements in the building provisions of the code, and any work or material applied to the same in violation of any of the provisions shall be first removed from such building, and all material not in compliance with the building provisions of the code shall be removed from the premises.

Revocation of permits. The Building Official shall revoke any permit whenever there has been a false statement, or misrepresentation in the application as to a material fact on which the permit was based, or whenever the permit was issued in error and conditions are such that a permit should not have been issued.

Soil inspection. Shall be made after excavation for structures is complete and trenches for footings, column pads, spread footings, or other types of footings are ready for concrete. No concrete is to be poured without this inspection.

Statement of occupancy. Written certification by the Department that a structure conforms to the requirements of the code.

Stopping work. Whenever in the opinion of the Building Department by reason of defective or illegal work in violation of a provision or requirement of the Code, the continuance of a building operation is contrary to public welfare, he shall order, in writing, all further work to be stopped and may require suspension of all work until the condition in violation has been corrected.

Stop work order. The Building Official may issue a stop work order with respect to any building, structure, lot or parcel upon a finding that work on any building or structure is being prosecuted contrary to the provision of the Building Code or the Ordinances in force.

Structural damage. Loosening, twisting, warping, cracking, distortion, or breaking of any piece, or of any fastening or joint, in a structural assembly with loss of sustaining capacity of the assembly. The following shall not be deemed to constitute structural damage; small crack in reinforced concrete perpendicular to the reinforcing bars; or deformation of material when a structural assembly is under applied load which increases as such load increases, but which disappears when such load is removed.

Surety bond. Before any building permit is issued the applicant shall produce evidence that he has filed with, and had approved by, the Department of Streets, a surety bond in an amount stated in the Code protecting the City against any and all damages that may arise to the public ways upon which such building abuts, and to the City and to any person, in consequence or by reason of, the proposed operation to be authorized by such permit, or by reason of any obstruction or occupation of any public ways in and about such building operations.

Time limitation of permit. Any building permit issued shall become invalid unless the work authorized by it shall be commenced within six (6) months of its date of issue, or if work authorized by it is suspended or abandoned for a period of one (1) year.

Time limits. If, after a building or other required permit shall have been granted, the operations called for by such permit shall not be begun within six (6) months after the date thereof, such permit shall be void and no operations thereunder shall be begun or completed until an extended permit shall be taken out by the owner or his agent, Two (2) extensions only shall be granted and if work is not begun within eighteen (18) months after the date of issuance of the original permit, all rights under the permit shall thereupon terminate by limitation. Where, under authority of a permit, or extended permit, work has begun and has been abandoned for a continuous or cumulative period of twelve (12) months, all rights under such permit shall thereupon terminate by limitation.

To place. The putting of a building or structure in a particular situation, whether this is by original construction or erection or by moving a building or structure to a particular situation.

Unlawful continuance. A person shall not continue any work in or about a building, structure, lot or parcel of land after having been served with a stop order. A person who shall be found guilty of such unlawful continuance shall be punished as set forth in the Code.

Unsafe building. Building which is a hazard to the safety or health of the occupants of the building, or of others, because it is structurally unsafe, or is unstable, or is unsanitary, or is inadequately provided with exit facilities, or is a fire hazard, or is inadequately maintained, or is provided with unsafe building service equipment, or because it is otherwise dangerous to its occupants or to public health or safety.

Unsafe structures. Any structure or portion thereof declared unsafe by a proper authority may be restored to a safe condition.

Use of water. Before the Building Department issues a permit, as provided in the Code, the applicant shall pay in advance for the water to be used in the work under such permit in accordance with the provision of the Code.

Valuation or value. As applied to a building, means the estimated cost to replace the building in kind, at date of estimate.

Value. As pertaining to a building or other structure, the estimated, present cost to replace the building or structure in kind but not including the value or cost of the foundation or land.

Value. Value of a building shall be the estimated cost to replace the building in kind, at the date of the estimate.

Value or valuation. Of a building shall be the estimated cost to replace the building in kind, at date of estimate.

Wrecking bond. Before any permit is issued granting authority to wreck a building or structure for which such permit is required, the person engaged in the work of wrecking the same shall file with the city clerk a bond with sureties to be approved by the city comptroller to indemnify, keep and save harmless the city against any loss, cost, damage, expense, judgment, or disability of any kind whatsoever which the city may suffer, or which may accrue against, be charged to or be recovered from said city, or any of its officials from or by reason or on account of accidents to persons or property during any such wrecking operations, and from or by reason or on account of anything done under or by virtue of any permit granted for any such wrecking operations.

Wrecking by government authorities. The administrator of Public Works of the United States or such other authority as may be created by acts of Congress with power to co-operate with the city in the making of public improvements, the department of public works, the department of streets and electricity, and the fire department may engage in the work of wrecking of buildings and structures, and in such cases where any of these agencies make application for a permit to wreck buildings or structures, the Building Department shall issue such permit without the fee and shall not require the filing of a bond with sureties as provided in the Code.

Wrecking permit. Before proceeding with the wrecking or tearing down of any building or other structure, a permit for such wrecking or tearing down shall first be obtained by the owner or his agent from the Building Department and it shall be unlawful to proceed with the wrecking or tearing down of any building or structure or any structural part of such building or structure unless such permit shall first have been obtained. Application for such permit shall be made by the owner, or his agent, to the Building Department who shall issue said permit upon such application and the payment of the fee.

Written notice. The Building Official may, by written notice to the owner or developer of the property or holder of building permits for such property, at the address reflected by the records, specifically list the violations found and require the person to whom such notice is sent to show cause before the Council why a "stop work order" requiring such person to cease and desist from the to correct such violations should not be issued.

Written notice. Shall be considered to have been served if delivered in person to the individual or to the parties intended, or if delivered at or sent by certifed mail to the last business address known to the party giving the notice.

Written notice. Shall be any notice delivered or served in writing, either in person to the individual or to the parties intended, or forwarded by registered mail to the last known address of the party to be served.

Written notice. Shall be considered to have been served if delivered at or sent by registered mail to the last business address known to the party giving the notice.

1-H PLANS, SPECIFICATIONS, SURVEYS, PLATS AND MAPS

Accepted engineering practice. Accepted engineering practice shall be in compliance with the building provisions of the code or, in the absence of such provision, with the standards of governmental bureaus or authoritative technical organizations.

Accepted engineering practice. Conforming to approved published standards, and conforming to a standard, method or test proven by performance to be adequate for an intended purpose.

Accepted engineering practice. That which conforms to principles, tests, regulations, specifications, or standards presently accepted or approved by the technical or scientific organizations or authorities listed in the various appendices to the code.

Accepted engineering practice. That which conforms to accepted principles, tests or standards of nationally recognized technical or scientific authorities.

Accepted engineering practice. The use of materials and equipment of a quality meeting the requirements and standards of authoritative agencies recognized as such by building officials, practicing architects and engineers and a manner of design or method of construction having an established record of satisfactory performance and commonly recognized as such by building officials, practicing architects and engineers.

Approval of changes on drawings. When it is necessary to change approved heating and ventilating drawings or specifications, revised drawings shall be approved before installation is commenced.

Approval of drawings and specifications. Where heating, ventilating and air conditioning equipment is required, complete drawings, specifications, and data sheets shall be submitted to the department of Building and human relations for approval. Approval shall be obtained before affected work is commenced and all work shall be executed according to the approved drawings and specifications.

Approval of plans. All drawings and plans for the construction, erection, addition to, or alterations of any building or other structure, for which a permit is required shall first be presented to the Building department for examination and approval as to proper use of building and premises and as to compliance in all other respects with the Zoning ordinance and shall thereafter be presented to the Board of Health, the Department of Smoke Inspection, Fire Department, Department of Water and Sewers, Department of Streets and Sanitation, Department of Boiler Inspection, and Department of Public Works for submission to the proper official of these departments and bureaus for their examination and approval with regard to such provisions of the code, as are within the duty of such office to enforce, and after said drawings and plans have been examined and passed upon, the same shall be returned to the Building Department where they shall be taken up for examination and approval by Commissioner of Buildings.

Approved drawings kept at building. A complete set of approved drawings shall be kept available at the job site.

Architect. Person who is legally qualified to practice the profession of architecture.

Architect. Person licensed under the provisions of the statute and entitled thereby to conduct a practice of architecture in the state of registration.

Architect. Person holding a certificate and is registered pursuant to the state statutes.

Architect. Within the meaning of the Code, shall be deemed to be a duly registered and licensed architect.

Architect or registered architect. A person technically qualified and professionally licensed by the State to practice architecture.

Complete plans. Plans consisting of floor plans, elevations, sections, structural plans and details, building service equipment plans and details, and such graphic representations, diagrams, and delineations as are necessary in the design of a building to show its size, construction, location, design, design live loads, unit stresses used in its structural design, foundation data, and other pertinent information as is necessary to determine that the building, when constructed in compliance with said plans, will comply with the code of the city pertaining to location, safety and health.

Complete plans. Includes general plan, detailed plans and specifications, and engineering reports.

Compliance approval. Approval as being in compliance with the provisions of the building code.

Compliance required. No building, structure or land shall be used, and no building, structure or part thereof shall be erected, converted, altered, moved, demolished or removed, except as in conformity with the ordinance.

Comprehensive development plan. Includes a comprehensive set of plans, specifications and measures for the private and/or public development of an industrial park, cluster development, apartment project, shopping center or other planned development permitted in the ordinance. The development plan shall include a site plan showing the location of streets, pedestrian ways, rail lines, utility systems, landscaped areas, parcel lines, building areas, exits and entrances to be provided; any restrictions to be included in the sale or lease of land, for parking, building location, property maintenance, sign control and any other protective measures; schedule for the development of streets, grading, utility installations, rail facilities, docking facilities or other improvements to be provided for the project area and occupants thereof; and a statement of intent to proceed and financial capability of the developer or sponsor.

Comprehensive development plan. The allocation of land areas to the several varieties of physical development, present and future, of the regulated area, the same having been prepared in accordance with the principles of comprehensive planning or having been developed through the approval of subdivisions previously submitted, wherever such plan exists and has been officially adopted by the Planning Commission and recorded; wherever the term "Development Plan" is used, it shall have the same meaning as the term "Master Plan" or "Comprehensive Plan."

Comprehensive plan. Includes charts, maps, descriptive matter officially adopted by a planning commission or governing body showing among other things recommendations for the most appropriate use of land; for the most desirable density of population; for a system of thoroughfares; parkways and recreation areas; for the general location and extent of facilities for water, sewer, light and power; for the general location, character and extent of community facilities.

Controlled construction. The construction of a building or structure or a specific part thereof which has been designated and erected under the supervision of a licensed or registered engineer or architect using controlled materials as defined in compliance with accepted engineering practice under the procedure of the Code.

Design plan. Plan prepared by the City for implementing components of the General Plan and may include, but is not limited to the design, bulk, use, height, location and arrangements of buildings in respect to streets, open spaces, other structures and natural features.

Designing engineer. Registered, professional engineer who is duly and regularly registered by the State.

Detailed mechanical plans. Plans showing each feature of the sewer or sewage works design and including elevations, grades, profiles, sections, all special designs and structures, typical and special manholes, lamp holes, flush tanks, intersections and stream crossings, pumping stations, catch basins, storm water inlets and all like appurtenances.

Development plan. Dimensioned presentation of the proposed development of a specified parcel of land which reflects thereon the location of buildings, easements, parking arrangement, public access and street pattern and other similar features.

Drainage plan. Plan showing all proposed and existing facilities to collect and convey surface drainage, described by grades, contours and other topographical data.

Elevation. Scale drawing of the front, rear, and sides of a building or structure.

Engineer. Person deemed to be a duly registered and licensed engineer in conformance with the State laws.

Engineering practice and standards. Includes, design and quality, use and installation of all materials and equipment or systems, and shall conform to the standards of accepted authoritative agencies and practices as listed in the code.

Engineer's report. A comprehensive report describing the project, the basis of the design, together with design data, and all other pertinent data necessary to present an accurate understanding of the work to be undertaken and the reasons for the same.

Final development plan. Final plan prepared by a developer based on the approved preliminary plan of a proposed development or development area which consists of detailed drawings, specifications, cost estimates and agreements for the construction of the site improvements and buildings for the proposed development or development area.

General development plan. Description of the development proposed within a particular Planned Community District includes map showing the location and arrangement of all proposed uses, and a written statement of the general regulations proposed to govern them.

General plan. Plan and statement of the objectives and recommendations for the general location and extent of desirable future land development, community facilities and street plans for the City, duly adopted or officially accepted.

Grid survey (entire site). Elevations on a rectangular grid section and at points where definite breaks in grade occur. Grid lines 50 ft. or 100 ft. unless otherwise specified and extended 10 feet beyond all property lines.

Landscape architect. Professional landscape architect licensed by the State.

Map. A drawing showing geographic, topographic or other physical features of the land.

Master Plan. Comprehensive plan including graphic and written proposals indicating the general location for streets, alleys, parks, schools, public buildings and all physical development of the City, and includes any unit or part of such plan, and any amendment to such plan or parts thereof. Such plan or part thereof may or may not be adopted by the Planning Board.

Mechanical Drawings, specifications and data. All drawings and specifications and associated calculations and data for the installation of heating, ventilating and air conditioning systems shall be designed and prepared to satisfy the requirements of the code. All drawings, specifications, calculations and associated design data shall be submitted for review and approval under the provisions of the code and shall be sealed or stamped by an engineer or architect registered in accordance with the laws of the state.

Metes and bounds survey. System of a land survey and description based on starting at a known reference point and tracing the boundary lines around an area.

Minimum building plans. Plans shall include drawings of floor plans of all habitable floors and the basement or foundation plan and such drawings shall clearly indicate sizes and spacings of all supporting members, sizes of rooms, glass areas, sizes of all footings and reinforcing, thickness of basement or foundation walls and the reinforcing, and all floor slabs with or without reinforcing, exterior and interior wall construction, sizes and spacing of all framing members, ceiling heights, and parapet wall heights.

Plan. Horizontal cross-section of a structure at any level showing room arrangement, location of doors, windows, etc. Also may show the site surrounding the building and objects thereon.

Plan. Drawing of a proposed design or of work to be performed.

Plans. Include drawings, specifications, and data submitted to the chief enforcement official in accordance with the Building Code for approval.

Plans and specifications. Plans for all buildings and structures other than one-and-two-family and multi-family dwellings, which are designed for human occupancy shall designate the number of occupants to be accommodated in the various rooms and spaces and when means of artificial lighting and ventilation are required, the application shall include sufficient details and description of the mechanical system to be installed as herein required or as specified in the Code.

Plans and specifications. One or more drawings or documents indicating and describing the amount, arrangement, kind and quality of the materials to be used for the construction of a building or structure.

Plat. Map, plan or layout of a Township, City, Village, Section or Subdivision, or any part thereof, including the boundaries of individual properties.

Plat. Map of a lot, parcel, subdivision or development area on which the lines of each element are shown by accurate distances and bearings.

Plat. Map or plan of a subdivision or land development.

Plat. Sketch, map or survey of a lot(s), tract or parcel of land including lot lines, street rights-of-way, and easements with the dimensions of these features inscribed thereon.

Plat. Map, plan or chart of a City, Town, Section or Sub-division, indicating the location and boundaries of individual properties.

Plat required. All applications for building permits shall be accompanied by a plat in duplicate of a dimensioned sketch or to-scale plan signed by the owner or his authorized agent, showing the actual dimensions of the lot to be built upon, the location and size of the building or structure to be erected, the location of adjoining or surrounding buildings or structures, and such other information as may be required by the Building Inspector, which is necessary to provide for the enforcement of the Ordinance.

Preliminary development plan. Drawing prepared by a developer, which may include explanatory exhibits and text, submitted to the designated authority for the purpose of study of a proposed development of land, or a preliminary plan of land use of a development area which, if approved by the designated authority, provides the basis for proceeding with the preparation of the final plan of a development or development area.

Preliminary subdivision plan. Complete and exact subdivision plan to define property rights and proposed streets and other improvements presented for purposes of securing preliminary approval.

Professional engineer. Means a person holding a certificate of registration in the State.

Professional engineer or architect. Individual technically and legally qualified and registered to practice the profession of engineering or architecture.

Record plan. Exact copy of the approved final plat, reproducible of standard size prepared for necessary signatures and recording with the County Recorder of Deeds.

Registered architect or registered engineer. Architect or professional engineer registered to practice as such in the State.

Registered contractor. Any contractor duly registered by the State in accordance with the State registration laws.

Seal required. Plans and specifications must bear the seal and signature of a registered architect or engineer in conformity with and when required by the statutes of the State.

Site map for project. Drawings shall include a site map drawn to scale, adequately dimensioned, clearly showing the exact location of all structures existing or to be constructed. When a private water supply or sewage disposal system is necessary, the site map shall show the location of proposed well, septic tank and disposal field in addition to existing wells, septic tanks, sewer lines, drains, sewage disposal fields, seepage pits, privies and cesspools within 100 feet of the dwelling.

Specifications. Technically describing the quality, kind, and grade of material and equipment may be required if not clearly shown on drawings.

Stamped plans on job. The stamped set of plans and specifications issued to the applicant shall be kept at the site of the construction or work, and shall be available to the authorized representative of the Building Department. There shall be no deviation from the stamped or approved application, plans, or specifications without official approval.

Surveyor. Duly registered and licensed surveyor or Civil Engineer.

System of rectangular surveys. Land survey system based on geographical coordinates of longitude and latitude originally established by acts of congress to survey the lands of public domain and is now used in 30 states.

Working drawing. Scale drawing showing dimensions, elevations, sections, and construction details.

1-1 PROJECT PREPARATIONS

Alteration. Any addition, or change or modification of a building, or the service equipment thereof, that affects safety or health and that is not classified as a minor alteration or ordinary repair. The moving of a building from one location or position to another shall be deemed an alteration.

Approved material, equipment and methods. Approved by the building official or by a recognized authoritative agency.

Approved rules. The legally adopted rules of the building official or of a recognized authoritative agency.

Beginning of construction. Incorporation of labor and materials on the site, tract or lot where a building or structure is proposed to be constructed; the incorporation of labor and material within the walls of a building where repairs or remodeling are proposed to be made; the incorporation of labor and materials at the site, lot or parcel where land is to be used for purposes other than construction of a building.

Blasting cap. Means a cap or detonator with wires attached for exploding the same by means of electricity.

Blasting powder. Means an explosive substance composed of sulphur, charcoal and sodium nitrate, specially prepared for the purpose of blasting.

Catch platform. A platform or other construction projecting from the face of a building, supported therefrom, and used to intercept the fall of objects and to protect individuals and property from falling debris.

Construction. Any or all work or operations necessary or incidental to the erection, demolition, assembling, installing, or equipping of buildings, or any alterations and operations incidental thereto. The term "construction" shall include land clearing, grading, excavating and filling. It shall also mean the finished product of any such work or operations.

Construction equipment. Includes such construction equipment and machinery as tools, derricks, cranes, hoists, scaffolds, platforms, runways, ladders, dozers, drilling equipment, and all material handling equipment safeguards and protective devices used in construction operations.

Construction equipment. The construction machinery, tools, derricks, hoists, scaffolds, platforms, runways, ladders and all material handling equipment safeguards and protective devices used in construction operations.

Construction operation. Includes erection, alteration, repair, renovation, demolition or removal of any building or structure, and the excavation, filling, grading and regulation of lots in connection therewith.

Construction operation. The erection, alteration, repair, renovation, demolition or removal of any building or structure; and the excavation, filling, grading and regulation of lots in connection therewith, in accordance with the Code.

Construction shed. A temporary building incidental to the construction of another building for which a permit has been issued and which is removed within 30 days after the completion of the building for which the permit was issued.

Construction site security. The owner, contractor, or responsible party constructing a new building, shall maintain security measures as deemed necessary or as required by the Building department safety superintendent to control vandalism, fires and other mischievous and deliberate acts of destruction.

Constructor. Person who contracts with the owner of a project for the work thereon and includes an owner who contracts with more than one person for the work on a project or undertakes the work on a project or any part thereof.

Contractor. A licensed person, company, or corporation who are registered and in good standing in the records for performance; who are financially able to provide labor, materials and equipment to erect, alter and demolish buildings or structures or to provide such services to an owner in accordance with the requirements of the Licensing Act.

Controlled construction. Construction of a building or structure or a specific part thereof which has been designed and erected under the supervision of a licensed or registered architect or engineer using controlled materials as herein defined in compliance with accepted engineering practice under the requirements of the code.

Controlled materials. Materials which have been determined by testing and experience to have the strength, durability, fire resistance ratings or other qualities in excess of that most commonly used as recognized in accepted engineering practices and meeting the requirements of the standard specifications of nationally approved testing agencies.

Controlled materials. Materials which are scientifically selected, graded, proportioned, and tested to produce specified results.

Controlled materials. Materials which are certified by an accredited authoritative agency as meeting accepted engineering standards for quality and as provided in the Code.

Demolition. The dismantling or razing of all or part of a building, including all operations incidental thereto.

Glare and heat. Any operation producing intense glare or heat shall be performed within an enclosure so as to completely obscure and shield such operation from direct view from any point along the lot line, except during the period of construction of the facilities to be used and occupied.

Loading. It shall be unlawful to load any structure, temporary support, scaffolding, sidewalk bridge or sidewalk shed or any other device or construction equipment during the construction or demolition of any building or structure in excess of its safe working capacity as provided in the Code for allowable loads and working stresses.

Material. An established size, quality, composition, or strength, or with respect to an established size, quality, composition or strength.

Material platform hoist. A power or manually operated suspended platform conveyance operating in guide rails for the exclusive raising or lowering of materials, which is operated and controlled from a point outside the conveyance.

Quality of materials. All materials, assemblies, construction, and equipment shall conform to the regulations of the Code, and shall conform to generally accepted standards with respect to strength, durability, corrosion resistance, fire resistance, and other qualities recognized under those standards. All test specimens and construction shall be truly representative of the material, workmanship, and details to be used in actual practice.

Runway. Any aisle or walkway constructed or maintained as a temporary passageway for pedestrians or vehicles.

Safety during construction. Construction, within the scope of this Code, shall be performed in such a manner that the workmen and public shall be protected from injury, and adjoining property shall be protected from damage, by the use of scaffolding, underpinning or other approved methods.

Safety during demolition. Safe and sanitary conditions shall be provided where demolition and wrecking operations are being carried on. Work shall be done in such a manner that hazard from fire, possibility of injury, danger to health, and conditions which may constitute a public nuisance will be minimized, in conformity with generally accepted standards.

Scaffold. Any elevated platform which is used for supporting workmen, materials or both.

Sidewalk shed. A construction over a public sidewalk, used to protect pedestrians from falling objects.

1-J SAFETY REGULATIONS

Approved. Sanctioned, endorsed, accredited, certified, or accepted as satisfactory by a duly constituted and nationally recognized authority or agency.

Authorized person. A person approved or assigned by the employer to perform a specific type of duty or duties or to be at a specific location or locations at the jobsite.

Building. Includes structures of all kinds during the course of construction, regardless of the purposes for which they are intended and whether such construction be below or above the level of the ground.

Competent person. One who is capable of identifying existing and predictable hazards in the surroundings or working conditions which are unsanitary, hazardous, or dangerous to employees, and who has authorization to take prompt corrective measures to eliminate them.

Construction elevator. Includes any means used to hoist persons or material of any kind on a building under course of construction, when operated by any power other than muscular power.

Defect. Any characteristic or condition which tends to weaken or reduce the strength of the tool, object or structure of which it is a part.

Employee. Every person who is required or directed by any employer, to engage in any employment, or to go to work or be at any time in any place of employment.

Employer. Shall also include every person having direction, management, control, or custody of any employment, place of employment, or any employee.

Employment. Includes the carrying on of any trade, enterprise, project, industry, business, occupation or work, including all excavation, demolition and construction work, or any process or operation in any way related thereto, in which any person is engaged or permitted to work for hire except household domestic service.

Insurer. Includes the State Compensation Insurance Fund and any private company, corporation, mutual association, reciprocal or interinsurance exchange authorized under the laws of the State to insure employers against liability for compensation and any employer to whom a certificate of consent to selfinsure has been issued.

Local order. Any ordinance, order, rule, or determination of the governing body of any county, city, district, or other public or quasi-public corporation, or an order or direction of any public official, board, or department upon any matter over which the division has jurisdiction.

Place of employment. Any place, and the premises appurtenant thereto, where employment is carried on, except a place the safety jurisdiction over which is vested by law in any state or federal agency other than the division.

Safe. As applied to a building, means free from danger or hazard to the life, safety, health, or welfare of persons occupying or frequenting it, or of the public, and from danger of settlement, movement, disintegration, or collapse, whether such danger arises from the method or materials of its construction or from equipment installed therein for the purpose of lighting, heating, the transmission or utilization of electric current, or from its location or otherwise.

Safe and safety. As applied to an employment or a place of employment mean such freedom from danger to the life or safety of employees as the nature of the employment reasonably permits.

Safety device and safeguard. Shall be given a broad interpretation so as to include any practicable method of mitigating or preventing a specific danger, including the danger of exposure to potentially injurious levels of ionizing radiation or potentially injurious quanities of radioactive materials.

Sanitary. As applied to a building, means free from danger or hazard to the health of persons occupying or frequenting it or to that of the public, if such danger arises from the method or materials of its construction or from any equipment installed therein for the purpose of lighting, heating, ventilating, or plumbing.

Serious hazard. A hazard of considerable consequence to safety or health through the design, location, construction, or equipment of a building, or the condition thereof, which hazard has been established through experience to be of certain or probable consequence, or which can be determined to be, or which is obviously such a hazard.

1-K STANDARDS, TESTING, LISTINGS AND RATINGS

Acceptability. Compliance with applicable provisions of generally accepted standards, except as otherwise prescribed in the Code, shall constitute compliance with the Code. Deviations from applicable provisions of generally accepted standards, when it shall have been conclusively proved that such deviations meet the provisions of generally accepted standards, when it shall have been conclusively proved that such deviations meet the performance requirements of the Code, shall constitute compliance with the Code.

Accepted. Pertaining to standards and practices. Refers to nationally accepted standards and practices as listed in the Rules and Regulations of the code.

Accepted engineering practice. That which conforms to accepted principles, tests or standards of nationally recognized technical or scientific authorities.

Accredited authoritative agencies. Agencies or organizations establishing accepted standards.

Accredited authoritative agency. An agency qualified under the code to perform a task or to render a service.

Approved. As to materials and types of construction, refers to approval by the Director as the result of investigation and tests conducted by him, or by reason of accepted principles or tests by national authorities, technical or scientific organizations.

Approved agency. An establishment or testing agency recognized and approved by the Building Officials to conduct tests on materials and products and issue reports on the results of such tests.

Approved agency. An established and recognized agency regularly engaged in conducting tests or furnishing inspection services, when such agency has been approved by the Building Official.

Approved agency. An established and recognized agency regularly engaged in conducting tests or furnishing inspection services, when such agency has been approved by the Director.

Approved fabricator. An established and qualified person, firm, or corporation approved by the Building Official pursuant to applicable Section of the Code.

Approved material. Shall mean any material, product, devise, assembly or mode of construction that has been submitted to investigation of the Research Committee for the Building Officials and determined suitable for use under the provisions of the Code.

Approved material, equipment and methods. Approved by the building official or by a recognized authoritive agency.

Approved rating methods. Ratings of fire-resistive assemblies shall be determined by one of the following methods: Test by approved testing laboratories. Typical examples as listed in the Code in lieu of approved test. Approved method of calculation in lieu of approved test.

Approved standards. The intent of the Code is to safeguard the public against dangerous and hazardous, as well as nuisance, conditions arising from the construction, repair, alteration, remodeling, moving, operation; and demolition of buildings and structures, electrical, gas and mechanical installations, including moving stairs, air conditioning, mechanical refrigeration, attic ventilation, boiler, unfired pressure vessels, and smoke abatement, and the operation and maintenance of power equipment and apparatus; and to establish reasonable standards whereby the foregoing activities may be carried on in a standard and approved manner.

Approved testing laboratories. Fire rating tests conducted according to listed standards shall be acceptable if conducted by the recognized testing laboratory for referenced test.

Approved test or approved analysis. Test or analysis conducted and evaluated by an accredited authoritative agency under the standard for test or analysis provided for the particular purpose as described in the code.

Check tests. When there is reasonable doubt as to the design capacity of any structural unit or assembly, the Building Official may require that check tests be made of the assembled unit and its connections or he shall accept certified reports of such tests from accredited testing authorities conducted in accordance with the approved rules.

Controlled Materials. Materials which are certified by an accredited authoritative agency as meeting accepted engineering standards for quality and as provided in the Code.

Durability and endurance tests. Whenever required by the Building Official or specified in the Code, the material or construction shall be subjected to sustained and repetitive loading to determine its resistance to fatigue, and to tests for durability and weather resisrance.

Finish standards. The National Bureau of Standards of the United States Department of Commerce have prepared product standards for the finishes used with builders' hardware.

Fire endurance. Measure of the elapsed time during which a material or assembly continues to exhibit fire resistance under specified conditions of test and performance, as applied to elements of buildings it shall be measured by the methods and the criteria defined in ASTM E-119 and ASTM E-152.

Fire protection rating. Time in hours or fraction thereof that a closure, window assembly, or glass block assembly will withstand the passage of flame when exposed to fire under specified conditions of test and performance criteria or as otherwise prescribed in the code.

Fire-protection rating. The time in hours or fractions thereof that an opening protective and its assembly will withstand fire exposure as determined by a fire test made in conformity with specified standards of the Code.

Fireresistance. That property of materials or their assemblies which prevents or retards the passage of excessive heat, hot gases or flames under conditions of use.

Fire resistance rating. The time in hours or major fractional parts thereof that a material, construction or assembly will withstand fire exposure, as determined in a fire test acceptable to Federal agencies.

Fire resistance rating. The time in hours by which materials will resist fire as determined by tests conducted in accordance with recognized national standards.

Fire resistance rating. The time in hours or fractions thereof that a material or assembly of materials will withstand the passage of flame and the transmission of heat when exposed to fire under specified conditions of test and performance criteria or as determined by extension or interpretation of information derived therefrom as prescribed in the Code.

Fire resistance rating. The time in hours that materials, construction, or assembly of materials will withstand fire exposure as determined in a fire test made in conformity with the "Standard Method of Fire Tests of Building Construction Materials," ASTM E119, and is referred to herein as "1 hour construction", "2 hour construction" "3 hour construction," or "4 hour construction," or "3/4 hour construction."

Fire resistance rating. Time in hours that the material or construction will withstand the standard fire exposure as determined by a fire test made in conformity with "Standard Methods of Fire Testing of Building and Materials" ASTM E-119.

Fire resistance rating. The degree of fire resistance of a fabricated unit or assembly of units of construction, by the standard fire test expressed in hours or fractions of an hour.

Fire resistance rating. The measured time in hours or fraction of an hour that a material or construction will withstand fire exposure as determined by tests conducted in conformity to approved standards.

Fire resistance rating. The time in hours, that materials or assemblies have withstood a fire exposure as established in accordance with the test procedures of Standard Methods of Fire Tests of Building Construction Materials.

Fire-resistance rating. The time in hours or fractions thereof that materials or their assemblies will withstand fire exposure as determined by a fire test made in conformity with a specified standard.

Fire-resistance rating. Time in hours or parts thereof that a material, construction, or assembly will withstand fire exposure, as determined in a fire test made in conformity with generally accepted standards, or as determined by extension or interpretation of information derived therefrom.

Fireresistance rating. The time in hours or fractions thereof that materials or their assemblies will resist fire explosure as determined by fire tests conducted in compliance with recognized standards.

Fire-resistive classification. Time in hours during which a material or assembly continues to exhibit resistance under conditions of tests and performance as specified in ASTM E-119, ASTM E-152, and ASTM E-163.

Fire-resistive rating. Time in hours that the material or construction will withstand the standard fire exposure as determined by a fire test in conformity with ASTM E-119.

Fire spread rating. Numerical value assigned to a material tested in accordance with ASTM E-84.

Flameresistance. The property of materials or combinations of component materials which restricts the spread of flame as determined by the flameresistance tests specified in the Code.

Flame spread. The propagation of flame over a surface.

Flame spread classification. Comparative rating of the measure of flame spread on a surface of a material or assembly as determined under conditions of tests and performances as specified in ASTM E-84.

Flame spread rating. The measurement of the comparative rate or propagation of flame over the surface of a material as determined by a fire test made in accordance with a specified standard in the Code.

Flame spread rating. That rating as determined by tests conducted in accordance with ASTM Standard E84 and shall be accepted by the Building Official if reported by the Underwriters Laboratories, Inc., or other recognized testing laboratory.

Flame spread rating. The measurement of flame spread on the surface of materials or their assemblies as determined by tests conducted in compliance with conforming standards.

Flame spread rating. Measurement of flame spread on the surface of materials or their assemblies as determined by accepted standard tests.

Flame spread rating. The measurement of flame spread on the surface of materials or their assemblies as determined by tests conducted in compliance with recognized standards.

Flame spread rating. That numerical value assigned to a material tested in accordance with ASTM E84 (Steiner Tunnel Test).

Flame spread rating. Rating obtained for the surface flame spread of interior finish tested according to the "Test for Surface Burning Characteristics of Building Materials." ASTM E84, in which asbestos cement board rates zero on the scale and red oak lumber 100.

Flame spread rating. The degree of flame resistance of materials used for interior finish and trim or for decorative purposes determined by the rate of flame spread in the standard tunnel test.

Flame spread rating. An index or classification indicating the extent of spread-of-flame on the surface of a material or an assembly of materials as determined in a standard fire test as prescribed in the Code.

Flame spread rating. A number of classifications indicating a comparative measure derived from observations made during the progress of the boundary of a zone of flame under conditions complying with recognized national standards.

Flame-spread rating. The measurement of flame spread on the surface of materials on their assemblies as determined by tests conducted in conformity with a generally accepted standard.

Flame spread rating. The flame spread rating of materials as determined by the Method of Test of Surface Burning Characteristics of Building Materials, NFPA NO. 255, ASTM E84, Underwriter's Laboratories, Inc., Standard. Such materials are listed in the Underwriter's Laboratories, Inc., Building Materials List under the heading "Hazard Classification (Fire)."

Gage. Designation for "U.S. Standard Gage" when applied to thickness of metal sheets or diameter of wires except where some other designation is specifically mentioned; also, a measure of pressure.

Grade material. Established size, quality, composition, or strength, or with respect to an established size, composition, quality, or strength.

Listed. Noted, as approved after test or inspection, in publications of UL, or other approved agencies.

Listed. Refers to appliances and accessories which are shown in a list published by an approved nationally recognized testing agencies, qualified and equipped for experimental testing and maintaining, and adequate periodic inspection of current production of listed models and whose listing states either that the appliance or accessory complies with nationally recognized safety requirements or has been tested and found safe for use in a specified manner.

Material standards. All building units used in wall, partition and floor construction and for fireproofing or other insulation purposes shall comply with the applicable standards listed in the Code.

New materials. All new building materials, equipment, appliances, systems or methods of construction not provided for in the Code, and any material of questioned suitability proposed for use in the construction of a building or structure, shall be subjected to the tests prescribed in the Code and in the approved rules to determine its character, quality and limitations of use.

Performance standard. Criterion established to control the dust, smoke, fire and explosive hazards, glare and heat, noise, odor, toxic and noxious matter, vibrations and other conditions created by or inherent in uses of land or buildings.

Qualification tests. Standard tests performed by national testing agencies, on materials and equipment to evaluate their proposed use.

Quality control. Inspection, testing and analysis of materials used in the manufacturing of products and the inspection of materials used in construction of buildings during the construction period.

Quality control. System whereby the manufacturer assures that materials, methods, workmanship, and final product meet the requirements of a standard.

Recognized testing laboratories. Equipment listed as satisfactory for specific purposes by nationally recognized testing laboratories (such as the Underwriters' Laboratories, Inc., the Factory Mutual Laboratories, and the American Gas Association), when installed and used for the purposes intended, will be accepted as approved, unless specifically prohibited elsewhere in the Code.

Ringlemann chart. The standard published by the U.S. Bureau of Mines to determine the density of smoke. It shall constitute the standard of the bureau of industrial hygiene and air pollution control in determining the density of smoke as hereinafter set forth in the Ordinance.

Safe. Applied to a building, means free from danger of hazard to the life, safety, health, or welfare of persons occupying or frequenting it, or of persons occupying or frequenting it, or of the public, and from danger of settlement, movement, disintegration, or collapse, whether such danger arises from the method or materials of its construction or from equipment installed therein for the purpose of lighting, heating, the transmission or utilization of electric current, or from its location or otherwise.

Smoke developed rating. The smoke developed rating of materials as determined by the Method of Test of Surface Burning Characteristics of Building Materials, NFPA NO. 255, ASTM E84, Underwriter's Laboratories, Inc. Standard. Such materials are listed in Underwriter's Laboratories, Inc., Building Materials List under the heading "Hazard Classification (Fire)."

Standard. Test, measure, model or example of quantity, extent or quality.

Standard fire test. Fire test formulated under the procedure of the American Standards Association as "American Standard." this "American Standard" is the "Standard Methods of Fire Tests of Building Construction and Materials" of the American Society for Testing Materials (ASTM Designation E119).

Standard fire test. The standard controlled furnace test formulated under the procedure of the American Society for Testing and Materials and designated as ASTM E119.

Standard fire test. Fire test formulated under the procedure of the American National Standards Institute, as the American Standard.

Strength tests. To determine the safe uniformly distributed working load, when not capable of design by accepted engineering analysis, or to check the adequacy of the structural design of an assembly when there is reasonable doubt as to its strength or stability, every system of construction, sub-assembly or assembled unit and its connections shall be subjected to strength tests prescribed in the Code, or to such other tests acceptable to the Building Official.

Structural analysis. The safe load for any structural member or system of construction shall be determined by accepted engineering analysis except as provided in the Code.

Test load. When approved by test, every structural assembly shall sustain without failure minimum superimposed loads equal to two and one-half (2-1/2) times the required live load; and under the approved working load, the deflection shall not exceed the limits prescribed in the Code.

Test methods. Test methods shall be in accordance with those specified in the Code for the material or method of construction in question. Where no appropriate test method is prescribed in the Code the test procedure shall be determined by the Building Department.

Tests. All structural steel. cast steel and cast iron shall be tested in accordance with the Code specifications when deemed necessary by the Director, and copies of such tests shall be filed in the office of the Director. No structural steel, cast steel and cast iron shall be used in any building or structure which does not comply with the above requirements or for which no test results have been filed with the Director. All such tests shall be made by competent testing laboratories at the expense of the owner.

Tests. Test methods, used to determine the specific requirements for physical properties of roll roofing, roofing felt, or felt membrane given in the Code shall be those methods set forth in the "Tentative Methods of Testing Felted and Woven Fabrics Saturated with Bituminous Substances for Use in Waterproofing and Roofing." (A.S.T.M. Standards).

Tests. Tests of materials shall be made in accordance with the standard specifications of the American Society for Testing Materials as such Standard Specifications are noted in the code.

Test procedure. Materials shall pass the test procedure of ASTM E136 for defining noncombustibility of elementary materials when exposed to a furnace temperature of 1,382 degree F. for a minimum period of 5 minutes, and do not cause a temperature rise of the surface or interior thermocouples in excess of 54 degrees F, above the furnace air temperature at the begining of the test and which do not flame after an explosure of 30 seconds.

Tests of materials and systems of construction. Where certain systems of construction, qualities of materials or tests are referred to or regulated by the Code, tests may be required by the Director. Such tests shall be made at the expense of the owner or his agent by a testing laboratory or other organization approved for the purpose by the Director. Certified copies of all test reports shall be filed with the Director for his approval before the systems of construction or materials are used. The owner or his agent shall notify the Director of the time and location of all such tests so that he may be present.

Used materials. The use of all second-hand materials which meet the minimum requirements of the Code for new materials shall be permitted.

2. Article 2

SITE WORK 2-A THROUGH 2-G

2-A DRAINAGE

Building house drain. That part of the lowest piping of a drainage system that receives the discharge from the soil, waste, and other drainage pipes and conveys it to the building house sewer by gravity. The building house drain shall be considered to extend 5 feet outside the exterior wall of the building.

Building house drain (combined). A building house drain that conveys storm water in combination with sewage of other drainage.

Building house drain (storm). That part of the lowest piping of a storm drainage system that receives clear drainage from leaders, surface runoff, ground water, subsurface water, condensate, cooling water, or other similar storm or clear drainage and conveys it to the building house storm sewer by gravity. The building house storm drain shall be considered to extend 5 feet outside the exterior wall of the building.

Building house storm sewer. That part of the horizontal piping of a storm drainage system that extends from the building house storm drain to the public storm sewer, combined sewer, or other point of disposal.

Building sub-house drain. That portion of a house drainage system that cannot drain by gravity into the building house sewer.

Combined building drain. Drain that is intended to conduct sewage and storm water.

Combined building sewer. Sewer that is intended to conduct sewage and storm water.

Drainage system. Assembly of pipes, fittings, traps and appurtenances that is used to convey sewage, clear-water waste or storm water to a public sewer or a private sewage disposal system but does not include subsoil drainage pipes.

Drainage system. All the piping within public or private premises, which conveys sewage, rain water, or other liquid wastes to a legal point of disposal, but shall not include the mains of public sewer system or private or public sewage-treatment or disposal plant.

Drywell. Covered pit with open-jointed lining or covered pit filled with coarse aggregate through which drainage from roofs, basement floors, foundation drain tile, or areaways may seep or leach into the surrounding soil.

Sand interceptor. Interceptor in a drain, primarily intended to intercept sand and earth.

Storm building drain. Building drain that may conduct only storm water or clear-water waste.

Storm building sewer. Building sewer that may conduct only storm water or clear-water waste.

Storm drainage system. Drainage system or a part of a drainage system that conveys only storm water or clearwater waste.

Storm sewer. Sewer that is installed to convey storm water.

Storm sewer. A sewer used for conveying rain water, surface water, condensate, cooling water, or similar clear liquid wastes which do not contain organic materials or compounds subject to decomposition.

Storm sewer. A sewer which carries storm and surface waters and drainage, but excludes sewage and polluted industrial wastes.

Storm system. That part of the drainage system which conveys storm water to a provided street sewer when a street sewer is not available, it shall be discharged as directed by the chief plumbing inspector.

Storm water. Water that is discharged from a surface as a result of rainfall or snowfall.

Storm water drain. Any pipe, or drain, which receives the discharge of rain water from buildings or premises (which may include the discharge of seepage or ground water) and shall not enter a sanitary sewer.

Subsoil drain. That part of a drainage system which conveys the subsoil, ground or seepage water from the foot of walls or below the cellar bottom under buildings.

Subsoil drainage pipe. Pipe that is installed underground to intercept and convey groundwater.

Yard drain. That part of the horizontal piping and its branches which conveys the surface drainage from areas courts and yards.

2-B MARINE REFERENCES

Backshore. That part of a beach which is usually dry, being reached only by the highest tides, and, by extension, a narrow strip of relatively flat coast bordering the sea.

Beach line. Is that side of a lot facing the beach.

Bulkhead line. Official line established by the code and enforced by the City Council to limit the extent of bulkheads at coastal lines.

Bulkhead line. Legally established property line and boundary of a riverfront lot, or the extent to which a riverfront lot can be filled.

Depth slope line. A line drawn between two (2) fixed depth points in a straight line from the lowest depth to the highest depth.

Dune line. The highest point of an existing dune or line of dunes, or in the event there is no existing dune or "dune line" in any respective location, shall be the highest point within one hundred fifty (150) feet of mean low water. In determining said high point, the elevation survey completed by the Department of Natural Resources and on file in the office of the City Engineer.

Elevation slope line. A line drawn between two (2) fixed elevation points in a straight line from the lowest elevation to the highest elevation.

Groin. Structure extending into a body of water designed to trap literal drift or retard erosion of the shore.

Hard beach. Portion of a beach especially prepared with a hard surface extending into the water, employed for purpose of loading or unloading into ships or vessels.

Seawall line. Wall situated so as to resist the encroachments of the river or sea. A seawall may or may not be the property line.

Sediment. The resulting residue from erosion.

Shoreline. A straight or smoothly curving line which, on tidal waters, follows the general configuration of the mean high water line (1.09 feet above mean sea level as determined by U.S. Coast and Geodetic Survey datum); and which, on non-tidal waters is determined by the annual average water level. Small boat slips and other minor indentions shall be construed as lying landward of the shoreline and are considered upland when measuring required yards or computing the lot area of waterfront property.

2-C SOILS AND EARTHWORK

Backfill. Replacement of excavated earth into a pit or trench or against a foundation wall.

Backfill. To place earth or selected fill material in an excavated void.

Backfill. Material used in refilling an excavation, such as for a foundation or subterranean pipe.

Bearing materials. Soil or compacted fill on which the foundations are supported.

Bedrock. The solid undisturbed rock in place either at the ground surface or beneath surficial deposits of natural soil or fill.

Bench. A bench is a relatively level step excavated into each material on which fill is to be placed.

Clay. Fine-grained inorganic soil which is plastic when wet and brittle when dry.

Clay mineral. Naturally occurring inorganic material (usually crystalline) found in soils and other earthy deposits, the particles being of clay size; that is, not greater than 0.002 mm in diameter.

Claypan. A horizon of accumulation or a stratum of dense compact and relatively impervious clay. Claypan is not cemented, but is hard when dry, and plastic or stiff when wet. Its presence, like that of a true hardpan, may interfere with water movement.

Compacted fill. Usually mixtures of sand, gravel and predominately granular materials, crushed stone or noncorrosive slag compacted in layers in accordance with accepted engineering practice. Other materials having the same characteristics as sand, gravel, crushed stone or noncorrosive slag may be approved by the authority as compacted fills.

Compacted fill density. The density of the soil as it is compacted into a man made fill.

Compaction. The volume change produced artificially by momentary application such as rolling, tamping, or vibration.

Compression. The volume change produced by application of a static external load.

Consolidation. Volume change that is achieved with the passage of time.

Deep excavations. Whenever an excavation is made more than 4 feet below adjacent ground surface.

Deep excavations. Whenever an excavation is made to a depth of more than four (4) feet below adjacent ground surfaces, the person who causes such excavation to be made, if afforded the necessary license to enter the adjoining premises, shall preserve and protect from injury at all times and at his own expense such adjoining structure or premises which may be affected by the excavation. If the necessary license is not afforded, it shall then be the duty of the owner of the adjoining premises to make his building or structure safe by installing proper underpinning or foundations or otherwise; and such owner, if it be necessary for the prosecution of his work, shall be granted the necessary license to enter the premises where the excavation or demolition is contemplated.

Depth of excavation. Depth is measured from the elevation of the street grade nearest to the point of excavation.

Dry density. Term normally used for expressing the unit weight of soil. The dry density is computed from the wet density and the water content data.

Earth Materials. Shall mean natural soil, bedrock and fill.

Elevation. Distance above or below a prescribed datum established by the city, county, state, or a federal agency from an established reference marker or monument.

Excavating machinery. Any power-driven machinery used to excavate or dig into the ground or used to level or grade any ground site.

Excavation. The breaking of the ground, except common household gardening and normal ground care.

Excavation or removal of soil. No person, firm or corporation shall strip, excavate or otherwise remove soil for sale or for any other purpose without first obtaining a permit from the Building department. The provisions of this paragraph shall not be construed to prohibit excavation or grading incidental to the construction or alteration of a building or structure on the premises for which a permit has been issued as required by the code.

Existing grade. Shall mean the grade prior to grading.

Expansion. Opposite of consolidation. The amount of expansion depends on the type of clay mineral and the availability of water, and is a function of time, confining load, initial density, and initial water content.

Fill. Soil and other materials deposited at a site, or secured from the site, to be used to change the grades on the site.

Filling. Depositing or dumping of any matter including septic tank effluent onto, or into the ground, except common household gardening and ground care.

Finish grade. Shall mean the grade after the completion of grading.

Frost action. The heaving of subgrades due to formation of ice lenses and the subsequent loss of stability on thawing. The freezing of the pore water in saturated fine-grained soils will decrease the density of the mass by expansion.

Frost action. That phenomenon occurring in wet soils which results in volume increase or the build-up of stresses when subjected to freezing.

Frostline. Depth of frost penetration in soil. This depth varies in different parts of the country. Footings should be placed below this depth to prevent movement.

Gradation. A descriptive term which refers to the distribution and size of grains in a soil.

Grade. Shall mean the vertical location of the ground surface.

Grading. Shall mean the excavating and filling of land for altering the contours of the ground to prescribed grades for the purpose of erecting buildings or structures thereon or other use thereof, and shall not include the processing of material for use on another site.

Hard clay. Clay when fresh, cannot be molded by pressure of the fingers and which when dry requires a pick for removal.

Hardpan. Highly compacted soil which requires a pick or special equipment for removal.

Hardpan. A cemented (indurated) or hardened soil horizon. This horizon which may have any texture is compacted or cemented by iron oxide, silica, organic matter or other substance.

Hard rock. Rock which requires drilling and blasting for removal.

Heave. Volume change produced by frost action or expansive soils.

Liquid limit. That water content expressed as a percentage of the dry weight of soil at which the soil first shows a small but definite shearing strength as the water content is reduced.

Loosening or scarifying. The operation opposite to that of compaction.

Medium Clay. Clay which, when fresh, requires substantial pressure of the fingers for molding and which requires spading for removal.

Mica. A small flake of metamorphic rock which gives the appearance of a glistening fish scale.

Mica schist. A metamorphic rock consisting of naturally cemented, closely spaced, approximately parallel layers of scale-like flakes. The rock, if hard, sound and massive requires large explosive charges for removal.

Natural density (in place). The unit weight of a soil, expressed in pounds per cubic foot, as it exists in a natural deposit at any particular time.

Natural soil. Shall mean naturally occurring surficial deposits overlying bedrock.

Ordinary fill. Usually comprised of clay, sand, gravel and rocks artificially deposited at a site. Some specifications allow 5% of the total weight of the fill to be noncorrosive slag.

Ordinary gravel. Gravel containing particles up to 1-1/2 inches in size.

Organic soil. A soil composed mainly of organic matter on a volume basis. (Twenty per cent or more organic matter by weight.)

Peat. Soil material consisting primarily of raw undecayed or slightly decomposed organic matter.

Perlite. Acid, igneous, glassy rock of the composition of obsidian, expanded by heating and divided into small spherical bodies by the tension developed by its contraction on cooling.

Permanent water level. Shall mean "sea level" unless special conditions exist. If special conditions exist, the term "permanent water level" shall mean such lower level as the Building Official in his opinion may deem to represent the permanent water level.

Permeability. The state of water movement in soil is called percolation; the measure of it is called permeability; and the factor relating permeability to unit conditions of control is called coefficient of permeability.

Plastic limit. That water content expressed as a percentage of the dry weight of soil at which the soil mass ceases to be plastic and becomes brittle, as determined by a procedure for rolling the soil mass into threads one-eight inch in diameter. The plastic limit is determined by reducing the water content of the soil mass.

Plasticity index. The difference between the liquid and plastic limits and represents the range of moisture within which soil is plastic.

Porosity. The percentage of space in the soil mass not occupied by the solids with respect to the total volume of the mass.

Porosity and void ratio. The measures of the state or condition of a soil structure.

Positive drainage. Sufficient slope to drain surface water away from buildings without ponding.

Quagmire. Saturated area with a surface of soft mud, or, at best, a surface providing a shaky and precarious footing.

Quicksand. Loose, yielding, wet sand which offers no support to heavy objects. The upward flow of the water has a velocity that eliminates contact pressures between the sand grains, and causes the sandwater mass to behave like a fluid.

Quicksand. Submerged, saturated sand into which a heavy object easily sinks. Sand is held in a very loose, unstable packing such that a shock moves the grains to a smaller bulk volume. Lack of bearing power may be due to seepage pressure of water percolating through the sand in an upward direction or it may be due to inherent instability of the structure of the sand, unaided by seepage pressure.

Quicksand. Any sand rendered unstable by an upward flow of ground water. Condition rather than a material. Any sand can be made quick by an upward flow through it of sufficient velocity.

Rebound. Opposite of compression. The normal rebound phenomenon which occurs on release of a compressive load.

Rock. Mineral matter of various composition found in natural resources.

Sedimentary rock. Rock formed by deposit of small particles of soil and consolidated over eons of time, which when sound, cannot be removed by a pick and usually requires blasting for removal.

Sensitive clay. A type of clay, when tested in an undisturbed condition, has substantial strength, but when disturbed and remolded loses its strength.

Shale. Laminated clay or silt compressed by earth over-burden. Unlike slate, it splits along its bedding planes.

Shale. Soft, laminated, easily split, fine-grained, sedimentary rock which is susceptible to weathering when exposed.

Shallow excavations. Wherever an excavation is made to a depth less than four (4) feet below the curb, the owner of a neighboring building or structure excavation, shall preserve and protect from injury and shall support his building or structure by the necessary underpinning or foundation. If necessary for the purpose, he shall be afforded a license to enter the premises where the excavation is contemplated.

Shrinkage. The volume change produced by capillary stresses during the drying of the soil.

Sod removal. Commercial removal of soil filled with the roots of grass or herbs.

Soft clay. Clay which, when fresh, can be molded with a slight pressure of the fingers.

Soft rock. Characteristic which allows rock to be easily penetrated by using a pick.

Soil. Softer matter mostly inorganic composing part of the surface of the earth in distinction from the firm rock; including gravel, clay, loam and the like, and filling materials of similar nature.

Soil. Inorganic material other than water in its natural state and location.

Soil. A natural body developed from weathered minerals and decaying organic matter, covering the earth in a thin layer. It is a natural medium on the surface of the earth in which plants may grow.

Soil consistency. The physical properties of soils affected by water content.

Soil consistency tests. The four stages, or states are recognized for describing the consistency of a soil. These are: (1) the liquid state, (2) the plastic state, (3) the semi-solid state, and (4) the solid state.

Soil cover (ground cover). Light roll roofing or plastic used on the ground of crawl spaces to minimize moisture permeation of the area.

Soil engineer. Shall mean a civil engineer experienced and knowledgeable in soil engineering.

Soil engineering. Shall mean the application of the principles of soil mechanics in the investigation, evaluation and design of civil works involving the use of earth materials and the inspection and testing of the construction thereof, by a Soil Engineer.

Soil engineering report. A written opinion concerning the engineering properties of earth materials as related to the proposed work.

Soil or soils. Earth or earth stratum condition found at and below the point of bearing of building, footing and foundations.

Soil removal. Removal of any kind of soil or earth matter, including top soil, sand or other type of soil matter or combination thereof, except common household gardening or ground care.

Sound rock. Rock having no visible cracks or seams.

Specific gravity. The ratio between the unit weight of a substance and the unit weight of water at 4°C.

Subgrade. Elevation established to receive top surfacing or finishing materials.

Swell. The opposite of shrinkage. To increase in bulk.

Terrace. A terrace is a relatively level step constructed in a graded slope for drainage and maintenance purposes.

Void ratio. The ratio of the space not occupied by the solid particles to the volume of the solid particles.

Water table. The upper surface of free ground water in a zone of saturation except when separated from an underlying body of ground water by unsaturated material.

Wet density. The unit weight of the solid particles and the contained moisture expressed in pounds per cubic foot.

2-D SUBSTRUCTURES

Batter pile. Pile which is installed at an angle to the vertical; a raking pile.

Bored pile. Concrete pile, with or without a casing, cast-in-place in a hole previously bored in soil or rock.

Caisson or pier foundations. Foundation system where the building structure is supported upon a system of holes (usually lined) bored in the earth to a strata which will provide adequate support of design loads and filled with concrete. Borings, 2 feet or larger to permit bottom inspection are usually considered caissons while borings less than 2 feet in diameter are considered piers. Where ground water is a problem, the pneumatic-caisson method is employed.

Caisson pile. Cast-in-place pile made by driving a tube, excavating it, and filling the cavity with concrete.

Casing. Large metal pipe or tubing inserted into the ground before the pouring of concrete caissons, to keep the surrounding soil from collapsing into the concrete pour.

Casing-off. The elimination of the frictional forces between a portion of a pile and the surrounding soil by use of a sleeve between the pile and the soil.

Cast-in-place pile. Shall mean a concrete pile cast in place without forms.

Cast-in-place pile. A concrete pile concreted either with or without a casing in its permanent location, as distinguished from a precast pile.

Composite pile. Pile made up of different materials, usually concrete and wood, or steel fastened together end to end, to form a single pile.

Concrete filled steel pipe piles. 1. Steel pipe. Steel pipe piles shall conform to A.S.T.M. "Standard Specification for Welded and Seamless Pipe Piles." If it is desired to use pipe of other materials, satisfactory substantiating data must be submitted. 2. Concrete. The concrete used in concrete filled steel pipe piles shall have an ultimate compressive strength (f'.) of not less than 2,500 pounds per square inch.

Concrete pile. Precast reinforced or prestressed concrete pile driven into the ground by a pile driver or otherwise placed.

Driving to refusal. Inability to drive a pile further under a hammer or approved adequate weight after the tube has been completely washed and blown at the bottom, and before filling with concrete.

Footing. That portion of the foundation of a structure which spreads and transmits load directly to the piles, or to the soil or supporting grillage.

Footing. Shall mean that portion of a structure or substructure which spreads and transmits loads directly to the soil or other bearing media.

Footing. That portion of the foundation of a structure which spreads and transmits loads directly to the soil or as a cap on the top of piles.

Footing. Shall mean a structural unit used to distribute loads to the bearing materials.

Footing. The spreading course at the base or bottom of a foundation wall, column, or pier intended to assume and distribute the total imposed load over a minimum required area of the accepted bearing soil.

Footing. The projecting base of a foundation which transmits the building load to the ground.

Footing. The spreading course of the base or bottom of a foundation wall, column or pier.

Footing. That portion of a foundation of a building which distributes and transmits to the ground the loads resulting from the building.

Footing. A foundation element consisting of an enlargement of a foundation pier or foundation wall, wherein the soil materials along the sides of and underlying the element may be visually inspected prior to and during its construction.

Footing or foundation. The spreading course at the base or bottom of a foundation wall, column or pier.

Foundation. Construction, below or partly below grade, which provides support for exterior walls or other structural parts of the building.

Foundation. System or arrangement of foundation units through which the loads from a building are transferred to the supporting soil or rock.

Foundation. That portion of a building or structure of which the chief purpose or use is to transmit the weight of the building or structure to the earth; the support of the lowest portions of columns, walls, piers or other vertical members.

Foundation. A wall or pier below first floor serving as support for a wall, pier, column, or other structural parts of a building.

Foundation. The supporting portion of a building below the floor construction nearest the finished grade of the ground adjoining the building, and includes the footings.

Foundation. Construction, below or partly below grade, which provides support for exterior walls or other structural parts of the building.

Foundation. The supporting portion of a structure below the first-floor construction, or below grade, including footings.

Foundation. The part of the structure on which the superstructure rests. It includes all construction which transmits the loads of the superstructure to the earth.

Foundation. The wall or pier below the first floor upon which the superstructure rests.

Foundation. Shall mean the construction of a building or a structure below the lowest floor of the building or structure, including the footing, and which directly transmits the loads to the supporting soil or other bearing media.

Foundation. Material or materials through which the load of a structure is transmitted to the earth.

Foundation (building). A construction that transfers building loads to the supporting soil.

Foundation pier. A foundation element consisting of a column embedded into the soil below the lowest floor to the top of a footing or pile cap. Where a pier bears directly on the soil without intermediate footings or pile caps, the entire length of the column below the lowest floor level shall be considered as a foundation pier.

Foundation piers. Sub-surface structures for the support of columns, above-ground piers, or other concentrated foundation loads, extending from the surface of the ground downward to suitable bearing.

Foundation unit. One of the structural members of the foundation of a building such as a footing, slab, raft, pile, pier, foundation wall or retaining wall.

Foundation wall. A wall extending below grade.

Foundation wall. Wall below the first floor extending below the adjacent ground level and serving as support for a wall, pier, column or other structural part of a building.

Foundation wall. A wall, below or partly below grade, providing support for the exterior or other structural parts of a building.

Foundation wall. All walls and piers built to serve as supports for walls, piers, columns, girders, parts or beams, etc.

Foundation wall. Shall mean any wall or pier built below the curb level or the nearest tier of beams to the curb, which serves as a support for walls, piers, columns or other structural parts of a structure.

Foundation wall. A wall below or partly below grade serving as a support for a structural member.

Foundation wall. A wall below the floor nearest grade serving as a support for a wall, pier, column or other structural part of a building or structure.

Foundation wall. A supporting wall below the floor construction nearest the finished grade of the ground adjoining the building. For this definition each wall of a building is considered separately.

Friction pile. Load-bearing pile which receives its principal vertical support from skin friction between the surface of the buried pile and the surrounding soil.

Lagging (pile). Pieces of timber or other material attached to the sides of piles to increase resistance to penetration through soil.

Metal cased pile. Shall mean a concrete pile cast in place in a previously driven metal casing.

Pedestal footing. A column footing which projects less than one-half its depth from the faces of the column on all sides and the maximum depth of which is three times its least width.

Pedestal pile. Cast-in-place concrete pile constructed so that concrete is forced out into a widened bulb or pedestal shape at the foot of the pipe which forms the pile.

Pile. Shall mean a column inserted into the ground which transmits loads directly to the soil or other bearing media.

Pile. A structural element introduced into the ground to transmit loads to lower strata and of such construction that the material underlying the base of the unit or along the sides cannot be visually inspected.

Pile. Slender timber, concrete, or steel structural element, driven, jetted, or otherwise embedded on end in the ground for the purpose of supporting a load or of compacting the soil.

Pile bent. Two or more piles driven in a row transverse to the long dimension of the structure and fastened together by capping and (sometimes) bracing.

Pile cap. (A). Structural member placed on, and usually fastened to, the top of a pile or a group of piles and used to transmit loads into the pile or group of piles and in the case of a group to connect them into a bent; also known as a rider cap or girder; also a masonry, timber, or concrete footing resting on a group of piles. (B). Metal cap or helmet temporarily fitted over the head of a precast pile to protect it during driving; some form of shock-absorbing material is often incorporated.

Pile cap. A construction encasing the heads of one or more piles which transfers loads to the pile or piles.

Pile driver. Any mechanism that is used to hammer or drive any type of piling into the ground.

Pipe column. Column made of steel pipe; often filled with concrete.

Pipe pile. Steel cylinder, usually between 10 and 24 in. (250 and 600 mm) in diameter, generally driven with open ends to firm bearing and then excavated and filled with concrete; this pile may consist of several sections from 5 to 40 ft (1.5 to 8 m) long joined by special fittings such as cast-steel sleeves and is sometimes used with its lower end closed by a conical steel shoe.

Pole footing. A type of construction in which a pole embedded in the ground and extending upward to form a column is used for both column and footing.

Precast pile. Reinforced concrete pile manufactured in a casting plant or at the site but not in its final position.

Raft foundation. Continuous slab of concrete, usually reinforced, laid over soft ground or where heavy loads must be supported to form a foundation.

Raking pile. Pile which is installed at a slight angle to the vertical. Also called a batter pile.

Retaining wall. Any wall used to resist the lateral displacement of any material including liquids.

Retaining wall. Designed to resist lateral pressure.

Retaining wall. Wall used to resist the lateral displacement of any material.

Retaining wall. Wall designated to prevent lateral movement of liquids, soil or fill.

Retaining wall. Wall used to resist laterally imposed pressures.

Retaining wall. Wall used to resist the lateral displacement of liquid, granular or other materials.

Sheet pile. Pile in the form of a plank driven in close contact or interlocking with others to provide a tight wall to resist the lateral pressure of water, adjacent earth, or other materials; may be tongued and grooved if made of timber or concrete and interlocking if made of metal.

Sheet piling. Construction consisting of materials installed into soil to create a retaining wall.

Stepped footing. A step-like support consisting of prisms of concrete of progressively diminishing lateral dimensions superimposed on each other to distribute the load of a column or wall to the subgrade.

Underpinning. Construction required to extend a foundation to a lower soil strata.

Wing pile. Bearing pile, usually of concrete, widened in the upper portion to form part of a sheet pile wall.

2-E WASTE, SEWAGE AND DISPOSAL

B.O.D. (Biochemical oxygen demand). An abbreviation denoting biochemical oxygen demand, means the quantity of oxygen utilized in the biochemical oxidation of organic matter under standard laboratory procedure in 5 days at 20 degrees centigrade, expressed in parts per million by weight.

Building house drain (sanitary). A building house drain that carries sewage only.

Building house sewer. That part of the horizontal piping of a drainage system that extends from the end of the building house drain and that receives the discharge of the building house drain and conveys it to a public sewer, private sewer, individual sewage-disposal system, or other point of disposal.

Building house sewer (combined). A building house sewer that conveys sewage in combination with storm water and other clear water wastes.

Building house sewer (sanitary). A building house sewer that carries sewage only.

Cesspool. A lined excavation in the ground which receives the discharge of a drainage system or part thereof, so designed as to retain the organic matter and solids discharging therein, but permitting the liquids to seep through the bottom and sides.

Cesspool. Covered pit with open-jointed lining in its bottom portions into which raw sewage is discharged, the liquid portion of the sewage being disposed of by seeping or leaching into the surrounding porous soil, and the solids or sludge being retained in the pit to undergo partial decomposition before occassional or intermittent removal.

Cesspool. Covered pit with open-jointed lining into which raw sewage is discharged.

Cesspool. A lined and covered excavation in the ground which receives the discharge of domestic sewage or other organic wastes from a drainage system, so designed as to retain the organic matter and solids, but permitting the liquids to seep through the bottom and sides.

Cesspool. Receptacle in the ground which receives crude sewage and is so constructed that the organic portion of such sewage is retained while the liquid portion seeps through its walls or bottom.

Cesspool. A tank with loose joints built in the ground to retain household sewage and permit the waste water to leach through the loose joint into the ground.

Collection and disposal of sewage. All sewage shall be collected and disposed of in properly constructed and managed sewers, treatment facilities, septic tanks, chemical toilets, privies, or by other methods approved by the authority.

Combined sewer. A sewer receiving both surface runoff and sewage.

Combined sewer system. System which carries both sanitary sewage and/or industrial wastes and storm water or drainage.

Domestic wastes. Means liquid wastes from the noncommercial preparation, cooking or handling of food, or containing human excrement and similar matter discharged from the sanitary facilities of dwellings, commercial buildings, industrial facilities and institutions, and other places where humans congregate.

Ground garbage. Means the residue from the preparation, cooking or dispensing of food that has been shredded to such degree that all particles will be carried freely in suspension under the flow conditions normally prevailing in public sewers with no particle greater than one-half inch in any dimension.

House sewer. The extension from the building drain to the public sewer or other place of disposal.

Industrial wastes. Means those particular liquid or other wastes resulting from any process of industry, agriculture, manufacture, trade or business or the development of any natural resource.

Industrial wastes. The liquid wastes from industrial processes as distinct from sanitary sewage.

Industrial wastewater. Means the liquid wastes resulting from the processes employed in industrial, manufacturing, trade or business establishments, as distinct from domestic wastes.

Leaching well or cesspool. Any pit or receptacle having porous walls which permits seepage into the ground.

Leaching well or pit. Pit or receptacle having porous walls which permit the contents to seep into the ground.

Natural outlet. Any outlet into a watercourse, pond, ditch, lake or other body of surface or ground water.

Parts per million. Shall be a weight to weight ratio as the parts per million value multiplied by the factor 8.345 shall be equivalent to pounds per million gallons of water.

Percolation test. Determination of the suitability of an area for subsoil effluent disposal by testing for the rate at which the undisturbed soil in an excavated pit of standard size will absorb water per unit of surface area.

pH. An abbreviation, means the logarithm of the reciprocal of the weight of hydrogen ions in grams per liter of solution.

Private sewage disposal. When waterclosets or other plumbing fixtures are installed in buildings which are not located within a reasonable distance of a sewer, suitable provisions shall be made for disposing of the building sewage by some method of sewage treatment and disposal satisfactory to the administrative authority having jurisdiction.

Private sewer. A sewer privately built by a person and connected to a public sewer.

Private sewer. Sewer privately owned on private property connecting a building drain or drains to a main sewer.

Private sewer. A sewer privately owned and controlled by public authority only to the extent provided by law.

Public sewage disposal. All plumbing fixtures installed in buildings intended for human habitation, occupancy or use in premises abutting on a street, alley or easement in which there is a public sewer shall be connected to such sewer.

Public sewer. A sewer in which all owners of abutting properties have equal rights, and which is controlled by the city.

Sanitary building drain. Building drain that may conduct sewage and clear-water waste, but not storm water.

Sanitary building sewer. Building sewer that may conduct sewage and clear-water waste, but not storm water.

Sanitary drainage system. Drainage system that may conduct sewage or clear-water waste, and includes a combined building drain and combined building sewer.

Sanitary sewage. Means a combination of water-carried wastes from residences, business buildings, institutions and industrial plants, (other than industrial wastes from such plants), together with such groundwaters, surface waters or stormwaters as may be present.

Sanitary sewer. Sewer that may conduct sewage or clear-water waste but not storm water.

Sanitary sewer. Any sanitary sewer owned, operated and maintained by the city and available for public use for the disposal of sewage or a sanitary system approved by the Division of Health of the City.

Sanitary sewer. A sewer which carries sewage and to which storm, surface and ground waters are not intentionally admitted.

Sanitary sewer facility. Public sanitary sewer facility, or a comparable common or package sanitary sewer facility approved by the authorities.

Seepage pit. A lined excavation in the ground which receives the discharge of a septic tank so designed as to permit the effluent from the septic tank to seep through its bottom and sides.

Seepage well or pit. A covered pit with open jointed lining through which the septic tank effluent it receives may seep or leach into the surrounding porous soil.

Septic tank. Watertight receptacle which receives sewage.

Septic tank. A water tight reservoir or tank which receives sewage and by sedinentation and bacterial action effects a process of clarification and partial purification.

Septic tank. A watertight receptacle which receives the discharge of a drainage system or part thereof, and is designed and constructed so as to separate solids from the liquid, digest organic matter through a period of detention, and allow the liquids to discharge into the soil outside of the tank through a system of open-joint or perforated piping, or disposal pit.

Septic tank. A horizontal continuous flow sedimentation tank through which sewage is allowed to flow slowly to permit suspended matter to settle to the bottom where it is retained until an aerobic decomposition is established, resulting in the changing of organic matter into liquid and gaseous substances.

Sewage. Any liquid waste containing animal and vegetable wastes in solution; and may include liquids from laboratories or industrial institutions containing minerals in solution.

Sewage. Liquid waste that contains animal, mineral or vegetable matter.

Sewage. Any liquid waste containing animal or vegetable matter in suspension or solution and may include liquids containing minerals in solution such as from laboratories or industrial establishments.

Sewage. Waste from a flush toilet, bath, sink, lavatory, dishwashing or laundry machine, or the water-carried waste from any other fixture or equipment or machine.

Sewage. Any liquid waste which includes human excreta, wastes from sink, lavatory, bathtub, shower, laundry, and any other liquid waste of organic or chemical nature, either singularly or in any combination thereof.

Sewage. Any liquid waste containing animal or vegetable matter in suspension or solution, and may include liquids containing chemicals in solution.

Sewage. A combination of the water-carried wastes from residences, business buildings, public buildings, institutions and industrial establishments, together with such ground, surface and storm water as may be present.

Sewage disposal system. A system for the disposal of sewage by means of a septic tank, cesspool, or mechanical treatment, all designed for use apart from a public sewer to serve a single establishment, building, or development.

Sewage ejector. A mechanical device used to pump or eject sewage.

Sewage facility. Any sewer, sewage system, sewage treatment works or part thereof designed, intended or constructed for the collection, treatment or disposal of liquid waste, including industrial waste.

Sewage system. Means all service mains and intercepting sewers and structures by which sewage or industrial waste is collected, transported, treated and disposed of. "Sewage system" does not include plumbing inside or in connection with buildings served or sewer laterals from a building or structure to city owned mains.

Sewage tank. Sump that is air tight except for the vent required and that receives the discharge of sewage from a subdrain.

Sewage treatment tank. A water-tight receptacle so constructed as to promote the separation and decomposition of sewage.

Sewer. A pipe or conduit for carrying sewage.

Sludge. Solid waste matter that settles to the bottom of a septic tank.

Sump pit. A tank or pit that receives clear liquid wastes that do not contain organic materials or compounds subject to decomposition, located below the normal grade of the gravity system and that must be emptied by mechanical means.

Sump pump. A mechanical device used to pump the liquid waste from a sump pit into the gravity drainage system.

Suspended solids. Means the total suspended matter that floats on the surface of, or is suspended in, water, wastewater, or other liquids, a high percentage of which is removable by laboratory filtering. Measurement of quantities of suspended solids shall be made in accordance with procedures set forth in the Code.

Wastewater. Means the liquid and water-carried industrial or domestic wastes from dwellings, commercial buildings, industrial facilities, and institutions, together with any groundwater, surface water, and stormwater that may be present, whether treated or untreated, which is discharged into or permitted to enter the system.

2-F WELLS

Abandoned well. A well whose use has been discontinued.

Abandoned well. A well whose use has been discontinued, or which is in such a state of disrepair that continued use for obtaining groundwater or other useful purpose is impracticable.

Aquifer. A geologic formation, group of such formations, or a part of such a formation that is water bearing.

Artesian well. A well, tapping a confined or artesian aquifer.

Casing. Pipe inserted in water wells to prevent the sides from collapsing.

Construction of wells. All acts necessary to construct wells for any intended purpose or use, including the location and excavation of the well; placement of casings, screens and fittings; development and testing.

Drill (drilling). All acts necessary to the construction of water well with power equipment including the sealing of unused water well holes.

Ground water resources. The State finds that improperly constructed, operated, maintained, or abandoned wells can adversely affect the public health and the ground-water resources of the State. Consistent with the duty to safeguard the public welfare, safety, health and to protect and beneficially develop the ground-water resources of the State, it is declared to be the policy of the State to require that the location, construction, repair and abandonment of wells, and the installation of pumps and pumping equipment conform to such reasonable requirements as may be necessary to protect the public, welfare, safety, health and ground-water resources.

Groundwater. Water of underground streams, channels, artesian basins, reservoirs, lakes and other water under the surface of the ground whether percolating or otherwise.

Installation of pumps and pumping equipment. The procedure employed in the placement and preparation for operation of pumps and pumping equipment, including all construction involved in making entrances to the well and establishing seals.

Nonpotable mineralized water. Brackish, saline, or other water containing minerals of such quantity or type as to render the water unsafe, harmful or generally unsuitable for human consumption and general use.

Pitless adapter. A threaded or welded device which provides underground connection between the well casing and the buried piping, and which provides ready access to the drop pipe and any working parts within the well casing in a manner to protect the well from contamination.

Polluted water. Water containing organic or other contaminants of such type and quantity as to render it unsafe, harmful or unsuitable for human consumption and general use.

Pressure tanks. Closed water storage containers constructed so as to operate under normal water system pressures.

Private well. A private well is permitted as a source of water when a public water facility is not available to the premise.

Pump. Any manufactured device designed to either raise the water from the well, or to discharge the water through a distribution system, or both.

Pump room or well room. Any enclosed structure, either above or below grade, that houses the pump, top of the well, any suction line, or any combination thereof.

Pumps and pumping equipment. Any equipment or materials utilized or intended for use in withdrawing or obtaining groundwater including well seals.

Repair. Work involved in deepening, reaming, sealing, installing or changing casing depths, perforating, screening, or cleaning, acidizing or redevelopment of a well excavation, or any other work which results in breaking or opening the well seal.

Rig permit and permit. A permit to operate a water well drilling rig required by the Code.

Water supply well. Any well intended or usable as a source of water supply.

Water supply well. Any well intended or usable as a source of water supply, but not to include a well constructed by an individual on land which is owned or leased by him, appurtenant to a single-family dwelling, and intended for domestic use (including household purpose, farm livestock or gardens).

Water well. Any excavation that is machine drilled cored, bored, washed, driven, jetted when the intended use of such excavation is for the location, diversion, artificial recharge acquisition of groundwater, but such term does not include excavation made for the purpose of obtaining or prospecting for oil, natural gas, minerals or products of mining or quarrying or for inserting media to repressure oil or natural gas or other products.

Water well contractor. Any person who contracts to machine drill, alter or repair any water well.

Water well drilling rig. The power machinery used in drilling a well.

Well. Any excavation that is cored, bored, drilled, jetted, dug or otherwise constructed for the purpose of locating, testing or withdrawing groundwater for potable consumption.

Well. A pipe or conduit installed in the ground for the purpose of providing a method of collection of ground water.

Well seal (sanitary well cap). A device or method used to protect a well casing or water system from the entrance of any external contaminant at the point of entrance into the casing of any pipe or pipes, electric conduits, or water level measuring equipment necessary to the proper functioning of the water system.

2. Article 3

CONCRETE WORK 3-A THROUGH 3-C

3-A CONCRETE — GENERAL

Air-entraining portland cement. The product obtained by pulverizing clinker consisting essentially of hydraulic calcium silicates, with which there has been interground an air-entraining addition.

Aluminate concrete. Concrete made with calcium-aluminate cement; used primarily where high-early-strength or refractory or corrosion-resistant concrete is required.

Architectural concrete. Concrete which will be permanently exposed to view and which therefore requires special care in selection of the concrete materials, forming, placing, and finishing to obtain the desired architectural appearance.

Architectural concrete. Concrete which is cast in a form to produce figures, designs, or textures so as to create an ornamental building surface. Usually used to replace stone masonry.

Average concrete. The term "average concrete" shall mean concrete mixed in accordance with the provisions of the specifications and any other applicable section of the Code.

Backfill concrete. Non-structural concrete used to correct over-excavation, or fill excavated pockets in rock, or to prepare a surface to receive structural concrete.

Beam clamp. Any of various types of tying or fastening units used to hold the sides of beam forms.

Boron-loaded concrete. High-density concrete including a boron-containing admixture or aggregate, such as mineral colemanite, boron frits, or boron metal alloys, to act as a neutron attenuator.

Bush-hammer finish. A finish on concrete obtained by means of a bush-hammer.

Centering. Specialized falsework used in the construction of arches, shells, and space structures, or any continuous structure where the entire falsework is lowered (struck or decentered) as a unit to avoid the introduction of injurious stress in any part of the structure.

Concrete. Mixture of portalnd cement, fine and course aggregates and water.

Concrete. A mixture of portland cement, fine aggregate, coarse aggregate and water.

Concrete. A mixture of cement, aggregates and water, of such proportions and manipulation as to meet specific requirements.

Concrete batch plant. Includes such plants as portland cement concrete plant, transit concrete mixing plant, sand, gravel and cement mixing plants and soil cement mixing plants.

Composite beams. The term "Composite Beam" shall apply to any rolled or fabricated steel floor beam entirely encased in a poured concrete haunch at least (4) inches wider, at its narrowest point than the flange of the beam, supporting a concrete slab on each side without openings adjacent to the beam.

Compression test. Test made on a test specimen of concrete to determine the compressive stress. Compression tests of concrete are made on cyclinders 6 inches in diameter and 12 inches in height.

Consistency. Relative plasticity of freshly mixed concrete or mortar.

Controlled concrete. Concrete mixed in accordance with the requirements of the code.

Controlled concrete. Concrete work on the site shall be inspected by a competent registered architect or engineer, preferabily the one responsible for its design, or by a qualified testing or inspection agency, who shall keep a record which shall indicate the quality and quantity of concrete materials; mixing, placing, and curing of the concrete; the placing of reinforcing steel; the sequence of erection in connection with precast members; all test samples taken; and the general progress of the work. The records shall be available to the building department for inspection during the progress of the work and for 2 years thereafter and shall be preserved by the architect or engineer for that purpose.

Darby. Hand operated straightedge, used in the early stages of a concrete floor pour, preceeding supplemental floating and finishing.

Dry packed concrete. Concrete placed by dry packing.

Early strength. Strength of concrete usually as developed at various times during the first 72 hrs after placement.

Falsework. The temporary structure erected to support work in the process of construction; composed of shoring or vertical posting, formwork for beams and slabs, and lateral bracing.

Fat concrete. Concrete containing a relatively large amount of plastic and cohesive mortar.

Float. Tool, usually of wood, aluminum, or magnesium, used in finishing operations to provide a finish to a fresh concrete surface.

Flying forms. Large mechanically handled sections of formwork; frequently includes supporting truss, beam, or scaffolding units completely unitized. Term usually applies to floor framing system.

Form. A temporary structure or mold for the support of concrete while it is setting and gaining sufficient strength to be self-supporting.

Form anchors. Form anchors are devices used in the securing of formwork to previously placed concrete of adequate strength. The devices normally are embedded in the concrete during placement. Actual load carrying capacity of the anchors depends on their shape and material, the strength and type of concrete in which they are embedded, the area of contact between concrete and anchor, and the depth of embedment and location in the member.

Form hangers. Form hangers are devices used to support formwork loads from a structural steel or precast concrete framework.

Formwork. Total system of support for freshly placed concrete including the mold or sheathing which contacts the concrete as well as all supporting members, hardware, and necessary bracing.

Ganged forms. Prefabricated panels joined to make a larger unit for convenience in erecting, stripping, and reusing.

Granolithic concrete. Concrete suitable for use as a wearing surface finish to floors, made with specially selected aggregate of suitable hardness, surface texture, and particle shape.

Heat-resistant concrete. Any concrete which will not disintegrate when exposed to constant or cyclic heating at any temperature below which a ceramic bond is formed.

High-density concrete. Concrete of exceptionally high density, usually obtained by use of heavyweight aggregates, used especially for radiation shielding.

Initial setting time. The time required for a freshly mixed concrete to achieve initial set.

Insulating concrete. Concrete having low thermal conductivity; used as thermal insulation.

Isolation joint. Joint placed to separate concrete into individual structural elements or from adjacent surfaces.

Jack. A mechanical device used to apply force to prestressing tendons, adjust elevation of forms or form supports, and raise objects small distances.

Jack shore. Telescoping, or otherwise adjustable, single-post metal shore.

Kickouts. Accidental release or failure of a shore or brace.

Low-density concrete. Concrete having an oven-dry unit weight of less than 50 pcf (800 kg/m^3).

Mold oil. Mineral oil that is applied to the interior surface of a clean mold, before casting concrete or mortar therein, to facilitate removal of the mold after the concrete or mortar has hardened.

Monolithic. Monolithic concrete poured in a continuous process so there are no separations or joints.

No-fines concrete. A concrete mixture containing little or no fine aggregate.

No-slump concrete. Concrete with a slump of 1/4 in. (6mm) or less.

Oiled and edge sealed. Surfaces of concrete forms lightly coated with a special oil and the edges sealed.

Permanent shores. The original shores that remain in place without being disturbed during and after formwork removal, and without permitting the new concrete to support its own weight and additional construction loads above.

Plain concrete. Concrete without reinforcement, or reinforced for shrinkage or temperature changes.

Plain concrete. Cast in place without metal reinforcement or reinforced only for shrinkage or temperature changes.

Plain concrete. Concrete without reinforcement or with only minimum reinforcement required to meet shrinkage or temperature stresses.

Plain concrete. A concrete without reinforcing or reinforced for shrinkage or temperature changes only.

Polymer-cement concrete. A mixture of water, hydraulic cement, aggregate, and a monomer or polymer; polymerized in place when a monomer is used.

Polymer concrete. Concrete in which an organic polymer serves as the binder, also known as resin concrete; sometimes erroneously employed to designate hydraulic cement mortars or concretes in which part or all of the mixing water is replaced by an aqueous dispersion of a thermoplastic copolymer.

Post. Vertical formwork member used as a brace; also shore, prop, jack.

Post shore or Pole shore. Individual vertical member used to support loads.

Quality assurance. A system of procedures for selecting the levels of quality required for a project or portion thereof to perform the functions intended, and assuring that these levels are obtained.

Raker. A sloping brace of metal or wood used in formwork as a shore head.

Reshores. Shores placed firmly under a stripped concrete slab or structural member where the original formwork has been removed, thus requiring the new slab or structural member to support its own weight and construction loads posted to it. Such reshores are provided to transfer additional construction loads to other slabs or members and/or to impede deflection due to creep which might otherwise occur.

Scaffolding. A temporary structure for the support of deck forms, cartways, or workmen, or a combination of these such as an elevated platform for supporting workmen, tools, and materials; adjustable metal scaffolding is frequently adapted for shoring in concrete work.

Shore head. Wood or metal horizontal member placed on and fastened to vertical shoring member.

Shores. Vertical or inclined support members designed to carry the weight of formwork, concrete, and construction loads above.

Side form spacers. A side form spacer is a device which maintains the desired distance between a vertical form and reinforcing bars.

Slump test. Test conducted to measure the consistency of freshly mixed concrete in a molded specimen immediately after removal of the mold.

Straightedge. Rigid, straight piece of wood or metal used to strikeoff or screed a concrete surface to proper grade, or to check the planeness of a finished grade.

Tamping. The operation of compacting freshly placed concrete by repeated blows or penetrations with a tamping device.

T-concrete beam. Beam whose cross section resembles a "T." Several T-Beams side-by-side, if acting as a unit, form a floor slab.

Thickened edge slab. Type of combination concrete floor slab foundation where the slab is constructed integrally with the foundation wall.

Transit-mixed concrete. Concrete, the mixing of which is wholly or principally accomplished in a truck mixer.

Tremie concrete. Subaqueous concrete placed by means of a tremic.

Vacuum concrete. Concrete from which water and entrapped air are extracted by a vacuum process before hardening occurs.

Vermiculite concrete. Concrete in which the aggregate consists exfoliated vermiculite.

Vibrated concrete. Concrete compacted by vibration during and after placing.

Vibrator. An oscillating machine used to agitate fresh concrete so as to eliminate gross voids, including entrapped air but not entrained air, and produce intimate contact with form surfaces and embedded materials.

Volumetric batch plant. Bituminous concrete mixing plant that proportions aggregate and bituminous constituents into the mix by volumetrically measured batches.

Waler. Horizontal timbers used in concrete form construction to brace the section.

3-B CONCRETE MATERIALS

Accelerator. A substance which, when added to concrete, mortar, or grout, increases the rate of hydration of the hydraulic cement, shortens the time of setting, or increases the rate of hardening of strength development.

Aggregate. All the materials used in the manufacture of concrete or plaster except water and the bonding agents (cement, lime, plaster). May include sand, gravel, cinders, rock, slag, etc.

Aggregate. An inert material with suitable physical characteristics and composition to form when combined with a cementitious material, a conglomerate having required physical properties, such as concrete, mortar or plaster.

Air entrainment. The occlusion of air in the form of minute bubbles (generally smaller than 1 mm) during the mixing of concrete or mortar.

Barrel (of cement). A quantity of portland cement; 376 lb (4 bags).

Bentonite. A clay composed principally of minerals of the montmorillonite group, characterized by high adsorption and very large volume change with wetting or drying.

Blast-furnace slag. The nonmetallic product, consisting essentially of silicates and aluminosilicates of calcium and other bases, that is developed in a molten condition simultaneously with iron in a blast furnace.

Boulder gravel. Gravel containing particles up to 8 inches in size.

Cement. Cement for mortar shall be Types I, II or III Portland Cement as specified in A.S.T.M. Designation C150 or Types I-A, or III-A air-entraining Portland Cement as specified in A.S.T.M.

Coarse aggregate. Includes, crushed stone, gravel, blast furnace slag, or other approved inert materials of similar characteristics, or combinations thereof, having hard, strong, durable pieces, free from adherent coatings.

Coarse gravel. Gravel containing particles up to 3 inches in size.

Coarse sand. Sand of which more than 50% by weight is retained on a number 20 mesh sieve.

Curing compound. A liquid that can be applied as a coating to the surface of newly placed concrete to retard the loss of water or, in the case of pigmented compounds, also to reflect heat so as to provide an opportunity for the concrete to develop its properties in a favorable temperature and moisture environment.

Dense-graded aggregate. Aggregates graded to produce low void content and maximum weight when compacted.

Dry-shake. A dry mixture of cement and fine aggregate, which is distributed evenly on an unformed surface after water has largely disappeared following the strike-off, and then worked in by floating.

Dusting. The development of a powdered material at the surface of hardened concrete.

Expanded shale (clay or shale). Lightweight vesicular aggregate obtained by firing suitable raw materials in a kiln or on a sintering grate under controlled conditions.

Expansive cement (general). A cement which when mixed with water forms a paste that, after setting, tends to increase in volume to a significantly greater degree than portland cement paste; used to compensate for volume decrease due to shrinkage or to induce tensile stress in reinforcement.

Fine aggregate. Includes, fine granular material, passing a no. 4 screen, such as natural sand as found in the earth, broken rock, or other inert materials having similar physical characteristics.

Fine sand. Sand which at least 50% by weight passes a Number 60 mesh sieve.

Flash coat. A light coat of shotcrete used to cover minor blemishes on a concrete surface.

Graded standard sand. Ottawa sand accurately graded between the U.S. Standard No. 30 (600 μm) and No. 100 (150 μm) sieves for use in the testing of cements.

Gravel. Mixture of mineral grains of which more than thirty (30) percent of weight by a sample is retained by a Number 4 mesh sieve.

Hardener. 1. A chemical (including certain fluosilicates or sodium silicate) applied to concrete floors to reduce wear and dusting. 2. In a two-component adhesive or coating, the chemical component which causes the resin component to cure.

Hydraulic cement. A cement that sets and hardens by chemical interaction with water and that is capable of doing so under water.

Medium sand. Sand of which at least 50% by weight passes a No. 20 mesh sieve and more than 50% by weight is retained on a No. 60 mesh sieve.

Mix. The act or process of mixing; also mixture of materials, such as mortar or concrete.

Neat cement grout. A fluid mixture of hydraulic cement and water, with or without admixture; also the hardened equivalent of such mixture.

Ottawa sand. Silica sand produced by processing of material obtained by hydraulic mining of massive orthoquartzite situated in deposits near Ottawa, Illinois, composed almost entirely of naturally rounded grains of nearly pure quartz; used in mortars for testing of hydraulic cement.

Perlite. A volcanic glass having a perlitic structure, usually having a higher water content than obsidian; when expanded by heating, used as an insulating material and as a lightweight aggregate in concretes, mortars, and plasters.

Pozzolan. Siliceous or siliceous and aluminous material which in itself possesses little or no cementitious value, but will in finely divided form and in the presence of moisture, chemically react with calcium hydroxide at ordinary temperatures to form compounds possessing cementitious properties.

Proportioning. Selection of proportions of ingredients for mortar or concrete to make the most economical use of available materials to produce mortar or concrete of the required properties.

Reactive aggregate. Aggregate containing substances capable of reacting chemically with the products of solution or hydration of the portland cement in concrete or mortar under ordinary conditions of exposure, resulting in some cases in harmful expansion, cracking, or staining.

Sand. Mixture of mineral grains which has no cohesion when dry and of which at least seventy (70) percent by weight passes a Number 4 mesh sieve and of which not more than fifteen (15) percent by weight passes through a Number 200 mesh sieve.

Shotcrete. Mortar or concrete pneumatically projected at high velocity onto a surface; also known as air-blown mortar; also pneumatically applied mortar or concrete, sprayed mortar and gunned concrete.

Vermiculite. A micaceous mineral, or hydrous silicate, derived generally from the alteration of some kinds of mica which expand when heated. Used in the expanded form as a lightweight aggregate.

Water. Used in mortar, grout, or masonry work shall be clean and free from injurious amounts of oil, acid, alkali, organic matter, or other harmful substances.

Water-cement ratio. The ratio of the amount of water, exclusive only of that absorbed by the aggregates, to the amount of cement in a concrete or mortar mixture; preferably stated as a decimal by weight.

Workability. That property of freshly mixed concrete which determines the ease and homogencity with which it can be mixed, placed, compacted, and finished.

3-C REINFORCED CONCRETE

Aggregate. Inert material which is mixed with Portland Cement and water to produce concrete.

Beam-and-slab floor. A reinforced concrete floor system in which the floor slab is supported by beams of reinforced concrete.

Beam-column. A structural member which is subjected to forces producing significant amounts of both bending and compression simultaneously.

Bond beam Continuous structural member having the same thickness as the wall of which it is a part of and which is designed and constructed to provide lateral stability to the wall.

Bond beam. A continuous beam, usually of reinforced concrete, but sometimes of reinforced brick or concrete block placed in masonry walls to tie them together and add lateral stability. It also distributes concentrated vertical loads along the wall.

Bonded reinforcement. Reinforcement bonded throughout its length to the surrounding concrete.

Column. An upright compression member, the length of which exceeds (3) times its least lateral dimension.

Column capital. An enlargement of the upper end of a reinforced concrete column designed and built to act as a unit with the column and flatslab.

Column strip. A portion of a flat slab panel one-half (1/2) panel in width occupying the two quarter (1/4) panel areas outside of the middle strip, and extending through the panel in the direction in which bending moments are being considered.

Combination column. A column in which a structural steel section is designed to carry the principal part of the load, is wrapped with wire and encased in concrete of such quality that some additional load may be allowed.

Composite column. A column in which a steel or cast-iron section is completely encased in concrete containing reinforcement of spiral reinforcement and longitudinal bars.

Concrete. A mixture of Portland Cement, fine aggregate, coarse aggregate and water.

Concrete. A mixture of cement, aggregates and water, of such proportions and manipulation as to meet specific requirements.

Deformed bar. Reinforcing bars with closely spaced shoulders, lugs or projections formed integrally with the bar during rolling so as to firmly engage the surrounding concrete. Wire mesh with welded intersections not farther apart than twelve inches in the direction of the principal reinforcement, and with cross wires not smaller than No. 10 W. & M. Gage may be rated as a deformed bar.

Deformed reinforcement. Metal bars, wire, or fabric with a manufactured pattern of surface ridges which provide a locking anchorage with surrounding concrete.

Diagonal band. Group of bars, covering a width approximately 0.4 symmetrical with respect to the diagonal running from corner to corner of the panel of a flat slab.

Drop-in-beam. A simple beam, usually supported by cantilever arms, with joints so arranged that it is installed by lowering into position.

Drop panel. The thickened structural portion of a flat slab in the area surrounding column, column capital, or bracket, in order to reduce the intensity of stresses.

Dropped panel. The structural portion of a flat slab which is thickened throughout an area surrounding the column capital.

Edge-bar reinforcement. Tension steel sometimes used to strengthen otherwise inadequate edges in a slab, without resorting to edge thickening.

Effective area of concrete. The area of a section which lies between the controid of the tensile reinforcement and the compression face of a slab or beam.

Effective area of reinforcement. The area obtained by multiplying the right cross sectional area of the metal reinforcement by the cosine of the angle between its direction and that for which the effectiveness of the reinforcement is to be determined.

End-anchored reinforcement. Reinforcement, in concrete, provided at its ends with anchorage capable of transmitting the tensioning forces to the concrete.

Expanded metal lath. A metal network, often used as reinforcement in concrete or mortar construction, formed by suitably stamping or cutting sheet metal and stretching it to form open meshes, usually of diamond shape.

Field bending. Bending of reinforcing bars on the job.

Flat plate. Flat slab without column capitals or drop panels.

Flat slab. Concrete slab reinforced in two (2) or more directions, generally without beams or girders to transfer the loads to supporting columns.

Ground wire. Small-gage high-strength steel wire used to establish line and grade as in shotcrete work; also called Alignment wire or Screed wire.

Heavy-edge reinforcement. Wire fabric reinforcement, for highway pavement slabs, having one to four edge wires heavier than the other longitudinal wires.

Indented wire. Wire having machine-made surface indentations intended to improve bond; depending on type of wire, may be used for either concrete reinforcement or pretensioning tendons.

Lift slab. Construction system in which the floor and roof slabs are cast one on top of the other at ground level and are then jacked into position and fastened to the columns.

Main reinforcement. Steel reinforcement designed to resist stresses resulting from design loads and moments, as opposed to reinforcement intended to resist secondary stresses.

Mechanical bond. Physical interlock between cement paste and aggregate, or between concrete and reinforcement (specifically, the sliding resistance of an embedded bar and not the adhesive resistance).

Middle strip. Portion of a flat slab panel, one-half panel in width, symmetrical with respect to the panel center line and extending through the panel in the direction in which bending moments are being considered.

Negative reinforcement. Steel reinforcement for negative moment.

Positive reinforcement. Shall mean, reinforcement so placed as to resist tensile stress due to positive bending moment.

Post-tensioning. Method of prestressing reinforced concrete in which the reinforcement is tensioned after the concrete has hardened.

Precast concrete. Concrete structural components which are not formed and poured in place in the structure, but are cast separately either at a separate location or on site.

Prestressed concrete. Concrete in which there have been introduced internal stresses of such magnitude and distribution that the stresses resulting from the service loads are counteracted to a desired degree. In reinforced concrete, the prestress is commonly introduced by tensioning the reinforcement.

Ratio of reinforcement. The term "ratio of reinforcement" shall mean ratio of the effective area of the reinforcement cut by a section of a beam or slab to the effective area of the concrete at that section.

Rebar. Abbreviation for "reinforcing bar."

Rectangular Direction. The term shall mean a direction parallel to a side of a flat slab panel.

Reinforced concrete. Concrete in which reinforcement other than that provided for shrinkage or temperature changes is combined in such manner that the two materials act together in resisting forces.

Reinforced concrete. Concrete in which reinforcement is embedded in such a manner that the concrete and steel act together in resisting forces.

Reinforced concrete. Concrete in which metal is embedded in such a manner that the two materials act together in resisting stresses.

Reinforced concrete. Concrete in which steel rods or mesh are embedded to increase strength.

Reinforced concrete. Concrete in which reinforcement other than that provided for shrinkage or temperature changes is embedded in such a manner that the two materials act together in resisting forces.

Reinforcing. Steel rods or metal fabric placed in concrete slabs, beams, columns or masonry to increase their strength.

Reinforcing. To strengthen. Steel rods or mesh are embedded in concrete to increase the strength in tension..

Reinforcing steel. Shall conform to the physical and chemical requirements for metal reinforcement in concrete, as specific in the Code.

Spandrel beam. Beam in the perimeter of a building, spanning between columns and usually supporting floors or roofs.

Splice. Connection of one reinforcing bar to another by lapping, welding, mechanical couplers, or other means; connection of welded wire fabric by lapping.

Stirrup. Reinforcement used to resist shear and diagonal tension stresses in a structual member; typically a steel bar bent into a U or box shape and installed perpendicular to or at an angle to the longitudinal reinforcement, and properly anchored; lateral reinforcement formed of individual units, open or closed, or of continuously wound reinforcement.

Tilt-up. Method of concrete construction in which members are cast horizontally at a location adjacent to their eventual position, and tilted into place after removal of molds.

Tilt-up construction. Method of construction where concrete wall sections are cast horizontally on the ground and tilted or lifted into position.

Transverse reinforcement. Reinforcement at right angles to the principal axis of a member.

Unbonded reinforcement. Reinforcement not bonded throughout its length to the surrounding concrete.

2. Article 4

MASONRY WORK 4-A THROUGH 4-D

4-A MASONRY — GENERAL

Approved masonry. Means masonry constructed of brick, stone, concrete, hollow-block, solid block, or other material approved after test, or a combination of these materials.

Ashlar masonry. The term "ashlar masonry" shall mean masonry of natural or manufactured stone rectangular units larger in size than brick having sawed, dressed or squared beds, and the joints of which are laid in mortar with proper bond.

Ashlar masonry. Is masonry composed of properly bonded rectangular units of burned clay or shale, or natural or cast stone, larger in size than brick, having sawed, dressed, or squared beds, and joints laid in mortar.

Ashlar masonry. Masonry composed of bonded, rectangular units, larger in size than brick, with sawed, dressed or squared beds and mortar joints.

Backing. Part of a masonry wall behind the exterior facing.

Bed. Horizontal surfaces on which masonry units are laid on a wall.

Belt. Course of masonry projecting from the face of the wall.

Blocking. Usually course of masonry crowning the top of the wall.

Bond. The physical connection of masonry units by the addition of a cementious or an adhesive material and lapping them in such an order as to accomplish a unit of single construction of a wall.

Bond. The adherence of one construction element to another. In masonry construction, the arrangement of bricks or concrete block, especially the arrangement of vertical joints.

Bonder (header.) A masonry unit which ties 2 or more wythes (leaves) of the wall together by overlapping.

Bonding. The facing and backing of cavity walls shall be bonded with 3/16in. diameter metal unit ties or the equivalent or/with the equivalent of metal reinforcement having #9 inch longitudinal rods and #9 gauge cross wires. Metal ties shall be of corrosion resistant metal or coated with a corrosion-resistant metal, or other approved protective coating.

Brick Masonry. The term "Brick Masonry" as used in the Code, shall be defined to mean masonry of burned clay or shale brick, sand-lime brick or concrete brick as specified in the Code.

Brick veneer. Facing of brick laid against the frame or rough masonry wall construction.

Brick veneer. A non-load-bearing single tier of brick applied to a wall of other materials.

Cavity wall masonry. Type of construction made with brick, structural clay tile or hollow concrete masonry units or any combination of such units in which facing and backing are completely separated except for the metal ties which serve as bonding. Type A, B, or C mortar shall be used in cavity walls having a nominal thickness of ten inches (10").

Chase. Groove or shaft in a masonry wall provided for the accommodation of pipes, ducts, or conduits.

Coping. Capping of masonry walls with a material which will protect the joints of the wall, provide a watershed and present a finished appearance to the wall.

Corbel. Masonry courses which have been stepped out to form a ledge for the support of structural members.

Corbels. Built only into solid masonry walls twelve inches (12") or more in thickness. The projection for each course in such corbel shall not exceed one inch (1"), and the maximum projection shall not exceed one-third the total thickness of the wall when used to support structural members and not more than six inches (6") when used to support a chimney built into the wall. The top course of all corbels shall be a header course.

Coursed rubble. Masonry composed of roughly shaped stones fitting approximately on level beds of mortar and well bonded.

Dry masonry. Masonry units that are not separated or held in place by mortar.

Exterior masonry. Building of exterior masonry construction if all enclosing walls are constructed of masonry or reinforced concrete with fire-resistive ratings as set forth in the code.

Facing. Outer wythe in a masonry wall or the exterior surface of a structure.

Grout. Mortar of pouring consistency.

Grout. Mortar into which sufficient additional water has been incorporated to produce a creamy consistency and permit the mixture to flow readily.

Grout. Mortar made of such consistency by the addition of water that it will just flow into the joints and cavities of the masonary work and fill them solid.

Grouted masonry. Solid masonry in which the outside and inside tiers or wythes are laid in full head and bed joints of mortar and all interior joints are filled with grout made of mortar of the same type as that used in laying the outer tiers or wythes.

Grout lift. An increment of grout height within the total pour; a pour may consist of one or more lifts.

Grout pour. The total height of masonry wall to be poured prior to the erection of additional masonry. A pour will consist of one or more lifts.

Gypsum masonry. Form of construction made with gypsum blocks or tile in which the units are laid and set in gypsum mortar. No gypsum masonry shall be used in any bearing wall or in any location where the gypsum will be directly exposed to weather or where subject to frequent or continuous wetting.

Header. Course of brick in which the masonry units are laid perpendicular to the face of the wall to tie 2 tiers of brick together.

Jack arch. One having horizontal or nearly horizontal upper and lower surfaces. Also called FLAT or STRAIGHT ARCH.

Joint. Space between masonry units—usually filled with mortar.

Jointer. Tool used for smoothing or indenting the surface of a mortar joint.

Joint grouting and pointing. Joints should be carefully packed with a mortar composed of one part Portland Cement, 1/4 part lime, and three parts of clean sand, mixed to buttery consistency. In pointing care shall be exercised to avoid deposits of excess mortar along the perimeter edges of stones. Deposits should be completely removed at once with a cellulose sponge to avoid staining of the material. Stones should be finally finished by rubbing in circular motion with dry sharp sand and small gunny sacks. Pointing of joints shall be done the same day the material is laid.

Keystone. Last wedge-shaped stone placed in the crown of an arch regarded as binding the whole.

Masonry. A built-up construction or combination of building units of such materials as clay, shale, concrete, glass, gypsum, or stone, set in mortar.

Masonry. Construction composed of separate units such as brick, block, hollow tile, stone or approved similar units or a combination thereof, laid up or built unit by unit and bonded by approved manner.

Masonry. Construction consisting of units made of clay, shale, concrete, glass, gypsum, stone or similar materials set in mortar.

Masonry. Includes but is not limited to architectural terra cotta, brick, and other solid masonry units of clay or shale, concrete masonry units, glazed building units, gypsum tile or block, plain concrete, stone, structural clay tile, structural glass block, or other similar building units or materials, or combination of same, bonded together with mortar.

Masonry. A form of construction composed of stone, brick, concrete, gypsum, hollow clay tile, concrete block or tile, or other similar building units or materials or combination of these materials laid up unit by unit and set in mortar.

Masonry. A built-up construction or combination of building units of such materials as clay, shale, concrete, glass, gypsum or stone set in mortar, or plain concrete.

Masonry. Construction of assembled units of stone, brick, concrete, gypsum or other similar incombustible materials separated from one another and held in place by mortar.

Masonry. Stone, brick, structural clay tile, concrete masonry units, gypsum tile or block, structural glass block or other similar building units or materials or a combination of same bonded together with mortar. Masonry also includes plain concrete.

Masonry. A construction of units of such materials as clay, shale, concrete, glass, gypsum, or stone, set in mortar. Plain concrete, because of its structural similarity to masonry, is often considered masonry.

Masonry. A construction of units of such materials as clay, shale, concrete, glass, gypsum, or stone, set in mortar, including plain concrete, but excluding reinforced concrete.

Masonry. Construction composed of shaped or molded units, usually small enough to be handled by one man and composed of stone, ceramic brick or tile, concrete, glass, adobe, or the like; sometimes used to designate cast-in-place concrete.

Masonry. A built-up construction or combination of building units or materials of clay, shale, concrete, glass, gypsum, stone or other approved units bonded together with mortar or monolithic concrete. Reinforced concrete is not classed as masonry.

Masonry. A dimension that may vary from actual masonry dimensions by the thickness of a mortar joint, but not to exceed one-half (1/2) inch.

Masonry. That form of construction, composed of stone, brick, concrete, gypsum, hollow clay tile, concrete block or tile, or other similar building units or materials, or a combination of these materials, laid up unit by unit and set in mortar. For the purpose of the Code plain monolithic concrete shall be considered as Masonry.

Masonry bond. A bond obtained from the arrangement of masonry units and the use of mortar and without other fastening devices.

Masonry bond. Bond obtained from the arrangement of the units and the use of mortar, epoxy or other methods or materials.

Masonry filler unit. Masonry unit used to fill in between joists or beams to provide a platform for a cast-in-place concrete slab.

Masonry mortar. Mortar used in masonry structures.

Masonry of hollow units. Masonry consisting wholly or in part of hollow masonry units laid contiguously in mortar.

Masonry (solid). Masonry built without hollow spaces.

Masonry unit. A construction unit in masonry.

Masonry veneer. Masonry veneer on wood frame structures shall be securely attached to the backing by corrosion-resistant corrugated metal ties, not less than No. 22 ga. in thickness and 7/8 inches in width or equivalent. One tie shall be used for at least each 2 square feet of wall area and the distance between ties shall not exceed 24 inches or by No. 13 ga, metal ties or equivalent located 36 inches horizontally and 18 inches vertically.

Mortar. Plastic mixture of cementitious material, fine aggregate and water.

Mortar. A pasty mixture of cement, lime, sand and water, used as a bonding agent for brick, stone or other masonry units.

Mortar. A plastic mixture of approved cementitious materials, fine aggregates and water used to bond masonry or other structural units.

Nominal dimension. Dimension that may vary from an actual masonry dimension by the thickness of a mortar joint, but not more than one-half (1/2) inch.

Parge. To coat with plaster, particularly foundation walls and rough masonry.

Parging. Thin coating of mortar applied to the inside surface of a masonry wall. Used on the exterior face of below grade walls as waterproofing or used to smooth a rough masonry wall.

Pointing. Process of removing deteriorated mortar from masonry and replacing it with new mortar, also the final patching, filling or finishing of mortar joints in new masonry work.

Quarry. Open, surface excavated pit, for the mining and removal of granite, marble, slate, limestone and similar stones used for the building industry.

Quarry. Opening made in an outcrop of rock with the purpose of obtaining stone for commercial purposes.

Quarry. Location of an operation where a natural deposit of stone is removed from the ground.

Quarrying. Removal of rock which has value because of its physical characteristics.

Queen-closure. Half-brick, usually of face brick quality, made by cutting a whole brick in half, lengthwise.

Quoin. Masonry design at exterior vertical outside corners, usually projecting slightly from the face of the wall.

Rake. In masonry it is a method of laying brick courses in an angular designed fashion.

Raked joint. Type of brick joint in which the mortar while fresh is removed or raked out to a specified depth.

Random rubble. Masonry composed of roughly shaped stone, laid in mortar without regularity of coursing, but fitting together to form well-defined joints.

Reinforced filled masonry cell construction. In walls or hollow unit masonry, structural members may be built by filling continuous cores or spaces with concrete or grout in which reinforcement is embedded. Such members may be designed as specified for reinforced brick masonry in the Code. The area of such core walls in contact with the fill, and of the face shells of units containing such cores not exceeding the length of one unit, may be included in the computation of the effective areas of the section. In such walls the required horizontal steel may be concentrated in bond beams and at the tops and bottoms of walls and openings. The minimum steel required shall be calculated on the gross area of the wall.

Reinforced grouted brick masonry. Reinforced brick masonry in which the continuous collar joint or cavity is filled with grout and steel reinforcement as the wall is built.

Reinforced masonry. Unit masonry in which reinforcement is embedded in such a manner that the two materials act together in resisting forces.

Rough or ordinary rubble. Masonry composed of unsquared or field stones laid without regularity of coursing but well bonded.

Rubble masonry. Masonry composed of roughly shaped stones.

Running bond. Repetitive arrangement in masonry construction by which units of not more than seven (7) successive courses overlaps units of the next lower course by the length of a half unit, the seven (7) successive courses being followed by a course of header units.

Solid masonry. Masonry consisting of stone, brick, sand-lime or concrete brick, or other solid masonry units, or a combination of these materials laid contiguously with the spaces between the units filled with mortar, or monolithic concrete.

Solid masonry. Masonry consisting of solid masonry units laid contiguously with the mortar joints. For the purposes of the Code, plain concrete shall be deemed to be solid masonry.

Solid masonry. Masonry built without hollow spaces or cavities. A wall with not more than 25% cavities or voids shall be considered solid masonry.

Solid masonry. Masonry consisting of solid masonry units laid continguously with the joints between the units filled with mortar or an unreinforced concrete.

Solid masonry. Masonry built without hollow spaces.

Solid masonry. Masonry consisting of solid masonry units laid contiguously with the joints between the units filled with mortar, or consisting of plain concrete.

Solid masonry. Masonry consisting wholly of solid masonry units laid contiguously in mortar.

Veneer. Nonstructural facing attached to a wall or other structural assembly for the purpose of providing ornamentation, protection, or insulation.

Veneer. Facing of any material attached to a wall for the purpose of providing protection, insulation or ornamentation, and not forming an integral part of the wall or contributing to load-bearing or stability.

Veneer. A facing of brick, stone, glass, concrete, tile or similar material attached to a wall for the purpose of providing ornamentation, protection, or insulation but not counted as adding strength to the wall.

Veneer. A facing of brick, stone, concrete, tile, metal, wood, glass or the like attached to a wall. Veneer shall not be calculated as contributing to the strength of a wall.

Veneer. Thin sheets of wood.

Veneered. The surfacing of a wall in which the veneering of brick, stone, concrete or tile is provided for the purpose of ornamentation, protection or insulation, but which is not bonded to the backing in such a manner as to be counted on as adding strength to the wall.

Wythe. The partition between two (2) chimney flues in the same stack. Also the inner and outer walls of a cavity wall.

Wythe. Inner or outer part of a cavity wall/or each continuous vertical section of a wall one masonry unit in thickness.

4-B MASONRY BUTTRESS, PIERS AND PILASTERS

Buttress. Masonry projecting from a wall to provide additional strength against the thrust of a roof or vault.

Buttress. Structural projection which is an integral part of a wall, primarily to provide resistance to lateral forces.

Buttress. An abutting pier or brace which strengthens or supports a wall, by opposing the horizontal forces.

Buttress. A projecting part of a wall built integrally therewith to furnish lateral stability which is supported on proper foundations.

Buttress. The term "buttress" shall mean a masonry structure built against and bonded into a wall.

Buttress. A projecting part of a masonry wall built integrally therewith to furnish lateral stability which is supported on proper foundations.

Flying buttress. Detached buttress or pier of masonry at some distance from a wall connected to the wall by an arch or part of an arch.

Pier. Isolated, above ground column of masonry; a bearing wall not bonded at the sides into associated masonry shall be considered a pier when its horizontal dimension measured at right angles to the thickness does not exceed four times its thickness.

Pier. An isolated column of masonry.

Pier. A vertical body of masonry used as a column, the portion of a masonry wall between thinner portions or between openings when the horizontal dimension parallel to the wall does not exceed four times the thickness.

Pier. An isolated column of masonry. A bearing wall not bonded at the sides into associated masonry and whose length does not exceed four times its thickness.

Piers. An isolated column of masonry. A bearing wall not bonded at the sides into associated masonry shall be considered a pier when its horizontal dimension measured at right angles to the thickness does not exceed four times its thickness. The least dimension shall not be less than 1/30 of the span, in inches, and the height shall not exceed 10 times the least dimension for solid or grouted masonry piers or 6 times the least dimension for hollow masonry piers.

Pilaster. An unreinforced masonry section bonded to the adjoining wall by one of the following mehtods: by the use of pilaster blocks, by alternate course bond of masonry with adjoining wall.

Pilaster. A part of a masonry or reinforced concrete wall projecting four (4) inches or more from one (1) or both faces and generally extending for the full height of the wall.

Pilaster. Column usually formed of the same material, and integral with, but projecting from, a wall.

Pilaster. Pier forming part of a masonry or concrete wall, partially projecting therefrom and bonded thereto.

Pilasters. The least dimensions in inches for pilasters carrying beams, trusses, and girders shall be not less than 1/40 the span and the height shall not exceed 12 times the least dimensions for solid or hollow masonry. The dimension of pilasters used for lateral stability only, shall be no less than 4 inches greater in thickness than the principal wall nor less than 16 inches in length.

4-C MASONRY MATERIALS

Architectural terra cotta. Is plain or ornamented hard-burned building units, larger in size than brick, consisting of mixtures of plastic clays, fusible materials, and grog, and having a glazed or unglazed ceramic finish.

Architectural terra cotta. Plain or ornamental hard-burned plastic clay units, larger in size than brick, with glazed or unglazed ceramic finish.

Ashlar. Masonry composed of squared stones; one pattern of masonry construction.

Ashlar. Wall construction of cut stone. Also refers to the square cut stones used in ashlar masonry work.

Ashlar. Facing of thin slabs of stone or terra cotta, which covers the rough brick, or concrete block and structural steel in the exterior walls of a building.

Ashlar facing. The term "ashlar facing" shall mean facing composed of solid rectangular units of burnt clay or shale, natural or manufactured stone, larger in size than brick, with sawed, dressed or squared beds, and joints laid in mortar and used in facing masonry walls.

Ashlar facing. Is a facing composed of solid rectangular units of burned clay or shale, or natural or cast stone, larger in size than brick, having sawed, dressed, or squared beds, and joints laid in mortar.

Ashlar facing. Facing of solid rectangular units larger in size than brick of burned clay or shale, natural or cast stone, with sawed, dressed and squared beds and mortar joints.

Backup. The lower cost material in a masonry wall which is covered by a facing of more expensive and ornamental material such as face brick, stone, marble, metal panels, etc.

Bonder. A masonry unit which overlaps two (2) or more adjacent wythes of masonry to bind or tie them together.

Bonder (header). A masonry unit which overlaps two (2) or more adjacent wythes of masonry to bind or tie them together.

Brick. The term "Brick" as used in the Code shall be defined to mean a structural unit of burned shale or clay, sand-lime or concrete, usually solid and about eight inches by three and three-quarters inches by two and one-half inches (8" × 3-3/4" × 2-1/2") in size.

Brick. The term "brick" shall mean a structural unit of burned clay or shale, formed while plastic into a rectangular prism, usually solid and approximately eight inches by three and three-quarters inches by two and one-quarter inches in size, the net cross-sectional area of which shall be at least seventy-five percent of the gross cross-sectional area. Similar structural units made of other substances, such as lime and sand, cement and suitable aggregates or fire clay, shall be considered as brick within the meaning of the Code.

Brick. Means a solid masonry unit, usually solid, having a shape approximating a rectangular prism not larger than twelve (12) inches by four (4) inches by four (4) inches. A brick may be made of burned clay or shale, of lime and sand, of cement and suitable aggregates, or of fire clay or other approved materials.

Brick. Is a material of construction in small, regular units, formed from inorganic substances and hardened in a shape approximately 2-1/4 × 3-3/4 × 8 inches or 2-1/3 × 3-5/8 by 7-5/8 inches or 3-5/8 × 3-5/8 × 11-5/8 inches in size, the net cross-sectional area of which in any plane parallel to the bearing surface is not less than 75 percent of its gross cross-sectional area measured in the same plane, except that brick used in radial brick chimneys may have a net cross-sectional area 65 percent of the gross cross-sectional area measured in the same manner.

Brick. A solid masonry unit having a shape approximating a rectangular prism, not larger than 12 × 4 inches. A brick may be made of burned clay or shale, of lime and sand, of cement and suitable aggregates, or of fire clay or other approved materials.

Brick. A solid masonry unit having a shape of a rectangular prism, made of burned clay or shale, of fire clay or mixtures thereof, of lime and sand, of suitable aggregates, or of other approved materials.

Brick. A masonry unit having a shape approximately a rectangular prism, usually not larger than 12" × 4" × 4", made of burned clay or shale, of fire or clay mixtures thereof, or of lime and sand, or cement and suitable aggregates, or other approved materials.

Brick. A masonry unit having a shape approximately a rectangular prism, made from burned clay, shale or mixture thereof.

Brick. A solid masonry unit not larger than 16 × 4 × 8 inches.

Brick (clay or shale). A solid masonry unit of clay or shale, usually formed into a rectangular prism while plastic and burned or fired in a kiln.

Brick made from clay or shale. Building brick of clay or shale shall be of a quality at least equal to that required by A.S.T.M. Designation C62 or C216. When in contact with the ground, brick shall be of at least Grade MW. Where severe frost action occurs in the presence of moisture, brick shall be at least Grade S.W.

Brick made from sand-lime. Building brick made from sand-lime shall be of a quality at least equal to that required by A.S.T.M. Designation C73. When in contact with the ground, brick shall be of at least Grade MW. Where severe frost action occurs in the presence of moisture, brick shall be at least Grade SW.

Brick veneer. A non-loadbearing single tier of brick applied to a wall of other materials.

Building brick. Building brick is a masonry unit, not less than 75 percent solid, having a shape approximating a rectangular prism and usually not larger than 4 inches by 4 inches by 12 inches. Brick may be made of burned clay or shale or mixtures thereof, of lime and sand or Portland Cement and suitable aggregates.

Building bricks in structure. All building brick shall be free from cracks, laminations and other defects or deficiencies which may interfere with proper laying of the brick or impair the strength or permanence of the structure.

Cast building stones. Shall conform to A.C.I. Designation 704. Every concrete unit more than 18 inches in any direction shall conform to the requirements for concrete in the Code.

Cast stone. Precast building stone manufactured from portland cement concrete and used as a trim, veneer, or facing on or in buildings or structures.

Cast stone. Precast concrete material made to resemble stone.

Cast stone. Shall be of a quality at least equal to that required by A.C.I "Specifications for Cast Stone". All cast stone shall be branded with a permanent identification mark of the manufacturer which shall be registered with the Director.

Cast stone identification. Each piece of cast stone shall be marked or otherwise identified with a permanent identification symbol of the manufacturer which shall be registered with the Building department.

Clay building brick. All building brick made of burned clay or shale shall conform to the requirements of ASTM C-62.

Clay masonry unit. A building unit larger in size than a brick composed of burned clay, shale, fireclay or mixtures thereof.

Concrete brick. Building brick of concrete shall be of a quality at least equal to that required by A.S.T.M. Designation C55.

Concrete brick. Approved concrete brick for use when exposed to freezing in the presence of moisture, shall have a minimum compressive strength of 3500 psi and when used as a backup in exterior walls or for general interior construction shall have a compressive strength of not less than 1250 psi, or shall meet the requirements of ASTM C-55.

Concrete masonry unit. Unit made from cement and suitable aggregates such as sand, gravel, crushed stone, cinders, burned clay, shale, or blast-furnance slag.

Concrete masonry unit. A building unit or block larger in size than 12 × 4 × 4 inches made of cement and suitable aggregates.

Concrete masonry units. Concrete masonry units shall be of a quality at least equal to that required by A.S.T.M. Designation C90 or C145, when used for bearing walls or piers, or when in contact with ground or exposed to the weather; or equal to A.S.T.M. Designation C129, when used for non-bearing purposes and not exposed to the weather. Solid units subject to the action of weather or soil shall be Grade A.

Concrete products. Products such as bricks, blocks or other units made of cement, aggregates and water.

Concrete sand-lime building brick. Brick made from sand-lime shall conform to requirements of the standard specifications of ASTM C-73.

Glass block. Hollow structural glass block laid as masonry for translucent effect in wall construction.

Glass masonry. The term "Glass Masonry" as used in the Code shall be defined to mean masonry of hollow glass blocks as specified in the Code.

Glazed building units. Shall conform to the structural requirements for building brick of clay or shale, and glazed structural tile shall conform to the structural requirements for structural clay tile.

Glazed building units. Glazed building units shall conform to requirements of A.S.T.M. "Tentative Specifications for Glazed Masonry Units", except that the requirements for finish shall not apply to salt-glazed building units.

Grade M.W. brick. Used where exposed to temperatures below freezing, but where brick are not likely to be permeated with water or where a moderate degree of resistance to frost action is permissible. (M.W.—Moderate Weather)

Grade N.W. brick. Used for backup or for interior construction exposed for use where no frost action occurs. (N.W.—Normal Weather)

Gypsum units. Gypsum partition tile or block shall be of a quality at least equal to that required by A.S.T.M. Designation C52.

Hollow clay tile. For exterior walls and bearing walls shall meet the requirements for the "5-15" clay tile given in the A.S.T.M. "Standard Specifications and Tests for Structural Clay Load Bearing Wall Tile". The exterior shell of such tile shall be not less than three-fourths of an inch (3/4") thick, except that a tolerance of one-sixteenth of an inch (1/16") will be permitted in such shell thickness. Hollow clay tile for non-bearing partitlons, fire protection and furring, shall meet the requirements of A.S.T.M. "Standard Specifications for Structural Clay Non-Load-Bearing Tile". Hollow Clay Tile for floor construction shall meet with the requirements of the "Standard Specifications for Structural Clay Floor Tile," (A.S.T.M.).

Hollow masonry unit. A masonry unit whose net cross-sectional area in any plane parallel to the bearing surface is less than 75 per cent of its gross cross-sectional area measured in the same plane.

Identification of brick. All building brick shall be of distinctive design or appearance, or so marked so that the manufacturer can be identified.

Jumbo brick. Generic term indicating a brick larger in size than the standard. Some producers use this term to describe oversize brick of specific dimensions manufactured by them.

Kiln. Large furnace used for baking drying or burning fire brick or refractories, or for calcining ores or other substances.

Kiln run. Brick or tile from one klin which have not been sorted or graded for size or color variation.

King closer. Brick cut diagonally to have one 2-in., end and one full width end.

Marble veneers. Class "A" marble shall be hard, sound marble and free of any unsound lines. It shall be of uniform thickness and sizes.

Masonry. The following materials, when laid up in mortar, shall be classed as masonry, and when used in any building, shall conform to the minimum requirement of the Code:

a. Brick
b. Concrete Block and Tile
c. Gypsum Block and Tile
d. Hollow Clay Tile
e. Stone
f. Concrete—Plain
g. Gypsum—Plain, poured
h. Structural Glass Blocks
i. Glazed Building Units

Masonry of hollow units. The term "Masonry of Hollow Units" as used in the Code shall be defined to mean masonry of hollow tile, concrete tile or blocks or gypsum tile or blocks as specified in the Code.

Masonry units. Masonry unit whose net cross-sectional area in any plane parallel to the bearing surface is less than 75% of its gross cross-sectional area measured in the same plane.

Quarry run. Stones taken directly from the quarry, and having no sawing, dressing or other finishing performed on them.

Quarry stone. Rock broken out of its natural stratification by drilling and splitting.

Second hand masonry materials. May be used in masonry when such materials conform to the provisions of the Code for corresponding new materials; provided that such second hand materials are sound and free from defects which would impair their suitability for re-use and have been cleaned of old mortar and other adherent coatings which would prevent proper assembly or bond, and their use has been approved by the Building Department.

Solid masonry unit. Masonry unit whose net cross-sectional area in every plane parallel to the bearing surface is seventy-five percent (75%) or more of its gross cross-sectional area measured in the same plane. Net cross-sectional area shall be taken as the gross cross-sectional area minus the area of the cores or cellular space. Gross cross-sectional area shall be determined to the outside of the scoring, but the cross-sectional area of the grooves shall not be considered as part of the area of the coring and shall not be deducted from the gross cross-sectional area to obtain the net cross-sectional area.

Solid masonry units. Masonry unit whose net cross-sectional area in every plane parallel to the bearing surface is seventy-five (75) percent or more of its gross cross-sectional area measured in the same plane, in accordance with Code.

Solid masonry units. Units in which the voids do not exceed 25 percent of the cross-sectional area at any plane parallel to the bearing surface.

Stone. Natural stone shall be sound, clean, and in conformity with other provisions of the Code.

Structural clay tile. A hollow masonry unit composed of burned clay, shale, fireclay or mixtures thereof and having parallel cells.

Structural clay tile. Hollow building unit made from burned clay, shale, fire clay or admixtures thereof.

Structural clay tile. Shall be of a quality at least to equal that required by A.S.T.M. Designation C34 Grade LB when used for bearing walls or piers, or Grade Designation C56, when used for interior non-bearing purposes; or equal, to A.S.T.M. Designation C57, when used for floor construction.

Structural glass block. Structural glass block shall have unglazed surfaces to allow adhesion on all mortared faces.

S.W. grade brick. Brick which may be frozen when permeated with water. (S.W.—Severe Weather)

Terra cotta. Architecturally hard-burned clay, usually molded into shapes for ornamentation of structural surfaces.

Terra cotta. Hard-burned clay, usually molded into shapes for ornamentation of structural surfaces.

Unburned clay bricks. Unburned clay brick shall conform to the requirements specified in the Code.

4-D MASONRY WALLS AND PARTITIONS

Brick cavity wall. A wall in which a space is left between inner and outer tiers of brick. The space may be filled with insulation.

Brick masonry bearing walls. Shall have a thickness of at least 1/20 of their unsupported height or width, whichever is shorter. In addition, the thickness of such bearing walls shall be not less than 6 inches for walls 10 feet or less in height and the minimum thickness shall be increased 1 inch for each successive 10 feet or fraction thereof in height.

Brick veneer wall. Usually used to describe a wall made up of brick veneer applied over wood framing.

Cavity wall. A wall built of masonry units or of plain concrete, or a combination of these materials, so arranged as to provide an air space within the wall, and in which the inner and outer parts of the wall are tied together with metal ties.

Cavity wall. A wall built of masonry units or of plain concrete, or a combination of these materials, arranged to provide an air space within the wall, and in which the inner and outer parts of the wall are tied together with approved metal ties.

Cavity wall. Masonry type of construction consisting of brick, hollow concrete masonry units, structural clay tile units, or any combination of these units in which facing and backing are completely separated except for the metal ties which serve as a bond. Mortar with strength required by code shall be used for cavity walls having a nominal thickness of 10 inches. Cavity walls 10 inches in thickness shall not exceed 25 feet in height, and in no case shall any cavity wall exceed 35 feet in height.

Cavity wall. Load-bearing and non-load-bearing walls of the cavity type may be built of solid or hollow masonry units or combinations thereof subject to the following requirements as well as other applicable requirements of the code. The description of a cavity wall is determined by its nominal out-to-out dimension. For allowable unit stresses in masonry, see the code. In computing the unit stresses, the effective cross sectional area of the cavity walls shall be taken as the gross sectional area minus the area of the cavity.

Cavity wall. A hollow wall in which the inner and outer wythes are tied together with metal ties in horizontal joints.

Cavity wall. A wall built of masonry units or of plain concrete, or a combination of these materials, so arranged as to provide an air space within the wall and in which the facing and backing of the wall are tied together with masonry metal ties or bonded together with masonry or concrete.

Cavity wall. A wall built of masonry units or of concrete, or a combination of these materials, so arranged as to provide an air space, 2 to 4 inches wide within the wall, and in which the facing and backing of the wall are tied together with metal or approved ties.

Cavity wall. Wall in which the inner and outer wythes are separated by an air space, but tied together with noncorrosive metal ties.

Cavity wall. Wall built of masonry units or of masonry and concrete units, arranged to provide an air space within the wall, and in which the inner and outer parts of the wall are tied together with approved metal ties.

Cavity wall. Wall built of a combination of an inner wall of wood studs and an outer wall of masonry units. Both walls separated by an air space and tied together with approved metal ties.

Composite wall. A wall built of a combination of two (2) or more masonry units of different materials bonded together, one forming the back-up and the other the facing elements.

Faced wall. A wall in which the masonry facing and the backing are of different materials and are so bonded as to exert a common reaction under load.

Faced wall. A wall in which the masonry facing and backing are so bonded as to exert common action under load. A facing unit shall be not less than nominal 4 inch in wall thickness.

Faced wall. Wall in which the masonry facing and backing are so bonded as to exert common action under load, as required by the Code.

Hollow bonded wall. Wall built of masonry units with or without any air space within the wall, and in which the facing and backing of the wall are bonded together with masonry units.

Hollow bonded wall. Wall built of masonry units with or without any air space within the wall, and in which the facing and backing of the wall are bonded together with required mortar.

Hollow masonry wall. Wall built of masonry units so arranged as to provide an air space within the wall, and in which the inner and outer parts of the wall are bonded together with masonry units.

Hollow unit masonry construction. That type of construction made with structural clay tile or hollow concrete masonry units in which the units are all laid and set in mortar. Specified type of mortar listed in the code for approved stresses shall be used in such construction except that interior non-bearing masonry of hollow units may be laid up in gypsum mortar.

Hollow wall. A wall built of masonry units so arranged as to provide an air space within the wall, and in which the facing and backing of the wall are bonded together with masonry units.

Hollow wall. Wall built of masonry units laid in and so constructed as to provide an air space within the wall. When hollow walls are built in 2 or more vertical separated wythes, these wythes shall be bonded together so as to exert common action under load.

Hollow wall. A wall in which the masonry units are arranged to form an air space between wythes.

Hollow wall. A wall built of masonry units arranged to provide an air space within the wall, and in which the backing and the facing of the wall are bonded together with approved mortar.

Hollow wall of masonry. A wall built of masonry units so constructed as to provide an air space within the wall, and in which the inner and outer parts of the wall are bonded together with masonry units, using the specified mortar.

Hollow wall of masonry. A wall built of masonry units so arranged as to provide an air space within the wall, and in which the inner and outer parts of the wall are bonded together with masonry units or steel.

Masonry bonded hollow wall. Wall built of masonry units installed as to provide an air space within the wall, and in which the inner and outer wythes of the wall are tied together with masonry units properly set.

Masonry crosswalls or pilasters. Construction which may be omitted on hollow concrete masonry bearing walls 12 inches or more in thickness where such walls are supported horizontally by floors or roofs at heights not exceeding 18 times the wall thickness.

Masonry wall. Bearing or non-bearing wall of hollow or solid masonry units.

Masonry walls below grade. Masonry walls which are in contact with the soil shall be of sufficient strength and thickness to resist the lateral pressure from the adjacent earth and to support their vertical loads without exceeding the allowable stresses. The minimum thickness for masonry walls below grade shall be 4 inches greater than the required thickness for the walls of the supported structures except that 12 inch walls will be accepted for buildings not more than 2 stories in height if substantial lateral support consisting of masonry walls, offsets or pilasters are provided at intervals of not to exceed 20 feet.

Minimum thickness of masonry bearing walls. Walls may be decreased except for walls below grade, and the height or length to thickness ratio may be increased when data is submitted to the Building Department, which may take under consideration a justification in the reduction of the requirements specified in the Code.

Nonbearing masonry walls. All exterior nonbearing walls, if constructed with one wythe of brick to the weather may be backed with S.W. or M.W. classified clay or shale brick, concrete masonry units or clay tile conforming to the requirements of the Code. If such walls are built of concrete masonry units or clay tile, such units shall conform to the requirements of the Code.

Solid masonry. Masonry consisting of solid masonry units laid contiguously in mortar, or consisting of plain concrete.

Solid masonry. Units in which the voids do not exceed 25 percent of the cross-sectional area at any plane parallel to the bearing surface.

Solid masonry. Masonry consisting of solid masonry units laid continuously with the joints between the units filled with mortar, or consisting of plain concrete.

Stack bond. Non-load-bearing walls, or wythes thereof, laid in stack bond or otherwise with inadequate longitudinal bond, shall be tied and reinforced as required in the Code; except that for interior non-load-bearing partitions the maximum spacing of joint reinforcement shall be 24 inches.

Stiffened masonry walls. Where solid masonry bearing walls are stiffened at distances not greater than 12 feet apart by masonry cross walls or by reinforced concrete floors, they may be of 12 inch thickness for the uppermost 50 feet, measured downward from the top of the wall, and shall be increased 4 inches in thickness for each successive 50 feet or fraction thereof.

Stone walls. Rough or random or coursed rubbled stone walls shall be 4 inches thicker than is required by the Code, but in no case less than 16 inches thick.

Veneered wall. Wall having a facing which is attached to the backing but not so bonded as to exert common action under load.

Veneered wall. Wall in which the facing does not support vertical loads.

Veneered wall. A wall having a masonry facing which is attached to the backing but not so as to exert common action under load.

Veneered wall. A wall with a masonry face which is attached to but not bonded to the body of the wall.

Veneered wall. Wall having a facing which is not attached and bonded to the backing so as to form an integral part of the wall for purposes of load bearing stability.

Veneered wall. A wall having a facing of masonry or other material securely attached to the backing, but not bonded so as to exert a common reaction under load.

Veneered wall. Wall having a facing of masonry or other weather-resisting noncombustible materials securely attached to the backing, but not so bonded as to exert common action under load.

Walls below grade. Masonry walls which are in contact with the soil shall be of sufficient strength and thickness to resist the lateral pressure from the adjacent earth and to support their vertical loads without exceeding the allowable stresses. The minimum thickness for masonry walls below grade shall be 4 inches greater than the required thickness for the walls of the supported structures except that 12 inch walls will be accepted for buildings not more than 2 stories in height if substantial lateral support consisting of masonry walls, offsets or pilasters are provided at intervals of not to exceed 20 feet.

2. Article 5

METALS 5-A THROUGH 5-C

5-A STEEL PLATE

Cold-pressing quality plate. Plate made of soft steel suitable for bending and forming at ordinary temperatures.

Drawing quality plate. Plate produced from low-carbon steel suitable for drawing into identified forms.

Firebox quality plate. Plate used for pressure vessels exposed to fire or radiant heat and subject to mechanical and thermal stresses.

Flange quality plate. Plate used for pressure vessels and similar installations but not exposed to fire or radiant heat.

Forging quality plate. Plate intended for forging, heat treating, or similar purposes requiring uniformity of composition and freedom from injurious defects.

Hot-pressing quality plate. Plate used for ordinary hot pressing, flanging, or bending work.

Marine quality plate. Plate used for pressure vessels and combustion chambers of marine boilers.

Structural quality plate. Plate used for bridges, buildings, and miscellaneous structures.

5-B STRUCTURAL

Assembly. Fitting together and/or attaching by any means, including brazing, welding, bonding, riveting, bolting and screwing.

Bay (part of structure). Space between 2 adjacent piers or mullions or between 2 adjacent lines of columns.

Bay (part of structure). The wall space between two columns; the whole space between column centers.

Beam. Structural member transversely supporting a load.

Beam. A horizontal load-bearing structural member, transmitting superimposed vertical loads to walls or columns.

Beam. A primary structural member, supporting secondary structural members, floor, roof, joists and the like.

Bearing. That portion of a structural member resting on supports.

Bearing. That portion of a beam, truss, or other structural member that rests on the supports.

Bolster. Short horizontal structural member resting on top of a column for the support of other structural members.

Bond. Binding together of individual parts of a structure with frictional forces in order that the parts act as a single unit.

Box girder. A girder having a hollow cross-section similar to that of a rectangular box.

Box system. Is a structural system without a complete vertical load-carrying space frame. In this system the required lateral forces are resisted by shear walls.

Brace. Any inclined structural member introduced in a truss or frame to provide stiffness.

Brace. Usually an inclined member used to form a triangle in a framework to stiffen the frame.

Brace. Any minor member designed to steady or stiffen a major member of a structure.

Bridging. Diagonal or cross bracing between joists to resist twisting.

Cast iron. Cast iron used in buildings or structures shall be of such quality as to conform to the "Standard Specifications for Gray Iron Castings", (A.S.T.M.)

Cast iron construction. Construction which utilizes cast iron for structural purposes. The cast iron product shall be of good foundry mixture producing a clean, tough, gray iron free from serious blowholes, cinder spots, or cold shuts, and shall have minimum tensile strength of 20,000 psi.

Cast Steel. Cast steel used in buildings or structures shall be of such quality as to conform to the "Standard Specifications for Carbon Steel Castings" (A.S.T.M.)

Cellular steel deck. Structural floor system, consisting of 2 layers of sheet metal shaped to form cells and welded together. Cells serve as electrical raceways.

Cellular steel floor. Construction shall consist of sheet or strip steel formed into an integrated system of parallel steel beams which combine the function of load-bearing members and a continuous deck spanning between main supporting girders, beams or walls.

Clear span. Distance between inside faces of supports.

Corrosion. Chemical action which causes gradual destruction of a metal.

Corrosion resistant material. Material that maintains its original surface characteristics under prolonged exposure and use.

Decking. Fabricated metal forms installed over other structural members to which other materials can be installed.

Floor assembly. Combination of materials providing the horizontal separation between stories including the ceiling, floor and horizontal structural members supporting the floor but excluding those primary structural members which serve as part of the structural frame.

Formed steel construction. Construction composed of sheet or strip steel, formed into structural panels, decks, studs, joists, and other fabricated structural members.

Girder. Horizontal structural piece or pieces which support the ends of floor beams or joists, or carrying walls over openings, carrying its loads in concentrations.

Girder. Horizontal beam used to support interior walls or joists. Most wood frame houses have a lengthwise centerline girder that supports the floor joists and plywood subflooring.

Girt. Horizontal framing member to aid in providing rigidity to columns and support for siding or sheathing.

Grade beam. Structural beam placed at or near ground level usually performing the functions of a foundation.

Jackhammer. Any power-driven device which causes any metal object to strike any other metal object for the purpose of riveting, bolting or fastening together any two (2) or more objects.

Lally column. Concrete filled steel pipe used as a column.

Light gage steel construction. Construction composed of sheet or strip steel, less than 3/16 inch thick, formed into structural panels, decks, studs, joists, and other structural members; formerly called Formed Steel Construction.

Light gage steel member. A steel structural member cold formed to shape from sheet or strip steel less than three-sixteenths (3/16) inches thick, which is used for load carrying purposes.

Light-gauge cold-formed steel structural member. Structural member cold formed to shape from sheet or strip steel and used for load-carrying purposes in buildings.

Lintel. Beam or girder placed over an opening in a wall, which supports the wall structure above.

Lintel. Horizontal structural member that supports the load over an opening such as a door or window.

Lintel. A beam in a masonry wall supporting the masonry above an opening.

Lintel. A beam or girder placed over an opening in a wall which supports the wall construction above and other transfer loads.

Lintel. A horizontal framing member carrying a load over a wall opening; a header.

Lintel. A structural member providing support for masonry above an opening in a wall or partition.

Lintel. Beam to support the weight of a wall or partition above an opening.

Low bay. Industrial area where the ceiling or truss height is relatively low, usually under 20 feet.

Non-corrodible metal. Metal which, under the conditions of its use, may reasonably be expected, without unusual or excessive maintenance, to serve its purpose throughout the probable life of the structure in which it is used as determined by the Building Department.

Open-web joists. Open web load-carrying structural members and assemblies in buildings and structures, including the accessory parts and materials included in the fabrication and construction of the assemblies.

Open web joists. Prefabricated metal, parallel chord trusses.

Open web steel joist. A steel made of hot or cold formed sections, strip or sheet steel, riveted or welded together, or by expanding, which is used for supporting floors and roofs between girders, trusses, beams or walls in buildings or structures.

Open web steel joists. Shall be designed as a truss, solid web steel joists as a beam. Deck or top slabs over steel joists shall not be assumed to carry any part of the compression stress in the steel joists.

Pedestal. Upright compression member, the height of which does not exceed three times its least lateral dimension.

Pipe. Steel pipe for steel pipe columns shall be of such quality as to conform to "Standard Specifications for Welded and Seamless Steel Pipe". (A.S.T.M.), and shall be a medium carbon steel manufactured by the open hearth or electric furnace process.

Positive bending moment. Shall mean, that moment the intensity of which is least at or near the supports.

Pounds per square foot (p.s.f). Measure of loads distributed over a square foot of surface. Used in determining allowable spans for subfloor and roof sheathing and floor framing.

Pounds per square inch (p.s.i.) Measure of loads distributed over a square inch of surface.

Primary member. A member of the structural frame of a building used as a column, or a grillage beam, or to support masonry walls or partitions; including trusses, isolated lintels spanning an opening of eight (8) feet or more and any other structural member required to brace a column or a truss.

Primary members. Structural member upon which a large portion of a structure, including at least one other structural member, is dependent for support.

Purlin. Horizontal framing members supporting the rafters or spanning between trusses to support the roof covering.

Riveted construction. All parts of riveted members shall be well pinned or bolted and rigidly held together while riveting. Drifting done during assembling shall not distort the metal or enlarge the holes.

Rivet steel. Rivet steel shall conform to the "Standard Specifications for Structural Rivet Steel" (A.S.T.M.).

Secondary member. Any member of the structural framework other than a primary member, including filling-in beams of the floor system.

Secondary members. Any structrual member which is not a primary member.

Shelf angles. Includes similar members which support facing materials only and which material is not more than 4-1/2 inches in thickness, need not be fireproofed where such members are part of an assembled lintel or spandrel beam. The major member of such lintel or spandrel beam shall be fireproofed when and as required and shall support the backing to which the facing material shall be securely bonded or anchored. Where a wall is supported by a lintel or spandrel beam which is required to be fireproofed, its stability shall not be dependent upon a shelf angle or similar member which is not fireproofed as required for the major member.

Space frame. Three-dimensional structural system composed of interconnected members, other than bearing walls, laterally supported so as to function as a complete self-contained unit with or without the aid of horizontal diaphragms or floor bracing systems.

Space frame-ductile moment resisting. A space frame-moment resisting, complying with the requirements for a ductile moment resisting space frame as specified in the Code.

Space frame-moment resisting. A vertical load carrying space frame in which members and joints are capable of resisting design lateral forces by bending moments.

Space frame-vertical load-carrying. A space frame designed to carry all vertical loads.

Span. Distance between structural supports, such as walls, columns, piers, beams, girders, and trusses.

Span. The clear horizontal distance between two supports.

Steel joist. Any approved form of open webbed beam or truss nominally 24 inches or less in depth, produced directly by rolling, cold-forming, or pressing or fabricating from rolled, cold-formed or pressed shapes by welding, riveting or expanding.

Steel joist. Any secondary steel member of a building or structure made of hot or cold-formed solid or openweb sections, or riveted or welded bar, strip or sheet steel members or slotted and expanded or otherwise deformed rolled sections.

Steel joist. Any secondary steel member of a building or structure made of hot or cold-formed solid or openweb sections, or riveted or welded bar, strip or sheet members or slotted and expanded or open-web sections, or riveted or welded bar, strip or sheet steel members or slotted and expanded or otherwise deformed rolled sections.

Structural alteration. Any change in the supporting members of a building or structure, including bearing walls, partitions, columns, beams, girders or similar parts of a building or structure, and any substantial change in the roof of a building.

Structural alterations. Any change in the supporting members of a building, such as footings, bearing walls or partitions, columns, beams or girders, or any substantial change in the roof or in the exterior walls, excepting such repair as may be required for the safety of the building.

Structural aluminum. Aluminum alloys used in structural members and assemblies in buildings and structures, and the accessory parts and materials included in the fabrication and construction of the assemblies.

Structural feature. Any part of a structure which is designed for or indicative of the intent to accommodate any given use.

Structural frame. All vertical load supporting members, other than bearing walls; all primary horizontal load-supporting members rigidly connected thereto; and all other primary members essential to the stability of the structural frame.

Structural member. Columns, studs, posts, bearing walls or partitions, trusses, girders, arches, beams, joists, rafters, roofing, flooring or other supporting devices which are essential in carrying vertical, horizontal or torque forces to bearing materials upon which a structure rests.

Structural steel. Structural steel shall conform to the "Standard Specifications for Structural Steel for Bridges and Buildings" (A.S.T.M.).

Structural steel. Steel structural members and assemblies and structures, and the accessory parts and materials included in the fabrication and construction of the assemblies, with the exception of open-web steel joists, members formed of flat rolled sheet or strip, and light-gauge steel construction.

Structural steel member. Any primary or secondary member of a building or structure consisting of a rolled steel structural shape other than formed steel, light gauge steel or steel joist members.

Structural steel member. A structural member other than a light-gage steel member or an open web steel joist, made of structural steel shapes or plates, including steel pipe columns.

Towers of trussed construction. Shall be designed and constructed to withstand wind pressures applied to the projected exposed areas as specified in the Code, and multiplied by the shape factors specified in the Code.

Truss. A frame or jointed structure designed to act as a beam of long span, while each member is usually subjected to longitudinal stress only, either tension or compression.

Truss. A structural framework composed of a series of members so arranged and fastened together that external loads applied at the joints will cause only direct stress in the members.

Truss. A rigid, open web structural member designed and engineered to carry roof or floor loads.

Truss. Complete or redundant framed structural unit composed of structural members connected at their intersections, in which, if loads are applied at their intersections, the stress in each member is in the direction of the length of the member.

Wall framing. Shall include columns, studs, beams, girders, lintels and girts.

5-C WELDING AND CUTTING

Arc welding. A group of welding processes wherein coalescence is produced by heating with an electric arc or arcs, with or without the application of pressure and with or without the use of filler metal. Pressure as herein used refers to pressure necessary to the welding process.

Bell-welded. Furnace-welded pipe produced in individual lengths, from cut-length skelp, having its longitudinal butt joint forge welded by the mechanical pressure developed in drawing the furnace-heated skelp through a cone-shaped die (welding bell) which serves as a combined forming and welding die.

Brazed. Joined by hard solder.

Braze welding. Method of welding whereby a groove, fillet, plug or slot weld is made using nonferrous filler metal having a melting point below that of the base metals, but above 800°F but lower than that of the base metals joined. The filler metal is distributed between the closely fitted surfaces of the joint by capillary action.

Buttweld. The term buttweld shall mean a weld in a butt joint. The term "groove weld" shall mean a weld made in the groove between two members to be joined. The size of a groove weld shall be expressed in terms of joint penetration or depth of chamfering plus the root penetration.

Butt weld joint. Welded pipe joint made with ends of the two pipes butting each other, the weld being around the periphery.

Butt weld pipe. Pipe welded along a seam butted edge to edge and not scarfed or lapped.

Cup weld. Pipe weld where one pipe is expanded on the end to allow the entrance of the end of the other pipe. The weld is then circumferential at the end of the expanded pipe.

Filler metal. Arc-Welding electrodes shall conform to the requirements of the "Specifications for Mild Steel Arc-Welding Electrodes" of the American Welding Society.

Fillet weld. Weld of approximately triangular cross-section joining 2 surfaces approximately at right angles to each other in a lap joint, tee joint or corner joint.

Fuel gas. Includes acetylene, hydrogen, natural gas, LP-Gas, methylacetyle-propadiene, stabilized and other liquefied and nonliquefied flammable gases which are stable because of their composition or because of the conditions of storage and utilization stipulated in the code.

Fusion weld. Joining metals by fusion, using oxyacetylene or electric arc.

Gas welding. Group of welding processes wherein coalescence is produced by heating with a gas flame of flames, with or without the application of pressure, and with or without the use of filler metal.

High pressure oxygen manifold. A manifold connecting oxygen containers having a DOT service pressure exceeding 200 psig.

Joint. The location where two or more members are to be or have been fastened together mechanically or by brazing or welding.

Low pressure acetylene. Acetylene at a pressure not exceeding 1 psig.

Low pressure oxygen manifold. A manifold connecting oxygen containers having DOT service pressure not exceeding 200 psig.

Machine. A device in which one or more torches using fuel gas and oxygen are incorporated.

Manifold. An assembly of pipe and fittings for connecting two or more cylinders for the purpose of supplying gas to a piping system or directly to a consuming device.

Medium pressure acetylene. Acetylene at pressures exceeding 1 psig but not exceeding 15 psig.

Oxygen cutting. Group of cutting processes wherein the severing of metal is effected by means of the chemical reaction of oxygen with the base metal at elevated temperatures. In the case of oxidation-resistant metals the reaction is facilitated by the use of a flux.

Pipe. Rigid conduit.

Piping. Pipe or tubing or both for any purpose and made of materials acceptable under the code.

Portable outlet header. An assembly of piping and fittings used for service-outlet purposes which is connected to the permanent service piping by means of hose or other non-rigid conductors.

Sleeve weld. Joint made by butting two pipes together and welding a sleeve over the outside.

Socket weld. Joint made by use of a socket weld fitting which has a prepared female end or socket for insertion of the pipe to which it is welded.

Stabilized methylacetylene-propadiene. A mixture of gases which, in the liquid phase, contains not more than 68 mole percent of the compounds methylacetylene and propadiene in combination, and which contains at least 6 mole percent butanes and at least 18 mole percent other saturated hydrocarbon dilutents. The mixture shall contain not more than 10 mole percent propylene nor more than 2 mole percent butadiene.

Station outlet. The point at which gas is withdrawn from the service piping system.

Weld dimensions. The term "weld dimensions" shall be expressed in terms of their size and length.

Welded construction. Surfaces to be welded shall be free from loose scale, slag, rust, grease, paint and any other foreign material, except that mill scale which withstands vigorous wire brushing, may remain. A light film of linseed oil may be disregarded. Joint surfaces shall be free from fins and tears. Preparation of edges by gas cutting shall, wherever practicable, be done with a mechanically guided torch. Parts to be fillet welded shall be brought in as close contact as practicable and in no event shall be separated more than 3/16 inch. If the separation is 1/16 inch or greater, the size of the fillet welds shall be increased by the amount of the separation. The separation between faying surfaces of lap joints shall not exceed 1/16 inch. The fit of joints at contact surfaces which are not completely sealed by welds, shall be close enough to exclude water after painting.

Welding. Union of two (2) or more metal members produced by the application of a welding process.

Weld length. The term "weld length" shall mean the unbroken length of the full cross-section of the weld exclusive of the length of any craters.

Welds, butt, groove, fillet, length and dimensions. (a) The term "butt weld" shall mean a weld in a butt joint. The term "groove weld" shall mean a weld made in the groove between two members to be joined. The size of a groove weld shall be expressed in terms of joint penetration or depth of chamfering plus the root penetration. (b) The term "fillet weld" shall mean a weld of approximately triangular cross-section joining two surfaces approximately at right angles to each other in a lap joint, tee joint or corner joint. The size of an equal leg fillet weld shall be expressed in terms of leg length of the largest isosceles right-triangle which can be inscribed within the fillet-weld cross-section. The size of an unequal leg fillet-weld shall be expressed in terms of the leg lengths of the largest right-triangle which can be inscribed within the fillet-weld cross-section. (c) The term "weld length" shall mean the unbroken length of the full cross-section of the weld exclusive of the length of any craters. (d) The term "weld dimensions" shall be expressed in terms of their size and length.

2. Article 6

WOOD AND PLASTICS 6-A THROUGH 6-H

6-A PLASTIC MATERIALS

Approved combustible plastic. A plastic material more than one-twentieth (1/20) inches in thickness which burns at a rate of not more than two and one-half (2-1/2) inches per minute when subjected to the ASTM standard test for flammability of plastic in sheets of six-hundredths (0.06) inch thickness.

Approved plastic. Any plastic material which meets the requirements for a specific use as described in the code.

Approved plastic. Any plastic which when tested for flammability in accordance with "Standard Method of Test for Flammability of PLASTICS over .050 Inch in Thickness," in sheets .060 Inch in Thickness, does not burn at a rate exceeding 2-1/2 inches per minute.

Glass fiber reinforced plastic. Plastic reinforced with glass fiber having not less than twenty (20) percent of glass fibers by weight.

Laminated plastic. Decorative high-pressure laminated plastic for facings, tops, splashed, wainscot, walls and caps, with plastic or metal trim applicable to these items such as: decorative high-pressure laminated plastic bonded to proper backing; backing sheet or sealer if required; sink rings with cutouts for same; metal or selfedge trim as required.

Light-diffusing system. A suspended construction consisting in whole or in part of lenses, panels, grids or baffles suspended below lighting fixtures.

Plastic. A pliable and impressionable material of synthetic nature, capable of being formed to a desired shape under heat or pressure, or both, and processed into parts or articles through molding, extrusion, casting, laminating, and machining operations.

Plastic. Material that contains as an essential ingredient, an organic substance of large molecular weight, is solid in its finished state, and which at some stage in its manufacture or processing into finished articles, can be shaped by flow.

Plastic glazing. Sheets of plastic material, or an assembly of plastic face sheets bonded to a noncombustible core, glazed or mounted in sash or frames, for use as light transmitting media.

Plastic glazing. Material glazed or set in frame or sash and not held by mechanical fasteners which pass through the glazing material.

Plastic light-diffusing ceiling. Installation of plastic panels suspended below lighting fixtures for the purpose of diffusing light.

Plastic panel in a wall. One or more plastic sheets, each of single thickness, not individually glazed in sash or frames, or an assembly composed of plastic face sheets bonded to a noncombustible core, used as a light-transmitting media in an exterior wall.

Plastic roof panel. One or more plastic sheets each of single thickness not individually glazed in sash or frames, or an assembly composed of plastic face sheets bonded to a noncombustible core, used as a light-transmitting media in the building roof.

Plastic roof panels. Approved plastic materials which are mechanically fastened to structural members or to structural panels or sheathing and which are used as light-transmitting media in roofs.

Plastic wall panel. Approved plastic materials which are mechanically fastened to structural members or to structural panels or sheathing and which are used as light-transmitting media in exterior walls.

Pyroxylin plastic. Any nitro-cellulose product or compound soluble in a volatile, flammable liquid, including such substances as celluloid, pyroxylin, fiberloid and other cellulose nitrates (other than nitro-cellulose film) which are susceptible to explosion from rapid ignition of the gases emitted therefrom.

Reinforced thermosetting plastic. Thermosetting plastic reinforced with a glass fiber mat having not less than 1-1/2 ounces of glass fiber per square foot.

Slow burning plastic. A plastic having a rate of combustion within the limits of a specified standard of the Code.

Thermoplastic material. Solid plastic material which is capable of being repeatedly softened by increase of temperature and hardened by decrease of temperature.

Thermoplastics. Plastics which become soft under heat and pressure or when reheated, and become dense and hard when cooled.

Thermosetting material. A solid plastic material which is capable of being changed into a substantially non-reformable product when cured under the application of heat or pressure.

Thermosetting plastics. Plastics which become permanently rigid on application of heat, above a critical temperature, and cannot be softened by reheating.

6-B WOOD — FINISHED/MANUFACTURED

Apron. The flat member of the inside trim of a window placed against the wall immediately beneath the stool. Function is to cover the rough edge of the interior wall finish.

Architrave. The group of mouldings above and on both sides of a door or other opening.

Aris. The meeting of two surfaces producing an angle.

Astragal. Small semicircular moulding, plain or ornamented.

Back band. The outside member of a door or window casing.

Baseboard. A finish board around the bottom of interior walls.

Base moulding. Moulding used to trim the upper edge of the interior baseboard.

Base or baseboard. Board installed against the wall around a room, next to the floor to provide a proper finish between the floor and the plaster or plasterboard.

Base shoe. Moulding used next to the floor on the interior baseboard, sometimes called a carpet strip.

Batten. A narrow strip of wood to cover a joint between boards, or to simulate a covered joint for architectural purposes.

Bead. Small moulding, semi-circular in section.

Bed. A moulding used to cover the joint between the plancier and frieze; also used as a base moulding upon heavy work, and sometimes as a member of a cornice.

Bed moulding. Moulding in an angle, as between an overhanging cornice or eaves of a building.

Blank. Piece of wood cut to a size from which the manufactured article is finished.

Brick mould. The outside casing of a frame intended for installation in a masonry or brick veneer wall.

Cabinet. Shop or job built unit, may include combinations of drawers, shelves and hinged doors.

Cabinet work. Refinished shop assembled cabinet work such as cases, counters, desks, cabinets.

Casing. The trimming around a door or window opening, either outside or inside, or the finished lumber around a post or beam, etc.

Clapboard. Exterior wood siding having one edge thicker than the other and laid so that the thick butt overlaps the thin edge of the board below.

Corner boards. Boards use for trimming external corners of a frame house to receive the ends of the wood siding.

Crown moulding. Large moulding usually used on a cornice or in the intersecting surfaces of the ceiling and the wall.

Cuts. Grading term used where the piece is to be cut up for further manufacture, as distinguished from other lumber which is intended for use as a piece.

Dovetail. Tenon shaped like dove's spread tail, fitting into corresponding mortise to form a joint.

Dowel. Cylindrical wooden pin used for holding two pieces of wood together.

Dowel joint. Made by boring the pieces involved to receive dowels, which are set in glue and assembled under pressure.

Dressed and matched. Board or planks machined in such a manner that there is a groove on one edge and a corresponding tongue on the other.

Dressed lumber. Lumber that has been surfaced to attain smoothness of a surface and uniformity of size on one side (SIS) two sides (S2S), one edge (SIE), two edges (S2E), or any combination of these.

Dressed size. The actual cross-sectional dimensions of lumber after planing. Such dimensions are usually from 1/4 to 1/2 inch less than nominal thickness and width, depending on the grading rule and size of lumber.

Drip cap. Moulded wood member for use on the top of outside door and window casings to shed water.

Drop siding. A pattern of boarding which is rebated and overlapped, and used to cover the exterior of buildings.

Edge-grained lumber. Lumber that has been sawed so that the wide surfaces extend approximately at right angles to the annual growth.

Edge treatment. Edge finishing method used, such as banding with wood or plastic, or filling with putty or spackling.

Exterior trim. Includes cornices, mouldings and other ornamental shop fabricated shapes attached to the outside face of exterior walls, doors, windows, porches and bay windows.

Face side. That side of a piece which shows the best quality.

Facings. Dressed boards, with or without mouldings, used in exposed places.

Factory or shop lumber. Lumber intended to be cut up for use in further manufacture and graded on the basis of the percentage of the area which will produce cuttings of a given quality and size.

Finish. As applied to lumber grades, this term refers to the upper grades suitable for natural or stained finishes.

Floor sleepers. Wood embedded in or laid directly on a concrete slab that is in direct contact with the ground shall be treated with an approved pressure preservative treatment. Sleepers are shimed level to accept a finish wood floor.

Floor sleepers. Unless heartwood of a durable species is used, wood floor sleepers or other wood embedded in or laid on masonry or concrete that is in direct contact with the ground shall first be treated with an approved pressure preservative treatment. Pieces shall be machined and cut before treatment whenever possible; when cutting after treatment is unavoidable, the cut surfaces shall be given two brush coats of a suitable preservative.

Furring. Strips of wood or lathing channels fastened to a wall surface or around structural members to provide the backup for nailing or wood screws for fastening other materials.

Furring. Construction applied in widely spaced strips to flat surfaces such as walls, ceilings or partitions and to which a finished surface is attached, creating an air space between the 2 surfaces.

Glue-laminates. Lumber composed of an assembly of wood laminations bonded with adhesives in which laminations are too thick to be classed as veneers.

Grounds. Strips of wood, of the same thickness as the lath and plaster, that are attached to walls before the plastering is done. Used around windows, doors, and other openings as a plaster stop and in other places for the purpose of attaching baseboards or other trim.

Interior trim. Materials generally not exceeding 12 inches in width around openings or on walls or ceilings, including but not limited to casings, aprons, baseboards, picture mouldings applied for decorative purposes.

Interior trim. Materials twelve (12) inches or less in width, applied to interior surfaces other than floors.

Interior trim. Worked lumber used for finishing the interior of buildings.

Kerf. Channel or groove cut by a saw or other tool.

Kerf. The space from which metal has been removed by a cutting process.

Kerf. The space which was occupied by the material removed during cutting.

Kerf. Cut made by the saw in solid lumber on the back face to prevent buckling and curving.

Millwork. Generally all building materials made of finished wood and manufactured in millwork plants and planing mills; doors, window and door frames, sashes, porch work, mantels, panel work, stairways, special wood-work and cabinetry. Excluded are finished dressed four sides, siding, or partition, which are items of yard lumber.

Millwork. Archictectural woodwork, cabinetry and related items manufactured in millwork plants.

Mortise. A notch or hole cut in a piece of wood or other material designed to receive a projecting part, called the tenon, of another piece of material for the purpose of joining the two.

Mortise and tenon joint. A projection, machined on one piece, snugly fits into a rectangular shaped recessed opening machined in a second piece, and is secured under pressure with an adhesive.

Planing-mill products. Products worked to pattern, usually in strip form, such as flooring, ceiling, and siding.

Plywood. Laminated wood construction consisting of an uneven number of individual sheets of wood placed so that the grain in adjacent sheets are approximately perpendicular to each other and with the sheets bonded together with water-resistant adhesives having greater strength, when dried, than the wood.

Plywood. A built up board of laminated veneers.

Quarter-sawed. The grain pattern that is produced when wood is cut so that the annular rings are at an angle of 45 degrees or less with the board surface.

Rabbet. A longitudinal channel, groove or recess cut out of the edge or face of any wood member, especially one intended to receive another member.

Rake. Trim members that run parallel to the roof slope and from the finish between wall and roof.

Ripping. In woodworking, the sawing or splitting of wood with the grain.

Scribing. Fitting woodwork to an irregular surface.

Sleeper. Strip of material installed in a concrete floor and to which top layers of flooring can be fastened.

Sleepers. A non-structural timber, board, or metal strip laid on the ground or a basic floor to support and provide a component to which the finish floor may be fastened.

Stool. Flat, narrow shelf forming the top member of the interior trim at the bottom of a window.

Subfloor. Boards or sheet material laid on joists over which a finished floor is to be laid.

Tenon. The end of a piece of lumber formed to fit into a mortise.

Terms (Abbreviations)

A.D.	Air-dried
a.l.	All lengths
B&S	Beam and stringer
bd ft	Board foot
bm	Board (foot) measure
btr.	Better
clr.	Clear
CM	Center matched—that is, the tongue and groove
com.	Common
cu ft	Cubic foot
D&H	Dressed and headed—that is, dressed one or two sides and worked to tongue and groove joints on both the edge and the ends.
D&M	Dressed and matched—that is, dressed one or two sides and tongued and grooved on the edges. The match may be center or standard.
D&SM	Dressed one or two sides and standard matched
D2S&CM	Dressed two sides and center matched
D2S&M	Dressed two sides and center or standard matched
D2S&SM	Dressed two sides and standard matched
dim.	Dimension
ECM	Ends center matched
E.G.	Edge grain
EM	Ends matched, either center or standard
ESM	Ends standard matched
fbm	Feet board measure
F.G.	Flat grain
ft	Foot or feet
Hdwd.	Hardwood
Hrtwd.	Heartwood
in.	Inch or inches
J&P	Joist and plank
k.d.	Knocked-down
K.D.	Kiln-dried
kip	Kilo-pound (1,000 pounds)
lb	Pound or pounds
lbr.	Lumber
l.c.l.	Less carload lots
lgth.	Length
lin ft	Linear foot; that is, 12 in.
Mbm	Thousand (feet) board measure
No.	Number
o.c.	On centers
P&T	Post and timber
res.	Resawed
rip.	Ripped
r.l.	Random lengths
Sap.	Sapwood
Sftwd.	Softwood
SM	Standard matched
Sq.E.	Square edge
S1E	Surfaced one edge
S2E	Surfaced two edges
S1S	Surfaced one side
S2S	Surfaced two sides
S1S1E	Surfaced one side and one edge
S2S1E	Surfaced two sides and one edge
S1S2E	Surfaced one side and two edges
S4S	Surfaced four sides
S4SCS	Surfaced four sides with a calking seam on each edge
S&CM	Surfaced one or two sides and center matched

S&M	Surfaced and matched—that is, surfaced on one or two sides and tongued and grooved on the edges—with the match either center or standard.
S&SM	Surfaced one or two sides and standard matched
S2S&CM	Surfaced two sides and center matched
S2S&M	Surfaced two sides and (center or standard) matched
S2S&SM	Surfaced two sides and standard matched
T&G	Tongued and grooved
Tbrs.	Timbers
V.G.	Vertical grain
wt	Weight
yd	Yard or yards

Tongue and groove joint. (T&G) Any joint made by one member with a projecting tongue fitting into another member with a matching groove.

Trim. Finish materials in a building, such as mouldings, applied around openings or at the floor and ceiling of rooms.

Trimmer. Beam or joist to which a header is nailed in framing for a chimney, stairway, or other opening.

Wallboard. Woodpulp, gypsum or other materials made into large rigid sheets that may be fastened to the frame of a building to provide a surface for other materials.

Worked lumber. Lumber that has been run through a matching machine, sticker, or moulder.

6-C WOOD — GENERAL

Acclimatization. Storing lumber in a building for a short period to acquire the desired interior temperature before installation.

Air dried. Dried by exposure to air, usually in a yard, without artificial heat (lumber).

Air dried lumber. Lumber that has been piled in yards or sheds for any length of time.

Air-dry. The condition reached by lumber seasoned under natural atmospheric conditions.

Air-dry. Seasoned or dry lumber in open-air, by placing on sticks or platforms leaving openings for air circulation. No artificial heat is used.

Approximate shrinkage. Tolerance allowed for normal variance from green to dry.

Blind nailing. Nailing accomplished in such a manner that the nailheads are not visible on the finished face of the work.

Board foot. The equivalent of a board (1) foot square and (1) inch thick.

Board measure. Standard basis of measuring lumber. The board foot is the unit of measurement.

Braces. Pieces fitted and firmly fastened to two others at any angle in order to strengthen the angle thus treated.

Brashness. Condition of wood characterized by low resistance to shock and by abrupt failure across the grain without splintering.

Brashness. Condition that causes some pieces of wood to be relatively low in shock resistance for the species and, when broken in flexure, to fail abruptly without splintering at comparitively small deflection.

Break joints. To arrange joints so that they do not come directly under or over the joints of adjoining pieces, as in shingling, siding, etc.

Buck. Rough framing around any type of openings on which a finished frame is to be installed.

Butt joint. Place where two pieces of wood are joined together end to end.

Butt joint. The junction where the ends of 2 timbers or other members meet in a square-cut joint.

Case-hardening. Defect produced through too severe drying of lumber owing to the exterior drying out while the interior of the piece is still moist.

Casein glue. Adhesive with a suitable mould inhibitor. The standard dry-use adhesive of the lamination industry.

Common. Term applied to a grade of lumber containing numerous defects which render it unsuitable for high-class finish, but suitable for general construction.

Compression failure. Deformations of the fibers due to excessive compression along the grain either in direct and compression or in bending.

Compression failures. Minute ridges formed by crumpling or buckling of the cells, resulting from excessive compression stresses along the grain.

Corner braces. Diagonal braces built into the framing of wood framed houses to reinforce the corners.

Countersink. To make a cavity for the reception of a metal plate or the head of a screw or bolt so that it shall not project beyond the face of the work.

Cross break. Separation of the wood across the grain. Such breaks may be due to internal stresses resulting from non-uniform longitudinal shrinkage or to external forces.

Dimension. All yard lumber except boards, strips and timbers; that is, yard lumber from 2 inches to, but not including, 5 inches thick, and of any width.

Dimensional stabilization. Reduction through special treatment in swelling and shrinking of wood, caused by changes in its moisture content with changes in relative humidity.

Dimension stock. Square or flat stock usually in pieces smaller than the minimum sizes admitted by standard lumber grades that is rough, dressed, green, or dry, and cut to the approximate dimension required for the various products of woodworking factories.

Direct nailing. Nailing perpendicular to the initial surface or to the junction of the pieces joined.

Dry kiln. Drying oven where green lumber is placed to be dried to a certain moisture content by the use of mechanical heat.

Durability. General term for permanence or lastingness, frequently used to refer to the degree of resistance of a species or of an individual piece of wood to decay.

Eased edges. Slightly rounded surfacing on pieces of lumber to remove sharp corners.

Edge joint. Place where 2 pieces of wood are joined together edge to edge.

Edgejoint. A joint between two pieces of wood glued edge to edge and in the direction of the grain.

Fibre board. Building board made from fibrous material such as wood pulp.

Finger joint. A joint consisting of a series of fingers, precision machined on the ends of two pieces to be joined, which mesh and are firmly held together by an adhesive.

Fished. An end butt splice strengthened by pieces nailed on the sides.

Fish-joint. Pieces are joined butt end to end, and are connected by wooden or metal plates on each side and secured to the pieces joined.

Foxtail wedging. A peculiar make of mortising, in which the end of the tenon is notched beyond the mortise, and is split and a wedge inserted, which, being forcibly driven in, enlarges the tenon and renders the joint firm and immovable.

Gage. Tool used by carpenters; to strike a line parallel to the edge of a board.

Glued-laminated wood. Several pieces of wood, glued together to act as one.

Glue joint. A joint held together with glue.

Glue line. Thin layer of glue between two surfaces.

Grade (Lumber). The classification of lumber in regard to strength and utility in accordance with the grading rules of an approved lumber grading agency.

Grading rules. Lumber is manufactured in a sawmill to grading rules of regional associations. As it leaves the saw, the wood is unseasoned. though it is usually partly air-seasoned when it reaches the job.

Ground. A strip of wood assisting the plasterer in making a straight wall and in giving a place to which the finish of the room may be nailed.

Halved joint. A joint made by cutting half the wood away from each piece so as to bring the sides flush.

Hardboard. Manufactured flat homogeneous panels from inter-felted lignocellulosic fibers consolidated under heat and pressure.

Hardboard. Fibrous-felted, homogeneous panel made from lignocellulose fibers consolidated under heat and pressure in a hot press to density not less than thirty-one (31) pounds per cubic foot conforming to the Code.

House joint. A joint in which a piece is grooved to receive the piece which is to form the other part of the joint.

Housing. A groove or trench in a piece of wood made for the insertion of a second piece.

Kerf. Cut made by the saw in solid lumber on the back face to prevent buckling and curving.

Kiln. Heated chamber for drying lumber, veneer, and other wood products.

Kiln-dried. Lumber which has been seasoned in a dry-kiln, usually, though not necessarily, to a lower moisture content than that of air-seasoned lumber.

Laminated wood. An assembly of wood built up of plies or laminations that have been joined either with glue or with mechanical fastenings.

Laminating. Process of bonding laminations together with adhesive.

Lap joint. A joint of two pieces lapping over each other.

Lap siding. Boards used to cover the sides of buildings, the lower edge of one board being lapped over the upper edge of the board below.

Level. A term describing the position of a line or plane when parallel to the surface of still water, an instrument or tool used in testing for horizontal and vertical surfaces, and in determining differences of elevation.

Lumber. Lumber is the product of the saw and planing mill that is not further manufactured than by the processes of sawing, resawing, passing lengthwise through a standard planing mill, cross-cutting to length, and working.

Marine glue. A form of glue resisting the action of water, and containing rubber, shellac and oil.

Matched lumber. Lumber edge dressed and shaped for a tongue-and-groove joint when pieces are laid edge to edge or end to end.

Matching or tonguing and grooving. The method used in cutting the edges of a board to make a tongue on one edge and a groove on the other.

Mill run. All the lumber produced in a mill, without reference to grade.

Miter. A moulding returned upon itself at right angles.

Moisture proofing. The process of making wood resistant to change in moisture content, especially to entrance of moisture.

Moisture-resistant adhesive. Adhesives whose bonds are sufficiently durable to withstand occassional exposure to wet and damp conditions.

Mortise. The hole which is to receive a tenon, or any hole cut into or through a piece by a chisel; generally of rectangular shape.

Nominal dimension. The dimension of lumber corresponding approximately to the size before dressing to actual size.

Nominal measure. In worked lumber, the dimensions of the rough board before dressing.

Nominal size. As applied to timber or lumber, the rough-sawed commercial size by which it is known and sold in the market.

Nominal size (lumber). The commercial size designation of width and depth, in standard sawn lumber and glued-laminated lumber grades; somewhat larger than the standard net size of dressed lumber, in accordance with the Standard for sawn lumber and the Standard for structural glued-laminated timber.

On-center spacing. Distance from the center of one member to the center of the adjacent member, as in the spacing of studs, joists, rafters, sleepers, or nailing in special earthquake requirements.

Oven-dry. The term applied to wood which does not continue to lose moisture after an interval of time in an oven at 100° centigrade.

Oven dry wood. Wood dried to constant weight in an oven at temperature above that of boiling water.

Plumb cut. Any cut made in a vertical plane.

Precision end trimmed lumber. Trimmed square and smooth on both ends to uniform length.

Prefabricated. Fabricated on or off the site before incorporation into a building or structure.

Premium grade. Term meaning "the best of its grade." A better selection from a given grade, the top of a grade.

Quarter sawn—quarter cut. Lumber cut in a radial direction that is, at right angles to the direction of the annual rings.

Rabbetted joint. A lapping joint machined on the meeting portions of the pieces to be connected.

Round timber. Lumber used in the original round form, such as poles, piling, and mine timbers.

Run. The length of the horizontal projection of a piece such as a rafter when in position.

Scarfed. A timber a spliced by cutting various shapes of shoulders, or jogs, which fit each other.

Scarfing. A joint between two pieces of wood which allows them to be spliced lengthwise.

Scribing. Fitting woodwork to an irregular surface.

Scribing. The marking of a piece of wood to provide for the fitting of one of its surfaces to the irregular surface of another.

Seasoned lumber. Lumber which has been air-dried for at least sixty (60) days, or which has at the time of installation in the structure reached a moisture content approximately equal to that which it will eventually contain to service.

Where green or recently cut lumber is used, tabulated bolt values shall be reduced one-third.

Sheathing paper. The paper used under siding or shingles to insulate the house.

Sills. The horizontal timbers of a house which either rest upon the masonry foundations or in the absence of such, form the foundations.

Spline. A rectangular strip of wood which is substituted for the tongue.

Splines. Used usually with double tongue and groove decking members to join the butt ends together where pieces are not end matched.

Square. Tool used by mechanics to obtain accuracy; a term applied to a surface including 100 square feet.

Standard matched. Matched lumber that is tongued and grooved to correspond with a specification.

Stub tenon. A short tenon intended for insertion in a plow or groove.

Threshold. The beveled piece attached to the floor over which the door swings.

Tongue. A projection on the edge of a board machined to fit into a groove in the adjacent piece.

Workability. The degree of ease and smoothness of cut obtainable in a material.

Yard lumber. Lumber of all sizes and patterns that is intended for general building purposes.

6-D WOOD — GROWTH

Advanced Decay. The older stage of decay in which the destruction is readily recognized because the wood has become punky, soft and spongy, stringy, ring-shaked, pitted or crumbly.

Annual ring. The ring seen on transverse section of a piece of wood caused by contrasting spring-wood and summer-wood and denoting one-year's growth of wood.

Bark. Outside layer of a tree, composed of a living, inner bark called "phloem" and an outer bark of dead tissue.

Birdseye. Small central spot with wood fibers arranged around it so as to give the appearance of an eye.

Black knot. Results from a dead branch which the wood growth of the tree has surrounded.

Bleeding. An exudation of resin, gum, creosote or other substance in lumber.

Blue Stain. Bluish or grayish discoloration of the sapwood caused by the growth of moldlike fungi on the surface and in the interior of the piece, made possible by the same conditions that favor the growth of other fungi.

Borer holes. Voids made by wood-boring insects, such as grubs or worms.

Branch knots. Two or more divergent knots sawed length-wise and tapering toward the pith at a common point.

Brown stain. Dark, often chocolate brown, discoloration found in the sapwood of some softwoods stored under unfavourable seasoning conditions, it is caused by a fungus.

Burl. Swirl or twist in the grain of the wood, which usually occurs near a knot but does not contain a knot.

Burl. Hardy wood excresence on a tree, more or less rounded in form, usually resulting from the entwined growth of a cluster of adventitious buds. Such burls are the source of the highly figured burl veneers used for purely ornamental purposes.

Cambium. The layer of tissue just beneath the bark from which the new wood and bark cells of each year's growth develop.

Cell. General term for the minute units of wood structure, including fibers, vessels, and other elements of diverse structure and functions.

Cellulose. The carbohydrate that is the principal constituent of wood and forms the framework of the cells.

Check. Lengthwise separation of wood fibers, usually extending across the rings of annual growth caused chiefly by strains produced in seasoning.

Checks. Small splits running parallel to grain of wood, caused chiefly by strains produced in seasoning.

Clear. Lumber and timbers selected for lack of imperfections and knot holes, desirable for finsished interior beams and paneling for construction.

Clear wood. Wood member free from splits, checks, shakes, knots, or other characteristics.

Close grain. Lumber with an average of approximately 6 and not more than 30 annual rings per inch on either one end or the other of the piece.

Coarse grain. Wood with wide and conspicuous annual rings in which there is considerable difference between springwood and summerwood, and also used to designate wood with large pores.

Compreg. Wood in which the cell walls have been impregnated with synthetic resin and compressed to give it reduced swelling and shrinking characteristics and increased density and strength properties.

Crotch swirl. Heartwood at base of the tree where grain is extremely distorted and irregular.

Curly grain. Grain which is undulating without crossing.

Decay. Decomposition of wood by fungi.

Decay. Disintegration of wood substance due to the action of wood-destroying fungi.

Decayed knot. Knot which is softer than the surrounding wood because of its advanced decay.

Decayed. (Knot) A knot which is softer than the surrounding wood, and which contains advanced decay.

Defect. Any irregularity occurring in or on the wood that may lower its strength or appearance.

Delignification. Removal of part or all of the lignin from wood by a chemical treatment.

Dense grain. Pieces averaging six or more annual rings with addition 1/3 or more. Summer wood on one end or other of the piece.

Dip-grained wood. Wood which has single waves or undulations of the fibers, such as occur around knots and pitch pockets.

Dote, doze and rot. Terms synonymous with decay and are any form of decay which may be evident as a discoloration or a softening of the wood.

Dry rot. Decay caused by certain fungi, which are peculiarly adapted in regard to supplying their own moisture requirements.

Encased knot. Knot whose annual rings of growth are not intergrown and homogeneous with those of the surrounding wood; the encasement may be partial or complete and may be composed of other pith or bark.

Figure. The pattern produced in a wood surface by irregular coloration and by annual growth rings, rays, knots, and such deviations from regular grain as interlocking and wavy grain.

Firm knot. Solid across its face but may contain incipient decay.

Fixed knot. Will retain its place in dry lumber under ordinary conditions, but can be moved under pressure though not easily pushed out.

Flat grain. Lumber is a piece or pieces sawn approximately parallel to the annual growth rings so that all or some of the rings form an angle of less than 45 degrees with the surface of the piece.

Grain. The direction of wood elements (fibres) that determines the direction of cleavage and designates the angle of the growth rings in relation to the axis of the board.

Green lumber. Unseasoned or wet lumber, in which free water still remains within the cells.

Green timber. Green is the condition of timber as taken from a living tree. Immediately upon being sawed from the tree, lumber begins to lose moisture and otherwise change its condition.

Growth ring. Growth layer put on a tree in a single year, including springwood and summerwood.

Growth rings. Concentric bands of distinctive coloration.

Heart. The portion of the tree contained within the sapwood. It is sometimes used to mean the pith.

Heart center. Is the pith or center core of the log.

Heart center decay. Is a localized decay developing along the pith in some species and is readily identifiable and easily detected by visual inspection.

Heart check. A check starting near the pith and extending toward, but not to the surface of a piece.

Heart shake. When pith or center of the log has grain separation, causing a weakening of its structural property.

Heartwood of a durable species. Includes the heartwood of tidewater red cypress, cedar, or redwood or other approved decay resistant wood.

Impreg. Synthetic resin-treated wood made so as to reduce materially its swelling and shrinking.

Incipient decay. Early stage of decay in which the disintegration has not proceeded far enough to soften or otherwise impair the hardness of the wood perceptibly.

Inter-grown knot. Knot whose growth rings are partially or completely intergrown on one or more faces with the growth rings of the surrounding wood.

Interlocking grain. Wood in which the fibers are inclined in one direction in a number of rings of annual growth, then gradually reverse, and are inclined in an opposite direction in succeeding growth rings, then reverse again.

Knot cluster. Is two or more knots grouped together as a unit with the fibers of the wood deflected around the entire unit.

Knots. Are formed when the part of a branch or limb embedded in a tree is cut through.

Large knot. A knot more than 1-1/2 inch in diameter.

Large pitch pocket. A pocket over 3/8 inch in width and over 4 inches in length, or over 1/8 inch in width and over 8 inches in length.

Lignin. A principal constituent of wood, second only to cellulose. It incrusts the cell walls and cements the cells together.

Loose knot. A knot which is not held tightly in place by growth or position and which cannot be relied upon to remain in place.

Loosened grain. The separation or raising of fiber along the rings as they run out on a flat-sawn surface.

Medium grain. Average of four or more annual rings per inch on either one end or the other of the piece using radial line basis of measurement.

Medium knot. Is one over 3/4 inch, but not more than 1-1/2 inches in diameter.

Medullary rays. Cellular tissues which usually run continuously from the pith to the bark, particularly prominent in quarter-cut oak.

Mineral streaks. Slight discoloration inherent in certain species.

Mould or mildew. A superficial fungus growth usually appearing in the form of a woolly or fussy coating on varying colour.

Open defects in wood. Include but are not limited to checks, knots, wormholes, or other defects interrupting the smooth continuity of the surface.

Open knot in wood. Opening where a portion of the wood substance of the knot has dropped out, or where cross-checks have occurred to present an opening.

Oval knot. Is a knot cut at slightly more than right angles to the length of the knot (limb).

Peck. Is channeled or pitted areas or pockets as sometimes found in cedar and cypress. Wood tissue between pecky areas remains unaffected in appearance and strength. All further growth of the fungus causing peckiness ceases after the trees are felled.

Pitch. Accumulation of resin in wood. This accumulation may be in the form of pitch pockets, pitch streaks and pitch seams.

Pitch streak. A well defined accumulation of pitch at one point in the piece.

Pith. The small, soft core occurring in the structural center of a log.

Pith knot. A sound knot except that it contains a pith hole not more than one-quarter inch in diameter.

Pocket rot. Typical decay that appears in the form of a hole, pocket, or area of soft rot usually surrounded by apparently sound wood.

Raised grain. Is an unevenness between springwood and summerwood on the surface of dressed lumber.

Rate of growth. The rate at which a tree has laid on wood, measured radially in the trunk or in the lumber cut from the trunk.

Rays. Strips of cells extending radially within a tree and varying in height from a few cells in some species to 4 inches or more in oak.

Resin passage. Intercullular passages or ducts that contain and transmit resinous materials.

Ring annual growth. The growth layer put on in a single growth year.

Rings. Rings are those circular markings around the center of a tree section which are produced by the contrast in density, hardness, color, etc.

Ring shake. Occurs between the growth rings to partially or wholly encircle the pith.

Round knot. A knot cut at approximately right angle to its long axis so that the exposed section is round or oval.

Sapwood. The living wood of pale color near the outside of a log. Under most conditions, sapwood is more susceptible to decay then heartwood.

Sapwood. The outer zone of wood, next to the bark. In the living tree it contains some living cells (the heartwood contains none), as well as dead and dying cells.

Shakes. Defects originating in the living tree due to frost, wind or other causes, or occuring through injury in felling, which later show in the manufactured lumber, most commonly as partial or complete separation between growth-rings.

Silver grain. A condition where there are conspicuous medullary rays in quarter-sawn wood.

Single knot. One occuring by itself, with the fibers of the wood deflected around it.

Softwood. General term for trees that have needle-like or scale-like leaves and bear cones. Also the wood produced by such trees. The term has no reference to the actual hardness of the wood.

Sound knot. A knot solid across its face, as hard as the surrounding wood, showing no indication of decay.

Sound wood. Wood free from any form of decay, incipient or advanced.

Split. Separation of wood fibers.

Splits. Separations of wood fiber running parallel with the grain.

Swirls. Irregular grain in wood usually surrounding knots or crotches.

Tight knot. A knot so fixed by growth or position that it will retain its place in the piece.

Timber. A broad term including standing trees and certain products cut from them, including lumber 5 inches or larger in least nominal dimension.

Tyloses. Masses of cells appearing somewhat like froth in the pores of some hardwoods, notably white oak and black locust.

Typical or advanced decay. The stage of decay in which the disintegration is readily recognized because the wood has become punky, soft, spongy, stringy, pitted, or crumbly.

Unsound knot. Usually contains decay.

Vessel. Wood cells of comparatively large diameter that have open ends that are set one above the other, forming continuous tubes.

Virgin growth. Original growth of mature trees.

Wane. Is bark or absence of wood or bark on an edge.

Warp. Any deviation from a true or plane surface, including bow, crook, cup and twist or any combination thereof.

Wavy-grained. Wood in which the fibers collectively take the form of waves or undulations.

White specks. Are small white pits or spots in wood caused by the fungus "Fomes pini."

6-E WOOD — STRUCTURAL

Balloon Frame. A framing system in which studs and corner posts extend from sill to top plate and upper story floor joists are carried on ledgers or girts let into the studs.

Balloon Frame. Light timber construction in which the exterior walls consist of studs that are either continuous through floors or interrupted only by thickness of plates.

Balloon frame construction. Type of framing for two-story buildings where both studs and first floor joists rest on the anchored sill. The second floor joists bear on a 1 × 4 inch ribbon strip which has been set into the inside edges of the studs.

Balloon framing. System of framing a building in which all vertical structural elements of the bearing walls and partitions consist of single pieces extending from the top of the sole-plate to the roofplate and to which all floor joists are framed.

Beam. Girders, rafters, and purlins acting as a structural member, supported at two ends.

Beams and stringers. Lumber having a minimum rectangular cross section, 5 inches by 8 inches, graded with respect to its strength when loaded on the narrow face.

Bearing. Point of support on a wall, beam or structural member.

Bracing. In buildings of one and two stories the corner post shall be the equivalent of not less than three (3) pieces of two by four (2 × 4″) inch studs. Bracing shall be accomplished by diagonal wood sheathing or plywood panels, or other sheathing specified in Code applied vertically in panels of not less than four by eight (4 × 8′) feet in area in compliance with approved nailing complying with the Code. Ledger members used to support joists shall not be less than two by four (2 × 4″) inch nominal side adequately nailed with 16d nails to doubled header joists.

Bridging. Floor and flat-roof joists and beams shall be securely bridged at intervals not exceeding 8 feet between bridging or between bridging and bearing, either with diagonal wood bridging not less than 1 inch × 3 inches fitted and double nailed at each end, or with approved rigid metal bridging, or with solid wood bridging, or with other approved bridging providing adequate stiffening; except that when the required joist depth is more than 6 times the breadth, joists shall be bridged at intervals not exceeding 6 times the joist depth. Solid blocking shall be placed between joists at all joist supports whenever the joists are not otherwise laterally braced or fastened. Where the required depth of rafters is more than 6 times the breadth, the rafters shall be bridged as required for floor joists.

Bridging. Short wood or metal braces or struts installed crosswise between joists to hold them in line.

Bridging. Small wood or metal members that are inserted in a diagonal position between the joists acting both as tension and compression members for the purpose of bracing the joists, and to create a truss action and also spreading the floor loads.

Bridging. Diagonal pieces fitted in pairs from the bottom of one floor joist to the top of adjacent joists, and crossed to distribute the floor load; sometimes pieces of width equal to the joists and fitted neatly between them.

Built-up members. Structural members, the sections of which are composed of combinations of sawn lumber or plywood, in which all parts are bonded or joined together with glue, bolts, nails, metal clips or other similar fastening.

Built up timber. Timber made of several pieces nailed, bolted or lag screwed together, forming one of larger dimensioned piece.

Camber. Convexity of a truss or beam to offset weight or pressure which might result in its becoming concave.

Ceiling joists. Structural members directly above a room to which the finished ceiling materials are fastened.

Chords. Top and bottom members of a Truss.

Collar Beam. Board usually 1 × 4 inches, 1 × 6 inches or 1 × 8 inches fastened to a pair of rafters in a horizontal position at some desired location between the plate and ridge of roof.

Columns. A support, square, rectangular, or cylindrical in section, for roofs, floors, etc., composed of base, shaft, and capital.

Combination frame. Combination of the principal features of the full and balloon frames.

Common rafters. Those which run square with the plate and extend to the ridge.

Cripple-rafter. Those which cut between valley and hip rafters.

Cross-wall construction. A type of construction in which floor and roof loads are carried entirely on walls or bearing partitions running across a building.

Dead load. Refers to type and weight of floor or roofing materials in relation to psf (pounds per square foot).

Decking. Prepared wood surfacing material installed over rafters or joists to which other finish materials can be applied.

Frame construction. Wood framing which conforms to all requirements of frame construction as described in building codes.

Framing. The rough timber structure of a building, including interior and exterior walls, floor, roof, and ceilings.

Girder. Large sized beam used as a main structural member, normally for the support of other beams.

Girt. (Ribband) The horizontal member of the walls of a full or combination frame house which supports the floor joists or is flush with the top of the joists.

Glued built-up members. Structural elements, the sections of which are composed of built-up lumber, plywood or plywood in combination with lumber; all parts bonded together with adhesives.

Glued built-up sections. Structural elements consisting of wood, plywood, or combinations of the two in which the grain is not parallel and in which all pieces are bonded together with glue.

Glue joint. Place where two pieces of wood are joined together by means of glue.

Glue laminated structural lumber. Lumber consisting of laminations in which the grain of all laminations is approximately parallel and where all laminations are bonded together with glue.

Grade (lumber). The classification of lumber in regard to strength and utility. A mechanical means of grading may be accepted when approved by the Building Department.

Grade-stress. A lumber grade defined in such terms that a definite working stress may be assigned to it.

Hardness. Hardness represents the resistance of wood to wear and marring.

Header. A short joist supporting tail beams and framed between trimmer joists; the piece of wood or finish over an opening; a lintel.

Header. Cross member installed between studs or joists to support openings, such as for stairways, chimneys, doors and windows.

Heavy timber construction. Construction composed of planks or laminated floors supported by beams or girders. Exterior walls may be frame, masonry, or metal.

Heel of a rafter. The end or foot that rests on the wall plate.

Hip rafter. Those extending from the outside angle of the plates toward the apex of the roof.

Impact bending. In the impact bending test, a hammer of given weight is dropped upon a beam from successively increased heights until complete rupture occurs.

Jack rafter. Those square with the plate and intersecting the hip rafter.

Joist. One of a group of closely spaced beams.

Joist. One of several parallel beams carrying a floor or ceiling, sometimes acting both as ceiling joist and rafter.

Joist. Supporting member. A solid loadbearing board set on edge, and set evenly on centers.

Joist lumber. Lumber of rectangular cross sections, from 2 inches up to, but not including, 5 inches thick, and 4 inches or more wide, graded with respect to its strength in bending when loaded either on the narrow face (joist) or on the wide face (plank). If 5 inches or more thick.

Joistplate. Plate at top of masonry walls supporting rafter or roof joists and ceiling framing lumber.

Joists and planks. Means lumber of a rectangular cross section 2 inches to less than 5 inches thick and 4 or more inches wide, graded with respect to its strength in bending, when loaded either on the narrow face as a joist, or on the wide face as a plank.

Joist spacing. Distances (specified by building codes) between joists. Usually specified as o.c. (on center), which is the distance from the center of one joist to the center of the next joist.

Kiln. Heated chamber for drying lumber, veneer, and other wood products.

Kiln-dried. Lumber which has been seasoned in a dry-kiln, usually, though not necessarily, to a lower moisture content than that of air-seasoned lumber.

Kiln-dried. Dried in a kiln with the use of artificial heat.

Knee brace. A corner brace, fastened at an angle from wall stud to rafter, stiffening a wood or steel frame to prevent angular movement.

Lamella. A short piece of lumber used in the construction of the network arches that form a lamella roof.

Lamella roof structure. Arched roof-framing structure identified by the diamond shaped arrangements of the pieces of plank from which it is formed.

Lintel (header). The piece of construction or finish, stone, wood, or metal, which is over an opening; a header.

Live load. Weight bearing term for floors, pitched roofs or level roofs in considering people, partitions, snow, rain and wind loads.

Load bearing. Term used in reference to one structural member of a building which supports another structural member, or loading bearing walk supporting structural loads.

Lumber. Product of the saw and planing mill, not further manufactured than by sawing, resawing, passing lengthwise through a standard planing machine, cross-cutting to length, and matching.

Lumber grade. Division of sawed lumber into quality classes with respect to its physical and mechanical properties as defined in published lumber manufacturer's standard grading rules.

Maximum crushing strength. The maximum crushing strength is the maximum stress sustained by a compression specimen having a ratio of length to least dimension of less than 11 under a load slowly applied parallel to the grain.

Mechanical properties. Mechanical properties are those properties of wood which enable it to resist deformations, loads, shocks or forces.

Modulus of elasticity. The modulus of elasticity of wood is a measure of its stiffness or rigidity.

Modulus of rupture. The modulus of rupture is a measure of the ability of a wood beam to support a slowly applied load for a short time.

Moisture content. Moisture content is the weight of water contained in the wood expressed as a percentage of the weight of the oven dry wood.

Nominal size. The commercial size designation of width, and depth, in standard sawn lumber and glued laminated lumber grades; somewhat larger than standard net size of dressed lumber.

Nominal size lumber. Commercial size designation of width and depth in standard lumber grades somewhat larger than the standard net size of dressed lumber.

Nominal thickness lumber. Full "designated" thickness. For example, a nominal 2 inch x 4 inch stud may be 1-1/2 inch x 3-1/2 inch when dry. It is a commercial size designation, subject to acceptable tolerances.

On center. Measurement of spacing joists, timbers, beams, purlins in a building from center to center of one next to it.

Physical properties. Physical properties, as the term is ordinarily used, are those properties of wood which have to do with its structure, such as density, cell arrangements, fiber length, etc.

Plank. Lumber having a rectangular cross-section in which the width is four (4) inches or more and the thickness is not less than two (2) inches, nor more than five (5) inches.

Plate. Horizontal framing members which provide the anchorage and bearing for floor, ceiling, and roof framing.

Plate. Horizontal dimension lumber member placed on top and/or bottom of the exterior wall studs to tie them together and to support the joists and rafters.

Plate cut. The cut in a rafter which rests upon the plate; sometimes called the seat cut.

Plate line. That part of the wall that supports the rafters.

Plates. The top horizontal piece on the walls of a frame building upon which the roof rests.

Post. Column not more than one (1) story high.

Posts. Lumber of square or approximately square cross section, 5 × 5 and larger, graded primarily for use as posts or columns carrying longitudinal load but adapted for miscellaneous uses in which strength in bending is not especially important.

Posts and timbers. Lumber of square or approximately square cross section, 5 by 5 inches and larger, graded primarily for use as posts or columns carrying axial loads, but adapted for miscellaneous uses in which strength in bending is not especially important.

Purlin. A timber supporting several rafters at one or more points, or the roof sheeting directly.

Rafter. Lumber used in the framing of a roof on a house to support the sheathing and roofing materials. Flat roof members are called joists.

Rafter or joist plate. Plate at top of masonry wall supportiny rafter or roof joist and ceiling framing.

Rafters. Supporting members immediately beneath the roof sheathing.

Rafters. Sloping roof joists. Roof framing lumber.

Ribbon. Narrow board let into the studding to add support to joists.

Ridge beam. Top horizontal member of a sloping roof, against which the ends of the rafters are fixed or supported.

Ridge board. Board placed on edge at the ridge of the roof to support the upper ends of the rafters.

Ridge framing. Term to denote the members used in a wood structure that comprise the frame or form of a ridge that makes up the roof of the building. Usually two chords mitered forming the roof peak and supporting web members with a bottom chord.

Rough dimension. Term applied to lumber indicating an actual measurement which is equal to or larger than that usually specified.

Rough lumber. Lumber that has not been dressed but has been sawed, edged, and trimmed at least to the extent of showing saw marks in the wood on the four longitudinal surfaces of each piece for its over-all length.

Rough stock. Lumber which has been sawed, edged and trimmed, but not dressed. Will vary in thickness and width owing to unavoidable variations in sawing, difference in shrinkage etc.

Seat cut or plate cut. The cut at the bottom end of a rafter to allow it to fit upon the plate.

Seat of a rafter. The horizontal cut upon the bottom end of a rafter which rests upon the top of the plate.

Shear. Shear is the name of the stress which tends to keep two adjoining planes or surfaces of a body from sliding, one on the other, under the influence of two equal and parallel forces acting in opposite directions.

Shearing strength parallel to grain. Shearing strength is a measure of the ability of timber to resist slipping of one part upon another along the grain.

Sheathing. Sheets of wall or roof construction which are attached directly to studs or rafters.

Sill. The lowest horizontal framing member of a structure, resting on the ground or on a foundation. Also, the lowest horizontal member of a window or door casing.

Sill. Lowest member of the frame of a structure, resting on the foundation and supporting the uprights of the frame.

Sill plate. The plate on top of a foundation wall which supports floor framing.

Sleeper. Timber laid on the ground to support a floor joist.

Span. The distance between the bearings of a timber or arch.

Specific gravity. Specific gravity is the ratio of the weight of a given volume of wood to that of an equal volume of water at a standard temperature.

Static bending. The fiber stress at proportional limit in static bending or flexure is the computed stress in the wood specimen at which the strain (or deflection) becomes no longer proportional to the stress (or load).

Strength of wood. Term in its broader sense includes all the properties of wood that enable it to resist different forces or loads.

Strength properties. In a broad sense of the term "strength" implies all those properties that fit a material to resist forces.

Strength ratio of wood. Represents the percentage of remaining strength left in commercial grades after making allowances for the effect on an unseasoned piece of the permitted growth characteristics, such as knots, cross grain, and shakes.

Stress. Stress is distributed force. Fiber stress is the distributed force tending to compress, tear apart, or change the relative position of the wood fibers.

Stress-grade lumber. Lumber 2 or more inches thick and 4 or more inches wide which has been graded for strength by a lumber grading or inspection bureau or other agency or individual recognized as being competent, according to the principles outlined in "Guide to the Grading of Structural Timbers and Determining of Working Stresses", and miscellaneous publications and supplements thereto of the U.S. Department of Agriculture, and the basic provisions for selection and inspection given in Part IV of the American Lumber Standards published by the National Bureau of Standards, U.S. Department of Commerce.

Stress grades. Lumber grades having assigned working stress and modulus of elasticity values in accordance with accepted basic principles of strength grading, and the provisions of the Code.

Stringer. Long horizontal timber in a structure supporting a floor.

Structural insulating board. A structural insulating material, made principally from wood, cane, or other vegetable fibers and preformed into a rigid, fibrous, insulating board, lath, or plank, and used principally in building construction.

Structural lumber. Lumber that is 2 inches or more thick and 4 inches or more wide, intended for use where working stresses are required.

Structural lumber. Consists of lumber classifications known as "beams and stringers", "joists and planks," and "posts and timbers," to each grade of which is assigned proper allowable unit stresses.

Structural timber. By the term "Structural Timber" is understood such products of wood in which the strength of the timber is the controlling element in its selection and use.

Stud. An upright beam in the framework of a building.

Studding. The framework of a partition or the wall of a house; usually referred to as 2 by 4's.

Studs. One of a group of closely spaced wood or metal vertical structural members used in the construction of walls or partitions to provide lateral stiffness and, in the case of bearing walls and partitions, to support vertical loads.

Studs. A series of vertical wall or partition framing members spaced not more than 24 inches o.c.

Studs. A series of slender wood or metal structural members placed as supporting elements in walls and partitions.

Studs. Generally, 2 × 4's used as the basic vertical framing members of walls.

Tail beam. A relatively short beam or joist supported in a wall on one end and by a header on the other.

Tensile strength perpendicular to grain. Tensile strength perpendicular to the grain is a measure of the resistance of wood to forces acting across the grain that tend to split a member.

Tie beam (collar beam). Beam so situated that it ties the principal rafters of a roof together and prevents them from thrusting the plate out of line.

Timber connectors. Rings, grids, plates, or dowels of metal or wood set in adjoining members, usually in pre-cut grooves or holes, to fasten the members together in conjunction with bolts.

Trimmer. The beam or floor joist into which a header is framed.

Truss. Lumber fabricated to span a large area, eliminating intermediate supporting members and thus providing free spanned open areas.

Trussed rafter. Truss where the chord members are also serving as rafters and ceiling joists and are subject to bending stress in addition to direct stress.

Valley rafters. Those extending from an inside angle of the plates toward the ridge or center line of the house.

Wall plate. Horizontal member anchored to a masonry wall to which other structural elements may be attached.

Wall plate. Plate at top or bottom of wall or partition framing. Further defined as top plate, at top, and sole plate, at bottom.

Wood structural members. Each wood structural member shall be of sufficient size to carry the design loads without exceeding the allowable unit stress specified in the Code. Adequate bracing and bridging to resist wind and other lateral forces shall be provided.

Work in bending to maximum load. Work to maximum load in static bending represents the ability of the timber to absorb shock with some permanent deformation and more or less injury to the timber.

6-F WOOD — TREATED LUMBER

Coal tar creosote. Distallate of coal tar produced entirely by high-temperature carbonization of bituminous coal. Heavier than water and has a continuous boiling range of at least 125°C.

Creosote. Creosote, creosote solutions, and oil-borne preservatives are relatively insoluble in water, so are usually recommended where a preservative is required to remain stable under varied conditions of use. Creosote, or creosote coaltar solutions are the only preservatives recommended for use in structures subject to attack by marine borers. Creosote treatment is best for tidewater structures.

Empty-cell process. Any process for impregnating wood with preservatives or chemicals, in which air is imprisoned in the wood under the pressure of the entering preservative; the air expands when the pressure is released and drives out part of the injected preservative.

Fireproofing. The process of making wood resistant to fire to a degree that it is difficult to ignite and will not support its own combustion.

Fire retardant lumber. Lumber treated by pressure impregnation to reduce combustibility.

Fire retardant lumber. Lumber so treated by a recognized impregnation process so as to reduce its combustibility or surface flame spread.

Fire retardant lumber. A wood so treated by a recognized pressure impregnation process as to make it less combustible as required.

Fire retardant lumber. Wood treated by a recognized pressure impregnation process or system as to make it less combustible as required by the code.

Fire retardants. Are water-soluble salts which increase the fire resistance of wood so that it will not support its own combustion, and will cease to burn if the source of the heat is removed.

Fire retardant treated lumber. The Building Official shall accept treated wood as meeting the requirements of the Code only when each piece bears a label certifying that adequate tests and examinations have been made to insure that the wood has been treated to meet the requirements of Underwriters Laboratories, Inc., for flame spread not over 25 and no evidence of significant progressive combustion in 30 minute test duration (ASTM E84). The label to be accepted for the purpose of the Code shall have been applied by an accredited authoritative agency adequately equipped and competent to conduct the required tests and to evaluate the results thereof.

Fire retardant treated wood. Fire-retardant wood includes lumber or plywood that has been treated with a fire-retardant chemical to provide classifications (flame-spread FSC) and fuel contributed (FCC) of 25 or less by ASTM method E84, shows no progressive combustion during 30 minutes of fire exposure by this method, and is so labeled. Fire-retardant wood for decorative and interior finish purposes provides reduced flame-spread classification (FSC) by ASTM method E84 as specified by the Code for materials used in the particular applications.

Fire retardant treated wood. Lumber or plywood impregnated with chemicals and when tested in accordance with the Code Standards for a period of 30 minutes shall have a flame-spread of not over 25 and show no evidence of progressive combustion. The fire-retardant properties shall not be considered permanent where exposed to the weather. All material shall bear identification showing fire performance rating issued by an approved agency having a re-examination service.

Fire retardant treated wood. Wood or wood products that has had its surface-burning characteristics such as flame spread, rate of fuel contribution and density of smoke developed, reduced by impregnation with fire retardant chemicals.

Full-cell process. Any process for impregnating wood with preservatives or chemicals in which a vacuum is drawn to remove air from the wood so that the cells may be filled with the preservative.

Kyanising. The process of impregnating timber with a solution of corrosive sublimate as a preservative.

Preservative. Any substance that will, for a reasonable length of time, prevent the action of wood destroying fungi, borers of various kinds, and similar destructive agents, when the wood has been properly treated.

Preservative treated wood. Wood treated by a recognized pressure impregnation process to increase its durability.

Preservative treatment (treated material). Unless otherwise noted, is impregnation under pressure with a wood preservative. Wood preservative is any suitable substance that is toxic to fungi, insects, borers, and other living wood-destroying organisms.

Protective treatments. Timbers exposed to attack by wood destroying organisms, decay, or damage by fire should be pressure-treated. Interior woodwork, except in termite infested regions, is seldom given a preservative treatment.

Salts preservatives. Are carried into the wood by a solution, with water or volatile solvents.

Sapwood preservative. Takes preservative treatment more readily than heartwood, and is equally durable when treated.

Treated woods. Pressure treatment or the term "pressure impregnated with an approved preservative" is that treatment of wood which is in accordance with standards of the American Wood Preservers Association.

Water repellents. Wood preservatives, with water resistant properties.

6-G BUILDERS' HARDWARE

Anchor bolts. Bolts used for anchoring materials to each other.

Anchor bolts. Bolts which secure wood plates to the construction.

Anchors. Special metal devices or bolts used to fasten timbers to masonry or concrete.

Bolt. A metal bar which, when actuated, is projected (or "thrown") either horizontally or vertically into a retaining member, such as a strike plate, to prevent a door from moving or opening.

Bolt projection (or bolt throw). The distance from the edge of the door, at the bolt center line, to the farthest point on the bolt in the projected position, when subjected to end pressure.

Builders' finishing hardware. Hardware which includes manufactured devices for supporting doors, including cabinet work devices, and hardware, and miscellaneous other finishing hardware.

Clamp. Mechanical device used to hold two or more pieces together.

Clasp Nail. Square section cut nail whose head has two pointed projections that sink into the wood.

Component. As distinguished from a part, is a subassembly which combines with other components to make up a total door assembly. The prime components of a door assembly include:

door, lock, hinges, jamb/wall, jamb/strike and wall.

Cylinder. The cylindrical subassembly of a lock containing the cylinder core, tumbler mechanism and the keyway. A double cylinder lock is one which has a key-actuated cylinder on both the exterior and interior of the door.

Cylinder core (or cylinder plug). The central part of a cylinder containing the keyway, which is rotated by the key to operate the lock mechanism.

Deadbolt. Lock bolt which does not have a spring action as opposed to a latch bolt, which does. The deadbolt must be actuated by a key and/or knob or thumb turn and when projected becomes locked against return by end pressure.

Dead latch (or deadlocking latch bolt). Spring-actuated latch bolt having a beveled end and incorporating a plunger which, when depressed, automatically lock the projected latch bolt against return by end pressure.

Door assembly. Unit composed of a group of parts or components which make up a closure for an opening to control passageway through a wall. For the purposes of this standard, a door assembly consists of the following parts: door; hinges; locking device or devices; operation contacts (such as handles, knobs, push plates); miscellaneous hardware and closers; the frame, including the head and jambs plus the anchorage devices to the surrounding wall and a portion of the surrounding wall extending 36 inches (900 mm) from each side of the jambs and 16 inches (400 mm) above the head.

Double style hanger. Designed for hanging joists that are aligned on each side of a girder or beam.

Drift bolts. Drift bolts or drift pins can be used to fasten heavy timbers together.

Hardware. Required to complete all of the rough carpentry installations, such as all types and sizes of nails, screws, lag screws, bolts, and all other rough hardware.

Jamb. Vertical member of a door frame to which the door is secured.

Jamb/strike. That component of a door assembly which receives and holds secure the extended lock bolt; the strike and jamb used together are considered a unit.

Jamb/wall. That component of a door assembly to which a door is attached and secured; the wall and jamb, used together, are considered a unit.

Joist hangers. Metal hangers for joists, headers, and other framed-in wood members shall be of approved type, shall provide adequate and firm support, and shall be punched to permit spiking to the supporting and supported members.

Joist Hangers. Metal preshaped hangers with a minimum 2 inch seat fastened to the support by nailing.

Joist-purlin hanger. One piece metal prefabricated shaped seat hanger with tapered throat for better installation fastened to the support by nailing.

Key-in-knob. Lockset having the key cylinder and other lock mechanisms such as a push or turn button contained in the knobs.

Lag screw. Type of screw required in construction.

Latch (or latch bolt). Beveled, spring-actuated bolt, which may or may not have a deadlocking device.

Lock front. Outer plate through which the locking bolt projects and which is usually flush with the edge of the door.

Lock (or lockset). Keyed device (complete with cylinder, latch or deadbolt mechanism, and trim such as knobs, levers, thumb turns, escutcheons, etc.) for securing a door in a closed position against forced entry. For the purposes of the Standard, a lock does not include the strike plate.

Part. As distinguished from component, is a unit (or subassembly) which combines with other units to make up a component.

Strike. Metal plate attached to, or mortised into, a door or door jamb to receive and to hold a projected latch bolt and/or deadbolt in order to secure the door to the jamb.

Swinging door. A stile (side) hinged door.

6-H FIXED WOOD LADDERS

Cage. A cage is a guard that may be referred to as a cage or basket guard, which is an enclosure that is fastened to the side rails of the fixed ladder or to the structure to encircle the climbing space of the ladder for the safety of the person who must climb the ladder.

Cleats. Cleats are ladder crosspieces of rectangular cross section placed on edge on which a person may step in ascending or descending.

Fastenings. A fastening is a device to attach a ladder to a structure, building, or equipment. Fixed, hinged, bearing, or slide-type fastenings may be used.

Fixed ladder. A fixed ladder is a ladder permanently attached to a structure, building, or equipment. Ladders referred to in the code shall be construed to be fixed ladders.

Grab bars. Grab bars are individual handholds placed adjacent to or as an extension above ladders for the purpose of providing access beyond the limits of the ladder.

Individual-rung ladder. An individual-rung ladder is a fixed ladder, each rung of which is individually attached to a structure, building, or equipment.

Ladder. A ladder is an appliance usually consisting of two side rails joined at regular intervals by crosspieces called steps, rungs, or cleats, on which a person may step in ascending or descending.

Ladder safety device. A ladder safety device is any device, other than a cage or well, designed to eliminate or reduce the possibility of accidental falls and which may incorporate such features as life belts, friction brakes, and sliding attachments.

Pitch. Pitch is the included angle between the horizontal and the ladder, measured on the opposite side of the ladder from the climbing side.

Railings. Railings when referred to in this section shall be any one or a combination of those railings defined in the Code.

Rail ladder. A rail ladder is a fixed ladder consisting of side rails joined at regular intervals by rungs or cleats and fastened in full length or in sections to a building, structure, or equipment.

Rungs. Rungs are ladder crosspieces on which a person may step in ascending or descending.

Side-step ladder. A side-step ladder is one from which a man getting off at the top must step sideways from the ladder in order to reach the landing.

Steps. Steps are the flat crosspieces of a ladder on which a person may step in ascending or descending.

Through ladder. A through ladder is one from which a man getting off at the top must step through the ladder in order to reach the landing.

Well. A well is a permanent complete enclosure around a fixed ladder, which is attached to the walls of the well. Proper clearances for a well will give the person who must climb the ladder the same protection as a cage.

2. Article 7

THERMAL AND MOISTURE PROTECTION 7-A THROUGH 7-C

7-A INSULATION

Blankets and batts. Flexible, fibrous type of insulation.

Board. Rigid or semi-rigid insulation formed into sections, rectangular both in plan and cross section.

Fiber glass insulations. Made from extra fine and long, glass fibers.

Foil faced fiber glass. Fiber glass rolls and batts faced with an aluminum foil vapor barrier.

Functions of insulation. Thermal insulation is used in a building to retard the transfer of heat from the warm side to the cold.

Insulation. Any material used to obstruct the passage of sound, heat, vibration, or electricity from one place to another.

Insulation. Substance which is a nonconductor. In building construction.

Insulation. Any material used to absorb, restrict, obstruct the passage of sound, heat, vibration, or electricity from one area to another.

Insulating materials and adhesives. For pipes, ducts, plenums and other components of heating, air handling, and cooking exhaust systems in hotels shall be noncombustible or shall have fire hazard classification ratings not exceeding flame spread—25; fuel contributed—35; and smoke developed—50; or shall be of other approved composition.

Loose-fill type insulation. Comes in bales or bags and may be poured or blown into spaces to be insulated.

Metallic insulation. Made of very thin tin plate or copper or aluminum sheets. It is also made of aluminum foil on the surface of rigid fiberboard or plasterboard.

Mineral insulation. Made from rock, slag, and glass.

Plastic insulation. Is made from polystyrene and rubber.

Reflective insulation. Sheet material with one or both surfaces of comparatively low heat emissivity, such as aluminum foil.

Rigid insulation board. Structural building board made of coarse wood or cane fiber.

Rubber insulation. Made from synthetic rubber containing cells filled with nitrogen.

Thermal insulation. Any material high in resistance to heat transmission that, when placed in the walls, ceiling, or floors of a structure, will reduce the rate of heat flow.

Vegetable or natural insulation. Made from processed wood, sugar cane, corn stalks, and certain grasses.

7-B ROOFS AND ROOFING

Abrasion resistance. Ability of the membrane to resist mechanical abrasion such as foot traffic and wind blown particles which tend to progressively remove materials from its surface.

Adhesion. Ability of the membrane to remain adhered during its service life to the substrate or to itself.

Aggregate. Crushed stone or water-worn gravel used for surfacing or ballasting.

Base sheet. One layer of felt or combination sheet secured to the deck over which may be applied additional felts, a cap sheet, organic or inorganic fiber shingles, smooth coating, or mineral aggregate.

Bitumen. Generic term applied to amorphous, semi-solid mixtures of predominantly hydrocarbons in viscous or solid form, derived from coal or petroleum. This term is normally used to describe either coal tar pitch or asphalt.

Bond. Adhesive and cohesive forces holding two roofing components in contact.

Boston ridge. Method of applying asphalt or wood shingles as a finish at the ridge or hips of a roof.

Built-up roof. Roofing application consisting of 3 to 5 layers of rag felt saturated with coal tar, pitch, or asphalt. The top is usually finished with crushed slag or gravel. Generally used on flat or very low pitched roofs.

Built-up roof covering. Two or more layers of roofing consisting of a base sheet, felts and cap sheet, mineral aggregate, smooth coating, or similar surfacing material.

Cant strip. Wedge or triangular shaped piece of lumber, compressed composition wood, precast lightweight masonry material used where the horizontal roof surface meets the vertical surface of a masonry wall or parapet.

Cap sheet. Roofing made of organic or inorganic fibers, saturated and coated on both sides with a bituminous compound, surfaced with mineral granules, mica, talc, ilminite, inorganic fibers, or similar materials.

Cementing. Solidly mopped application of asphalt, cold liquid asphalt compound, coal tar pitch, or other approved cementing material.

Class A roof coverings. Roof coverings which are effective against severe fire exposures (meeting the 3 methods for fire tests of class A roof coverings—ASTM E-108) and possess no flying brand hazard.

Class B roof coverings. Roof coverings which are effective against moderate fire exposures (meeting the 3 methods for fire tests of class B roof coverings—ASTM E-108) and possess no flying brand hazard.

Class C roof coverings. Roof coverings which are effective against light fire exposures (meeting the 3 methods for fire tests of class C roof coverings—ASTM E-108) and possess no flying brand hazard.

Closed valley. Roof valley where the shingles of the intersecting slopes are continuous without open space.

Coal tar. Tar derived from the destructive distillation of coal during the conversion of coal into coke.

Combination sheet. Glass fiber felt integrally attached to kraft paper.

Corrosion-resistant. Any nonferrous metal, or any metal having an unbroken surfacing of nonferrous metal, or steel with not less than 10 percent chromium or with not less than 0.20 percent copper.

Creep. Permanent deformation of a roofing material or roof system caused by the movement of the roof membrane that results from a combination of thermal stresses and mechanical loading.

Dead loads. Non-moving roof top loads, such as mechanical equipment, air conditioning units and the roof system itself.

Deck. The structural surface to which the roofing or waterproofing system (including insulation) is applied.

Delamination. Separation of the plies in a membrane system or separation of any laminated materials in composite form.

Elastomer. A macromolecular material that returns rapidly to its approximate initial dimensions and shape, after substantial deformation by a low level stress and the release of that stress.

Elastomeric. Term used to describe the elastic, rubber-like properties of a material.

Felt. Matted organic or inorganic fibers, saturated with bituminous compound.

Felt (non-bituminous saturated). Matted asbestos fibers with binder for use with wood shingle and wood shake assemblies as defined in the Code.

Fire retardant roof coverings. Roof coverings shall be classified on the basis of protection provided against fire originating outside the building or structure on which they have been installed.

Flammability. Ability of the membrane to resist combustion and spreading of the flame.

Flashing. The system used to seal the edges of a membrane at walls, expansion joints, drains, gravel stops, terminations and other areas where the membrane is interrupted. Usually a base flashing covers the edges of the membrane while a counter flashing is used to shield the upper edges of the base flashing.

Flashing. Sheet metal or other impervious material used in roof and wall construction to protect a building from water seepage.

Flashing cement. A plastic mixture of bitumen and asbestos reinforced with inorganic reinforcing fibers and a solvent to soften the material for hand troweling.

Flood coat. The top layer of bitumen in a mineral aggregate surfaced built-up roof assembly into which the aggregate is embedded.

Gable. The vertical triangular end of a building from the eaves to the apex of the roof.

Gambrel. Symmetrical roof with two different pitches or slopes on each side.

Glass fiber felt. Glass fiber sheet coated on both sides with bituminous compound.

Impact resistance. Ability of the membrane to resist hail and falling objects without puncturing.

Inorganic. Being or composed of material other than hydrocarbons and their derivatives; not of plant or animal origin.

Interlayment. Layer of felt or nonbituminous saturated asbestos felt not less than 18 inches wide, shingled between each course of roof covering.

Interlocking roofing tiles. Individual units typically of clay or concrete possessing matching ribbed or interlocking vertical side joints that restrict lateral movement and water penetration.

Live loads. Moving or non-permanent loads such as wind, snow, ice, rain or portable equipment.

Mansard roof. Roof which has a slope of 15 degrees and not over 30 degrees from the vertical.

Mansard roof. A roof with two slopes on all four sides. The lower slope is very steep, and the upper one almost flat.

Membrane. Thin sheet or film of waterproof material installed in such a method or procedure using other materials if necessary to prevent the movement of moisture through a floor, roof or wall.

Membrane. Continuous flexible (or semi-flexible) roof covering that forms the water control element of a roofing system. It is normally assembled on site from single or multiple plies of material, e.g. polyvinyl chloride roofing in single ply and bituminous felt roofing in multiple ply.

Metal roofing. Metal shingles or sheets for application on solid roof surfaces, and corrugated or otherwise shaped metal sheets or sections for application on solid roof surfaces or roof frameworks.

Mineral surfaced roofing. Felt or fabric saturated with bitumen, coated on one or both sides with a bituminous coating and surfaced on its weather side with mineral granules.

Non-nailable deck. Any deck which is incapable of retaining an approved fastener.

Open valley. Roof valley where the shingles of the intersecting slopes leave an open space covered by metal flashing.

Organic. Material composed of hydrocarbons or their derivatives; matter of plant or animal origin.

Prepared roofing. Any manufactured or processed roofing material other than untreated wood shingles and shakes as distinguished from built-up coverings.

Protected membrane. Roofing membrane with insulation and protective surfacing or ballasting on top; also called inverted or upside down roof.

Ridge. Peak of a double-pitched roof.

Ridge. Horizontal line at the junction of the top edges of two sloping roof surfaces. Rafters are nailed at the ridge board.

Rise. Height of a roof rising in horizontal distance (run) from the outside face of a wall supporting the rafters or trusses to the ridge of the roof. In stairs, the perpendicular height of a step or flight of steps.

Roll roofing. Roofing material laid and overlapped from roll material.

Roof. Roof slab or deck with its supporting members, no including vertical supports.

Roof. Cover of a building, including the slab or deck with its supporting members, with the exception of the vertical supporting members such as columns and walls.

Roof. Structural cover of a building with a slope range bearing from horizontal to a maximum of 60 degrees to the horizontal.

Roof construction. Combination of materials providing cover for a building including the ceiling (if any) beneath the roof, the roof decking, roofing and all horizontal or sloping structural members supporting the roof only, but excluding columns and other vertical load supporting members.

Roof covering. Covering applied to the roof for weather resistance, fireresitance or appearance.

Roof covering. Material applied to roof surfaces for protection against the elements. Roof insulation shall not be deemed to be a roof covering.

Roof covering. The covering applied to the roof for weather resistance fireresistance or appearance.

Roofing. Covering applied to a roof surface for the purpose of providing protection from the weather. Roofing does not include insulation.

Roofing square. 100 square feet of roofing surface.

Roofing system. An assembly of interacting roofing and metal materials and methods to weatherproof the exposed surface areas of a structure.

Roofs. Flat roof is one that has 4 inches or less fall per foot; a pitch roof is one that has a rise of more than 4 inches per foot.

Roof sheathing. Boards or sheet material fastened to the roof rafters on which the shingles or other roof covering is laid.

Roof structure. Structure erected above the roof of any part of a building supporting or enclosing a tank enclosing a stairway, elevator, machinery or other equipment, and not used for living quarters, recreational areas, or other purposes involving the more or less continuous presence of persons within the structure. When the roof structure or structures exceed 50% of the roof area, it shall be considered an additional story.

Roof structure. A structure above the roof of any building, enclosing a stairway, tank, elevator machinery, or ventilating apparatus, or such part of a shaft as extends above the roof.

Roof structure. A structure above the roof of any part of a building enclosing a stairway, tank, elevator machinery or ventilating apparatus, or such part of a shaft as extends above the roof, and not housing, living, or recreational facilities.

Roof structure. A structure above the roof of any part of a building, enclosing a stairway, tank, elevator machinery or ventilating apparatus, or such part of a shaft as extends above the roof and not housing, living or recreational accommodations.

Rooftop equipment. Any equipment located on the exterior roof of a building and/or structure. Such equipment shall be installed on a platform and enclosed or screened with incombustible material as approved by the Building Official. All rooftop equipment over 200 pounds in weight shall be engineered for roof loading, wind loads and stability.

Run. In reference to roofs, the horizontal distance from the face of a wall to the ridge of the roof.

Sawtooth roof. A roof with serrated cross-section whose shorter, vertical sides have fenestration for light and ventilation.

Scupper. An opening or receptacle usually in the side of walls so that precipitation falling on the roof surface is conducted away from the building.

Single ply. A nominal description of roofing membranes completely installed in one application effort. The single ply membrane may be homogeneous or composite in nature.

Slip sheet. Sheet material placed between two layers of a roofing system to assure that there is no adhesion between them.

Slope. The tangent of the angle between the roof surface and the horizontal normally measured in inches per foot.

Smooth surfaced roofing. Felt or fabric saturated with bitumen coated on one or both sides with a bituminous coating and surfaced with fine mineral surfacing.

Spot-cementing. Discontinuous application of asphalt, cold liquid asphalt compound, coal tar pitch or other approved cementing material.

Underlay. One or more layers of felt applied as required for a base sheet, over which finish roofing is applied.

Underlayment. One or more layers of felt or nonbituminous saturated asbestos felt over which finish roofing is applied.

Valley. Internal angle formed by the junction of two sloping sides of a roof.

Valley. Line of intersection of two roof slopes, where their drainage combines.

Weatherability. The ability of the membrane to resist weathering; i.e., degradation due to sun, rain, wind, etc.

Weather-exposed surfaces. All surfaces of walls, ceilings, floors, roofs, soffits and similar surfaces exposed to the weather.

Weight. The manufacturer's shipping weight in pounds per 100 square feet of roof covering.

Wide-selvage asphalt roll roofing. Surfaced with mineral granules.

7-C WATERPROOFING

Asbestos. Mineral of fibrous crystalline structure composed, chemically, of silicates of lime and magnesia, and alumina.

Asbestos felt. Sheets made of asbestos shreds.

Asphalt. Derivative obtained from the distillation of crude petroleum.

Asphalt cement. Fluxed or unfluxed asphaltic material.

Asphalt mastic. Refined asphalt, particularly that obtained from bituminous rocks.

Bitumen. Any asphalt or coal-tar product used in the application of waterproofing and dampproofing.

Bituminous grouts. Suitable for waterproofing above or below ground level as protective coatings.

Built-up-membranes. Consisting of several plies of treated felt cemented with asphalt or coal-tar-pitch.

Coal-tar pitch. Derivative obtained from the distillation of bituminous coal tar.

Continuous membranes. Essential to prevent penetration of moisture under hydrostatic head.

Creosote primer. Refined coal tar creosote oil having liquid properties.

Dampproofing. Treatment of a surface or structure to retard passage of liquid water.

Emulsion. Asphalt or coal-tar pitch which has been rendered liquid by suspension of asphalt or pitch particles in water, usually with the aid of a small quantity of an emulsifying agent.

Fabric. Woven cotton or glass fiber cloth saturated with asphalt or coal tar and used in waterproofing.

Felt. Rag, asbestos or wood-fiber felt saturated with asphalt or coal tar and used in waterproofing.

Grout. Mixture of cement and water or cement, sand and water or thinner consistency than mortar.

Grouting. Process of injecting grout or mortar to fill small holes and seems in and around subsurface structures.

Hydrostatic pressure. Foundations, walls or other portions of structures exposed to hydrostatic pressure.

Integral compound. Material incorporated in mortar or concrete.

Iron (powder). Cast iron or pig iron in powder form.

Low carbon tars. Tars containing a low percentage of free carbon.

Membrane. Thin layer or layers of bituminous material with or without fabric reinforcement.

Membrane system. System of applying a combination of elastic, membranous waterproofing materials.

Membrane waterproofing. Three or more hot-applied coatings and layers of a combination of compatible bituminous-type materials.

Oil asphalts. Artificial oil pitches or asphaltic cements produced as a residuum from asphaltic petroleum.

Pitch. Resin from pine tar.

Resin. Dried and hardened pitch from pine and similar trees.

Surface coating. Compound applied to a masonry surface for dampproofing or waterproofing purposes.

Tar. Bitumen which yields pitch upon fractional distillation and which is produced as a distillate by the destructive distillation of bitumens pyrobitumens, or organic material.

Vapor barrier. Material used to retard the flow of vapor or moisture into walls and thus to prevent condensation within them.

Waterproofing. Treatment of a surface or structure, which prevents the passage of water.

Water repellent. Property of a waterproofing material which hinders or prevents it miscibility with water.

2. Article 8

DOORS AND WINDOWS 8-A THROUGH 8-D

8-A DOORS

Automatic collapsible revolving door. A door which is designed, supported and constructed so that the wings will release and fold back in the direction of egress under pressure exerted by persons under panic conditions, providing a legal passageway on both sides of the door pivot.

Batten door. One made of sheathing, secured by strips of board, put crossways, and nailed with clinched nails.

Bi-parting door. A door which slides vertically and consists of two or more sections or pairs of sections that open away from each other and are so interconnected that two or more sections operate simultaneously.

Combination door. One with interchangeable storm sash and screen insert.

Combination doors. Door with glazed and screen replacement sections to provide summer and winter use.

Corestock. Stock which is used to form the inner portion of a solid door. The core stock is covered on both sides by the type of wood paneling or facing desired.

Doors required in a dwelling. Each door opening which provides an entrance to the dwelling, bedrooms, or bathroom or toilet compartment shall be provided with a door.

Dutch door. A door divided into two parts, each part separately hinged, and separately locked. When used in residences, at the kitchen, the lower part has a shelf for serving purposes.

Exterior door. One for installation in an entrance which provides access from the outside to the inside of any building.

Filler. Material used in slab door construction that fills the area between the core framing.

Flush door. (Slab door), one of composite construction having flush surfaces. It may be prepared to take glass.

French door. One consisting all or in large part of divided glass openings.

Hollow core. A type of slab door construction in which the core filler consists of a spaced ribbing or grid fabricated to support the facing surfaces.

Interior door jamb. The surrounding case into which, and out of, a door closes and opens. It consists of two upright pieces, called jambs, and a head piece, fitted together and rabbeted.

Louvred door. One with one or more sections of angled slats providing ventilation.

Panel door. One constructed of stiles, rails, and one or more panels of wood or glass.

Plank door. One constructed of a number of solid individual planks, joined vertically and assembled with horizontal blind battens or splines. It may be prepared to take glass.

Solid core. A term pertaining to slab door construction wherein solid built-up wood extends the full width and height of the door.

8-B FIRE DOORS

Automatic. As applied to a fire door or other opening protective means normally held in open position and automatically closed by a releasing device actuated by abnormal high temperatures, or by a pre-determined rate of rise in temperature.

Automatic. As applied to a door, window or protection for an opening shall mean that the means of protection is so constructed and arranged that it will close when subjected to a predetermined temperature or rate of temperature rise.

Automatic-closing. As applied to opening protectives such as fire doors, shall mean normally held in an open position and automatically closing upon the action of a heat actuated releasing or operating mechanism.

Automatic doors, shutters and windows. The term "automatic," as applied to fire doors, fire shutters, fire windows and other opening protectives, shall mean doors, shutters, windows and other opening protectives which are normally held in an open position and which close automatically upon the action of some heat actuated releasing mechanism.

Automatic fire door. A fire door or other opening protective so constructed and arranged so that if open, it shall close when subjected to: A predetermined temperature, or a predetermined rate of temperature rise, or smoke or other products of combustion.

Closing device (fire door). Closing device which will close the door, and be adequate to latch and/or hold hinged or sliding door in a closed position.

Door holders. Intended for use with swinging, sliding, or rolling doors or fire doors, to hold doors in the open position under normal usage and also to release fire doors under fire conditions. Also intended for use with a suitable door closer and automatic operating devices.

Fire door. Door and its assembly, so constructed and assembled in place as to give the specified protection against the passage of fire.

Fire door. Fire-resistive door assembly, including frame and hardware, which under standard test conditions, meets the fire protective requirements for the location in which it is to be used.

Fire door assembly. The assembly of a fire door and its accessories, including all hardware and closing devices and their anchors; and the door frame, when required, and its anchors.

Fusible link. Device consisting of 2 pieces of brass or other suitable metal connected by solder or other metal fusible at a moderate temperature, arranged to release in the presence of fire or excessive heat; or any equivalent approved device.

Fusible links. Devices used in connection with automatic closing devices for doors and windows, and other automatic devices requiring fusible links.

Self-closing. A device which will maintain the door in a closed position.

Self-closing. Equipped with an approved device which will insure closing after having been opened.

Self-closing. As applied to a fire door or other opening protective, means normally closed and equipped with an approved device which will insure closing after having been opened for use.

Self-closing. Normally kept in a closed position by some mechanical device and closing automatically after having been opened.

Self-closing fire door. A door which is normally kept in the closed position and which if opened is returned to closing position by a closing device.

Self-closing fire door. A fire door, which when actuated by a fire or smoke detector system, a fusible link or other device, the closing mechanism will automatically close the door.

Smoke and fire door holder. Fire door holders that keep the door open by automatic attraction. When a fire is detected the control panel automatically releases the doors by interrupting the current flow to the electromagnet.

Smoke stop doors. Where installed to meet the requirements of the Code, shall be of metal, metal covered, or approved treated wood construction with clear wired glass panels, except that in buildings not over two stories in height and not required by other sections of the Code, to be of fire-resistive construction, smoke stop doors may be of ordinary solid wood type not less than 1-3/8" thick with clear wired glass panels. Such doors shall be of self-closing, single swinging type and may be either single or double. They shall close the opening completely with only such clearance as is reasonably necessary for proper operation.

8-C GLASS AND GLAZING

American National Standards Institute (ANSI). Document ANSI Z97.1-(latest issue) provides specifications and methods of test for safety of glazing materials used in buildings.

Annealed glass. Glass that has been subjected to a slow, controlled cooling process during manufacture to control residual stresses so that it can be cut or subjected to other fabrication.

Bite. The amount of overlap between the stop and the panel or light.

Clips. (Wire) Wire spring devices to hold glass in rabbeted sash, without stops, and face glazed.

Exterior glazed. Glass set from the exterior of the building.

Federal specifications. The following Federal Specifications establish the quality standards for glass: (a) DD-G-451 sets the quality characteristics as well as the thickness and dimensional tolerances of flat glass products; (b) DD-G-1403 provides standards for tempered glass, heat strengthened glass and spandrel glass; (c) DD-M-411 prescribes the standards for mirrors and mirror frames.

Glass. A hard, brittle, amorphous substance produced by fusion, usually consisting of mutually dissolved silica and silicates that also contains soda and lime. Glass may be manufactured transparent, translucent, or opaque.

Glass area. The gross glass within a sash, door, or exterior opening, and may include the area of small vertical or horizontal muntins and division bars.

Glass door. A door which is composed of a single glass panel, usually tempered glass, which is not set in a frame.

Glazed door. A glass panel of any size which is installed in a wood or metal frame as part of the door assembly.

Glazing. The securing of glass or plastics in prepared openings in windows, door panels, partitions and decorative items.

Glazing materials. Includes plastics, glass, annealed glass, organic-coated glass, tempered glass, laminated glass, wired glass; or combinations of any of these.

Interior glazed. Glass which has been installed from the interior of the building.

Interior stop. The removable molding or bead that holds the glass panel or light in place.

Jalousie door. A door having the opening glazed with operable, overlapping glass louvers.

Laminated glass. Glazing material composed of two or more pieces of glass, each piece being either tempered glass, heat strengthened glass, annealed glass or wired glass, bonded to an intervening layer or layers of resilient plastic material.

Leaded glass. A decorative composite glazing material made of individual pieces of glass whose circumference is enclosed by durable metal such as lead or zinc.

Mirror. A treated, polished or smooth glazing material that reflects images.

Organic-coated glass. A glazing material consisting of a piece of glass and bonded on one or both sides with an applied polymeric coating, sheeting or film.

Permanent label. A permanently legible identification visible after installation of the glazing material, which cannot be removed or destroyed.

Plastic. A sheet of organic glazing material.

Plate glass. Glass which is manufactured in a continuous ribbon and cut into large sheets. Both surfaces are ground and polished. Glass is produced in thicknesses from one-eighth (1/8) inch to one and one-quarter (1-1/4) inches.

Tempered glass. A piece of glass specially heat treated or chemically treated, that after it has been produced in its final form cannot be cut or drilled, ground or polished after treatment without fracture. When fractured at any point, if highly tempered, the entire piece will disintegrate into small particles.

Transparent glass door. Doors containing transparent glass in a ratio of eighty (80) percent or more to the total area of the door.

Wired glass. Glass, not less than one-quarter (1/4) inch thick, in which a mesh structure of wire is embedded and completely covered.

8-D WINDOWS

Approved attachments for window cleaner safety belts. To which belts may be fastened at each end. Said attachments shall be permanent devices that shall be firmly attached to the window frame, or to the building proper, and so designed that a standard safety belt may be attached thereto.

Bay window. A window structure which projects from a wall. Technically, it has its own foundation. If cantilevered, it would be an ariel window, however, this is not common usage.

Bay window. A rectangular, curved, or polygonal window, supported on a foundation extending beyond the main wall of the building.

Bay window. A rectangular, curved or polygonal window, projecting beyond the exterior wall of the building.

Bay window. A window projecting beyond the wall line of the building and extending down to the foundations.

Blind stop. Rectangular moulding, used in the assembly of a window frame.

Casement. Window hinged vertically, swinging open from either side.

Casement. A window, the sash of which is hinged vertically to a frame.

Casement. Frames of wood or metal enclosing part or all of the sash which may be opened by means of hinges affixed to the vertical edges, and can swing in or out from the frame.

Casement. A window sash which is made to open by turning on hinges affixed to its vertical edges.

Casement frames and sash. Frames of wood or metal, enclosing part or all of the sash which may be opened by means of hinges affixed to the vertical edges.

Center-hung sash. A sash hung on its centers so that it swings on a horizontal axis.

Check rails. Meeting rails which are sufficiently thicker than the other members of pair of sash so that they fill the opening between the two sash made by the parting strip in the frame.

Checkrails. Meeting rails sufficiently thicker than a window to fill the opening between the top and bottom sash made by the parting stop in the frame.

Clerestory. The part of a building rising clear of the roofs or other parts and whose walls contain windows for lighting the interior of the building.

Clerestory windows. Series or band of windows installed above the primary roof deck.

Dormer window. Substantially vertical window and its enclosing structure erected as an appendage to a sloping roof.

Double hung window. A window having top and bottom sashes, each balanced by springs or weights to be capable of vertical movement with relatively little effort in its own grooves.

Double window. A window arranged with double sashes enclosing air, intended to act as a sound and heat insulator.

Fenestration. The arrangement of windows in a building.

Fenestration. Any opening or arrangement of openings for the admission of daylight and ventilation.

Fire window. A window, frame, sash, and glazing which will successfully resist fire to the degree required by, and which has been constructed and installed in accordance with the requirements of the National Board of Fire Underwriters, and Fire Underwriters Inc.

Fire window. A window and its assembly so constructed and assembled in place as to give protection against exposure fires.

Fire window. A window frame with sash and glazing having a fire resistive rating of three-quarters (3/4) of an hour in accordance with the rules of the board.

Fire window. A window and its assembly, so constructed and assembled in place as to give specified protection against the passage of fire.

Fire window. A window constructed and glazed to prevent the passage of fire.

Fire window. Window assembly, including frame, wired glass and hardware, which under the standard test method meets the fire protective requirements for the location in which it is to be used. ASTM E-163.

Fire window. A window constructed and glazed to give protection against the passage of fire.

Fire window assembly. Fire window includes glass, frame, hardware and anchors constructed and glazed to give protection against the passage of flame.

Fixed sash. A sash permanently fixed in a solid frame.

French window. A glazed casement, serving as both window and door.

Jalousie. Adjustable glass louver. Also refers to doors or windows containing jalousies.

Jalousie. Louvered glass window, stationary or operable.

Meeting rail. The bottom rail of the upper sash of a double-hung window. Sometimes called the check rail.

Meeting rails. The rails of a pair of sash that meet when the sash are closed.

Mullion. The construction between the openings of a window frame to accommodate two or more windows.

Muntin. The vertical member between two panels of the same piece of panel work. The vertical sash-bars separating the different panels of glass.

Oriel bay window. Extension of a room suspended or projected beyond the general wall line of an enclosed structure in one or more vertically aligned spaces but not extending to the foundations, whereby floor space is gained in the room.

Oriel window. A window projecting beyond and suspended from the wall of a building, or cantilevered therefrom.

Oriel window. A window which projects from the main line of an enclosing wall of a building and is carried on brackets or corbels.

Pulley stile. The member of a window frame which contains the pulleys and between which the edges of the sash slide.

Sash. The framework which holds the glass in a window.

Sash. Single frame containing one or more panes of glass.

Sash balance. Device, usually operated with a spring designed to counter-balance window sash. Use of sash balances eliminates the need for sash weights, pulleys, and sash cord.

Window. Glazed opening including glazed doors in a building which open upon a yard, court, or recess from a court or a vent shaft open and unobstructed to the sky.

Window assembly. Unit composed of a group of parts or components that make up a closure for an opening in a wall to control light, air and other elements, and which normally includes: glazed sash, hinges or pivots, sash lock, sash operator, window frame, miscellaneous hardware and the anchorage between the window and the wall.

Window (bay). A window projecting beyond the wall of a building and extending down to or below the ground.

Window (bay). A rectangular, curbed or polygonal window supported on a foundation extending beyond the main wall of the building.

Window cleaning. Where the tops of windows to be cleaned are more than twenty (20) feet above the floor, ground, flat roof, balcony, or permanent platform, fastening and latching means shall be provided to protect the window cleaners.

Window (dormer). A substantially vertical window and its enclosing structure erected as an appendage to a sloping roof.

Window (oriel). A window projecting beyond and suspended from the wall of a building, or cantilevered therefrom.

Window (oriel). A window which projects from the main line of an enclosing wall of a building and is carried on brackets or corbels.

Window (show). A window in which goods or wares are displayed for sale or advertising purposes.

2. Article 9

FINISHES 9-A THROUGH 9-C

9-A FLOORING

Asphalt tile. A resilient floor covering laid in mastic, is available in several colors. Standard size is 9″ × 9″, also comes in several other sizes. Asphalt is normally used only in the darker colors, the lighter colors having a resin base.

Block flooring. Form of parquetry assembled into square or rectangular blocks at the mill, installed in mastic or by nailing. Also used as industrial flooring laid in common-brick pattern where heavy wear or great loading requires strength.

Flagstone. Flat stones, pre-cut, and used for walks, steps, floors, and similar areas.

Floor finish. Material used for the finished floor. Floor finish is not deemed to be part of either interior finish or interior trim.

Mosaic. A decoration in which small pieces of glass, stone or other materials are laid in mortar to form a design.

Parquet. Flooring laid in geometrical designs with small pieces of wood.

Parquet flooring. Prefabricated and finished wood blocks laid in patterns.

Quarry tile. Prefabricated, hard-burned, unglazed ceramic tile.

Terrazzo. Flooring surface of marble chips in cement. After the cement has hardened, the floor is ground and polished to expose the marble chips.

Tile. A ceramic surface unit, usually relatively thin in relation to facial area, made from clay or a mixture of clay and other ceramic materials, called the body of the tile, having either "glazed" or "unglazed" face and fired above red heat in the course of manufacture to a temperature sufficiently high to produce specific physical properties and characteristics.

9-B LATH AND PLASTER

Back plastering. Plaster applied to one face of a lath system following application and subsequent hardening of plaster applied to the opposite face.

Base coat. Any plaster coat or coats applied prior to application of the finished coat.

Bond plaster. Specifically formulated gypsum plaster designed as first coat application over monolithic concrete.

Brown-coat. As applied to plastering, means the second coat of plaster applied.

Brown-coat. The second in three-coat plaster application.

Brown-out. To complete application of basecoat plaster.

Clip. Wire or sheet-metal device used to attach various types of lath to supports or to secure adjacent lath sheets.

Contact. As applied to ceiling construction, means that the lath is attached in direct contact with the construction above, without the use of runner channels or furring.

Contact ceiling. Means that the lath is attached in direct contact with the construction above, without the use of main runners or cross furring.

Corner bead. Strip of preformed galvanized sheet iron, usually combined with narrow strips of metal lath attached to each wing and placed on exterior corners of walls before plastering is started.

Corner bead. Rigid formed unit or shape used at projecting or external angles to define and reinforce the corners of interior surfaces.

Cornerite. Shaped reinforcing unit of expanded metal or wire fabric used for angle reinforcing and having minimum outstanding legs of not less than 2 inches.

Corner reinforcement. Metal reinforcement for plaster at re-entrant corners to provide continuity between two intersecting planes; or concrete reinforcement used at wall intersections or near corners of square or rectangular openings in walls, slabs, or beams.

Corrosion-resistant materials. Materials that are inherently rust-resistant or materials to which an approved rust-resistive coating has been applied either before or after forming or fabrication.

Craze cracks. Fine, random cracks or fissures caused by shrinkage which may appear in a surface of plaster.

Cross furring. Furring members which are attached at right angles to the underside of main runners or structural supports, in a ceiling assembly, for the support of the lath.

Cross furring. As applied to metal construction, means the furring members which are attached at right angles to the underside of main runners or other structural supports.

Dash-bond coat. Thick slurry of portland cement, sand, and water flicked on surfaces with a paddle or brush to provide a base for subsequent portland cement plaster coats; sometimes used as a final finish on plaster.

Drywall. Any finish material applied to the interior wall or partition in a dry state as opposed to the application of water mixed plaster. Drywall could be plywood, fiberboard, plasterboard; usually used as the term for plasterboard.

External corner reinforcement. A shaped reinforcing unit for external corner reinforcement for portland cement plaster formed to insure mechanical bond and a solid plaster corner.

Finishing-coat. White-coat, putty-coat, sand-finish, or acoustical plaster, as applied to plastering, means the third or last coat applied.

Furred ceiling. Cross furring used for the support of the lath which is attached directly to the structural members of the building.

Furring. Used in vertical construction, means spacer elements used to maintain a space between the structural elements and the lath supports.

Furring. Strips of wood or metal fastened to a wall or other surface to even it, to form an air space, to give appearance of greater thickness, or for the application of an interior finish such as plaster.

Interior stucco. Regional term designating a finish plaster for walls and ceilings finishing smooth or textured. It is a mechanically blended compound of Keene's cement, lime (Type "S") and inert fine aggregate. Color pigment may be added to produce integrally colored interior stucco.

Interior surfaces. Surfaces other than weather-exposed surfaces.

Keene's cement. Cement composed of finely ground, anhydrous, calcined gypsum, the set of which is accelerated by the addition of other materials.

Keene's cement. Quick-setting, white, hard-finish plaster which produces a wall of extreme durability. It is made by soaking plaster of paris in a solution of alum or borax and cream of tarter.

Key. The grip or mechanical bond of one coat of plaster to another coat or to a plaster base. It may be accomplished physically by the penetration of wet mortar or crystals into paper fibers, perforations, scoring irregularities, or by the embedment of the lath.

Lath. Any material used as a base for plaster including gypsum lath, wire and metal lath.

Lime. Specifically, calcium oxide (CaO). A general term for the various chemical and physical forms of quicklime, hydrated lime, and hydraulic hydrated lime.

Main runners. The runners which are attached to or suspended from the construction above for the support of cross furring.

Moist curing. Any method employed to retain sufficient moisture for hydration of portland cement plaster.

Plaster. Cementitious material or combination of cementitious material and aggregate that, when mixed with a suitable amount of water, forms a plastic mass or paste which when applied to a surface, adheres to it and subsequently hardens, preserving in a rigid state the form or texture imposed during the period of plasticity; also the placed and hardened mixture.

Portland cement plaster A mixture of portland cement, or portland cement and lime and aggregate and other approved materials as specified in the Code.

Scratch-coat. As applied to plastering, means the first coat applied.

Scratch-coat. First coat of plaster, which is scratched or scored after it is applied to the lath to provide a bond for second coat.

Screeds. Strips of plaster of the desired coat thickness laid on a surface to serve as guides for plastering the intervals between them. Also the intermediate leveling strips in concrete slabs.

Stripping. Flat reinforcing units of expanded metal glass fiber or wire fabric not less than 3 inches wide to be installed as required over joints of gypsum lath.

Stucco. Outside plaster made with Portland cement as its base.

Suspended. As applied to ceiling construction, means that the furring members are suspended below the structural members of the building.

Suspended ceiling. Main runners and cross furring that are suspended below the structural members of the building.

Tie wire. Wire for securing together metal framing or supports, for tying metal and wire lath and gypsum lath and wallboard together and for securing accessories.

9-C PAINTS AND PAINTING

Acrylic resin. Ingredient of water-base (latex) paints. Synthetic resin with excellent weathering characteristics. Acrylics may be colorless and transparent or pigmented.

Blistering. Coating failure common to paints, varnishes, lacquers, and related formulations. It is characterized by the formation of local or scattered blisters varying in size from small pimples to large patches.

Bloom. Surface coating failure associated with high gloss paints, varnishes, lacquers, and related formulations. It is characterized by the formation of a surface haze which lowers the original specular gloss, imparting a dull or semigloss appearance to the coating.

Bodied linseed oil. Oil that has been thickened in viscosity by suitable processing with heat or chemicals. Bodied oils are obtainable in a great range in viscosity from a little greater than that of raw oil to just short of a jellied condition.

Chalking. Surface coating failure common to paints, varnishes, lacquers, and related formulations, particularly when exposed to exterior environmental weathering. It is characterized by the formation of a chalklike powder on the surface attributed to film deterioration by the blue and ultra-violet wavelengths of the sun's radiant energy.

Checking. Surface coating failure of paints, varnishes, lacquers, and related formulations, characterized by the formation of small surface breaks in the coating which do not penetrate to the underlying surface. If the underlying surface is visible, crackling is the term used to denote these breaks.

Checking. Fissures that appear with age in many exterior paint coatings, at first superficial, but which in time may penetrate entirely through the coating.

Filler. Material used for filling nail holes, checks, cracks or other blemishes in the surfaces of wood before the application of paint, varnish, lacquer or other finishes or coatings.

Finishing room. Room in which flammable finishes are applied to objects, products, and materials by spraying or dipping, or in which highly flammable finishes are applied to objects, products, and materials by brushing, flow coating, or roller coating.

Interior finish. Materials applied directly to walls, ceilings, and other exposed interior surfaces of a building (except floors) for acoustical correction, surface insulation, decorative treatment, or similar purposes, including but not limited to, veneer, wainscoting, and paneling. Surfaces finished of wallpaper or other similar materials not more than 1/28th inch thick having no greater fire hazard than wallpaper shall not be considered interior finish.

Japan. Solutions of metallic salts in drying oils, or varnishes containing asphalt and opaque pigments.

Kettle-boiled oil. A union or blending of linseed oil and driers by boiling, in an open kettle.

Lead based coatings. Any paint, lacquer, or other applied liquid surface coatings, and putty, which contain a quantity of lead more than six-hundredths of one-percent (0.06 of 1%) by weight of its non-volatile content.

Painter. Person, firm, or corporation handling and mixing paints, varnishes, oils, or other flammable liquids to be used, in painting and finishing by himself or his employees on premises other than those occupied by him.

Prime coat. First coat of paint, an undercoat, to prepare the surface for finish coats.

2. Article 10

WALLS AND PARTITIONS 10-A AND 10-B

10-A PARTITIONS

Bearing. As applied to a wall or partition, shall mean supporting any vertical load in addition to its own weight.

Bearing partition. Partition which supports any vertical load in addition to its own weight.

Bearing partition. A partition used to support loads other than its own weight.

Bearing partitions. Masonry bearing partitions shall be supported either vertically or horizontally (whichever distance is the lesser) at right angles to the face of the wall at intervals not exceeding twenty-four (24) times the wall thickness for hollow masonry units when laid in Type M, S or N mortar, Gypsum partition tile or block shall not be used in bearing walls.

Fire partition. A partition of construction which subdivides a building to restrict the spread of fire or to provide areas of refuge, but is not necessarily continuous through all stories nor extended through the roof, and which has a fire-resistance rating as required by the Code.

Fire partition. A vertical, horizontal, or other construction having the required fire resistance rating to provide a fire barrier between adjoining rooms or spaces within a building, building section or fire area.

Fire partition. The term "fire partition" shall mean a partition providing for the purpose of protecting life by furnishing an area of exit or refuge, and having a fire resistive rating of at least three hours.

Fire partition. Partition of construction which subdivides a building to restrict the spread of fire or to provide areas of refuge as required by Code.

Fire partition. Partition shall have a fire rating of at least 2 hours. They shall be constructed of approved masonry or reinforced concrete, or other approved forms of construction of incombustible materials.

Fire partition. Partition providing for the purpose of protecting life by furnishing an area of exit or refuge, and having a fire resistive rating of at least 3 hours.

Fire partition. Partition of construction which subdivides a building to restrict the spread of fire or to provide areas of refuge, but is not necessarily continuous through all stories nor extended through the roof, and which has a fire-resistance rating as required by the code for this use.

Fire partition. A partition which subdivides a building to restrict the spread of fire or to provide areas of refuge, but is not necessarily continuous through all stories nor extending through the roof and has a fire-resistance rating of at least two (2) hours.

Fire partition. A partition having a fire resistance rating required by the Code.

Fire partition. A partition which subdivides the floor area of a building or structure to provide an area of refuge or to resist the spread of fire, including but not limited to stairway, elevator, and public hallway enclosures.

Fire partition. A partition which subdivides a story of a building to provide an area of refuge or to restrict the spread of fire.

Fire partition openings. Fire Partitions shall have no openings other than required door openings, or properly protected duct openings.

Fireproof partition. Partition, other than a fire partition, provided for the purpose of restricting the spread of fire, and having a fire resistive rating of at least 1 hour.

Fireresistive partition. A partition other than a fire partition which is required to subdivide the floor area of a fireresistive building for the purpose of restricting the spread of fire.

Nonbearing. As applied to a wall or partition, shall mean one that supports no vertical load other than its own weight.

Nonbearing partition. Partition which supports no load other than its own weight.

Nonbearing partition. Interior nonbearing wall one (1) story or less in height.

Non-bearing partitions. Non-bearing partitions shall be supported either vertically or horizontally (whichever distance is the lesser) at right angles to the face of the wall at intervals not exceeding forty-five (45) times the nominal wall thickness exclusive of plaster.

Office type partition. Permanent or movable partition in which the height does not extend within 18 inches of the finished ceiling, or one in which 1/3rd or more of the surface area of the partition consists of glass or other translucent or transparent material.

Partition. An interior nonload-bearing wall one story or less in height, supporting no vertical load other than its own weight.

Partition. Any interior wall in a building.

Partition. An interior bearing or nonbearing wall not over one (1) story in height, the chief function of which is to separate rooms, corridors, or other spaces.

Partition. A minor interior wall used to subdivide a floor area.

Partition. An interior nonbearing vertical element serving to enclose or divide an area, room or space.

Partition. A non-bearing interior wall one story or less in height.

Partition. Vertical separating construction between rooms or spaces.

Partition. An interior wall.

Partition. A wall whose primary function is to divide space within a building or structure.

Partition. An interior wall, other than folding or portable that subdivides spaces within any story, attic or basement of a building.

Partition (non-rated). Any partition not higher than three fourths of the ceiling height of the room in which it is located or, which has one-fourth of its height in plain glass or openings, or which is constructed entirely of incombustible materials.

Partition (permanent). Any partition not classed as a non-rated partition.

Rated partition or wall. Rated partition or wall is a vertical construction having the required fire resistance as set forth in the code to provide a fire barrier between adjoining rooms or spaces within a building or structure.

Smoke stop. Partition in corridors, or between spaces, to retard the passage of smoke, with any opening in such partition protected by a door equipped with a self-closing device.

10-B WALLS

Apron wall. That portion of a skeleton wall below the sill of a window.

Apron wall. The term "apron wall" shall mean that part of a panel wall between the window sill and the support of the panel wall.

Apron wall. A wall which supports any vertical load in addition to its own weight.

Area separation wall. Any wall other than an exterior wall, which extends the full height of a building and through the roof. Such walls may be bearing walls or self-supporting only.

Bearing wall. Any wall meeting either of the following classifications: 1. Any metal or wood stud wall which supports more than 100 pounds per lineal foot of superimposed load. 2. Any masonry or concrete wall which supports more than 200 pounds per lineal foot superimposed load, or any such wall supporting its own weight for more than one story.

Bearing wall. A wall which supports upper floor or roof loads.

Bearing wall. A wall which supports any load other than its own weight.

Bearing wall. A wall which supports any vertical load in addition to its own weight.

Bearing wall. Any wall which carries any load other than its own weight.

Bearing wall. A wall which supports a floor, roof or other vertical load in addition to its own weight.

Bearing wall. A wall which supports in addition to its own weight, another load having a vertical component.

Bearing wall. A wall which supports more than 100 pounds per foot of ceiling load or supports any other vertical loads except its own weight; or, any wall self-supporting for more than 20 feet or any wall more than one story in height.

Bearing wall. A wall supporting any vertical load in addition to its own weight.

Blank wall. A wall constructed without openings.

Brick cavity wall. A wall in which a space is left between inner and outer tiers of brick. The space may be filled with insulation.

Brick veneer wall. Usually used to describe a wall made up of brick veneer applied over wood framing.

Building division. Wall used for separation between 2 buildings on the same property identical in construction to a party wall.

Cavity wall. A wall built of masonry units or of plain concrete, or a combination of these materials so arranged as to provide an air space within the wall, and in which the inner and outer wythes of the wall are tied together with metal ties.

Cavity wall. A hollow wall unit of masonry units so arranged as to provide a continuous air space within the wall (with or without insulating material), and in which the inner and outer wythes of the wall are tied together with metal ties.

Cavity wall. A wall built of masonry units or of plain concrete, or a combination of these materials, arranged to provide an air space within the wall, and in which the inner and outer parts of the wall are tied together with metal ties.

Common wall. Wall which separates adjacent dwelling units within an apartment building.

Common wall. Single wall used jointly by two buildings.

Composite wall. A wall built of a combination of (2) or more masonry units of different materials bonded together, one forming the back-up and the other the facing elements.

Curtain wall. Non-bearing wall supported by the structural frame of a building.

Curtain wall. A wall, usually nonbearing, between piers or columns.

Curtain wall. Any exterior non-bearing wall between columns or piers which is not supported by beams or girders at each story.

Curtain wall. An exterior non-bearing wall built between columns or piers which may not be supported at every story.

Curtain wall. A non-bearing enclosure wall not supported at each story.

Curtain wall. The term "curtain wall" shall mean a non-bearing wall built between piers or columns for the enclosure of the structure, but not supported at each story.

Curtain wall. An exterior nonbearing wall.

Curtain wall. A non-bearing wall built between piers or columns for the enclosure of the structure, but not supported at each story.

Curtain wall. A non-bearing wall between columns or piers and which is not supported by girders or beams, but is supported on the ground.

Curtain wall. Exterior, non-bearing wall more than one story high and not supported at each floor level, which is laterally stayed by masonry piers or by the frame of the building.

Curtain wall. Non-bearing wall between columns or piers which is not supported by girders or beams.

Curtain wall. Non-bearing wall in skelton construction, anchored to columns, piers, or floors, but not necessarily built between columns or piers.

Curtain wall. A nonbearing wall between columns or piers that is not supported at each story.

Curtain wall. An exterior nonload-bearing wall not wholly supported at each story. Such walls may be anchored to columns, spandrel beams, floors, or bearing walls, but not necessarily built between structural members.

Division wall. Any interior wall or partition in a building.

Division wall. Wall dividing a building into separate areas.

Division wall. Wall used to divide the floor area of a building or structure into separate parts for fire protection, for different uses, for restricted occupancy, or other purposes specified in the code.

Division wall. Any interior wall in a building or structure.

Enclosure wall. Interior wall, bearing or nonbearing, which encloses a stairway, elevator shaft or other vertical opening.

Enclosure wall. An exterior, non-bearing wall in skeleton construction, anchored to columns, piers, or floors, but not necessarily built between columns or piers.

Exposed building face. That part of the exterior wall of a building which faces one direction and is located between ground level and the ceiling of its top story, or where a building is divided into fire compartments, the exterior wall of a fire compartment which faces in one direction.

Exterior wall. Any outer enclosing wall of a building.

Exterior wall. A wall located on the perimeter of the area under the roof of a building. Open spaces under the perimeter of the roof shall be presumed to be openings in the exterior wall of the building.

Exterior wall. Wall, bearing or nonbearing, which is used as an enclosing wall for a building, but which is not necessarily suitable for use as a party wall or a fire wall.

Exterior wall. Any wall or element of a wall, or any member or group of members, which defines the exterior boundaries or courts of a building and which has a slope of 60 degrees or greater with the horizontal plane.

External wall. The outer wall or vertical enclosure of a building other than a party wall.

Facade. Usually applying to the face of a building having the principal entrance; exterior face of a building, often applied to the important face.

Faced wall. A wall in which the masonry facing and backing are so bonded as to exert common action under load.

Facing walls. Walls opposite to and parallel with one another and wall lines or wall lines extended of opposite walls intersection at angles of less than sixty-five (65) degrees.

Fire division. Wall extending from the lowest floor level to or through the roof to restrict the spread of fire.

Fire division. Construction having a fire resistance rating required by the code for the purpose of separating occupancies.

Fire division. Interior means of required separation of one part of a floor area from another part, together with required fire-resistive floor construction, to form a complete fire barrier between adjoining areas and floor areas above or below in the same building.

Fire division. Portion of a building so separated from the rest by separations that it may be erected to the maximum height and area allowed for its principal occupancy and type of construction, independently of adjoining occupancies; portion of a building separated from the rest by fire walls; fire division may not be larger than a maximum unit of occupancy and may be further limited by the application of requirements of the building code.

Fire division. The interior means of separation of one part of a floor area from another part together with fire resistive floor construction, to form a complete fire barrier between adjoining or superimposed floor areas in the same building or structure.

Fire division. The interior vertical means of separation of one part of a building from another part together with fire-resistive floor construction to form a continuous fire barrier between adjoining or superimposed floor areas in one or more stories.

Fire division. A building is considered to be located in a fire division when, due to segregation by open space, fire walls or other means of protection, a fire therein, under normal conditions, would burn itself out without spreading to buildings or combustible materials outside the fire division, and in which a fire originating in buildings or combustible materials outside such fire division would burn itself out without spreading to a building in the fire division.

Fire wall. A wall having adequate fire resistance and structural stability under fire conditions to accomplish the purpose of completely subdividing the building or completely separating adjoining buildings to resist the spread of fire.

Fire wall. A wall constructed in accordance with the Code for the purpose of subdividing buildings to restrict the spread of fire.

Fire wall. A wall of non-combustible material having fire resistance as follows: storage occupancy, four (4) hours, business occupancy, three (3) hours, all other occupancies, two (2) hours. Fire walls shall subdivide a building or separate buildings to restrict the spread of fire and which starts at the foundation and extends continuously through all the stories to and above the roof, except where the roof is fireproof or semi-fireproof and the wall is carried up tightly against the under side of the roof slab. (NOTE: Fire walls must include stability under extreme fire conditions as well as resist the effects of fire. It is therefore necessary that a fire wall be considered as a special device requiring careful consideration of the use of materials and the workmanship.

Fire wall. A wall which subdivides a building or separates buildings to restrict the spread of fire, and which starts at the foundation and extends continuously through all stories to and above the roof, except where the roof is fireproof and the wall is carried up tightly against the underside of the roof slab. Fire walls shall be built of brick, concrete, or other approved materials or assemblies of materials, and shall have a fire-resistive rating of not less than four (4) hours, and shall meet all other requirements for structural stability and thickness set forth in the Code for walls of various materials.

Fire wall. Shall mean a wall provided primarily for the purpose of resisting the passage of fire from one structure to another, and having a fire resistive rating of at least four hours.

Fire wall. A wall separating two fire divisions of a building.

Fire wall. A wall of noncombustible materials having a required fire resistance rating, the purpose of which is to divide a structure into floor areas, none of which exceeds the maximum permitted by the Code.

Fire wall. A wall having a specified fire resistance rating and adequate structural stability for the purpose of subdividing a building, or of separating buildings, to restrict the spread of fire.

Fire wall. A wall of incombustible construction which subdivides a building or separates buildings to restrict the spread of fire and which starts at the foundations and extends continuously through all stories to and above the roof, except where the roof is of fireproof or fireresistive construction and the wall is carried up tightly against the underside of the roof slab.

Fire wall. A wall of noncombustible materials which subdivides a building into separate fire areas to restrict the spread of heat, hot gases, or flames and which has the fire resistance rating as required by the Code.

Fire wall. A wall of masonry units or monolithic concrete which subdivides a building or separates buildings to restrict the spread of fire, which shall have not less than four-hour fire-resistive rating; and which starts at the foundation and extends continuously through all stories to and above the roof, except where the roof is fireproof or semi-fireproof, and the wall is carried up tightly against the under side of the roof slab.

Fire wall. A wall of noncombustible material having a fire-resistance rating of not less than 4 hours, and having sufficient structural stability under fire conditions to allow collapse of construction on either side without collapse of the wall. Fire walls shall start at the foundation and extend continuously through all stories to and above the roof, except where the roof is of fire-resistive construction and the wall is carried up tightly against the under side of the roof slab.

Fire wall. A wall of non-combustible material having fire resistance as follows: storage occupancy-four (4) hours, business occupancy—three (3) hours, and all other occupancies two (2) hours. Fire walls shall subdivide a building or separate buildings to restrict the spread of fire and which starts at the foundation and extends continuously through all stories to and above the roof, except where the roof is fireproof or semifireproof and the wall is carried up tightly against the underside of the roof slab.

Fire wall. A wall of noncombustible construction, with qualities of fire resistance and structural stability, which completely subdivides a building into fire areas, and which resists the spread of fire.

Fire wall. A fireresistive wall, having protected openings, which restricts the spread of fire and extends continuously from the foundation to or through the roof.

Fire wall. Wall constructed so as to give protection against the passage of fire.

Fire wall. Interior wall which completely subdivides a building into limited fire areas in all stories or which separates 2 or more buildings to restrict the spread of fire; and which is supported on a foundation and extends continuously through all stories to and above the roof except buildings of fireproof construction.

Firewall. A wall of incombustible construction which subdivides a building or separates buildings to restrict the spread of fire and which starts at the foundation and extends continuously through all stories to, and three feet above the roof, except where the roof is of fireproof or fire-resistive construction and the wall is carried up tightly against the underside of the roof slab. Firewalls shall be of approved masonry or reinforced concrete and shall have a fire resistance rating of at least four hours.

Firewall. A type of fire separation of noncombustible construction which subdivides a building or separates adjoining buildings to resist the spread of fire and which has a fire-resistance rating as prescribed in the Code and has structural stability to remain intact under fire conditions for the required fire-rated time.

Firewall. A wall with qualities of fire resistance and structural stability which subdivides a building into fire areas, and which resists the spread of fire.

Hollow wall. A wall built of masonry units so arranged as to provide an air space within the wall, and in which the facing and backing of the wall are bonded together with masonry units.

Hollow wall. A wall built of solid or hollow masonry units so arranged as to provide an air space within the wall between the inner and outer wythes.

Hollow wall. The term "hollow wall" shall mean a wall built of solid masonry units so arranged as to provide an air space within the wall.

Hollow wall of masonry. A wall built of masonry units so arranged as to provide an air space within the wall, and in which the inner and outer parts of the wall are bonded together with masonry units or steel.

Interior nonbearing walls. Walls may be built of materials conforming with the requirements of the code, or of gypsum block or other approved materials.

Interior wall. Wall entirely surrounded by the exterior walls of a building.

Interior wall. Wall either bearing or non-bearing, other than exterior, fire or party walls.

Nonbearing exterior wall. Wall which supports no vertical load other than its own weight.

Nonbearing wall. Wall which supports no vertical load other than its own weight.

Nonbearing wall. A wall which supports no other load than its own weight.

Non-bearing wall. A wall which does not support vertical load other than its own weight.

Non-bearing wall. The term "non-bearing wall" shall mean any wall which carries no load other than its own weight.

Non-load-bearing walls. Non-load-bearing walls shall be anchored to each other at intersections and to supporting masonry by means of masonry bond or corrosion-resistant corrugated metal ties or equivalent. Corrugated metal ties shall be not less than 7/8 inches wide and No. 22 gauge in thickness and shall be located at vertical intervals not more than 16 inches on center or shall be equivalent to the foregoing.

Panel wall. Non-bearing wall in skeleton construction built between columns or piers and wholly supported at each story. Window and other openings shall be included in the wall dimensions.

Panel wall. Synonymous with "Enclosure Wall."

Panel wall. A non-bearing exterior wall not over one story high, or supported at each floor level.

Panel wall. A nonbearing wall supported at each story by a skeleton frame.

Panel wall. An exterior nonbearing wall in skeleton construction.

Panel wall. A nonbearing wall built between columns in skeleton construction and wholly supported at each story.

Parapet wall. That part of any wall entirely above the roof line.

Parapet wall. Free standing portion of a wall above the roof.

Parapet wall. That portion of a wall extending above the roof.

Parapet wall. The portion of a wall which projects above the roof line.

Parapet wall. Part of a wall which extends less than a story above a floor or roof of a structure.

Parapet wall. Extension of a wall above the roof level.

Party wall. A wall on an interior lot line used or adapted for joint service between two buildings or structures.

Party wall. Walls used for separation between 2 buildings on the property line between adjoining buildings.

Party wall. A fire wall on an interior lot line used or adopted for use as part of two (2) structures.

Party wall. Wall used or adapted for joint service between two (2) buildings.

Party wall. Wall constructed in the same manner as a "Fire Wall," and used, or built to be used, as a separation of two (2) or more buildings; also, a wall constructed as above and built upon the dividing line between adjoining premises for their common use, extending to and above the roof, except where the roof is of fireproof or fire-resistive construction and the wall carried-tightly against the underside of the roof slab.

Party wall. A wall between 2 buildings of different ownership, and used or adapted for use by both buildings.

Party wall. A wall used or adapted for use in common as a part of two buildings.

Party wall. Wall used jointly by two (2) parties under easement agreement, erected upon a line separating two (2) parcels of land, each of which is a separate real estate entity.

Party walls. A wall on an interior lot line to be used or adapted for joint service between two (2) buildings.

Proscenium wall. Means the wall which separates the stage section of a building from the auditoruim.

Retaining wall. Any wall used to resist the lateral displacement of any material.

Room separation wall. A room separation wall is defined as a wall or partition located between adjacent rooms or areas occupied by persons or material. It shall include separating walls between adjacent schoolrooms, hospital rooms, storerooms, service rooms and any combination of such rooms adjacent to one another.

Separation wall. Any wall forming a separation between occupancies.

Shear wall. A wall which resists horizontal forces applied in the plane of the wall.

Shear wall. Wall designed to resist lateral forces parallel to the wall. Braced frames subjected primarily to axial stresses shall be considered as shear walls for the purpose of this definition.

Single-wythe wall. A wall containing only one masonry unit in wall thickness.

Skeleton or panel wall. A nonbearing wall supported at each story on a skeleton frame.

Spandrel wall. That portion of a skeleton wall above the head of a window or door.

Spandrel wall. That part of a panel wall above the window and below the apron wall.

Spandrel wall. Portion of an exterior wall between top of one opening and bottom of another opening in the story directly above.

Spandrel wall. That part of a wall above the top of a window in one story and below the sill of the window in the story above.

Solid masonry wall. A wall built of concrete masonry units laid contiguously with joints between units filled with mortar or grout.

Veneered wall. A wall having a facing of masonry or other weather-resisting noncombustible materials securely attached to the backing, but not so bonded as to exert common action under load.

Veneered wall. A wall having a facing which is not attached and bonded to the backing so as to form an integral part of the wall for purposes of load bearing stability.

Vertical compartment. That portion of the building bounded by exterior walls and the required two (2) hour fire rated and smoke barrier interior wall assembly which divides the building in a vertical plane.

Wall. Structural element which is vertical or within thirty (30) degrees of vertical, serving to enclose space, form a division, or support superimposed weight.

Wall. Shall mean any structure or device required by the ordinance for screening purposes forming a physical barrier, which is so constructed that fifty percent or more of the vertical surface is closed and prevents the passage of light, air and vision through said surface in a horizontal plane. This shall include concrete, concrete block, wood or other materials that are solids and are so assembled as to form a screen. Where a solid wall is specified, one hundred percent of the vertical surface shall be closed, except for approved gates or other access ways. Where a masonry wall is specified, said wall shall be concrete block, brick, stone or other similar material and one hundred percent of the vertical surface shall be closed, except for approved gates or other access ways.

Wall. A vertical member of a structure whose horizontal dimension measured at right angles to the thickness exceeds three (3) times its thickness.

Wall (bearing). A wall which supports any vertical load in addition to its own weight.

Wall (bearing). A wall which supports any load other than its own weight.

Wall (bearing). A wall supporting any vertical load including to its own weight.

Wall (blank). A wall without openings.

Wall (cavity). A wall built of masonry units or of plain concrete, or a combination of these materials, so arranged as to provide an air space within the wall, and in which the inner and outer parts of the wall are tied together with metal ties.

Wall, (curtain). A non-bearing wall between columns or piers and which is not supported by girders or beams, but is supported on the ground.

Wall (curtain). Any exterior non-bearing wall between columns or piers which is not supported by beams or girders at each story.

Wall (division). Any interior wall in a building.

Wall (enclosure). An exterior non-bearing wall in skeleton construction, anchored to columns, piers or floors, but not necessarily built between columns or piers.

Wall (exterior). A wall, bearing or non-bearing which is used as an enclosing wall for a building, but which is not necessarily suitable for use as a Party Wall or Fire Wall.

Wall (external). The outer wall or vertical enclosure of a building other than a party wall.

Wall (faced). A wall in which the masonry facing and backing are so bonded as to exert common action under load.

Wall (fire). A wall of incombustible construction which subdivides a building or separates buildings to restrict the spread of fire and which starts at the foundation and extends continuously through all stories to and above the roof, except where the roof is of fireproof or fire-resistive construction and the wall is carried up tightly against the underside of the roof slab.

Wall (fire). A wall of masonry units or monolithic concrete which subdivides a building or separates buildings to restrict the spread of fire, which shall have not less than four-hour fire-resisting rating; and which starts at the foundation and extends continuously through all stories to and above the roof, except where the roof is fireproof or semi-fireproof, and the wall is carried up tightly against the under side of the roof slab.

Wall (foundation). A wall below the first floor extending below the adjacent ground level and serving as support for a wall, pier, column or other structural part of a building.

Wall (foundation). All walls and piers built to serve as supports for walls, piers, columns, girders, parts or beams etc.

Wall (front). That wall of the building facing a public street, alley, or public approved place. Where a lot abuts on more than one street, walls facing on any street shall be considered a front wall under the provisions of the Code.

Wall (height). The vertical distance from the foundation wall or other immediate support of such wall to the top of the wall.

Wall (interior). A wall entirely surrounded by the exterior walls of the building.

Wall (non-bearing). One which supports no load other than its own weight.

Wall (non-bearing). Is a wall which supports no vertical load other than its own weight.

Wall (non-bearing). A wall which supports no load other than its own weight.

Wall of hollow masonry. A wall built of masonry units so arranged as to provide an air space within the wall, and in which the inner and outer parts of the wall are bonded together with masonry units or steel.

Wall (panel). A non-bearing wall in skeleton or framed construction, built between columns or piers and wholly supported at each story.

Wall (parapet). That part of any wall entirely above the roof line.

Wall (party). A wall used for joint service between two (2) buildings.

Wall (party). A wall constructed in the same manner as a "Fire Wall," and used, or built to be used, as a separation of 2 or more buildings; also, a wall constructed as above and built upon the dividing line between adjoining premises for their common use, extending to and above the roof, except where the roof is of fireproof or fire-resistive construction and the wall carried tightly against the underside of the roof slab.

Wall (partition). Any interior wall in a building.

Wall (retaining). Any wall built to resist lateral pressure.

Wall (retaining). Any wall used to resist the lateral displacement of any material.

Walls. The upright members forming the lateral enclosure of a building or portion thereof.

Wall (veneered). A wall having a facing which is not attached and bonded to the backing so as to form an integral part of the wall for purposes of load bearing and stability.

2. Article 11

EQUIPMENT 11-A

11-A FOOD SERVICE EQUIPMENT

Commercial cooking appliance. Includes ranges, ovens, broilers, and other miscellaneous cooking appliances of the types designed for use in restaurants, hotel kitchens and similar commercial establishments.

Commercial cooking hood. Hood for the collection of cooking odors, smoke, steam or vapors from commercial food heat-processing equipment.

Commercial food heat-processing equipment. Equipment used in a food establishment for heat-processing food or utensils and which produces steam, vapors, smoke or odors which are required to be removed through a local exhaust ventilation system.

Cooking appliances. All ranges, ovens, food boilers, upright broilers, charcoal broilers, charbroilers, griddles, deep-fat fryers or similar appliances used to heat, cook or process food for human or animal consumption.

Corrosion-resistant material. Material which maintains its original surface characteristics under prolonged influence of the food, cleaning compounds, and sanitizing solutions which may contact it.

Disposables. Term applied to plates, cups, saucers, and utensils that are made of material such as paper or plastic and designed for one-time use.

Equipment. Equipment used for the preparation, storage and serving of food including but not limited to the following with all appurtenances such as blenders, counter freezers, counters, dishwashing machines, hoods, meatblocks, meat grinders, ice-making machines, ranges, stoves, steamtables, baking ovens, refrigerators, slicers and all other similar items used in the preparation and serving of food.

Equipment. All stoves, ranges, barbeque facilities, hoods, meat blocks, tables, counters, refrigerators, sinks, dishwashing machines, steam tables and similar items other than utensils used in the operation of a food service establishment.

Equipment. Includes but is not limited to such equipment and appurtenances as stoves, ranges, hoods, counters, freezers, meatblocks, refrigerators, coolers, sinks, ice-making machines, dishwashers, steamtables, blenders, meat grinders, meat slicers, and other such similar items, other than utensils, used in the operation of the restaurant.

Equipment. All aparatus, utensils and articles used in the preparation, conveyance and service of food and beverages.

Kitchen equipment. Equipment used or intended to be used in connection with the preparation of food including: Kitchen sink, or other device of equipment used as a kitchen sink. Range, stove, hot plate, broiler, burner, heater, or other device or equipment used for the cooking or warming of food. Refrigerator and cabinets used for food storage or utensils or dish storage in connection with the preparation or service of food, when installed within the same room, recess, or space as other kitchen equipment.

Kitchenware. All multi-use utensils other than tableware used in the storage, preparation, conveying or serving of food.

New. As used in reference to new restaurants or new equipment, applies to restaurants constructed or machines or equipment installed after the effective date of the act.

Range. Cooking appliance consisting of an oven or ovens, along with additional top or side burners for other cooking purposes.

Utensils. Includes any kitchenware, tableware, glasses, cutlery, containers, implements, or other equipment with which food or beverage comes in contact during storage, display, preparation, serving, or through use by an employee or consumer.

Vending machine. Any self-service device offered for public use, which, upon insertion of a coin, coins, or token, or by other means, dispenses unit servings of food or beverage, either in bulk or in package, without the necesity of replenishing the device between each vending operation, but not including devices dispensing peanuts, wrapped candy, gum, or ice exclusively.

2. Article 12

EXITING AND TRAVEL EXITING
12-A THROUGH 12-H

12-A CORRIDORS

Common corridor. Enclosed passageway serving a single story and with access to and from individual apartments offices, or rooms, and leading to egresses (exitways), public hallways or stairways. Walls of said corridor shall extend from the floor to the under side of the floor or roof above and shall be permanently affixed thereto. Covering of said walls shall be made of rated materials and shall extend their entire height on both sides.

Corridor. Enclosed passageway.

Corridor. Passageway or hallway which provides a common way of travel to an exit. Dead-end corridor is one which provides only one direction of travel to an exit.

Corridor. Enclosed passageway leading to a means of legal egress.

Corridor. Passageway or hallway which provides a common way of travel to an exit or to another passageway leading to an exit.

Corridor. Synonymous with passageway, hallway and similar designations for horizontal travel from or through sub-divided areas.

Corridor. Enclosed passageway in a building for public ingress and egress to and from dwelling units, rooms or other areas and leading to a lobby, foyer or exit discharge.

Corridor. Horizontal passageway for use of occupants of a story to reach an exitway.

Corridor. Passage-way which gives communication between the various parts of a building.

Corridor or hallway. Substantially horizontal passageway located between walls, partitions, or similar enclosures in a building, designed for and devoted to access to exit ways, rooms, or other enclosed areas in the building.

Corridor-required exit. Fire-rated enclosure beginning at the end point of maximum allowable exit distance and continuing to the exit discharge door.

Dead-end areas in corridors. Or other interior exit in hotels, and in additions thereafter made to hotel, shall not exceed twenty (20) feet in length.

Dead-end corridor. Corridor or portion of a corridor with the exit therefrom at one end only.

Dead end (exit). A portion of a corridor in which the travel to an exit is in one direction only.

Exit corridor. Any corridor or passageway used as an integral part of the exit system, that portion of a corridor or passageway which exceeds the allowable distance of travel to an exit, becomes an exit corridor or passageway.

Exterior corridor. Substantially horizontal unenclosed passageway projecting from the wall of a building and protected by parapet or railings on open sides.

Public corridor. Corridor that provides access to exit from individually rented rooms, suites of rooms or dwelling units.

Public corridor. Enclosed public passageway with access to and from individual apartments, offices, or rooms leading to a public hallway or to the exitways.

Public corridor or public hallway. Corridor or hallway used by the public.

Skip corridor floor. Floor or level in a multistory apartment building not containing a public corridor with required exitways, but having access thereto by means of a stairway leading up or down to an adjacent floor or level provided with a public corridor leading to fire tower stairs or exits.

12-B EXITS, EXITWAYS AND EXITING

Access exit. That part of a means of egress within a floor area that provides access to an exit serving the floor area.

Balcony used as an exterior exit. A landing or porch projecting from the wall of a building, and which serves as a required means of egress. The long side shall be at least 50 per cent open, and the open area above the guardrail shall be so distributed as to prevent the accumulation of smoke or toxic gases.

Basement exits. All basements having habitable rooms shall have an exit providing a means of egress directly to the outdoors.

Building exits. All buildings, except an accessory building, shall be provided with at least 2 independent exitways providing safe and continuous means of egress to a street or to a ground level open space. At least one of these exitways shall be a minimum of 36 inches in width. Exitways shall be no farther apart than 40 feet.

Directional signs. Signs reading TO EXIT and TO STAIRWAY or similar designation with an arrow indicating the direction shall be placed in locations where the direction of travel to reach the nearest exit is not immediately apparent and near all elevators or escalators where in event of fires, persons accustomed to use only the elevators or escalators would have to use a stairway or other alternate exit, unless such stairway or alternate exit is near enough so that the way to reach it is unmistakable.

Doorway. Opening, such as a door or gate closure, in a wall or partition, or a similar opening, provided for ingress or egress.

Entrance. Confined passageway immediately adjacent to the door through which people enter a building.

Exit. A passageway from a portion of a building to the exterior ground surface including every intervening doorway, passageway, stairway or ramp.

Exit. Way of departure from the interior of a building or structure to the open air outside at the ground level and shall comprise vertical and horizontal means of travel, such as doorways, stairways, escalators, ramps, and enclosed passageways.

Exit. Continuous and unobstructed means of egress to a public way, and shall include intervening doorways, corridors, ramps, stairways, smoke-proof towers, horizontal exits, and exterior courts.

Exit. Means of egress; a way out.

Exit. That portion of a means of egress which is separated from all other spaces of the building or structure by construction or equipment as required in the code to provide a protected way of travel to the exit discharge.

Exit. A confined passageway immediately adjacent to the door through which people leave a building.

Exit. Way of departure from the interior of a building or structure, providing a means by which occupants may proceed in reasonable safety from a room or space to a street or to a way or open space which provides a safe access to a street. Exit which comprises all elements for the escape of occupants, including vertical and horizontal means of travel such as doorways, foyers, stairways, escalators, ramps, corridors, passageways, and fire escapes.

Exit. Continuous and unobstructed means of egress to a public way, and which includes intervening doors, doorway, corridors, exterior exit balconies, ramps, stairways, smokeproof enclosures, horizontal exists, exit passageways, exit courts, and yards.

Exit. Way of departure from a living unit to the exterior at street or grade level, including doorways, corridors, stairways, ramps and other elements necessary for egress or escape.

Exit. Way of departure from the interior of a building or structure, to the exterior at street or grade, including doorways, passageways, hallways, corridors, stairways, ramps, fire escapes, and all other elements necessary for egress or escape.

Exit. That portion of a means of egress which is separated from the area of the building from which escape is to be made, by walls, floors, doors or other means which provide the protected path necessary for the occupants to proceed with reasonable safety to the exterior of the building.

Exit. That point which opens directly into a safe dispersal area or public way. All measurements are to be made to that point when determining the permissible distance of travel.

Exit. Means of egress from a room, corridor, stairway, lobby, or other area in a building or structure, means the doorway, stairway, horizontal exit, or other way for departure, which affords a conforming way for immediate egress from the confines of such areas to a conforming exit way which leads to outside of the confines of the building or structure and to the ground level, or which affords a conforming way for immediate egress to outside of the confines of the building or structure at the ground level.

Exit. Required doorway in a structure forming part of an exitway.

Exit. Means of egress from a building or structure including outside exits, vertical exits, horizontal exits, and exit connection.

Exit. That part of a means of egress that leads from the floor area it serves, including any doorway leading directly from a floor area to a public thoroughfare or to an approved open space.

Exit. That portion of the way of departure from the interior of a building or structure to the exterior at street, or grade level accessible to a street.

Exit access. That portion of a means of egress which leads to an entrance to an exit.

Exit capacity. Number of persons an exit or exit way of a given type and width is considered as accommodating for egress.

Exit connection. Includes doorways, aisles, corridors, foyers, lobbies and other horizontal means of exit leading to a vertical exit, a horizontal exit or an outside exit.

Exit court. A yard, court, or inner court providing egress to a public way for one or more required exits.

Exit court. Exterior unoccupied space which is open to the sky for its entire area, located on the same lot with a theater or other assembly building which it serves exclusively as an unobstructed exit to the street or other public space.

Exit court. Exterior court providing a pathway for public egress from an exit to a public thoroughfare.

Exit court. Yard or court providing egress to a public way for one or more required exits.

Exit discharge. That portion of a means of egress between the termination of an exit and a public way.

Exit doorway. Doorway opening directly to the exterior, to a horizontal exit, to an exit stairway, or to a similar place of safety.

Exit doorway. Doorway leading into an Exitway or to a street or to an open place giving safe access to a street.

Exit facilities. All required exits and exit ways; doors, gates, and similar devices used in exit ways; and the hardware in connection therewith; exit signs; and all other ways and devices used and required to effect the rapidity or safety of egress.

Exit facility. Exit facilities permitted for use in Exitways are: Interior Exit Stairways, Fire Towers, Horizontal Exits, Exterior exit stairways, Exit Ramps, Exit Hallways and Exit Doorways, Exitways from the main entrance floor shall discharge directly to a street or an open space which gives safe access to a street.

Exit from a building. Doorway or other means of egress from the building opening upon a street or upon an open space with unobstructed access to a street.

Exit from a room. Doorway or other means of egress from the room on the way toward an exit from the building.

Exit from a story. Stairway, ramp, ladder or other means of egress from the story on the way toward an exit from the building.

Exit way. Exit doorway, or such doorways and connecting corridors, hallways, passageways, and stairways, through which persons may pass from a room or area in a building to a street, or to an open space which provides safe access to a street.

Exit way. Way of departure from a room or other enclosed area of a building or structure, to the outside of the confines of the building or structure and to the ground level. The exit way includes all aisles, doorways, corridors, stairway, fire escapes and other ways of egress which constitute the way of departure.

Exitway. Exit doorway or doorways, or such doorways together with connecting hallways or stairways, either interior or exterior, or fire escapes, designed to provide means by which individuals may proceed safely from a room or space to a street or to an open space which provides safe access to a street.

Exitway. Any required means of direct egress in either a horizontal or vertical direction leading to the exterior grade level.

Exitway. Required means of departure in a structure leading to an open space. The exitway may consist of any combination of doorways, corridors, stairs, ramps, escalators, smoke-proof towers, balconies, vestibules and fire escapes.

Exitway. Exit doorway or doorways, or such doorways together with connecting hallways or stairways, either interior or exterior, or fire escapes, designed to provide means by which individuals may proceed safely from a room or space to a street or to an open space or court communicating with a street.

Exitway. Every required means of egress and the public hallways leading thereto.

Exitway. The necessary combination of "Exit Facilities" through which persons may proceed safely in case of emergency from any floor of a building to the main entrance floor or to a street or an open space which provides safe access to a street; provided that Exitways from the main entrance floor shall discharge directly to a street or an open space which gives safe access to a street.

Exitway. The exit doorway or doorways, or such doorways together with connecting hallways or stairways, either interior or exterior, or fire escapes, designed to provide means by which individuals may proceed safely from a room or space to a street, or to an open space which provides safe access to a street, or to a court which leads to a street.

Exitway. That portion of a means of egress which is separated from all other spaces of a building or structure by construction or equipment as required in the Code to provide a protected way of travel to the exitway discharge.

Exity way or Exit. Applied to a doorway, corridor, fire escape, lobby, or other such facility; or to a door, gate or similar device; or to hardware, lighting, signs, or similar items of equipment; or to construction, means a facility, device, item of equipment, or construction which constitutes a part of the exit or exit way, or which affects the rapidity or safety of egress through an exit or exit way.

Fire hazard. The obstruction to or of fire escapes, ladders which may be used as escapes, stairways, aisles, exits, windows, passageways, or halls, likely in the event of fire to interfere with the operation of the fire department or of the safety and ready egress of occupants.

Ground floor exit. Means of egress from a floor level less than one story above or below grade level and so situated that such exit leads directly from such floor level to a street or to an exterior open space leading to a street.

Path of exit. Continuous series of doorways, connecting rooms, corridors, passages, stairways, ramps and the like, which leads from any exit from a room through an exit from the building.

Private. As applied to an exit way, toilet room, or other part of a building, means that such exit way or other part of the building is an adjunct to not more than one room or one suite of rooms.

Public. As applied to an exit way, toilet room, or other part of a building, means that such exit way or other part of the building is subject to common use by those who occupy or enter the building and includes those parts of the building which are not included within the meaning of "private."

Public occupancy. As applied to an exit way, toilet room, or other part of a building, means that such exit way, toilet room, or other part of the building is subject to common use by those who occupy or enter the building, except as required for the separation of the sexes.

Ramp. Sloping walkway providing access to and from floors at different elevations.

Remote. In reference to two or more exits, removed or distant from one another in such manner that a person in any place served by such exits may choose either of two directions in a path toward an exit and in such manner that a single fire could not, in its early stages, block both paths toward an exit.

Required exit. Required means of egress.

Required means of exit. Continuous means of egress to the outside and consist of vertical exits, horizontal exits and outside exits together with the exit connections leading thereto, and arranged, located and constructed as required by Code.

Supplemental vertical exit. An enclosed stair, ramp or escalator providing means of egress to an area of refuge at another level nearer to the street floor.

Unit of width. Required width of a path of travel either horizontally or vertically, for one (1) person or a single line of persons to exit from a building or from any of its parts. All units of width shall be unobstructed by railings and (not more than eight inches by doors when in a full open position in corridors).

Vertical exit. Means of egress used for ascension or descension between two (2) or more floors, or other levels, and shall include approved exterior stairways, automatic (moving) stairways, fire escapes, ramps, stairways, and smokeproof stair towers.

Vertical exit. Means of egress used to ascend or descend between 2 or more levels, including, but not limited to stairways, smoke-proof towers, ramps, escalators, elevators, and fire escapes.

Vertical exitway. Means of exit between two (2) or more levels of a structure including horizontal passageways connecting upper and lower portions of the same exitway.

Vestibule. Enclosed space, with doors or opening protectives, to provide protected passage between the exterior and interior of a building, or between spaces within a building.

12-C HALL OR HALLWAYS

Common hallway. A common corridor or space separately enclosed which provides any of the following in any story: Common access to the required exitways of the building, or Common access for more than (1) tenant, or Common access for more than (30) persons.

Dead end. When a hallway, corridor or other space is so arranged that a person therein is able to travel in one direction only in order to reach an exit.

Dead end pockets or hallways. Exits and exit access shall be so arranged that dead end pockets or hallways in excess of twenty (20) feet in depth shall not occur.

Grade hallway—grade lobby—grade passageway. An enclosed hallway, exitway or corridor connecting a required exit to a street or to an open space or court communicating with a street.

Hallway. The enclosed passageway leading to a stairway or other required exit, which provides common access to rooms or exitways in the same story in a building.

Hallway. Enclosed hall or corridor leading to a stairway, fire tower or other required exit.

Hallway. Corridor for the passage of people.

Public hall. Hall, corridor, or passageway, not within an apartment, which may be used by an occupant of such building.

Public hallway. Public corridor or space separately enclosed which provides common access to all the exitways of the building in any story.

Public hallway. Public corridor or space separately enclosed or providing common access to all the exitways in any story of a building. The Grade Hallway is the public hallway on the grade floor providing direct access to the street or other open public space from the stairways and elevators.

Public hallway. Corridor or hallway leading directly to a stairway, fire tower or other required exit, within a story of a structure which story is occupied by more than one tenant or lessee.

Public hallway. A hallway, corridor, passageway, vestibule, stairway, landing, or platform in an apartment house, hotel, or apartment hotel; but not within any apartment, guest room or suite of rooms.

12-D HORIZONTAL EXIT

Horizontal exit. Exitway consisting of protected openings through or around a fire wall, exterior wall, party wall or fire partition connecting 2 adjacent floor areas of the required size and exit facilities which furnish an area of refuge and escape complying with the provisions of the code.

Horizontal exit. Means of passage from one building into another building occupied by the same tenant, or from one section of a building into another section of the same building occupied by the same tenant, through a separation wall having a minimum fire resistance of one hour.

Horizontal exit. Means of egress through or around a fire wall or approved fire separation, or by means of a bridge, to a floor area of refuge in the same building, or in a separate building, the floor area having clear floor space sufficient to contain the total number of occupants of both of the connected spaces, allowing not less than 3 square feet of floor area per person.

Horizontal exit. Horizontal passageway or doorway through a fire division wall leading into another building or into another fire division of the same structure. The horizontal exit may serve as a required exit only when it meets all the requirements of the code.

Horizontal exit. That type of exit connecting 2 floors areas at substantially the same level by means of a doorway, vestibule, bridge or balcony, such floor areas being located either in different buildings or located in the same building and fully separated from each other by a firewall.

Horizontal exit. Protected opening through or around a fire wall, connecting 2 adjacent floor areas, each of which furnishes an area of refuge, and from each of which required exits lead to legal open spaces.

Horizontal exit. Exit through or around a fire wall, party wall, or exterior wall, by means of a door, bridge, balcony, or vestibule, which provides access to a floor area protected from fire, in the area evacuated by the fire wall, party wall, or exterior wall, with all openings therein protected with approved opening protectives.

Horizontal exit. Protected opening through a two-hour fire partition through or over or around a fire wall or a bridge connecting two buildings.

Horizontal exit. Means of egress through one or more protected openings connecting 2 adjacent fire areas with the required size of access and egress facilities to safely accommodate the number of persons under consideration for refuge and escape.

Horizontal exit. Opening with fire doors passing through or around fire walls, exterior walls, or fire divisions and connecting two (2) adjacent floor areas.

Horizontal exit. Way of passage from one building into another building on approximately the same level, or is a way of passage through or around a wall constructed as specified in the Building Code for a two-hour occupancy separation and which completely divides a floor into two or more separate areas so as to establish an area or refuge affording safety from fire or smoke coming from the area from which escape is made.

Horizontal exit. Connection of any two floor areas, whether in the same structure or not, by means of a vestibule, or by an open air balcony or bridge, or through a fire partition or fire wall.

Horizontal exit. Way of passage on the same level, from one building or fire area to an area of refuge in another building, or if in the same building, separated by a fire wall, party wall, or fire partition affording fire safety.

Horizontal exitway. One or more protected openings through or around a fire wall or fire partition, or one or more bridges connecting two buildings. When horizontal exits are employed, provisions shall be made to eventually reach grade level.

12-E MEANS OF EGRESS

Egresses. Free and unobstructed ways by doors, corridors, foyers, and stairways leading as directly as possible to, and discharging into a street, public way or right of way, an open space leading to a street, or into an exit court or passageway leading into a street or into an approved open space with access to a street.

Egress (means of). A continuous path of travel from any point in a building or structure to the open air outside at ground level and consists of two separate and distinct parts: (1) the exit access, and (2) the exit. A means of egress comprises the vertical and horizontal means of travel and may include the room space, doorway, corridor, hallway, passageway, stairs, ramp, lobby, fire escape, escalator, and other paths of travel.

Foyer. A foyer, lobby, corridor or passageway, one (1) or more in combination, adjacent to the auditorium of a theatre or assembly hall at the level of the main floor or a balcony thereof and into which one (1) or more exitways therefrom open, in the path of normal egress from the building.

Means of egress. A continuous and unobstructed path of travel from any point in a building or structure to a public space and consists of (3) separate and distinct parts:

The exitway access, the exitway, and the exitway discharge; a means of egress comprises the vertical and horizontal means of travel and shall include intervening room spaces, door, hallways, corridors, passageways, balconies, ramps, stairs, enclosures, lobbies, escalators, horizontal exits, courts and yards.

Means of egress. Is measured in units of exit width of 22 inches. Fractions of a unit shall not be counted, except that 12 inches added to one or more full units shall be counted as 1/2 a unit of exit width.

Means of egress. A continuous path of travel provided by a doorway, hallway, corridor, exterior passageway, balcony, lobby, stair, ramp or other egress facility, or combination thereof, for the escape of persons from any point in a building, floor area, room, or contained open space to a public thoroughfare or other approved open space. Means of egress includes exits and access to exits.

Means of egress. The continuous and unobstructed way of exit travel from any point in a building or structure to a public way and consists of three (3) separate and distinct parts: the way of exit access, the exit, and the way of exit discharge. A means of egress comprises the vertical and horizontal ways of travel and shall include intervening room spaces, doorways, hallways, corridors, passageways, balconies, ramps, stairs, enclosures, lobbies, escalators, horizontal exits and yards.

Means of egress. Continuous path of travel from any point in a building or structure to the open air outside at ground level and consists of 2 separate and distinct parts: the exit access, and the exit. A means of egress comprises the vertical and horizontal means of travel and may include the room space, doorway, corridor, hallway, stairs, passageway, ramp, lobby, fire escape, escalator, and other paths of travel.

Means of egress. Any means for vacating a room area, premises, or building, including exit doors, exterior and interior stairways, fire escapes, slide escapes, smokeproof fire towers, horizontal exits, ramps, moving stairways, and exit passageways.

Required means of egress. The exit required by the Code or deemed necessary by the Building Official to provide safe egress. Doors, stairways, ramps, and other means and devices for intercommunication which are not part of a means of egress required by the Code, are not "Required Means of Egress" and are subject to regulation and restrictions only to the extent specifically defined in the Code.

12-F PASSAGEWAY

Exit passageway. Enclosed means of egress connecting a required exit or exit court with a public way.

Exit passageway. Enclosed hallway, or corridor connecting a required exit to a street.

Exit passageway. Fire-resistive passageway connecting a means of egress with a street or with an open space leading to a street.

Passageway. The term "passageway" shall mean an enclosed passage or corridor connecting a stairway, fire tower or elevator with a street or open space communicating with a street.

Passageway. An enclosed hallway or corridor connecting a required exit to a street or other open space connecting with a street.

Passageway. An enclosed hallway or corridor connecting a required exit to a street, or open place leading to a street.

Passageway. Enclosed hallway or corridor connecting a required means of exit to a street, or other open space adjoining a street when such required exit does not lead directly to a street.

Passageway. Nonhabitable space which serves as a means of travel to or from other enclosed areas.

Passageway. Continuous way, of required width, kept clear for use as an exit, whether enclosed or not.

Passageway. Enclosed hallway, exitway or corridor connecting a required exit to a street or to an open space or court communicating with a street.

Passageway. Enclosed passage or corridor connecting a stairway, fire tower or elevator with a street or open space communicating with a street.

Passageways and corridors. Safe and continuous passageways, aisles or corridors leading to exits and so arranged as to provide convenient access to exits for every occupant shall be maintained at all times on all floors and in all buildings. The minimum width of any passageway, aisle or corridor shall be 3 feet at the narrowest point, and doors swinging into such passageway shall not restrict the effective width of any point during their swing to less than the minimum width herein required by the Code.

Protected passageway. Exit passage such as from interior stairs to the outside of the building or tunnels discharging to the outside of the building.

12-G VERTICAL TRAVEL

Access Stair. A stair between two floors, which does not serve as a required exit.

Baluster. Small pillar or column of various shapes or forms, supporting a rail, used in a balustrade.

Baluster. Small spindles or members forming the main part of a railing for a stairway or balcony, fastened between a bottom and a top rail.

Baluster. A small individual column in a balustrade or railing.

Balustrade. Protective barrier that acts as a guard around openings in floors or at the open sides of stairs, landings, balconies, mezzanines, galleries, raised walkways, or other locations to prevent accidental falls from one level to another. Such barrier may or may not have openings through it.

Box stairs. Built between walls, and usually with no support except the wall strings.

Closed and open stairway. May have one closed side and one side partially open, or both sides may be partially open.

Closed stairway. Constructed between two walls, and generally used for rear or attic stairs.

Enclosed stairway. Stairway separated by fire-resistive partitions and walls, and fire-resistive doors and frames from the remainder of the building.

Enclosed stairway. Interior stairway, enclosed on all sides with fire-resistant construction, with means of access from the floor area of each story it serves protected by approved self-closing doors and with separately enclosed direct exit or exit passageway to the street at the grade floor, which does not require passage into the floor area of any story for access to any higher or lower story.

Exterior stair. A stair open to the outdoor air, that serves as a required exit.

Exterior stairway. Stairway on the outside of a building but not including fire escapes.

Exterior stairways and fire escapes. Outside stairways, fire tower balconies, fire escapes or other means of egress shall project not more than 4 feet beyond the face of the wall.

Fire escape. An exitway at least one (1) story in height erected on the outside of a structure.

Fire escape. Exterior vertical exit used primarily as an emergency means of egress.

Fire tower. An interior stairway constructed and arranged as described and provided in the code.

Flight. Portion of stairs between successive landings.

Flight of stairs. Combination of consecutive treads and risers between one floor or landing level and the next.

Flight of steps. Series of steps between successive landings or between a landing and a floor.

Guard. Vertical barrier erected along exposed edges such as of stairways, balconies, etc.

Guardrail. Rail secured to up-rights and erected along the exposed sides and ends of platforms.

Handrail. Includes a bar, pipe, extruded section, or wood, or similar materials used to provide support for a person or persons when using a stairway.

Headroom. Minimum vertical clearance between a tread nosing in a stairway and the open end of the wellhole, and is specified in building Codes.

Interior stair. A stair within a building, that serves as a required exit.

Interior stairway. Stairway within a building.

Landing. Platform in a flight of stairs between 2 stories, or a platform at the end of a flight of stairs.

Landing. Platform between flights of stairs or at the termination of a flight of stairs.

Monumental stair. Stair whose steps have a rise of not more than 6-1/2 inches, and which have a run of not less than 12 inches; which is, for the most part, unenclosed; and which is more elaborate in architectural design than an ordinary stair.

Moving stairway. Power driven inclined continuous stairway used for raising and lowering passengers.

Newel. Principal post at the foot of a staircase; also the central support of a winding flight of stairs.

Newel post. Post at which the railing terminates at each floor level.

Nosing. That part of a stair tread which projects beyond the riser.

Open stairway. One or both sides exposed. Exposed side guarded with a railing.

Outside stairs. Includes stairs in which at least one side is open to the outside air.

Platform. An extended step or landing breaking continuous run of stairs.

Platform stairway. Constructed when it is necessary to change the direction of the stairs at a given point between a lower level and the next floor level.

Private stairway. A stairway serving one (1) tenant only and not for general use.

Railing. Protective bar placed at a convenient distance above the stairs for a handhold.

Rise. Distance from floor to floor.

Rise. Vertical distance from the top of a tread to the top of the next higher tread.

Riser. Vertical part of a step between two successive treads or between a tread and a landing or floor.

Rough stairways. Used for access to basements or cellars, and is usually constructed at a steeper angle than are major stairs between levels.

Run. Total length of stairs in a horizontal plane, including landings.

Slidescape. Straight or spiral chute erected on the interior or exterior of a building which is designed as a means of human egress with direct exit to the street or other public space.

Smokeproof tower. Interior stairway accessible at each story at and above grade through exterior balconies or ventilated vestibules.

Smokeproof tower (fire tower). An interior enclosed stairway, with access from the floor area of the building either through outside balconies or ventilated fireproof vestibules opening on a street or yard or open court, and with a separately enclosed direct exit or exit passageway to the street at the grade floor.

Stair carriage. Stringer for steps on stairs.

Staircase. Entire stair structure, including stair stringers, balusters, railing or handrail assembly, moldings, brackets, and panels.

Stair landing. Platform between flights of stairs or at the termination of a flight of stairs.

Stair or stairs. A series of steps without an intervening platform or landing.

Stair rise. Vertical distance from the top of one stair tread to the top of the next tread.

Stair rise. The vertical distance between successive treads or steps measured always from the same relative position thereon.

Stair rise. Vertical distance from the top of one stair tread to the top of the one next above.

Stair riser. Upright face of a step.

Stair run. Horizontal distance between the same face of two successive risers.

Stair tread. Horizontal distance from the face of the riser to the nosing.

Stair tread. The horizontal top surface of a step, including the run and the nosing.

Stairway. One or more flights of steps and the necessary landings connecting them to form a continuous and uninterrupted passage from one story to another in a building.

Stairway. A series of four or more risers with appurtenant treads.

Stairway. One or more flights of stairs and any landings or platforms connected therewith to form a continuous passage from one floor to another.

Stairway. One or more flights of stairs and the necessary landings and platforms connecting them to form a continuous and uninterrupted passage from one story to the next.

Stairway. Two or more risers.

Stairway. Two or more risers shall constitute a stairway.

Stairway. One or more flights of stairs and the necessary landings and platforms connecting them to form a continuous and uninterrupted passage from one floor to another.

Stairway. One or more flights of stairs and the necessary landings and platforms connecting them, to form a continuous and uninterrupted passage from one story to another in a building or structure.

Stairway. One or more flights of stairs and the necessary landings and platforms connecting them to form a continuous and uninterrupted passage from one floor to another. A flight of stairs, for the purposes of the Code, must have at least (2) risers.

Stairway headroom. All stairs shall have unobstructed headroom of at least 6-1/2 feet measured vertically above the tread in line with the face of the riser.

Stairway risers and treads. In straight treads the run of the treads shall be not less than 10 inches wide and the rise shall be not more than 7-1/2 inches high and tread shall be so proportioned to riser that the sum of the height of 2 risers in inches, and the width of one tread in inches is not less than 24 nor more than 25 inches. The width of tread including nosing shall be not less than 10-1/2 inches.

Stairway width. All stairways (except basement and grade landings) shall be not less than 3 feet in width in the clear. A handrail shall be provided on the open portion of the stairs which are not adjoining a wall or partition and must extend 18 inches beyond the top and bottom steps.

Stair winders. Where winders are used the width of the tread at a point one foot from the center of rail or newel shall be not less than 8 inches, and at a point 2 feet from the center of the rail or newel the width of the tread shall be not more than 12 inches.

Stair winders. Curving or angular alignment obtained by treads that are wider at one end than at the other.

Stringer. Lumber supporting member for a series of cross members. Frequently applied to stair supports.

String or stringer. A timber or other support for cross members. In stairs, the support on which the stair treads rest; also string-board.

Total rise. Vertical distance from finish floor to finish floor.

Total run. Overall horizontal distance occupied by a stairway is called the total run.

Tread. Horizontal part of a step between two (2) successive risers in a flight of stairs.

Tread width. Horizontal distance from front to back of tread including nosing when used.

Winder. A step in a winding stairway.

Winders. Are any stairway steps which have variations in the width of the treads.

Winders. Are any stairway steps which have variations in the width of the treads of more than three-fourths inches per one foot of stair width.

Winder stair. Treads of non-uniform width.

Winding stair. Stairs with winder treads where the angle between the longitudinal edges or sides exceed ten (10) degrees.

12-H GENERAL EXITING SAFETY

Occupancy. The purpose for which a building, or part thereof, is used or intended to be used.

Occupancy. The purpose for which a building is used or intended to be used. Change of occupancy is not intended to include change of tenants or proprietors.

Occupancy. The purpose for which a building, or part thereof, is used or intended to be used. The term shall also include the building, room or enclosed space that houses such use.

Occupancy. The purpose for which a building is used or intended to be used. The term shall also include the building or room housing such use. Change of occupancy is not intended to include change of tenants or proprietors.

Occupancy. As used in the Code pertains to and is the purpose which a building is used or intended to be used. Change of occupancy not intended to include change of tenants or proprietors.

Occupancy. The purpose or activity for which a building or space is used or is designed or intended to be used.

Occupancy. Means the purpose for which a building is used, according to the classification of buildings or as defined in the Code.

Occupancy classification. The various use groups as classified in the building Code.

Occupancy classification. The purpose for which a building is used or intended to be used. The term shall also include the building or room housing such use.

Occupancy group. The category in which a building or space is classified by the provisions of the Code, based on its occupancy or use.

Occupancy load. The number of individuals normally occupying the building or part thereof, for which the exit facilities have been designed.

Occupancy sprinkler system. An automatic sprinkler system servicing a use group in a building enclosed by construction assemblies as required by the Code.

Occupant. The individual, partnership or corporation that has the use of or occupies any business building or a part or fraction thereof, whether the actual owner or tenant. In the case of vacant business buildings or any vacant portion of a business building the owner, agent or other person having custody of the building shall have the responsiblity of an occupant of a building.

Occupant. Any person using or having actual possession of any non-residential premises.

Occupant load. The number of occupants of a space, floor or building for whom exit facilities shall be provided.

Occupant load. The total number of persons that may occupy a building or portion thereof at any one time.

Occupiable room. A room or enclosed space designed for human occupancy in which individuals congregate for amusement, educational, or similar purposes or in which occupants are engaged at labor; and which is equipped with means of egress, light, and ventilation facilities meeting the requirements of the Code.

Occupiable room. A room or space, other than a habitable room designed for human occupany or use, in which persons may remain for a period of time for rest, amusement, treatment, education, dining, shopping, or other similar purposes, or in which occupants are engaged at work.

Occupied. As applied to a building shall be construed as though followed by the words "or intended, arranged or designed to be occupied."

Occupied. Occupied shall mean, with regard to buildings, structures, or portions thereof, the possession by physical presence of materials, persons or objects, or an area that is designed or intended for utilization.

Occupied. Refers to any room or enclosure used by one or more persons for other than incidental maintenance.

Occupied. Lived, or intended, arranged or designed to be used.

Occupied space. The term "occupied space" shall mean any room or space in which any person normally does or is required to live, work or remain for any period of time.

2. Article 13

SPECIAL CONSTRUCTION 13-A THROUGH 13-D

13-A INCINERATORS

Charging chute (incinerator). An enclosed vertical passage through which refuse is fed to an incinerator.

Charging gate (incinerator). A gate in an incinerator used to control the flow of combustion gases into the charging chute and the entry of refuse into the combustion chamber.

Commercial and industrial incinerator. One designed to burn waste matter incidental to any class of occupancy.

Flue-fed apartment-type incinerator. An incinerator having a chimney which also serves as a charging chute from one or more floors above the incinerator.

Garbage. The refuse or residue of animal or vegetable matter which has been used as food for man or beast and all refuse animal or vegetable matter. It shall include fowl manure, decayed vegetables, fruit or any condemned food.

Incinerator. Combustible apparatus designed for high temperature operation in which solid, semi-solid, liquid, or gaseous combustible wastes are ignited and burned efficiently and from which the solid residues contain little or no combustible material.

Incinerator. An incinerator is a device, using heat, for the reduction of garbage, refuse, or other waste materials.

Incinerator. An appliance or combustion chamber used for burning rubbish, garbage and other combustible waste material.

Incinerator, (domestic gas-fired). A domestic appliance used to reduce combustible refuse material to ashes and which is manufactured, sold and installed as a complete unit.

Paper. Newspapers, periodicals, cardboard and all wastepaper.

Refuse. All putrescible and non-putrescible solid wastes, except body wastes, including, but not limited to: garbage, rubbish, ashes, street cleaning, dead animals, abandoned automobiles, and solid market and industrial wastes.

Residential incinerator. An incinerator used by not more than 3 families, gas-fired, with a charging compartment of not over 5 cubic feet.

Rubbish. Non-putrescible solid wastes consisting of both combustible and noncombustible wastes, such as paper, wrapping, cigarettes, cardboard, tin cans, yard clippings, leaves, wood, glass crockery, and similar materials.

13-B PREFABRICATED CONSTRUCTION

Building component. Any sub-system, sub-assembly, or other system designed for use in, or as a part of, a structure, including but not limited to: structural, electrical, mechanical, fire protection, and plumbing systems, and other systems affecting health and safety.

Building system. Plans, specifications and documentation for a system of manufactured building or for a type or a system of building components, including but not limited to, structural, electrical, mechanical, fire protection and plumbing systems, and including such variations thereof as are specifically permitted by regulation, and which variations are submitted as part of the building system or amendment thereof.

Closed construction. Any building, component, assembly or system manufactured in such a manner that all portions cannot be readily inspected at the installation site without disassembly, damage to, or destruction thereof.

Installation. The process of affixing or assembling and affixing, manufactured buildings or building components on the building site, or to an existing building.

Label. Approved device affixed to a manufactured building or building component, by an approved agency, evidencing code compliance.

Manufactured building. Any building which is of closed construction and which is made or assembled in manufacturing facilities, on or off the building site, for installation, or assembly and installation, on the building site. Manufactured building may also mean, at the option of the manufacturer, any building of open construction, made or assembled in manufacturing facilities away from the building site, for installation, or assembly and installation, on the building site.

Open construction. Any building, component, assembly or system manufactured in such a manner that all portions can be readily inspected at the installation site without disassembly, damage to, or destruction thereof.

Portable cabana. Any prefabricated cabana which is designed to be readily assembled and disassembled and adapted to ready transportation from place to place.

Portable grandstand. Assembly of prefabricated units, readily erected, dismantled and transported, and used or intended for use as movable, permanent or temporary support of audiences.

Prefabricated. Fabricated prior to installation or erection.

Prefabricated. Construction materials or assembled units fabricated prior to erection or installation in a building or structure.

Prefabricated. Fabricated prior to erection or installation in a building.

Prefabricated. Fabricated prior to erection or installation at a building site.

Prefabricated assembly. A building unit, the parts of which have been built up or assembled prior to incorporation in a building.

Prefabricated building. The completely assembled and erected building or structure, including the service equipment, of which the structural parts consist of prefabricated individual units or subassemblies using ordinary or controlled materials; and in which the service equipment may be either prefabricated or at-site construction.

Prefabricated subassembly. A built-up combination of several structural elements designed and fabricated as an assembled section of wall, ceiling, floor or roof to be incorporated into the structure by field erection of two (2) or more such subassemblies.

Prefabricated unit. A built-up section forming an individual structural element of the building, such as a beam, girder, plank, strut, column or truss, the integrated parts of which are prefabricated prior to incorporation into the structure, including the necessary means for erection and connection at the site to complete the structural frame.

Prefabricated unit service equipment. A prefabricated assembly of mechanical units, fixtures, and accessories comprising a complete service unit of mechanical equipment, including bathroom and kitchen plumbing assemblies, unit heating and air-conditioning systems and loop-wiring assemblies of electric circuits.

13-C MEMBRANE STRUCTURES

Air-inflated structure. A building where the shape of the structure is maintained by air pressurization of cells or tubes to form a barrel vault over the usable area. Occupants of such a structure do not occupy the pressurized area used to support the structure.

Air-supported structure. A building wherein the shape of the structure is attained by air pressure and occupants of the structure are within the elevated pressure area. Air-supported structures are of two basic types: 1. Single skin—Where there is only the single outer skin and the air pressure is directly against that skin. 2. Double skin—Similar to a single skin, but with an attached liner which is separated from the outer skin and provides an air space which serves for insulation, acoustic, aesthetic or similar purposes.

Assemblage tent. A tent used or intended for use as a place of assemblage. Large Tent means a tent designed and intended for any use for occupancy by 10 or more persons. Small Tent means a tent designed and intended for any use for occupancy by less than 10 persons.

Cable-restrained air-supported structure. One in which the uplift is resisted by cables or webbing which are anchored to either foundations or dead men. Reinforcing cable or webbing may be attached by various methods to the membrane or may be an integral part of the membrane. This is not a cable-supported structure.

Cable structure. A nonpressurized structure in which a mast and cable system provides support and tension to the membrane weather barrier and the membrane imparts structural stability to the structure.

Flame retardant or flame resistant. Fabric or material resistant to flame or fire to the extent that it will successfully withstand standard flame resistance tests adopted and promulgated by the State Fire Marshal. Membrane is a thin, flexible, impervious material capable of being supported by an air pressure of 1.5 inches of water column.

Frame-covered structure. A nonpressurized building wherein the structure is composed of a rigid framework to support tensioned membrane which provides the weather barrier.

Membrane. A thin, flexible, impervious material capable of being supported by an air pressure of 1.5 inches of water column.

Noncombustible membrane structure. A membrane structure in which the membrane and all component parts of the structure are noncombustible as defined by the code.

Tent. A shelter, structure or enclosure made of fabric or similar pliable material which derives its support from mechanical means such as poles, ropes, cables, stakes, or similar devices.

Tent. Any structure, enclosure or shelter constructed of canvas or pliable material supported by any manner except by air or the contents it protects.

13-D FALLOUT SHELTERS

Dual-use fallout shelter. A dual-use fallout shelter is a fallout shelter having a normal, routine use and occupancy as well as an emergency use as a fallout shelter.

Fallout shelter. A building, structure, or other real property, or an area or portion thereof, constructed, altered or improved to afford protection against harmful radiation resulting from radioactive fallout, including such plumbing, heating, electrical, ventilating conditioning, filtrating and refrigeration equipment and other mechanical additions or installations, if any, as may be an integral part thereof.

Fallout shelter. A structure, room or space that protects its occupants from fallout gamma radiation, with a protection factor of at least 40.

Fallout shelter. A fallout shelter is any room, structure, or space designated as such and providing its occupants with protection at a minimum protection factor of forty (40) from gamma radiation from fallout from a nuclear explosion as determined by a Qualified Fallout Shelter Analyst certified by the Office of Civil Defense. Area used for storage of shelter supplies need not have a protection factor of 40.

Protection factor. A factor used to express the relation between the amount of fallout gamma radiation that would be received by an unprotected person and the amount that would be received by one in a shelter. For example, an occupant of a shelter with a PF of 40 would be exposed to a dose rate 1/40 (or 2-1/2%) of the rate to which he would be exposed if his location were unprotected.

Protection factor. A factor used to express the relation between the amount of fallout gamma radiation that would be received by an unprotected person and the amount that would be received by one in a shelter.

Single purpose fallout shelter. A single purpose fallout shelter is one having no use or occupancy except as a fallout shelter.

Unit of egress width. A unit of egress width is 22 inches.

2. Article 14

CONVEYING SYSTEMS 14-A THROUGH 14-I

14-A CONVEYORS AND ESCALATORS

Conveyors. A system of machinery and manual or mechanized devices other than elevator and dumbwaiter equipment consisting of belts, chains, rollers, buckets, aprons, slides and chutes and other miscellaneous equipment for hoisting, lowering and transporting materials and merchandise in packages or in bulk in any direction in a building or structure.

Escalator. A power-drive, inclined, continuous stairway used for raising or lowering passengers.

Escalator. A moving stairway for transporting passengers from one level to another.

Escalator. A moving, inclined stairway for passengers.

Moving stairs. All moving stairs, to be acceptable under the code, must be designed to safely carry the load, must be marked with the manufacturer's name, the load and the speed; must not exceed a greater than 30° angle from horizontal; must be 24 inches to 48 inches wide between hand rails. Hand rails must move at the same speed as stairway, and the stair well must have solid, smooth sides, with no paneling. Treads and landings must be non-slip type. All moving stairs must have mechanical brakes, emergency stop buttons, speed governors and other safety devices, as required by the "Safety Code for Elevators, Dumbwaiters and Escalators". ANSI-A17.1 Moving stairs may be used as required stairways if moving in the direction of egress, or if equipped with stop devices at head of each flight, provided they are properly enclosed. When not used as required stairway, enclosure shall be required only at upper landing.

Moving stairway (escalator). A power operated moving inclined continuous stairway used for raising and lowering individuals on single horizontal treads.

Moving stairway (escalator). A power driven, inclined, continuous stairway used for raising and lowering passengers.

Moving walk. A type of passenger-carrying device on which passengers stand or walk, and in which the passenger-carrying surface remains parallel to its directions of motion and is uninterrupted.

14-B CRANES AND DERRICKS

A-frame derrick. A derrick in which the boom is hinged from a cross member between the bottom ends of two upright members spread apart at the lower ends and joined at the top; the boom point secured to the junction of the side members, and the side members are braced or guyed from this junction point.

Angle indicator (boom). An accessory which measures the angle of the boom to the horizontal.

Axis of rotation. The vertical axis around which the crane superstructure rotates.

Axle. The shaft or spindle with which or about which a wheel rotates. On truck and wheel mounted cranes it refers to an automotive type of axle assembly including housing, gearing, differential, bearings and mounting appurtenances.

Axle (bogie). Two or more automotive type axles mounted in tandem in a frame so as to divide the load between the axles and permit vertical oscillation of the wheels.

Base (mounting). The base or carrier on which the rotating superstructure is mounted such as a truck, crawler or platform.

Basket derrick. A derrick without a boom, similar to a gin pole with its base supported by ropes attached to corner posts or other parts of the structure. The base is at a lower elevation than its supports. The location of the base of a basket derrick can be changed by varying the length of the rope supports.

Boom. A timber or metal section or strut. The heel (lower end) is affixed to a base, carriage or support, and the upper end supports a cable and sheaves where the load is lifted by means of wire rope and a hook.

Boom angle. The angle between the longitudinal centerline of the boom and the horizontal. The boom longitudinal centerline is a straight line between the boom foot pin (heel pin) centerline and boom point sheave pin centerline.

Boom harness. The block and sheave arrangement on the boom point to which the topping lift cable is reeved for lowering and raising the boom.

Boom hoist. A hoist drum and rope reeving system used to raise and lower the boom.

Boom point. The outward end of the top section of the boom.

Boom stop. A device used to limit the angle of the boom at the highest position.

Brake. A device used for retarding or stopping motion by friction or power means.

Breast derrick. A derrick with a boom. The mast consists of two side members spread farther apart at the base than at the top and tied together at top and bottom by rigid members. The mast is prevented from tipping forward by guys connected to its top. The load is raised and lowered by ropes through a sheave or block secured to the top crosspiece.

Cab. A housing which covers the rotating superstructure machinery and/or operator's station.

Cableway. The power operated system for moving loads in a generally horizontal direction in which the loads are conveyed on an overhead cable, track or carriage.

Chicago boom derrick. A boom which is attached to a structure, an outside upright member of the structure serving as the mast, and the boom being stepped in a fixed socket clamped to the upright. The derrick is complete with load, boom, and boom point swing line falls.

Climber crane. A crane erected upon and supported by a building or other structure which may be raised or lowered to different floors or levels of the building or structure.

Clutch. A friction, electromagnetic, hydraulic, pneumatic, or positive mechanical device for engagement or disengagement of power.

Counterweight. Weight used to supplement the weight of the machine in providing stability for lifting working loads.

Crane. A power operated machine for lifting or lowering a load and moving it horizontally which utilizes wire rope and in which the hoisting mechanism is an integral part of the machine.

Crawler crane. A crane consisting of a rotating superstructure with power plant operating machinery and boom, mounted on a base, equipped with crawler treads for travel.

Derrick. An apartus consisting of a mast of equivalent members held at the top by guys or braces, with or without a boom, for use with a hoisting mechanism and operating rope, for lifting or lowering a load and moving it horizontally.

Drum. The cylindrical members around which ropes are wound for raising and lowering the load or boom.

Dynamic (loading). Loads introduced into the machine or its components by forces in motion.

Folding boom. A boom constructed of hinged sections which is articulate in a folding manner and may be folded for storage or transit.

Gantry (A-frame). A structural frame, extending above the super-structure of a mobile crane, to which the boom support ropes are reeved.

Gin pole derrick. A derrick without a boom. Its guys are so arranged from its top to permit leaning the mast in any direction. The load is raised and lowered by ropes reeved through sheaves or blocks at the top of the mast.

Gudgeon pin. A pin connecting the mast cap to the mast, allowing rotation of the mast.

Guy. A rope used to steady or secure the mast or other members in the desired position.

Guy derrick. A fixed derrick consisting of a mast capable of being rotated, supported in a vertical position by guys, and a boom whose bottom end is hinged or pivoted to move in a vertical plane with a reeved rope between the head of the mast and the boom point for raising and lowering the boom, and a reeved rope from the boom point for raising and lowering the load.

Hoisting machine. A power operated machine used for lifting or lowering a load, utilizing a drum and a wire rope, excluding elevators. This shall include but not be limited to a crane, derrick and cableway.

Hydraulic boom. A boom which is operated by means of a hydraulic system.

Jib. An extension attached to the boom point to provide added boom length for lifting specified loads. The jib may be in line with the boom or offset to various angles.

Lay. That distance measured along a cable in which one strand makes a complete revolution around the cable axis.

Load (working). The external load, in pounds, applied to the crane or derrick, including the weight of auxiliary load attaching equipment such as load blocks, shackles and slings.

Load block (lower). The assembly of hook or shackle, swivel, sheaves, pins, and frame suspended by the hoisting ropes.

Load block (upper). The assembly of hook or shackle, swivel, sheaves, pins and frame suspended from the boom point.

Load hoist. A hoist drum and rope reeving system used for hoisting and lowering loads.

Load ratings. Maximum loads that may be lifted by a crane or derrick at various angles and positions as approved by the Department of Safety.

Mast. The upright member of a derrick.

Mobile crane. A crawler crane; a truck crane; or a wheel mounted crane.

Outriggers. Extendable or fixed metal arms, attached to the mounting base, which rests on supports at the outer ends.

Reeving. A rope system in which the rope travels around drums and sheaves.

Rope. Refers to wire rope unless otherwise specified in the Code.

Shearleg derrick. A derrick without a boom. The mast, wide at the bottom and narrow at the top, is hinged at the bottom and has its top secured by a multiple reeved guy to permit handling loads at various radii by means of load tackle suspended from the mast top.

Side loading. A load applied at an angle to the vertical plane of the boom.

Sill. A member connecting the foot block and stiffleg of a member connecting the lower ends of a double member mast.

Standing (guy) rope. A supporting rope which maintains a constant distance between the points of attachment to the two components connected by the rope.

Stiffleg derrick. A derick similar to a guy derrick except that the mast is supported or held in place by two or more stiff members, called stifflegs which are capable of resisting either tensile or compressive forces. Sills are generally provided to connect the lower ends of the stifflegs to the foot of the mast.

Structural competence. The ability of the machine and its components to withstand the stresses imposed by applied loads.

Superstructure. The rotating upper frame structure of the machine and the operating machinery mounted thereon.

Swing. Rotation of the superstructure for movement of loads in a horizontal direction above the axis of rotation.

Swing mechanism. The machinery involved in providing rotation of the superstructure.

Tackle. An assembly of ropes and sheaves arranged for hoisting and pulling.

Telescopic boom. A boom constructed of sections of diminishing cross section in which the sections fit within each other. The boom may be extended in a manner similar to a telescope.

Tower crane. A crane in which a boom, swinging jib or other structural member is mounted upon a vertical mast or tower.

Transit. The moving or transporting of a crane from one job site to another.

Travel. The function of the machine moving from one location to another, on a job site.

Travel mechanism. The machinery involved in providing travel.

Truck crane. A crane consisting of a rotating superstructure with power plant, operating machinery and boom, mounted on an automotive truck equipped with a power plant for travel.

Truck mounted tower crane. A tower crane which is mounted on a truck or similar carrier for travel or transit.

Wheel base. Distance between centers of front and rear axles. For a multiple axle assembly the axle center wheel base measurement is taken as the midpoint of the assembly.

Wheel mounted crane (wagone crane). A crane consisting of a rotating superstructure with power plant, operating machinery and boom, mounted on a base or platform equipped with axles and rubber tired wheels for travel. The base is usually propelled by the engine in the superstructure, but it may be equipped with a separate engine controlled from the superstructure. Its function is to hoist and swing loads at various radii.

Whipline (auxiliary hoist). A separate hoist rope system of lighter load capacity and higher speed than provided by the main hoist.

Winch head. A power driven spool for handling of loads by means of friction between fiber or wire rope and spool.

14-C DUMBWAITERS

Dumb-waiter. A hoisting or lowering mechanism for moving materials or products in a substantially vertical direction between two or more floors of a building or structure.

Dumb-waiter. Hoisting and lowering mechanism equipped with a car, which moves in a substantially vertical direction, the floor area which does not exceed 9 square feet, whose compartment height does not exceed 4 feet, the capacity of which does not exceed 500 pounds, and which is used exclusively for carrying materials.

Dumbwaiter. A hoisting and lowering mechanism equipped with a car that moves in guides in a substantially vertical direction, the floor area of which does not exceed 9 sq. ft., whose total inside height whether or not provided with fixed or movable shelves does not exceed 4 ft., the capacity of which does not exceed 500 lbs., and that is used exclusively for carrying materials.

Dumbwaiter. A hoisting and lowering mechanism with a car of limited capacity and size which moves in guides in a substantially vertical direction and is used exclusively for carrying material.

Dumbwaiter. A special form of freight elevator, power driven or manually operated, dimensions not exceeding 9 square feet in horizontal section, which lands not less than 2 feet above the floor.

Hoistway enclosure. The fixed structure, consisting of vertical walls or partitions, which isolates the hoistway from all other parts of the building or from an adjacent hoistway and in which the hoistway doors and door assemblies are installed.

Undercounter type (dumbwaiter). A dumbwaiter which has its top terminal landing located underneath a counter.

14-D WINDOW CLEANING EQUIPMENT

Anchor. The fitting, fastened to the window frame or wall, to which the belt terminal is attached. Anchor, double head. An anchor having two heads. Anchor, single head. An anchor having one head.

Belt terminal. That part of the window cleaner's safety belt which is fastened to the terminal strap to be attached to the anchor during the operation of window cleaning.

Building. Any building more than one story in height or having window sills more than 12 feet above grade, which is a place of employment.

Davit (fixed). A davit designed to remain at a fixed location.

Davit (mobile). A davit designed to be used in association with a roof car.

Davit (portable). A davit designed to be moved manually from socket to socket.

Davits. Suspension devices at roof level which are capable of bringing the work platform of the suspended scaffold or boatswain chair onto the roof or into a working position.

Davit socket (pivoted). An anchoring device that pivots inboard from the building face and transfers loads imposed by the davit to the roof structure or parapet.

Fixed anchorage. A secure point of attachment (not part of the work surface) for safety lines or lifelines that is capable of supporting a minimum dead weight of 5400 pounds.

From the inside. From a position in which all of the window cleaner's body except one arm and shoulder shall be on the interior side of the line of the window frame and with both feet on the floor.

From the outside. From a position in which more of the window cleaner's body than one arm and shoulder is outside of the line of the window frame.

Grade. The ground, the floor, the sidewalk, the roof, or any other approximately level solid surface of sufficient area and having sufficient structural strength to be considered as a safe place to work.

Machine screw or bolt. A screw or bolt used to install anchors on metal window frames.

Outrigger beam. A suspension device, located at the roof level, which is not capable of bringing the work platform of the suspended scaffold or boatswain's chair onto the roof.

Outrigger beam (fixed). An outrigger beam designed to remain at a fixed location.

Outrigger beam (mobile). An outrigger beam designed to be used in association with a roof car.

Outrigger beam (portable). An outrigger beam designed to be moved manually from one location to another.

Safety device (approved). Those devices acceptable to the Department of Safety for use in window cleaning operations shall be in compliance with the ANSI (American National Standards Institute) standards specified in the Code.

Safety factor. The ratio of the stress at the yield point of the material to the allowable unit stress.

Scaffold. The complete scaffold structure including the work platform and all supporting members.

Scaffold (permanent). A scaffold that is designed for a specific building and is used on that building only.

Scaffold (rolling). A fixed-height or extensible self-supporting scaffold that can be manually moved into place.

Scaffold suspended (manually operated). A scaffold suspended from above by ropes or cables and rigged with manually operated pulley blocks or winches or equivalent means so that the work platform elevation is easily adjustable.

Scaffold suspended (power driven). Any suspended scaffold equipped with one or more power units (not manually powered) for raising or lowering the scaffold platform.

14-E ELEVATOR EQUIPMENT

Alteration. Any change or addition to the equipment other than ordinary repairs or replacements.

Annunciator car. An electrical device in the car which indicates visually the landings at which an elevator landing signal registering device has been actuated.

Automatic operation. As applied to an elevator, shall mean operation whereby the starting of the car is effected in response to the momentary actuation of operating devices at the landing, and/or of operating devices in the car identified with the landings, and/or in response to an automatic starting mechanism, and whereby the car is stopped automatically at the landings.

Automatic operation. Operation wherein the starting of the elevator car is effected in response to the momentary actuation of operating devices at the landing, and/or of operating devices in the car identified with the landings, and/or in response to an automatic starting mechanism, and wherein the car is stopped automatically at the landings.

Auxiliary rope fastening device. A device attached to the car or counterweight or to the overhead dead-end rope-hitch support which will function automatically to support the car or counterweight in case the regular wire-rope fastening fails at the point of connection to the car or counterweight or at the overhead dead-end hitch.

Bi-parting door. A vertically sliding or a horizontally sliding door, consisting of two or more sections so arranged that the sections or groups of sections open away from each other and so interconnected that all sections operate simultaneously.

Blind hoistway. The portion of a hoistway which passes floors or other landings at which no normal landing entrances are provided.

Bottom car clearance. The clear vertical distance from the pit floor to the lowest structural or mechanical part, equipment or device installed beneath the car platform, except guide shoes or rollers, safety jaw assemblies and platform aprons or guards, when the car rests on its fully compressed buffers.

Bottom elevator counterweight runby. The distance between the counterweight buffer striker plate and the striking surface of the counterweight buffer when the car floor is level with the top terminal landing.

Bottom terminal landing. The lowest landing served by the elevator which is equipped with a hoistway door and hoistway door locking device which permits egress from the hoistway side.

Buffer. A device designed to stop a descending car or counterweight beyond its normal limit of travel by storing or by absorbing and dissipating the kinetic energy of the car or counterweight.

Buffer. A device designed to stop an elevator car or counter-weight when it has passed its normal range of travel.

Bumper. A device, other than an oil or spring buffer, designed to stop a descending car or counterweight beyond its normal limit of travel by absorbing the impact.

Cableway. A power-operated system for moving loads in a generally horizontal direction, in which the loads are conveyed on an overhead cable, track or carriage.

Car door or gate. As applied to an elevator, shall mean the sliding portion of the car that closes the opening, giving access to the car.

Car door or gate electric contact. An electrical device, the function of which is to prevent operation of the driving machine by the normal operating device unless the car door or gate is in the closed position.

Car door or gate power closer. A device or assembly of devices which closes a manually opened car door or gate by power other than by hand, gravity, springs or the movement of the car.

Car door or gate switch. As applied to an elevator, shall mean an electrical device, the function of which is to prevent operation of the driving machine by the normal operating device unless the car door or gate is in the closed position.

Car (elevator). The load-carrying unit including its platform, car frame, enclosure and car door or gate.

Car enclosure. The top and the walls of the car resting on and attached to the car platform.

Car frame (sling). The supporting frame to which the car platform, upper and lower sets of guide shoes, car safety, and the hoisting ropes or hoisting-rope sheaves, or the plunger of a direct plunger elevator are attached.

Car or counterweight safety. A mechanical device attached to the car frame or to an auxillary frame, or to the counterweight frame, to stop and hold the car or counterweight in case of predetermined overspeed or free fall or if the hoisting ropes slacken.

Car or hoistway door or gate. The movable portion of the car or hoistway entrance which closes the opening providing access to the car or to the landing.

Car platform. The structure which forms the floor of the car and which directly supports the load.

Car-switch automatic floor-stop operation. Operation in which the stop is initiated by the operator from within the car with a definite reference to the landing at which it is desired to stop, after which the slowing down and stopping of the elevator is effected automatically.

Car-switch operation. Operation wherein the movement and direction of travel of the car are directly and solely under the control of the operator by means of a manually operated car switch or of continuous-pressure buttons in the car.

Car-switch operation. Operation of an elevator wherein the movement and direction of travel of the car is under the control of a manually operated car switch or of continuous-pressure buttons in the car.

Certified. A certification by a testing laboratory, a professional engineer, a manufacturer or a contractor that a device or an assembly conforms to the requirements of the Code.

Closed shaft. A shaft enclosed at the top.

Closed shaft. Shaft used for vertical transportation which is closed at the top.

Compensating-rope sheave switch. A device which automatically causes the electric power to be removed from the elevator driving machine motor and brake when the compensating sheave approaches its upper or lower limit of travel.

Continuous-pressure operation. Operation by means of buttons or switches in the car and at the landings, any one of which may be used to control the movement of the car as long as the button or switch is manually maintained in the actuating position.

Control. The system governing the starting, stopping, direction of motion, acceleration, speed and retardation of the moving member.

Controller. A device or group of devices which serves to control in a predetermined manner the apparatus to which it is connected.

Door or gate closer. A device which closes a hoistway door or a car door or gate by means of a spring or gravity.

Door or gate. (manually operated). A door or gate which is opened and closed by hand.

Door or gate power-operator. A device or assembly of devices which opens a hoistway door and/or a car door or gate by power other than by hand, gravity, springs or the movement of the car; and which closes them by power other than by hand gravity or the movement of the car.

Door or gate—self-closing door or gate. A manually opened hoistway door or a car door or gate which closes when released.

Elevator and dumbwaiter entrance. The protective assembly which closes the hoistway enclosure openings normally used for loading and unloading.

Elevator automatic dispatching devices. A device, the principal function of which is to either: Operate a signal in the car to indicate when the car should leave a designated landing, or Actuate its starting mechanism when the car is at a designated landing.

Elevator automatic signal transfer device. A device by means of which a signal registered in a car is automatically transferred to the next car following, in case the first car passes a floor for which a signal has been registered without making a stop.

Elevator car flash signal device. One providing a signal light in the car, which is illuminated when the car approaches the landings at which a landing signal registering device has been actuated.

Elevator car leveling device. Any mechanism which will, either automatically or under the control of the operator, move the car within the leveling zone toward the landing only, and automatically stop it at the landing.

Elevator construction or alteration. Before proceeding with the construction, installation, or alteration of any elevator or mechanical equipment used in the raising and lowering of any stage curtain, dumbwaiter, escalator, platform lift, stage or orchestra floor, or amusement device or apparatus, the owner, contractor or the agent for the building shall obtain a permit for the work.

Elevator landing. That portion of a floor, balcony, or platform used to receive and discharge passengers of freight.

Elevator landing signal registering device. A button or other device, located at the elevator landing, which when actuated by a waiting passenger, causes a stop signal to be registered in the car.

Elevator or dumbwaiter hoistway. Includes a shaftway, hatchway, wellhole, or any other vertical opening or space without any interferences from pipes or ducts or structural piers, or pilasters, in which an elevator or a dumbwaiter can operate as designed.

Elevator or dumbwaiter hoistway. A shaftway for the travel of one or more elevators or dumbwaiters. It includes the pit and terminates at the underside of the overhead machinery space floor or grating, or at the underside of the roof where the hoistway does not penetrate the roof.

Elevator parking device. An electrical or mechanical device, the function of which is to permit the opening from the landing side of the hoistway door at any landing when the car is within the landing zone of that landing. The device may also be used to close the door.

Elevator pit. That portion of a hoistway extending from the threshold level of the lowest landing door to the floor at the bottom of the hoistway.

Elevator repairs. All work necessary to maintain present elevator equipment in a safe and serviceable condition and to adjust or replace defective broken or worn parts, with parts made of equivalent material, strength and design, and only where the replacing part performs the same function as the replaced part.

Elevator separate signal system. One consisting of buttons or other devices located at the landings, which when actuated by a waiting passenger illuminate a flash signal or operate an annuciator in the car indicating floors at which stops are to be made.

Elevator signal transfer device. A manually operated switch, located in the car, by means of which the operator can transfer a signal to the next car approaching in the same direction, when he desires to pass a floor at which a signal has been registered in the car.

Elevator truck zone. The limited distance above an elevator landing within which the truck-zoning device permits movement of the elevator car.

Elevator truck-zoning device. A device which will permit the operator in the car to move a freight elevator within the truck zone with the car door or gate and a hoistway door open.

Emergency interlock release switch. As applied to an elevator, shall mean a device to make inoperative, in case of emergency, door or gate electric contacts or door interlocks.

Emergency stop switch. A device located in the car which, when manually operated, causes the electric power to be removed from the driving-machine motor and brake of an electric elevator or from the electrically operated valves and/or pump motor of a hydraulic elevator.

Emergency terminal speed limiting device. A device which automatically reduces the speed as a car approaches a terminal landing, independently of the functioning of the operating device, and the normal-terminal stopping device if the latter fail to slow down the car as intended.

Existing elevator installation. Any equipment covered by the related section of the Code which was installed prior to the effective date of the Code or for which an application for permit to install.

Existing installation. One for which, prior to the effective date of the Code: All work of installation was completed, or the plans and specifications were filed with the enforcing authority and work begun not later than twelve (12) months after the approval of such plans and specifications.

Final terminal stopping device. A device which automatically causes the power to be removed from an electric elevator or dumbwaiter driving-machine motor and brake, or from a hydraulic elevator or dumbwaiter machine, independent of the functioning of the normal-terminal stopping device, the operating device or any emergency terminal speed limiting device, after the car has passed a terminal landing.

Generator-field control. A system of control which is accomplished by the use of an individual generator for each elevator or dumbwaiter wherein the voltage applied to the driving-machine motor is adjusted by varying the strength and direction of the generator field.

Hoistway. An enclosed or partly enclosed shaft used for the travel of an elevator, dumbwaiter, platform or bucket.

Hoistway. Vertical opening, space, or shaftway in which an elevator or dumbwaiter is installed.

Hoistway. Enclosed shaft in which one or more elevators is designed for vertical travel.

Hoistway access switch. A switch, located at a landing, the function of which is to permit operation of the car with the hoistway door at this landing and the car door or gate open, in order to permit access to the top of the car or to the pit.

Hoistway door. As applied to an elevator, shall mean the hinged or sliding portion of a hoistway enclosure which closes the opening giving access to a landing.

Hoistway-door combination mechanical lock and electric contact. A combination mechanical and electrical device the two related, but entirely independent, functions of which are: To prevent operation of the driving-machine by the normal operating device unless the hoistway door is in the closed position, and to lock the hoistway door in the closed position and prevent it from being opened from the landing side unless the car is within the landing zone.

Hoistway door interlock. A device used to prevent the operation of the driving machine of an elevator by the normal operating device unless the hoistway door is locked in the operating device unless the hoistway door is locked in the closed position, and also used to prevent the opening of the hoistway door from the landing side unless the car is within the landing zone and is either stopped or being stopped.

Hoistway door interlock. A device having two related and interdependent functions which are: To prevent the operation of the driving-machine by the normal operating device unless the hoistway door is locked in the closed position, and to prevent the opening of the hoistway door from the landing side unless the car is within the landing zone and is either stopped or being stopped.

Hoistway-door electric contact. An electrical device, the function of which is to prevent operation of the driving-machine by the normal operating device unless the hoistway door is in the closed position.

Hoistway-door interlock retiring cam device. A hoistway-door interlock retiring-cam device is a device which consists of a retractable cam with its actuating mechanism and which is entirely independent of the car-door or hoistway-door and power-operator.

Hoistway-door or gate locking device. A device which secures a hoistway door or gate in the closed position and prevents it from being opened from the landing side except under certain specified conditions.

Hoistway-gate separate mechanical lock. A mechanical device, the function of which is to lock a hoistway gate in the closed position after the car leaves a landing and prevent the gate from being opened from the landing side unless the car is within the landing zone.

Hoistway-unit system. A series of hoistway-door interlocks, hoistway-door electric contacts or hoistway-door combination mechanical locks and electric contacts, or a combination thereof, the function of which is to prevent operation of the driving machine by the normal operating device unless all hoistway-doors are in the closed position and, where so required by the Code, are locked in the closed position.

Horizontal slide type entrance. An entrance in which the panel(s) or door(s) slides horizontally.

Installation. A complete elevator, dumbwaiter, escalator, private residence inclined lift or moving walk including its hoistway, hoistway enclosures and related construction, and all machinery and equipment necessary for its operation.

Installation placed out of service. An elevator or dumbwaiter whose suspension ropes have been removed, whose car and counterweight rest at the bottom of the hoistway and whose hoistway doors are permanently boarded up or barricaded on the hoistway side or an escalator whose power feed lines have been disconnected and whose top and bottom entrances have been permanently boarded up or barricaded.

Landing zone. A zone extending from a point 18 inches above the landing.

Leveling zone. The limited distance above or below an elevator landing within which the leveling device is permitted to cause movement of the car toward the landing.

Multiple hoistway. A hoistway for more than one elevator or dumbwaiter.

Multi-voltage control. A system of control which is accomplished by impressing successively on the armature of the driving-machine motor a number of substantially fixed voltages such as may be obtained from multi-commutator generators common to a group of elevators.

New installation. Any installation not classified as an existing installation by definition, or an existing elevator, dumbwaiter, escalator, private residence inclined lift or moving walk moved to a new location subsequent to the effective date of the Code.

Non-stop switch-elevator. A switch, which when operated, will prevent the elevator from making registered landing stops.

Normal terminal stopping device. A device or devices to slow down and stop an elevator or dumbwaiter car automatically at or near a terminal landing independently of the functioning of the operating device.

Oil buffer. As applied to an elevator, shall mean a buffer using oil as a medium which absorbs and dissipates the kinetic energy of a descending car or counterweight.

Oil buffer. A buffer using oil as a medium which absorbs and dissipates the kinetic energy of the descending car or counterweight.

Oil buffer stroke. The oil-displacing movement of the buffer plunger or piston, excluding the travel of the buffer-plunger accelerating device.

One-way automatic leveling device. A device which corrects the car level only in case of under-run of the car, but will not maintain the level during loading and unloading.

Operating device. The car switch, push button, lever or other manual device used to actuate the control.

Operation. The method of actuating the control.

Overhead structure. All of the structural members, platforms, etc., supporting the elevator machinery, sheaves and equipment at the top of the hoistway.

Overslung car frame. A car frame to which the hoisting-rope fastenings or hoisting-rope sheaves are attached to the crosshead or top member of the car frame.

Position indicator. A device that indicates the position of the elevator car in the hoistway. It is called a hall position indicator when placed at a landing or a car position indicator when placed in the car.

Power-operated door or gate. A hoistway door or a car door or gate which is opened and closed by a door or gate power operator.

Pre-register operation. Operation in which signals to stop are registered in advance by buttons in the car and at the landings. At the proper point in the car travel, the operator in the car is notified by a signal, visual, audible, or otherwise, to initiate the stop, after which the landing stop is automatic.

Rated load. The load which the elevator, dumbwaiter, escalator, or private residence inclined lift is designed and installed to lift at the rated speed.

Rated speed. The speed at which the elevator, dumbwaiter, escalator, or inclined lift is designed to operate under.

Rheostatic control. A system of control which is accomplished by varying resistance and/or reactance in the armature and/or field circuit of the driving-machine motor.

Safety. As used with elevators, a device attached to the car or counterweight which applies sufficient gripping or clamping action to the guide rails to retard and stop the elevator car or counter-weight when it has passed its designed rate of speed in the downward direction.

Safety bulkhead. A closure at the bottom of the cylinder located above the cylinder head and provided with an orifice for controlling the loss of fluid in the event of cylinder head failure.

Safety (car on counterweight). A mechanical device attached to an elevator car frame or to an auxilliary frame, or to the counterweight frame, to stop and hold the car or counterweight in case of predetermined overspeed or free fall, or if the hoisting ropes slacken.

Semi-automatic gate. A gate which is opened manually and which closes automatically as the car leaves the landing.

Signal operation. Operation by means of single buttons or switches (or both) in the car, and up-or-down direction buttons (or both) at the landings, by which predetermined landing stops may be set up or registered for an elevator or for a group of elevators.

Single automatic operation. Automatic operation by means of one button in the car for each landing served and one button at each landing, so arranged that if any car or landing button has been actuated, the actuation of any other car or landing operating button will have no effect on the operation of the car until the response to the first button has been completed.

Single hoistway. A hoistway for a single elevator or dumbwaiter.

Single speed alternating current control. A control for a driving-machine induction motor which is arranged to run at a single speed.

Slack-rope switch. A device which automatically causes the electric power to be removed from the elevator driving-machine motor and brake when the hoisting ropes of a winding-drum machine become slack.

Special equipment. Any permanently or semi-permanently located device, manually or power operated, used for moving or lifting materials or persons, but not considered as an elevator, dumbwaiter, escalator, moving walk or amusement device. Special equipment shall include but shall not be limited to: ski lifts, cable cars, aerial passenger tramways and devices, inclined at an angle for carrying one (1) or two (2) persons but which are not included under the term elevator or escalator; manhoists; lift bridges; elevators which are used only for handling building materials and workmen during construction; stage and orchestra lifts; belt; bucket scoop, roller or similarly inclined or vertical freight conveyors; telescopic ash hoists; tiering or piling machines; skip hoists and wharf ramps.

Spring buffer. A buffer utilizing a spring to cushion the impact force of the descending car or counterweight.

Spring buffer load rating. The load required to compress the spring an amount equal to its stroke.

Spring buffer stroke. The distance the contact end of the spring can move under a compressive load until all coils are essentially in contact.

Starters control panel (elevator.) An assembly of devices by means of which the starter may control the manner in which an elevator or group of elevators function.

Stopping device elevator landing. A button or other device, located at an elevator landing, which when actuated, causes the elevator to stop at that floor.

Swing type entrance. An entrance in which the panel(s) of door(s) swings around vertical hinges.

Sub-post car frame. A car frame all of whose members are located below the car platform.

Suspension rope equalizer. A device installed on an elevator car or counterweight to equalize automatically the tensions in the hoisting wire ropes.

Terminal stopping device-machine (final stop-motion switch). A final-terminal stopping device operated directly by the driving machine.

Top car clearance. The shortest vertical distance between the top of the car crosshead, or between the top of the car where no crosshead is provided, and the nearest part of the overhead structure or any other obstruction when the car floor is level with the top terminal landing.

Top counterweight clearance. The shortest vertical distance between any part of the counterweight structure and the nearest part of the overhead structure or any other obstruction when the car floor is level with the bottom terminal landing.

Top direct-plunger hydraulic elevator runby. The distance the elevator car can run above its top terminal landing before the plunger strikes its mechanical stop.

Top terminal landing. The highest landing served by the elevator which is equipped with a hoistway door and hoistway door locking device which permits egress from the hoistway side.

Transom. A panel or panels used to close a hoistway enclosure opening above a hoistway entrance.

Travel rise. The vertical distance between the bottom terminal landing of an elevator, dumbwaiter, escalator, or a private residence inclined lift.

Traveling cable. A cable made up of electric conductors, which provides electrical connection between an elevator or dumbwaiter car and fixed outlet in the hoistway.

Two-speed alternating current control. A control for a two-speed driving-machine induction motor which is arranged to run at two different synchronous speeds by connecting the motor windings so as to obtain different numbers of poles.

Two-way automatic maintaining leveling device. A device which corrects the car level on both under-run and overun, and maintains the level during loading and unloading.

Two-way automatic non-maintaining leveling device. A device which corrects the car level on both under-run and over-run, but will not maintain the level during loading and unloading.

Underslung car frame. A car frame to which the hoisting-rope fastenings or hoisting-rope sheaves are attached at or below the car platform.

Vertical slide type entrance. An entrance in which the panel(s) or door(s) slides vertically.

Waiting-passenger indicator. An indicator which shows at which landings and for which direction elevator hall stop-or-signal calls have been registered and are unanswered.

Weatherproof. So constructed or protected that exposure to the weather will not interfere with its successful operation.

Working pressure. The pressure measured at the cylinder of a hydraulic elevator when lifting car and its rated load at rated speed, or with Class C2 loading when leveling up with maximum static load.

14-F ELEVATOR MACHINE TYPES

Belt-drive machine. As applied to an elevator, shall mean an indirect-drive machine having a single belt or multiple belts as the connecting means.

Chain-drive machine. As applied to an elevator, shall mean an indirect drive machine having a chain as the connecting means.

Direct-drive machine. An electric driving machine the motor of which is directly connected mechanically to the driving sheave, drum, or shaft without the use of belts or chains either with or without intermediate gears.

Direct-plunger driving machine. One in which the energy is applied by a plunger or piston directly attached to the car frame or platform and which operates in a cylinder under hydraulic pressure; it includes the cylinder and plunger or piston.

Driving machine. The power unit which applies the energy necessary to raise and lower an elevator or dumbwaiter car or to drive an escalator, a private residence inclined lift or a moving walk.

Electric driving machine. One where the energy is applied by an electric motor. It includes the motor and brake and the driving sheave or drum together with its connecting gearing, belt or chain if any.

Geared-drive machine. A direct-drive machine in which the energy is transmitted from the motor to the driving sheave, drum, or shaft through gearing.

Gearless-traction machine. A traction machine, without intermediate gearing, which has the traction sheave and the brake-drum mounted directly on the motor shaft.

Hydraulic driving machine. One in which the energy is applied by means of a liquid under pressure in a cylinder equipped with a plunger or piston.

Indirect-drive machine. A electric driving machine, the motor of which is connected indirectly to the driving sheave, drum or shaft by means of a belt or chain through intermediate gears.

Roped-hydraulic driving machine. One in which the energy is applied by a piston, connected to the car with wire ropes, which operates in a cylinder under hydraulic pressure. It includes the cylinder, the piston, and multiplying sheaves if any and their guides.

Screw machine. An electric driving machine, the motor of which raises and lowers a vertical screw through a nut, with or without suitable gearing, and in which the upper end of the screw is connected directly to the car frame or platform. The machine may be of direct or indirect drive type.

Traction machine. A direct-drive machine in which the motion of a car is obtained through friction between the suspension ropes and a traction sheave.

Winding-drum machine. A geared-drive machine in which the hoisting ropes are fastened to and wind on a drum.

Worm-geared machine. A direct-drive machine in which the energy from the motor is transmitted to the driving sheave or drum through worm gearing.

14-G ELEVATOR TYPES

Direct-plunger elevator. A hydraulic elevator having a plunger or piston directly attached to the car frame or platform.

Electric elevator. A power elevator where the energy is applied by means of an electric driving-machine.

Electro-hydraulic elevator. A direct-plunger elevator where liquid is pumped under pressure directly into the cylinder by a pump driven by an electric motor.

Elevator. A hoisting and lowering mechanism equipped with a car or platform which moves in guides for the transportation of individuals or freight in a substantially vertical direction through successive floors or levels of a building or structure.

Elevator. A hoisting and lowering mechanism equipped with a car or platform which moves in guides for the transportation of individuals or freight in a substantially vertical direction.

Elevator. A hoisting and lowering mechanism equipped with a car or platform which moves in guides in a substantially vertical direction, and which serves two or more floors of a building or structure.

Elevator (residential). Residential elevators shall mean any elevator, man-lift or power operated hoisting equipment for humans.

Freight elevator. An elevator primarily use for carrying freight and on which only the operator and the persons necessary for loading and unloading and employees having special permission of the Superintendent of Buildings are permitted to ride.

Freight elevator. An elevator for the transportation of freight and only such individuals as are necessary for its safe operation for the handling of the freight.

Freight elevator. An elevator primarily used for carrying freight and on which only the operator and the persons necessary for unloading and loading the freight are permitted to ride.

Gravity elevator. An elevator utilizing gravity to move the car.

Hand elevator. A freight elevator that is driven by manual power.

Hand elevator. An elevator utilizing manual energy to move the car.

Hand elevator. An elevator that is driven by manual power and made safe by control devices.

Hand elevator. Elevator operated by manual power.

Hydraulic elevator. Elevator in which the motion of the vertical action is obtained by liquid under pressure.

Hydraulic elevator. A power elevator in which the motion of the car is obtained through the application of energy from liquid under pressure.

Hydraulic elevator. A power elevator in which the motion of the car is obtained through the application of force from liquid under pressure.

Hydraulic elevator. A power elevator where the energy is applied, by means of a liquid under pressure, in a cylinder equipped with a plunger or piston.

Inclinator. An inclinator is an inclined power lift to be used for humans in residences.

Passenger elevator. An elevator used primarily to carry persons other than the operator and persons necessary for loading and unloading.

Passenger elevator. An elevator for the transportation of individuals.

Power elevator. An elevator utilizing energy other than gravitational or manual to move the car.

Power elevator. An elevator in which the motion of the car is obtained through the application of energy other than by hand or gravity.

Power elevator. An elevator in which the motion of the car is obtained through the application of force other than by hand or gravity.

Private residence elevator. A power passenger electric elevator, installed in a private residence, and which has a rated load not in excess of seven hundred (700) pounds, a rated speed not in excess of forty (40) feet per minute, a net inside platform area not in excess of twelve (12) square feet, and a rise not in excess of fifty (50) feet.

Roped hydraulic elevator. A hydraulic elevator having its piston connected to the car with wire ropes.

Sidewalk elevator. Any shaft or opening through the surface of a street utilized for mechanical conveyance of goods, together with an approved cover for such shaft or opening.

Sidewalk elevator. A freight elevator which has a hatch opening wholly outside of the building or structure, and with no openings into the building at its upper terminal landing.

Sidewalk elevator. A freight elevator which operates between a sidewalk or other area exterior to the building and floor levels inside the building below such area, which has no landing opening into the building at its upper limit of travel and which is not used to carry automobiles.

Sidewalk elevator. Freight elevator with the hatch opening located partially or wholly outside of a building and with no opening into the building at the upper terminal landing of the elevator.

Sidewalk elevator. An elevator of the freight type for carrying material exclusive of automobiles and operating between a landing in a sidewalk or other area exterior to a building and floors below the sidewalk or grade level.

Sidewalk elevator. A freight elevator that operates between a sidewalk or other area outside of a building and floor levels inside the building below such area, which has no landing opening into the building at its upper limit of travel.

14-H HOISTS, LIFTS AND RAMPS

Auto lift. A mechanized device for raising motor vehicles above the ground or grade level but not through successive floors of the building or structure for maintenance or repair purposes.

Automatic lift. A vehicle-lifting device, the purpose of which is to raise an entire vehicle to provide accessibility for under-chassis service.

Automotive lift. A fixed mechanical device for raising an entire motor vehicle above the floor level, but not through successive floors of the building or structure.

Console lift. A section of the floor area of a theater or auditorium that can be raised and lowered.

Inclined lift. A hoisting and lowering mechanism equipped with a car or platform that moves in guides, installed at a degree not substantially vertical, that is designed to carry passengers and freight.

Industrial lift. A hoisting and lowering mechanism of a nonportable power-operated type for raising or lowering material vertically, operating entirely within one story of a building.

Industrial lift (material lift). A non-portable power operated raising or lowering device for transporting freight vertically.

Industrial lift (material lift). A non-portable power operated by raising or lowering device for transporting freight vertically, operating entirely within one (1) story of the building or structure.

Loading ramp. A hinged, non-portable device, either mechanical or hydraulic, hand or power operated, used for spanning gaps or adjusting heights between loading surface and carrier or between loading surface and unloading surface.

Manlift. Device consisting of a power driven endless belt provided with metal steps or platforms and handholds attached to the belt for the transportation of persons in a vertical position through successive floors or levels of a building or structure.

Material lift. A power-operated raising or lowering device for transporting freight vertically, operating entirely within one (1) story of a building or structure.

Material platform hoist. A power or manually operated suspended platform conveyance operating in guide rails for the exclusive raising and lowering of materials, or equipment, which is operated and controlled from a point outside the conveyance.

Miscellaneous hoisting and elevating equipment. All power operated hoisting and elevating equipment for raising, lowering and moving persons or merchandise from one level to another such as, but not limited to, inclined elevators, cranes, slings and hooks, tiering and piling machines not permanently located in a fixed position, mine elevators, skip hoists for blast furnaces, stage and orchestra lifts, lift-bridges and temporary builders' hoists and similar equipment.

Private residence inclined lift. A power passenger lift, installed on a stairway in a private residence, for raising and lowering persons from one floor to another.

Special hoisting and conveying equipment. Manually or power-operated hoisting, lowering or conveying mechanisms, other than elevators, moving stairways or dumbwaiters for the transport of persons or freight in a vertical, inclined or horizontal direction on one floor or in successive floors.

Workmen's hoist. A hoisting and lowering mechanism equipped with a car that moves in guides in a substantially vertical direction and that is used primarily for raising and lowering workmen to the working levels.

14-1 MOVING WALKS

Belt pallet type moving walk. A moving walk with a series of connected and power-driven pallets to which a continuous belt treadway is fastened.

Belt type moving walk. A moving walk with a power-driven continuous belt treadway.

Edge supported belt type moving walk. A moving walk with the treadway supported throughout its width by a succession of rollers.

Moving walk. A passenger-carrying device on which persons stand or walk, and in which the passenger-carrying surface remains parallel to its direction of motion and is uninterrupted.

Moving walk. A type of passenger-carrying device on which passengers stand or walk.

Moving walk. A type of passenger-carrying device on which passengers stand or walk, and in which the passenger-carrying surface remains parallel to its direction of motion and is uninterrupted, through its designated distance.

Moving walk landing. The stationary area at the entrance to or exit from a moving walk or moving walk system.

Moving walk slope. The angle which the treadway makes with the horizontal.

Moving walk system. A series of moving walks in end to end or side by side relationship with no landings between treadways.

Moving walk threshold comb. The toothed portion of a threshold plate designed to mesh with a grooved treadway surface.

Moving walk threshold plate. That portion of the landing adjacent to the treadway consisting of one or more stationary or slightly movable plates.

Moving walk width. The width of a moving walk is the exposed width of the treadway.

Pallet-moving walk. One of a series of rigid platforms which together form an articulated treadway or the support for a continuous treadway.

Pallet type moving walk. A moving walk with a series of connected and power-driven pallets which together constitute the treadway.

Slider-bed type moving walk. A moving walk with the treadway sliding upon a supporting surface.

2. Article 15

MECHANICAL 15-A THROUGH 15-M

15-A AIR CONDITIONING

Air conditioner. Means one or more factory made assemblies which include an evaporator or cooling coil and an electrically driven compressor and condenser combination, and may include a heating function.

Air conditioning. The process of treating air so as to control simultaneously its temperature, humidity, cleanliness and distribution to meet the requirements of the space to be conditioned.

Air conditioning. The process by which the temperature, humidity, movement and quality of air in buildings and structures used for human occupancy are controlled and maintained to secure health and comfort.

Air conditioning. The process by which the temperature, humidity, movement and quality of air in buildings and structure are controlled and maintained.

Air conditioning. The process of treating air so as to control simultaneously its temperature, humidity, cleanliness, and distribution to meet the comfort requirements of the occupants of the conditioned space. The system may be designed for summer air conditioning or for winter air conditioning or for both.

Air conditioning. Is the process of treating air to control simultaneously its temperature, humidity, cleanliness and distribution to meet the requirements of the conditioned space.

Air conditioning. The process of treating air so as to control simultaneously its temperature, humidity, cleanliness and distribution to meet the requirements of the conditioned space. The classifications of air conditioning are as follows: Winter air conditioning which shall include the proper distribution of the cleaned, humidified, and heated air to and within the spaces to be conditioned. Summer air conditioning which shall include the proper distribution of the cleaned, dehumidifited, and cooled air to and within the spaces to be conditioned.

Air-conditioning or comfort-cooling equipment. All of that equipment intended or installed for the purpose of processing the treatment of air so as to control simultaneously its temperature, humidity, cleanliness and distribution to meet the requirements of the conditioned space.

Air conditioning or refrigeration contractor. Any person, firm or corporation engaging in the business of installing, constructing, reconstructing, replacing, altering, repairing or servicing any air conditioning or refrigeration system or apparatus.

Air conditioning system. System of a building which provides conditioned air for comfort cooling by the lowering of temperature, requiring a total of more than 15 motor horse power or a total or more than 15 tons of mechanical refrigeration, in a single or multiple units, and air distribution ducts.

Air conditioning system or apparatus. A system or apparatus which ventilates, heats, and humidifies in winter and/or cools and humidifies in summer the space under consideration and provides the desired degree of air motion and cleanliness.

Air duct. A tube or conduit, or an enclosed space or corridor within a wall of structure used for conveying air.

Air filter units. Both washable and throw-away types used for removal of dust and other airborne particles from air circulated mechanically in equipment and systems.

Air quantity. The quantity of air used to ventilate a given space during period of occupancy shall always be sufficient to maintain the standards of air temperature, air quality, air motion and air distribution.

Air supply. Means the supply and distribution of the air required for heating, ventilating, and air conditioning.

Apparatus-heating and cooling. A device which utilizes fuel or other forms of energy to produce heat, refrigeration or air conditioning.

Central air conditioner. Means an air conditioner which is not a room air conditioner.

Conditioned space. Means space, within a building, which is provided with a positive heat supply or a positive method of cooling, either of which has a connected output capacity in excess of ten Btu/hr. per square foot.

Damper. Any device which when installed will restrict, retard or direct the flow of air in any duct, or the products or combustion in any heat producing equipment, its vent connector, vent or chimney therefrom.

Damper. Movable device in a duct or at a face grille which can regulate the flow of air.

Duct. Means any tube, pipe, conduit, flue, or passageway used or intended to be used for the conveyance of air, gases, vapors, or entrained materials as caused by the operation of a heating, ventilating, or air conditioning system.

Duct systems. All ducts, duct fittings and plenums assembled to form a continuous passageway for the transmission of air.

Major apparatus. Central air-handling equipment supplying more than one occupancy or rooms and heat-producing equiment generating heat for the heating and ventilating systems.

Outdoor opening. Means an opening to the outside air.

Outlet. Opening, the purpose of which is to deliver air into any space to provide heating, ventilation, or air conditioning.

Outlet or supply opening. Means an opening the sole purpose of which is to deliver air into any space to provide heat, ventilation, or air conditioning.

Outside air. Air that is taken from outside the building and is free from contamination of any kind in proportions detrimental to the health or comfort of the persons exposed to it.

Outside intake. Includes the ducts and outdoor openings through which outside air is admitted to a ventilating, air conditioning or heating system.

Packaged terminal air conditioner. Means a room air conditioner consisting of a factory-selected combination of heating and cooling components, assemblies or sections, intended to serve an individual room or zone and constructed in a manner which complies with the definition contained in the Standard for Packaged Terminal Air Conditioners approved by the Air-Conditioning and Refrigeration Institute in 1976, known as ARI-76.

Plug in air conditioning appliance. A complete factory tested air conditioning unit or a window air conditioning unit, in a suitable frame or enclosure which is fabricated and shipped in one complete assembly in a ready to operate condition, requiring no refrigerant containing parts to be connected in the field and requiring no field expertise to place it in operation.

Return. Any opening, the sole purpose of which is to remove air from any space being heated, ventilated or air conditioned.

Return duct. Duct for conveying air from a space being heated, ventilated or air-conditioned back to the heating, ventilating or air-conditioning appliance.

Room air conditioner. Means a factory encased air conditioner designed as a unit for mounting in a window or through a wall, or as a console. It is designed for delivery of conditioned air to an enclosed space without ducts. "Room air conditioner" includes packaged terminal air conditioners.

Warm air all-year air conditioning system. Includes a mechanical warm-air heating plant, together with such other devices and such automatic controls as will secure the simultaneous control of the temperature, motion, humidity, and a reduction in the dust and odor content, of the air employed in the ventilation of rooms. This includes both warming and humidifying in winter and cooling and dehumidifying in summer.

Warm air winter air conditioning system. Includes a mechanical warm air heating plant, together with such other devices and such automatic controls as will secure the simultaneous control of the temperature, motion, humidity, and a reduction in the dust and odor content, of the air employed in the ventilation of rooms, but not provided with such devices and automatic controls as will provide for cooling and dehumidifying in summer.

15-B CHIMNEYS, FLUES AND VENTS

Appliance flue. The flue passages within an appliance.

Barometric damper. Is a device placed in a smoke pipe, or in a liquid—or solid-fuel-fired device or appliance connected to a smoke pipe, which is so designed, arranged, and installed as to effectively limit the stack action of the chimney, flue, or smokestack to a predetermined maximum draft.

Barometric draft regulator. A device with functions to maintain a desired draft in an appliance by automatically reducing the chimney draft to the desired value.

Chimney. A vertical masonry or reinforced concrete shaft enclosing one or more flues designed for the purpose of removing the products of combustion of solid, liquid or gas fuel to the outside atmosphere.

Chimney. A vertical shaft of masonry, reinforced concrete, or other approved non-combustible, heat resisting material enclosing one or more flues, for the purpose of removing products of combustion from solid, liquid, or gas fuel.

Chimney. Primarily a vertical shaft enclosing at least one flue for conducting flue gases to the outside atmosphere.

Chimney. Primarily vertical enclosure containing one or more passageways.

Chimney. Part of a building which contains a flue or flues for transmitting products of combustion from a furnace, fireplace, or boiler or any appliance to the outer air.

Chimney. Vertical shaft enclosing one or more flues for conveying flue gases to the outside atmosphere.

Chimney. Vertical structure of masonry with one or more flues in which smoke or the products of combustion are conducted upward for disposal in the open air at a height above the ground.

Chimney. A structure which is primarily vertical and encloses one or more flues for the removal of the flue gases to the atmosphere.

Chimney. A vertical enclosure containing one or more flues used to remove hot gases from burning fuel, refuse, or from industrial processes.

Chimney connector. The pipe which connects a fuel-burning appliance to a chimney.

Chimney connector. A pipe or metal breeching that connects combustion equipment to a chimney.

Chimney draft. Available natural draft of the chimney, measured at or near the base of the chimney.

Chimney (factory-built). A chimney manufactured at a location other than the building site and composed of listed factory-built components assembled in accordance with the terms of the listing to form the completed chimney.

Chimney flue. The passage in a chimney for conveying the flue gases to the outside atmosphere.

Chimney flue. Conduit for conveying the flue gases delivered into it by a flue or vent connector to the outer air.

Chimney (high-heat industrial appliance-type). A factory-built, masonry or metal chimney suitable for removing the products of combustion from fuel-burning high-heat appliances producing combustion gases in excess of 2000°F. measured at the appliance flue outlet.

Chimney liner. A vent pipe or flue inserted within a type A flue or vent for the purpose of minimizing condensation of flue products and preventing such condensation from contact with the interior of the type A flue or vent in which it is inserted.

Chimney liner. A lining material of fire clay or other approved material that meets the requirements of ASTM C27.

Chimney liner (metallic). A vent pipe or chimney liner inserted within a chimney for the purpose of minimizing condensation of flue products and preventing such condensation from contact with the interior of the type A flue or vent in which it is inserted.

Chimney liner. (non metallic). A lining material of fire clay or other approved material that meets the requirements of the Building Code.

Chimney (low-heat industrial appliance-type). A factory-built, masonry or metal chimney suitable for removing the products of combustion from fuel-burning low-heat appliances producing combustion gases not in excess of 1000°F. under normal operating conditions but capable of producing combustion gases of 1400°F. during intermittent forced firing for periods up to one hour.

All temperatures are measured at the appliance flue outlet.

Chimney (medium-heat industrial appliance-type). A factory-built, masonry or metal chimney suitable for removing the products of combustion from fuel-burning medium-heat appliances producing combustion gases not in excess of 2000°F. measured at the appliance flue outlet.

Chimney (metal). A field-erected chimney of metal.

Chimney or vent. A conduit or passageway, vertical or nearly so for conveying flue gases to the outer air.

Chimney (residential appliance-type). A factory-built or masonry chimney suitable for removing products of combustion from residential-type appliances producing combustion gases not in excess of 1000°F. measured at the appliance flue outlet.

Chimneys. A vertical shaft enclosing one or more passageways for the removal of combustion products.

Clearance. The distance between a heat-producing appliance, chimney, chimney connector, vent, vent connector, or plenum, and other surfaces.

Draft hood. A device built into a gas appliance, or made a part of the vent connector from an appliance, which is designed to: 1. Assure the ready escape of gases in the event of no draft, back draft, or stoppage beyond the draft hood. 2. Prevent a back draft from entering the appliance.

Draft hood. A device placed in and made part of the vent connector from an appliance, or in the appliance itself, which is designed to; (1) To insure the ready escape of the products of combustion in the event of escape of the products of combustion in the event of no-draft, back-draft or stoppage beyond the drafthood; (2) Prevent a back-draft from entering the appliances; (3) Neutralize the effect of stack action of the chimney flue upon the operations of the appliance.

Draft regulator. A device which functions to maintain a desired draft in the appliance by automatically reducing the draft to the desired value.

Duct. A tube, pipe, conduit or continuous enclosed passageway used for the conveying of air, gases or vapors.

Exterior chimney. A chimney built outside the walls of a building but receiving lateral support from the exterior walls of the building.

Factory-built chimney. Chimney that is factory made, listed by an accredited authoritative testing agency, for venting gas appliances, gas incinerators, and solid or liquid fuel burning appliances.

Factory-built chimney. A chimney composed of listed factory-built components assembled in accordance with the terms of listing to form the completed chimney.

Factory-built chimney. Chimney consisting entirely of factory made parts, each designed to be assembled with the other without requiring field fabrication.

Fireplace. A hearth, fire chamber, or similarly prepared place and a chimney.

Flue. An enclosed passageway in a chimney or smokestack used for the removal of the products of combustion.

Flue. Conduit for conveying the products of combustion or fumes to the outer air.

Flue. Conduit or pipe, vertical or nearly so in direction, designed to convey all the products of combustion to the outside atmosphere.

Flue. The general term for the passage and conduits through which flue gases pass from the combustion chamber to the outer air.

Flue. Primarily vertical passageway used for the purpose of removing products of combustion and suitable to serve devices or appliances using any type of fuel.

Flue or vent. A conduit or passageway, vertical or nearly so, for conveying flue gases to the outer air.

Flue or vent connector. The pipe connecting an appliance with the flue or vent. This corresponds to the smoke pipe used with solid or liquid fuels.

Forced and induced draft fuel burning appliances. Fuel burning appliances listed as exhausting low temperature flue gases and listed for use with type "L" venting systems.

Gasvent. Enclosed passage used for removal to the outer air of products of combustion from gas-fired equpment only.

Gas vent connector. Pipe connecting a gas fired appliance with the vent pipe.

Gas vents. Factory-built vent piping and vent fittings listed by a nationally recognized testing agency, that are assembled and used in accordance with the terms of their listings, for conveying flue gases to the outside atmosphere.

Gas vents—type "B". Listed factory-made gas vents for venting listed or approved appliances, equipped to burn only gas, except those specifically listed for use with chimneys only.

Gas vents—type "B-W". Listed factory-made gas vents for venting listed or approved gasfired vented recessed heaters.

Gas vents—type "C". Vents constructed of sheet copper not less than No. 24 U.S. Standard Gage or Galvanized Steel of not less than No. 20 U.S. Standard Gage, or other approved noncombustible corrosion-resistant materials.

Gas vents—type "L". Low-temperature, venting systems. A venting system consisting of listed factory made piping and fittings for use with fuel burning appliances listed as exhausting low temperature flue gases and approved for use with a type "L" venting system.

High-heat appliance type chimney. Factory-built, masonry or metal chimney suitable for removing products of combustion from fuel-burning high-heat appliances producing combustion gases in excess of 2000° F., measured at the appliance flue outlet.

Hood. A canopy or similar device connected to a duct for the removal of heat, fumes or gases.

Low-heat appliance type chimney. Factory-built, masonry or metal chimney suitable for removing the products of combustion from fuel-burning low-heat appliances producing combustion gases not in excess of 1000° F., under normal operating conditions, but capable of producing combustion gases of 1400° F., during intermittent forced firing for periods up to one hour. All temperatures are measured at the appliance flue outlet.

Masonry chimney. Field constructed chimney built in accordance with nationally recognized standards and the requirements of the code.

Masonry chimney. Chimney of solid masonry units, bricks, stones, listed masonry units or reinforced concrete, lined with suitable flue liners.

Masonry chimney. A field constructed chimney built in accordance with nationally recognized Code or Standards.

Masonry chimney. A chimney of masonry units, bricks, stones or listed masonry chimney units lined with approved flue liners. For the purpose of the code masonry chimneys shall include reinforced concrete chimneys.

Masonry or concrete chimney. A field constructed chimney of brick, stone, concrete, or masonry units.

Medium-heat appliance type chimney. Factory-built, masonry or metal chimney suitable for removing the products of combustion from fuel-burning medium-heat appliances producing combustion gases not in excess of 2000°F., measured at the appliance flue outlet.

Metal chimney. Field constructed single-walled chimney of ferrous metal.

Metal chimney. Chimney fabricated of metal of adequate thickness, galvanized or painted unless suitably treated corrosion-resistant, properly welded or riveted and built in accordance with nationally recognized standards and codes.

Metal chimney. A chimney constructed of metal with a minimum thickness not less than that of No. 10 Manufacturers' Standard gage steel sheet.

Metal chimney. A chimney made of metal of adequate thickness, galvanized or painted unless suitably corrosion resistant, properly welded or riveted and built in accordance with nationally recognized Codes or Standards.

Metal chimney or smoke stack. Being a field constructed single walled chimney.

Residential appliance type chimney. A factory-built or masonry chimney suitable for removing products of combustion from residential-type appliances producing combustion gases not in excess of 1000° F., measured at the appliance flue outlet.

Smoke pipe. Flue, approximately horizontal, of metal or other material, in which smoke or the products of combustion are conducted from a furnace to a chimney or stack.

Smoke pipe. A pipe or breeching which is primarily horizontal and which connects a heating appliance to a flue.

Smoke stack. Vertical flue of metal or reinforced concrete, whether or not lined with masonry or other protective material, in which smoke, or the products of combustion, are conducted upward for disposal in the open air at a height above the ground.

Smokestack. Enclosed passage, primarily vertical, used for removal to the outer air of products of combustion of any fuel.

Smokestack. A vertical flue constructed of metal (lined or unlined), not insulated, to which is connected one or more smoke pipes.

Stack. A smokestack, chimney, flue, duct, exhaust pipe or other conveyor for carrying into the open air pollutants in any physical state from any source.

Stack. Any structure or part thereof which contains a flue or flues for the discharge of gases.

Vent. A passageway, vertical or nearly so, for removing vent gases in to the outer air.

Vent connector. (vent connector pipe.) That portion of the vent system which connects the gas appliance to the gas vent or chimney.

Vented appliance. An indirect-fired appliance provided with a flue collar to accommodate a venting system for conveying flue gases to the outer air.

Vented decorative appliance. A vented appliance whose only function lies in the esthetic effect of the flames.

Vented wall furnace. A vented heating appliance designed for incorporation in, or permanent attachment to, a wall, floor, ceiling or partition, and arranged to furnish heated air by gravity or by a fan. This definition shall not include floor furnaces, unit heaters and room heaters.

Vent gases. Products of combustion from fuel burning appliances plus excess air, plus any diluted air in the venting system above a draft hood or draft regulator.

Venting. The removal of combustion products as well as noxious or toxic fumes.

Venting collar. The outlet opening of an appliance provided for connection of the vent system.

Venting system. A continuous open passageway from the flue collar of draft hood of a fuel burning appliance to the outside atmosphere for the purpose of removing flue gases.

Venting system. The vent or chimney and its connectors assembled to form a continuous open passageway from an appliance to the outside atmosphere for the purpose of removing products of combustion. This definition also shall include the venting assembly which is an integral part of an appliance.

Venting system-gravity type. A system which depends entirely on the heat from the fuel being used to provide the draft required to vent an appliance.

Venting system-power type. A system which depends on a mechanical device to provide a positive draft within the venting system.

Vent pipe. As applied to a heating appliance, a pipe for the removal of products of combustion from a gas-fired appliance.

Vent system. A continuous open passageway from the flue collar or draft hood of a fuel burning appliance to the outside atmosphere for the purpose of removing products of combustion.

Vent system. Gas vent or chimney and vent connector, if used, assembled to form a continuous open passageway from the gas appliance to the outside atmosphere for the purpose of removing vent gases.

15-C HEATING FUELS — EQUIPMENT AND ACCESSORIES

Appliance. A device using gas as a fuel to produce heat, light, power or refrigeration.

Appliance accessory. Is a unit designed to be attached to a gas appliance, such as pilot lights, regulators, safety devices, control valves, and relief valves.

Automatic control. A device or devices installed on an appliance to accomplish, without manual attention, a complete turn-on or shut-off of gas to the main burner or burners, or to regulate the supply of gas to the main burner or burners.

Compressed gas. Any material or mixture having in the container an absolute pressure exceeding 40 psia at 70 F, or regardless of the pressure at 70 F, having an absolute pressure exceeding 104 psia at 130 F.

Container appurtenances. Items connected to container openings needed to make a container a gastight entity. These include, but are not limited to, safety relief devices; shutoff, backflow check, excess flow check and internal valves; liquid level gages; pressure gages and plugs.

Conversion burner. A gas burner accessory or device designed to supply gaseous fuel to and properly burn same within the combustion chamber of a boiler, furnace, or other device originally designed to utilize another fuel.

Conversion range oil burner. An oil burner designed to burn kerosene, range oil or similar light fuels. The burner is intended primarily for installation only in a stove or range, a portion or all of which originally was designed for the utilization of solid fuel and which is flue-connected.

Directly connected. Physically connected in such a way that water or gas cannot escape from the connection.

Drip. Short piece of pipe, capped at one end, and connected at the other end to low points in fuel lines for the purpose of trapping condensation formed in the fuel lines.

Dual fuel burning. A gas burner firing into the same combustion zone into which another fuel is utilized.

Fixed liquid level gage. A type of liquid level gage using a relatively small positive shutoff valve and designed to indicate when the liquid level in a container being filled reaches the point at which this gage or its connecting tube communicates with the interior of the container.

Fixed maximum liquid level gage. A fixed liquid level gage which indicates the liquid level at which the container is filled to its maximum permitted filling density.

Flammable liquid. Any liquid having a closed cup flash point below 140 F and having a vapor pressure not exceeding 40 psia at 100 F.

Flexible appliance connectors. Only approved double-walled flexible appliance connectors shall be used.

Flexible tubing. A gas conduit other than that formed by a continuous one-piece metal tube.

Float gage. A gage constructed with a float inside the container resting on the liquid surface which transmits its position through suitable leverage to a pointer and dial outside the container indicating the liquid level. Normally the motion is transmitted magnetically through a nonmagnetic plate so that no LP-Gas is released to the atmosphere.

Fuel burning equipment. Any furnace, boiler, water heater, device, mechanism, stoker, burner, stack, structure, oven, stove, kiln, still or other apparatus, or a group or collection of such units used in the process of burning fuel, refuse or other combustible material.

Fuel line. Any gas piping system extending from the meter to outlets supplying appliances or other gas burning equipment.

Fuel oil. Kerosene or any hydrocarbon oil conforming to nationally recognized standards and having a flash point not less than 100°F.

Fuel oil. Kerosene or any hydrocarbon oil specified in the standard in the code, having a flash point not less than 100°F.

Fuel oil. Oil commonly used as a fuel, of grades commonly numbered 1, 2, 4, 5 and 6 and having the requirements shown in Table 1 of Commercial Standard CS 12 (Latest Edition) published by the United States Department of Commerce.

Fuel oil. A liquid mixture or compound derived from petroleum which does not emit flammable vapor below a temperature of one hundred and twenty-five (125) degrees F. in a Tag closed-cup tester.

Fuel oil. Any liquid used as fuel and having a flashpoint of not less than 100 degrees Fahrenheit.

Fumes. Air-borne colloidal systems which are formed by chemical reactions or physical processes, such as, but not limited to combustion, distillation, sublimation, calcination or condensation.

Gas. Liquefied Petroleum Gas in either the liquid or vapor state. The more specific terms "liquid LP-Gas", or "vapor LP-Gas" are normally used for clarity.

Gas. Natural gas, manufactured gas or a mixture of the two gases, which may be used to produce light, heat, power or refrigeration.

Gas. Formless fluid which occupies space and which can be changed to a liquid or solid state only by increased pressure with decreased or controlled temperature or by decreased temperature with increased or controlled pressure.

Gas appliance. Fixture or apparatus manufactured and designed to use natural, manfactured, or mixed gas as a fuel medium for developing light, heat, or power.

Gas building house line. Piping system extending from the outlet of the utility meter to the inlet connection of equipment.

Gas building service line. Pipe which brings gas from the main to the meter.

Gas cleaning equipment. A device or process designed for removing particulate matter from the gas or air in which it is entrained.

Gas distribution piping. All piping from the house side of the gas meter piping the distributes gas supplied by a public utility to all fixtures and apparatus used for illumination or fuel in any building.

Gases. Escape of or emission of any gas which is injurious or destructive or explosive shall be unlawful and may be summarily caused to be abated, except as required in the provision of essential services.

Gasfitting. The art of installing, repairing or altering pipes, fittings, fixtures and other apparatus for distributing gas for heat, light, power or other purposes; the system of pipes, fittings, fixtures and other apparatus for distributing gas for heat, light, power or other purposes.

Gas meter (piping). The piping from the shutoff valve inside the building to the outlet of the meter.

Gas service. A pipe conveying as to a building from the distribution system of a public service corporation.

Gas service piping. The supply pipe from the street main through the building wall and including the stopcock or shutoff valve inside the building.

Hand fired fuel burning equipment. Any fuel burning equipment in which fresh fuel is manually thrown directly on the hot fuel bed.

Handling of solid fuel. Includes but is not limited to its transport by water on boats, barges, car ferries and motor vehicle ferries; its transport by land, by railroad, truck or trailer; its transfer from water transport to land transport and vice versa; its transfer to and from storage bins, silos, hoppers or piles; and its transfer to or from the equipment in which it is processed or burned.

Heating and cooking appliance. An oil-fired appliance not intended for central heating purposes. Appliances include kerosene stoves, oil stoves and conversion range oil burners.

House gas piping. Gas pipes and fittings installed on any premises or in any building or other structure on the outlet side of the gas meter, or extending from gas pipes anywhere beyond that location, and ending as capped or plugged outlets ready to connect with gas fixtures or gas appliances; but shall not include the connection of gas fixtures or gas appliances or gas meters, or any portion of the gas service piping from the street mains.

Internal valve. A primary shutoff valve for containers which has adequate means of actuation and which is constructed in such a manner that its seat is inside the container and that damage to parts exterior to the container or mating flange will not prevent effective seating of the valve.

Journeyman gasfitter. Any person, engaging in the trade of installing, constructing, or repairing gas piping, gas appliances or other apparatus using gas as a fuel.

Kerosene. Oil or liquid product of petroleum which does not emit a flammable vapor below a temperature of 115 degrees F., when tested in a Tag closed-cup tester.

Licensed gasfitter. Individual having a license from the Department of Safety and Permits to do gasfitting work.

Liquefied petroleum gas equipment. All containers, apparatus, piping (not including utility distribution piping systems), and equipment pertinent to the storage and handling of liquefied petroleum gas. Gas consuming appliances is not to be considered as being liquefied petroleum gas equipment.

Liquefied petroleum gas (gases) or LPG (LP-gas). Includes any material composed predominantly of any of the following hydrocarbons or mixtures of them: Propane, propylene, butanes (normal butane of isobutane) and butylenes. When reference is made to liquefied petroelum gas it refers to liquefied petroleum gases in either the liquid of gaseous state.

Liquefied petroleum gas (LP gas). Any material which is composed predominantly of the following hydrocarbons, or mixtures of them: Propane, propylene, butane (normal butane or isobutane) and butylenes.

Liquefied petroleum (LP) gas-air mixture. Liquefied petroleum gases distributed at relatively low pressures and normal atmospheric temperatures which have been diluted with air to produce desired heating value and utilization characteristics.

Liquefied petroleum (LP) gases. Fuel gases, including commercial propane (predominantly propane and/or proplene) or commercial butane (predominantly butane, isobutane and/or butylene). Liquefied petroleum gases supplied in cylinders and used without liquid vaporizers must be either commercial propane or commercial butane as defined above to avoid major variations in heating value of the gas as it is released from the cylinder.

LP-gas. Liquefied petroleum gas, any one of several petroleum products such as "butane" or "propane" stored under pressure as a liquid and vaporized and burned as gas.

LP-gas. Material composed predominantly of any of the following hydrocarbons, or mixtures of them: Propane, propylene, normal butane, isobutane and butylenes.

LP-gas system. An assembly consisting of one or more containers with a means for conveying LP-Gas from the container(s) to dispensing or consuming devices (either continuously or intermittently) and which incorporates components intended to achieve control of quantity, flow, pressure, or state (either liquid or vapor).

Main burner. A device or group of devices essentially forming an integral unit for the final conveyance of gas or a mixture of gas and air to the combustion zone, and on which combustion takes place to accomplish the function for which the appliance is designed.

Main burner control valve. A valve (or cock) of the plug and barrel type designed for use with gas, operated manually to control or shut off the supply of gas.

Manual shut-off valve. A valve (or cock) of the plug and barrel type designed for use with gas, operated manually to control or shut off the supply of gas.

Manufactured gas. Fuel existing in a gaseous state at standard conditions, having a heating value of between 500 and 600 B.T.U. per cubic foot.

Measured gas. Gas which has passed through and the volume of which has been measured by a meter, or gas which has been otherwise measured such as by weight or volume.

Mechanically fired fuel burning equipment. Any device by means of which fresh fuel is mechanically fired from outside the furnace into the zone of combustion, the same being actuated by automatic control.

Meter. The instrument installed to measure the volume of gas delivered through it.

Natural gas. A fuel existing in a gaseous state at standard conditions, having a heating value of between 1,000 and 1,100 B.T.U. per cubic foot.

Oil burner. A device for burning oil in heating appliances such as boilers, furnaces, water heaters, ranges and the like. A burner of this type may be furnished with or without a primary safety control; and it may be a pressure atomizing gun type, a horizontal or vertical rotary type, or a mechanical or natural draft vaporizing type.

Oil burner. Any device used for the purpose of burning or preparing to burn fuel oil and having a tank or container with a capacity of ten or more gallons connected thereto.

Oil burning equipment. An oil burner of any type together with its tank, piping, wiring, controls and related devices and shall include all oil burners, oil-fired units, and heating and cooking appliances but exclude those exempted in the code.

Oil burning equipment. Oil burning equipment shall be considered as consisting of oil burners and all equipment connected thereto, including every internal and external tank, pumping apparatus, piping, wiring and all accessories thereto.

Oil-fired unit. A heating appliance equipped with one or more oil burners and all the necessary safety controls, electrical equipment and related accessories, devices and equipment manufactured for assembly as a complete unit. This definition does not include kerosene stoves or oil stoves.

Outlet. The end of a particular branch of a fuel line at which an appliance is connected, or is to be connected.

Pilot light. Flame from a small burner having control independent from the main burner or burners which is utilized to ignite the main burner or burners of an appliance, or the flame from the main burner or burners of an appliance resulting from the diversion of a limited amount of gas through a bypass when the main gas supply is shut off, which flame is utilized to maintain ignition of the main burner or burners.

Protective equipment. Safety devices designed to forestall the development of a hazardous or undesirable condition in the fuel supply piping, in the gas equipment, in the medium being treated, or in the combustion products.

Semi-rigid tubing. Metallic conduit sufficiently flexible to permit bending so as to eliminate the use of elbows, generally connected to the appliance and house gas piping by means of flare or compression couplings.

Special gas fired commercial and industrial equipment. Shall consist of boilers for other than central heating plants, various types of ovens, processing equipment, sterilizing equipment, gas illuminated fixtures, clothes dryers other than domestic type for private residences, and other commercial and industrial gas fired equipment.

Tanks for flammable liquids. Any person desiring to install a tank or tanks for the storage of any flammable liquids, corrosive liquids, or oxidizing materials, highly toxic materials, or hazardous chemicals in liquid form, as provided in the Code, shall first obtain a permit from the Commissioner of Buildings, provided however, that no permit shall be required for an aggregate capacity of tanks, or one hundred twenty (120) gallons or less for Class 1, Class II and Class III flammable liquids. The aplication for the permit shall be made by the owner or his agent as required by the Code.

Valve (shutoff). A valve used to shut off either individual equipment (valve to be located in the piping system and readily accessible and operable by the consumer) or the entire piping system (valve to be located between the meter or source of supply and the piping system.)

Valve (tamperproof). Valves designed and constructed to minimize the possibility of the removal of the core of the valve accidentally or willfully with ordinary household tools.

Vaporizer. A device for converting liquid LP-Gas to vapor by means other than atmospheric heat transfer through the surface of the container.

Vaporizer-burner. The combination of a vaporizer with a burner into a unit dependent upon part of the heat generated by the burner as the source of heat to vaporize the liquid so that it may be burned as a vapor.

Vent connector. That portion of the venting system which connects the gas appliance to the gas vent, chimney or single-wall metal pipe.

Venting. Removal of combustion products as well as noxious or toxic process fumes to the outer air by means of roof openings, natural draft chimneys, flue stacks, or mechanical exhaust systems.

Vent pipe. As applied to heating, means a pipe for removing products of combustion from gas appliances.

15-D HEATING — GENERAL

Alteration of a heating system. A change, addition or modification of an existing heating system, or of any part thereof; replacement of an existing furnace with a new furnace; substitution of another furnace for an existing furnace; moving of an existing furnace, and conversion of a furnace to the use of fuel other than that previously used.

Apparatus—heating and cooling. A device which utilizes fuel or other forms of energy to produce heat, refrigeration or air conditioning.

Appurtenance. Is an accessory or adjunct to a gas appliance, intended to be used in connection with it.

Atmospheric tank. Means a storage tank which has been designed to operate at pressures from atmospheric through 0.5 psig.

Auxiliary heating equipment. That equipment used in connection with combustion equipment, such as stokers, burners, draft regulations, etc.

Baseboard convector. Means a low type of convector normally installed at the floor of the space to be heated.

Baseboard heating. Heating in which the heating element, usually an electric resistance or forced hot water, is located at the base of the wall.

Boiler. A closed heating appliance intended to supply hot water or steam for space heating, processing or power purposes.

Boiler. Means a closed vessel in which liquid is heated or vaporized, under pressure or vacuum, by the direct application of heat from the combustion of fuels, or from electrical or nuclear energy.

Boiler. "Boiler" shall mean a closed vessel intended (a) for use in heating water or other liquids; (b) for generating steam or other vapors under pressure or vacuum by the direct application of heat from combustible fuels, electricity, or atomic energy.

Boiler. Appliance intended to supply hot water or steam for space heating, processing or power purposes.

Boiler and furnace room. Every boiler or furnace room, including the breeching and fuel room, in places of outdoor assembly, shall be enclosed with a 2-hour fire-resistive enclosure or better and all interior openings in walls forming such enclosures shall be protected by self-closing, fire-resistive doors. Gas-fired appliances for heating water shall be installed in a boiler or furnace room. Chimneys shall be constructed in conformity with the requirements of the Code.

Boiler burning fuel in suspension. Fuel burning device in which fuel is conditioned or pulverized previous to admitting the fuel into the furnace for combustion. The combustion process is completed with the fuel in suspension.

Boiler (high pressure). A closed vessel in which steam or other vapor (to be used externally to itself) is generated at a pressure of more than fifteen (15) psig by the direct application of heat.

Boiler (low pressure). A boiler operated at pressures not exceeding fifteen (15) psig steam or at water pressure not exceeding 160 psig and temperatures not exceeding 250 degrees F.

Boiler room. Any room containing a fuel-fired steam or hot water boiler or furnace.

Boilers. The size, number, and location of power or heating boilers to be installed shall be marked on the plans, and, except in single dwellings, shall be approved by the department for the inspection of steam boilers, unfired pressure vessels and cooling plants, and by the department of smoke inspection and abatement, before a permit is issued by the department of Buildings for the erection of such building.

Ceiling-type direct-fired unit heater. Includes direct-fired unit heaters which are suspended from the ceiling, mounted between uprights or on a wall or column brackets. Unit heaters under this definition are appliances having a common enclosure and placed within or adjacent to the space being heated.

Ceiling-type unit heater. Unit heater which is suspended from a ceiling, mounted between uprights, or mounted on a wall or on brackets.

Central furnace. Means a furnace in which air is distributed and returned through ducts either by gravity or a fan.

Central heating boilers and furnaces. Heating furnaces and boilers shall include warm air furnaces, floor mounted direct-fired unit heaters, hot water boilers, and steam boilers operating at not in excess of 15 psig, used for heating of buildings or structures.

Central heating plant. Comfort heating plant equipment installed in such a manner to supply ducts or pipes to areas other than the room in which the heating equipment is located.

Central heating plant. Equipment centrally located, designed to heat buildings in part or in whole by means of pipes or ducts, water, steam or air is used as a medium for conveying the heat so generated to radiators or registers to be liberated in those portions of the building to be heated.

Central heating system. A system whereby heat is furnished from a central source of supply by a heat producing mechanism or device which is completely separated from those parts of the dwelling to which heat is supplied.

Central warm air heating system. A system consisting of a heat exchanger with an outer casing or jacket, or an electric heating unit, connected to a supply system and a return system. It may be either gravity or mechanical circulation.

Closed hot water system. A forced hot water system in which the circulating water is completely enclosed, under pressure above atmospheric, and closed to the atmosphere.

Convector. Radiator for either hot water or steam heat with many radiation surfaces, such as fins, to increase contact with air moved either by natural or forced convection.

Direct-fired low static unit heater. Direct-fired suspended, self-contained automatically controlled, vented heating appliance, having integral means for circulation of air by means of a propeller fan or fans.

Direct-fired unit heater. Unit heater which uses liquid or gas fuel for heating of the heat emitting element.

Direct gas fired air heaters. A gas heating device in which gas is burned and in which the products of combustion are mixed with the air which is to be heated in passing through the heater. The term shall be taken to mean the unit and equipment from its outside air inlet to the exit where the heated air leaves the unit.

Direct heating system. System in a building which produces heat to raise the temperature of the space within the building for the purpose of human comfort in which electric heating elements, or products of combustion exchange heat either directly with the building supply air or indirectly through a heat exchanger and using an air distribution system of ducts.

Direct-return system. A hot water heating system in which the water, after it has passed through a heating unit, is returned to the boiler along a direct path so that the total distance traveled by the water is the shortest feasible, and so that there are considerable differences in the lengths of the several piping circuits composing the system.

Down-feed system. A steam heating system in which the supply mains are above the level of the heating units which they serve.

Duct. Tube, pipe, conduit, or passageway used to convey air, gases, or vapors for a designed distribution system.

Duct and vent material. Ducts and vents shall be constructed of aluminum, copper, monel metal, galvanized steel, cement-asbestos or other approved noncombustible, corrosion-resistive materials of adequate strength, durability and for the temperatures involved; and the seams shall be made secure and substantially air and gas tight.

Duct furnace. Suspended direct-fired heating appliance normally installed in air ducts. Air circulation is provided by a blower not furnished as part of the appliance.

Duct furnace. Furnace designed for insertion or installation in a duct or an air distribution system to supply warm air for heating. The duct furnace depends for air circulation on a blower which is not part of the furnace.

Duct systems. All ducts, duct fittings, plenums and fans assembled to form a continuous passageway for the distribution of air.

Electric boiler. Any heating apparatus which provides for heat transfer from an electric heating element to a liquid conduction medium such as water.

Expansion tank, closed system. An airtight tank which provides a means of pressurizing the system over a wide range of conditions.

Floor furnace. Completely self-contained unit furnace suspended from the floor of the space being heated, taking air for combustion from outside this space, and with means for observing flames and lighting the appliance from such space.

Floor furnace. Means a furnace installed under a floor with a single register, the central portion emitting hot air and the outer portion returning air to furnace casing. No ducts are used.

Forced air heating system. A central warm air heating system that is equipped with a fan or blower which provides the primary means for circulation of air.

Forced air type central furnace. A central furnace equipped with a fan or blower which provides the primary means for circulation of air.

Forced hot water heating system. A system in which water is heated in the boiler and is forced through the pipes by the action of a circulating pump.

Forced hot water system. Forced hot water system in which circulation is created by means of a pump, usually driven by an electric motor.

Forced warm air heating plant. A plant which shall consist of one or more warm air furnaces, enclosed within casings, together with necessary appurtenances thereto, consisting of warm air pipes and fittings, cold air or recirculating pipes, ducts, boxes and fittings, smoke pipes and fittings, registers, borders, faces and grilles, the same intended for means of moving or forcing warm air, the same intended for the heating of buildings in which they may be installed.

Furnace. Completely self-contained direct-fired, automatically controlled, vented appliance for heating air by transfer of heat of combustion through metal to the air and designed to supply heated air through ductwork to spaces remote from the appliance location.

Furnace. Suspended direct-fired heating appliance normally installed in air ducts. Air circulation is provided by a blower not furnished as part of the furnace.

Furnace. A chamber or enclosure in which any combustion process takes place.

Gravity heating system. A central warm air heating system through which air is circulated by gravity. It may also use an integral fan or blower to overcome the furnace resistance to air flow.

Gravity hot water heating system. A system in which water is heated in the boiler and as the water temperature rises, it flows out through supply pipes to the space distribution units, the cooled water flows downward to the return pipes to the boiler.

Gravity hot water systems. Gravity Hot water heating systems where circulation of water is due to the head created by the difference in density of the water between the supply and return risers.

Gravity low pressure steam heating system. One in which the condensate is returned to the boiler by gravity due to the static head of water in the return mains. The elevation of the boiler water line must consequently be sufficiently below the lowest heating units and steam main and dry return mains to permit the return of condensate by gravity. The water line difference must be sufficient to overcome the maximum pressure drop in the system and the operating pressure of the boiler when radiator and drip traps are used as in two-pipe vapor systems. This applies only to closed circuit systems, where the condensation is returned to the boiler. If the condensation is wasted, no water line difference is required.

Gravity or circulating type space heater. Vented, self-contained free standing or wall recessed heating appliance using liquid or gas fuels.

Gravity warm air heating plant. A plant which shall consist of one or more warm air furnaces, enclosed within casings, together with necessary appurtenances thereto, consisting of warm air pipes and fittings, cold air or recirculating pipes, ducts, boxes and fittings, smoke pipes and fittings, registers, borders, faces and grilles, the same intended for the heating of buildings, in which they may be installed.

Gravity warm air heating system. A warm air heating system in which the motive head producing flow depends on the difference in weight between the heated air leaving the casing and the cooler air entering the bottom of the casing.

Gravity warm air heating system. Warm air heating system in which the warm air is transported and delivered through pipes or ducts by natural stack action and without the use of any mechanical impeller.

Heater room. Space containing central heat producing or heat transfer equipment.

Heating. Heating system of a building, which requires the use of high or low pressure steam, vapor or hot water, including all piping, ducts, and mechanical equipment appurtenant thereto, within, adjacent to or connected with a building, for comfort heating.

Heating and hot water supply installation. A fuel burning installation used only for space heating or hot water supply.

Heating appliance. Any device designed or constructed for the generation of heat from solid, liquid or gaseous fuel or electricity to be used for the purposes other than heating a building or structure.

Heating boiler. Any boiler carrying not in excess of 15 pounds per square inch steam or 30 pounds per square inch water pressure.

Heating surface. All surfaces in contact with hot gases for the purpose of transferring the heat by conduction, radiation or convection.

Heating system. Any combination of building construction, machinery, devices or equipment, so proportioned, arranged, installed, operated, and maintained as to produce and deliver in place the required amount and character of heating service.

Heat producing appliance. An appliance or device used for the production of heat by the combustion of fuel.

Heat pump. Means a system in which refrigeration equipment is used in such a manner that heat is taken from a heat source and given up to the conditioned space when heating service is wanted, and is removed from the space and discharged to a heat sink when cooling and dehumidification is desired.

Heat pump. Device which transfers heat from a cooler reservoir to a hotter one, expending mechanical energy in the process, especially when the main purpose is to heat the hot reservoir rather than refrigerate the cold one.

High capacity heating equipment. Containing equipment having an individual or combined rated gross capacity of 1,000,000 Btu per hour or more, or capable of operating at more than 15 psi for steam or more than 30 psi or 250 degrees F., for hot water.

High pressure boiler. Closed vessel in which steam or other vapor, is generated at a pressure of more than 15 psi by the direct application of heat.

High pressure boiler. Means a boiler generating steam at a pressure exceeding 15 psig or heating water at a pressure exceeding 160 psig or a temperature exceeding 250 degrees F.

High static pressure type heating appliance. Direct-fired suspended or floor standing, self-contained, automatically controlled and vented, heating appliance having an integral means for circulation of air against 0.2 inch or greater static pressure.

Horse power. Boiler horse power and shall be figured as equivalent to the evaporation of 34-1/2 pounds of water per hour from and at 212 degrees F.

Hot water forced circulation system. A system that has a booster on the circulating pump installed to mechanically circulate the water in the system.

Hot water gravity system. A system that depends on the difference in the weight or difference in the density between the hot water and the cold water to create circulation within the piping system.

Hot water heating system. A heating system in which water is used as the medium by which heat is carried from the boiler to the heating units.

Humidifier. Device designed to accumulate and discharge water vapor into a confined space for the purpose of increasing or maintaining the relative humidity in an enclosure.

Hydronics. Science of heating and cooling with liquids.

Industrial furnaces and power boilers—stationary type. For the purpose of the standards, stationary-type industrial furnaces and power boilers shall be classified as low, medium or high heat appliances in accordance with their character and size and the temperatures developed in the portions thereof where substances or materials are heated for baking, drying, roasting, melting, vaporizing or other purposes.

Jacketed stove. Vented, self-contained free standing non-recessed heating appliance, using solid, liquid or gas fuels. The effective heating is dependent or a gravity flow of air circulation over the heat exchanger.

Limit control. A thermostatic device installed in the duct system to shut off the supply of heat at a predetermined temperature of the circulated air.

Limit control. A thermostatic or pressure device installed to shut off the supply of heat at a predetermined temperature.

Liquid piping system. Shall mean a system in which water or other liquid is used as the medium by which heat is carried through pipes from the supply source to or from the heating or cooling units.

Low capacity heating equipment. Containing equipment having a rated gross capacity of less than 250,000 Btu per hour, and operating at less than 15 psi for steam or less than 30 psi or 250 degrees F., for hot water.

Low pressure boiler. Boiler operated at pressures not exceeding 15 psi-steam or at water pressures not exceeding 160 psi and temperatures not exceeding 250° F.

Low pressure boiler. Means a boiler generating steam at a pressure not exceeding 15 psig, or heating water at a pressure not exceeding 160 psig or at a temperature not exceeding 250 degrees F.

Low pressure tank. Storage tank which has been designed to operate at pressures above 0.5 psig but not more than 15 psig.

Low static type unit heater. Direct-fired suspended, self-contained automatically controlled, vented heating appliance, having integral means for circulation of air by means of a propeller fan or fans.

Low-temperature gas-oil burning equipment. Fuel-burning appliances listed as exhausting low-temperature combustion products containing a minimum of excess air and listed for use with Type L Low-Temperature vent systems.

Mechanical return low pressure steam heating system. One in which the condensate flows to a receiver and is then forced into the boiler against the boiler pressure. The lowest parts of the supply side of the system must be kept sufficiently above the water line of the receiver to insure adequate drainage of water from the system, but the relative elevation of the boiler water line is unimportant in such cases except that the discharge head on the mechanical return device becomes greater as the height of the boiler water line above the pump increases.

Mechanical warm air furnace. A warm air furnace which is equipped with a fan to circulate the air.

Mechanical warm air heating plant. One or more warm air furnaces enclosed within casings, together with necessary appurtenances thereto, consisting of warm air supply pipes and fittings, cold air or recirculating pipes, ducts, boxes and fittings, smoke pipes, dampers and registers, grilles, fans, or blowers, the same being intended for heating the buildings in which they may be installed. The circulation of air within such a system shall be dependent upon the motive power furnished by a fan or blower, and the duct work in connection therewith shall be designed especially for such system. However, the incorporation of a booster fan, blower or any power driven device for the purpose of accelerating the air circulation in a gravity warm air heating plant shall be construed as changing the classification of such gravity system to a mechanical system.

Mechanical warm air heating system. A warm air heating system in which circulation of air is effected by a fan. Such a system may include air cleaning devices, such as removeable filters.

Metal housing for heating appliances. The burner of the appliance shall be enclosed with a metal housing so constructed that there will be no open flame and the burner housing shall be effectively guarded against personal contact. The arrangement shall be such that the shield will prevent any combustible material in the vicinity of the appliance from coming in contact with the flame or with the housing that encloses the burner.

Moderate capacity heating equipment. Containing equipment having an individual or combined rate gross capacity, from 250,000 to 1,000,000 Btu per hour, and operating at less than 15 psi for steam or less than 30 psi or 250 degrees F. for hot water.

Net rating of heating boiler. That net rating specified by the Institute of Boiler & Radiator Manufacturers for cast iron boilers and by the Steel Boiler Institute for Steel Boilers.

One pipe steam heating system. A system where the pipe which carries the steam to the distribution units also returns the condensed steam to the boiler.

One-pipe system. A steam heating system in which a single main serves the dual purpose of supplying steam to the heating unit and conveying condensation from it. Ordinarily to each heating unit there is but one connection which must serve as both the supply and return, although separate supply and return connections may be used.

One-pipe system. A hot water system in which the cooled water from the heating units is returned to the supply main. Consequently, the heating units farthest from the boiler are supplied with cooler water than those near the boiler in the same circuit.

One-pipe system. A one pipe system employs a single pipe main with special fittings installed at riser connections to the heating elements.

One-pipe systems. Those in which the flow of the steam supply to the radiation and the return of condensation flow are in opposition to each other.

Overhead system. Any steam or hot water system in which the supply main is above the heating unit. In a steam system the return must be below the heating units; in a hot water system the return may be above or below the heating units.

Panel heating. Means a method of space heating in which radiant heat is supplied by large heated areas of the room surface operating at low surface temperatures, 80 degrees to 140 degrees F.

Plenum. A compartment or chamber to which one or more ducts are connected as a part of an air distribution or exhaust system, other than the occupied space.

Prefabricated heating panels. Means a factory-built heating unit. Heat may be supplied by hot water, steam or electricity.

Pressure piping. Includes piping for power, refrigeration, hydraulic, liquefied petroleum gas, and heating piping as listed in the Code.

Pressure vessel. Storage tank or vessel which has been designed to operate at pressures above 15 psig.

Radiant heaters. Reflector heaters, gas steam radiators, gas warm air radiators, gas unit heaters, gas floor heaters, or floor furnaces.

Radiant heating system. A heating system in which only the heat radiated from panels is effective in providing the heating requirements.

Radiator. A heating unit exposed to view within the room or space to be heated, that emits heat to objects within visible range by radiation and to the surrounding air by convection.

Room or space heater. Space heaters are above the floor appliances for direct heating of the space in and adjacent to that which the appliance is located without heating pipes or ducts.

Space heater. Individual gas heating appliance, vented or unvented, and used for heating a room or rooms which is self-contained and connected to a gas outlet.

Space heater. (Gravity or circulating type). A vented self-contained free standing or wall recessed heating appliance using liquid or gas fuels.

Space heater. Space-heating appliance for heating the room or space within which it is located, without the use of ducts.

Space heating appliance. Appliance intended for the supplying of heat to a room or space directly, such as a space heater, fireplace or unit heater, or to rooms or spaces of a building through a heating system, such as a central furnace or boiler.

Space heating appliance. Any device designed and constructed for the generation of heat from electricity or burning gases, liquid or solid fuels and used for heating all or any portion of a building.

Standard air. Means air that is equivalent to dry air at 70 degrees F and 29.92 inches of mercury.

Steam. Any boiler, generator, pressure vessel, system, piping or equipment used for the purpose of heating or distributing steam for heating, power, and processing, operating at pressures of fifteen (15) psig or less, shall be classed as low pressure.

Steam heating system. A heating system in which heat is transferred from the boiler or other source of heat to the heating units by means of steam at, above, or below atmospheric pressure.

Steam heating system. The steam is generated in the boiler, and rises to the space radiators, convectors and coils, where it condenses and forms water and returns to the boiler.

Steam piping system. Shall mean a system in which steam is transferred from a source to a steam utilizing device at, above or below atmospheric pressure for a purpose other than for heating a building or other structure.

Tempered air. Air transferred from heated area of building.

Tempered air. Means treated air heated or cooled, compatible with the space to which it is being introduced.

Tempered outside air. Outside air heated before distribution.

Two pipe direct return system. A system in which the heating medium after it has passed through a heat exchanger unit, is returned to the boiler by the shortest direct path, resulting in considerable differences in the lengths of the several circuits composing the system.

Two pipe reversed returned system. A system in which the heating medium from each heat transfer unit is returned along paths arranged so that all circuits composing the system are of equal length.

Two pipe steam heating system. A system where the steam rises through a supply main to the distribution units, air in the system and the condensed steam are forced through thermostatic traps at the bottom outlets of the units to the return main. An air eliminator in the return main expels the air through a vent and allows the water to return the water to the boiler.

Two-pipe system. A steam or hot water heating system in which one pipe is used for the supply of the heating medium to the heating unit and another pipe for the return of the heating medium to the source of the heat supply. The essential feature of a two-pipe system is that each heating unit receives a direct supply of the heating medium which medium cannot have served a preceding heating unit.

Two-pipe systems. Those in which one pipe is used for the supply of steam to the radiator and another for the return of condensation.

Underfloor horizontal furnaces. Horizontal flow forced warm air devices located under the floor of the structure and connected to a duct system.

Unfired pressure vessel. A closed metal vessel which contains air, steam, gas or liquid pressure in excess of fifty (50) pounds per square inch gage which is supplied from an external source.

Unfired pressure vessels. Includes jacketed kettles, steam cookers, stills, digesters, compressed air tanks and other pressure vessels used for storage of gases under pressure exceeding fifteen (15) pounds per square inch.

Unfired pressure vessels. The term "unfired vessels" as used in the Code, shall be deemed to include jacketed kettles, steam cookers, stills, digesters, compressed air tanks and other pressure vessels used for storage of gases under pressure exceeding (15) pounds per square inch.

Unit heater. Heating appliance other than a floor-mounted space heater, which consists of a heat emitting element and fan contained within a common enclosure, designed and installed for delivery of warm air for space heating directly into the space in which or adjacent to which the appliance is located.

Unit heater. A suspended space heater with an integral air circulating fan.

Unit heater. An appliance which consists of an integral combination of heating element and fan within a common enclosure and which is located within or adjacent to the space to be heated.

Unit heater. Means a self contained space heating appliance usually used for heating the space in which it is installed and provided with a fan for circulating the heated air. They may be floor or wall mounted or suspended. Gas, oil, steam, circulated hot water and electricity may be used for furnishing the heat.

Up-feed system. A heating system in which the supply mains are below the level of the heating units which they serve.

Unit heater (infrared). Means a heater consisting of an element heated to a high temperature electrically or by burning fuel and the radiant heat so produced is directed by a reflector.

Vacuum heating system. A two-pipe steam heating system equipped with the necessary accessory apparatus which will permit operating the system below atmospheric pressure when desired.

Vapor heating system. A steam heating system which operates under pressures at or near atmospheric and which returns the condensation to the boiler or receiver by gravity. Vapor systems have thermostatic traps or other means of resistance on the return ends of the heating units for preventing steam from entering the return mains; they also have a pressure-equalizing and air-eliminating device at the end of the dry return.

Volume damper. Any device which when installed will restrict, retard or direct the flow of air, in any duct, or the products of combustion in any heat producing equipment, its vent connector, vent or chimney therefrom.

Wall furnace. Means a vertical furnace designed to be installed in or against a wall. Cool air enters at bottom and is discharged at the top. Air may be circulated by gravity or by a fan.

Wall heater. A unit heater which is supported from or recessed in the wall of the room or space to be heated.

Warm air furnace. A solid, liquid or gas-fire appliance for heating air to be distributed with or without duct systems to the space to be heated.

Warm-air furnaces. It shall be unlawful for any person to construct, replace or install any warm air heating furnace, with appurtenances, ducts, or register, without first obtaining a permit from the Building Department for such work, as provided by the Code.

Warm air heating system. A heating plant consisting of an air heating appliance from which the heated air can be distributed by means of ducts or pipes and shall include any accessory apparatus and equipment in connection therewith.

Warm air heating system. A warm air heating plant consists of a heating unit, such as a fuel burning furnace, enclosed in a casing, from which heated air is distributed to the various rooms of the building through ducts.

Warm air heating system. Heating system which utilizes warm air as a heat transmitting and distributive medium. A warm air heating system shall be deemed to include all devices and equipment used for heating and conditioning the air and for transmitting and distributing it, including the furnace; heat exchange; fan; filters; humidifiers; controls; supply, return and fresh air pipes, ducts and fittings; and the registers, grilles, faces, and other distributing openings and heating surfaces incidental thereto.

15-E MANUAL FIRE EXTINGUISHING EQUIPMENT

Class A fires. Extinguishers suitable for Class A fires should be identified by a triangle containing the letter "A" If colored, the triangle should be green.

Class B fires. Extinguishers suitable for Class B fires should be identified by a square containing the letter "B". If colored, the square should be red.

Class C fires. Extinguishers suitable for Class C fires should be identified by a circle containing the letter "C". If colored, the circle should be blue.

Class D fires. Extinguishers suitable for fires involving metals should be identified by a five-pointed star containing the letter "D". If colored the star should be yellow.

Class A hazards. Extinguishers for protecting Class A hazards shall be selected from among the following: water types, foam, loaded stream, and multipurpose dry chemical.

Class B hazards. Extinguishes for protection of Class B hazards shall be selected from the following; bromotrifluormethane (Halon 1301) bromochlorodifluoromethane (Halon 1211), carbon dioxide, and dry chemical types, foam, and loaded stream.

Class C hazards. Extinguishers for protection of Class C hazards shall be selected from the following: bromotrifluoromethane (Halon 1301), bromochlorodifluoromethane (Halon 1211), carbon dioxide, and dry chemical types.

Class D hazards. Extinguishers and extinguishing agents for the protection of Class D hazards shall be of types approved for use on the specific combustible-metal hazard.

Factory test pressure. The pressure at which the shell was tested at time of manufacture. The pressure is indicated on the nameplate.

Fire extinguisher. Portable device, the contents of which are for extinguishing a fire.

First-aid hose. Hose that is permanently attached to a standpipe outlet and provided primarily for use in fire-fighting by the occupants of the building.

First aid hose station. Hose connection with valve in a system of piping adequately supplied with water, hose and nozzel for use of the of a building in extinguishing a fire.

Gas cartridge or cylinder. Explellent gas is confined in in a separate pressure vessel until the operator releases it to pressurize the extinguisher shell.

Hand fire extinguishing equipment. Portable equipment intended for the control of small or incipient fires and designed for manual operation.

Hand propelled. The material is applied with scoop, pail or bucket.

Horizontal fire line. A fire line installed around the interior walls and columns of a building, pier or wharf, with hose outlets located so that every part of the floor area is within reach of at least one fire stream.

Inspection. A check to see that the extinguisher is fully charged, and operable. That it has not been tampered with or damaged.

Maintenance. A check to see that the extinguisher will operate safely and effectively.

Manual fire-extinguishing equipment. All hand operated auxiliary fire-extinguishing equipment of an approved type suitable to the occupational use of the building and installed in the corridors or other locations, visible and readily accessible to the occupants of the building in accordance with the requirements of the Building and Fire department.

Mechanically pumped. The operator provides expelling energy by means of a pump and the vessel containing the agent is not pressurized.

Mild steel shell. Except for stainless steel and steel used for compressed gas cylinders, all other steel shells are "mild steel" shells.

Recharging. The replacement or replenishment of the extinguishing agent.

Self-expelling. The agents have sufficient vapor pressure at normal operating temperatures to expel themselves.

Self-generating. Actuation causes gases to be generated that provide expellent energy.

Service pressure. The service pressure is the normal operating pressure as indicated on the gage and nameplate.

Stored pressure. The extinguishing material and expellent gas are kept in a single container.

15-F MECHANICAL REFRIGERATION

Absorber (adsorber). That part of the low side of an absorption system used for absorbing (adsorbing) vapor refrigerant.

Absorption system. A refrigerating system in which the gas evolved in the evaporator is taken up by an absorber or adsorber.

Absorption unit. An absorption refrigerating system which has been factory-assembled and tested in accordance to manufacturer's design specifications.

Air compressors. Includes all stationary power driven compressors.

Alteration of a refrigerating system. Any change involving an extension or addition to the system; a change in refrigerant from a refrigerant of one Group Classification to a refrigerant of another Group Classification; an increase in the quantity of refrigerant contained in the system; a change in the arrangement, type, or purpose of the original installation; a change in the size of the equipment utilized; relocation of a compressor, generator, condenser, condensing unit, or evaporator; or replacement of any equipment, part, or installation which is not in conformity with the provision of the Code.

Boiling point or boiling temperature. The temperature at which a fluid will change from a liquid to a gas. The boiling point will depend upon the pressure exerted on the surface of the liquid.

Brazed joint. For the purpose of the Code, is a gas-tight joint obtained by the joining of metal parts with alloys which melt at temperatures higher than 1000°F, but less than the melting temperatures of the joined parts.

Brine. Any liquid used for the transmission of heat without a change in its state, having no flash point or a flash point above 150°F.

Bypass. A means of circumventing an object. A connection around a coil for the purpose of reducing the capacity of the coil.

Central cooling plant. Comfort cooling equipment installed in such a manner to supply cooling by means of ducts or pipes to areas other than the room or space in which the equipment is located.

Central plant. A refrigeration system utilizing refrigerant containing components which are interconnected by piping in the field.

Central refrigerating plant. System utilizing refrigerant containing components which are interconnected by piping.

Centrifugal compressor. A non-positive displacement compressor which depends at least in part on centrifugal force for pressure use.

Commercial system. A system assembled or installed in a building used for business or commercial purposes.

Compressor. Equipment when used with or without accessories is used for compressing a given refrigerant vapor.

Compressor. A specific machine, with or without accessories, for compressing a given refrigerant vapor.

Compressor (refrigeration). A machine used for the purpose of compressing a refrigerant.

Compressor unit. A condensing unit less the condenser and liquid receiver.

Condenser. A vessel or arrangement of piping or tubing in which vaporized refrigerant is liquefied by the removal of heat.

Condensing unit. A specific refrigerating machine combination for a given refrigerant, consisting of one or more power-driven compressors, condensers, liquid receivers (when required), and the regularly furnished accessories.

Cooling tower. Tower designed to cool water by evaporation.

Cooling tower. Structure designed for use in the cooling of liquids used in the operation of a refrigeration, air conditioning system, or similar installations by exposure of the liquids to the open air.

Cooling tower. Fixture or structure used to cool water by vaporizing some of the water into the atmosphere.

Cooling unit. A self-contained refrigeration system.

Damper. A valve or vane, leaf or butterfly type for controlling the flow of air.

Dampers, face and bypass. A set of co-ordinated dampers, arranged to direct the air through an evaporator, around an evaporator, or partly through and partly around an evaporator in any desired proportion, in response to control demand.

Dehydrator. A device containing a desiccant for the purpose of removing moisture from the refrigerant.

Density. The weight per unit volume of a substance.

Depressor fork. A device which, when energized, moves to hold open the cylinder suction valves in certain compressor capacity modulation systems.

Desiccant. A chemical agent used for moisture removal.

Design temperature (indoor). The temperature to be maintained within the conditioned space.

Design temperature (outdoor). The outdoor temperature arbitrarily established as the maximum against which the system must be able to maintain the desired indoor conditions. For economic planning it is somewhat lower or higher than the actual maximum.

Direct refrigeration expansion. Refrigeration system or apparatus in which the evaporator is in direct contact with the refrigerated material or space or is located in air circulating passages communicating with such spaces.

Direct system. A system in which the evaporator is located in the material or space refrigerated or in the air circulating passages communicating with such space.

Equalizer. A pipe connection between two or more pieces of equipment made in such a way that the pressure in each piece is maintained equally.

Equalizer line. A line that equalizes the gas or oil pressure in two or more pieces of equipment.

Evaporation. The process of converting a liquid substance into a vapor or a change of state.

Evaporator. That part of the refrigerating system in which liquid refrigerant is vaporized to produce refrigeration.

Expansion valve. A valve designed to meter the flow of liquid refrigerant to an evaporator.

Face and bypass type of control. A device, usually a valve or damper, to divert the flow of air over the face of an extended surface evaporator or through a passage around the evaporator.

Fluid. Any substance which, in its normal state, is a liquid or gas.

Fusible plug. A safety device having an insert of low melting point alloy. At excessive temperature the alloy will melt and release the refrigerant.

Gas suction. The gas entering the suction side of the compressor.

Gas Tracer. A gas having a powerful odor. Sometimes used in small quantities with odorless refrigerants to give warning of a leak.

Gas velocity. The speed of the gas in the piping or equipment. Usually stated in feet per minute or feet per second.

Generator. Equipment with a means of heating used in an absorption system to drive refrigerant out of solution.

Generator. Any device equipped with a heating element used in the refrigerating system to increase the pressure of refrigerant in its gas or vapor state for the purpose of liquifying the refrigerant.

Heat of compression. Heat developed within a compressor when a gas is compressed as in a refrigeration system.

High side. Any portion of a refrigerating system under condenser pressure.

High side charging. The process of introducing liquid refrigerant into the high side of the refrigerating system. The acceptable manner for placing the refrigerant into the system.

Indirect open spray system. A system in which a liquid such as brine or water, cooled by an evaporator located in an enclosure external to a cooling chamber, is circulated to such cooling chamber and is sprayed therein.

Indirect refrigeration. Refrigerating system or apparatus in which liquid such as a brine or water cooled by the refrigerant, is circulated to the material or space refrigerated or is used to cool air so circulated.

Indirect system. A system in which a liquid, such as brine or water, cooled by the refrigerant, is circulated to the material or space refrigerated or is used to cool air supplied to such space.

Industrial system. A system used in the manufacture, processing, or storage of materials located in a building used exclusively for industrial purposes.

Limited charged system. A system in which, with the compressor idle, the internal volume and total refrigerant charge are such that the design working pressure will not be exceeded by complete evaporation of the refrigerant charge.

Liquid receiver. A vessel permanently connected to a system by inlet and outlet pipes for storage of a liquid refrigerant.

Low side charging. The process of introducing refrigerant into the low side of the system. Usually reserved for the addition of small amount of refrigerant after repairs.

Low sides. Term which refers to the parts of a refrigeration system under evaporator pressure.

Machinery. The refrigerating equipment forming a part of the refrigerating system including any or all of the following: compressor, condenser, generator, absorber (adsorber), liquid receiver, connecting pipe, or evaporator.

Manual shutoff valve. Hand-operated device to stop flow of liquids in piping system.

Multiple compressors. Two or more compressors installed in parallel.

Multiple dwelling system. A refrigerating system employing the direct system in which the refrigerant is delivered by a pressure imposing element to two (2) or more evaporators in separate refrigerators or refrigerated spaces located in rooms of separate tenants in multiple dwellings.

Multiple system. A system employing the direct system of refrigeration in which the refrigerant is delivered to two (2) or more evaporators in separately refrigerated spaces.

Multispeed. Usually refers to a machine designed to operate at more than one speed.

Portable cooling unit. A self-contained refrigerating system, usually not over three (3) horsepower rating. The unit is factory assembled and tested to meet manufacturer's specifications.

Positive displacement compressor. A compressor in which increase in vapor pressure is attained by changing the internal volume of the compression chamber.

Pressure-relief device. A pressure-actuated valve or rupture member designed to automatically relieve excessive pressure.

Pressure-relief valve. A pressure-actuated valve held closed by a spring or other means and designed to automatically relieve pressure in excess of its setting.

Reciprocating compressor. A positive displacement compressor with a piston or pistons moving in a straight line, but alternately in opposite directions.

Refrigerant. A substance used to produce refrigeration by absorbing heat in its expansion or vaporization.

Refrigerant. The medium used to produce cooling or refrigeration by the process of expansion or vaporization.

Refrigerant. A substance used to produce refrigeration by its expansion or vaporization.

Refrigerating system. A combination of parts in which a refrigerant is circulated for the purpose of extracting heat.

Refrigerating system. A combination of interconnected refrigerant-containing parts constituting one closed refrigerant circuit in which a refrigerant is circulated for the purpose of extracting heat.

Refrigeration. Process utilizing the application of external work to remove heat from a region of lower temperature and discharge it into a region at a higher temperature.

Refrigeration. The mechanical process of removing heat from the air in an enclosed space of a building or structure.

Refrigeration system or apparatus. Any system or apparatus which will remove heat from a region of lower temperature and discharge it into a region at a higher temperature. Such refrigeration system or apparatus may utilize, but shall not be necessarily limited to utilization of one of the following types of refrigeration cycles: vapor compression cycle, absorption cycle, water vacuum cycle, steam jet cycle or dense air cycle.

Refrigerator. Any room or space in which an evaporator or brine coil is located for the purpose of reducing or controlling the temperature below 50°F.

Remote system. A system in which the compressor or generator is located in a space other than the cabinet or fixture containing the evaporator.

Rupture member. A mechanical device that will rupture at a predetermined pressure to control automatically the compressor or maximum pressure of operation of the refrigerant.

Sealed absorption system. A unit for Group 2 refrigerants only in which all refrigerant-containing parts are made permanently tight by welding or brazing against refrigerant loss.

Self-contained. A term which means having all essential working parts to produce the desired results, except energy and control services and connections.

Self-contained system. A complete factory-made and factory-tested system in a suitable frame or enclosure which is fabricated and shipped in one or more sections and in which no refrigerant-containing parts are connected in the field other than by companion or block valves.

Single package. Complete factory-made and factory-tested refrigeration system in a suitable frame or enclosure which is fabricated and shipped in one or more sections and in which no refrigerant containing parts are connected in the field.

Thermostat. A device for controlling equipment in response to temperature change. A temperature sensitive controller.

Ton of refrigeration. Term used to describe that Unit for heat removal at the rate of twelve thousand (12,000) BTU per hour.

Ton of refrigeration. A unit of refrigeration capacity corresponding to the removal of 200 Btu per minute, 12,000 Btu per hour or 288,000 Btu per day. It is so named because it is equivalent in cooling effect to melting one ton of ice in 24 hours.

Unit compressor. A unit consisting of a compressor, motor, drive and frequently the essential compressor controls, all mounted on a common base.

Unit system. A self-contained system which has been assembled and tested prior to its installation and which is installed without connecting any refrigerant-containing parts. A unit system may include factory-assembled companion or block valves.

Unit system. A system which can be removed from the premises without disconnecting any refrgerant containing parts, water connections, or fixed electrical connections.

Vacuum. A reduction in pressure below atmospheric pressure. Usually stated in inches of mercury.

Vaccum pump. A pump for exhausting a system. A pump designed to produce a vacuum in a closed system or vessel.

Valve (purge). A valve through which noncondensable gases may be purged from the condenser or receiver.

Valve (receiver shutoff). A valve in the line which connects the condenser to the receiver. Usually located at the inlet to the receiver.

Valve (refrigerant drum). A shut-off valve to control the flow of refrigerant from the drum.

Valve (relief). A valve designed to relieve the pressure from a vessel or system, whenever the pressure exceeds the setting of the valve.

Valve (solenoid). A magnetically operated valve generally used to control the flow of liquid to an evaporator, but may be used wherever off-on control is permissible. A solenoid electrical winding controls the action of the valve.

Water (condensing). The water supplied to cool a water cooled condenser or condensing unit.

Water make-up. The water added to a cooling tower or evaporative condenser to replace that lost through evaporation or other causes.

15-G MECHANICAL VENTILATION

Air Filter. A screen which prohibits the passage of fine air-borne particles from infiltrating the air stream delivered to a space.

Air outlet. Any opening through which air is delivered to a space.

Attic ventilators. Openings provided in gable ends of roofs with louvers or the use of exhaust fans placed behind the louvers.

Auxiliary ventilating openings. The free area when louvres, dampers, or other devices are in position to deflect or diffuse the air currents in such a manner that there will be no objectionable drafts.

Blower. A fan unit used to force air under pressure into a space.

Blower. A fan used to force air under pressure into an affected area.

Duct. A fabricated conduit used to deliver air to a given space.

Duct. Means a tube, pipe, conduit, or passageway used to convey air, gases, or vapors.

Duct system. A continuous passageway for the transmission of air which, in addition to ducts, may include duct fittings, dampers, plenums, fans and accessory air handling equipment.

Duct (ventilation). A pipe, tube, conduit, or an enclosed space within a wall or structure, used for conveying the air.

Exhaust duct. Duct through which air is conveyed from a room or space to the outdoors.

Exhauster. A fan used to withdraw air from an affected area under suction.

Exhaust system. Means any combination of building construction, machinery, devices, or equipment, so proportioned, arranged, maintained and operated that gases, dusts, fumes, vitiated air, or other materials injurious to health are effectively withdrawn from the breathing zone of employees and frequenters, and disposed of in a proper manner.

Exhaust ventilation system. Any combination of construction, machinery, devices or equipment, designed and operated to remove harmful gases, dust, fumes or vitiated air, from the breathing zone of the persons using the space.

Exhaust ventilating system. Any combination of building construction, machinery, devices or equipment, designed and operated to remove harmful gases, dusts, fumes or vitiated air, from the breathing zone of employees and frequenters.

Fan. An assembly comprising blades or runners and housing or casings, and being either a blower or exhauster.

Fan. An assembly of blades equally spaced or runners installed in a housing or case, operated by a motor.

Fresh air. Outdoor air.

General ventilation. Ventilation in which air is supplied to or removed from any area.

Gravity exhaust ventilation. A process of removing air by natural means, the effectiveness depending on atmospheric condition, such as difference in relative density, difference in temperature or wind motion.

Local exhaust ventilation. Ventilation in which dusts, fumes, vapors, gases and mists are removed from the atmosphere near the sources of their generation.

Mechanical ventilation. The process of supplying or removing air by power-driven fans or blowers.

Mechanical ventilation. The process of supplying or removing air under pressure or suction of fans or other approved means for inducing air movement.

Mechanical ventilation. Supply and removal of air by power driven devices.

Mechanical ventilation. Ventilation by power-driven devices.

Mechanical ventilation. Ventilation by opening to outer air through windows, skylights, doors, louvers, or stacks with or without wind-driven devices.

Mechanical ventilation. Ventilation derived by means of fans or other mechanical devices.

Mechanical ventilation. The mechanical process for introducing fresh air or for providing changes of air in a building or structure.

Mechanical ventilation. Ventilation which depends upon the operation of power-driven equipment to remove air from or deliver air to the desire location or area.

Mechanical ventilation. The process of introducing outdoor air into, or removing vitiated air from a building by mechanical means. A mechanical ventilating system may include air heating, air cooling, or air conditioning components.

Mechanical ventilation. The process of supplying a mixture of tempered outside air and/or simultaneously removing contaminated air to the outside by power-driven fans or blowers.

Outside air. Air that is taken from outside the building and is free from contamination of any kind in porportions detrimental to the health or comfort of the persons exposed to it.

Plenum. An air compartment or chamber to which one (1) or more ducts are connected, and which forms part of an air distribution system.

Plenum. Air compartment, or chamber, to which one or more ducts, or grilles, are connected, which forms part of either the supply or return system, or houses or contains aid-handling equipment.

Plenum. Chamber associated with air-handling apparatus, for distributing the processed air from the apparatus (supply plenum) to the supply ducts, or for receiving air to be processed by the apparatus (return plenum).

Plenum. A chamber for distributing air from a furnace to supply ducts or for receiving air to be heated by the furnace.

Plenum. An air-compartment or chamber to which one or more ducts are connected and which forms a part of any system of air movement incorporated into the subject building.

Plenum. A chamber designed for use as a part of an air distribution system.

Plenum. A compartment or chamber to which one or more ducts are connected as a part of an air distribution or exhaust system, other than the occupied space.

Plenum chamber. Compartment or chamber to which one or more ducts are connected and which form a part of either the supply or return air system.

Plenum chamber. An air compartment or enclosed space to which one or more distributing air ducts are connected.

Plenum floor system. The use of space between the floor structural system, earth (separated by non-porous vapor barrier and sand or concrete), and foundation walls in a one-story building for the distribution of conditioned space.

Return (or exhaust opening). Any opening the sole purpose of which is to remove air from any space being heated, ventilated or air conditioned.

Recirculated air. The transfer of air from a space through the air-handling equipment and back to the space.

Tempered air. Air transferred from a heated or cooled area of a building.

Tempered outside air. Outside air heated or cooled before distribution.

Ventilated. Provided with a means to permit circulation of the air sufficiently to remove an excess of heat, fumes, or vapors.

Ventilating ceiling. A suspended ceiling containing many small apertures through which air, at low pressure, is forced downward from an overhead plenum dimensioned by the concealed space between suspended ceiling and the floor or roof above.

Ventilating hood. Device suspended above a heat-producing appliance and connected to a vent or flue in order to collect and remove heated gases.

Ventilating system (exhaust). Any combination of building construction, machinery, devices or equipment, designed and operated to remove harmful gases, dusts, fumes or vitiated air, from the breathing zone of employes and frequenters.

Ventilation. The process of supplying or removing air by natural or mechanical means to or from any space.

Ventilation. Supply and removal of air to and from any space by natural or mechanical means.

Ventilation. Supply and removal of air to and from any space by natural or mechanical means. Such air may or may not have to be conditioned.

Ventilation. Process of supplying or removing air by natural or mechanical means, to or from any space, based on Code requirements.

Ventilation. The supply and removal of air to and from any space.

Ventilation. The process or method of supplying air to or removing air from any space by natural or mechanical means.

Ventilation. The process of replacing foul air with fresh air by natural or mechanical means.

Ventilation. The process of supplying and removing air by natural or mechanical means to or from any space. Such air may or may not have been conditioned before distribution.

Ventilation. The process of supplying and/or removing air by natural or mechanical means to or from any space in a manner which will protect the health, safety and comfort of the occupants of that space. Such air may or may not have been conditioned, unless required by Code.

Ventilation. Defined as the providing and maintaining in rooms or spaces, by natural or mechanical means, air conditions which will protect the health and comfort of the occupants thereof.

Ventilation exhaust system. A complete device, including all hoods, ducts, fans, separators and receptacles when required, any other part necessary for the receptacles when required, and any other part necessary for the proper installation and operation tereof.

Ventilation exhaust system refuse receptacle. That part of the exhaust system into which dust or other material separated from the air is deposited.

Ventilation exhaust system separator. That part of an exhaust system in which the contaminant or entrained material is separated from the air which conveys it.

Ventilation fan. Means the machine which creates the movement of air in a mechanical system ventilation.

Ventilation (mechanical). The process of supplying or removing air by power-driven fans or blowers.

Ventilation of basements and cellars. Basement and cellars, and all rooms located therein except storage rooms shall be lighted and ventilated by windows in exterior walls having both a glass and ventilation area of not less than two (2) percent of the room or space floor area.

Ventilation or exhaust duct. Any pipe, flume or channel, forming a part of an exhaust or ventilation system, used to convey air, dusts, fumes, mists, vapors or gases.

Ventilation or exhaust hood. That part of a ventilation or exhaust system into which the contaminated air or dust, fume, mist vapor or gas first enters.

Ventilation system. A complete device, including all hoods, ducts, fans separators, and receptacles when required, and any other part necessary for the proper installation and operation thereof.

Ventilation systems. All equipment intended or installed for the purposes of supplying air to, or removing air from any room, space or equipment by gravity or mechanical means, and which is not a portion of any warm air heating system.

15-H NATURAL LIGHT AND VENTILATION

Court. An open, unobstructed space on the same lot with the building.

Court vent. Inner court solely for the light and ventilation obtained for rooms facing the court such as bathrooms, kitchens, public halls, corridors and stair halls.

Gravity exhaust ventilation. Process of removing air by natural means, the effectiveness depending on atmospheric condition, such as difference in relative density, difference in temperature or wind motion.

Infiltration. Leakage of air into a building through doors and windows and through the cracks around them.

Inner court. A court entirely surrounded by building walls or lot lines or any court not defined as a through or outer court.

Light and ventilation of habitable rooms. Total glass area of required windows in any habitable room shall be not less than one-tenth of the floor area of the rooms. Total area of the ventilating portion of required windows in any habitable room shall be not less than 4 percent of the floor area.

Light and ventilation of laundry and recreation rooms. Laundry and recreation rooms located above the basement shall be lighted by windows located in exterior walls having a glass area of not less than 10 percent of the floor area and having a ventilating area and having a ventilating area of not less than 2 percent of the floor area.

Natural ventilation. Ventilation by openings to outside air through windows, doors or other openings.

Natural ventilation. Ventilation by opening to outer air through windows, skylights, doors, louvers, or stacks with or without wind-driven devices.

Natural ventilation. By providing openings to outer air through windows, skylights, doors, louvers, or stacks without using power-driven devices.

Natural ventilation. Ventilation derived from open windows, doors, louvers or other openings opening directly to the outside.

Natural ventilation. Ventilation by use of the outer air obtained through window, skylight, door, transom or court.

Natural ventilation. Ventilation which depends upon natural air currents to provide air movement in the environment or occupied area.

Openings screened. Any openings placed in exterior walls of dwellings beneath the first floor level such as those used for ventilation shall be completely closed off with noncorrodible 16 mesh wire cloth and a substantial grille. All such materials shall have openings not more than 1/2 inch in size.

Open shaft. Shaft extending through the roof of a structure and open to the outer air at the top.

Outer court. A court open on one end to an open space not less than thirty (30) feet wide. The open space shall be a public way, yard or through court or any combination of two (2) or more such areas.

Rear court. A court extending the entire width of the lot and extending from the building to the rear of the lot.

Recess. An open area formed by the indentation of a wall of a court, such indentation having a depth of not more than one-half (1/2) its length.

Required window. Window which provides all or part of the required natural light and ventilation in the room or space it is located.

Shaft. An enclosed vertical or inclined space for the transmission of light, air, materials, or persons through one or more stories of a building or other structure, and connecting two or more openings in successive floors or in a floor or floors and a roof; but not including ducts forming an integral part of a heating or ventilating system, or of a blower or exhaust system.

Shaft. Vertical opening through a building for air, light, elevators, dumb-waiters, mechanical equipment, or similar purposes.

Shaft. A space enclosed with side walls and extending through two or more stories.

Shaft. An enclosure of a vertical opening in two or more stories.

Shaft. A vertical opening in a building extending through one or more stories and/or roof, other than an inner court.

Shaft. A vertical opening or passage through one or more floors of a building, or through a floor and the roof.

Shaft. A vertical opening extending through one or more stories of a building, for elevators, dumbwaiter, light, ventilation or similar purpose.

Shaft. A vertical opening through a building for elevators, dumbwaiter, light ventilation or similar purposes.

Shaft. An enclosed shaftway or space, extending vertically through one (1) or more stories of a building, connecting a series of two (2) or more openings in successive floors, or floors and roof.

Shaft. A vertical opening or passage through two or more floors of a building or through floors and roof.

Shaft. An enclosed space for the transmission of light, air materials or persons through one or more stories of a structure which connects a series of two or more openings in successive floors, or floors and roof, except as may be otherwise provided in the Code.

Shaft. Vertical opening or enclosed space extending through two or more floors of a building, or through a floor and roof.

Shaftway. A vertical or oblique passageway through one (1) or more floors of a structure.

Skylights. The area of the maximum opening to the outer air, provided that it does not exceed the area of the sashed openings to the outer air, or the area of the skylight well. If this area exceeds either the area of the sashed openings or the skylight well, the smaller area is the ventilating area.

Through court. A court open at both ends to two (2) open spaces, one of which shall not be less than thirty (30) feet wide, the opposite not less than ten (10) feet wide. The open spaces shall be public ways, yards, permanent easements, outer courts, or another court or a combination of two (2) or more such areas.

Transoms. The free area through the sashed opening if the transom swings through an arc of not less than sixty degrees. It is the same percentage of the free area as the maximum angle of the transom when open is to sixty degrees if the transom swings through an arc of less than sixty degrees.

Ventilating openings. In any room or space are hereby defined as apertures opening upon a public way, yard, court, public park, public waterway, or onto a roof of a building or structure in which the room or space is situated. They shall be windows, skylights, transoms, or other openings which are provided for ventilating purposes and which are equipped with adjustable louvres, dampers, or other devices to deflect or diffuse the air currents. French windows and doors shall be considered to be ventilating openings in living quarters.

Ventilation. The process of supplying or removing air by natural or mechanical means to or from any space. Such air may or may not have been conditioned.

Ventilation gravity system. Ventilation depending wholly upon relative air density.

Vent shaft. Court used only to ventilate or light a water closet, both, toilet or utility room or other service room.

15-1 PLUMBING FIXTURES AND ENCLOSURES

Accessibility of fixtures. All plumbing fixtures shall be installed and spaced so as to be reasonably accessible for their intended use and maintenance.

Accessible. "Accessible" when applied to a fixture connection, appliance or equipment means having access thereto, but which first may require the removal of an access panel, door or similar obstruction; "readily accessible" means direct access without the necessity of removing or moving any panel, door, or similar obstruction.

Bathroom. The enclosed space containing one or more bathtubs, showers, or both, and which shall also include toilets, lavatories, or fixtures, serving similar purposes.

Battery of fixtures. Any group of two or more similar adjacent fixtures which discharge into a common horizontal waste or soil branch.

Bedpan washer. A fixture designed to wash bedpans and to flush the contents into the soil drainage system. It may also provide for steaming the utensils with steam or hot water.

Bedpan washer hose. A device supplied with hot and cold water and located adjacent to a water closet or clinic sink to be used for cleansing bedpans.

Clinic sink or bedpan hopper. A sink designed primarily to receive wastes from bedpans provided with a flush rim, integral trap with a visible trap seal, having the same flushing and cleansing characteristics as a water closet.

Combination fixture. A combination fixture is a fixture combining one sink and tray or a two- or three-compartment sink or tray in one unit.

Construction of fixtures. Plumbing fixtures shall be made of smooth nonabsorbent material, free from concealed fouling surfaces, and the fixture enclosures shall be ventilated.

Fixture. Device, applicance, or equipment which is installed in a building as a functional part of the occupancy or use and which may be fastened to but not made a part of the building in which it is being installed.

Fixture. A receptacle, appliance, apparatus or other device that is used for discharging sewage or clear-water waste and includes a floor drain.

Fixture. A receptacle or device which is either permanently or temporarily connected to the water distribution system of the premises, and demands a supply of water therefrom, or it discharges used water, liquid-borne waste materials, or sewage either directly or indirectly to the drainage system of the premises, or which requires both a water supply connection and a discharge to the drainage system of the premises.

Fixture branch. A pipe connecting several fixtures.

Fixture drain. Pipe that connects a trap serving a fixture to another part of a drainage system.

Fixture drain. The drain from the trap of a fixture to the junction of that drain with any other drain pipe.

Fixture outlet pipe. Pipe that connects the waste opening of a fixture to the trap serving the fixture.

Fixtures. Shall include water closets, bathtubs, sitz tubs, catch basins, slop sinks, kitchen sinks, urinals, wash trays, wash basins, lavatories, pantry sinks, showers, drinking fountains, floor drains, laundry tubs and all other appliances requiring running water or connection to a sewer.

Fixture supply. The water supply pipe connecting a fixture to a branch water supply pipe or directly to a main water supply pipe.

Fixture unit. A fixture unit is a quantity in terms of which the load-producing effects on the plumbing system of different kinds of plumbing fixtures are expressed on some arbitrarily chosen scale.

Fixture unit. The fixture unit is the rate of discharge through a plumbing fixture of 7-1/2 gallons per minute. This is termed one fixture unit.

Fixture unit, drainage or d.f.u. A measure of the probable discharge into the drainage system by various types of plumbing fixtures. The drainage fixture unit value for a particular fixture depends on its volume rate of drainage discharge, on the time duration of a single drainage operation, and on the average time between successive operations.

Fixture unit, supply or s.f.u. A measure of the probable hydraulic demand on the water supply by various types of plumbing fixtures. The supply fixture unit value for a particular fixture depends on its volume rate of supply, on the time duration of a single supply operation, and on the average time between successive operations.

Lavatory. A basin or other similar vessel (that is fixed in place and is plumbed with water) used for washing the hands, arms, face and head.

Plumbing appliance. Any one of a special class of plumbing fixtures which is intended to perform a special plumbing function. Its operation or control or both may be dependent upon one or more energized components, such as motors, controls, heating elements, or pressure or temperature sensing elements. Such fixtures may operate automatically through one or more of the following actions: a time cycle, a temperature range, a pressure range, a measured volume or weight; or the fixture may be manually adjusted or controlled by the user or operator.

Plumbing fixture. A receptacle or device which is either permanently or temporarily connected to the water distribution system of the premises, and demands a supply of water therefrom or it discharges used water, liquid-borne waste materials, or sewage either directly or indirectly to the drainage system of the premises.

Plumbing fixtures. Receptacles intended to receive and discharge water, liquid or water-carried wastes into a drainage system with which they are connected.

Plumbing fixtures. Installed receptacles, devices, or appliances that are supplied with water or which receive or discharge liquids or liquid-borne wastes.

Plumbing fixtures. Plumbing fixtures are installed receptacles, devices or appliances which are supplied with water or which receive or discharge liquids or liquid-borne wastes, with or without discharge into the drainage system with which they may be directly or indirectly connected.

Receptor. An approved plumbing fixture or device of such material, shape and capacity as to adequately receive the discharge from indirect waste pipes, so constructed and located as to be readily cleaned.

Sanitary fixtures. Each dwelling unit abutting on a public sewer or with a private sewage disposal system shall have at least one (1) water-closet, one (1) lavatory, one (1) tub or shower bath and one (1) kitchen type sink. All other structures for human occupancy or use abutting on a sewer or with a private sewage disposal system shall have at least one (1) watercloset and one (1) fixture for cleansing purposes. In hotel and dormitory residential buildings, there shall be not less than one (1) toilet room for each sex containing not less than one (1) watercloset, one (1) lavatory and one (1) tub or shower bath for every six (6) occupants.

Soil pipe. Any pipe which conveys the discharge of water closets, with or without the discharges from other fixtures.

Sterilizer instrument. A nonpressure type fixture used for boiling for disinfecting instruments, utensils or other equipment which may be portable or connected to the plumbing system.

Toilet compartment. No watercloset shall be located in a room or compartment which is not properly lighted and ventilated.

Toilet facility. A fixture, maintained within a toilet room, which may be used for defecation or urination, or both.

Toilet room. Enclosed space, containing one or more water closets, which may also contain one or more lavatories, urinals, and other plumbing fixtures.

Urinal. A toilet facility which is used only for urination.

Washrooms. Enclosed space containing one (1) or more bathtubs, showers or both and which shall also include toilets, lavatories, or fixtures, serving similar purposes.

Water-closet. A toilet fixture.

Water-closet. A toilet facility (which may be used for both defecation and urination) in which the waste matter is removed by flushing with water.

Water closet bowls. Water closet bowls and traps shall be glazed vitreous earthenware made in one piece and of such form as to hold sufficient quantty of water, when filled to the trap overflow, to prevent fouling of surface, and shall be provided with integral flushing rims constructed so as to flush the entire interior of the bowl. The use of water closet bowls with side inlets, or of the valve-in-bowl type, are prohibited.

Water closet combination. Water closet bowls may be siphon-jet, washdown, reverse-trap, or blowout type with wall outlet. Water closet bowls and traps shall be made in one piece and shall be provided with integral flushing rims so constructed as to flush the entire interior of the bowl.

Water-closet compartment. An enclosed space containing one (1) or more toilets or one (1) or more urinals and other plumbing appliances.

Water-closet compartment. Enclosed space containing one or more toilets which may also contain (1) or more lavatories, urinals, and other plumbing fixtures.

15-J PLUMBING — GENERAL

Backwater traps. When there is a possibility that a plumbing drainage system will be subject to backflow of sewage, suitable provisions shall be made to prevent its overflow into the building.

Branch. The branch of any system of piping is the part of the system which extends horizontally at a slight grade with or without lateral or vertical arms from the main, to receive fixture outlets not directly connected to the main.

Branch vent. Vent pipe that is connected at its lower end to the junction of 2 or more vent pipes and is connected at its upper end either to a stack vent, vent stack or header, or is terminated in open air.

Brass and copper pipe. Brass and copper pipe shall be of standard weight iron pipe size and shall be of the grade known as containing sixty-seven per cent copper.

Building drain. The building drain is that part of the lowest horizontal piping of a drainage system which receives the discharge from soil, waste, and other drainage pipes inside the walls of any building and conveys the same to the building sewer.

Building drain. That part of the lowest horizontal piping that conducts sewage, clear-water waste or storm water to a building sewer.

Building sewer. Pipe that is connected to a building drain 3 feet outside of a wall of a building and that leads to a public sewer or private sewage disposal system.

Building sewer. That part of the horizontal piping of a drainage system extending from the building drain to its connection with the main sewer or septic tank and conveying the drainage of but one (1) building site.

Building subdrain. That part of a drainage system that cannot drain by gravity into the building sewer.

Building trap. Trap that is installed in a building drain or building sewer to prevent circulation of air between a drainage system and a public sewer.

Circuit vent. Vent pipe that is connected at its lower end to a branch and its upper end to a vent stack or terminates in open air.

Cleanout. Pipe fitting that is intended to provide access to a pipe to permit pipe cleaning.

Cleanouts. Provision shall be made in the drainage system for adequate cleanouts so arranged that all pipes may be readily cleaned.

Clear-water waste. Clear water that does not contain sewage or storm water.

Conductors or roof leaders. Pipes which carry storm or rain water from roofs of buildings.

Continuous vent. Pipe extending vertically above the soil waste branch.

Continuous vent. Vent pipe that is an extension of a vertical section of a branch or fixture drain.

Copper tubes for waste and vent lines. Copper tube when used in waste and vent lines shall be seamless, cold drawn commercially pure hard copper tubing of standard U.S. Government or A.S.T.M. Types.

Dead end. Pipe that is 2 feet or more in developed length and terminates with a closed fitting.

Developed length. Length along the center line of a pipe.

Diameter. The nominal diameter by which a pipe, fitting, trap or other item of similar use is commercially designated.

Drainage system. The drainage system shall be designed, constructed and maintained so as to guard against fouling and clogging.

Dual vent. Vent connected at the junction of 2 fixture drains and serving as a vent for both fixtures. Usually called a common vent.

Effective opening. Opening that has a cross-sectional area equal to the minimum area through which water is discharged at a discharge opening, control valve inlet or control valve seat of a water supply inlet to a fixture or device.

Fresh air inlet. Vent pipe that is connected at its lower end to a building drain and is extended through the wall of a building to terminate in outside air.

Header. Vent pipe that is installed to connect the upper end of one or more vent stacks or stack vents to a vent stack, stack vent or open air.

House drain. That part of the underground or lowest horizontal piping of a house drainage system which receives the discharge of all soil, waste and other drainage pipes inside the wall of any building and conveys the same to the house sewer 5 feet outside the building.

House sewer. That part of the horizontal piping of a house drainage system extending from the house drain to its connections with the main sewer, cesspool or septic tank, and conveying the drainage of but one building site.

Indirect connections. Proper protection shall be provided to prevent contamination of food, water, sterile goods and similar materials by backflow of sewage and if necessary the fixture, device or appliance shall not connect directly with the building drainage system.

Indirectly connected. Not directly connected to a drainage system.

Individual vent. Vent pipe that serves not more than one fixture.

Inside copper water tube. Copper tubing for inside water supply distribution system shall be seamless, cold drawn, commercially pure hard copper tubing of standard U.S. Government type 'K" or "L" except that copper tubing used for water supply under or in concrete slabs shall be soft or hard drawn Type "K" copper. On that part of inside water piping extending from the floor or wall to the plumbing fixture, a fixture supply of the type known as "Flexible" supplies may be used, such supplies to be of not less than one-quarter inch I.D. copper tubing.

Interceptor. A receptacle that is installed to prevent oil, grease, sand, or other materials from passing into a drainage system.

Interceptor. Receptacle designed and constructed to separate or intercept and prevent the passage of oil, volatile flammable liquid, grease, sand, or other material into the drainage system to which such receptacle is directly or indirectly connected.

Joint. Point at which two sections of pipe are fitted together.

Journeyman plumber. A person certified as such under the terms of the ordinance skilled in the practice of installing plumbing and drainage systems, who alters, repairs, dismantles and maintains plumbing and drainage systems or parts thereof, as an employee, but who does not furnish materials or supplies or contracts work.

Leader. Pipe that is installed to carry storm water from a roof to a storm building drain or sewer or other place of disposal.

Lead water service. Lead water service pipe shall be of best quality of not less weight per linear foot than defined in the Code and in accordance with Federal Specifications. Lead water supply pipe shall be AAA double extra strong.

Loop vent (as applying to plumbing systems). Vent pipe that is connected at its lower end to a branch and at its upper end to a stack vent or header or to a branch vent that is connected to a stack vent or header.

Main. Main of any system of horizontal, vertical, or continuous piping which is that part of such system that receives the wastes, vent or revents, from fixture outlets or traps, direct or through branch pipes.

Main. Construed as the main of any system of horizontal, vertical or continuous piping, means that part of such system which receives such wastes, vents or back vents from fixture outlets or traps directly or through branch pipes.

Main vent. Soil or waste stack serving 1 or more water closets, that is most distant from the building sewer, together with any vent pipe that joins the top of the stack to open air, except that in a building where fixtures are located only on the lowest story, main vent means the vertical soil or waste pipe that is most distant from the building sewer, together with any vent pipe, that connects the top of the pipe to open air.

Mild steel pipe. All steel pipe shall conform to the A.S.T.M. Standard Specifications for welded and seamless steel pipe (A120, A53 or better) and shall be galvanized.

Offset. In a line of piping, a combination of elbows or bends which brings one section of the pipe out of line with, but into line parallel to, another section.

Outside copper water tube. Copper tubing used for underground water supply or water service shall be "K" type only and shall be fully annealed.

Pipe dope. Standard compounds used as a pipe and fitting thread lubricant.

Plumber. A master plumber duly licensed by the division of inspections of the city.

Plumbing. The practice, materials, and fixtures used in the installation, maintenance, extension, and alteration of all piping, fixtures, appliances, equipment, and appurtenances in connection with any of the following: Sanitary drainage or storm drainage or storm drainage facilities, the venting system and the public or private water supply systems, within or adjacent to any building; also the practice and materials used in the installation, maintenance, extension, or alteration of storm water, liquid-waste, or sewerage, and water-supply systems of any premises and their connection with any point of public disposal or other acceptable terminal.

Plumbing. System of pipes, fixtures, apparatus and appurtenances, installed upon the premises, or in a building, to suply water thereto and to convey sewage or other waste therefrom.

Plumbing. The system of pipes, fixtures, apparatus and appurtenances, installed upon the premises, or in a building, to supply water thereto and to convey sewage or other waste therefrom, in accordance with the Code.

Plumbing. The profession, art or trade of, and all work done and all materials used in and for: introducing, maintaining and extending a supply of water, provide for the removal of sewage, waste and drainage through pipes or conduits, or any appurtenances thereof, in a building, lot, premises or establishment.

Plumbing. The installation of all interior plumbing work shall comply with the standards of the American Standard National Plumbing Code (Latest edition).

Plumbing. Includes gas pipes and provided gas burning equipment, heaters and tanks or boiler for hot water, waste pipes, water pipes, water closets, sinks, lavatories, furnace for steam heat and other heating appliances, bathtubs, showerbaths, catchbasins, drains, vents, water-cooled air conditioning system, and any other provided fixtures, together with the connection to the water, sewer or gas lines.

Plumbing. Art of installing, repairing or altering the pipes, fixtures and other apparatus for distributing the water supply and removing liquid and water-carried wastes; the system of pipes, fixtures, and other apparatus installed in buildings for distributing the water supply and for the disposal of liquid and water-carried wastes, including valves, traps and soil, waste and vent pipes; provided, that nothing herein contained shall include the work of steamfitting.

Plumbing. All of the following supplies, facilities, and equipment: gas pipes, gas-burning equipment, water pipes, garbage disposal units, waste pipes, water closets, sinks, lavatories, bathtubs, shower baths, catch basins, vents, and any other similar supplied fixtures, together with all connections to water, sewer, or gas lines, and water pipes and lines utilized in conjunction with air-conditioning equipment.

Plumbing system. The plumbing and drainage system of a building, which includes the pipes within the building distributing the city water supply, the plumbing fixtures and traps, the soil, waste and vent pipes, the house drain and human sewer, the storm water drain pipes with their devices, appurtenances and connections, and the air pipes with the plumbing and drainage system.

Plumbing system. Drainage system, a venting system and a water system.

Plumbing system. The water-supply and distribution pipes; plumbing fixtures and traps; soil, waste, and vent pipes; building house drains and building house sewers including their respective connections, devices, and appurtenances within the property lines of the premises; and water-treating or water-using equipment.

Plumbing systems appliance. Receptacle or equipment that receives or collects water, liquids or sewage and discharges water, liquids or sewage either directly or indirectly to a plumbing system.

Plumbing system tests. The plumbing system shall be subjected to such tests as will effectively disclose all leaks and defects in the work.

Relief vent. Vent pipe that is connected at its lower end to a nominally horizontal branch and at its upper end to a branch vent, header, stack vent or vent stack, or is terminated in open air.

Return offset. A double offset installed so as to return the pipe to its original alignment.

Revent pipe. A pipe which connects directly at or near the junction of an individual trap outlet with a waste or soil pipe underneath or back of a fixture and extends to a connection with the main or branch vent above the top of the fixture.

Riser. A water supply pipe which extends vertically one full story or more to convey water to branches or to a group of fixtures.

Roughing in. The installation of all parts of the plumbing system which can be completed prior to the installation of fixtures. This includes drainage, water supply, and vent piping, and the necessary fixture supports, or any fixtures that are built into the structure.

Sanitary piping. The piping of the plumbing system shall be of durable material, free from defective workmanship and so designed and installed as to give satisfactory service for the reasonable life expectancy of the building.

Sewage interceptors. No substances which will clog the pipes, produce explosive mixtures, or destroy the pipes or their joints shall be allowed to enter the building drainage system or interfere with sewage disposal processes.

Size and length. The given caliber or size of pipe is for a nominal internal diameter except that other than iron pipe size brass pipe is measured by outside diameter. The developed length of a pipe is its longitudinal length along the center line of pipe and fittings.

Size of pipe or tubing. Size of pipe or tubing, unless otherwise stated, is the size, which has been graded for strength by a lumber grading or inspection bureau or other agency or individual recognized by the Building Official as being competent and responsible.

Size of pipe or tubing. Unless otherwise specified is the nominal size by which pipe or tubing is commercially designated. Actual dimensions of the different kinds of pipe and tubing are given in the Code specification applying to its use.

Size of piping. Unless otherwise indicated means the nominal size by which a pipe, fitting, trap or other item is commercially designated.

Sizes and lengths. Referring to the given caliber or size of pipe, means a standard internal diameter except for brass pipe other than iron pipe size which may be outside the diameter; and, referring to the length of the pipe, means its developed length along its center lines.

Soil-or-waste pipe. Pipe in a sanitary drainage system.

Soil-or-waste stack. Vertical soil-or-waste pipe that passes through 1 or more stories, and includes any off set that is part of the stack.

Soil or waste vent. That part of the main soil or waste stack extending through the roof and above the highest fixture outlet.

Soil pipe. Any pipe which receives the discharge of water closets, with or without the discharge from other fixtures, and conveys the same to the house drain.

Soil stack. General term for the vertical main of a system of soil, waste, or vent piping.

Soil vent. A soil vent is that part of a stack which extends above the highest installed water closet.

Stack. As a general term, means any vertical line of soil waste or vent piping.

Stack. The general term for any vertical line of soil waste or vent piping as used in the Code.

Stack vent. Vertical vent pipe that is an extension of a soil-or-waste stack.

Structural integrity. Plumbing systems shall be installed so that structural members of the building are not impaired nor adjacent surfaces damaged through ordinary use of the plumbing fixtures.

Tailpiece. A connection used from outlet of fixture strainer to trap connection.

Tempered water. Water at a temperature of not less than 90 degrees and not more than 105 degrees.

Testing. Requires that the plumbing system be filled with water, air, smoke or as otherwise required by the Plumbing Code.

Trap. A fitting or device so constructed as to prevent the passage of air or gas through a pipe without materially affecting the flow of sewage or waste water through it.

Trap. Fitting or device that is designed to hold a liquid seal that will prevent the passage of gas but will not materially affect the flow of a liquid.

Trap. A fitting device so constructed as to retain a water seal to prevent the passage of air or gas, through a pipe, without materially affecting the free flow of sewage or waste water.

Trap. Pertaining to plumbing, is a fitting or device, so designed and constructed as to provide a liquid seal, which will prevent the passage or air or gas through it without materially affecting the flow of sewage or liquid wastes.

Trap arm. A trap arm is that portion of a fixture drain between a trap and its vent.

Trap (building). A building (house) trap is a device, fitting, or assembly of fittings installed in the building drain to prevent circulation of air between the drainage system of the building and the building sewer.

Trap dip. Lowest part of the upper interior surface of a trap.

Trap primers. A trap primer is a device or system of piping to maintain a water seal in a trap.

Trap (resealing). A resealing trap is a trap so constructed and installed as to retain a satisfactory seal when subjected to siphonic or aspiratory effects of wastes discharging through or past the branch, into which the trap is connected.

Trap seal. Vertical distance between the trap dip and the trap weir.

Trap seal. The trap seal is the vertical distance between the crown weir and the dip of the trap.

Trap seal. The maximum vertical depth of liquid that a trap will retain, measured between the crown weir and the dip of the trap.

Trap weir. Highest part of the lower interior surface of a trap.

Vent. A passageway, vertical or nearly so, for removing vent gases to the outer air.

Vent. A pipe installed to provide a flow of air to or from a drainage system or to provide a circulation of air within such systems to protect trap seals from siphonage and back pressure.

Venting system. All vent terminals shall extend to the outer venting system. All vent terminals shall extend to the outer air and shall be so installed as to minimize the possibilities of clogging and the return of foul air to the building.

Venting system. Assembly of pipes and fittings that connects a drainage system with outside air for circulation of air and the protection of trap seals in the drainage system.

Vent pipe. Any pipe provided to ventilate a building drainage system and to prevent trap siphonage and back pressure.

Vent pipe. Pipe that is a part of a venting system.

Vent pipes. Any pipe provided to ventilate a house drainage system and to prevent trap siphonage and back pressure.

Vent stack. Vertical vent pipe that is connected at its lower end to a soil-or-waste stack or to a building drain and is connected at its upper end either to a stack vent or header, or is terminated in open air.

Volume of a pipe. The measurement of the space within pipe walls.

Waste pipe. Any pipe which receives the discharge of any fixture, except water closets, and conveys the same to the house drain, soil or waste stacks.

Waste pipe and special waste. A waste pipe is any pipe which receives the discharge of any fixture, except water closets, and conveys the same to the building drain, soil or waste stacks. When such pipe does not connect directly with a building drain or soil stack it is termed a special waste.

Weather protection. Drainage and water piping shall be protected against freezing temperatures and water supply systems shall be installed to permit complete drainage when necessary.

Wet vent. A pipe connecting to side outlet fitting, venting one water closet, or where the vent of one fixture is used as the waste for other minor fixtures.

Wet vent. Soil-or-waste pipe that also serves as a vent pipe.

Wrought iron pipe. All genuine wrought iron pipe shall conform to the A.S.T.M. Standard Specifications for welded wrought iron pipe and shall be galvanized.

Yoke vent. Vent pipe that is connected at its lower end to a soil-or-waste stack and at its upper end to a vent stack or a branch vent that is connected to a vent stack.

15-K SPRINKLER SYSTEMS

Automatic dry pipe sprinkler system. A sprinkler system in which the piping up to the sprinkler heads is filled with air, either compressed or at atmospheric pressure, with the water supply controlled by a Type "A" or Type "B" dry pipe valve.

Automatic fire pump. A pump that maintains a required water pressure in a fire extinguishing system and which is actuated by a starting device adjusted to cause the pump to operate when the pressure in the system drops below a predetermined pressure, and to stop the pump when the pressure is restored.

Automatic sprinkler head. Is a device which will open automatically within a pre-determined temperature range because of fusing of a fusible element, or will operate automatically when controls are subjected to a predetermined accelerated rise in temperature.

Automatic sprinkler head. A device connected to a water supply system that opens automatically at a predetermined fixed temperature and disperses a stream or spray of water.

Automatic sprinklers. Is an arrangement of piping and sprinklers designed to operate automatically by the heat of the fire and to discharge water upon the fire.

Automatic sprinklers. System of piping supplied with water under pressure with devices for releasing under the influence of heat and the spraying of water on ceilings, walls and floors.

Automatic sprinkler system. Automatic sprinkler system shall be of a standard approved type, and installed to provide complete coverage of all portions of the building being protected.

Automatic sprinkler system. Is a sprinkler system which will operate automatically to discharge or diffuse a stream or spray of water through automatic sprinkler heads or through open sprinkler heads.

Automatic sprinkler system. An arrangement of piping and sprinklers designed to operate automatically by the heat of fire and to discharge water upon the fire.

Automatic water supply source. Water supplied through a gravity or pressure tank, or automatically operated fire pumps, or from a direct connection to an approved city water main.

Automatic wet pipe sprinkler system. A sprinkler system in which all piping and sprinkler heads are at all times filled with water under pressure which is immediately discharged when a sprinkler head operates, with the water continuing to flow until the system is shut off.

Deluge sprinkler system. An open head sprinkler system without water in the system piping, with the water supply controlled by an automatic valve operated by smoke or heat-responsive devices installed throughout the sprinklered area, and independent of the sprinkler heads.

Deluge system. A sprinkler system designed to deliver large quantities of water through open sprinkler heads, in which the water supply is controlled by a valve actuated by a thermostatic device on a predetermined temperature or rate of temperature rise.

Deluge system. A system employing open sprinklers attached to a piping system connected to a water supply through a valve; valve is opened by the operation of a heat responsive system installed in the same areas as the sprinklers. When this valve opens, water flows into the piping system and discharges from all sprinklers attached thereto.

Dry-pipe sprinkler system. Sprinkler system to which water is supplied upon operation of an automatic dry-pipe valve, actuated and operated automatically by release or application of air pressure, or actuated and operated by remote thermostatic control.

Dry pipe systems. A system employing automatic sprinklers attached to a piping system containing air under pressure, the release of which as from the opening of sprinklers permits the water pressure to open a valve known as a "dry-pipe valve." The water then flows into the piping system and out the opened sprinklers.

Dry pipe valve. A valve that automatically controls the water supply to a sprinkler system so that the system beyond the valve is normally maintained dry.

Freezing of systems. Systems, or portions of a system, subject to freezing shall be maintained dry, filled with antifreeze solution, heated by electric strip heater, or otherwise protected against freezing.

Heat banking area. Area of the upper portion of a story under the ceiling or roof, within draft curtains, or between walls or partitions or any combination thereof in which heat banks up to actuate sprinklers and or open smoke vents.

Limited water supply system. A system employing automatic sprinklers and conforming to the Code Standard, but supplied by a pressure tank of limited capacity.

Non-automatic sprinkler system. A sprinkler system in which all pipes are maintained dry and which is equipped with a siamese fire department connection in an approved location.

Nonautomatic sprinkler system. A sprinkler system in which all pipes and sprinkler heads are maintained dry and which is supplied with water through a fire department siamese connection.

Non-automatic sprinkler systems. When approved by the administrative official, a dry sprinkler system with fire department connection may be accepted in buildings and structures which involve low fire and life hazard and in which adequate heat is not provided, in place of an automatic sprinkler system. Such systems shall be provided with an approved automatic heat-actuated alarm with an outside alarm gong or connection to the fire department, or to the central station of an approved supervisory service.

One-source sprinkler system. An automatic sprinkler system which is supplied from one of the approved automatic sources of water supply.

Partial sprinkler system. An automatic sprinkler system consisting of a limited number of automatic sprinkler heads serviced from the building water supplies with one or more fire department siamese connections as required, for use in exitway facilities and isolated hazardous locations when approved by the building official.

Pre-action system. A system employing automatic sprinklers attached to a piping system containing air that may or may not be under pressure, with a supplemental heat responsive system of generally more sensitive characteristics than the automatic sprinklers themselves, installed in the same areas as the sprinklers; actuation of the heat responsive system, as from a fire, opens a valve which permits water to flow into the sprinkler piping system and to be discharged from any sprinklers which may be open.

Sprinkled. Equipped with an approved automatic sprinkler system properly maintained.

Sprinkler alarm. An apparatus constructed and installed so that a flow of water through the sprinkler system equal to, or greater than, that required for a single automatic sprinkler head will cause an alarm to be given.

Sprinkler alarm. A local alarm unit is an assembly of apparatus approved for the service and so constructed and installed that any flow of water from a sprinkler system equal to or greater than that from a single automatic sprinkler will result in an audible alarm signal on the premises.

Sprinklered. As applied to a structure shall mean equipped throughout with an approved system of automatic sprinklers.

Sprinklered. Provided with a fire sprinkler system which conforms to the requirements of the Code.

Sprinkler system. A complete automatic sprinkler system which is installed in compliance with generally accepted standards.

Sprinkler system. A sprinkler system for fire protection purposes by an integrated system of underground and overhead piping designed in accordance with fire protection engineering standards. The system includes a suitable water supply, such as a gravity tank, fire pump, reservoir or pressure tank and/or connection by underground piping to a city main. The portion of the sprinkler system above ground is a network of specially sized or hydraulically designed piping installed in a building, structure, or area, generally overhead, and to which sprinklers are connected in a systematic pattern. The system includes a controlling valve and a device for actuating an alarm when the system is in operation. The system is usually activated by heat from a fire and discharges water over the fire area.

Sprinkler system. System of piping connected to one or more approved sources of water supply and provided with distributing devices so located and arranged as to discharge or diffuse an effective stream or spray of water over the interior of a building, or part of a building, or on the exterior of a building. Such system may be set off by extraordinary heat or manually.

Sprinkler system. A system of piping connected to one or more acceptable sources of water supply, which system is provided with distributing devices so arranged and located as to discharge an effective spray over the interior of the building area.

Sprinkler system. A system of piping and appurtenances designed and installed in accordance with generally accepted standards so that heat from a fire will automatically cause water to be discharged over the fire area to extinguish it or prevent its further spread.

Sprinkler system. An approved system of automatic sprinklers throughout the building; unsprinklered means not so equipped.

Sprinkler system. A system of piping and sprinkler heads connected to one or more sources of water supply.

Sprinkler system—chemical. A system of automatic sprinklers controlled by thermostatic operating devices for the diffusion of approved fire-extinguishing chemicals or gases.

Sprinkler system—dry pipe. A system in which all pipes and sprinkler heads are filled with air under pressure and the water supply is controlled by an approved automatic dry-pipe valve in the event of fire, actuated either by the release of air or by thermostatic electric control.

Sprinkler system—thermostatic. An open or closed head sprinkler system operated through an auxillary thermostatic device which functions at a predetermined rate of temperature rise.

Sprinkler system—wet pipe. A system of automatic sprinklers in which all pipes are filled with water at all times.

Supervised sprinkler system. A system in which all water supply, valves and accessory equipment is provided with electrical contact devices to transmit signals to an outside central supervisory station.

Two-source system. An automatic sprinkler system which is supplied from a combination of any two of the approved automatic sources of water supply, or from two (2) pressure tanks, or by direct connections to the municipal water supply on two (2) streets in which the water mains are separately controlled.

Water curtain. A system of approved open or closed sprinkler heads or perforated pipes installed on the exterior of a building at eaves, cornices, window openings, and on mansard or peak roofs with water supply under manual control; or installed around openings in floors or walls of a building with water supply under thermostatic control.

Wet pipe systems. A system employing automatic sprinklers attached to a piping system containing water and connected to a water supply so that water discharges immediately from spinklers opened by a fire.

15-L STANDPIPE SYSTEMS

Automatic dry standpipe system. A standpipe system in which all piping is filled with air, either compressed or at atmospheric pressure. Water enters the system through a control valve actuated either automatically by the reduction of air pressure within the system or by the manual activation of a remote control located at each hose station.

Class I standpipe system. System for use by fire department and those trained in handling heavy fire streams (2-1/2 inch hose).

Class II standpipe system. System for use primarily by the building occupants until the arrival of the fire department (1-1/2 inch hose).

Class III standpipe system. System for use by either fire department and those trained in handling heavy hose streams or by the building occupants.

Combination standpipe. A standpipe fire line having a constant water supply available in addition to the Fire Department inlet connections and installed primarily for Fire Department use.

Cross-connection (fire extinguishing system). Piping between risers and siamese connections in a standpipe or sprinkler system.

Dry standpipe. Standpipe without fixed connection to a source of water and maintained without water in the piping.

Horizontal fire line. A fire line installed around the interior walls and column of a building, pier or wharf, with hose outlets located so that every part of the floor area is within reach of at least one fire stream.

Hose. Standpipes inside buildings and structures shall have not more than 100 feet of 1-1/2 inch diameter hose equipped with the proper accepted and approved nozzel and couplings at each outlet and hung in an approved rack or cabinet to meet the requirements of the code.

Manual fire pump. A pump that feeds water into a fire extinguishing system that must be started by either the building personnel or members of the fire department.

Nonautomatic standpipe system. A standpipe system in which all piping is maintained dry, and which is supplied with water through a fire department siamese connection.

Siamese connection. A fitting connected to a fire extinguishing system and installed on the outside of a building, with two hose inlets for use of the fire department, to furnish or supplement the water supply to the system.

Standpipe. Wet or dry fire line installed exclusively for the fighting of fire, extending from the lowest to the topmost story of a building or structure with hose outlets at every floor, equipped with reducing valves and designed to operate at required working pressures.

Standpipe. A wet or dry fire line installed exclusively for the fighting of fire, extending from the lowest to the topmost story of a building or structure with hose outlets at every floor equipped with reducing valves and designed to operate at required working pressures, and located in protected areas.

Standpipe (dry). A standpipe fire line without permanent or automatic water supply equipped with a siamese connection for use of the fire department.

Standpipe during building construction. Every building six stories or more in height shall be provided with one standpipe for temporary Fire Department use during construction, when topmost construction reaches six stories in height. Such a standpipe shall be provided with Fire Department inlet connections at an accessible location and located adjacent to usable stairs.

Standpipe-first-aid. An auxiliary vertical or horizontal fire line designed primarily for emergency use by the occupants of the building or by the private fire brigade before the arrival of the municipal fire department.

Standpipes (Interior wet standpipes). Every building three or more stories in height shall be provided with interior wet standpipes.

Standpipe system. Approved installation of piping and appurtenances, whereby all parts of a building can be quickly reached with an effective stream of water.

Standpipe system. The system of wet or dry piping, including the necessary appurtenances, within a building or structure.

Standpipe system. A system of piping, for fire-fighting purposes, consisting of connections to one or more sources of water supply, and serving one or more hose outlets.

Standpipe system. An approved system of piping, valves, water supply, siamese connections, hose, nozzles and outlets to enable water to be used for fire fighting purposes on upper floors of a building with sufficient pressure.

Standpipe system. A standpipe system is an arrangement of piping, valves, hose outlets and allied equipment installed in a building or structure with outlets located in such a manner that water can be discharged in streams or spray patterns through hose and nozzles, attached to such hose outlets, for the purpose of extinguishing a fire and so protecting a building or structure and its contents in addition to protecting the occupants. This is accomplished by connections to water supply systems or by pumps, tanks and other equipment necessary to provide an adequate supply of water to the hose outlets.

Standpipe system. An arrangement of piping, valves, hose outlets and allied equipment installed in a building or structure with outlets located in such a manner that water can be discharged through hose and nozzles.

Standpipe (wet). A standpipe fire line having a primary water supply constantly available at every hose outlet, or made available by opening the hose outlet or by automatic functioning of a control station.

Wet standpipe. A system of vertical pipes and interior outlets which is connected to a permanent water supply.

Wet standpipe. A standpipe connected to an open or automatically provided source of water with water available at all outlets at all times.

Wet standpipe. A standpipe fire line having a primary water supply constantly available at every hose outlet, or made available by opening the hose outlet or by automatic functioning of a central station.

Wet standpipe. A standpipe fire line having a primary water supply constantly available at every hose outlet.

Wet standpipe. Auxiliary fire line system with a constant water supply installed primarily for emergency fire use by the occupants of the building.

Wet standpipe system. A standpipe system in which all of the piping is filled with water under pressure, that is immediately discharge upon the opening of any hose valve.

Wet standpipe system. A system having supply valve open and water pressure maintained at all times.

Zone. A vertical division of a building fire standpipe system used to establish the water working pressures within the system and also to limit the pressure at the lowest hose outlet in the zone.

15-M WATER SUPPLY AND PIPING

Air gap. The vertical distance between the lowest point of a water supply outlet and the flood level rim of the fixture or device into which the outlet discharges.

Anti-siphonage. The drainage system shall be so designed as to provide an adequate circulation of air in all pipes without danger of siphonage, aspiration or forcing of trap seals under normal use.

Approved water supply. The term "approved water supply" means any water supply approved by, or under the public health supervision of, a public health agency of the State.

Auxiliary supply. The term "auxiliary supply" means any water supply on or available to the premises other than the public water supply.

Backflow. Flow of water or other liquids, mixtures or substances into the distributing pipes of a supply of potable water from any source other than its intended source and may be produced by the differential pressure existing between two systems either or both of which are at pressures greater than atmospheric.

Backflow connection. Any connection or condition that may permit backflow.

Backflow preventer. Device or a method to prevent backflow caused by gravity or back pressure into a potable water supply.

Backflow (water supply). A flow of water or other substances into distribution pipes of a potable water supply from any source other than the intended source.

Back-siphonage. Flow-back of water from a plumbing fixture or vessel or other source into a water supply pipe due to negative pressure in such pipe.

Branch distributing pipe. A pipe which is connected to a distributing pipe or riser pipe and conveys the water therefrom to the plumbing fixtures.

Branch supply pipe. A pipe which is connected to a principal supply pipe and conveys the water therefrom to the riser pipe or distributing pipe.

Critical level. Highest level to which a back-siphonage preventer, when subjected to a specified test, can be submerged before backflow begins.

Cross-connection. Any unprotected connection between any part of a water system used or intended to supply water for drinking purposes and any source or system containing water or substance that is not or cannot be approved as safe, wholesome, and potable for human consumption.

Cross-connection (potable water system). A physical connection or arrangement between two otherwise separate piping systems, one of which contains potable water, and the other of which contains water of questionable safety, or steam, gases, or chemicals whereby there can be a flow from one system to another.

Distributing pipe. A pipe which is connected to a riser pipe or a branch supply pipe and conveys the water therefrom to the branch distributing pipe.

Drinking water. Water provided or used for human consumption or for lavatory or culinary purposes.

Extension. The extension of a water main along a street, avenue or highway. An extension shall not include the water-service connection as defined in the Code.

Flood level rim. Top edge at which water can overflow from a fixture or device.

Flushing and service water supply. Buildings in which there are waterclosets and other plumbing fixtures shall be provided with a supply of water adequate in volume and pressure for flushing and other building service purposes.

Frost-proof closet. Water closet that has no water in the bowl and has a trap and water control valve that are designed for installation below the frost line.

Hot water. Any boiler, generator, pressure vessel, system, piping or equipment used for purpose of heating or distributing hot water for heating or processing, operating at pressures in excess of one hundred sixty (160) psig and/or temperatures in excess of two hundred fifty (250) degrees F., shall be classed as high pressure.

Hot water. Water at a temperature of not less than 120°F. All fixtures such as kitchen sink, lavatory basin, bathtub, shower and all other appliances requiring hot water shall be connected to the hot water and the cold water.

Hot water supply. Devices for heating and storing water in boilers or hot water tanks shall be so designed and installed as to prevent danger from explosion through overheating.

Indirect service water heater. Heater that derives its heat from a heating medium such as warm air, steam or hot water.

Individual water supply. A supply other than an approved public water supply which serves one or more families.

Main supply pipe. A pipe which is connected to the service pipe of any building, structure, or premises and conveys the water therefrom to the principal supply pipe.

Non-potable water. Water not safe for drinking, personal or culinary use.

Potable water. Water free from impurities present in amounts sufficient to cause disease or harmful physiological effects. Its bacteriological and chemical quality shall conform to the requirements of the Regulations of the State Department of Public Health.

Potable water. Safe for human consumption.

Potable water. Water free from impurities present in amounts sufficient to cause disease or harmful physiological effects and conforming in its bacteriological and chemical quality to the requirements of the public health service drinking water standards or the regulations of the public health authority having jurisdiction.

Potable water. Potable water is water which is satisfactory for drinking, culinary, and domestic purposes, and meets the requirements of the Director of Public Health.

Potable water supply. All premises intended for human habitation or occupancy shall be provided with an adequate supply of pure and wholesome potable water, neither connected with unsafe water supplies nor subject to the hazards of backflow or back-siphonage.

Principal supply pipes. Are such pipes which are the water supply arteries in buildings and structures. They are connected to the water supply main and convey the water therefrom to the pumps, tanks, filters, heaters, and other equipment together with all their appurtenances and to the branch supply pipes.

Quality of water. All premises intended for human habitation or occupancy shall be provided with potable water that has been declared safe for domestic use by the proper authorities.

Riser. A riser is a water-supply pipe which extends vertically one full story or more to convey water to branches or fixtures.

Riser. A water supply pipe which extends vertically one full story or more to convey water to branches or to a group of fixtures.

Riser pipe. A pipe which is installed perpendicular to the horizontal through the floors, stories, and other open spaces of buildings and structures and conveys the water from the main or branch supply pipes to the distributing pipes or branch distributing pipes.

Service water heater. Appliance intended for the heating of water for plumbing services as distinct from water for space heating.

Size of pipes. The pipes conveying water to fixtures shall be of sufficient size to supply the water at an adequate rate for use and flushing without undue noise or undue reduction of the pressure at other fixtures.

Storage type service water heater. Service water heater with an integral hot water storage tank.

Water distributing pipe. Pipes for plumbing systems shall be of lead, galvanized wrought iron or steel, brass, copper, cast iron, CPVC or polybutyelene (PB) plastics with brass, cast-iron, galvanized malleable-iron, copper or appropriate plastic fittings.

Water distributing pipes. The pipes which convey water from the service pipe to the plumbing fixtures.

Water-distribution piping. The pipes in a building or premises that convey water from the water service pipe to the plumbing fixtures and other water outlets.

Water heater. A device for the heating and storage of water to be used for other than heating or industrial purposes.

Water heater. Tank, system of coils or combination of both, containing water directly heated by gaseous, liquid, solid fuel, or electrical heating elements.

Water main. A water-supply pipe for public or community use.

Water seal. All fixtures directly connected to the drainage system shall be equipped with a water seal trap.

Water-service connection. The pipe serving a premises from the main into the premises to a point 3 feet beyond the meter, including the meter.

Water service pipe. The pipe from the water supply to the building being served.

Water service pipe. Pipe in a water system that conveys water from a public water main or a private water source to the inner side of the wall or floor through which the system enters the building.

Water service pipe. The pipe from the water main or other source of water supply to the building or structure served.

Water service pipe. The pipe from the water (street) main or other source of water supply to the building served.

Water (street) main. A water-supply pipe for public or community use controlled by public authority.

Water supply. Every living unit shall have available a supply of safe water obtained from a public or municipal water supply if available; or by a drilled, driven or dug well.

Water-supply system. A system for a building or premises includes the water-service pipe, the water-distributing pipes, and the necessary connecting pipes, fittings, and all appurtenances.

Water-supply system. The water-service pipe, the water-distribution piping, and all of the necessary connecting pipes, fittings, control valves, and appurtenances used for conveying water in a plumbing system.

Water-supply system. The "water-supply system" of a building, or premises, consists of the water-service pipe, the water-distributing pipes, and the necessary connecting pipes, fittings, control valves and all appurtenances in or adjacent to the building or premises.

Water-supply system. The water-supply system of a building or premises consists of the water-service pipe, the water-distributing pipes, standpipe system and the necessary connecting pipes, fittings, control valves, and all appurtenances in or on private property.

Water system. Assembly of pipes, fittings, control valves and appurtenances that conveys water from a public main or private water source to the water supply outlets of fixtures or devices, and includes a private water source.

2. Article 16

ELECTRICAL 16-A THROUGH 16-D

16-A AUTOMATIC DETECTION SYSTEMS

Automatic fire detecting system. A system which utilizes theremostatic or other approved detecting elements for the detection of fire and the automatic transmission of an alarm.

Combination detector. Device that either (a) responds to more than one of the fire phenomena classified in the Code or (b) employs more than one operating principle to sense one of these phenomena. Typical examples are (a) a combination of a heat detector with a smoke detector, or (b) a combination rate-of-rise and fixed temperature heat detector.

Duct detectors. The function of air duct smoke detectors is to detect smoke for the primary purpose of controlling blowers and dampers of air conditioning and ventilating systems in an attempt to prevent possible panic and damage from distribution of smoke and gaseous products.

Fire. Phenomenon which occurs when a substance upon reaching a critical temperature reacts chemically, as for example, with oxygen, to produce heat, flame, light, smoke, water vapor, carbon monoxide, carbon dioxide, or other products and effects.

Fire and smoke-detecting system. An approved installation of equipment which automatically actuates a fire alarm when the detecting element is exposed to fire, smoke or abnormal rise in temperature.

Fire detecting system. An approved installation of equipment which automatically actuates a fire alarm when the detecting element is exposed to fire or abnormal rise in temperature.

Fire detection system. System for automatically detecting the presence of smoke, particles of combustion, fire or abnormal heat and producing an alarm signal on the premises, at the fire station, or at a central alarm office.

Fixed temperature detector. Device which will respond when its operating element becomes heated to a predetermined level.

Flame. Column of gases, made luminous by heat, emanating from a burning substance. Flame from some substances (e.g., hydrogen) may not be visible to the unaided human eye.

Flame detector. A device which detects the infrared, or ultraviolet, or visible radiation produced by a fire.

Flame detector. A device which responds to the appearance of radiant energy visible to the human eye (approximately 4000 to 7700 Angstroms) or to radiant energy outside the range of human vision.

Flame detectors. Flame detectors are sensitive to glowing embers, coals, or actual flames, which radiate to the detectors energy of sufficient intensity and spectral quality to initiate action.

Flame flicker detector. A photoelectric flame detector including means to prevent response to visible light unless the observed light is modulated at a frequency characteristic of the flicker of a flame.

Heat. Added energy that causes substances to rise in temperature. Also, the energy liberated by a burning substance.

Heat-detector. Device which detects abnormally high temperature or rate-of-temperature rise.

Infrared detector. A device whose sensing element is responsive to radiant energy outside the range of human vision (above approximately 7700 Angstroms).

Ionization detector. Detector has a small amount of radioactive material which ionizes the air in the sensing chamber, thus rendering it conductive and permitting a current flow through the air between two charged electrodes. This gives the sensing chamber an effective electrical conductance. When smoke particles enter the ionization area, they decrease the conductance of the air by attaching themselves to the ions, causing a reduction in mobility. When the conductance is less than a predetermined level, the detector circuit responds.

Line-type detector. Device in which detection is continuous along a path. Typical examples are rate-of-rise pneumatic tubing detectors, projected beam smoke detectors, and heat sensitive cable.

Non-restorable detector. Device whose sensing element is designed to be destroyed by the process of detecting a fire.

Photoelectric beam-type detector. The photoelectric beam-type detector consists of a light source which is projected across the area to be protected into a photosensing cell. Smoke between the light source and the receiving photosensing cell reduces the light reaching the cell, causing actuation.

Photoelectric flame detector. Device whose sensing element is a photocell which either changes its electrical conductivity or produces an electrical potential when exposed to radiant energy.

Photoelectric spot-type detector. Contains a chamber with either overlapping or porous covers which prevent the entrance of outside sources of light but which allow the entry of smoke. The unit contains a light source and a special photosensitive cell in the darkened chamber. The cell is either placed in the darkened area of the chamber at an angle different from the light path or has the light blocked from it by a light stop or shield placed between the light source and the cell. With the admission of smoke particles, light strikes the particles and is scattered and reflected into the photosensitive cell. This causes the photosensing circuit to respond to the presence of smoke particles in the smoke chamber.

Pneumatic rate-of-rise tubing. Line-type detector comprising small diameter tubing, usually copper, which is installed on the ceiling or high on the walls throughout the protected area. The tubing is terminated in a detector unit, containing diaphragms and associated contacts set to actuate at a predetermined pressure. The system is sealed except for calibrated vents which compensate for normal changes in temperature.

Rate compensation detector. Device which will respond when the temperature of the air surrounding the device reaches a predetermined level, regardless of the rate of temperature rise.

Rate-of-rise detector. Rate-of-rise detector is a device which will respond when the temperature rises at a rate exceeding a predetermined amount.

Restorable detector. Device whose sensing element is not ordinarily destroyed by the process of detecting a fire. Restoration may be manual or automatic.

Self-restoring detector. Restorable detector whose sensing element is designed to be returned to normal automatically.

Smoke. Totality of the airborne visible or invisible particles of combustion.

Smoke detector. A device which detects the visible or invisible particles of combustion.

Smoke detector. Device which senses visible or invisible particles of combustion.

Smoke detector. A device installed in the plenum chamber or in the main supply air duct of an air-conditioning system to automatically shut off the blower and close a fire damper in the presence of smoke.

Smoke detector. An approved detector which senses visible or invisible particles of combustion. The detector shall bear a label or other identification issued by an approved testing agency having a service for inspection of materials and workmanship at the factory during fabrication and assembly.

Spot-type detector. Device whose detecting element is concentrated at a particular location. Typical examples are bimetallic detectors, fusible alloy detectors, certain pneumatic rate-of-rise detectors, certain smoke detectors and thermoelectric detectors.

Spot-type pneumatic rate-of-rise detector. Device consisting of an air chamber, diaphragm, contacts, and compensating vent in a single enclosure.

Thermoelectric effect detector. Device whose sensing element comprises a thermocouple or thermopile unit which produces an increase in electric potential in response to an increase in temperature. This potential is monitored by associated control equipment, and an alarm is initated when the potential increases at an abnormal rate.

Ultraviolet detector. Device whose sensing element is responsive to radiant energy outside the range of human vision (below approximately 4000 Angstroms).

16-B ELECTRIC — GENERAL

Appliance. Includes but is not limited to any device, machine, piece of apparatus or tool which utilizes or transmits electrical current or energy.

Appliance. An appliance is utilization equipment. It is generally other than industrial; normally built-standardized sizes and types; installed or connected as a unit to perform one or more functions, such as clothes washing, air conditioning, food mixing, cooking, etc.

Appliance (fixed). An appliance which is fastened or otherwise secured at a specific location.

Appliance (portable). An appliance which is actually moved, or can easily be moved from one place to another in normal use.

Appliance (stationary). An appliance which is not easily moved from one place to another in normal use.

Automatic. Automatic means self-acting, operating on its own mechanism, when actuated by some impersonal influence; as for example, a change in current strength, pressure, termperature or mechanical configuration.

Auto-transformer. A transformer in which part of the winding is common to both the primary and secondary circuits.

Bonding jumper. A reliable conductor to assure the required electrical continuity.

Branch circuit. That portion of a wiring system extending beyond the final overcurrent device protecting the circuit.

Busway. An approved assembly consisting of a sheet metal or equivalent insulated enclosure having a continuous ground, enclosing and supporting fixed rigid conductors.

BX. Electrical cable consisting of a flexible metal covering enclosing two or more wires.

Cabinet. An enclosure designed either for surface or flush mounting, and provided with a frame, mat or trim in which swinging doors are hung.

Cable. Factory assembled combination of any number of conductors with their required individual insulations, all within a protective outer sheath, with or without fill, as may be required for specific use.

Circuit. Any set of branch lighting conductors which have been extended from a distribution center, and which may be utilized for the transmission of electrical energy.

Circuit breaker. A device designed to open and close a circuit by non-automatic means, and to open a circuit automatically on a predetermined overload of current, without injury to itself when properly applied within its rating.

Conduit. Rigid metal piping, flexible metal piping, fibre duct conduit or other raceway approved as a wiring enclosure.

Convenience outlet. Plug-in receptacle, as well as the outlet box in which it is housed.

Cord. Type of flexible electrical conductor used in connection with an appliance.

Dead front. Switchboard or panelboard where there is no exposed portion of any current-carrying conductor, part, device or other equipment on the front of such boards.

Dead rear. Switchboard or panelboard where there is no exposed portion of any current-carrying conductor, part, device or other equipment on the rear of such board.

Demand factor. The demand factor of any system or part of a system, is the ratio of the maximum demand of the system, or part of a system, to the total connected load of the system, or of the part of the system under consideration.

Device. A unit of an electrical system which is intended to carry but not consume electrical energy.

Distribution center. Two or more sets of overcurrent devices located in a switchboard, or a panelboard, or in two or more grouped fused switches or circuit breaker cases.

Distribution panel. Panelboard containing fuses or circuit breakers and which distributes the current to various circuits.

Distribution system. System to provide overhead and/or underground wiring services (electrical or communications) to individual lots. This definition applies to the system as a whole, or any part thereof (cables, conduits or wires) used as feeders, primaries, secondaries or similarly designated conductor systems forming a part of such distribution system and not including transmission systems.

Electrical conduit. Metal pipes in which electrical wires or conductors are installed.

Electrical construction. Includes all work and materials used in the installing, maintaining and/or extending a system of electrical wiring for light, power or heat and all appurtenances, apparatus, or equipment used in connection therewith, inside or attached to any building or structure, lot or premises.

Electrical contractor. Any person engaged in the business of installing or altering by contract, electrical equipment for the utilization of electricity supplied for light, heat, or power, not including radio apparatus or equipment for wireless reception of sound and signals, conductors, and other equipment installed for or by public utilities including common carriers which are under state jurisdiction for use in their operation as public utilities. The term does not include employees employed by such contractor to do or supervise such work.

Electrical contractor. Any individual, partnership or corporation or association or other organizations who or which contacts for, proceeds with or employs others for the construction, installation, alteration, repair or addition to any electrical wiring, apparatus, or fixed electrical equipment for the furnishing of light, power, or heat in a building or structure within the limits of the City.

Electrical equipment. Includes but is not limited to conductors and apparatus installed for the utilization of electricity for light, heat, power, aural or visual communication or signal transmission or convenience.

Electrical equipment. All installations of electrical conductors, fittings, devices and fixtures within or on public and private buildings.

Electrically supervised. As applied to a control circuit, shall mean that in the event of interruption of the current supply or in the event of a break in the circuit, a specific signal will be given.

Electrical regulations. Includes but is not limited to the installation of all electrical wiring and electrical equipment provided for in the Code, which shall be done in strict accordance with the electrical regulations of the Code.

Electrical service equipment. The equipment located at point of entrance of supply conductors to a building which constitutes the main control of supply and means of cut-off of electricity, including circuit breaker, switches, fuses and electrical accessories, all in accordance with the Code.

Electrical service equipment. The equipment located at point of entrance of supply conductors to a building which constitutes the main control of supply and means of cut-off of electricity, including circuit breaker, switches, fuses and electrical accessories.

Electrical service equipment. The electrical equipment, located near the point of entrance of electric supply conductors to a building, which constitutes the main control of supply and means of cutoff of electricity for the building and includes switches, fuses, and electrical accessories and other appurtenances.

Electrical wiring. Includes wiring methods, electrical conductors, conduits, insulators, junction boxes, switches, outlets, gas tubes, and other devices used for the functioning and safety of an electrical installation.

Electric equipment standards. The materials, fittings, appliances, devices and other equipment listed in publications of inspected electrical equipment of the Underwriters' Laboratories, Inc., and other accredited authoritative agencies and testing organizations, and installed in accordance with the recommendations of the written approval of those authorities shall be accepted as meeting the requirements of the Code.

Electrician. Person who is engaged in the trade or business of electrical construction and who is qualified and has been registered under the terms of the code.

Electric installation standards. Conformity of installations of electric equipment to the applicable standards of the City Electrical Code, National Electrical Safety Code and other accepted Engineering standards listed in the Code, shall be prima facie evidence that such installations are reasonably safe for use in the service intended and in compliance with the provisions of the Code.

Feeder. Any conductors of a wiring system between the service equipment, or the generator switchboard of an isolated plant, and the branch circuit over-current device except those conductors which supply all the current consumed in an individual apartment or living unit in an apartment building or residential type occupancy.

Fished. A manner of installing concealed electric wiring in existing inaccessible hollow spaces of buildings with a minimum damage to the building finish.

Fitting. An accessory such as a locknut, bushing or other part of a wiring system which is intended primarily to perform a mechanical rather than an electrical function.

Flexible tubing. Conduit other than that formed by a continuous one-piece metal tube.

Full-connected Load. The total load or total current required for operation at maximum capacity of all appliances, motors, lighting, or other equipment.

Fuse. An overcurrent protective device with a circuit-opening fusible member directly heated and destroyed by a passage of a predetermined overcurrent through it.

General-use switch. A switch intended for use as a switch in general distribution and branch circuits. It is rated in amperes and is capable of interrupting its rated current at its rated voltage.

Ground. A conducting connection, whether intentional or accidental between an electrical circuit or equipment and earth or some conducting body which serves in place of the earth.

Grounded. Connected to earth or to some conducting body which serves in place of the earth.

Grounded conductor. A conductor intentionally grounded in any manner.

Grounding conductor. That conductor used to connect equipment, devices or wiring systems to the required ground.

Guarded. Covered, shielded, fenced, enclosed or otherwise protected, by means of suitable covers or casings, barriers, rails or screen, mats or platforms, to remove the liability of dangerous contact or approach by persons or objects to a point of danger.

Horsepower. Term as used in the code shall apply to electric motors and shall be determined by tables listed in the National Electrical Code NFPA 70.

Insulated. Covered and protected in an approved manner with suitable materials of the type and thickness adopted as standard requirements for the voltage and location of the particular material, conductor or part involved.

Interrupting capacity. The maximum current at its rated voltage that a device will open without injury to any parts other than replaceable fuses. The interrupting capacity shall be determined from the recognized standards for testing equipment.

Isolated. Not readily accessible to persons unless special means for access are used.

Isolated plant. A private electrical installation deriving energy from its own generator driven by a prime mover.

Isolating switch. A switch intended for isolating an electric circuit from the source of power. It has no interrupting rating and is intended to be operated only after the circuit has been opened by some other means.

Low-energy power circuit. A circuit which is not a remote control or signal circuit, but which has the power supply limited in accordance with the requirements of Class 2 remote control circuits.

Low-voltage. Any voltage not exceeding 600 volts.

Mains. Those conductors of a distribution system of a serving agency or isolated plant which are designed, intended, or used to supply electric energy to two or more separate and independent premises.

Manually operable. Designed and intended for operation by the hand directly to a handle, lever, push button or other suitable contrivance that is an integral part of the equipment.

Master electrican. Individual who possesses the necessary qualifications, training, experience, and technical knowledge to plan, layout, and supervise the installation of electrical wiring, apparatus or equipment for light, power or heat, and who is qualified to be licensed under the provisions of the code.

Master service. The service conductors and service equipment supplying a group of buildings under one management.

Meter. Device installed by the utility company for measuring the quantity of electric current used by the consumer.

Motor-circuit switch. A switch, rated in horsepower, capable of interrupting the maximum operating overload current of a motor of the same horsepower as the switch at the rated voltage.

Multi-outlet assembly. A type of surface raceway, designed to hold conductors and receptacles, assembled in the field or at the factory.

Neutral. Unfused current-carrying conductor, common to ground, designed and intended to carry the difference in loading of the associated conductors.

Open wiring. Conductors not in a wiring enclosure.

Outlet. A point on the wiring system at which current is taken to supply fixtures, lamps, heaters, motors and current consuming equipment generally.

Outlet. Electrical service connection by the way of a wired prong device.

Outside wiring. Electric wiring located outside of buildings, not including wiring for signs or any extensions of circuits supplying any load within the building.

Panelboard. A single panel, or a group, of panel units designed for assembly in the form of a single panel; including buses and with or without switches and/or automatic overcurrent protective devices for the control of light, heat, or power circuits of small individual as well as aggregate capacity; designed to be placed in a cabinet or cutout box placed in or against a wall or partition and accessible only from the front.

Portable appliance. An appliance capable of being readily moved where established practice or the conditions of use make it necessary or convenient to be moved while in use for it to be detached from its source of current by means of flexible cord and attachment plug.

Raceway. Any channel for holding wires, cables or bus-bars, which is designed expressly for, and used solely for, this purpose.

Raintight. So constructed or protected that exposure to a beating rain will not result in the entrance of water.

Readily accessible. Capable of being reached quickly for operation, renewal, or inspection, without requiring those to whom ready access is requisite to climb over or remove obstacles or to resort to portable ladders, chairs, etc.

Receptacle. A contact device designed, intended, and used solely for the transmission of electric energy from fixed circuit conductors to an attachment plug.

Receptacle outlet. An outlet equipped with one or more receptacles, not of the screw-shell type, or provided with one or more points of attachment within one foot or less, intended to receive attachment plug caps.

Remote-control circuit. Any electrical circuit which controls any other circuit through a relay or an equivalent device.

Rigid conduit. Rigid pipe used as a raceway and protective cover for electrical wiring.

Sealable equipment. Equipment enclosed in a case or cabinet that is provided with means for sealing or locking so that live parts cannot be made accessible without approval.

Service. The conductors and equipment for delivering energy from the electricity supply system to the wiring system of the premises served.

Service drop. That portion of overhead service conductors between the pole and the first point of attachment to the building or other structure.

Service drop. Overhead service conductors from the last pole or other aerial support to and including the splices, if any, connecting to the service-entrance conductors at the building or other structure.

Service entrance. System of service conductors, bringing electricity into the building, and service equipment, which controls and distributes it where needed in the building.

Service equipment. The necessary equipment, usually consisting of circuit-breaker or switch and fuses, and their accessories, located near point of entrance of supply conductors to a building and intended to constitute the main control and means of cutoff for the supply to that building.

Service feeder. Set of entrance conductors from the utility company's network to main building switches or switchboard.

Thermal cutout. An overcurrent protective device which contains a heater element in addition to and affecting a renewable fuse member which opens the circuit. It is not designed to open short circuits.

Underground service supply. The electric wiring installed partially or wholly underground between the mains of the serving agency from the connection at the vault, manhole or pole and a terminating point, such as a terminating pull box, a switchboard pull section or other terminating enclosure.

Underground service supply conductors. The conductors of the underground service supply.

Use. Assumed to be in use of.

Utilize. (As applied to electricity). To employ electricity for light, heat or power, other than for transmission purposes.

Vaportight. So enclosed as to resist the passage of vapor, as by the use of a gasket.

Ventilated. Provided with a means to permit circulation of the air sufficiently to remove an excess of heat, fumes or vapors.

Voltage (of a circuit). The greatest effective difference of potential between any two conductors of the circuit concerned. (On various systems such as three-phase 4-wire, single-phase 3-wire and 3-wire direct current there may be various circuits of various voltages.)

Voltage to ground. For grounded circuits, the voltage between the given conductor and that point or conductor of the circuit that is grounded; for ungrounded circuits, the greatest voltage between the given conductor and any other conductor of the circuit.

Voltage to ground. The voltage between any given live ungrounded part and any grounded part in the case of grounded circuits, or the greatest voltage existing in the circuit in the case of ungrounded circuits.

Watertight. An enclosure which moisture cannot enter.

Weatherproof. So constructed or protected that exposure to the weather will not interfere with its successful operation.

Wiring enclosure. Any raceway, cabinet, box, fitting, enclosed switchboard compartment or other case approved for the protection of electric wiring and permitted by the Code.

Wiring method. Type of wiring and equipment, and details of installation required or used in an electrical installation.

16-C LIGHTING — ILLUMINATION

Aeronautical beacon. Aeronautical ground light visible at all azimuths, either continuously or intermittently, to designate a particular point on the surface of the earth.

Aeronautical light. Luminous sign or signal, recognized by competent authority, which is established, maintained, exhibited, or operated as an aid to air navigation.

Approach lights. Configuration of aeronautical ground lights located in extension of a runway or channel to provide visual approach and landing guidance to pilots.

Commercial electric lamp posts. Posts which shall deemed to be street lamp posts on public property, not owned, operated, or maintained by the city.

Cornice lighting. Light sources shielded by a panel parallel to the wall and attached to the ceiling, and distributing light over the wall.

Course light. Aeronautical ground light, supplementing an airway beacon, used to indicate the direction of the airway and to identify by a coded signal the location of the airway beacon with which it is associated.

Cove lighting. Light sources shielded by a ledge or horizontal recess, and distributing light over the ceiling and upper wall.

Decorative lighting. Means superfluous light, not used as part of an advertising display, intended to increase the attractiveness or other incidental use.

Decorative street lighting equipment. Shall be deemed to be lamps attached to wires or structures which extend over any public property and shall not apply to lamps attached to commercial electric lamp posts, or to any sign, canopy, or structure authorized by the Code.

Diffused. Illumination by means of light which travels through a material other than the bulb or tubing necessary to enclose the light source so that the light is spread evenly over the surface of the diffusing material.

Direct illumination. Illumination by means of light which travels directly from its source to the viewer's eye.

Emergency lighting. Lighting designed to supply illumination essential to safety of life and property in the event of failure of the normal supply.

Emergency lighting system. Independent lighting system supplementing the general lighting system.

Floodlight. Projector designed for lighting a scene or object to a brightness considerably greater than its surroundings. It usually is capable of being pointed in any direction and is of weather proof construction.

General lighting. Lighting designed to provide a substantially uniform level of illumination throughout an area, exclusive of any provision for special local requirements.

Indirect illumination. Illumination by means of light cast upon an opaque surface from a concealed source.

Lampholder. A device intended to support an electric lamp mechanically and connect it electrically to circuit wires.

Light. Exterior lighting shall be so installed that the surface of the source of light shall not be visible from any bedroom window, and shall be so arranged as far as practical to reflect light away from any residential use.

Lighting. Electric illumination shall be maintained in accordance with the minimum values shown in the Code. The illumination shall be measured on a plane 30 inches above the floor.

Lighting fixture. An assembly having one or more lampholders therein or thereon or a lampholder used in lieu of such assembly, but shall not include strip, trough, border, sign, outline, festoons, or footlights or similar lighting equipment for which specific provisions are made in the code.

Lighting fixture (portable). A fixture placed upon a horizontal surface, but not on a wall or ceiling.

Lighting outlet. An outlet intended for the direct connection of a lampholder, a lighting fixture, a pendant cord terminating in a lampholder, or for the supply of a fluorescent fixture suspended directly below the outlet.

Lighting outlet. An outlet intended for the direct connection of a lampholder, a lighting fixture or a pendant cord terminating in a lampholder.

Lighting standard or lamp post. A support provided with necessary internal attachments for wiring and external attachments for bracket and luminaire.

Local lighting. The lighting used to provide illumination over a relatively small area or confined space without providing any significant general surrounding lighting.

Louver. A series of baffles used to shield a source from view at certain angles or to absorb unwanted light.

Louvered ceiling. A ceiling area lighting system comprising a wall-to-wall installation of multicell louvers shielding the light sources mounted above it.

Luminous ceiling. A ceiling area lighting system comprising a continuous surface of transmitting material of a diffusing or light controlling character with light sources mounted above it.

Luminous ceiling. Light-transmitting panels suspended below light sources and supported from the construction above.

Outline lighting. An arrangement of incandescent lamps or gaseous tubes to outline and call attention to certain features such as the shape of a building or the decoration of a window.

Pole. A standard support generally used where overhead lighting distribution circuits are employed.

Portable lighting. The lighting equipment designed for manual portability.

Range lights. Groups of color-coded boundary lights provided to indicate the direction and limits of a preferred landing path normally on an airport without runways but exceptionally on an airport with runways.

Runway lights. Aeronautical ground lights arranged along a runway indicating its direction or boundaries.

Searchlight. A projector which produces an approximately parallel beam and which can be aimed by controlled rotation about two mutually perpendicular axes.

Street lighting luminaire. A complete lighting device consisting of a light source together with its direct appurtenances such as globe, reflector, refractor, housing, and such support as is integral with the housing. The pole, post, or bracket is not considered a part of the luminaire.

Street lighting unit. The assembly of pole or post with bracket and luminaire.

Supplementary lighting. The lighting used to provide an additional quantity and quality of illumination which cannot readily be obtained by a general lighting system and which supplements the general lighting level, usually for specific work requirements.

Troffer. A long recessed lighting unit usually installed with the opening flush with the ceiling. The term is derived from "trough" and "coffer."

16-D SIGNALING AND WARNING SYSTEMS

Alarm. A signal indicating an emergency requiring immediate action to safeguard life or property from the hazards of fire, explosion or panic.

Alarm Service. The service required following the receipt of an alarm signal. In auxiliary systems the alarm service is primarily provided by the municipal fire department to which the transmission circuits are connected.

Alarm Service. The service required following the manual operation of a fire alarm box, the transmission of an alarm indicating the operation of protective equipment or systems, such as an alarm from waterflow in a sprinkler system, the discharge of carbon dioxide, the detection of smoke, the detection of excessive heat, or the transmission of an alarm from other protective systems.

Alarm signal. A signal indicating an emergency requiring immediate action, as an alarm for fire from a manual box, a waterflow alarm, or an alarm from an automatic fire alarm system.

Alarm signal. A signal indicating an emergency requiring immediate action, as an alarm for fire from a manual box, a waterflow alarm, an alarm for an automatic fire alarm system, or other emergency signals.

Annunciator. A unit containing two or more identified targets or indicator lamps in which each target or lamp indicates the circuit, condition or location to be annunciated.

Automatic. Providing a function without the necessity of human intervention.

Automatic fire alarm system. System which automatically detects a fire condition and actuates a fire alarm signal device.

Automatic smoke alarm service. Consists of equipment designed and installed, whereby abnormal smoke density in a specific area results in the transmission of a distinctive smoke alarm signal to a central supervising station.

Auxiliarized local system. A local system (NFPA 72A) that is connected to the municipal alarm facilities.

Auxiliarized proprietary system. A proprietary system (NFPA 72D) that is connected to the municipal alarm facilities.

Auxiliary alarm system. A connection to the municipal fire alarm system to transmit an alarm of fire to the fire department. Fire alarms from an auxiliary alarm system are received at municipal fire alarm headquarters on the same equipment and by the same alerting methods as alarms transmitted from municipal fire alarm boxes located on streets.

Auxiliary system. A fire alarm signal system wherein a "local system," is connected to the community fire department system through the community's fire department alarm circuits.

Central station system. Automatic sprinkler or fire alarm system in which all equipment is supervised by a central or proprietary station to which all alarm signals are transmitted and relayed to the fire department.

Class A fire alarm system. A closed circuit electrically supervised fire alarm system. System may be coded or selective code ringing type, a noncoded type with or without annunciators (automatic detection type), or combination fire and sprinkler alarm.

Class B fire alarm system. A fire alarm system which may consist of one or more bells, sirens, or other sounding devices electrically or mechanically operated so that all sounding devices will sound when any fire alarm station is operated.

Closed circuit. Fire alarm system so wired as to permit a continuous flow of electric current.

Closed circuit system. System in which all circuits between alarm sending stations, sounding devices, and annunciators are maintained as normally closed electrical circuits.

Coded closed circuit. Circuit used for an interior fire alarm signal system consisting of sending stations and signaling devices operated on supervised closed electrical circuits where rounds of automatically sounded coded signals are transmitted to indicate the floor or portion of the same from which the alarm was sent.

Coded or selective code ringing. Fire alarm system where each sending station, when operated, will cause sounding devices to signal a predetermined number of strokes, which is indicative of the location of the sending station.

Combination system. A local protective signaling system for fire alarm, supervisory or watchman service whose components may be used in whole or in part in common with a nonfire-emergency signaling system, such as a paging system, a musical program system, or a process monitoring service system, without degradation of or hazard to the protective signaling system.

Direct circuit auxiliary alarm system. An auxiliary alarm system connected by a municipally controlled individual circuit to the protected property, to interconnect the actuating devices and the municipal fire alarm switchboard.

Electrically supervised. A system so designed, installed and maintained that upon break or ground fault of its circuits, or a failure of its main operating current supply source, a trouble signal shall sound or the signal sounding devices of the system shall be activated.

Fire alarm. A system, automatic or manual, arranged to give a signal indicating a fire emergency.

Fire alarm—coded or selective code-ringing. A fire alarm system where each sending station, when operated, will cause sounding devices to signal a predetermined number of strokes, which indicates the location of the sending station.

Fire alarm—noncode system. A fire alarm system in which all sending stations when operated will cause sounding devices to sound the same signal.

Fire alarm system. Signaling system, usually electrically operated and actuated either automatically or manually for producing a signal at one or more locations which will warn building occupants of a fire.

Fire alarm system. An approved installation of equipment for sounding a fire alarm.

Interior fire alarm system. System installed within a building and provided and intended to warn the occupants of the building in the event of a fire by means of an audible signal transmitted from manual trip stations, signal stations, or boxes, or automatic fire detecting devices.

Local alarm system. A local system sounding an alarm as the result of the manual operation of a fire alarm box or of the operation of protection equipment or systems, such as water flowing in a sprinkler system, the discharge of carbon dioxide, the detection of smoke or the detection of heat.

Local energy auxiliary alarm system. An auxiliary alarm system which employs a locally complete arrangement of parts, initiating devices, relays, power supply, and associated components, to automatically trip a municipal transmitter or master box over electric circuits which are electrically isolated from the municipal system circuits.

Local supervisory system. A local system arranged to supervise the performance of watch patrols, or the operative condition of automatic sprinkler systems or of other systems for the protection of life and property against the fire hazard.

Local system. A fire alarm signal system in which the operation and alarm signals are on the protected premises and which is designed to warn the occupants of the premises.

Local system. A local system is one which produces a signal at the premises protected.

Manual fire alarm service. A manually actuated fire alarm service.

Manual fire alarm system. An interior alarm system composed of sending stations and signaling devices in a building, operated on an electric circuit, so arranged that the operation of any one station will ring all signals throughout the building or at one or more approved locations. Signals may be either non-coded or coded to indicate the floor area in which the signal originated and may be transmitted to an outside central station.

Master box. A municipal fire alarm box that may also be operated by remote means.

Mechanical fire alarm service. A manually actuated and operated mechanical device for sounding an alarm, and consists primarily of a vertical rod, tripping mechanism, spring, clapper, and bell, and wherein the alarm does not have continuity after actuation.

Municipal fire alarm box. A specially manufactured enclosure housing a transmitting device that can only be operated manually.

Non-coded closed circuit. Interior fire alarm signal system, consisting of sending stations and signaling devices operated on supervised closed electrical circuits so arranged that operation of any station will automatically sound the sounding devices throughout all portions of the building.

Non-coded fire alarm system. All sending stations when operated will cause sounding devices to sound the same signal.

Open circuit interior fire alarm system. System in which all circuits are maintained as electric circuits normally open at the alarm sending stations and which is operated by closing the circuit at any alarm sending station.

Presignal fire alarm system. Coded alarm shall originally be sounded only at certain designated location. Provision is made for sending a coded alarm on all sounding devices from the same sending station.

Pre-signal system. A coded closed circuit interior fire alarm signal system so arranged that the operation of any station will cause the sounding of the sounding devices located at a main telephone switchboard, office, custodian's room, chief engineer's office, or other place or places within a building, to notify a responsible person or persons that an alarm box has been operated and the location of such alarm box. Alarm stations of a pre-signal system shall be so equipped that operation of any alarm sending station by an authorized person in a special manner or by use of a special key shall cause the sounding of all sounding devices throughout the building.

Process monitoring alarm system. An alarm system used to supervise the functioning of a commercial process, such as manufacturing operations, heating or refrigerating systems temperature control, etc., when failure of the supervised process could result in fire or explosion endangering life or property.

Proprietary system. A fire alarm signal system supervised by competent and experienced personnel in a central supervising station at the property protected, and in which the personnel take such action as is required under rules established for their guidance.

Protective signaling systems. Electrically operated circuits, instruments, and devices, together with the necessary electrical energy, designed to transmit alarms and supervisory and trouble signals necessary for the protection of life and property. In a compressed or liquefied gas-type system, pressure-operated instruments and devices, together with the necessary compressed gas energy to accomplish the same purpose, are used.

Protective systems, equipment or apparatus. Automatic sprinklers, standpipes, carbon-dioxide systems, automatic covers, and other devices used for extinguishing fires and for controlling temperatures or other conditions dangerous to life or property.

Remote station system. A fire alarm system employing a direct connection between alarm signal initiating devices in protected premises and signal-indicating equipment in a remote station, such as fire or police headquarters, or a conforming central station.

Shunt auxiliary alarm system. An auxiliary alarm system electrically connected to an integral part of the municipal alarm system extending the municipal circuit into the protected property to interconnect the actuating devices, which, when operated, open the municipal circuit shunted around the trip coil of the municipal transmitter or master box, which is thereupon energized to start transmission, without any assistance whatsoever from a local source of energy.

Signaling equipment. Equipment intended to give an audible signal such as bells, horns, chimes, buzzers and the like.

Stations. Devices for initiating the signal. Stations are either manually operable or automatically operable by temperature changes or rate of rise of temperature.

Supervised fire alarm system. That accidental interruption of current flow will be signaled as prescribed by the regulations.

Supervisory service. The service required to assure performance of watch patrols and the operative condition of automatic sprinkler systems and of other systems for the protection of life and property.

Supervisory service. An agreed-upon service rendered upon receipt of a signal or non-receipt of a scheduled signal from devices or equipment being supervised, at a central station remote from the building, devices, or equipment being supervised.

Supervisory signal. A signal indicating the need of action in connection with the supervision of watchmen, sprinkler and other extinguishing systems or equipment, or with the maintenance features of other protective systems.

Trouble signal. A signal indicating trouble of any nature in a fire warning system such as a circuit break or ground occurring in the devices or wiring associated with a protective signaling system.

Trouble signal. A signal indicating trouble of any nature, such as a circuit break or ground, occurring in the devices or wiring associated with a protective signaling system. In a compressed or liquefied gas-type system, the trouble signal indicates trouble of any nature, such as a break or a leak in the tanks, devices, or tubing of the system.

Voice communication system. Electrically supervised communicating system provides the following: Functions, a two-way emergency communication system and a one-way address communication system.

Watchman's tour supervisory service. Consists of one or more watchman with specific routes having reporting and signaling stations and facilities for keeping a permanent record at a supervising location, of each time a signal transmitting station is operated.

APPENDIX

DEPARTMENT OF COMMERCE

HERBERT HOOVER, SECRETARY

A STANDARD STATE ZONING ENABLING ACT

UNDER WHICH MUNICIPALITIES MAY ADOPT ZONING REGULATIONS

BY

THE ADVISORY COMMITTEE ON ZONING

APPOINTED BY SECRETARY HOOVER

EDWARD M. BASSETT . . . . Counsel, Zoning Committee of New York.
Lawyer.

IRVING B. HIETT . . . . Ex-President, National Association of Real Estate Boards.
Realtor.

JOHN IHLDER . . . . Manager, Civic Development Department of the Chamber of Commerce of the United States.
Housing Consultant.

MORRIS KNOWLES . . . . From the Chamber of Commerce of the United States.
Consulting Engineer.

NELSON P. LEWIS . . . . From the National Conference on City Planning and National Municipal League; Past President, American City Planning Institute.
Municipal Engineer.

J. HORACE McFARLAND . . . . President, The American Civic Association.
Master Printer and Civic Investigator.

FREDERICK LAW OLMSTED . . . . . President, The American Society of Landscape Architects; Past President, American City Planning Institute.
Landscape Architect.

LAWRENCE VEILLER . . . . Secretary and Director, The National Housing Association.
Housing Expert.

JOHN M. GRIES

Chief, Division of Building and Housing, Bureau of Standards, Department of Commerce

PRICE 5 CENTS

SOLD ONLY BY THE SUPERINTENDENT OF DOCUMENTS
GOVERNMENT PRINTING OFFICE, WASHINGTON, D. C.

WASHINGTON
GOVERNMENT PRINTING OFFICE
1924

PREFACE.

A standard State zoning enabling act, under which municipalities may adopt zoning regulations, was first issued in mimeographed form in August, 1922. A revised edition was made public in the same form in January, 1923, and since then a few minor changes have been made. The act is now printed in order to make it available to a larger number of people.

The circulation of the standard act in its preliminary form was not confined to those directly interested in the drafting of State zoning legislation. Calls for it were received from persons in all sections of the country who desired to use it on account of its general bearing on the legal and social aspects of zoning.

CONTENTS.

78055°—24

FOREWORD.

By Herbert Hoover.

The importance of this standard State zoning enabling act can not well be overemphasized. When the advisory committee on zoning was formed in the Department of Commerce, the proposal to frame it received unanimous support from the public-spirited organizations represented on the committee, and other groups interested in zoning. The urgency of the need for such a standard act was at once demonstrated, when, within a year of its issuance, 11 States passed zoning enabling acts which were modeled either wholly or partly after it. Similar acts have been introduced in 4 other States, with the prospect of more to follow.

The discovery that it is practical by city zoning to carry out reasonable neighborly agreements as to the use of land has made an almost instant appeal to the American people. When the advisory committee on zoning was formed in the Department of Commerce in September, 1921, only 48 cities and towns, with less than 11,000,000 inhabitants, had adopted zoning ordinances. By the end of 1923, a little more than two years later, zoning was in effect in 218 municipalities, with more than 22,000,000 inhabitants, and new ones are being added to the list each month.

In this rapid movement the fundamental legal basis on which zoning rests can not be overlooked. Several of our States, fortunately, already have zoning enabling acts that have stood the test in their own courts. This standard act endeavors to provide, so far as it is practicable to foresee, that proper zoning can be undertaken under it without injustice and without violating property rights. The committee did not make it public until it had given it the most exacting and painstaking study in relation to existing State acts and court decisions and with reference to zoning as it has been practiced and found successful in cities and towns throughout the country. Practical zoners who have been associated with a majority of zoned cities were consulted for their opinions, and the committee itself represents the professional, commercial, and civic societies most interested in zoning problems.

The drafting of the act has required very large effort, and the members of the advisory committee on zoning, particularly those who served on the subcommittee on standard law, merit the gratitude of the people of the United States for the thoroughness with which they executed their task.

February 15, 1924.

A STANDARD STATE ZONING ENABLING ACT UNDER WHICH MUNICIPALITIES MAY ADOPT ZONING REGULATIONS.

EXPLANATORY NOTES IN GENERAL.

1. *An enabling act is advisable in all cases.*—There is a tendency in some States to assume that the powers contained in home rule charters are sufficient to enable a municipality to undertake zoning. This is often a mistaken belief, and some zoning ordinances have been set aside because the municipality had not been granted the specific power to do that which zoning implies.

2. *Constitutional amendments not required.*—No amendment to the State constitution, as a rule, is necessary. Zoning is undertaken under the police power and is well within the powers granted to the legislature by the constitutions of the various States.

3. *Modify this standard act as little as possible.*—It has been prepared with a full knowledge of the decisions of the courts in every case in which zoning acts have been under review. A safe course to follow is to make only those changes necessary to have the act conform to local legislative customs and modes of expression.

4. *Adding new words and phrases.*—Especial caution is given to beware of adding additional words and phrases which, as a rule, restrict the meaning, from the legal point of view.

5. *Do not try to consolidate sections.*—It is natural to try to shorten the act by consolidating sections. This may defeat one of the purposes of the act, namely, of keeping the language of the statute as simple and concise as possible. It is much better to have an act broken up into a number of sections, provided they are properly drawn, than to have one or two, or a few long involved sections. While it is recognized that some of the sections in the standard act could be combined, it is put purposely in its present form.

6. *Title and enacting clause necessary.*—No title of the act and no enacting clause have been included. These are purposely omitted, as the custom varies in almost every State. The act should, of course, be preceded by the appropriate title and enacting clause in accordance with the local legislative custom.

7. *Definitions.*—No definitions are included. The terms used in the act are so commonly understood that definitions are unnecessary. Definitions are generally a source of danger. They give to words a restricted meaning. No difficulty will be found with the operation of the act because of the absence of such definitions.

8. *Validity of one section affecting other sections.*—Some States have included in the enabling act a declaration to the effect that the finding void or unconstitutional by the courts of one section or provision shall not affect the rest of the act. This is so well accepted a principle of legal interpretation that it seems unnecessary to include it in the act. If any State desires to have it included, it can be added without danger.

9. *No declaration that Act is not retroactive.*—Some laws contain a provision to the effect that "the powers by this Act conferred shall not be exercised so as to deprive the owner of any existing property of its use or maintenance for the purpose to which it is then lawfully devoted." While the almost universal practice is to make zoning ordinances nonretroactive, it is recognized that there may arise local conditions of a peculiar character that make it necessary and desirable to deal with some isolated case by means of a retroactive provision affecting that case only. For this reason it does not seem wise to debar the local legislative body from dealing with such a situation.

10. *The repeal clause.*—No repeal clause has been included in the act for the reason that the method of phrasing such a clause will vary in nearly every State. The local legislative custom as to repeal clauses should be followed.

11. *Date of taking effect.*—For similar reasons the act does not include any provision as to the date on which it will take effect. Here also the local legislative custom should be followed.

12. *Typical ordinances or local regulations.*—The department is now making a careful study of a group of typical zoning ordinances. It is expected that a supplementary publication on this subject will be forthcoming.

13. *Interim ordinances.*—After the local legislative authorities have the power to zone, they are nearly always pressed to bring immediate protection to certain threatened localities. Sometimes the authorities frame an ordinance to cover a few blocks, or only a part of the city; this is called piecemeal zoning. Its adoption is inadvisable and may lead to much litigation. Interim zoning, although undesirable, is not as objectionable as piecemeal zoning. Interim zoning, at least, has the advantage of applying to the whole city. For instance, an ordinance providing that, wherever three-fourths of the houses in a block are residential, then no new business structure or factory can be built in that block, is an illustration of interim zoning. The reason it is objectionable is because it is too general, not sufficiently adapted to the particular need of each street, and therefore likely to be arbitrary in many cases. In such case, if a new house is built or an old one destroyed, the legal protection of the district may be altered. In this sense the district is a "travel-

ing zone." As such, a district has no stability, and as the police power may be differently applied according to the acts of property owners, it is not looked upon with favor by the courts. To prevent this, the words " at the time of the passage of this ordinance " should be inserted. If it is deemed necessary to prohibit a nonconforming building because of the consents or protests of the property owners, the ordinance should always be phrased so as to prohibit the nonconforming use, *unless* the desired majority files written consents with the officials. In other words, a provision which conditions the permission to have a nonconforming use upon the consents of a majority of the property owners is void. If at all possible, the first zoning ordinance should be comprehensive.

A STANDARD STATE ZONING ENABLING ACT.

SECTION 1. GRANT OF POWER.—For the purpose of promoting health,[1] safety, morals, or[2] the general welfare[3] of the community, the legislative body[4] of cities and incorporated villages[5] is hereby empowered to regulate and restrict[6] the height, number of stories,[7]

---

[1] "*health*": It is to be noted that the word used is "health." not "public health," for the latter narrows the application. There are some things that relate to the health only of the people living in a given dwelling, such, for instance, as the size of yards, and have only a remote relation to public health. If the term "public health" were used, the act might be set aside in a given case where it would be possible to show that the particular provision in which legal action was being taken did not concern itself with the public health but only with health.

[2] "*or*": It should be noted that the word used is "or" and not the word "and." If the latter word were used, then it might be necessary to show to the satisfaction of the court that *all* four of the purposes mentioned were involved in a given case, viz, health, safety, morals, and general welfare. The use of the word "or" limits the application to any one of the four instead of to all of them.

[3] "*general welfare*": The main pillars on which the police power rests are these four, viz, health, safety, morals, and general welfare. It is wise, therefore, to limit the purposes of this enactment to these four. There may be danger in adding others, as "prosperity," "comfort," "convenience," "order," "growth of the city," etc., and nothing is to be gained thereby.

[4] "*legislative body*": This term is sufficiently understood to include all forms of government, including commission and city manager, as well as the older forms of government. Whatever form of government exists, there must be some local body performing legislative functions.

[5] "*cities and incorporated villages*": This phrase includes those municipalities which ordinarily will find it advantageous to be given zoning powers. In some States, where different forms of governmental provisions exist, it will be necessary to add those municipalities to the term "cities and incorporated villages;" in other States the word "town" or "borough" will probably need to be added. The term "cities and incorporated villages," however, will cover the normal situation.

[6] "*regulate and restrict*": This phrase is considered sufficiently all-embracing. Nothing will be gained by adding such terms as "exclude," "segregate," "limit," "determine."

[7] "*number of stories*": It is thought wise to add this to the term "height," as courts may construe this expression narrowly, as limited to a given number of feet only, and may hold that this does not give the power to limit the number of stories, provided the building in question came within the limitation of the number of feet imposed by the ordinance. It is obvious that the power to restrict the number of stories should be granted.

and size of buildings[8] and other structures,[9] the percentage of lot[10] that may be occupied, the size of yards, courts, and other open spaces,[11] the density of population,[12] and the location and use[13] of buildings, structures, and land for trade, industry, residence, or other purposes.[14, 15]

SEC. 2. DISTRICTS.—For any or all of said purposes the local legislative body may divide the municipality[16] into districts of such number, shape,[17] and area as may be deemed best suited to carry out the purposes of this Act; and within such districts it may regulate and restrict the erection, construction, reconstruction, alteration, repair,

---

[8] "*size of buildings*": The term "size" is a better expression to use than "bulk" or "area," for the reason that both "bulk" and "area" imply, to some extent, a regularity of outline that may not be involved in all cases, whereas "size" is sufficiently all-inclusive to cover all contingencies.

[9] "*other structures*": This phrase would include other structures which possibly might not be defined as "buildings," such as open sheds, billboards, fences, spite fences, etc., none of which can be strictly considered as "buildings," as commonly understood.

[10] "*percentage of lot*": This is a better method of expression than granting the power to limit "the area of the building," as has been done in some laws, for the latter expression does not imply a variation of the fraction of the lot built upon.

[11] "*other open spaces*": This is a catch-all expression and is necessary in view of the fact that "yards" and "courts" are not defined in the Act.

[12] "*density of population*": The power to regulate density of population is comparatively new in zoning practice. It is, however, highly desirable. Many different methods may be employed. For this reason the phrase "density of population" is a better phrase to use than one giving the power to "limit the number of people to the acre," as this is only *one* method of limiting density of population. It may be more desirable to limit the number of families to the acre or the number of families to a given house, etc. The expression "number of people to the acre" is therefore more limited in its meaning and describes only one way of reducing congestion of population, while the phrase "limiting density of population" is all-embracing. It is believed that, with proper restrictions, this provision will make possible the creation of one-family residence districts.

[13] "*use*": This term is broad enough to include all meanings desired.

[14] "*other purposes*": This is a catch-all phrase. It will include every use.

[15] Although the power to require open spaces allows the fixing of setback building lines, some recent acts contain a specific grant of that power. The establishment of setback lines is somewhat novel in zoning practice but is beginning to be employed. As it is in the minds of some people of doubtful legality and has not as yet been sustained by the courts, this power has not been included here. If it should be desired to grant such power, it can readily be done by adding at the end of this section the following words: "and may also establish setback building lines."

[16] "*municipality*": This term is sufficiently broad to include cities, towns, villages, boroughs, or whatever governmental unit may be involved.

[17] "*shape*": This permits districts of irregular outline, something that is quite necessary.

U. S. DEPARTMENT OF COMMERCE.

or use[18] of buildings, structures, or land. All such regulations shall be uniform for each class or kind of buildings throughout each district,[19] but the regulations in one district may differ[20] from those in other districts.

SEC. 3. PURPOSES IN VIEW.[21]—Such regulations shall be made in accordance with a comprehensive plan[22] and designed[23] to lessen congestion in the streets; to secure safety from fire, panic, and other dangers; to promote health and the general welfare; to provide adequate light and air; to prevent the overcrowding of land; to avoid undue concentration of population; to facilitate the adequate provision of transportation, water, sewerage, schools, parks, and other public requirements. Such regulations shall be made with reasonable consideration, among other things, to the character of the district and its peculiar suitability for particular uses,[24] and with a view to conserving the value of buildings[25] and encouraging the most appropriate use of land throughout such municipality.

---

[18] "*reconstruction, alteration, repair, or use*": All of these words are thought necessary, so as to allow no loophole for evasion of the law.

[19] "*uniform for each class or kind of buildings throughout each district*": This is important, not so much for legal reasons as because it gives notice to property owners that there shall be no improper discriminations, but that all in the same class shall be treated alike.

[20] "*may differ*": This is the essence of zoning, and without this express authority from the legislature to make different regulations in different districts, zoning might be of doubtful validity.

[21] "*Purposes in view*": This section should be clearly differentiated from the statement of purpose (under the police power) contained in the first sentence of section 1. *That* defined and limited the powers created by the legislature to the municipality under the police power. *This* section contains practically a direction from the legislative body as to the purposes in view in establishing a zoning ordinance and the manner in which the work of preparing such an ordinance shall be done. It may be said, in brief, to constitute the "atmosphere" under which the zoning is to be done.

[22] "*with a comprehensive plan*": This will prevent haphazard or piecemeal zoning. No zoning should be done without such a comprehensive study.

[23] "*and designed*": This is the statement of direction given by the legislature referred to in note 21. It has purposely been made to include many purposes. There are not the same dangers involved here that there are in adding to the statement of purposes under the police power, as set forth in the first sentence of section 1.

[24] "*peculiar suitability for particular uses*": This is a reassurance to property interests that zoning is to be done in a sane and practical way.

[25] "*conserving the value of buildings*": It should be noted that zoning is not intended to enhance the value of buildings, but to conserve that value—that is, to prevent depreciation of values such as come in "blighted districts," for instance—but it *is* to encourage the most appropriate use of land.

A STANDARD STATE ZONING ENABLING ACT.

SEC. 4. METHOD OF PROCEDURE.—The legislative body of such municipality shall provide for the manner[26] in which such regulations and restrictions and the boundaries of such districts shall be determined, established, and enforced, and from time to time amended, supplemented, or changed. However, no such regulation, restriction, or boundary shall become effective until after a public hearing[27] in relation thereto, at which parties in interest and citizens[28] shall have an opportunity to be heard. At least 15 days' notice[29] of the time and place of such hearing shall be published in an official paper, or a paper of general circulation, in such municipality.

SEC. 5. CHANGES.[30]—Such regulations, restrictions, and boundaries may from time to time be amended, supplemented, changed, modified, or repealed. In case, however, of a protest against such change,[31] signed by the owners of 20 per cent or more either of the area of the lots[32] included in such proposed change, or of those immediately

---

[26] "*provide for the manner*": In view of the great variety in the form of government that exists throughout the country, it is not thought wise to use the expression "provide by ordinance," for that method may be inappropriate in those communities that have commission government or city managers.

[27] "*after a public hearing*": It is thought wise to require by statute that there must be a public hearing before a zoning ordinance becomes effective. There should be, as a matter of policy, many such hearings.

[28] "*and citizens*": This permits any person to be heard, and not merely property owners whose property interests may be adversely affected by the proposed ordinance. It is right that every citizen should be able to make his voice heard and protest against any ordinance that might be detrimental to the best interests of the city.

[29] "*15 days' notice*": This requirement can be varied to conform to local custom. All that is important is that there should be due and proper notice and ample time for citizens to study the proposals and make their opposition manifest.

[30] "*Changes*": It is obvious that provision must be made for changing the regulations as conditions change or new conditions arise, otherwise zoning would be a "strait-jacket" and a detriment to a community instead of an asset.

[31] "*change*": This term, as here used, it is believed will be construed by the courts to include "amendments, supplements, modifications, and repeal," in view of the language which it follows. These words might be added after the word "change," but have been omitted for the sake of brevity. On the other hand, there must be stability for zoning ordinances if they are to be of value. For this reason the practice has been rather generally adopted of permitting ordinary routine changes to be adopted by a majority vote of the local legislative body, but requiring a three-fourths vote in the event of a protest from a substantial proportion of property owners whose interests are affected. This has proved in practice to be a sound procedure, and has tended to stabilize the ordinance.

[32] "*area of the lots*": Most laws heretofore enacted, based on the first enactment in New York City, have used ownership of feet frontage as the basis for this consent. This has given rise to many difficulties in practice,

adjacent[33] in the rear thereof[34] extending—— feet therefrom,[35] or of those directly opposite[36] thereto extending—— feet[35] from the street frontage of such opposite lots, such amendment shall not become effective except by the favorable vote of three-fourths of all the members[37] of the legislative body of such municipality. The provisions of the previous section relative to public hearings and official notice shall apply equally to all changes or amendments.

SEC. 6. ZONING COMMISSION.—In order to avail itself of the powers conferred by this act,[38] such legislative body shall appoint a commission,[39] to be known as the Zoning Commission, to recommend the boundaries of the various original districts and appropriate

---

especially with corner lots which have frontage on two streets and whose owners accordingly have had two votes to the single vote of the other property owners. In order to get rid of this unnecessarily complex method of determining solely the question of assent to a change in the ordinance, it is recommended that *area of the lots* included in the proposed change be used as the basis, instead of feet frontage. This will do away with the present unfair element of double voting and the unnecessary complications of the generally used method.

[33] "*or of those immediately adjacent*": There are three groups of property ownership, and if 20 per cent of *any one* of these object to the proposed change it will require a three-fourths vote of the legislative body before the change can become effective. These three are (1) the owners of the lots included in the change, (2) the owners of the lots immediately adjacent in the rear, and (3) the owners of the lots directly opposite.

[34] "*immediately adjacent in the rear thereof*": This phrase is necessary for precision; otherwise there will be doubt, and owners of lots in the rear but some distance away might claim the right to be included in the objection.

[35] "*extending —— feet therefrom*": There should be inserted in the Act the number of feet which is the prevailing lot depth in the municipalities of the state.

[36] "*directly opposite*": The same considerations apply to this phrase as to "immediately adjacent in the rear thereof."

[37] "*all the members*": It is important to use this expression, otherwise changes in the ordinance might be made by a three-fourth's vote of the members present at a given meeting.

[38] "*In order to avail itself of the powers conferred by this Act*": Without this phrase it would be necessary for the local legislative body forthwith to appoint a zoning commission, even though it was not desired to take up zoning at that time. This Act is an enabling act *empowering* action, not making it mandatory.

[39] "*shall appoint a commission*": Even though a committee of the local legislative body might be entirely competent to undertake the painstaking, careful, and prolonged detailed study that is ordinarily involved in the preparation of a zoning ordinance and map, the appointment of an outside body of representative citizens is most desirable as a means of securing that participation in and thorough understanding of the zoning ordinance which will insure its acceptance by the people of the particular municipality. One of the most important functions of such a commission is the holding of numerous conferences in all parts of the city with all classes of interests. No zoning ordinance should be adopted until such work has been done.

regulations to be enforced therein. Such commission shall make a preliminary report and hold public hearings thereon before submitting its final report, and such legislative body shall not hold its public hearings or take action until [40] it has received the final report of such commission. Where a city plan commission [41] already exists, it may be appointed [42] as the Zoning Commission.[43]

SEC. 7. BOARD OF ADJUSTMENT.—Such local legislative body may provide for the appointment of a Board of Adjustment, and in the regulations and restrictions adopted pursuant to the authority of this act may provide that the said Board of Adjustment may, in appropriate cases and subject to appropriate conditions and safeguards, make special exceptions to the terms of the ordinance in harmony with its general purpose and intent and in accordance with general or specific rules therein contained.

The Board of Adjustment shall consist of five members, each to be appointed for a term of three years [44] and removable for cause by the appointing authority upon written charges and after public hearing. Vacancies shall be filled for the unexpired term of any member whose term becomes vacant.

The board shall adopt rules in accordance with the provisions of any ordinance adopted pursuant to this act. Meetings of the board shall be held at the call of the chairman and at such other times as the board may determine. Such chairman, or in his absence the acting chairman, may administer oaths and compel the attendance of witnesses. All meetings of the board shall be open to the public. The

---

[40] "*shall not hold its public hearings or take action until*": This is a proper safeguard against hasty or ill-considered action. It should be carefully noted that this is in no sense a delegation of its powers by the local legislative body to the Zoning Commission. The legislative body may still reverse the recommendations of the Zoning Commission.

[41] "*city plan commission*": It is highly desirable that all zoning schemes should be worked out as an integral part of the city plan. For that reason the city plan commission, preferably, should be intrusted with the making of the zoning plan.

[42] "*may be appointed*": It should be noted that its appointment is not made mandatory, however, as sometimes there will be local reasons for desiring a separate body.

[43] "*Zoning Commission*": Some laws contain a provision to the effect that all changes in the ordinance shall be reported upon by the Zoning Commission before action on them can be taken by the legislative body. Such a provision has *not* been included here. In the first place, that involves continuing the Zoning Commission as a permanent body, which may not be desirable. In the second place, it is *before* a zoning ordinance is established that the necessity exists for that careful study and investigation which a Zoning Commission can so well perform. Amendments to the original ordinance do not, as a rule, require such comprehensive study and may be passed upon by the legislative body provided that proper notice and opportunity for the public to express its views have been given.

[44] "*each to be appointed for three years*": This can be altered to provide for overlapping terms, if desired.

board shall keep minutes of its proceedings, showing the vote of each member upon each question, or, if absent or failing to vote, indicating such fact, and shall keep records of its examinations and other official actions, all of which shall be immediately filed in the office of the board and shall be a public record.

Appeals to the Board of Adjustment may be taken by any person aggrieved or by any officer, department, board, or bureau of the municipality affected by any decision of the administrative officer. Such appeal shall be taken within a reasonable time, as provided by the rules of the board, by filing with the officer from whom the appeal is taken and with the Board of Adjustment a notice of appeal specifying the grounds thereof. The officer from whom the appeal is taken shall forthwith transmit to the board all the papers constituting the record upon which the action appealed from was taken.

An appeal stays all proceedings in furtherance of the action appealed from, unless the officer from whom the appeal is taken certifies to the Board of Adjustment after the notice of appeal shall have been filed with him that by reason of facts stated in the certificate a stay would, in his opinion, cause imminent peril to life or property. In such case proceedings shall not be stayed otherwise than by a restraining order which may be granted by the Board of Adjustment or by a court of record on application on notice to the officer from whom the appeal is taken and on due cause shown.

The Board of Adjustment shall fix a reasonable time for the hearing of the appeal, give public notice thereof, as well as due notice to the parties in interest, and decide the same within a reasonable time. Upon the hearing any party may appear in person or by agent or by attorney.

The Board of Adjustment shall have the following powers:

1. To hear and decide appeals where it is alleged there is error in any order, requirement, decision, or determination made by an administrative official in the enforcement of this act or of any ordinance adopted pursuant thereto.

2. To hear and decide special exceptions to the terms of the ordinance upon which such board is required to pass under such ordinance.

3. To authorize upon appeal in specific cases such variance from the terms of the ordinance as will not be contrary to the public interest, where, owing to special conditions, a literal enforcement of the provisions of the ordinance will result in unnecessary hardship, and so that the spirit of the ordinance shall be observed and substantial justice done.

In exercising the above-mentioned powers such board may, in conformity with the provisions of this act, reverse or affirm, wholly or partly, or may modify the order, requirement, decision, or determination appealed from and may make such order, requirement, decision, or determination as ought to be made, and to that end shall have all the powers of the officer from whom the appeal is taken.

The concurring vote of four members of the board shall be necessary to reverse any order, requirement, decision, or determination of any such administrative official, or to decide in favor of the applicant on any matter upon which it is required to pass under any such ordinance, or to effect any variation in such ordinance.

Any person or persons, jointly or severally, aggrieved by any decision of the Board of Adjustment, or any taxpayer, or any officer, department, board, or bureau of the municipality, may present to a court of record a petition, duly verified, setting forth that such decision is illegal, in whole or in part, specifying the grounds of the illegality. Such petition shall be presented to the court within 30 days after the filing of the decision in the office of the board.

Upon the presentation of such petition the court may allow a writ of certiorari directed to the Board of Adjustment to review such decision of the Board of Adjustment and shall prescribe therein the time within which a return thereto must be made and served upon the relator's attorney, which shall not be less than 10 days and may be extended by the court. The allowance of the writ shall not stay proceedings upon the decision appealed from, but the court may, on application, on notice to the board and on due cause shown, grant a restraining order.

The Board of Adjustment shall not be required to return the original papers acted upon by it, but it shall be sufficient to return certified or sworn copies thereof or of such portions thereof as may be called for by such writ. The return shall concisely set forth such other facts as may be pertinent and material to show the grounds of the decision appealed from and shall be verified.

If, upon the hearing, it shall appear to the court that testimony is necessary for the proper disposition of the matter, it may take evidence or appoint a referee to take such evidence as it may direct and report the same to the court with his findings of fact and conclusions of law, which shall constitute a part of the proceedings upon which the determination of the court shall be made. The court may reverse or affirm, wholly or partly, or may modify the decision brought up for review.

Costs shall not be allowed against the board unless it shall appear to the court that it acted with gross negligence, or in bad faith, or with malice in making the decision appealed from.

All issues in any proceeding under this section shall have preference over all other civil actions and proceedings.

SEC. 8. REMEDIES.[45]—In case any building or structure is erected, constructed, reconstructed, altered, repaired, converted, or main-

[45] "*Remedies*": This section is vital. Without it the local authorities, as a rule, will be powerless to do more than inflict a fine or penalty for violation

tained, or any building, structure, or land is used in violation of this act or of any ordinance or other regulation made under authority conferred hereby, the proper local authorities of the municipality, in addition to other remedies, may institute any appropriate action or proceedings[46] to prevent such unlawful erection, construction, reconstruction, alteration, repair, conversion, maintenance, or use, to restrain, correct, or abate such violation, to prevent the occupancy of said building, structure, or land, or to prevent any illegal act, conduct, business, or use in or about such premises.

SEC. 9. CONFLICT WITH OTHER LAWS.[47]—Wherever the regulations made under authority of this act require a greater width or size of yards, courts, or other open spaces, or require a lower height of building or less number of stories, or require a greater percentage of lot to be left unoccupied, or impose other higher standards than are required in any other statute or local ordinance or regulation, the provisions of the regulations made under authority of this act shall govern. Wherever the provisions of any other statute or local ordinance or regulation require a greater width or size of yards, courts, or other open spaces, or require a lower height of building or a less number of stories, or require a greater percentage of lot to be left unoccupied, or impose other higher standards than are required by the regulations made under authority of this act, the provisions of such statute or local ordinance or regulation shall govern.

---

of the zoning ordinance. It is obvious that a person desiring undue privileges will be glad to pay a few hundred dollars in fines or penalties if thereby he can obtain a privilege to build in a manner forbidden by law, or use his building in an unlawful manner, when he may profit thereby to the extent of many thousands of dollars. What is necessary is that the authorities shall be able to stop promptly the construction of an unlawful building before it is erected and restrain and prohibit an unlawful use.

[46] "*any appropriate action or proceedings*": Under the provisions of this section the local authorities may use any or all of the following methods in trying to bring about compliance with the law: They may sue the responsible person for a penalty in a civil suit; they may arrest the offender and put him in jail; they may stop the work in the case of a new building, and prevent its going on; they may prevent the occupancy of a building and keep it vacant until such time as the conditions complained of are remedied; they can evict the occupants of a building when the conditions are contrary to law, and prevent its reoccupancy until the conditions have been cured. All of these things the local authorities should be given power to do if zoning laws are to be effective.

[47] "*Conflict With Other Laws*": By this provision the community is always assured of the maintenance of the higher standard. Without a provision of this kind the later enactment would probably govern. This requirement is especially necessary in those States which now have or later may enact housing laws, as housing laws also contain requirements as to height of dwellings, size of yards, and other open spaces, etc.

Index—Part One

ZONING ORDINANCES

Index—Part Two

BUILDING CODES